Wm Beaumont

A PIONEER AMERICAN PHYSIOLOGIST

Washington University School of Medicine

William Beaumont

Chester Harding

WILLIAM BEAUMONT

A PIONEER AMERICAN PHYSIOLOGIST

A Newly Edited and Completely Reillustrated Edition of

Life and Letters of Dr. William Beaumont

By JESSE S. MYER, A.B., M.D.

with an Introduction by

Estelle Brodman, Ph.D.

to which is added

A Reprint of the First Edition of

Experiments and Observations on the Gastric Juice and the Physiology of Digestion

By WILLIAM BEAUMONT, M.D.

with an Introduction by

Robert Ely Shank, M.D.

1906 75 YEARS 1981

THE C. V. MOSBY COMPANY

ST. LOUIS · TORONTO · LONDON 1981

Printed in the United States of America

The C.V. Mosby Company
11830 Westline Industrial Drive, St. Louis, Missouri 63141

Library of Congress Cataloging in Publication Data

Main entry under title:

William Beaumont, a pioneer American physiologist.

Includes index.
1. Beaumont, William, 1785-1853. 2. Physiologists—United States—Biography. 3. Gastric juice. 4. Digestion. 5. St. Martin, Alexis, 1797?-1880. I. Myer, Jesse Shire, 1873-1913. Life and Letters of Dr. William Beaumont. 1981. II. Beaumont, William, 1785-1853. Experiments and observations on the gastric juice and the physiology of digestion. 1981. [DNLM: 1. Physiology—Biography. 2. Digestive system—Physiology. WZ 100 B379M]
QP26.B4W54 610'.92'4 [B] 81-2770
ISBN 0-8016-0535-0 AACR2

01/A/029

William Beaumont: A Pioneer American Physiologist
has been published on April 26, 1981, to commemorate the
seventy-fifth anniversary of the founding of
The C. V. Mosby Company.
The edition has been limited to
one thousand two hundred and fifty copies,
of which this is copy number

699

PUBLISHER'S ANNOUNCEMENT

THE C. V. MOSBY COMPANY was founded in 1906, the year when Dr. Charles Virgil Mosby first published Bernay's *Golden Rules of Surgery.* Since that time Mosby has enjoyed seventy-five years of steady growth and expansion, producing books and journals of outstanding quality and content.

Many Mosby books published over the years on laboratory medicine, orthopaedics, cancer research, pathology, and blood diseases continue to be revised and have become classics in their respective fields. In dentistry, books on prosthodontics, periodontics, oral anatomy, histology, and embryology have enjoyed similar success, becoming the standard works in those disciplines. Books for nurses and allied health personnel in the areas of pharmacology, anatomy, and physiology have also been leaders in the marketplace for many years, as have texts in biological science, physical education, and special education. With an active book list of 1326 titles in 1980, Mosby is truly a leading publisher for the health care professions throughout the world.

In January 1915 Mosby published its first professional journal, *The International Journal of Orthodontia,* still published under the title of the *American Journal of Orthodontics.* In October of that same year Mosby's second journal, the *Journal of Laboratory and Clinical Medicine* appeared; it was followed by the *American Journal of Obstetrics and Gynecology* in 1920 and the *American Heart Journal* in 1925. Mosby's list of journals has now reached eighteen, including eleven in medicine, three in dentistry, three in nursing, and one in emergency technology. The eighteenth journal, the *Journal of Emergency Nursing,* began publication in February 1981. Mosby journals now serve over 250,000 professionals in all aspects of health care.

In 1967 Mosby was acquired by the Times Mirror Company of Los Angeles, one of the nation's most successful and respected communications and media corporations.

As Mosby celebrates its seventy-fifth anniversary we also approach the one hundred fiftieth anniversary of Dr. William Beaumont's classic work, *Experiments and Observations on the Gastric Juice and the Physiology of Digestion.* We believe that honoring Dr. Beaumont's achievement is a fitting way to commemorate Mosby's anniversary.

Dr. Beaumont's book presented findings about the function of the stomach that were revelations to the medical world of 1833, although they are commonly understood today. His experiments are still recognized as the greatest single advancement in the understanding of digestion.

In 1912 Mosby published the first definitive biography of Dr. Beaumont, Dr. Jesse S. Myer's *Life and Letters of Dr. William Beaumont,* with an introduction by Sir William Osler. A second edition, published in 1939, featured an appreciation by Dr. Andrew Conway Ivy. Now, in 1981, we are proud to present a third edition—an Anniversary Edition—which includes several significant changes and additions. It focuses on Dr. Beaumont's work on digestion and omits the information on genealogy, early life, and the years after the conclusion of his experiments with St. Martin.

To enhance the reader's appreciation for Dr. Beaumont's work, we have reprinted the entire text of *Experiments and Observations* . . . , first published in Plattsburgh, New York, in 1833. We have added a reproduction of Chester Harding's color portrait of Dr. Beaumont and have included many new illustrations of original documents and artifacts from The William Beaumont Collection at the Washington University School of Medicine Library and the William Beaumont Papers at the University of Chicago. New contributions by Dr. Estelle Brodman and Dr. Robert Ely Shank also augment this edition.

We have received invaluable assistance and cooperation from these two contributors, who have both previously published articles on Dr. Beaumont. Dr. Brodman, Professor of Medical History and Head Librarian at the Washington University School of Medicine, has contributed a biographical study that gives personal insights into Dr. Beaumont as an individual. Dr. Brodman was responsible for cataloging and microfilming the more than 1000 documents in The William Beaumont Collection, a task completed in 1963 after the manuscripts had lain virtually untouched for nearly fifty years.

Dr. Shank, Professor and Head of the Department of Preventive Medicine, Washington University School of Medicine, has provided an analysis of Dr. Beaumont's work, bringing an updated medical perspective to this edition.

We are indebted to the St. Louis Medical Society and its executive director, Gordon R. Garrett, for allowing us access to documents and photographs in the Society's collection. We are also indebted to The Joseph Regenstein Library at The University of Chicago and its manuscript research specialist, Elizabeth Teleky, for permitting us to photograph personal artifacts in the Library's collection.

Dr. William Beaumont remains an eminent figure in the history of medicine and a prominent name in the City of St. Louis, where he resided for the last twenty years of his life.

The C. V. Mosby Company is pleased on this occasion to publish this volume.

CONTENTS

Wm Beaumont

A PIONEER AMERICAN PHYSIOLOGIST

A Newly Edited and
Completely Reillustrated Edition of

LIFE AND LETTERS OF

DR. WILLIAM BEAUMONT

By Jesse S. Myer, A.B., M.D.

with an Introduction by

Estelle Brodman, Ph.D.

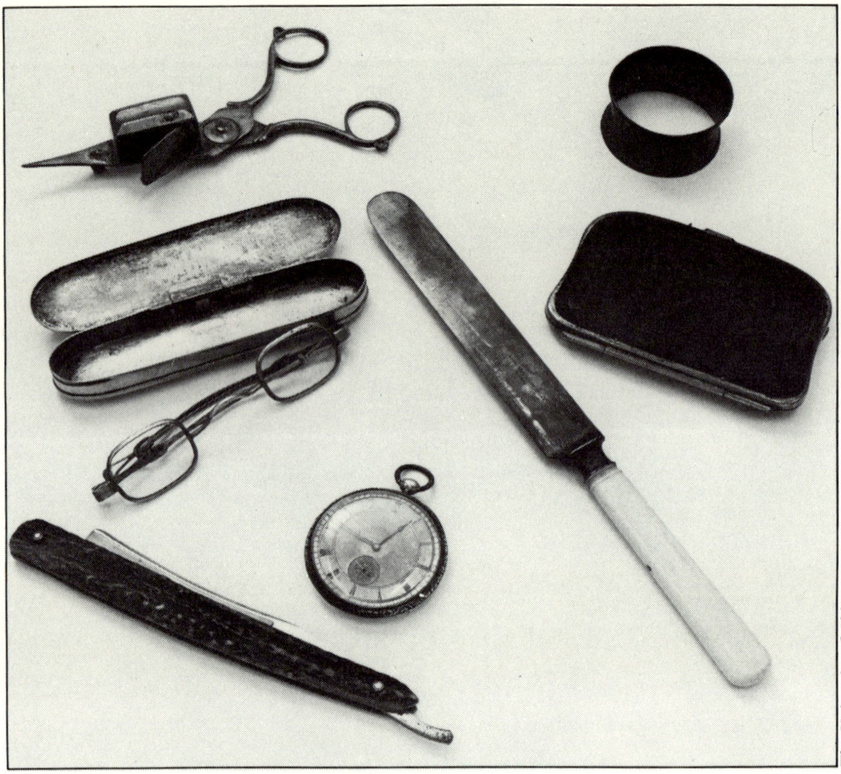

The University of Chicago

Beaumont family's personal effects: brass candle snuffer; inscribed napkin ring with flower and leaf design; Deborah's spectacles with metal case; bone-handled table knife, one of a set of thirteen; leather change purse; bone-handled straight razor; and gold-stem winding open-face watch.

WILLIAM BEAUMONT
THE INDIVIDUAL

Estelle Brodman St. Louis, January 1981

Over the years much has been written about the scientific and experimental work which William Beaumont undertook, especially his work on digestion with Alexis St. Martin as the human test tube for observations. But comparatively little has been published about Beaumont as a person—Beaumont's individuality—and since the individual attributes of a person explain in some degree what he did and how, it is instructive to look at Beaumont from a more personal viewpoint, perhaps, than has been common in the past. In line with this, it is interesting to note that the preface to the original edition of this biography, in June 1912, shows that Jesse Myer was aware of the necessity for portraying the man as well as the scientist. Explaining why he has quoted Beaumont verbatim throughout, Myer remarks, "Preserving Dr. Beaumont's diction gives the reader his mannerisms of speech, his unique and forceful modes of expression, and at the same time presents a concept of the man's traits and character which could not be conveyed by the words of another."[7,p.xiii] Yet throughout the biography, Myer presents nothing of a negative nature—probably due in part to his hero worship of Beaumont and in part to the mores of the time, which held "De mortuis, nihil nisi bonum"—"Of the dead say nothing but the good."

All of this is not meant to imply that terrible exposés of Beaumontian actions and feelings may yet appear; merely that, as Mommsen, the German historian of the nineteenth century noted, each generation must interpret history from its own point of view. Beaumont was definitely a figure of his time, molded by the conditions of his time, and he must be described in the light of that time, and not by the standards of the twentieth century.

Unless otherwise credited all quotations in this introduction are from manuscripts and printed material in The William Beaumont Collection at the Washington University School of Medicine Library.

Beaumont was preeminently a creature of the American frontier. Specifically impinging on him were (1) the conditions of that frontier, (2) the establishment of a medical corps in a newly founded Army, (3) the stance of the federal government on underwriting scientific research and publication, and (4) the general state of medical knowledge to that time. Each of these influenced Beaumont's behavior; in turn, he influenced many of them by what he did and said.

THE FRONTIER

The United States up to quite recent times was a land of great expanses and small populations. As Frederick Turner[13] pointed out in 1920, the pioneer life calls on each person to be a generalist instead of a specialist. With no one nearby to rely on, the pioneer must be a carpenter, bricklayer, farmer, butcher, ironmonger, weaver, educator, millwright, physician, and a host of other things. His very existence depends on his craftsmanship and his pragmatic knowledge, not his book-learning. As a consequence of this necessity, Beaumont throughout his life acquired information in many diverse fields—as a young man he was schoolteacher, farmer, storekeeper, student, military surgeon, and omniverous reader. His early notebooks show, by the items which he copied for his continuing use, the subjects and forms of literature and general knowledge that were of interest to him.[5]

His experiences as an army physician increased his self-reliance and his tendency to act forthrightly—his belief in the need to "do something" when faced with a difficult situation. This was in stark contrast with the Galenic principle of *vis medactrix naturae,* which held that the greatest good a physician could do was to let nature take its course. Thus his medical notebook of private cases shows that he did not spend time unnecessarily ruminating and philosophizing, but that he set immediately to do what he thought necessary. It is not surprising, therefore, that when Alexis St. Martin was presented to him in a very serious condition, he reported, "I applied the best means in my power and took the responsibility of his surgical treatment and medical attendance. . . ."[4] When he thought he detected malingering in Lieutenant Griswold in 1825, he prescribed an emetic-cathartic of heroic proportions to determine whether the lieutenant was following orders or was feigning illness.

Quarrels and litigation were commonplace in Beaumont's time.

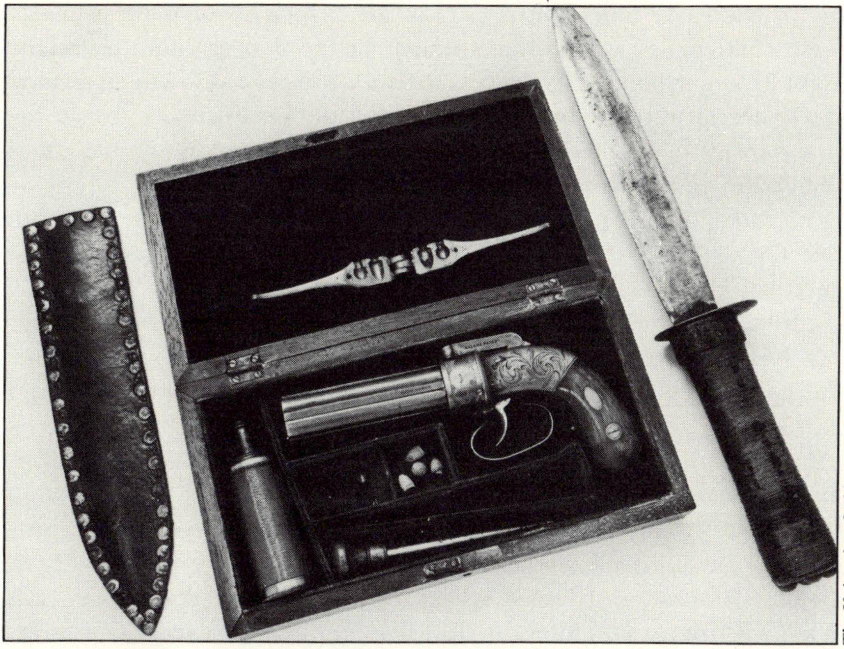

The University of Chicago

Beaumont's five-barrelled, automatically revolving derringer in mahogany case with round and elliptical lead bullets, gunpowder container, ramrod, and fulminate of mercury caps. On the cover of the gun case, the hinged and opened bullet mold for the manufacture of round and elliptical bullets, is also shown. Beaumont's Bowie knife and sheath are displayed on either side of the gun case.

We are accustomed to romanticize about the American frontier and the sturdy, self-reliant, liberty-loving men and women who inhabited it. However, the truer picture includes many negative elements described by visitors to the United States, such as Charles Dickens and Harriet Martineau—the slovenly appearance of the people and the careless use of their possessions, excessive brawling and vicious personal fights between individuals (shown, for example, by the turbulence in the St. Louis Medical Society until the Civil War), bribery in electioneering, and general bad manners among all classes. Sharp practices among land speculators were rampant, as Beaumont found out when he became embroiled in the Spanish land-grant litigation after he purchased property in St. Louis. Corruption among office holders was rife (as exemplified by Andrew Jackson's statement "To the victor belong the spoils" to explain his rewarding his followers from the government till).

In the early nineteenth century the principles of private liberty were carried to extraordinary limits. Jacksonian egalitarianism, the right of Everyman to have access to everything equally, was enshrined in the Declaration of Independence and the Constitution, but in the first part of the nineteenth century was not restricted by any feeling of obligation or personal discipline. The results were what might be expected: slander, denigrations of conduct and character, and a continual struggle of the weak against the powerful, the braggart, or the corrupt.

Beaumont was certainly caught up in this net. He was intimately involved in personal disputes and legal suits throughout his lifetime. He had difficulties with such diverse people as Lieutenants Griswold and Potter, whom he accused of malingering; G. H. Richards, who he thought had slandered him and whom he challenged to a duel after an exchange of broadsides; Surgeon General Lawson, who refused to allow Beaumont the personal freedoms that his predecessor, Surgeon General Lovell, had; and his partners and one-time protéges, George Johnson and James Sykes. In every case the controversy was filled with personal animosity and degenerated into acrimonious accusations about the motives and actions of the other person.

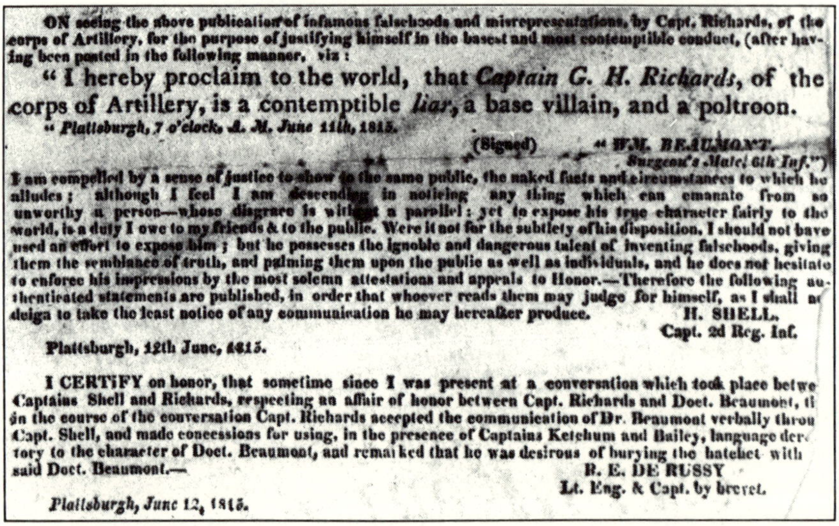

ON seeing the above publication of infamous falsehoods and misrepresentations, by Capt. Richards, of the corps of Artillery, for the purpose of justifying himself in the basest and most contemptible conduct, (after having been posted in the following manner, viz:

"I hereby proclaim to the world, that *Captain G. H. Richards*, of the corps of Artillery, is a contemptible *liar*, a base villain, and a poltroon.

"*Plattsburgh, 7 o'clock, A. M. June 11th, 1815.*

(Signed) "*WM. BEAUMONT,*
Surgeon's Mate, 6th Inf.")

I am compelled by a sense of justice to show to the same public, the naked facts and circumstances to which he alludes; although I feel I am descending in noticing any thing which can emanate from so unworthy a person—whose disgrace is without a parallel: yet to expose his true character fairly to the world, is a duty I owe to my friends & to the public. Were it not for the subtlety of his disposition, I should not have used an effort to expose him; but he possesses the ignoble and dangerous talent of inventing falsehoods, giving them the semblance of truth, and palming them upon the public as well as individuals, and he does not hesitate to enforce his impressions by the most solemn attestations and appeals to Honor.—Therefore the following authenticated statements are published, in order that whoever reads them may judge for himself, as I shall not deign to take the least notice of any communication he may hereafter produce.

H. SHELL,
Capt. 2d Reg. Inf.

Plattsburgh, 12th June, 1815.

I CERTIFY on honor, that sometime since I was present at a conversation which took place betwe[en] Captains Shell and Richards, respecting an affair of honor between Capt. Richards and Doct. Beaumont, t[hat] in the course of the conversation Capt. Richards accepted the communication of Dr. Beaumont verbally throu[gh] Capt. Shell, and made concessions for using, in the presence of Captains Ketchum and Bailey, language der[oga]tory to the character of Doct. Beaumont, and remarked that he was desirous of burying the hatchet with said Doct. Beaumont.—

R. E. DE RUSSY
Lt. Eng. & Capt. by brevet.

Plattsburgh, June 12, 1815.

Washington University School of Medicine

Part of a circular issued by Beaumont as the last word of a controversy with Captain G. H. Richards, of the corps of artillery, in June 1815.

Beaumont had particular difficulties with other physicians in St. Louis, including those who testified against him in the Darnes-Davis and Mary Dugan cases. And the strong words he used against the instigator of the suit against him are characteristic of Beaumont and the times in which he lived. He accused the instigator of attempting to "appeal to the fluctuating sympathies and seek and obtain the support of the weak, wilful, and wicked of the community, or the pusillanimous and prejudiced of the profession to sustain himself in his sinking condition." This reaction shows the emotional intensity that surrounded the suit. Beaumont and his partner Sykes were also accused by the members of the Medical Society of improper advertising, and had to agree to change the wording of their published notice. As a result, Beaumont's feelings about the other local physicians were so strong that when he was elected president of the Medical Society in 1840, his inaugural address was so condemnatory that throughout his tenure in office there never was a quorum present at any meeting of the society! And what he said of the local physicians to Dr. Combe still blisters the pages of the correspondence in the Beaumont collection.

The dislike Alexis St. Martin had of the handling he received from Beaumont is, of course, well-known. At least some of it was probably due to the conflict between Beaumont's concept of "servant" and Alexis' of free "engagé" and voyageur. This conflict continued to the end of their correspondence in the 1850s, when Beaumont was attempting to persuade Alexis to return to him for further experiments and for showing to groups.

The Medical Corps

Beaumont was also influenced by many other situations of his time and position, of course. As a member of the Army Medical Corps at a period when it was so new its traditions were just being formed, and when medical officers were in an anomalous position, often not accepted either as line officers or as staff, Beaumont found himself in difficulty at several times. The most important was his conflict with Surgeon General Lawson, which resulted in Beaumont's resignation from the army in a huff, but there were also other problems connected with the private practice in which he engaged, and with regular officers who acted as servants to Beaumont and were then repaid by him with accusations of malingering or cowardice.

The elements in the Lawson-Beaumont conflict are particularly

The University of Chicago

Brass plate engraved "Wm. Beaumont, Surg^{n} U. S. Ay." It was originally attached to his small leather-covered army trunk.

enlightening as to the feelings of each individual. Because the Surgeon General Joseph Lovell had seen the value of the unusual work which Beaumont was undertaking, he had indulgently granted Beaumont's special requests for almost a decade. When Lovell died, he was replaced by Lawson, who had a very different view of Beaumont and a much more rigid idea of the rules and regulations of an army. After all the years of special treatment Beaumont had received, it is not surprising that he had come to expect it—perhaps as his due. When, therefore, he was ordered by Lawson to procure a substitute at his own expense if he wished a furlough of 6 to 8 weeks to attend to private business in Green Bay, Wisconsin, Beaumont wrote a rather intemperate letter to the Surgeon General in which he also reported receipt of a rumor that he was to be transferred to Jefferson Barracks from St. Louis. He stated that he thought such an order would be unfair and a personal affront to him. Later, when Lawson did order Beaumont to Florida as a member of an examination board for medical officers, Beaumont wrote a series of angry letters to several people, and, finally, on December 20, 1839, forwarded to Lawson a letter of conditional resignation stating that if the order were rescinded, he would stay in the army; otherwise he was resigning. To his great surprise, the resignation was immediately accepted. This so flabbergasted him that he spent the rest of his life trying to get the acceptance removed. Obviously the likelihood of such a result had never occurred to him!

Science and the Federal Government

Accustomed as we are today to the large amounts of money which the federal government pours annually into research and development in scientific fields, it is sometimes difficult for us to realize that this interest dates back only to World War II, when the intimate relationship between such public outlays and the well-being of the citizens was first clearly demonstrated. Many in the early United States believed the less government the better, and thus scientific research and the publications from it were left to the private sector. The only exception to this rule seems to have been the expeditions to map and describe the country itself, of which the Lewis and Clark Expedition is a prime example. In other parts of the scientific field, however, the federal government resolutely stayed out.

Beaumont was the victim of this situation in several ways. Having taken Alexis St. Martin into his own home after the local political entity refused to pay further for his upkeep, Beaumont petitioned Congress for reimbursement. In spite of many supporting letters and documents, the petition was turned down by the Congress because of "the theoretical objection to the appropriation by Congress of money for any scientific or philanthropic purposes whatever," according to the report on the vote. Later, when he was publishing his book on the experiments conducted on Alexis, Beaumont once more asked the government for aid, this time requesting that 100 copies of the book be bought for the navy. This request, too, was turned down by the Secretary of the Navy. A third petition for compensation, at Senator Everett's suggestion, also had no success.

Medical Knowledge of the Time

As a medical officer in the army, Beaumont faced primarily three kinds of problems: (1) the prevalent agues or fevers, often malarial in origin, but occasionally added to by severe epidemics of cholera, (2) traumatic injuries due to military engagements and the general work conditions of the peacetime army on a frontier, and (3) the low level of sanitation and personal hygiene which has been the ever-present specter in all armies from the first recorded ones to the present. Since the understanding of the causes of the fevers, malaria, and cholera was either primitive or nonexistent, the physician was forced to treat the symptoms rather than the causes of the disease; this explains the multitude of remedies tried, whose rare successes were poorly understood. Only a few things were known: that some diseases were contagious; some were caused by unsanitary conditions; and some could recur in the same individual a number of times, while others conferred a degree of immunity.

In the early nineteenth century more was known, perhaps, about human anatomy than was true earlier. Medicine was struggling with the concept of cowpox vaccination to ward off smallpox; it was gradually working its way to a sounder knowledge of the physiology of the internal organs and of the circulatory system, and biochemistry was beginning to be an important study by the 1830s. However, treatment was not very different from that recommended by Sydenham, the Hunters, and others of the eighteenth century. The very books that Beaumont read and from which he made extracts were those of a century before his time.

Although the level of knowledge was relatively low, it should be pointed out that it enabled Beaumont to act without much of a feeling of inferiority—his knowledge of medicine was probably as good and as up-to-date as that of most other physicians of his time. Consequently, he could undertake experimental work, as well as practice, without enormous personal doubts. It must not be thought, however, that Beaumont was never without despair at his impotence in the face of disease. During a cholera epidemic, he wrote to a friend describing his treatment of cases of this disease. At the end he said:

> But if the subject has passed the verge into the cold blue stage of collapse no medical means, skill or human effort seems entirely impotent of relief [sic]. Relatives and friends shrink with sorrowful emotions, and doctors stand aghast in mute astonishment and mortification at the resistless rapidity of a disease so little understood and over which their most profound skill and judgment and professional efforts can have no salutary influence or effect—a disease which I conceive to be *sui generis* of atmospheric origin, sent by God over the face of the earth to scourge mankind—Test the wisdom and ability of the most learned and experienced of medical faculty of all conditions and to admonish us, minor masters of our Profession, of our almost total ignorance of all diseases or the best means of cure.

And this was written even after his experience as a physician during the great epidemics of cholera which had decimated the Indians in the Black Hawk uprising and had caused so many problems to General Scott's armies in the field against the Indians!

Beaumont the Student

One thing is certain. Just as the frontier engendered "the tumultuous and acrimonious debates which wracked the [Medical Society] organization in the mid-1840's,"[12] so the pioneer world tended to lead to self-study and the belief that learning is a lifelong experience. Educated formally only in the village school and through apprenticeship, Beaumont early acquired the habit of reading widely, copying portions of what he read into notebooks for further use, and pondering on what he read. As a truly "self-educated man, curious and eager to learn,"[5,p.ix] he did not have the feeling, so common in those educated at schools and universities, that learning required study halls and a faculty of lecturers, but understood that all experience teaches and all learning requires thoughtful consideration and contemplation. Thus, throughout his life, as his notebooks and records show, he examined critically the effects of his treatments on his patients and their diseases, and he reported both his successes and his failures critically and honestly.[2] When he was working on his experiments on digestion with Alexis St. Martin, he kept an index (a "synoptical index of various authors on digestion") of the writings he was able to see on the subjects which were confusing to him, and much of this was incorporated into the final draft of his book, which he submitted to Surgeon General Lovell for comments. Other quotations he copied, which have survived in the Beaumont collection at the Washington University School of Medicine Library, are from the *American Journal of the Medical Sciences,* from Jackson, the Harvard chemist, from Van Helmont, and from Lucretius; and are on nature, philosophy, science, and truth. Excerpts from Haller and Van Swietan, Brown and Cullen, Sydenham and Huxham show the breadth and depth of his medical readings. It is doubtful whether many American physicians of his time, school trained or apprentice trained, were better educated.

Nor did Beaumont ever believe that he knew all that was known in his field. For that reason he had no hesitation in requesting time from his army duties to go to Europe to study biochemistry from the masters in the field—a subject in which he felt (and rightly so) very deficient. When that furlough had to be canceled because of the Black Hawk War, he accepted the offer of the Surgeon General to spend some time in Washington, D. C., to read up on the subject of digestion and to talk to those who were more knowledgeable than he in chemistry. (It is thought by some that the books which Surgeon General Lovell procured for Beaumont's reading may be the basis for the present collections of the National Library of Medicine; though unfortunately there is only circumstantial evidence to prove this contention.)

And yet, in spite of all of this, there is a puzzling aspect to this seeming willingness of Beaumont's to acknowledge his lack of knowledge in this field. After Lovell had arranged for help for Beaumont from the eminent Virginia physiologist, Robley Dunglison, through Nicholas P. Trist, private secretary to Andrew Jackson, and after Dunglison had sent him several long letters with suggestions for further experiments (experiments which he faithfully carried out), Beaumont forwarded a letter to Dunglison which leads to some doubt about the motives he had in his work with Alexis. On January 25, 1833, he wrote:

> Dr. B. is much indebted, and will be ever grateful, for the aid and instructions received from Professor D., and is disposed to reciprocate every generous desire, mutually to afford subjects for Physiological investigation and improvement so far as he is capable. Dr. B. hopes his feelings may be perfectly appreciated—similar confidence he could repose in but few—but he feels well aware that by Professor Dunglison he will never be improperly anticipated in his intentions, sooner or later, to publish the experiments collectively by premature communication to any Periodical Publication of partial results, whereby impressions of an undue transfer merit or demerit from its original or collateral sources might be liable to be made upon the public mind. . . .

As might be expected, Dunglison was not very pleased by the inuendoes in this letter. Although he replied to Beaumont courteously, he wrote in quite a different vein in his reminiscences. "The suggestions, contained in this note," he observed, "although savouring somewhat of a want of delicacy, I need scarcely say were needless. . . ." He then went on to complain:

> It might have been anticipated, however, from Dr. Beaumont's sensitiveness as to his own rights, that he would have been extremely liberal as regarded those of others, and especially of one, of whose services he speaks so highly in the letters referred to above; and with such anticipations I confess, I was surprised, on the appearance of his *Experiments,* to find no allusion whatever to my having visited Washington, by request, in a Virginia winter from a distance of 100 miles; or to my having made a single suggestion to him, or been associated, in any manner, with his investigations. . . . Soon after the appearance of his book, I saw Dr. Beaumont in Baltimore. . . . I did not conceal from him my expectation of seeing in it a more detailed account of

> our joint labours, and reminded him that the only allusion to my being concerned in the matter was in [one] paragraph . . . when he replied that he did not think I would care to be associated with so humble an individual as himself.[10,p.52]

Since Beaumont referred to other work by Dunglison in very flattering terms in his book, it would appear that he wished to keep for himself alone all the glory of the experiments on digestion with Alexis St. Martin. The idea that others might have suggested the experiments which he performed was resolutely left out of the final work; and this is true whether those suggestions were from his revered mentor, Surgeon General Lovell, or from the doyen of American physiologists, Robley Dunglison.[1]

Perhaps the difficulty with Dunglison lay in the unconscious feeling of contempt which the latter early felt for Beaumont. Dunglison notes his view of Beaumont in his reminiscences: "Dr. Beaumont was neither by nature nor his stars well fitted for the investigation of any profound physiological subject. He was most persevering, however; and his zeal and industry compensated in some degree for his deficiency in ability."[10,p.51] It is possible Beaumont sensed this criticism and resented it.

Family Relationships

There is also presumptive evidence of some difficulties of personality in Beaumont's family relationships. His only son evinced problems in school, apparently changing schools several times, and then was in financial straits much of his life. An allusion to alcoholism may seem to be found in some of his father's letters, but nothing definite is known of this, and it may not have been true. Beaumont could not bear to think of his daughter Sarah marrying and leaving the household he had provided, with the result that when he received a proposal for her hand from Lieutenant J. H. La Motte, he refused to hand her the letter, in spite of his wife's belief that the marriage would be a good thing. Instead, he insisted that La Motte come to St. Louis to plead his case in person—knowing full well that La Motte was on active duty against the Seminole Indians in Florida and could not possibly travel to Missouri. Later Sarah was married to a man who committed suicide and left her and her daughter penniless.[8] Sarah in turn made her daughter Lucretia (Lily) miserable,[3] and Lily made an unhappy marriage and soon returned to her home divorced—in a society that did not condone such actions. Sarah Beaumont Keim,

The University of Chicago

Colored miniature of the Beaumont family. Mrs. Beaumont has black hair and the Doctor blue eyes and a full head of heavy auburn hair. The children are, from left to right, Lucretia ("Cush," b. 1827, at Ft. Howard, Green Bay, Wis.), Sarah ("Tasey," b. 1822, at Ft. Mackinac, 2 days after the famous accident to Alexis St. Martin), and Israel Green ("Bud," b. 1829, at Ft. Crawford, Prairie du Chien, Wis.). A son and namesake, William W., b. at Michilimackinac in 1824 died 2 years later at Plattsburgh, N. Y. Since Beaumont's book was published first in 1833, the miniature portrays the surviving members of the family at about the time that Beaumont, aged 48, had completed his work with St. Martin begun 8 years previously.

who died in 1913 at the age of 91, also muddied the historical record by promising the Beaumont papers to several institutions before Lily presented them to the Washington University School of Medicine in St. Louis.[9]

Throughout most of her life it was Mrs. Beaumont—Debby—who smoothed ruffled feathers and kept the family happy and made their home so pleasant that a number of the young officers of the army and many local inhabitants were often seen there, including the Robert E. Lees, Ethan Allan Hitchcock, the Kingsburys, the Cabannes, the Frelighs, and many others whose names are now associated with places in the city of St. Louis.[6]

Beaumont's high standard of conduct may account for some of his difficulties with people close to him. That Beaumont was somewhat priggish in his outlook and narrow in his view of right and wrong seems obvious. His early letters to his family before and during the War of 1812 (most of which are now at the Yale University Libraries) may be taken to exhibit the views of a very young man (although he was more than 25 years old at the time), but his advice to his younger brother Abel in 1809 and his condemnation of Federalism in 1811 go beyond what might have been expected of a young man following the conventions of letter writing common at that time. To Abel he wrote:

> You are now on a critical state of life—errors and improprieties will beset you on every side in spite of your precautionary efforts to evade them. . . . Think not too much of dancing scenes, but give your mind time to reflect on what you are and of what use you may be to yourself and to society by properly improving your time . . . Let virtue, truth and honor be your planetary guides; temperance, justice, and fortitude, and prudence your cardinal points; faith, hope, and charity your horizon; philanthropy, benevolence, friendship, and philosophy your atmosphere; and the elements of life will be smooth, transparent, and pleasant, gently gliding over you waving imagination like the eastern morning breeze across the swelling field of wheat.

Beaumont's family, poor but industrious, espoused the Republican cause, and Beaumont's deep-seated feelings about the conflict between the Federalists and the Republicans are revealed in his correspondence. Assuring his father that he had not turned Federalist, Beaumont wrote in 1811:

> That principle [of Republicanism] is the legitimate sentiment of every true American. It was hereditarily bestowed and carefully cultivated by an anxious father whose precepts and instructions have made an indelible impression on the heart of his son, who is farther than ever from having the name Federal

justly applied to his character or associated with his name. . . . Sooner might they remove the everlasting hills than bribe my integrity, make my faith waver, shake my belief, or divert my course from the pole star of Republicanism while reason holds her empire over the province of my intellect.

To his wife Debby, then in Plattsburgh with the children while he and Alexis were in Washington, Beaumont wrote letters of advice. She was exhorted to "examine Sarah's improvements day by day—every day to correct every little irregularity of her disposition; not to suffer Lucretia to be too much petted, but rather to check her natural vanity and affectation;" and to "coerce obedience from the young gentleman, by application of the rod when necessary."[8,p.20]

Conclusion

How, then, can we characterize Beaumont the individual? He was certainly a complex character with many traits, some of which seem contradictory. To his patients he apparently was always sympathetic and concerned. "In the sick-room he was a model of patience and kindness, his intuitive perceptions, guiding a pure benevolence, never failed to inspire confidence, and thus he belonged to that class of physicians whose very presence affords nature a sensible relief," his colleague, biographer, and eulogiser, Dr. T. Reyburn, remarked after Beaumont's death.[11] Yet he was anything but patient with those he thought in the wrong, whether they were his colleagues, his partners, or his superior, the Surgeon General. He was personally upright and reacted strongly when he thought someone else was not acting honorably.

He was hot tempered, as shown in his letters to Combe, Lawson, and Hitchcock, as well as in what he said and wrote to and about Alexis St. Martin.

He was willing to redress wrongs to another as well as to himself, as when he took the sick Alexis into his own household after the citizens of the Mackinac area refused to support him any longer as a pauper. Yet Beaumont seemed to see a conspiracy against him when his petitions for reimbursement were turned down.

It is difficult to judge from this distance whether Beaumont was as badly treated as he seemed to think. He was hard-of-hearing, and it is possible that this made it difficult for him to understand what people said, leading to suspicion in his dealings with others.

Beaumont's feelings about religion are really unknown, but certainly they were not institutionalized. He seems to have been a Deist, never enrolled in any formal church; yet his diary reveals that he did on occasion attend church services.

Beaumont was a man socially and economically unsure of himself, it might be argued, because he found it necessary at intervals to total up his assets. On one occasion he wrote a friend, "I have a very handsome, lucrative, and respectable private practice, reputation far above my merits, and professional popularity more than commensurate with my best practical skill or abilities." That he thought well of his abilities professionally, in spite of the last part of the preceding sentence is shown in his letter to the Surgeon General complaining about his transfer to Jefferson Barracks: "for any professional advantages I may be supposed to enjoy here [St. Louis] I am not indebted to the government, but to my character and exertions. These advantages . . . are not an incident to the Station, but are fruits of toil, I am sure, if not talent. . . ." As Reyburn put it, Beaumont was "a man who used well his resources, who never threw a legitimate opportunity away, and never stooped to avail himself of one that was unlawful." Certainly his request of Dunglison that he refrain from immediately publishing information of the case in which they were both working underlines this trait of never throwing a legitimate opportunity away; the request was equally lawful, if not politic.

In general, Beaumont's colleagues agreed with his evaluation of his professional abilities. "He was gifted with strong natural powers, which working upon an extensive experience in life, resulted in a species of natural sagacity, which . . . was something peculiar to him, and not to be attained by any course of study," Reyburn noted in his obituary. Dr. J. B. Johnson, faculty member at both of the medical schools which later formed the Washington University School of Medicine, remarked in his reminiscences that Beaumont was one of the best physicians in the city. The number of times other physicians called him in for consultation shows the generally high opinion they had of his professional knowledge.

All recognized Beaumont's zeal, industriousness, perseverance, and thoroughness, even though (as Dunglison felt) they considered him without scientific imagination. Most realized that he was impatient, of a suspicious nature, both sure and unsure of himself, more efficient in acting than in thinking, a good student, but difficult to get along with in many ways. He possessed "intelligence, industry, and integrity, essential qualifications for success," says the *Dictionary of American Biography,* and with this his contemporaries agreed.[14]

To his family Beaumont was a stern but loving paterfamilias, not uncommon in the nineteenth century; to his friends a good companion who lent them money and advice; to his patients a beloved physician; and to his colleagues a changeable person who could unpredictably express esteem or initiate a dispute. In sum, as V. C. Vaughan described him, "he possessed an excess of *fortiter in re* and was deficient in *suaviter in modo.*"

We must then conclude that William Beaumont, like most human beings, was not a simple soul, but had many traits that tended to contradict and negate each other, and he was greatly affected by his milieu. Just as the stomach is "not a fermenting vat, not a still, but a stomach, a stomach," so Beaumont was neither a saint nor a devil, but a man possessed of greatnesses as well as the flaws of a great man.

REFERENCES

1. Brodman, E.: Relations of William Beaumont and Joseph Lovell, Bull. Hist. Med. 38:127-132, 1964.
2. Brodman, E.: William Beaumont as a physician, Wisconsin Med. J. 66:272-276, 1976.
3. Carson, W. G. B.: Miss Lily Irwin, Bull. Missouri Hist. Soc. 18:231-238, 1962.
4. Medical Recorder 8:14, 1825.
5. Miller, G.: William Beaumont's formative years; two early notebooks, 1811-1821, New York, 1946, Schuman.
6. Myer, A.M.: The golden page; a biographical novel, Hicksville, N. Y., 1977, Exposition Press.
7. Myer, J.: Life and letters of Dr. William Beaumont, St. Louis, 1912, The C. V. Mosby Co.
8. Myer, Mrs. M. W.: Sarah Beaumont: her life and loves, Bull. Missouri Hist. Soc. 17:16-44, 1960.
9. Pizer, I. H.: Source materials and the library; the dispersion of the Beaumont papers, Bull. Med. Lib., A, 52:328-336, 1964.
10. Radbill, S. X.: Autobiographical ana of Robley Dunglison, M.D., Trans. Am. Philos. Soc., 1963, n.s. 53, pt. 8.
11. Reyburn, T.: William Beaumont, St. Louis Med. Surg. J. 12:127-139, 1854.
12. Soule, S. D.: Medicine in St. Louis. Our medical ancestors, 1764-1864, St. Louis, 1978 St. Louis Medical Service Bureau, p. 25.
13. Turner, F. J.: The frontier in American history, New York, 1920, Holt, Rinehart & Winston.
14. Vaughan, V. C.: William Beaumont. In Dictionary of American biography, vol. 2, New York, 1929, Charles Scribner's Sons.

INTRODUCTION

William Osler Oxford, January 1912

It is a great pleasure to introduce to the Profession a worthy study of William Beaumont, in whose work three generations have now been interested—as well as in "that old fistulous Alexis," the subject of his experiments. While the chief facts of his life are well known from fragmentary sketches that have appeared, the sources for a full biography have only recently been made available. The family papers have been placed at the disposal of Dr. Myer, to whom it has been a labor of love to present a picture of the first great American physiologist in its proper setting.

To the medical bibliographer there are few more treasured Americana than the brown-backed, poorly printed octavo volume of 280 pages with the imprint "Plattsburgh, Printed by F. P. Allen 1833." To have had the book printed at Philadelphia, and in better style, as Dr. Franklin Bache suggested, would not have been in keeping with the conditions in which the work was done—in the backwoods, in camps, at prairie garrisons—and it was characteristic of the man to overcome the difficulties of having the book published at a small army post of Northern New York.

To the physiologist Beaumont's experiments have always had an extraordinary interest. In the first place his scientific method is admirable. As has been remarked by Dr. Combe—"It would be difficult to point out any observer who excels him in devotion to truth, and freedom from the trammels of theory or prejudice. He tells plainly what he saw, and leaves every one to draw his own inferences; or, where he lays down conclusions, he does so with a degree of modesty and fairness of which few perhaps in his circumstances would have been capable." Beaumont himself remarks in the Preface: "I submit a body of facts which can not be invalidated. My opinions may be doubted, denied, or approved, according as they conflict or agree with the opinions of each individual who may read them, but their worth will be best determined by the foundation on which they rest—the incontrovertible facts."

His experiments settled finally the chemical nature of the digestive process so much discussed from the days of the fundamental experiments of Réaumur (1752) and Spallanzani (1783). To appreciate the importance of Beaumont's experiments we must read contemporary accounts of the nature of stomach digestion, in which the old views of concoction, putrefaction, trituration, fermentation and maceration are all considered. Even before Réaumur and Spallanzani shrewd guesses were made. I came across the other day the manuscript notes of lectures on medicine of one of my seventeenth century predecessors at Oxford, whose quaint description of digestion is worth quoting:

". . . This then ye meate being chaw'd in ye mouth receives ye first tincture from ye spittle, and going through ye oesophagus is imbued with ye same, till in ye fundus of ye stomach it lyes in a liquor of ye same nature but greater activity, & being encompassed about by ye coate of ye stomach, it lyes in that watry bed as in its proper menstrum: & receiving by little & little this piercing and searching liquor into all its parts and ferments, and so dissolves into a kind of mash, from whence ye nutrimentall parts of life of ye meate are now become easily separable from ye faeculent. Which being done, ye stomach having performed its proper worke, by ye compression of itselfe sends it forth through ye Pylorus into ye guts, leaving a small quantity still as ye remaining ferment, necessary for ye digestion of ye next meale. . . . That such a ferment belongs to ye stomach is clear, by ye stomachs of calves, which ye dayry-maides use for runnet in making cheese: and strong cheese is very good for digestion, because it has much of ye runnet, & so adds a connaturall ferment to ye stomach. This I conceive to be ye true Nature of digestion, by which it appeares to be no boyling or proper concoction, nor ye immediate effect of heate, but of an acidity, or subtle penetrating and fermenting quality of ye proper liquor of ye stomach. I conclude therefore ye Digestion is a Solution tending to Putrefaction made by ye way of fermentation."

Among other important observations may be mentioned the confirmation of the discovery by Prout of the presence of hydrochloric acid in the gastric juice; the recognition that the essential elements of the gastric juice and the mucus secretion were separate; the establishment by direct observation of the profound influence of mental disturbances on the secretion of the gastric juice and on digestion; the fuller and more accurate comparative study of digestion in the stomach with digestion outside the body; the rapid disappearance of water from the stomach through the pylorus; the first comprehensive and full study of the motions of the stomach; the study of the digestibility of different articles of diet, which remains today one of the most important contributions ever made to practical dietetics; the relation between the amount of food taken and the quantity of gastric juice secreted; and many other points, the true significance of which have not been recognized until the recent researches of Professor Pavlov.

The man was greater than his work; and it is well that the story of Beaumont's life should be told by a member of the Profession of the city of which for so many years he was a distinguished ornament. The pioneer physiologist of the United States and the first to make a contribution of enduring value, his work remains a model of patient, persevering research. The highest praise that we can give it is to say that his life fulfilled the ideal with which he set out, and which he so well expressed in the sentence: "Truth, like beauty, when 'unadorned, is adorned the most,' and, in prosecuting these experiments and inquiries, I believe I have been guided by its light."

Bronze medallion bearing the likeness of Beaumont issued at the Thirteenth International Physiological Congress, Boston, 1929.

A PRESENT-DAY APPRECIATION
OF BEAUMONT'S EXPERIMENTS ON ALEXIS ST. MARTIN

Andrew Conway Ivy CHICAGO, 1939

BEHIND EVERY EXPERIMENT is a man. The character and the mental attitude of the man bear on the character of his observations and their interpretation. Hence, it is appropriate, first, to depict the character of the man, Beaumont. This can best be done by using his own statements.

"I had opportunities for the examination of the interior of the stomach. . . ." "I have availed myself of the opportunity afforded . . ." "I submit a body of facts which cannot be invalidated." "The reader will perceive some slight seeming discrepancies . . .; but he will recollect that the human machine" is subject to some variation in the same and different individuals. "I am fully aware of the importance of the subject . . .; I am therefore willing to risk the censure or neglect of critics, if I may be permitted . . . to be the means, either directly or indirectly, of subserving the cause of truth, . . ." "I make no claim to originality in my opinions, . . ." "I consider myself but a humble inquirer after truth—a simple experimenter." "My opinions may be doubted, denied or approved . . .; but their worth will be best determined by the foundation on which they rest—the incontrovertible facts." "Truth, like beauty, when 'unadorned, is adorned the most,' and, in prosecuting these experiments and inquiries, I believe I have been guided by its light." More inspirational statements cannot be found in the writings of any other scientist.

It is no wonder that a man imbued with such a spirit made many observations all of which have been confirmed. Certain of his interpretations were in error, but it must be remembered that Beaumont worked with only a "gum-elastic tube," a magnifying glass, a scale, a thermometer, an incubator consisting either of a basin of water on a sand-bath or an armpit, a muslin filter, nutgall and his five senses. Moreover, in his day the science of physiology, particularly in the field of digestion, was but an embryo. Thus, certain of his interpretations were erroneous, not because of an error in observation, but because the observations have since assumed a new meaning which either was not or could not have been recognized by Beaumont. Yet, even today, one marvels at the high degree of accuracy of his fifty-one inferences, or conclusions.

In inference 24, Beaumont states that gastric juice "contains free muriatic acid and some other active chemical principles." In another place he states, "I had been long convinced of the existence of free muriatic acid in the gastric fluids. Indeed, it is quite obvious to the sense of taste . . ." Yet, in some of his classic experiments on artificial digestion he added muriatic acid and acetic acid to water. In experiment 104, he used "four drachms of a mixture of dilute muriatic and acetic acids, reduced with water to the flavor of the gastric fluid, as nearly as practicable—three parts of the muriatic to one part of the acetic. . . ." Does this mean that Beaumont was able to detect by taste a difference between the free and total acidity of gastric fluid, or was he misled by the odor of the gastric contents of the bibulous St. Martin?

Beaumont's scientific method of conducting experiments proved that something is present in gastric juice other than acid which facilitates digestion. He showed that gastric juice digested meat at body temperature much better than at room temperature and that gastric juice was a better digestant than dilute muriatic acid. If he had boiled gastric juice, he would certainly have discovered pepsin, and perhaps rennin. However, his experiments on artificial digestion were far in advance of his day, and it appears that Beaumont's suggestion led to the discovery of pepsin in 1836 by Schwann, who refers to Beaumont.

The following passages briefly outline Beaumont's observations and opinions, which render him partly responsible for promulgating three erroneous views. There is nothing wrong with the observations; in fact, they are classic; only their interpretation is misleading. In inference 25, Beaumont states that gastric juice "is never found free in the gastric cavity; but is always excited to discharge itself by the introduction of food, or other irritants." In inference 14, he states that "hunger is the effect of distension of the vessels that secrete the gastric juice." In his discussion (pages 232 and 233) he states, "On applying aliment to the internal coat of the stomach . . . the orifices, undoubtedly, of the gastric vessels, immediately throw out a quantity of the fluid, which mixes with the food . . . when the stomach is destitute of food or some other irritating substance, no such secretion can be found in it. And it is more than probable . . . that a large quantity of this fluid must be contained in appropriate vessels, during a fast, ready to obey the call of aliment. I would not be understood to say that the whole quantity necessary for an ordinary meal is eliminated from the blood, previous to the commencement of alimentation; but that enough is contained in the gastric vessels to produce the sensation of pain or hunger." He points out that hunger distress is almost immediately relieved when a little food causes the juice in the gastric vessels to be extruded. Regarding the secretion of the fasting stomach he says, "The gastric juice never appears to be accumulated in the cavity of the stomach while fasting; and is seldom, if ever, discharged from its proper secerning vessels, except when excited by the natural stimulus of aliment, mechanical irritation of tubes, or other excitants" (page 262). Yet, the protocols of many of his experiments show that he obtained from one-half to two and one-half ounces of acid fluid from the fasting stomach. His interpretation (page 283) was: "I infer that the fluid, in these experiments, was incited to discharge itself by the irritation of the tube used in extracting it." In experiment 63 (page 326) is found the following interesting observation: "There was no free fluid in the gastric cavity, until after the elastic tube was introduced, when it began slowly to distill from the end of the tube, drop by drop, perfectly transparent, and distinctly acid. I obtained about one drachm of this kind, and then gave him a mouthful of bread to eat. No sooner had he swallowed it, than the fluid commenced flowing more freely from the tube, and I obtained two drachms, less pure, however, with saliva and mucus mixed with it, and slightly tinged with yellow bile. The surface of the protruded portion of the villous coat at this time became covered with a limpid fluid, uniformly spread over its whole surface, distilling from myriads of very fine papillary points, and trickling down the sides." Later he "introduced the tube, and obtained about two drachms more of very pure gastric juice." It is clear that Beaumont believed that hunger distress was due to distention of the gastric glands with stored gastric juice, that the secretion he found in the empty stomach was due to the mechanically-excited extrusion of the stored secretion, that contact of food with the gastric mucosa caused secretion by the chemical action of the food on the mucosa, and that the sudden response of the gastric glands to swallowed food was due to the chemical stimulation of the food in contact with the gastric mucosa. *In retrospect,* if Beaumont had emptied the stomach completely and then had in some instances collected for a half-hour the continuous secretion of the stomach and then had manipulated the mucosa with his thermometer, he would have found no sudden discharge or mechanical stimulation of secretion. If he had distended the stomach, he probably would have found that mechanical distension, the only significant and adequate mechanical stimulus, would stimulate. Having found that gastric juice was not stored in the gastric glands, he would not have proposed his theory of hunger. Also, if he had told St. Martin to take the bread in his mouth, and chew but not swallow it, he would have discovered the psychic secretion of gastric juice, which he actually observed without realizing it. Further, if Beaumont had associated the pain in the pit of the stomach, which he produced when he caused St. Martin's stomach to contract by irritation with the thermometer bulb, with the pain of hunger, he would probably have recognized that hunger pain was due to gastric contractions. Beaumont's idea that the gastric juice secreted in fasting was stored in the gastric glands caused him to misinterpret hunger distress, to believe that the empty stomach does not secrete acid, and perhaps to miss the discovery of psychic secretion or the effect of nerves on secretion. In his day, of course, the concept of secretory nerves had not arrived. The existence of secretory nerves was proved by Ludwig in 1851, and the concept was prevalent when Bidder and Schmidt in 1852 first observed a psychic secretion of gastric juice.

Beaumont unequivocally distinguished between mechanical and chemical stimulation of the gastric glands. He discovered that certain foods act as chemical excitants of the gastric glands, and he demonstrated unequivocally that oily food is difficult to digest. He showed, but was not the first to do so, that bile aids the emulsification of fat. He knew that bile was alkaline, but suspected that pancreatic juice (page 299) was slightly acid. He believed that saliva has no digestive action, that it only lubricates. This view was probably based on the fact that he added saliva to gastric juice, the acidity of which inactivated the ptyalin, and being on the frontier, he was not acquainted with the discovery of Leuchs (1831) that saliva liquefies and decomposes cooked starches.

Most of Beaumont's experiments on the rate of digestion are classics. However, he could not distinguish between what we now call enzymatic digestion and chymification. He proved that vegetables, especially large pieces of raw or cooked vegetables, were more difficult of digestion, or rather chymification, than meats and farinaceous foods. Because he did not know of the digestive action of saliva on cooked starch he surmised, on the basis of filtration and sedimentation, that gastric juice dissolved starches. If chymification is substituted for digestion in Beaumont's statements regarding the changes the minutely divided vegetables underwent in his experiments, little objection can be made to his interpretations.

The movements of the pyloric portion of the stomach were remarkably well described by Beaumont when it is considered that he had to infer what was happening by feeling through the aid of his thermometer. In regard to the movements of the fundus of the stomach, he described a rotary motion of the contents, which he said was not "very exact," the contents being "intimately mixed and circulating promiscuously." His emphasis on the rotary motion (page 265) was probably due to the pathologic fixation of St. Martin's stomach. Yet, he observed that the movements of the fundus increase as digestion progresses; and, in his conclusions, he states correctly that "the movements of the stomach produce a constant churning of its contents" and that "these motions are in two directions—transversely and longitudinally."

In regard to the effect of emotions, overindulgence in food and drink, and febrile conditions on the appearance of the stomach, every modern gastroscopist would profit by reading Beaumont's descriptions of the gastric mucosa as affected by these conditions. In this connection a statement made by Beaumont should be quoted (page 348) for emphasis, "It is interesting to observe to what extent the stomach, perhaps the most important organ of the *animal* system, may become diseased, without manifesting any external symptoms of such disease, or any evident signs of functional aberration."

One truly does not know gastric physiology and pathology as one should until Beaumont has been read critically and *in toto.* It should be emphasized that many important observations are not mentioned in his list of fifty-one inferences. For example, among his observations, he records that hot, humid weather depresses gastric secretion, that the temperature of the pyloric antrum is higher than that of the fundus, that too much cooking hardens the food so that it is not easily digested, that too much condiment irritates the gastric mucosa, that the appearance of the tongue is an indicator of the healthy or unhealthy state of the stomach, that the mechanical action of a muslin bag containing food and prevented from passing the pylorus by means of a string irritates the gastric mucosa, that the cardiac sphincter visibly opens and closes after swallowing, and the idea that thirst "is felt in the mouth and fauces" and "the proximate cause may exist in an irritation, a kind of sub-inflammation of the mucous membranes of the mouth and fauces, the effect of the viscid state of the blood, and consequently impervious state of the secretory vessels of these membranes." These and many other important facts and ideas may be gleaned only by reading all of Beaumont.

Beaumont's observations not only provided many important facts but they suggested a new method of approach to the study of gastric physiology and directly and indirectly stimulated a tremendous amount of investigation.

At the meeting of the International Congress of Physiologists in Boston in 1929, each guest was presented with a medallion bearing the likeness of Beaumont and a facsimile of the original edition of his book. William Beaumont, the frontier physician and physiologist, on this occasion, was figuratively canonized as the patron saint of American Physiology. Modern American physiologists can pay no greater honor and respect to any man.

PREFACE

Jesse S. Myer St. Louis, June 1912

This biographic sketch of William Beaumont is presented on the hundredth anniversary of his entry into the practice of medicine. Just a century ago—June 9, 1812—he was granted a license by the Third Medical Society of the State of Vermont to practice "physic and surgery," and a decade later—June 6, 1822—destiny let fall in his path the opportunity which he recognized, grasped, and improved with a zeal and an unselfishness not excelled in the annals of medical science. Ever since that time the story of Dr. Beaumont and the case of Alexis St. Martin has had a peculiar fascination for persons in various walks of life—from the schoolboy, who gains his first knowledge of the incident from his *Essentials of Physiology,* to the scientist, who, by comparison of Beaumont's work with the most recent developments in the physiology of digestion, marvels at the thoroughness of this remarkable piece of research, conducted by a young army officer under the most discouraging circumstances.

Though there have been a number of brief biographic sketches of Beaumont presented in the form of addresses on various occasions, they have dealt with the man largely from the standpoint of his contribution to medical science, under the general assumption, no doubt, that little was to be learned of other phases and stages of his life. This impression was likewise shared by me until through my studies the various chapters of his life began to unfold, when my greatest difficulty was, not how to obtain sufficient material for his biography, but how to condense into a volume of reasonable size the mass of facts gathered from hitherto unexplored sources. The investigation once begun, hidden treasures of information were revealed without number.

The nucleus of this work was formed from the contents of two old chests found in possession of Mrs. Sarah Beaumont Keim, who at the age of four score years and ten still lives in the city of St. Louis, where her father, Dr. Beaumont, spent the last twenty years of his life and where his remains lie buried. To her my profound thanks are due for the great privilege of delving into those old records of the past. These strong boxes proved to be a veritable storehouse of biographic material in the form of documents, manuscripts, memoranda, diaries, letters, clippings, books, etc. Beaumont was so methodical throughout his whole life that he had treasured every scrap of paper which seemed to him of the slightest importance, and preserved not only all letters received, but also a rough draft in his handwriting of those sent by him, a fact which renders the material of unusual value and importance from an historic standpoint.

It would seem ostentatious on my part to enter into any extended recital of the difficulties encountered in deciphering, abstracting, and copying this vast material, or to enumerate the long list of inquiries that is was necessary for me to make in order to secure a complete chronological arrangement of the data. It is sufficient to say that in the preparation of this work several hundred letters were exchanged with many individuals in various sections of the United States in which Beaumont at some time lived and conducted his activities. Librarians, town clerks, postmasters, Beaumont descendants, and the descendants of those who through association seem to have exerted an influence on his character, received requests from me to aid in the investigation of such sources of information as were open to them, and almost invariably these appeals met with replies containing either the definite information desired, or the suggestion of sources from which it might be obtained. To these persons, without whose aid my work would have been incomplete, I wish to extend my sincere thanks, and at the same time express my regrets that space will not permit an enumeration of all their names and contributions. During the extensive correspondence, which proved so important in the evolution of this biography, I often recalled to mind Boswell's statement in his preface to *The Life of Samuel Johnson:* "I was sometimes obliged to run half over London in order to fix a date correctly; which, when I had accomplished, I well knew could obtain me no praise, though a failure would have been to my discredit."

It was interesting to see the data thus gathered from far and near fall into place as do the stones in an arch. Even replies of a negative character were of positive value, for the biographer gains a peculiar satisfaction from the knowledge that every promising thread of information was at least followed to its end. For example, it is of the greatest importance to have learned from officials that there are no records bearing on Beaumont to be found at Plattsburgh Barracks, his rendezvous in the War of 1812; at Fort Mackinac, where he had St. Martin under treatment from 1822 to 1825; at Jefferson Barracks and the arsenal at St. Louis, the scenes of his activities from 1835 to 1840; and among such archives as those of the Wisconsin Historical Library at Madison, the Kellogg Public Library at Green Bay, Wis., etc.

To Surgeon-General Torney and his capable associates in the Surgeon-General's office and in the Surgeon-General's Library I wish to extend my sincere thanks for many acts of kindness. Through their courtesy I gained personal access to archives including many volumes of old ledgers containing abstracts of letters received at and sent from the Surgeon-General's office from 1812 to 1840, the period during which Beaumont was an officer in the army, from which many important dates and interesting facts were gleaned. In addition to this, through the remarkable system maintained in his department, the Surgeon-General succeeded in bringing to light for me a number of most important letters, which had for years been stored in the basement of the Surgeon-General's Library. I can not say too much in recognition of the courtesy and substantial aid extended me by Lieutenant-Colonel J. R. Kean and every other member of this department with whom it was my good fortune to come in contact.

It gives me pleasure to say a word in commendation of the genial New England town clerks, loyal custodians of those priceless volumes—the New England town records—on which is based much of the early history of our country. I am especially indebted to them for the opportunities which enabled me to collect most of the material contained in the chapter on Genealogy, and I shall always recall with pleasant recollections the moments spent with them in an investigation of the old town records, dating back as far as 1680. This chapter was undertaken not only with a view to completeness, but also with a desire to determine, if possible, whether heredity played any part in accounting for William Beaumont's unusual powers of keen observation and tenacious perseverance.

Sir William Osler, whose Biographical Essays are an inspiration to every student of American medical history, has done me the honor to write an introduction to this book, for which favor I desire to express my grateful appreciation.

My thanks are due to Miss Helen S. Bixby, of Plattsburgh, N. Y., for her careful and successful search through the files of the Plattsburgh *Republican* from 1811 to 1835, the results of which are evident in the early chapters of this book. I am likewise indebted to Mrs. George Fuller Tuttle, of Plattsburgh, author of *Three Centuries in Champlain Valley,* for valuable data; to Mrs. C. C. Bishop, of Dorchester, Mass., for information concerning her grandfather, Dr. Beaumont's medical preceptor; to Mrs. Lydia Brown, of Hartford, Conn., for valuable letters throwing light on his early life; and to many others too numerous to mention here for aid in collecting material relating to the life of Dr. Beaumont. To those of my friends who read portions of my manuscript and made recommendations I wish to acknowledge indebtedness for the interest manifested.

Many interesting anecdotes concerning Dr. Beaumont have come to me in the course of this work, but, when compared with absolutely authentic data in my hands, I found so many discrepancies in them that I determined to discard all that could not be verified, and to confine my narrative to such statements as could be substantiated by documentary evidence.

As the same remark that has been applied to most men who have accomplished great deeds has also been made concerning Beaumont—that he was only a man of destiny, that mere accident brought to him a certain opportunity, that another in the same position would have accomplished similar results—I have added in the appendix a list of literature references and abstracts of cases of gastric fistulæ that occurred prior to the case of St. Martin. Though some of these cases fell into the hands of true scientists in large medical centers, where every facility was afforded for investigation, not one case was systematically utilized to increase our then meager knowledge of the physiology of digestion. No comment is therefore necessary to make conspicuous by comparison the indefatigable efforts of Dr. Beaumont in his research.

In conclusion it may be stated that it has not been my endeavor to write a popular biography, which so frequently necessitates sacrificing fact for style. I have therefore avoided altering the phraseology of Dr. Beaumont, and have presented intact such parts of his letters and papers as are used in this book. The peculiar composition, orthography, and capitalization of Dr. Beaumont's letters and memoranda have been followed in order to retain his characteristic style, and only so much of the punctuation has been corrected as seemed necessary to aid the reader in more readily comprehending the ideas intended to be conveyed. This desire to retain the writer's individuality has been followed also in the letters written by others to Dr. Beaumont. Preserving Dr. Beaumont's diction gives the reader his mannerisms of speech, his unique and forceful modes of expression, and at the same time presents a concept of the man's traits and character which could not possibly be conveyed by the words of another. In this I again observe Mr. Boswell's idea when he said: "Indeed I can not conceive a more perfect mode of writing any man's life, than not only relating all the most important events of it in their order, but interweaving what he privately wrote, and said, and thought; by which mankind are enabled as it were to see him live, and to 'live o'er each scene' with him, as he actually advanced through the several stages of his life. . . . And he will be seen as he really was, for I profess to write, not his panegyrick, which must be all praise, but his life; which, great and good as he was, must not be supposed to be entirely perfect. . . ."

The University of Chicago

William Beaumont was born in Lebanon, Connecticut, in this house on November 21, 1785.

LIFE AND LETTERS OF

DR. WILLIAM BEAUMONT

1822-1828

EARLY IN THE MONTH of June, 1822, Indians and voyageurs were returning to Mackinac with the results of their winter's catch. The little village had awakened from its long sleep, and the beach was again crowded with tents and wigwams and a seething mass of strange humanity. New arrivals of canoes and bateaux were being heralded, and friends who had been stationed far apart in the wilds of the North were familiarly greeting one another. Some were pitching tents in which to sleep when not otherwise engaged in carousing; newer arrivals were unpacking pelts, watching their appraisal by the officers of the fur company, and eagerly awaiting the figures that were to indicate the results of their winter's work; others, whose fate had already been decided, were engaged in games or watching the fight of two of the brigade bullies for the proverbial "black feather"; others still were crowding into the retail store of the American Fur Company in an effort to buy buckskin coats, moccasins, flannel shirts, and gaudy neck bands. It was in this little throng that a tragedy occurred on June 6th which was to leave its imprint on the pages of medical history for all time to come. A gun was accidentally discharged, and a young voyageur dropped to the floor, with a cavity in the left upper abdomen that would have admitted a man's fist. He proved to be a young French Canadian about 19 years of age, who had recently come down from Montreal, doubtless with one of the expeditions of Mr. Matthews.

Harvard Medical School

Alexis St. Martin was wounded in the basement of this house. It was then the retail store of the American Fur Company. The upper part of the old structure has been replaced. It stands at the foot of the hill below the fort and was entered by a door behind the lattice work.

Gurdon S. Hubbard[1] was an eyewitness of this accident, and the only one who has left any statement concerning the affair. He says:

The late Major John H. Kinzie[2] had charge of the American Fur Company's retail store at Michilimackinac. I was in the habit of assisting him occasionally when a press of customers needed extra clerks. The store comprised the ground floor near the foot of Fort Hill, on the corner of the street and the road leading up to the fort. The rear part of the store was underground, built of stone, which is still standing.

This St. Martin was at the time one of the American Fur Company's engagees, who, with quite a number of others, was in the store. One of the

[1]*Incidents and Events in the Life of Gurdon Saltonstall Hubbard,* collected from personal narrations and other sources, and arranged by his nephew, Henry E. Hamilton, 1888.

[2]Gurdon S. Hubbard and John Kinzie remained lifelong friends, and at the time of their death were among the most prominent citizens in the city of Chicago. Chicago was incorporated in 1835, with Mr. Hubbard and Mr. Kinzie as two of the five trustees. Mr. Hubbard was for many years the largest pork packer in the West, director of the Chicago branch of the State Bank of Illinois, and one of the organizers of the Chicago Board of Trade. The firm of Hubbard & Co., commission merchants, was for many years one of the largest in the country, and established the Eagle line of steamers between Buffalo and the Upper Lakes. After the formation of the Republican party he became one of Mr. Lincoln's warmest supporters for the presidency. (Biographical Sketch, Chicago Historical Society, 1907.)

party was holding a shotgun (not a musket), which was accidentally discharged, the whole charge entering St. Martin's body. The muzzle was not over three feet from him—I think not over two. The wadding entered, as well as pieces of his clothing; his shirt took fire; he fell, as we supposed, dead.

Dr. Beaumont, the surgeon of the fort, was immediately sent for, and reached the wounded man within a very short time—probably three minutes. We had just got him on a cot and were taking off some of his clothing.

After Dr. Beaumont had extracted part of the shot, pieces of clothing, and dressed his wound carefully, Robert Stewart and others assisting, he left him, remarking, "The man can't live thirty-six hours; I will come to see him by and by." In two or three hours he visited him again, expressing surprise at finding him doing better than he anticipated. The next day, I think, he resolved on a course of treatment, and brought down his instruments, getting out more shot and clothing, cutting off ragged ends of the wound, and made frequent visits, seeming very much interested, informing Mr. Stewart in my presence that he thought he could save him.

As soon as the man could be moved he was taken to the fort hospital, where Dr. Beaumont could give him better attention. About this time, if I am not greatly mistaken, the doctor announced that he was treating his patient with a view to experimenting on his stomach, being satisfied of his recovery. You know the result.

I knew Dr. Beaumont very well. The experiment of introducing food into the stomach through the orifice, purposely kept open and healed with that object, was conceived by the doctor very soon after the first examination.

Beaumont's own version of the accident was summed up in the following statement:

Being then stationed at the garrison on that island, and no citizen within 300 miles, I was immediately called to his relief, and when I arrived at this place 15 or 20 minutes after the accident found him senseless and apparently in a moribund state. To endeavor to resuscitate and relieve his sufferings being the first impulse, I applied the best means in my power and took the responsibility of his surgical treatment and medical attendance. . . .

The most complete description of the wound received by St. Martin, the treatment and progress of his case, is exhibited in Beaumont's hospital records and bedside notes, extending over a period of two years. This record was found in a journal of various cases under his observation from 1822 to 1824, and bears the title, "Sundry Cases of Practice—San Martin, St. Andre, Sheridan, etc., etc." All of the patients received the same careful consideration as did St. Martin. St. Andre's case was one of "great swelling of his belly," found to be a "confirmed case of ascites." The case of Sheridan was that of a deserter from the post of Mackinac, who "got lost and wandered about for seven days without either food, drink, fire, or bed clothes, the thermometer ranging between 20 degrees above and 14 degrees below zero," while the case of St. Martin was to him at this time simply one of "wounded stomach," calling for his best efforts to bring about a complete cure.

It is interesting to note that he had considerable difficulty in understanding the lad's patois, as is evidenced by various attempts to get his name correctly. The record, kept over a period of two years, is a model of thoroughness and completeness that would be a credit to our modern-day hospitals, and is worthy of being reproduced here in full:

> Alex Samata, St. Martin, San Maten, a Canadian lad about 19 years old,[3] hardy, robust and healthy, was accidentally shot by the unlucky discharge of a gun on the 6th of June, 1822. The whole charge, consisting of powder and duck shot, was received in the left side at not more than 2 or 3 feet distance from the muzzle of the piece, in a posterior direction, obliquely forward and outwards, carrying away by its force the integuments more than the size of the palm of a man's hand; blowing off and fracturing the 6th rib from about the middle anteriorily, fracturing the 5th, Rupturing the lower portion of the left lobe of the Lungs, and lacerating the Stomach by a spicula of the rib that was blown through it[s] coat, Lodging the charge, wadding, fire in among the fractured ribs and lacerated muscles and integuments, and burning the clothing and flesh to a crisp. I was called to him immediately after the accident. Found a portion of the Lungs as large as a turkey's egg protruding through the external wound, lacerated and burnt, and below this another protrusion resembling a portion of the Stomach, what at first view I could not believe possible to be that organ in that situation with the subject surviving, but on closer examination I found it to be actually the Stomach, with a puncture in the protruding portion large enough to receive my forefinger, and through which a portion of his food that he had taken for breakfast had come out and lodged among his apparel. In this dilemma I considered any attempt to save his life entirely useless. But as I had ever considered it a duty to use every means in my power to preserve life when called to administer relief, I proceeded to cleanse the wound and give it a superficial dressing, not believing it possible for him to survive twenty minutes. On attempting to reduce the protruding portions, I found the Lung was prevented from returning by the sharp point of the fractured rib, over which its membrane had caught fast, but by raising up the Lung with the front of the forefinger of my left hand I clipped off with my penknife, in my right hand, the sharp point of the rib,

[3]According to the statement of Rev. Mr. Chicoine, curé of St. Thomas ("William Beaumont, a Pioneer American Physiologist," *Journal American Medical Association,* November 15, 1902, Osler), St. Martin died in 1880 at 83 years of age. If this be true, he was born in 1797, and must have been 25 years old when the accident occurred.

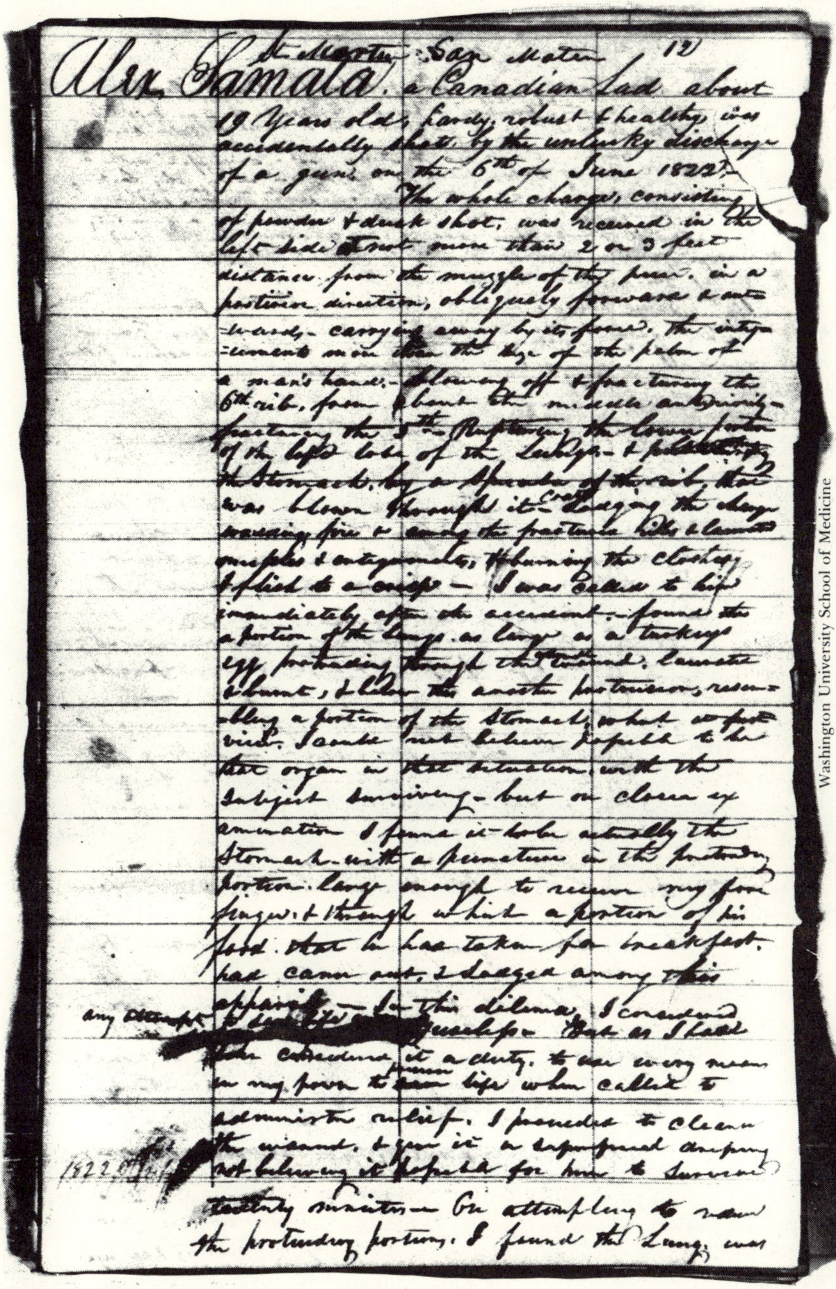

The first of seventeen pages of the elaborate hospital record of St. Martin's case kept by Beaumont.

which enabled me to return the Lung into the cavity of the Thorax, but could not retain it there on the least efforts of the patient to cough, which were frequent.

After giving the wound a superficial dressing, the patient was moved to a more convenient place, and in about an hour I attended to dressing the wound more thoroughly, not supposing it probable for him to survive the operation of extracting the fractured spicula of bones and other extraneous substances, but to the utter astonishment of every one he bore it without a struggle or without sinking.

The University of Chicago

The one-story frame hospital at the old fort to which Alexis was removed and where he lay during his long convalescence.

After taking away the fragments of the ribs, old flannel, wad and the principal charge of shot, all driven together under the skin and into the muscles, and replacing the lungs and stomach as much as practicable, I applied to the wound the carbonated fermenting poultice, composed of flour, hot water, charcoal, and yeast, changing once every 8, 10, or 12 hours, according to the quicker or slower process of fermentation, keeping the parts around constantly bathed with a solution of muriate or ammonia in Spirits and vinegar. This was done with an intention to excite local reaction as soon as possible upon the surface and occasional sloughing of contused, lacerated and burnt muscles and integuments, which had the desired effect in less than 84 hours, with assistance of the Camphorated Aq. Amon. Acet. given internally in liberal quantities. Under the above treatment a lively reaction commenced in about 24 hours, accompanied with strong arterial action and high inflammatory symptoms of the system generally, more specially of violent pneumonia and inflammation of the Lungs, with great dyspnoea and distressing cough. At the commencement of those symptoms I opened a vein and took 12 or 14 oz. of blood from the arm. Gave a mild cathartic. The bleeding abated the action and gave relief. The cathc. had no effect, having escaped from the stomach through the wound. I continued the Caphd. Aq. Acetat. every hour for the first 72 internally and the carbonated poultice and wash externally, omitting the muriate. The fever continued for 8 or ten days, running into the Typhoid type and the wound becoming very fetid. Nothing passed his bowels after the 2d day, and they became impervious and inactive, scarcely to be excited by stimulating injections. From the 2d day till the 10th nothing passed, no reaction from his bowels at all, everything he took into his stomach was either absorbed or made its exit at the wound externally.

About the 5th day a partial sloughing took place in the wound and the febrile symptoms abated. The protruded portion of the lungs and the small lacerated piece of the stomach also sloughed off, and left the puncture of the Stomach plain to be seen, and large enough to admit my forefinger its whole length directly into the cavity of the stomach, and a passage into the cavity of the Thorax half as large as my fist, exposing to plain view the action of the left lobe of the Lungs, and admitting at every respiration full escape of air, bloody mucus, froth, etc.

About the 10th day a more extensive sloughing took place. The febrile symptoms all subsided, and the whole surface of the wound put on a healthy, granulating appearance. The fractures of the ribs commenced exfoliating, and nature kindly performing what human foresight viewed as hopeless and professional skill might calculate upon with dubious odds. All that entered his stomach came out again at the wound for 12 or 15 days, and the only means of sustaining him was by nutritious injection given per anus until all was sloughed, and compresses and adhesive strap could be applied to stop the orifice in the stomach and retain the food and drinks.

A lucky and perhaps the only circumstance to which his miraculous survival can be attributed was the protruded portion of the Stomach, instead of falling back into the cavity of the abdomen to its natural position, adhered by the first intention to the intercostal muscles, and by that means retained the orifice in the wounded stomach in contact with the external wound, and afforded a free passage out and a fair opportunity to apply the dressings. The carbon poultice was continued constantly until the sloughing was complete and the granulating process established. They were afterwards occasionally applied as a corrective when the wound was becoming ill conditioned or languid. The Aq. Am. Acetat. was continued for several weeks, in proportion to the febrile symptoms or fetid condition of the wound.

No sickness or peculiar irritability of the Stomach was ever experienced, not even nausea, during the whole time; and after 3 weeks the appetite regular and healthy, alvine evacuation became regular, and all the functions of the system seemed as regular and healthy as in perfect health, excepting the wounded parts.

Cicatrization and contraction of the external wound commenced about the 5th week, and continued gradually and almost uninterruptedly. The Stomach at the wounded part became more and more firmly attached to the intercostals by its external coats, but showed not the least disposition to close its puncture by granulations forming from its own lacerated coats any more than is in the anus or mouth. By applying the nitrate of silver to the edges of the wounded muscles of the stomach, I could extend the attachments by its external and cellular coats more firmly to the intercostals or external integuments, which seemed rather to enlarge than contract the orifice in the Stomach, bringing it more and more external as its adhesions to the external wound increased, resembling in its appearance (all but the Sphincter) a natural anus, with a slight prolapse every time I removed the dressings, and the contents of the Stomach would run out fairly in proportion to the quantity received. If the Stomach happened to be empty when I dressed it, a prolapse and partial invertion of the Stomach would follow the removal of the compresses of lints from the wound, unless prevented by the application of my thumb, finger, or something else to its orifice while the dressing was reapplying. Frequently upon removing the dressings, when they had become looser than usual by some derangement of the bandages, the stomach would be inverted and the inner coats protruded through the orifice large as a hen's egg. No difficulty occurred in reducing it; a gentle pressure with the thumb or finger upon the protruded portion would quickly return it to its place without giving the least pain and almost without sensation. Nitrate of silver, applied to the lips of the wound and upon the inner coats of the Stomach so as to produce sloughing, occasioned less sensation than when applied to the most common fungus or ulcer upon the surface of the body or limbs, a conclusive evidence in my opinion that the Stomach is not so exquisitely sensible an organ as is represented by anatomists and Physiologists in general.

About the 6th or 7th week exfoliation from the fractured ribs and the separation of the ribs from the cartilagenous ends began to take place; spiculae of bones and fragments of old cloth and shot also were working out from among the muscles and integuments. The 6th rib, that was worst injured and blown off entirely in the first place, was also abraded of its periostium for about three inches back of the fracture towards the Spine and became carious at its fractured extremity. So that I was obliged to amputate it about midway between sternum and spine, which I did by dissecting around, separating and retracting the intercostals to the sound portion of the rib, and then sawing it off by introducing between the ribs a very narrow, short saw, which I had made for the occasion. In this operation I succeeded admirably, beyond my most sanguine expectations, taking the rib off smoothly without injuring any parts whatever. The granulations shot immediately out and formed soundly over the amputated end. About half of the interior edge of the other rib exfoliated longitudinally from about the center to the Sternum, and then the healthy granulation formed soundly over the other part and continued so.

After removing all the exfoliations and extraneous substances that were to be found about the wound, my next object was to contract the external wound and close up the puncture in the stomach if practicable, which I attempted by drawing the edges of the wound as near together as I could by adhesive straps laid on in radiative form, the circumference of the external wound being at least 12 or 14 inches, the orifice in the Stomach about in the center. To retain the food and drink as much as possible, I kept to the orifice a firm compress of lint, fitted to the shape and size of the puncture, and confined by the straps. Under these dressings and management cicatrization went on rapidly, his health improving, and all functions of the system regular. Digestion was as completely performed as in the most healthy person in the vicinity (and I could even see it go on every time I dressed the wound). I kept the granulating surface duly stimulated by applying sometimes Cincona pulv., sometimes Mirc. precip. Rub., and sometimes Nit. Silver, as the state of the granulation indicated.

After trying every means within my power to close the puncture of the Stomach by exciting adhesions between the lips of the wound of its own proper coats, without the least appearance of success, I gave over trying, convinced that the Stomach of itself will not close a puncture in its coats by granulations, and the only alternative left seemed to be to draw the external wound together as fast as cicatrization would form and contracting as much as possible the orifice in the Stomach, and make the granulations from the intercostal muscles and integuments shoot across the form over and close it that way. But to this method there seemed an insuperable difficulty, for, unless there be kept constantly upon the orifice a firm plug of lint compound, all the contents of the Stomach flow out and the patient must die for want of aliment, and this lint, intersepting, prevents the granulation from forming across.

The lacerated portion of the lungs sloughed off and digested away, leaving a surface suppurating in the lobe of the lungs large as the concave surface of a teacup, from which continued to issue much purulent matter for two or three months until it became completely filled up with healthy granulations and cicatrized over externally, with the lower portion of the left lobe firmly adhering to the Pleura costalis. Four months after the injury an abscess formed about two inches below the wound, anteriorly, over the cartilaginous ends of the 1st and 2d false ribs, very painful and extremely sore, producing a violent symptomatic fever, checking the process of cicatrization, rendering the granulation languid and pale, and the wound ill-conditioned and unfavorable to the prospects of recovery. After applying emolient poultices for several days, the swelling pointed externally, and I punctured and laid it open with the bistoury and director for about 2 or 3 inches. It discharged copiously very fetid, purulent matter for the first 2 days. On the 3d I could feel with the probe a small extraneous substance, which in the course of 3 or 4 days, by the use of the soap plaster and compresses, proved to be a shot and a small portion of the wad. After the exit of these I could introduce a common pocket-case silver probe nearly its whole length in the longitudinal direction of the ribs, and a great soreness and pain extended from the opening in the abscess upon the track of the cartilaginous ends of the false ribs to the spine, with a copious discharge from a long fistulous sinus. In the course of about 5 or 6 days [appeared] the cartilaginous end of a rib about an inch long; soon after this followed some small spicula of bone. The discharge, soreness and inflammation continued in the same direction. In about 6 or 7 days longer came away another similar cartilage about an inch and a half long, and in about the same length of time another 2 inches, and so continued to come away every 5 or 6 days, increasing in length in about the same proportion until five had made their exit through the same passage. The last was about three inches long, and seemed to be separated from the last false rib, as the soreness terminated at that point, and after which the sinus commenced closing, the discharge diminished, and the soreness subsided from that point forward with regular progression. The discharge, pain and irritation during the 4 or 5 weeks all those cartilages were working out reduced the strength of the patient very much, induced a general febrile habit, and stopped the healthy healing process of the original wound. Directly after exit of the last mentioned cartilage an inflammation appeared at the lower end of the sternum, about over the ensiform cartilage, from the anterior end of the original wound, extremely irritable and very painful. By the use of emolient poultice a few days it terminated in a large abscess, which I punctured and laid open an inch or two with the bistoury. About half a pint of very offensive matter discharged from this, and in a few days followed a cartilaginous substance about 3 inches long, after which the inflammation subsided. In a day or two after this came away another small cartilage and the discharge abated. To support the patient's strength under all these debilitating incidents, I gave him the diluted muriatic acid and wine, which very much improved his health and increased his strength.

It is now going on the 7th month since the injury was received, and the orifice in the stomach is still visible and but little contracted. The integuments are all cicatrized, smooth to within the circumference of a half eagle, immediately around the wound in the Stomach. His health daily improving, his spirits good, his appetite regular, his sleep refreshing, and all the functions of the system natural and healthy.

DEC. 2, 1822. Can the puncture of the Stomach be successfully closed by mechanical means until the granulations have time to form over and across it?

DEC. 3d. Omitted dressing wound myself, though dressed by the nurse.

DEC. 4th. On dressing today found a portion of the new cutis vera at the anterior extremity of the cicatrix raised up and separated from the pectoral muscle by the matter forced back from the sinus above in consequence of an obstruction in its usual outlet. On puncturing and laying it open about an inch and a half, found a small piece of cartilage loose and took it away and a large piece which seemed to be separating from the Sternum.

DEC. 5th. Dressed it today. Found it in better condition than yesterday. Inflammation and soreness subsiding, discharge less.

DEC. 6th. Dressed today. Still doing well.

DEC. 7th. Dressed. Still improving.

DEC. 8th. Dressed. Still improving.

DEC. 9th. Dressed. Original wound diminishing, sinuses discharge less, condition better.

DEC. 10th. Dressed. Condition of the wound, sinuses, etc., improving. Keep them dressed with cincona pulv., dry lint and adhesive straps, bathing the parts around with comphorated Spts. and water, and continue the muriatic acid.

DEC. 11th. Dressed. Improving, in healthy condition.

DEC. 12th. Dressed. Sores lessening in circumference and filling up with granulations. All healthy, except the one in front of the Sternum, which seems to have something extraneous behind yet—a piece of cartilage no doubt.

DEC. 13th. Dressed. To facilitate the closing of the puncture of the stomach, and remove the impediments to the granulations, which the constant pressure of compresses necessary to retain the food in the Stomach from day to day cause, I made fast a piece of lint to a ligature, just large enough to pass through the orifice into the Stomach, and then with the end of the probe pushed it on the inside, and, suspended by the ligature, drew it up against the inside of orifice, so as to stop the food, etc., from flowing out upon the inner side, and by this means suffering the granulations from the edges of the wound to contract nearer each other around the small ligature.

DEC. 14th. Dressed in the same manner as yesterday.

DEC. 15. Dressed as yesterday, excepting the introduction of a small silver canula introduced instead of the ligature and lint.

DEC. 16. Dressed as yesterday. Find the canula will not answer the purpose so well as the plug of lint suspended by the string.

DEC. 17th. Removed the canula and dressed as usual.

DEC. 22d. The above manner of dressing continued without any variation since the 17th. Continued the wash around the wound of Camphorated Spirits, water and vinegar, injecting frequently into the sinuses the same wash more diluted.

DEC. 23d. Added to the Muriatic acid, diluted, about one-third of the tincture of ______, to be taken three times a day in doses of a half teaspoonful. The discharge still purulent and healthy, and diminishing in quantity. The orifice in the Stomach remaining about the same as it has been for 3 months, tho the wounds of the muscles continue to contract and are fast cicatrizing upon the Stomach. The food still makes its exit whenever the dressings are removed.

JANY. 3d, 1823. Dressed every day since the 23d last mo., condition continuing nearly the same until, 3 or 4 days, an inflammatory redness appeared around the lower part of the Sternum, with considerable heat and some swelling, and today appeared another cartilage, which I took out—about one inch and a half long.

JAN. 6th. Swelling and inflammation subsiding. Took away at the opening of the sinus another cartilage 2 and half inch in length.

JAN. 10th. Since extracting the last cartilage the sinus has closed and the discharge ceased, excepting a very little from the sternal end, where some small pieces of diseased cartilage still remain undetached and keep up a small discharge.

JAN. 19th. Small piece of cartilage came from the ulcer on the sternum; other sinuses all closed; orifice in the Stomach about the same as ever.

FEB. 1st. Dressed every day since last date; all the ulcers and sinuses closed sound; the patient in perfect health; the orifice in the stomach in Statu quo.

MARCH 10th. Dressed daily since last date; much in same condition; orifice in the stomach a little diminished.

MARCH 25th. Dressed daily since last date; no alterative in condition; orifice in the stomach the same.

APL. 1st. Dressed daily; continues in same condition; general health good.

MAY 30th. Continued daily dressing as usual since last date; no essential variation in the wound; health good until within a few days past. Having complained of pain in the head, nausea, a cathartic of Rhei and sulphur administered, it is presumed, as never medicine was before administered to man since the creation of the world—to wit, by pouring it in through the ribs at the puncture into the stomach. I administered it in the form of dry powder. It occasioned a slight nausea in less than 10 minutes and operated briskly as a cathc. in less than two hours.

MAY 31st. Feels relieved by the operation of the medicine given yesterday. Administered in the same manner today 11 oz. manna, 1 oz. magnesia.

The County refusing any further assistance to the patient (who has become a pauper from his misfortune), I took him into my own family from mere motives of charity and a disposition to save his life, or at least to make him comfortable, where he has continued improving in health and condition, and is now able to perform any kind of labour from the whitling of a stick to the chopping of Logs, and is as healthy, active and strong as he ever was in his life, or any man in Mackinac, with the apperture of the Stomach in much the same condition as it was at the last mentioned date. June 1, 1824.

In a memorandum made in the preparation of a memorial to Congress, I find a more exhaustive statement concerning the attitude of the community toward poor Alexis:

Being a poor boy, without relation or friends at the time of the accident, he became a pauper upon the Town and County of Mackinac. But their charity fund being small, soon became exhausted, and, the people not being able to raise any more, he was left entirely destitute of every means of comfort, and even of a place to rest his wounded body or sustain his suffering, debilitated nature, and in this helpless condition the *Benevolent* authorities of the Town were about to pack him off in an open batteau to his native place, Montreal, L. C., a distance of nearly 2,000 miles. Repeated remonstrances against such an inhuman disposal of a person in his condition being of no avail in arresting the execution of the design, and knowing that his life must inevitably be sacrificed in such an attempt, I resolved to rescue him from the pending misery and death. I took him into my own family in April, 1823, at a time when he was helpless, sick and suffering under the debilitating effects of his wounds, naked and destitude of everything, but pain, a little "breath of life," and a wounded body. In this condition he remained with me, gradually improving, for a year or two, when he became able to walk about and help himself a little, but unable to provide for his own necessities. During this time I nursed him, fed him, clothed him, lodged him and furnished him with every comfort, and dressed his wounds daily and for the most part twice a day.

Let us not forget, in this connection, that the salary of an assistant surgeon at this period was the munificent sum of $40 a month and two to four rations daily, and that Beaumont, in addition to supplying out of this sum his own needs and those of Alexis, had a family to support.

The inhuman attitude of the citizens of Mackinac on this occasion was doubtless responsible for the sarcastic soliloquy on Charity, which appears in his journal, though not immediately following the case of St. Martin. It was his custom, in fact, to jot down his ideas wherever and whenever they occurred to him; not infrequently, therefore, we find kindred thoughts widely separated from one another, the relationship remaining none the less certain.

CHARITY.

Happy are they that die in the poor-house of this place, and three times doubly happy are the people of the Borough who reluctantly contribute to their wretched support while living upon the public charity! Were I reduced to the necessity of existing upon the charity of this Borough, I would commit suicide without scruple or hesitation, from principles of justice to poor human nature, from no other motive than to rescue her from the opprobrious ignominy of suffering a fellow being to perish in distress for want of comfortable and timely assistance! It is truth, abhorent to human nature and too melancholy to be recorded, that, in my opinion, the public officers of the Borough, whose duty it is to provide relief for the distressed, would sooner pay a round sum for the extinction of life of a pauper than to make an exertion or take any trouble to procure the necessary assistance.

"May the Lord deliver us from evil," than which a greater could never befall a human being than to become dependent upon the charity and benevolence of the people for relief in time of distress, for this is the Kingdom of Satan, the power of Tyrants, the Benevolence of Turkey and the Glory of Heathens! Amen! & Amen!

No one can doubt, after a careful perusal of St. Martin's case history, that every possible effort was made to close the wound com-

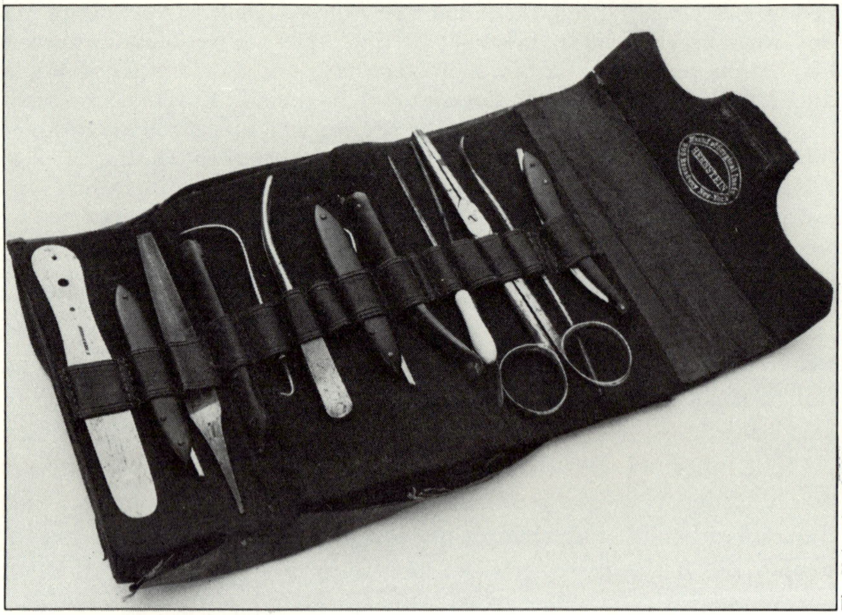

The University of Chicago

Beaumont's surgical pocket kit showing five differently bladed surgical "jack knives"; spatula; file; dissecting forceps; two directors, one round, the other "cutting" and grooved with white bone handle; spud and aneurysm needle without handles; and one bullet forceps.

pletely, and that Mr. Hubbard was laboring under a misapprehension when he stated that the tract was purposely kept open with a view to conducting experiments upon the stomach. Beaumont even states that, when all other means failed, he suggested incising the lips of the wound and "bringing them together with sutures, an operation to which the patient would not submit."

The idea of carrying on a series of experiments on the stomach of St. Martin seems not to have occurred to him until early in 1825, just prior to his leaving Mackinac. He now began to realize the great importance of the case that had so strangely fallen into his hands, and to appreciate the duty that he owed to humanity as a result of the incident, for about this time we find in one of the manuscripts devoted to a description of the case the following:

He will drink a quart of water or eat a dish of soup, and then, by removing the dressings and compress, can immediately throw it out through the wound. On removing the dressings I frequently find the stomach inverted to the size and about the shape of a half blown rose, yet he complains of no pain, and it will return itself, or is easily reduced by gentle pressure.

When he lies on the opposite side I can look directly into the cavity of the Stomach, and almost see the process of digestion. I can pour in water with a funnel, or put in food with a spoon, and draw them out again with a syphon. I have frequently suspended flesh, raw and wasted, and other substances into the perforation to ascertain the length of time required to digest each; and at one time used a tent of raw beef, instead of lint, to stop the orifice, and found that in less than five hours it was completely digested off, as smooth and even as if it had been cut with a knife.

It now beings to dawn upon him that an unusual opportunity is offered him, and he comments as follows:

This case affords an excellent opportunity for experimenting upon the gastric fluids and process of digestion. It would give no pain, nor cause the least uneasiness, to extract a gill of fluid every two or three days, for it frequently flows out spontaneously in considerable quantities. Various kinds of digestible substances might be introduced into the stomach, and then easily examined during the whole process of digestion. I may, therefore, be able hereafter to give some interesting experiments on these subjects.

During all this time Alexis' life had not been a bed of roses. His two years of convalescence had been anything but uneventful, and now, that he had practically recovered his strength and usefulness, his trials were not at an end. It is true that his misfortune had made him famous for all time to come, that his name was being heralded throughout the medical world, that every "Child's Health Primer," "Essentials of Physiology," and "Elementary History of the United States" was to contain for generations an account of his passive contribution to science, but all this was in no sense adequate compensation for past and future trials. Now he would have to submit to having all manner of digestible and indigestible objects poked into the orifice in the stomach, would have to fast for hours, lie in certain positions for interminable periods, carry small bottles about in his armpits, and all this for the sake of science, in which he had not the slightest interest. Not only this, but the finger of curiosity was being pointed at him from every side as "the man with a lid on his stomach." It is said that he was often provoked to resent the jeers in hot blood, and not infrequently made threats to the Doctor of "giving up his job."[4]

[4]*Early Mackinac,* by Meade C. Williams.

Hon. Chase Osborn[5] calls attention to an interesting little story by Mrs. Mary Hartwell Catherwood,[6] in which she describes a conflict between Etienne St. Martin, a brother of Alexis, also in the employ of the American Fur Company, and Charlé Charette in the court yard of the old warehouse and barracks.[7] While the fight was primarily due to jealousies aroused by the fickleness of Tite Laboise, the wife of Charlé, and her inclination to dance with Etienne at the Mackinac balls, it was finally precipitated by Charlé's sarcastic reference to "these St. Martins without stomachs" and "the little lard-eater from Canada, brother to a man with a stomach having no lid." Charlé was the wearer of the "black feather" in his brigade of a dozen boats, and had no trouble putting the little voyageur on his back in the presence of a crowd of officers and employees of the American Fur Company that had stopped long enough to watch the sport. Etienne had at least attempted to vindicate his family's name, and arose from his humiliating position with the threat that "if they do not let my brother's stomach alone, I will have to kill the whole brigade."[8]

In the fall of 1824 Beaumont sent a complete report of Alexis' case to Surgeon-General Lovell[9] for his approval and correction, with the suggestion that it be published in some reputable medical journal, and in response to this communication received the following reply from the Surgeon-General on November 9th:

---

[5]Mr. Osborn referred to this story in his address at the dedication of the monument erected to Beaumont at Mackinac. I am indebted to him for courtesies in this connection.

[6]*Mackinac and Lake Stories—The Black Feather,* by Mrs. Mary Hartwell Catherwood.

[7]The John Jacob Astor House is still well preserved and utilized as a summer hotel. It was built for the storing of furs and the housing of men during the summer gatherings.

[8]I was unable to learn through correspondence with Miss Hazel Catherwood whether the above story was based on fact or fiction.

[9]Dr. Joseph Lovell was the first incumbent of the high office of "Surgeon-General." which was tendered him in 1818 in recognition of his skill as a surgeon and his splendid qualifications as executive and organizer while hospital surgeon. To him belongs much credit for the excellent foundation upon which the present high efficiency of the medical corps of the army has been built. In spite of the efforts of Dr. James Tilton, his predecessor, the medical service was still in a rather chaotic state, and Dr. Lovell at once set to work to reorganize the corps along the lines of its present organization. Among the problems that he had to solve were the system of admission upon examination, the question of private practice on the part of military surgeons, their choice of stations and choice of quarters, the numerical increase of the corps, the abolition of the whisky ration, etc. He died on October 17, 1836, shortly after the death of his wife, and was mourned by the medical corps as a "fearless and diligent officer, an honest and high-minded gentleman, and an unselfish and disinterested philanthropist." (*Journal of the Association of Military Surgeons,* vol. XIV, No. 5.)

Headquarters, Department of the Army, Office of the Surgeon-General

Joseph Lovell, Surgeon-General of the United States Army, 1818-36.

I have received your letter enclosing your valuable communication of the case of wounded stomach. The cure is a full demonstration of the wonderful powers of nature and highly creditable to yourself. Agreeably to your suggestion, I shall send it to the *Medical Recorder* for publication, altering it only so far as may be necessary to introduce a part of your letter relative to the present state of your patient.

I will endeavor to send you some book of experiments on the gastric liquor, which will be somewhat of a guide to you in making your observations, which may be done with perfect ease and safety. It is stated, for example, that if several articles of food be taken into the stomach, that it would digest all of *one kind* first, then all of a second, and so on, and that this is the cause of the bad effects of a variety of food at the same meal. Suppose a man eat beef, potatoes, fish, cabbage, and pudding, it is expected that he will first digest all the beef, the others in the meantime remaining untouched; then all the pudding, then all the potatoes, and lastly the cabbage. Now, it is thought if he eat a dozen articles, by the time the stomach has disposed of eight or ten, it will become exhausted, and the rest will be left to ferment and produce indigestion and its consequent evils. Could you make experiments to ascertain this, and also the digestibility of various articles? This alone would afford a most valuable paper for publication. I should be happy to receive an account of any experiments or observations you may make, and they will no doubt be very acceptable to the public.

The article did appear in the *Medical Recorder*[10] early in 1825, but through an oversight on the part of someone was published as "A Case of Wounded Stomach, by Joseph Lovell, Surgeon-General, U.S.A." It is not stated how the error occurred, but toward the end of the same year (No. 4, page 840) this correction was inserted in the journal:

A mistake was made in stating that the case of wounded stomach produced by the penetration of a ball was communicated by Dr. Lovell for the January number of this year. It should have been mentioned as reported by Dr. William Beaumont, and also that it occurred at the post of Mackinac. We should be obliged to Dr. B. for any further particulars of the case, which was considered as highly interesting, instructive and well drawn up.

The Medical Society of Michigan Territory was the first to take recognition of Beaumont's article, making him an honorary member of the organization, as shown by the following extract from the proceedings of that organization:

---

[10] *The Medical Recorder of Original Papers and Intelligence in Medicine and Surgery,* conducted by Samuel Calhoun, assisted by an association of physicians in Philadelphia, New York, Baltimore, and Norfolk. 1825, vol. VIII, No. 1, article III, page 14.

DETROIT, March 3d, 1825.

Doct. William Beaumont, United States Army, Michilimackinac.

SIR: It is with much pleasure that I hereby transmit to you an extract from the minutes of the Medical Society of this Territory, at a meeting held at the home of Capt. Woodworth, in the city of Detroit, on Monday, 7th ultimo:

"Dr. William Beaumont, of the United States Army, duly proposed by Dr. Pitcher, and unanimously elected by Ballot an honorary member of this Society."

Whereupon it was ordered that the Secretary be directed to inform Dr. Beaumont of his election as aforesaid.

I remain, Sir, with much respect your most ob. Servant,

JOHN S. WHITING,

Secretary of the Medical Society of the Territory of Michigan.

Beaumont now began to feel keenly his complete isolation, and realized the great difficulties that confronted him in conducting experiments under the conditions which prevailed at Mackinac. Not only had he no facilities for research, but there was not even a physician within a radius of many hundred miles from whom he could receive suggestions, or with whom he could exchange ideas—a factor of no little importance even in this day of unlimited facilities for research. He felt that some eastern post might offer him better facilities for taking advantage of the rare opportunity that had so unexpectedly fallen to his lot, and in all probability requested a transfer of the Surgeon-General, for I find in the old files of the Surgeon-General's office an order of February 2, 1825, to this effect: "When the troops at Mackinac shall be removed to Green Bay . . . you will be ordered to Ft. Niagara to the commanding officer for duty there." On May 10th, however, another order came, no doubt as a great disappointment to him, stating that, "as the order for the removal of the troops from Mackinac has been countermanded, it will, of course, be impracticable for you to leave the post at present." Thereupon he commenced at Mackinac in a desultory way his first series of gastric experiments, but a solution to the problem of a transfer was evidently soon found by the Surgeon-General, for he was shortly ordered to Fort Niagara, and in June we find him actively engaged in the hospital there. This being no great distance from Plattsburgh, the girlhood home of his wife, Mrs. Beaumont and the children[11] went on to that place to await future developments.

[11]Sarah, aged 2 years, and William W., aged 3 months.

In the meantime he continued to prosecute his experiments until the month of August, when they came to a sudden termination through a most unexpected occurrence. In July Beaumont had requested a furlough of two months, with the evident intention of first visiting his family at Plattsburgh and then exhibiting St. Martin to some of the prominent scientists of that day. He received a communication from the War Department July 22d, granting a furlough for two months, "provided a citizen can be obtained to attend the post at the compensation authorized by the army regulation." All necessary arrangements were completed, and he left Niagara, accompanied by Alexis, stopping first at Burlington, Vermont, and then going on to Plattsburgh. They had been here but a very short time, when the close proximity of the Canadian border and the scenes of his boyhood proved an irresistible temptation to St. Martin, and he took "French leave" of his benefactor. This possibility seems not to have occurred to Beaumont, in spite of his knowledge of human nature in general and the wily ways of the voyageur in particular. Naturally he was broken-hearted over his loss, and made every possible effort to learn of the whereabouts of his truant, but at the expiration of his furlough he had obtained no clew concerning him, and was compelled to return to his duties at Niagara broken in spirit, but seeking consolation in the thought that his protege might fall into the hands of some one better fitted than himself for the important work of clearing up the knotty problems of gastric digestion.

Undaunted by this experience, he decided to give to the world the result of the few observations that he had been able to make, and sent a series of experiments to Mr. Webster, publisher of the *Medical Recorder,* saying:

> Enclosed I send you the result of a few experiments carefully made upon the gastric fluids of the Stomach of Alexis St. Martin, the circumstances of which were communicated by Dr. Jos. Lovell, Surgeon Genl. U.S.A., and published in the 29th number of the *Med. Recorder* for Jany., 1825, and, if you think them worthy to be published, they are entirely at your disposal.[12]
>
> I regret very much that it is not in my power to offer more varied and satisfactory results, but, unfortunately for me (tho fortunately perhaps for some more capable person than myself to make experiments), he has absconded and gone to Canada, at the very time I was commencing a number of more interesting and important experiments upon the process of digestion

[12]Published in the *Medical Recorder,* January, 1826, No. 33, page 94.

and power of the gastric liquors, and I very much fear I shall not be able to recover possession of him again. He was unwilling to be experimented upon, though it caused him but little pain or distress. At the time he left me he was much in the same condition as when my report was made to the Surgn. Genl. in Sept., 1824, and by him subsequently communicated to you.

N.B.—I have other interesting cases, of some practical importance perhaps, recorded on my journal of practice, which I should have no objections to communicate if I could bring myself to a belief that they would be acceptable and well recd. by the public.

This series of experiments numbered four, in which he noted the length of time required for the digestion of various articles of food suspended in the stomach by a silk string, which he could introduce and withdraw at pleasure, thus noting the changes in these substances from time to time. He determined "the natural warmth of the stomach" by introducing the glass portion of a thermometer, marking the height of the mercury on the glass and then placing it on the graduated scale. He compared the length of time of digestion of certain articles in the stomach and in vials containing gastric juice at the body temperature, agitated from time to time to imitate the gastric peristalsis. He found that the digestion took place more rapidly in the stomach than in the vial after the first half hour, that some meats digested more slowly than others because the texture was closer, etc. Even these incomplete experiments assured him that the old theories of maceration and putrefaction were insufficient to explain the processes that he had observed, and that the gastric secretion possessed innate solvent powers.

But now his dreams and ambitions of aiding in the solution of the important problems of digestion were a thing of the past, and his whole attention was again claimed by the routine military duties at Fort Niagara. One incident especially in his life at this frontier post is well worthy of careful consideration.

A trial by court-martial, which took place at this time, and in which Beaumont figured as the chief witness, excited general interest in army circles, and especially among members of the medical staff. It seems that on the morning of June 21, 1825, Lieutenant Griswold reported himself sick and was bled by Dr. Beaumont and given medicine, with the result that he recovered promptly from the symptoms of which he was complaining. He returned to his post, complaining of no further indisposition, but continued to neglect his duties at the expense of two of his fellow-officers, Lieutenants Russell and Morris. Their patience was finally exhausted, and they complained to the commanding officer on July 3d of Griswold's evasion of duty, whereupon he was informed that he must either do his duty or report to the surgeon. "Reduced to this dilemma, Lieutenant Griswold had but two alternatives left—either to admit the imposition and go on duty, or successfully persevere in the game of deception he had been playing; too willfully obstinate for the first, and, thinking himself capable perhaps of the last, he renewed his efforts and made an unsuccessful attempt." He therefore chose to return to the surgeon, complaining of his arm where he had been bled. Though Beaumont could find nothing, he took him on the sick list with a view to observing him carefully. Griswold then began complaining of all manner of "nondescript sensations." Beaumont made numerous efforts to locate a cause for the lieutenant's complaints, but, not succeeding, came to the conclusion that he was a malingerer. In this connection he wrote:

> Resolved never to be made the tacit medium of deception, nor the convenient organ of official falsehood, I determined neither to let the case pass unnoticed or waive my duty of making a correct report to the commanding officer. Viewing the case as novel and unprecedented, artfully calculated to evade proof and requiring more than ordinary means and management for detection, I consulted my duty to government and my professional character only, and at once resolved upon the course to be pursued, fully aware of the delicacy and difficulties of deciding judiciously upon the first case of feigned sickness in an officer that had ever occurred within the sphere of my official duty. . . . Under these circumstances and impressions, I found I had an extremely delicate and disagreeable duty to perform, and, to demonstrate the correctness of my belief and enable me to form a final opinion, I resolved to prescribe a powerful dose of medicine for him, in the effect of which, if taken as directed, there could be no mistake. I did so. Prepared and left with him, with directions, an emetic-cathartic of 15 or 20 grains of calomel and 5 or 6 grains of tartar emetic. Whether he took the medicine as directed, I am unable to say, not staying to administer or see him take it. I visited him at this quarters frequently in the course of the day, but discovered no effects of medicine upon him. I found him at each of my visits during the day sitting at his table or window, reading or writing, appearing very comfortable and undisturbed. In the afternoon of the same day, about 3 or 4 o'clock, just after a shower, while the atmosphere was yet cloudy and damp, and the ground wet, I saw him out of his Quarters, walking about the garrison, with no other protection from the weather than his usual apparel, which occasioned me to remark, "If Mr. Griswold had taken medicine as I directed him that morning, he would have swelled jaws and a sore mouth for his imprudence." I visited

him the next morning, saw no appearance of the operation of the medicine, neither did he complain of any soreness of his throat or mouth. . . . Being fully convinced from the preceding circumstances and observations that he was not then sick, neither had been since the 25th of June, I immediately left his room, without making any suggestions to or receiving any intimations from him of his ability or inability to do his duty, and without any remarks struck him off my report, marked him fit for duty, and reported him to the commanding officer. . . . Having conscientiously discharged my duty to government and Lieutenant Griswold on the morning of the 8th of July, I had no more to do with his official conduct. Whether he went to the office and saw or was informed of my report, or, judging from my last visit, that his situation was not tenable with impunity, is best known to himself. Sufficient be it for the occasion, he resumed his salutiferous faculties, dressed himself, and went on duty immediately after I had sent in my morning report, and never after, while at this post, exhibited the least indisposition to my knowledge or belief until the time of his trial.

Griswold was placed under arrest, tried by court-martial, and found guilty, "to be dismissed from the service," in spite of the fact that the very officers who had complained of his shirking his duties, and who knew all of the circumstances of the case, came to the rescue of their comrade finally, and endeavored to give him immunity through favorable evidence. The sentence of the court-martial was disapproved by President Adams, and was reversed by the "Old Man Eloquent" in General Order No. 9, issued February 18, 1826.[13] The President, in submitting his disapproval, stated:

It is at least very questionable whether the finding of the Court is warranted by the evidence in the face of the record. With regard to the only charges which, even if proved, could justify the severity of the Court's final sentence—namely, the third and fourth—the evidence before the Court was contradictory. Two unimpeached witnesses testify to facts which, if credited, show that the unfitness for duty by indisposition of Lieutenant Griswold at the time specified in the charges was real and not pretended. The facts are indeed otherwise stated by Assistant-Surgeon Beaumont, but the finding of the accused guilty by the Court on these charges appears to rest more upon the professional opinion of the surgeon than upon his statement of facts, shaken as that must be by the testimony of Lieutenant Russell and Lieutenant Morris.

The testimony of Brevet-Major Thompson and that of Assistant-Surgeon Beaumont both bear internal marks of excited feelings, impairing their credibility. The former was properly excluded by the Court from acting as prosecutor for misconduct pending the trial, and Surgeon Beaumont states an

[13]Files of the War Department at Washington, room 254. By courtesy of the Adjutant-General.

experiment as having been made by himself upon Lieutenant Griswold of more than doubtful propriety in the relations of a medical adviser to his patient. A medicine of violent operation, administered by a physician to a man whom he believes to be in full health, but who is taking his professional advice, is a very improper test of the sincerity of the patient's complaints, and the avowal of it as of a transaction justifiable in itself discloses a mind warped by ill-will, or insensible to its own relative duties.

It is needless to describe the effect of these strictures on the sensibilities of Dr. Beaumont. Pride and honor were two of his chief qualifications and his constant boast. Through this sweeping order of the highest tribunal of the land he felt that both had been levelled in the dust. His duty was clear, and he determined at once on his course of action in the matter. He first made a number of applications to the President for a court of inquiry to determine whether the facts in the case justified the President in such treatment as he had meted out to him, but, not succeeding in this, prepared a public reply to the "President's animadversions," evidently without the slightest regard for the results that it might have on his future career. When certain that he was in the right, no thought of self ever deterred him from giving expression to his opinion, no matter how lofty the position of him who had given offense. The following note, yellow with age, was found in the files of the War Department:

Eight weeks have elapsed, and my repeated applications for a court of inquiry not yet noticed by the proper authority, I consider further delay injustice to my own character, and am most imperiously required, by every principle of honour and feeling of manliness, to give circulation to the following statement coextensively with the president's animadversions upon my conduct, as promulgated in his order of February 18, 1826, with a view to correct erroneous impressions and remove any undue prejudice that may have existed in consequence of said order.

The statement, which follows in part, was issued at Fort Niagara, New York, May 8, 1826, in the form of a circular[14] of about four thousand words, and was executed with all the fearlessness and courage of one convinced that he was pleading in the cause of justice.

Considering my official reputation and moral character egregiously and unjustly assailed under General Order of February 18, 1826, and believing the Heads of Departments to have been imposed upon by misrepresentations and total perversions of truth, I consider it due to myself and the Govern-

---

[14]Files of the War Department at Washington, room 254.

ment, and demanded by justice, to give an explanation, with a view to correct the abuse and expose the imposition.

Silence under such circumstances would be cowardice, and submission meanness indeed, suffering falsehood and error to preponderate over truth and correctness, to the total subversion of good order and correct discipline, and entire destruction of military pride and laudable ambition.

In submitting the following statements, I am governed by the inflexible principles of truth, incited by a lively sense of self-respect, the good of the service and regard for the Medical Department, disavowing any other feeling of prejudice, excitement or ill-will than those inseparably connected with just abhorrence of falsehood, dissimulation and roguery.

Equal rights in a free Government are due to all; justice in a Military Department is at the disposal of a few. The first is my birthright as an American, the last is due me as an officer of the Army upon the common principles of honor and rectitude. I claim them both upon the fair and equitable grounds that every man (military in particular) is entitled to a candid hearing and careful investigation before being censured or condemned for supposed offences and misconduct. I applied for a Court of Inquiry, on the receipt of the Order, to investigate my conduct. Eight weeks have elapsed, and I am without an answer. My feelings are not callous; I have yet sense of honor sufficient not to rest easy under the imputation of baseness and discredit; neither is my mind so warped by ill-will, or so insensible to its own dignity, as to seek sly and silent revenge for its satisfaction. My reputation is dear to me; the wide spreading influence of that Order cannot be anticipated with indifference.

I ask a candid and patient perusal of this statement, every part of which I will substantiate if an opportunity be afforded me. Single and unsupported, save by truth and conscious rectitude, I ask but to be heard—then judged by the merits of the case. I seek not the aid of whining sympathy or boisterous advocates. I offer no borrowed documental influence to support my cause; by the merits or demerits of my own conduct I wish to rise or fall in the estimation of Government and the Heads of Departments.

It is presumed the Officers of the Army generally have seen and read the President's remarks under General Order No. 9 for 1826, in which my testimony before a Court Martial for the trial of Lt. E. B. Griswold is severely censured, my character for truth and candor impeached, and my Disposition and Professional judgment implicated. Whether the inferences drawn from the face of the records of the Court be correct and warranted by the evidence thereon exhibited, I presume not to offer an opinion, but that the final conclusions are as unjust as they are severe, I am able and ready to substantiate.

The character of the Court, the nature of the charges, the universal findings, and corresponding sentence of that Court are sufficient justifications of my conduct, and ample protection to my character from general censure. . . .

Whether the plan adopted, either in a moral or professional point of view, be justifiable or not, I leave for medical men and candid judges to decide. It was salutary, and had the intended effect of returning Lt. Griswold to his duty without prejudice to his health or constitution; neither is it of very great moment with me whether a successful experiment be of less or more than doubtful propriety that speedily restores a soldier from the sick report to the effective service of the government, be he private, non com. or commissioned officer; neither do I think it of very great consequence whether it be done secundum artem, secundum naturam, or terrorem, provided it be well done.

It may not be amiss here to remark that, so far from having administered a medicine of violent operation to a man whom I then believed (and have ever since) to be in full health, I neither required Lt. Griswold to take, believed he did take or had any intention of taking the medicine left with him, but, on the contrary, believed he studied to deceive by pretending to have taken it and then representing its effects. It was impossible to prove he did not feel those strange, indescribable sensations, but I know he could not disguise the effects of the medicine if taken. With this view I prescribed the calomel and emetic tartar. . . .

And no consideration can ever warp my mind from its fixed principles of acting honestly and independently in the discharge of its relative duties. Should I again, an hundred times, be placed in a similar situation, I would do as in this case I have done; fearless of censure, reproach, or temporary discredit, regarding far more the sanctity of an oath than either the applause or disapprobation of the highest earthly tribunal, considering Integrity, Faithfulness, and Fidelity to my God, my Conscience and my Country paramount to every other consideration.

My testimony was confined to facts that transpired between the 2d and 9th of July, and no other; was made upon the firm basis of truth, accurately identified with time and circumstances mentioned in the 3d and 4th charges, with a perfect recollection of facts, corroborated by the Morning Reports of the Garrison, needing not the aid of fortuitous events and equivocal memory to substantiate it. . . .

Treacherous indeed must have been their memories who could not have readily distinguished between the healthful enjoyments and festive associations of a convivial scene on the 4th of July and the more grave and sober acts of sympathetic kindness bestowed upon a friend on the 22d and 23d of June, and still more truant and pitiable his recollection who, after having officially complained of specific grievances and impositions on 3d of July, should forget that time and the occurrence of the very cause which induced him to make that complaint—viz., Lt. Griswold's evasion of duty between the 26th and 30th of June.

Yet, from the strangely defective recollection and extraordinary extraneous testimony, made to preponderate (tho it did not even refer to the facts or time specified in said charges), my professional opinion was shaken, my conduct as a witness and my disposition as a medical adviser severely animadverted upon by the President of the United States, by whose Veto and a few waves of his pen the sentence of an honourable Court-Martial is made void, and my reputation stigmatized throughout the army, without prospects of extenuation or retrieve, but by the aid of lenient charity, or through the more honorable medium of official reconsideration. I have respectfully claimed the last.

His verbose circular evidently received the scant consideration on the part of the chief executive that he anticipated it would,[15] but it served to convey his ideas on the subject to his friends in the army and operated as a safety valve for his pent-up feelings. Shortly after this episode he received orders "to be in readiness to accompany the troops from Sacketts Harbor and Niagara to Green Bay, and to have all medical supplies under his charge carefully packed for transportation." He replied that he would be in readiness at the appointed time, and requested permission "to prescribe for the citizens of Green Bay whenever necessity required it, they being entirely dependent upon army surgeons for medical aid and assistance." While there was a paragraph in the army regulations at this time forbidding medical officers of the corps to engage in private practice, permission was usually granted post surgeons on application to render professional aid to citizens under such conditions as those which prevailed at Green Bay. This regulation had been in existence since 1814, but was not enforced until Lovell was made Surgeon-General. He afterward explained, in response to Surgeon McMahon's inquiry, that the enforcement of this regulation was not intended to prevent acts of humanity on the part of surgeons, but simply "to prevent neglect of duty by entering extremely into it [private practice], as well as an improper application of public property, which often occurred." Beaumont's detachment soon[16] got under way, and arrived at Fort Howard early in June. On the 18th of this month he wrote to Washington: "Due regard for the comforts of the sick placed under my charge requires that a timely representation of their destitute condition be made, with a view to its future amelioration." He found the hospital department "entirely destitute of every means of comfort, save medicine," and asked that the hospital equipment at Sacketts Harbor be transferred to Green Bay, which was promptly granted.

---

[15]It is interesting to note that, in spite of this episode, Beaumont was commissioned "surgeon" by President Adams on November 28, 1826.

[16]His last communication from Niagara to the War Department is under the date of May 8, 1826. He evidently left his family to follow later, for I find recorded in the family Bible the death of his little son, William, at Plattsburgh on October 14, 1826.

The life at Green Bay, Michigan Territory, was not new to him, for much the same conditions prevailed here as at Mackinac—even the same traditions. The place was founded in 1634 by Jean Nicollet when he was sent out by Champlain to discover a short route to China. Marquette, Joliet, Allouez, and Tonti had played important roles here as they had at Mackinac. At the close of the French and Indian war the settlement fell into the hands of the English, and so remained until after the War of 1812. The fur companies for many decades had branch houses at La Baye Verte, as it was called by the French, and the brigades sent out in the fall from Mackinac to the Mississippi country passed here and filled the air with chansons of the voyageur. At the beginning of the War of 1812 Colonel Dickson came here, after aiding in the capture of Mackinac, with his Indian allies, among whom was the famous chief Black Hawk. This point was the center of an Indian country of many tribes—the Menominees, Winnebagoes, Foxes, Chippewas, Sauks, and others—and in the immediate vicinity were more than 3,000 savages and as many more belonging to the adjoining tribes. When peace was declared with England and Americans took possession, these savages were in an ugly mood as the result of their great losses from the war—their crops had failed, they had lost their cattle, and the fur trade had been destroyed. They were prepared to lay all of the blame at the door of their new masters. There was scant chance, therefore, of maintaining peace among them under such conditions. In view of the ugly mood of the Indians and the importance of Green Bay as a trading center, it was deemed advisable to place a garrison at this point, and in 1816 Fort Howard was built just across the river, together with a government trading post and an Indian agency. The fort was inclosed in a stockade of timber thirty feet high, inclosing the usual type of barracks, blockhouses, and officers' quarters, except those of the surgeon and quartermaster, which were built outside of the inclosure. Many years elapsed before the Indians could be reconciled to their new condition, and many acts of violence were committed by them. On one occasion the post surgeon, W. S. Madison, was ambushed and killed by a Chippewa Indian. In 1825 Lewis Cass, governor of Michigan Territory, succeeded in making at Prairie du Chien a treaty of peace with the Sioux, Sauks, Chippewas, Foxes, Winnebagoes, and Menominees, but within a year the treaty had been broken, and the Winnebagoes, who throughout had been more or less discontented, had found a *casus belli.*

Upon Beaumont's arrival here there were already undercurrents and mutterings of an Indian outbreak. The fomentation continued until the spring of 1827, when the tribe began committing outrages on the whites, and found sufficient courage to wipe out a family of seven. With this taste of blood, Chief Red Bird and his warriors started on the war path. Detachments of soldiers were promptly sent out from Forts Crawford, Snelling, and Howard, while Governor Cass himself proceeded to St. Louis to obtain aid from Jefferson Barracks. With this array of troops practically surrounding the savages, they soon realized the futility of further opposition, and Red Bird and his band made a dramatic surrender to Major Whistler's forces, to which Beaumont was attached. The Indian chief approached, carrying a flag of truce, singing his death knell, and casting a handful of dust over his shoulder to indicate that he gave up his life as freely. His only request was that he and his braves should not be put in chains. He was taken prisoner and sent to Prairie du Chien, where he died later. Two of his accomplices were condemned to be hung, but were pardoned by President Adams. This put an end, for a time at least, to Indian outrages in this section.

On returning to the fort from the interesting scenes that he had just witnessed, Beaumont entered into correspondence[17] with Dr. Lovell concerning the question of vaccine and vaccination, which from an historical standpoint should be of some interest. On September 18, 1827, he wrote:

> Sense of duty and common benevolence urge me most respectfully to request a fresh supply of vaccine infection to be forwarded to this post as soon as practicable. Accident alone has deprived me at this time of any efficient matter. The importance as well as necessity of having it on hand in a country like this, subject at all times to the incursion of the smallpox from the continual passing and repassing of wandering Indians and Canadian voyageurs, will doubtless be duly appreciated. I verily hope that some fresh and genuine infection be forthcoming immediately, so as to reach me before the close of the fall navigation of the lakes.

To this the Surgeon-General responded on October 5th: "Agreeably to your request, I herewith inclose you a crust of vaccine virus, the only one on hand, and in a few days another supply will be forwarded." On January 31, 1828, Beaumont acknowledged the receipt of these packages, saying:

---

[17]Files in Surgeon-General's office.

I have the honour to acknowledge the receipt of your two communications, containing each a Vaccine crust—the one dated October 25th this day, and that of Novem. 3d on the 11th instant, per express Via Chicago. I am apprehensive that, from some cause or other, the virus is all destroyed, inasmuch as immediately upon receiving the first package (which, by the by, was the last forwarded from the Department) I vaccinated a number of young and healthy subjects, none of which took effect. I repeated it several times, with the same result; no appearance of the disease could be produced by it.

I have this day vaccinated several with the last received (first forwarded), but do not expect any effects from it. Does frost destroy the virus? If so, it must have been destroyed in coming from Detroit to this place by express, the weather being extremely cold. And as it will be impossible to send any to this Post in the winter season, therefore I respectfully request that some more may be forwarded soon as the season will permit, so as to arrive here by the first vessel from Detroit in the spring.[18]

Fearing that the vaccine was not being properly administered, Dr. Lovell sent another crust, with directions for using: "The introduction of a small portion of crust softened with a little warm water, after allowing the puncture to bleed pretty freely, has been found most effectual in securing success; or making half a dozen pretty deep incisions with the edge of the lancet and crossing them in the same way, thus [indicating the customary cross incisions], and covering them well with the matter."

About this same time glad tidings came concerning the whereabouts of the truant Alexis. Practically two years had elapsed since he took his unceremonious departure, and from that time nothing had been heard of his whereabouts. Soon after his disappearance, Dr. Beaumont communicated with his friends in the employ of the American Fur Company, urging them to be constantly on the outlook for St. Martin, and to report to him immediately on learning his rendezvous. W. W. Matthews, whose annual duty it was to go into the interior of Canada for the purpose of employing voyageurs, had just returned to Mackinac from such an expedition, and wrote Dr. Beaumont:

[18]During the winter months Green Bay was practically cut off from the world. Mail was delivered from Detroit by a soldier only twice in six months. It is said that a private route was established between Fort Howard and Fort Wayne, a distance of three hundred miles, after which mail was delivered about once a month. Moses Hardwick, a sturdy frontiersman, carried the mail from Detroit to Green Bay from 1817 to 1824, and each time, especially in the winter, was compelled to fight his way through packs of wolves, burdened with mail pouch and loaded musket. *(Historic Green Bay.)*

While in Canada last winter I succeeded in finding your ungrateful boy, Alexis St. Martin. He is married, and lives about twelve miles back from Berthier at a place called LaChalaupe. He is poor and miserable beyond description, and his wound is worse than when he left you. The distance from Montreal to where he lives is fifty-seven miles. I did all I could to bring him up, but could not succeed, but my endeavors cost me $14. I will be obliged if you will let me know by return of boat whether I shall do anything more to get him back, and how I shall get my money back, as the company will not allow it to me.

The money was evidently promptly sent to Mr. Matthews, together with an authorization to get hold of St. Martin if possible, for a few days later another letter came:

I received your favor of the 15th inst., inclosing $14, the amount of my disbursements in finding Alexis St. Martin. There will be no difficulty in getting him back at any reasonable price, providing you will employ his wife (!) also. He is miserably poor, and will remain so while he lives unless he comes back to you. I think you had better write to me in New York, requesting me to engage Alexis and his wife to come up and remain with you one or two years. I will bring them up to Mackinac in the company's boats, and charge as low as possible. Dr. Caldwell saw him when he was in [Canada?] last, and wishes very much to get him to remain with him.

While this news encouraged Dr. Beaumont to hope that he might again have Alexis with him for the purpose of conducting further experiments, much time elapsed before any more information was forthcoming. In the meantime he received orders to join the Fifth Regiment at Fort Crawford, also in Michigan Territory. He left Fort Howard with some reluctance, for he had formed many friendships in the garrison and in the little village across the river. In after years he paid several visits to Green Bay to look after property which he owned, and to visit relatives whom he had induced to come here.[19] It is said that many years after he took his departure from this post the old French settlers were proud to recall the days of intimate relationship with "Le bon doc-teur who mak' de mer-a-cle." His good wife, too, through her domestic and social tendencies, had made herself much beloved among both citizens and soldiers, and her house was the rendezvous of those who sought the best and highest in the social life of the military post. It is said that she was fond of such modern classics as Scott, Cooper, Irving, and Byron, and that her greatest

---

[19]Quite a colony of Plattsburgh people emigrated to Green Bay in 1834.

delight was to read Shakespeare to her little family group and a circle of intimate friends while they busied themselves with needle and thread. Her renditions of the author's plays were of rare excellence, in spite of the fact that, Quaker-like, she had never entered a theater. Such were their congenial surroundings when the Beaumont family left Fort Howard. On August 10, 1826, he reported their arrival at the new post on the upper Mississippi.[20]

[20]A daughter, Lucretia, was born at Fort Howard on July 26, 1827.

1828-1833

ALMOST TWO YEARS had elapsed since receiving the news of Alexis' whereabouts from Mr. Matthews, and in the meantime the officers of the fur company had been making every possible effort to bring him back to his benefactor. They were unremitting in their zeal to see him again serving the man to whom he owed so much, and toward whom he had proven himself so ungrateful. Without the efforts of Ramsay Crooks, W. W. Matthews, and William Morrison, St. Martin's case would have been lost to science. To them, therefore, belongs much credit, which, so far as I know, has never been accorded them, except by Beaumont himself, for this important chapter in the history of physiology of digestion.

During Alexis' absence of four years he had married and had become the father of two children. He had worked hard to support his family as a voyageur in the employ of the Hudson Bay Fur Company, and was sent by them into the Indian country of Canada, where he suffered many privations. In spite of all this, he had remained robust and well. Dr. Beaumont's comment on Alexis' condition at this time was as follows:

> After considerable difficulty, and at great expense to me, they succeeded in engaging him and transported him from lower Canada, with his wife and two children, to me at Fort Crawford, Prairie du Chien, Upper Mississippi, a distance of nearly two thousand miles, in August, 1829. His stomach and side were in a similar condition as when he left me in 1825. The aperture was open and his health good.
>
> He now entered my service, and I commenced another series of experiments on the stomach and gastric fluids, and continued them interruptedly until March, 1831. During this time, in the intervals of experimenting, he performed all the duties of a common servant, chopping wood, carrying burthens, etc., with little or no suffering or inconvenience from his wound. He laboured constantly, became the father of more children, and enjoyed as good health and as much vigour as men in general. He subsisted on crude food in abundant quantities, except when on prescribed diet for particular experiment purposes and under special observance.

Beaumont states elsewhere that he gave him at this time "800 livres annually, clothed and subsisted him and his family."

Now that his dream was again realized, Dr. Beaumont lost no time in beginning anew his experiments on the stomach of St. Martin. With no laboratory equipment but a thermometer, a few open-mouth vials, and a sand bath, he conducted at this isolated military post a series of observations on digestion which have not been excelled even under more favorable conditions and surroundings. How exact these observations were can best be determined by an examination of his manuscript. From a comparison of his old manuscript and notes with his publication, one finds that these investigations were made "in the true spirit of enquiry," without any "particular hypothesis to support," and that he "honestly recorded the result of each experiment exactly as it occurred." Between December 6, 1829, and April 9, 1831, there were recorded fifty-six experiments. He took up the train of thought just where he had left off four years before, with a series "to ascertain the variation in temperature in the interior of the stomach under different circumstances and conditions of the system and vicissitudes of the atmosphere." Eight experiments were recorded by him on this particular point, of which No. 1 is a fair example:

DEC. 6, 1829. At 9 o'clock, A.M., I introduced the glass tube of a Thermometer (Fahrenheit's) through the artificial opening into the stomach, in a healthy and empty condition, nearly the whole length of the stem. In six or eight minutes the mercury became stationary at 98°. Weather cloudy, damp, and almost raining; ground wet, muddy and thawing. Wind S. and mild. Thermometer, in a North exposure, 63°. Commenced raining at 11 o'clock, A.M., and continued all day, with oppressive atmosphere.

He concluded from this series that "the variations of the atmosphere produce effects on the temperature of the stomach, a dry atmosphere increasing and a humid one diminishing it; that the ordinary temperature of the stomach is about 100° F.; that there is probably some difference of temperature in different regions of the stomach, it being higher at the pyloric than at the splenic end." Two of these experiments showed that the heat of the stomach was increased during the active period of digestion, but subsequent ones showed the temperature to be the same in its full and empty state.

He now started out "to ascertain whether the gastric juice began accumulating in the stomach during periods of fasting, or even from the immediate or direct influence of hunger." There were fourteen of these experiments, of which No. 12 is an average:

MARCH 13, 1830. At 10 o'clock, A.M., stomach empty, introduced tube, but was unable to obtain any gastric juice. On the application of a few crumbs of bread to the inner surface of the stomach, the juice began slowly to accumulate and flow through the tube. The crumbs of bread adhered to the mucous coat, soon became soft, and began to dissolve and digest. On viewing the villous membrane before applying the bread crumbs, the mucous coat and subjacent follicles only could be observed; but immediately afterwards small, sharp papillae and minute lucid points, situated in the interstices of and less than the mucous follicles, became visible, from which exuded a clear, transparent liquor. It then began to run through the tube.

These observations led him to the conclusion that "the stomach contains no gastric juice in the free state where aliment is not present," that "any digestible or irritating substances, when applied to the internal coat, excites the action of the gastric vessels," that "even the application of the tube used in withdrawing the fluid was sufficient to excite its flow." He took issue with those who maintained that the gastric juice was discharged into the stomach during a fast. The result of one of these experiments caused him to believe that "in febrile diathesis very little or no gastric juice is secreted," and that for this reason food should be withheld in febrile complaints.

The majority of his subsequent investigations "were instituted with the view of ascertaining the relative difference between natural and artificial digestion; to demonstrate the performance of digestion out of the stomach with the gastric juice, and also the continuation of the natural process when taken out during the period of chymification," as indicated, for example, in the following experiment, No. 24:

DEC. 14, 1829. At 1 o'clock, P.M., I took one and a half ounces of gastric juice fresh from the stomach, after eighteen hours' fasting, into an open mouthed vial, put into it twelve drachms recently salted beef (boiled), and placed it in a basin of water on a sand bath, and kept it at about 100° (Fahrenheit), with frequent gentle agitation. Digestion commenced in a short time on the surface of the meat, and progressed in that manner uniformly for about six hours, when its solvent action seemed to cease. The meat was at this time nearly half dissolved; the texture of the central portion considerably loosened and tender, resembling the same kind of aliment when ejected, partly digested, from the stomach hours after being swallowed, as frequently seen in cases of indigestion.

The vial continuing in the same situation, its contents varied considerable in their sensible qualities in twenty-four hours; the digested portion separated into a reddish brown precipitate and whey-coloured fluid.

I now separated the undigested from the chymous portion by filtration through thin muslin. When squeezed dry, it weighed five drachms, two scruples, and eight grains, which, deducted from the twelve drachms of meat put in at first, leaves six drachms and twelve grains digested in twelve fluid-drachms of gastric juice.

This experiment was conducted with as much precision and integrity of observation as possible, with the temperature of the digesting bath kept as near blood heat as was practicable to regulate and continue artificial warmth, the Thermometer varying during the time from 90° to 100°.

While "it took twelve drachms of gastric juice to digest six drachms and twelve grains of aliment," he felt that no definite conclusion should be drawn from this experiment on account of the great variation in the secretion under different circumstances. He expresses the opinion "that different kinds of diet require different proportions of gastric juice for their solution."

In experiment No. 25 he found "that a quantity of aliment taken out of the stomach twenty minutes after having been eaten had a sufficient amount of gastric juice to insure its perfect digestion when placed on the bath." In experiment No. 26 he demonstrated the "comparative digestibility of animal and vegetable diet," and showed "that vegetables are less easily disposed of by the gastric organs than animal or farinaceous substances." In subsequent experiments he demonstrated the coagulability of milk by the gastric juice, and its importance in connection with the digestion of milk; the digestion of different kinds of meat, the importance of meat in the action of the gastric juice, and showed that when gastric secretion into which meat had been placed was kept in a cool place, digestion ceased, but began again when transferred to the bath. In another experiment he calls attention to "the effect of violent passion on the digestive apparatus." The necessity of a perfect comminution of the articles of diet was cleverly demonstrated in experiment No. 34, as follows:

MARCH 14. At 8 o'clock 15 mins. introduced two ounces of rare roasted beef, suspended by a string, into the stomach, and at the same time put one drachm of the same kind of meat into twelve drachms of gastric juice, contained in a vial, and put it into his bosom. The piece in his stomach, examined every hour till 12 o'clock, M., exhibited an uniform, but very slow process of digestion, confined entirely to the surface of the meat. In four hours about half of it only was dissolved and gone. That in the bosom, at the same time, digested still slower, owing probably to the circumstances that the fluid in the vial had been taken out when the stomach was in a morbid condition, and had been permitted to get cold, even to the freezing point. This last circumstance, however, was probably of less importance than the other. The meat in the stomach was too much confined by the string, was not permitted to move about freely in the gastric fluids by the natural motions of the stomach, and consequently did not digest so fast as it otherwise would have done. Another circumstance or two may also have contributed to interrupt the progress of digestion, such as anger and impatience, which were manifested by the subject during this experiment.

His conclusions following experiment No. 35 contain some interesting statements.

In this experiment it took two hours for the digestion of a meal of bread and milk, something shorter than the usual time for the disposal of an ordinary meal. For those who have healthy and unsophisticated stomachs, milk appears to be one of the best articles of diet we possess. It is less stimulating than flesh, and more nutritious than vegetables. For persons who are disposed to pyrexial complaints, and who are not obliged to perform hard and exhausting labour, it is the most appropriate diet. But the stomach is a creature of habit. It can become accustomed to any kind of diet; and sudden changes are liable to derange its healthy actions. To those accustomed to what is called high living, such as strong meats, strong drinks, and high seasoned food of all kinds, the transition to a milk diet, which contains a considerably lowered stimulation, would probably be an imprudent change. When necessary, the change should be so gradual that the stomach should by degrees become accommodated to it.

Throughout the entire series of experiments he was suspending various articles of diet in the stomach, noting carefully the time required for the digestion of each, and concluded this lot of experiments with a series of ten, intended to demonstrate, if possible, "what effects were produced by the bile and pancreatic juice when added to chyme," making the following notation:

Not being able to procure human bile in a pure state, I obtained some beef's gall, and for pancreatic juice substituted diluted muriatic acid (one scruple acid to six ounces water). I was induced to use this acid from a resemblance observed between its taste and that of the pancreatic juice, and not being able to obtain any of that fluid at the time. These experiments are certainly very imperfect, but, such as they are, I submit them to the public. They may tend to pave the way to more perfect experiments on these fluids.

It is needless to say that he arrived at no satisfactory conclusions concerning the bile and pancreatic juice. In summing up this entire series of fifty-six experiments, he says:

The preceding Experiments, I think, plainly demonstrate the solvent properties of the gastric juice. When aliment is submitted to it out of the stomach, its operation is rather slower than when the process of digestion is assisted by the natural warmth and motion of that organ. One reason probably is the difficulty of maintaining a bath at the exact necessary temperature, and another one may present itself in the impossibility of perfectly imitating the motions of the stomach. With all these disadvantages, however, chyme formed in this way presents the same uniform, sensible appearance as that which is formed in the stomach by natural process.

That the cold gastric juice should not act at all, or but very imperfectly, on aliment is no proof, in my opinion, that it does not possess solvent powers, even on the admission that it was a debatable question. There are but few chemical combinations that do not require caloric to effect their operations, and none perhaps that are not facilitated by it. Some, and indeed many of them, require an intense heat. I am under the impression, though I have never fairly tested the truth of it, that gastric juice would, in a sufficient length of time, act on aliment in a cold state. But I am not anxious to contend for any extraordinary or unnecessary powers of this fluid, nor is it necessary to prove that it will act on cold substances or in cold situations. It is perfectly manifest that its operation is that of a chemical agent, that it dissolves aliment out of the stomach when the warmth and motions of that organ are imitated, and that it changes the various and heterogeneous articles submitted to its action to an uniform, homogeneous semi-fluid, varying, however, slightly in color and consistence, according to the aliment used.

While in the midst of these interesting observations, Alexis again determined to go to Canada, but this time not without the consent of his benefactor. Dr. Beaumont, referring to this event, says:

In the spring of 1831 circumstances made it expedient for him to return with his family from Prairie du Chien to Lower Canada again. I relinquished his engagements to me for the time on a promise that he would return when required, and gave him an outfit for himself, wife, and children. They started in an open canoe via the Mississippi, passing by St. Louis, Mo., ascended the Ohio river, then crossed the state of Ohio to the lakes, and descended Lake Erie, Ontario, and the River St. Lawrence to Montreal, where they arrived in June.

Unlike the previous occasion on which St. Martin took his departure, Dr. Beaumont felt certain that he would be able to obtain his services again for further experimentation. He felt, however, that he had gone about as far as he could without the assistance of others, and determined to get in touch, if possible, with men trained in the art of investigation, and especially on the subject of chemistry. He was aware that the best chemists, in fact the only physiological chemists, were then in Europe, and so he applied for a furlough of one year, to begin in September, 1831, for the purpose of going abroad with

Alexis. He was gratified to receive on May 23, 1831, the following communication from Surgeon-General Lovell:

> I have just returned to Washington after an absence of several weeks, and submitted your application for a furlough for one year to the secretary of war, who has directed it to be given at such time as you may wish. Nearly every officer of the department being on duty, and but one at a post, with but two exceptions, it will take some time to relieve you, but I will endeavor to do it by the time you mention—that is, "on or before the last of September, next." I should be much pleased to see Alexis on your route to the East.

Shortly after this, "Special Order No. 78" was sent out from the Adjutant-General's office by order of Major-General Macomb to the effect that "Assistant-Surgeon Elwers will repair to Prairie du Chien,

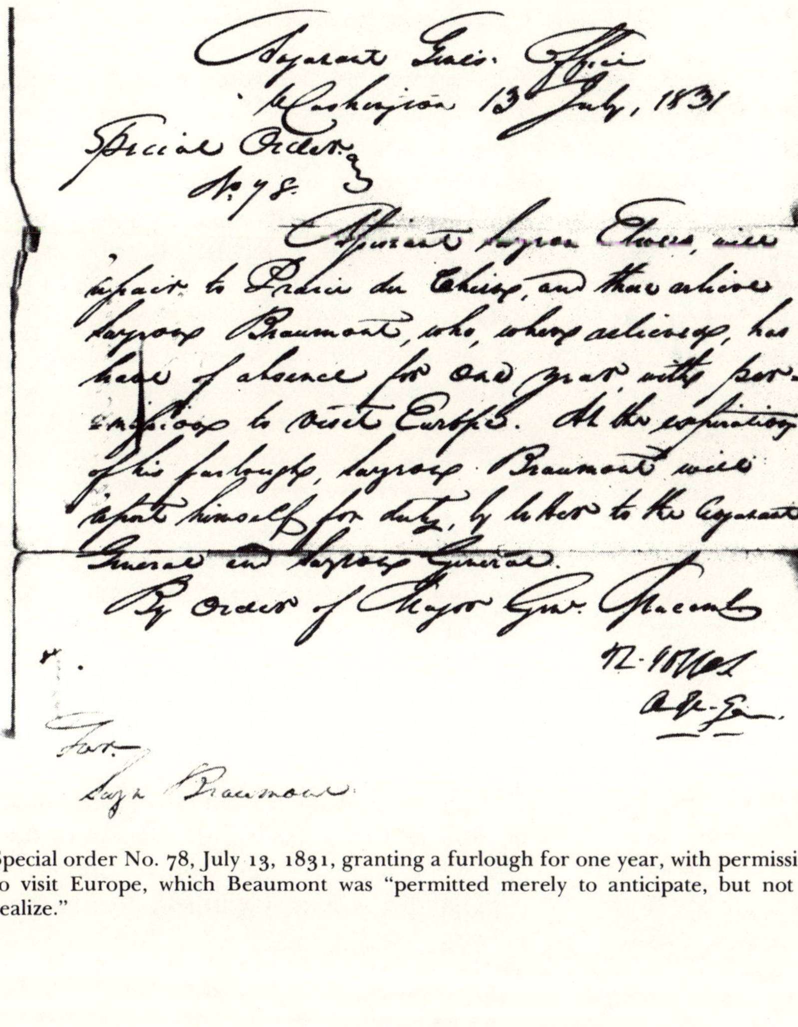

Adjutant Genl's Office
Washington 13th July, 1831
Special Orders
No. 78.

Assistant Surgeon Elwers, will repair to Prairie du Chien, and there relieve Surgeon Beaumont, who, when relieved, has leave of absence for one year, with permission to visit Europe. At the expiration of his furlough, Surgeon Beaumont will report himself for duty, by letter to the Adjutant General and Surgeon General.

By order of Major Genl. Macomb
R. Jones
Adj. Genl.

For
Surgn Beaumont.

Washington University School of Medicine

Special order No. 78, July 13, 1831, granting a furlough for one year, with permission to visit Europe, which Beaumont was "permitted merely to anticipate, but not to realize."

and there relieve Surgeon Beaumont, who, when relieved, has leave of absence for one year, with permission to visit Europe. At the expiration of his furlough, Surgeon Beaumont will report himself for duty by letter to the Adjutant-General and Surgeon-General."

Beaumont was doubtless elated over the fact that he would now be able, through the aid of the European scientists, to fathom some of the problems of digestion, which up to this time had remained insurmountable. Everything was in readiness, therefore, for his departure, when, to his great disappointment, there was issued by order of the same authority on October 24, 1831, "Special Order No. 127":

> The public service will not permit the absence of Surgeon Beaumont from duty at this time; accordingly the indulgence intended to have been accorded to him in Special Order No. 78 is recalled until otherwise directed.

The "public service," in all probability, had reference to an Indian outbreak which was brewing about this time. Black Hawk, a noted chief of the Sac and Fox Indians, who had fought on the side of the British in the War of 1812, now began to resent what he considered the unlawful encroachment of the whites on his territory, and refused to submit to a treaty made by the other chiefs of his tribe in the previous year at Prairie du Chien, ceding all of the lands east of the Mississippi to the whites. During the year 1831 the savages were pretty completely humbled by General Gaines, but in the spring of 1832 they renewed their activities, and the Black Hawk Indian war began in reality. The scenes of their outrages were in upper Illinois and Michigan. Therefore all of the regular troops from Forts Howard, Crawford, and Dearborn were centered at Fort Dearborn, with a view to crushing them as quickly as possible. It was feared that other tribes, especially the Winnebagoes, might join the belligerent Sacs and Foxes, but through much diplomacy and maneuvering such an alliance was prevented. The war continued with vigor through the spring and summer of 1832, but in two battles fought on July 21st and August 2d the Indians were so thoroughly whipped and routed that a complete cessation of hostilities was brought about, and Beaumont returned with his regiment to Prairie du Chien.

It was anticipated in the beginning of the hostilities that the forces in the West would not be sufficient to cope with the Indians, and troops were ordered to the scene from Fortress Monroe and New York harbor under the command of General Winfield Scott. Seven companies of these troops left Buffalo for Detroit on July 3d on the steamer "Henry Clay." On the following day a man became suddenly ill on board with symptoms of Asiatic cholera, and for the first time in the history of the country the disease was introduced here. Just after their arrival at Detroit a second case occurred, and, in spite of every effort to prevent the spread of the disease, cases multiplied rapidly, and within ten days thirty-four deaths had occurred. Soldiers deserted in the panic which ensued and spread the disease broadcast. It was transmitted to Chicago by steamboats from Detroit, and there the victims were numbered by the hundreds. Troops marching from Chicago to the Mississippi, to aid those at Fort Crawford in the warfare against the Indians, carried the dreaded disease to Prairie du Chien and other points along the upper Mississippi. The conflict with the Indians was over before these troops reached the scene of battle, but they had met even a more fatal foe. Thus Dr. Beaumont's detention at Fort Crawford gave him a wide experience in two campaigns—that against the Indians and that against the cholera. In fact, the Black Hawk war was afterward known in army circles as the "cholera campaign."[1] His experience with the disease was evidently quite large, and he manifested considerable interest in the study of it. Some of his views on the subject are unique, to say the least. The following thoughts were found scribbled on a fragment of paper, and were probably written about this time:

> The Greater proportional numbers of deaths in the cholera epidemics are, in my opinion, caused more by fright and presentiment of death than from the fatal tendency of violence of the disease.
>
> Like a snake watching and writhing for its prey, the Cholera lurks unseen through the pestiferous and malarious atmosphere, stealing upon human beings and thickly populated places, and gluts its cadaverous appetite more by the fears and dread it occasions than by its otherwise naturally fatal effects upon human life.
>
> The consternation and terror excited in community when the cholera makes its appearance is like that extraordinary instinct which compels animals and birds to thrust themselves into the open mouth of the serpent, whose almost only power to destroy consists in the fatal charm or terrific influence of the most distressing fear and dread which it is capable of inflicting on its intended victim.
>
> The general panic, like a phantom, pursues the public imagination; community becomes alarmed; individuals cannot abstain from the most distressing foreboding and sickening apprehensions of an attack and fatal termina-

[1]Brown's *Medical Department of the United States Army.*

tion; some fly from the first scenes of its occurrence, with a view of escaping its ravages; but from the effects of their fears, like the charm'd bird, flying in the utmost distress and agony from bush to bush and constantly approximating the dreaded destruction, they only approach nearer and nearer to the fatal influence of the dreaded disease, and finally thrust themselves resignedly into the gaping jaws of death, as if to rescue their minds or imaginations from the most agonizing pains of fear and dread.

So long as the general panic be unexcited and individuals remain unalarmed, and reason has control of intellect, the disease is little or no more fatal than many other epidemics. But when community becomes alarmed, and the public imagination dwells upon its fatal influence, and individuals constantly forbode death and general destruction, great fatality ensues, thousands sacrifice their rational faculties to their fears and become an easy prey to this fatally dreaded disease.

In this connection it will not be amiss to present some notes jotted down by him on the "History of the Intermittent Fever As It Prevailed at Prairie du Chien in the Summer and Fall of 1830," in which he shows a very clear insight into some of the causative factors of malaria, though unaware of its true cause.

In the month of July a few irregular intermittents made their appearance, both among the soldiers and inhabitants of the place, in company with mild remittent Dysenteries and Diarrhoeas. The weather during this month was extremely hot and dry, with almost constant south winds. The Mississippi river had been very high during the months of May and June, so as to overflow much of the prairie, but subsided rapidly in July, leaving numerous stagnant ponds and marshes exposed to the excessive heat of the sun. About the first of August the river rose rapidly again, so as to overflow its borders considerably, and fill the ponds and marshes again, but soon fell more rapidly than before, the hot and dry weather continuing, without any rain. Frequent heavy fogs came over early in the mornings during the forepart of the month, with occasional sudden changes of the hot, dry wind from the south to the west and north, continuing for a few hours only very chilly and damp.

In the first week in August Intermittents began to be more frequent and severe, but quite obscure in their type at first, appearing very commonly in the form of Remittents, accompanied with Dysenteric symptoms; but in a few days would put on their true character of regular Intermittents—at first of the Tertian type, but in the third week assumed the Quotidian type, generally with severity of attacks and obstinacy of symptoms; and cases increasing rapidly from 10 to 20 daily.

Early in June he had again been permitted to anticipate a furlough for the purpose of visiting Europe, and his hopes were again raised by "Special Order No. 79" from the Adjutant-General's office:

A furlough of six months, with permission to visit Europe, is hereby granted Surgeon Beaumont, to take effect as soon as his services can be dispensed with.

With the Indian war now ended and the cholera epidemic under control, the public service no longer required Dr. Beaumont's presence at Fort Crawford, so that on August 22, 1832, Colonel Zachary Taylor, then in command of the United States forces at Fort Crawford, issued the final order which permitted his departure from the Northwest:

The services of Surgeon Beaumont being no longer necessary here, he has permission to avail himself of the furlough which has been granted him for six months by the proper authority. Dr. Beaumont will accompany and attend the wounded militia as far as St. Louis who were left here in the hospitals and who are going down in the steamboat William Wallace to that place.

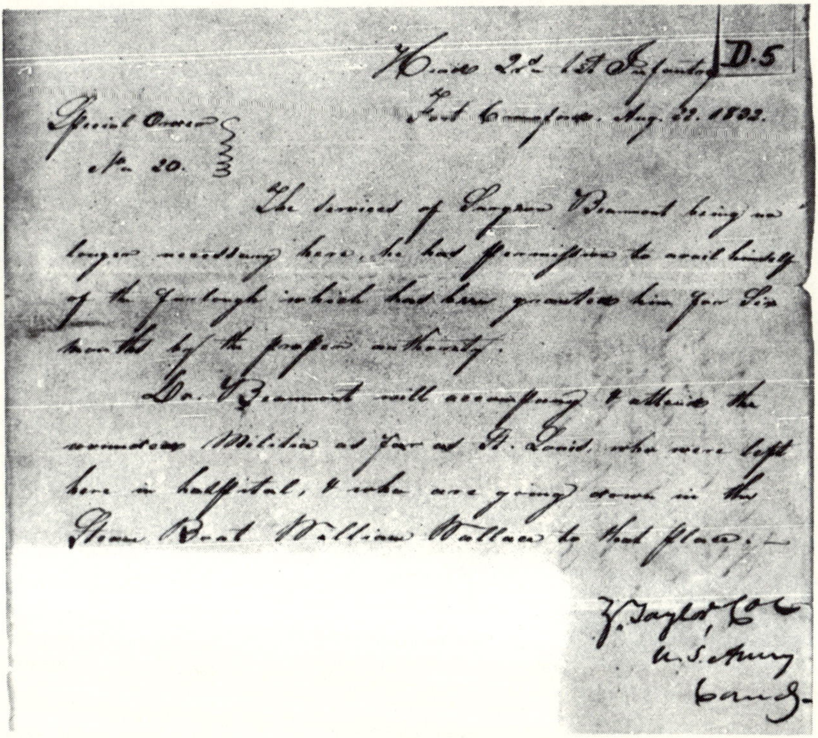
D.5

Head Qrs 1st Infantry
Fort Crawford. Aug. 22. 1832.

Special Orders
No. 20.

The services of Surgeon Beaumont being no longer necessary here, he has permission to avail himself of the furlough which has been granted him for six months by the proper authority.

Dr. Beaumont will accompany & attend the wounded Militia as far as St. Louis, who were left here in hospital, & who are going down in the Steam Boat William Wallace to that place. —

Z. Taylor Col.
U.S. Army
Comdg.

Washington University School of Medicine

Special order No. 20, August 22, 1832, signed by Zachary Taylor, colonel in the United States Army, granting Beaumont a six month furlough.

He evidently left the fort on the following day with the full intention of spending his furlough abroad, for we find among his papers a letter of introduction, dated August 24, 1832, from James G. Soulard, then of Galena, Illinois, whom he met on board the "William Wallace," to his uncle, Mr. Benjamin A. Soulard, of Paris:

I have just returned from a short trip, and on the same boat was my friend, Dr. Beaumont, a distinguished surgeon in the United States Army. He told me that he was about to leave for Europe and that he should pass some time in Paris. He is taking with him a man who is said to have recovered from the most extraordinary wound ever known up to the present time. I take pleasure in availing myself of this opportunity of introducing him to you as a very eminent man, worthy of your confidence. I foresee, however, that considerable difficulty will arise when you meet, as he does not speak any French and you speak no English, but I trust that you will find some means of communicating with each other, as he is a lovable and dignified man possessed of great knowledge.

After escorting the wounded militia safely to St. Louis, he promptly made his way toward Plattsburgh with his family, and in September reported himself there, "in quest of his man, St. Martin."

The home-coming of the Beaumonts on this occasion is well described by their niece,[2] then a mere child, living in Plattsburgh. She describes the home of her childhood to which the Beaumonts came:

A homely, bizarre bit of architecture, gable fronting on the turn of Margaret into Broad street. . . . No coat of paint in our remembrance had ever marred the gray coloring of its woody outside fibre, to which time had already given a soft, velvety tinge. . . . The year 1834[3] closed up our residence in this ancient landmark, yet deserves another note from one of its inmates, as it brought back to the home fold, from a far off western military post, Fort Crawford, Dr. William Beaumont, his interesting wife and trio of young children.

Many years had come and gone since the last visit of Mrs. Beaumont to her Lake Champlain home, and the event seemed like some jubilant holiday to all concerned. The trio had but recently passed through the fearful ordeal of an Indian raid at Prairie du Chien, called the "Black Hawk War" in military parlance, and, although of brief duration, the raid had its horrors outside the fort. These laid strong hold on the nervous system of the madame, by nature the most cheery, brave, and buoyant among heroic women. But sight of tomahawk, scalping knife, and bleeding scalps carried outside by painted

[2]"A Flash From Memory," by Mrs. Elizabeth Smith Martin in the Plattsburgh *Republican* of June 6, 1896.

[3]Should be 1832.

fiends wrought mischief on her gentle and sympathetic temperament. Like her young relative a brief term of years before, she at last found happy retreat among kith and kin, who also shared in the glad jubilation. To this day the writer cannot solve the problem as to the how and where this increased domestic population found stow away accommodation in the small, low browed domicil. Yet there never seemed lack of room, enough and to spare, while ever from morning to night fall good cheer and boncomradeship kept carnival.

Dr. Beaumont's greatest pleasure on this occasion was found in Alexis' prompt appearance on the scene according to promise. The Doctor lost no time in again beginning his investigations, but before doing so he endeavored to force Alexis' loyalty through the medium of a written agreement. On October 16, 1832, the following legal contract was drawn up between the two parties, the like of which has never been duplicated in history:

Articles of agreement and Covenant, indented, made, concluded and agreed upon at Plattsburgh, in the County of Clinton and State of New York, the Nineteenth day of October, in the year of our Lord one thousand eight hundred and thirty-two, between William Beaumont, Surgeon in the Army of the United States of America, of the one part, and Alexis St. Martin, Laborer, of Berthier, in the Province of Lower Canada, of the other part, as follows, to wit:

The said Alexis St. Martin, for the consideration herein mentioned, doth covenant, promise and agree to and with the said William Beaumont, his heirs, executors, administrators and assigns, by these presents in manner following—that is to say, that he, the said Alexis, shall and will for and during the full term of one year, to begin and to be accounted from the date of these presents, serve, abide, and continue with the said William Beaumont, wherever he shall go or travel, or reside in any part of the world, his covenant Servant, and diligently and faithfully, and according to the utmost of his power, skill and knowledge, exercise and employ himself in and do and perform such service and business matters and things whatsoever as the said William shall from time to time order, direct, and appoint to and for the most profit and advantage of the said William, and likewise be just and true and faithful to the said William in all things and in all respects.

And the said Alexis, for the consideration herein after mentioned, further specially covenants and agrees with said William that he, the said Alexis, will at all times during said term, when thereto directed or required by said William, submit to, assist and promote by all means in his power such Physiological or Medical experiments as the said William shall direct or cause to be made on or in the Stomach of him, the said Alexis, either through or by the means of the aperture or opening thereto in the side of him, the said Alexis, or otherwise, and will obey, suffer and comply with all reasonable and proper orders or experiments of the said William in relation thereto, and in relation to the exhibiting and showing of his said Stomach, and the powers and properties thereof, and of the appurtenances, and powers, properties, situation and state of the contents thereof. It being intended and understood both by said William and said Alexis that the facilities and means afforded by the wounds of the said Alexis in his side and stomach shall be reasonably and properly used and exhibited at all times upon the request or direction of said William for the purposes of science and scientific improvements, the furtherance of knowledge in regard to the power, properties and capacity of the human Stomach.

And in consideration of the premises, and of the several matters and things by the said Alexis to be performed, suffered and done as aforesaid, according to the true intent and meaning of the premises, and on condition that the said Alexis shall and does perform the same on his part, according to the true intent and meaning thereof, and not otherwise, the said William Beaumont doth for himself, his heirs, executors and administrators covenant, promise and agree to and with the said Alexis by these presents that the said William shall and will at all times during said term find and provide unto and for the said Alexis suitable, convenient rooms or house when with and in the service of the said William, and also defray the necessary expenses, and furnish the said Alexis good, suitable and sufficient subsistance, washing, lodging and wearing apparel when journeying with and at the request and directions of the said William. And also well and truly pay, or cause to be paid, unto the said Alexis, his executors or administrators, the just and full sum of one hundred and fifty dollars lawful money of the United States of America in Manner following, to wit: the sum of forty dollars, parcel thereof, to be paid to said Alexis at or within one day after the execution of these presents, and the residue thereof, being one hundred and ten dollars, to be paid on personal application to said William, his executors or administrators at the expiration of the said term, which will be one year from the date hereof.

In witness whereof, as well the said Beaumont as the said Alexis St. Martin have hereunto set their respective hands and seals, the day and year first herein written, in the presence of each other and in the presence of Jonathan Douglas Woodward, Esquire, the subscribing notary Public.

J. DOUGLAS WOODWARD. WM. BEAUMONT.
THOMAS GREEN. his
BENJ. J. MOORES. ALEXIS X ST. MARTIN.
mark.

UNITED STATES OF AMERICA,
STATE OF NEW YORK,
CLINTON COUNTY.

I, the subscriber, a Notary Public in and for said State, duly Commissioned and sworn and authorized in all respects to act as such Notary Public, do hereby certify and attest that the William Beaumont and Alexis St. Martin, who are both personally known to me, executed and signed and delivered the above instrument in writing in my presence and in the presence of each other,

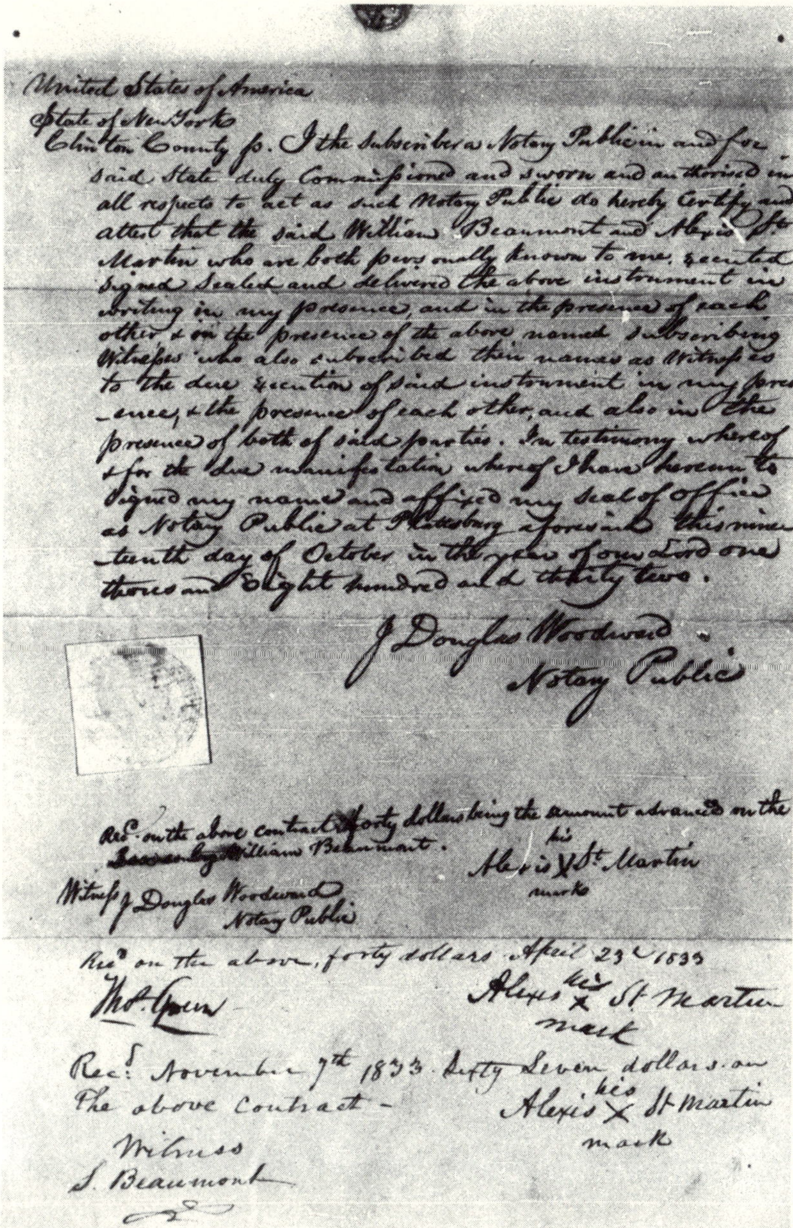
United States of America
State of New York
Clinton County ss. I the subscriber a Notary Public in and for said State duly Commissioned and sworn and authorised in all respects to act as such Notary Public do hereby Certify and attest that the said William Beaumont and Alexis St. Martin who are both personally known to me, executed signed sealed and delivered the above instrument in writing in my presence, and in the presence of each other & in the presence of the above named subscribing Witnesses who also subscribed their names as Witnesses to the due execution of said instrument in my presence, & the presence of each other, and also in the presence of both of said parties. In testimony whereof & for the due manifestation whereof I have hereunto signed my name and affixed my seal of office as Notary Public at Plattsburg aforesaid this nineteenth day of October in the year of our Lord one thousand Eight hundred and thirty two.

J Douglas Woodward
Notary Public

Recd on the above contract forty dollars being the amount advanced on the ~~[illegible]~~ William Beaumont.
Alexis his X mark St Martin
Witness J Douglas Woodward
Notary Public

Recd on the above, forty dollars April 23d 1833
[illegible]
Alexis his X mark St Martin

Recd November 7th 1833 Sixty Seven dollars on the above Contract —
Alexis his X mark St Martin
Witness
S. Beaumont

Washington University School of Medicine

The last page of a contract dated October 19, 1832, between Beaumont and St. Martin, granting the use of the latter's stomach for experimental purpose. At the bottom are seen St. Martin's acknowledgments of payments, amounting to $147.

and in the presence of the above named subscribing witnesses, who also subscribed their names as witnesses to the due execution of said instrument in my presence and in the presence of each other, and also in the presence of both of said parties.

In testimony whereof and for the due manifestation whereof, I have hereunto signed my name and affixed my seal of office as Notary Public, at Plattsburgh aforesaid, this Nineteenth day of October, in the year of our Lord one thousand Eight hundred and thirty-two.

J. DOUGLAS WOODWARD, Notary Public.

On taking his departure from Plattsburgh with Alexis, with the intention, no doubt, of going abroad, he left a little missive for his wife which he entitled "Pill Cochie's[4] parting admonition to his dear old wife Debh:"

Take good care of thyself and the children; use every means to preserve their health and thine own; be neighborly and sociable with all thy good friends; eat their good dinners and drink their good tea when you have a fair chance. Be not melancholy or much depressed at my absence; drive evil forebodings from thy mind, and anticipate my happy return in the Spring. Make all around you happy by the exercise of your natural vivacity and cheerfulness of disposition.

Bestow a goodly share of your care and attention upon Sarah; you can scarcely be too vigilant in the daily observance of the improvement of her mind and morals, in the correction of every little irregularity of her disposition. Excite her emulation to excel in virtue and usefulness; Examine her improvements from day to day—every day. Suffer not too much indulgence. But, above all, guard well her associations, and watch the communions she holds with herself and her playmates.

Suffer not Lucretia to be too much petted, but rather check her natural vanity and affection. Don't let the "young gentleman" get the upper hand of you, but apply the rod effectually when necessary to coerce obedience to your authority.

If Dr. Beaumont really had any intention of devoting his six months' furlough to a sojourn abroad, he seems to have decided soon after reaching Washington not to do so. It is said that he never quite forgave his superiors for curtailing his leave of absence, for he felt that the actual time that he could devote to his work abroad would now be much too short when the time required in going and coming had been subtracted. He determined, therefore, to spend the re-

[4]A pet name of Indian derivation, used by his wife. Mrs. Keim informs me that it was her father's custom, upon leaving home, to place loving missives for the various members of the family where they would be readily found.

mainder of his furlough in Washington, where he could have at least every facility that this country offered for conducting his experiments, and more especially could have access to books that were necessary to enable him to acquaint himself thoroughly with the literature on the physiology of digestion. That he did obtain a thorough knowledge of the literature is evidenced by the vast amount of manu-

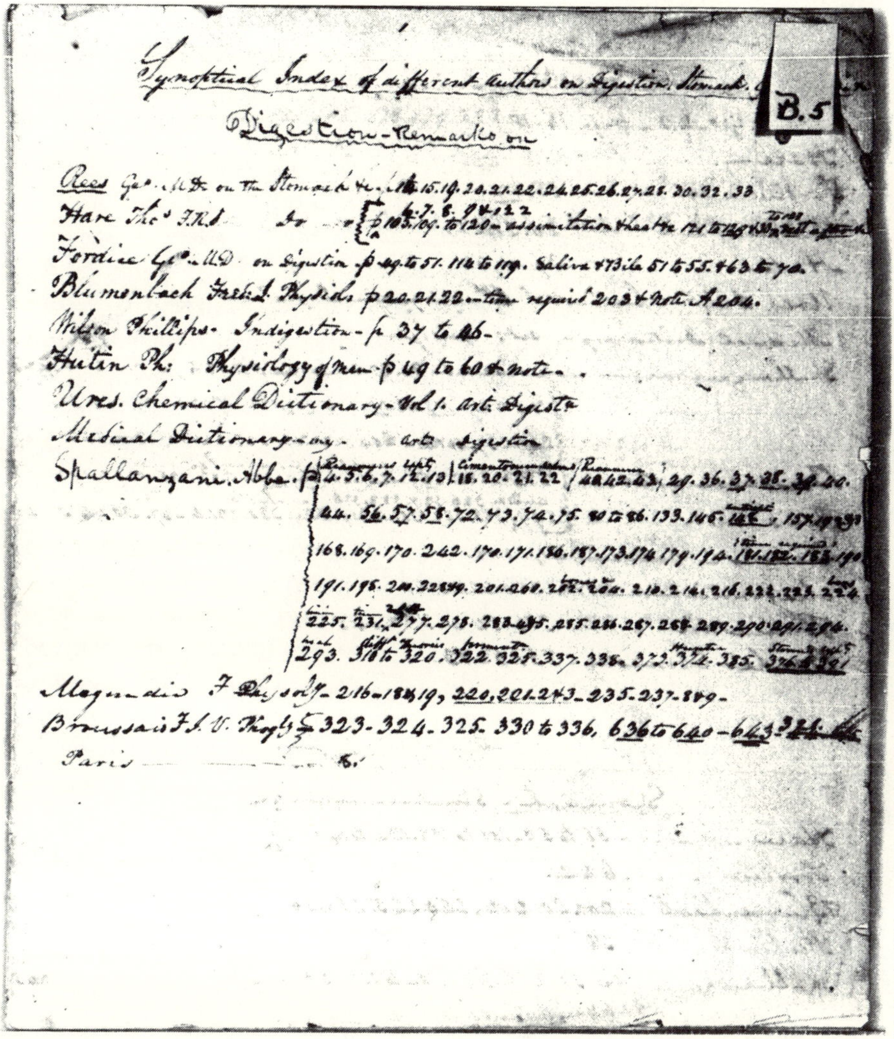

Washington University School of Medicine

Beaumont's synoptical index of various authors on digestion, revealing the great care and precision with which he worked.

script found in his own handwriting, in which he had copied page after page from the works of various authorities of those and previous days. While reading the works of Spallanzani, Philip, Magendie, Broussais, Leuret and Lassaigne, Gmelin and Tiedemann, and others, he copied verbatim that which impressed him because of its conformity with his own observations, as well as that with which he wished to take issue. And as he copied, he would intersperse the subject matter with parenthetical remarks, such as "true," "verily so," "certainly," "right," "even so," "so in Alexis," "this description exactly answers to the appearance of the mucous coats of my Frenchman," "your humble servant believes that the partial solution of food, etc.," "more time would have completed the solution, and more patience and closer observation would have confirmed the fact, no doubt," "so says Stevens of Edinburgh," "this last is hard to be believed, and the potato and parsnip difficult to be understood." His systematic and accurate methods are well shown in a "Synoptical Index of Different Authors on Digestion, Stomach, Gastric Juice, Etc.," which he kept much as one keeps a card index system today.

Life in Washington was much to his liking after years of trial and toil in military garrisons on the wild frontiers, surrounded by savages who were constantly haunting the forts in paint and feathers, and not infrequently parading with the scalps of their victims mounted on poles. How different at the capital—four months more of leisure to spend just as he wished, unhampered by routine duties, surrounded by books and men of note in medicine, law, and diplomacy, and with Alexis St. Martin at his beck and call. His work had already received recognition from such men as Edward Everett, Thomas Hart Benton, Lewis Cass, and others, and he received many invitations to dine and commune with the country's most illustrious men. Dr. Lovell was now able to render much assistance to him, and did so freely. There is evidence that they discussed every phase of the work together, and that the Surgeon-General gave him much valuable advice. Through him Alexis was made sergeant of a "detachment of orderlies stationed at the War Department in Washington, commanded by Lieutenant S. Cooper," and received payment for his services at the rate of $12 per month, an allowance of $2.50 per month for clothing, and ten cents per day for subsistence. This was done in order to make lighter the burdens of Dr. Beaumont, on whom Alexis was constantly making demands beyond the terms of their agreement. The "descriptive Role of Sergeant Alexis St. Martin" states that he is 28 years old, has dark eyes, dark hair, dark complexion, and is five feet five inches high, that he was born in Berthier, Lower Canada, that his occupation is that of laborer, and that he enlisted on December 1, 1832, for a period of five years.

As soon as all of these details were accomplished, Dr. Beaumont started on his third series of observations, and between December 1, 1832, and March 1, 1833, recorded 116 experiments, some of which were simply in confirmation of those that had gone before—the testing of the temperature of the stomach when full, when fasting, when exercising, and when resting; the length of time required to digest various food substances, etc. Some experiments disproved the old theory of maceration, and in others he made an effort to imitate the gastric juice. Experiment No. 104, for instance, in which he tests the powers of a solution of muriatic and acetic acids, which at least simulated gastric juice in taste, revealed to him that "the gastric juice contains some principles inappreciable to the senses or to chemical tests." He found that "when there are indications of disease on the coats of the stomach and on the tongue, digestion is consequently protracted." Alexis, it seems, was very fond of alcohol, and at times drank immoderately. Accordingly, therefore, we find in connection with one of the experiments such a statement as, "the diseased appearance of the stomach at this examination was probably the effect of intoxication the day before." While Dr. Beaumont had found that the "sense of hunger resides in the stomach, and is as well allayed by putting the food directly into the stomach as when the previous steps have been gone through with," there is no indication that Alexis was ever satisfied to receive his whisky in this manner, but took the position that this was one appetite which could not be appeased by pouring it in through the opening in the stomach. It will be remembered, however, that Dr. Beaumont had also found that the "sense of taste is very essential to the proper digestion" in that it stimulated the flow of gastric juice. It is interesting to note that in many of these experiments in which the Doctor wished to have the vials kept at the body temperature, they were carried about by the hour in the axilla, evidently with the idea that there were certain forces in animal heat or body temperature which could not be obtained on the water bath.

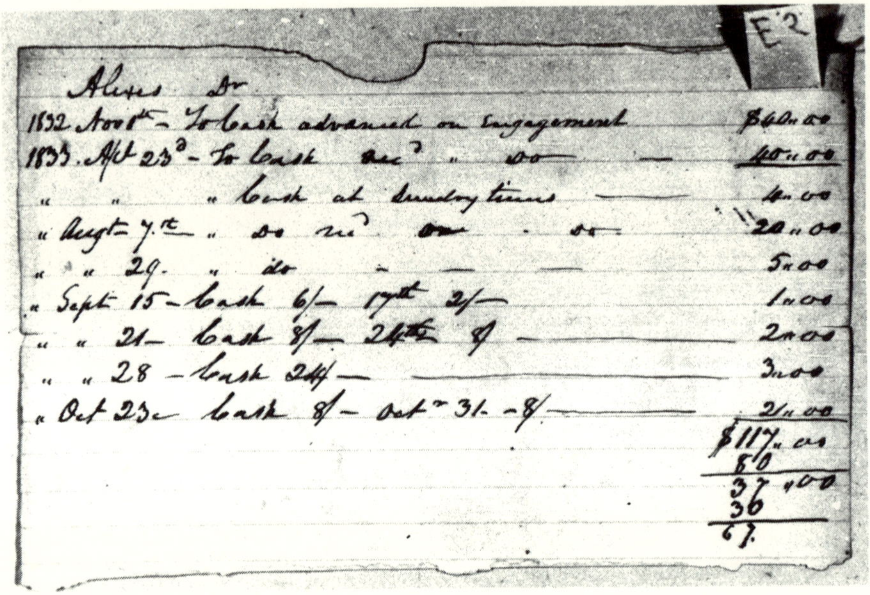

Alexis Dr

1832 Nov 1st – To Cash advanced on engagement $40.00

1833. Apl 23d – To Cash rec'd " do — 40.00

" " " Cash at sundry times — 4.00

" Augt 7th " do rec'd on do 20.00

" " 29 " do — 5.00

" Sept 15 – Cash 6/– 17th 2/– 1.00

" " 21 – Cash 8/– 24th 8/ — 2.00

" " 28 – Cash 24/ — 3.00

" Oct 23 Cash 8/– Oct 31 – 8/ — 2.00

$117.00

80

37.00

36

67.

Beaumont's cash account of sums paid St. Martin from November 1, 1832, through October 23, 1833.

1833

WHILE IN THE MIDST of these researches he succeeded in enlisting the assistance of two of America's leading scientists, Robley Dunglison, professor of physiology in the medical department of the University of Virginia, and Benjamin Silliman, professor of chemistry in Yale University. The correspondence with these two gentlemen is of great historic interest, and I am glad to be able to present here, in full, copies of all the letters. Not infrequently one side of such correspondence is brought to light after some years, but seldom indeed is it possible to present the letters interchanged by two individuals after a period of more than seventy-five years. For this opportunity we have again to thank Dr. Beaumont's remarkable traits of accuracy and method. Whenever he wrote a letter of some importance, he first scribbled a rough draft, interlining, erasing, and adding until the composition suited him, and then he would make a copy, filing away the rough draft for reference. This rendered his letters exceedingly difficult to decipher, and much time and care were required in the process. I have endeavored to reproduce them practically as I found them, and take this opportunity to state that allowance should be made for composition, use of capitals, punctuation, etc., in the letters of Dr. Beaumont, since these were not the exact copies of those which were forwarded.

Dr. Lovell had evidently written Professor Dunglison[1] on behalf of Dr. Beaumont, and received the following reply:

---

[1]Robley Dunglison came to the University of Virginia in 1824 from England at 26 years of age, crossing the ocean in the "Competitor." The voyage required four months, six weeks of which were spent beating about the English Channel. He was practically the founder of the medical department, and at different times served as secretary and chairman of the faculty. Dr. Dunglison soon took a high rank as teacher and writer, and was Mr. Jefferson's favorite physician. In 1833 he went to the University of Maryland as professor of therapeutics, and later to Jefferson College to take the chair of the institutes of medicine. He died April 1, 1869, having made many important contributions to medicine. *(Archives of University of Virginia.)*

UNIVERSITY OF VIRGINIA, January 12, 1833.

For Dr. Lovell, Surgeon-General,
Washington City.

DEAR SIR: My friend Mr. Trist will explain to you the causes that have prevented me from witnessing and assisting at the experiments instituted by Dr. Beaumont and yourself on the interesting and rare case which is now and has been engaging your attention, as well as from expressing in person my high sense of the honor you do me by desiring that I should be in any manner associated with you. I had seen a detail of certain experiments by yourself and Dr. Beaumont on the subject of this case several years ago,[2] and had fully intended to notice them in a work on Human Physiology recently published by me, but the Journal was mislaid, and I could not refer to it in time. One of these experiments, if I recollect rightly, was on artificial digestion, and the result showed that a piece of corned beef experiences the same changes as when it was inserted through the wound into the interior of the Stomach; and another, that the vegetable substances underwent chymefication in the stomach more speedily and thoroughly than different kinds of meat which were passed in at the same time. Both experiments were interesting as confirming the views of Spallanzani and the results of the experiments made at the Hospital La Charite of Paris on a female with a fistulous opening in the stomach, and those obtained by Helm, of Vienna, on two similar cases, as regards the effect of the juice contained in the Stomach in the solution of alimentary substances.

It would have been additionally instructive to have witnessed the effect of saliva only on the same Aliments out of the body for the purpose of deducing whether that fluid possesses the sole agency in digestion, as supposed by Montegre, or merely an adjuvant, as presumed by the best Physiologists; or what are the changes effected by it on the Aliment compared with those that result from the action of the compound fluid (formed of the various secretions from the supra-diaphramatic portion of the alimentary canal and of the stomach itself) met with in the interior of that organ, and which was the solvent in your experiments.

The results of your second experiment impressed me forcibly, being somewhat at variance with the inferences of Gosse, Montegre, Magendie, and others from their experiments, to which reference is made in the first volume, page 955, of my Physiology, regarding the comparative digestibility of animal and vegetable substances. Since the period at which your first experiments were made you have, doubtless, instituted others, which may have led you to confirm or disprove your first obtained result.

It would be signally interesting to me to learn the comparative digestibility, as tested by this individual, of the great chemical divisions of Aliments—amylaceous, mucilagenous, saccharine, aciduous, oily and fatty, caseous, gelatinous, and fibrinous—taking, as example of each, Starch (Arrow Root,

[2]*Medical Recorder,* containing first short series of experiments made at Mackinac.

University of Virginia Medical Center

Robley Dunglison, first Professor of Medicine at the University of Virginia.

sago, ordinary wheaten starch), Mucilage or Gum (Gum Arabic), Sugar, Acidulous fruits (with and without the skins), Butter or Suet, Cheese (mild and pungent), Gelatine (Isinglass), Albumen (fluid and concrete, as in the raw and boiled white of egg), and Fibrine (formed by repeatedly macerating the slices of muscular flesh in water under 150° Fahrenheit), as well as of the individual articles forming these divisions when compared with each other. These experiments might be made either in the stomach or artificially, and it might be instructive to adopt both courses.

As regards the nature of the fluids met with in the stomach of fasting Animals, and to which, collectively, the term gastric juice has been applied, experiments exhibit great discrepancy. It would be gratifying to me to learn the general result of your Researches. Did Litmus Paper indicate the presence of any free acid or alkali? If acid, could it be discovered, by burning, whether the muriatic or Acetic was the one in question, or did the Fumes of Ammonia indicate the existence of either, or a solution of Nitrate of Silver that of the former? Did the fluid obtained from the stomach, whilst fasting, deprive putrid substances of their septic characters? Did it remove the flavor of certain aliments, as of wine? Did you examine it with the microscope—that least satisfactory of all methods for investigating the nature of animal fluids—and, if so, what were the appearances? Did it always coagulate milk? It would afford me great satisfaction also to learn the effect of subjecting pure Fibrine, Albumen, or Gelatine to the action of this fluid out of the stomach, and to see how far they experience mutations in their sensible and chemical properties. When the experiment is made on pure Fibrine, the Albumen contained in the gastric fluid may be precipitated by a solution of the Bichloride of Mercury. The precipitate is a compound of the salt of albumen, in the proportion of about one of the former to three or four of the latter, so that by drying the precipitate the quantity of Albumen in the fluid can be easily determined. If the fibrinous solution be now evaporated at a moderate heat until it forms a thick mass, and concentrated acetic be added, with the assistance of heat a tremulous jelly is formed, which is completely dissolved by the addition of warm water, provided the mass be simply fibrine; but if the Fibrine has experienced changes during the process of artificial digestion or chymefication, different results will be obtained.

In like manner, if the artificial digestion has been accomplished on Albumen, provided the Albumen has experienced no conversion, the solution of Bichloride of Mercury will precipitate it, and the quantity of Albumen so precipitated may be compared with that subjected to the process of digestion.

Lastly, if Gelatine has been employed, the quantity remaining after digestion may be approximated by precipitating it with Tan, prepared by infusing an ounce of Gall nuts in a pint of water. The quantity of Gelatine in the precipitate may be roughly appreciated by considering that there are somewhat less than two parts of Tan to three of Gelatine.

In the case of the patient at La Charite, the food during its conversion into Chyme appeared to have acquired an increase of its gelatine and a substance in appearance fibrinous, but others have asserted that gelatine has not been met with in the chyme, which is scarcely comprehensible where gelatine has been the Aliment, as the conversion must have been total.

You will pardon me for the length of the preceding detail, every topic of which has probably suggested itself to you already. It will afford me great pleasure to learn any facts—which the Case has taught you and Dr. Beaumont—of a physiological character. I do not abandon the hope of being in Washington in the course of a few days. At present the roads from their roughness are almost impassible, but, if the frosty weather continues, they will be in order probably by the commencement of next week. The whole journey has now to be performed by land, and our Virginia Roads are proverbial for their badness. Should I, however, be prevented from visiting Washington, I may perhaps be permitted to request an account of your experiments and observations, in order that I may make all of them in a second edition of my work on Physiology, should one be demanded, or communicating them in your name to the American Philosophical Society, or both, should such be your desire.

I am with best respect,

Obediently yours,

ROBLEY DUNGLISON.

Conditions "under foot" improved within a few days, as Professor Dunglison prophesied, and he was soon able to make the trip from Charlottesville to Washington over the proverbially bad Virginia roads. He had the opportunity, therefore, to satisfy himself in person as to many of the points suggested in his letter, and to give Dr. Beaumont some valuable suggestions in the further conduct of his experiments. A careful perusal of the experiments made between January 24th (experiment No. 70) up to the end of this series shows that Beaumont carried out practically all of the suggestions offered him by Dunglison. The microscopic examinations were also made during Professor Dunglison's stay in Washington, and were conducted by him and Captain H. Smith, of the army. He concluded, however, that the microscopical examination of the gastric juice afforded very little information on the subject of digestion. What Dr. Beaumont most desired of Dunglison at this time was a careful chemical analysis in order to determine, if possible, those "solvent ingredients" which up to this time had baffled the chemists who had attempted the analysis of the gastric juice of both the human being and lower animals.

On January 25th Beaumont left a bottle of gastric juice, doubtless at Professor Dunglison's hotel on the day of his departure, with the following note, in which he thanks him profusely for his assistance, and in a subtle way requests Professor Dunglison not to anticipate him in the publication of any of the experiments that he had seen while in Washington:

Dr. Beaumont takes pleasure in acknowledging his high regard and thankfulness to Professor Dunglison for his kind and valuable assistance and lively interest taken in the prosecution of his Gastric Experiments, and hopes amply to be able to remunerate him for his generous exertions by affording him a more full and satisfactory view of the result of his observations when he shall have had time properly to collate and classify them.

The Bottle of Gastric juice is cheerfully submitted for chemical analysis, with a strong hope and expectation that Professor D. will succeed in obtaining very important and satisfactory results and communicate these in detail to Dr. Beaumont so soon as practically convenient.

Dr. B. is much indebted, and will be ever grateful, for the aid and instructions received from Professor D., and is disposed to reciprocate every generous desire, mutually to afford subjects for Physiological investigation and improvement so far as he is capable. Dr. B. hopes his feelings may be perfectly appreciated—similar confidence he could repose in but few—but he feels well aware that by Professor Dunglison he will never be improperly anticipated in his intentions, sooner or later, to publish the experiments collectively by premature communication to any Periodical Publication of partial results, whereby impressions of an undue transfer merit or demerit [?] from its original or collateral sources might be liable to be made upon the public mind. A simple notice of the case and the intention of publication may not be improper, and is not objectionable.

All communications and suggestions from Professor D. on this subject will be thankfully received, and gratefully, if not scientifically, appreciated and attended to.

WASHINGTON, D. C., 1 O.K., P. M., Friday, Jan. 25, 1833.

To this Dr. Dunglison promptly and courteously replied, putting at rest any fears entertained by Dr. Beaumont as to his publishing any part of the work.

UNIVERSITY OF VIRGINIA, January 29, 1833.

Dr. Beaumont, United States Service,
Washington City.

DEAR SIR: I received your letter so late that I was unable to answer it before I left Washington. I thank you very much for your Bottle of the gastric fluid, which I will examine at leisure, and be most happy to communicate the results as early as I conveniently can, as a trifling return for your kindness in permitting me to examine the case and to be associated with you in your experiments.

With regard to the publication of any results, you may rely upon me implicitly. I have but one desire in the prosecution of my profession—by teaching and by practice to benefit my fellow-men, and that can always be done with due credit without forestalling my coadjutors in the field of science, or arrogating to myself merit to which I may be but secondarily entitled. The praiseworthy manner in which you have entered your experiments demands my thanks, as well as those of every lover of science, and I have no doubt that when complete they will add largely to our amount of useful knowledge.

I will with pleasure afford you every facility in the communication of the results of your labors to the public, and shall be quite satisfied to perform a secondary part. Certainly I shall not anticipate you. The field is yours, and I wish you, for your own sake, to tell it to the best account. By the bye, do me the favor, at your leisure, to experiment artificially with the saliva, acidulated with the phosphoric the muriatic and acetic acids, and to note the results.[3] They will be valuable as connected with the question on the agents of digestion raised by Montegre, and referred to at page 589 of my Physiology.

It will afford me sincere pleasure to hear from you at any time, and to find that your labors are duly appreciated by the liberal and enlightened members of our Profession, who unfortunately form but a small portion of the mass.

Tomorrow I hope to be able to institute experiments for detecting with accuracy the nature of the acid which gives character to the gastric fluid. The sensible properties and the rough chemical experiments made at Washington lead me to believe that it is the muriatic. This we shall fix with certainty, as well as the other great ingredients.

Believe me, dear sir, with great respect and esteem,

Yours truly,

ROBLEY DUNGLISON.

Without awaiting a reply to this letter, Dunglison sent another on February 6th, reporting the preliminary results of his chemical tests:

UNIVERSITY OF VIRGINIA, February 6, 1833.

To Dr. Beaumont, United States Service,
Washington City.

MY DEAR SIR: Since I last wrote to you, my friend and colleague, Professor Emmet and myself have examined the Bottle of Gastric fluid which I brought with me from Washington, and we have found it to contain free muriatic and Acetic acid, Phosphates and muriates, with bases of Potassa, Soda, Magnesia, and Lime, and an animal matter soluble in cold water, but insoluble in hot. We were satisfied, you will recollect, in Washington that free muriatic Acid was present, but I had no conception it existed to the amount met with in our experiments here. We distilled the gastric fluid when the free

---

[3]*Beaumont's Experiments,* No. 96, Plattsburgh edition, 1833.

acids passed over, the salts and animal matter remaining in the retort. The quantity of Chloride of Silver thrown down on the addition of the Nitrate of Silver was surprising, but unfortunately the quantity of juice we possessed proved insufficient to estimate the precise amount of the ingredients, a matter, however, which I conceive to be of no great importance. Should you be able, however to accumulate for me half a pound of the Juice, and give it to Mr. Trist to send on to me, we may be capable of proceeding still farther. I am satisfied, however. I think it probable that the Acetic Acid is the great disinfecting agent. If you will take a drop or two of the strong Acetic Acid (Radical vinegar), and add it even to a wineglass of water, you will find it disinfects a putrid substance with great rapidity. Allow me to suggest to you to try a portion of animal substance (masticated by Alexis), half in a mixture about as sour as the real juice and the other half in the gastric juice, and observe the character of the solution effected in each.[4] Suppose you take a wineglass and afterwards precipitate it by Tan.

By the way, in the number of the *American Journal of Medical Science* just issued you will see a Review of my *Physiology* in which the writer accords with Dr. Jackson and others on the subject of stomachal digestion. There is a systematic obstinacy occasionally amongst Physiologists which requires time as well as evidence for its downfall, and one of the first things to be learned is to disinfect the mind of all bias, which is extremely difficult. Had I not been associated with you in our professional experiments and with my friend, Professor Emmett, in the chemical, I have no doubt it could have been said that I had determined to find certain results, and accordingly I had found them. You will, I am sure, do me the justice of saying that I endeavored to observe fairly.[5]

Your publication, which I hope will make its appearance soon, will upset, or ought to upset, this delusion on the subject of digestion, and I will back it so far as my feeble support can do so. Permit me to recommend to you, as a vehicle for publication, the *American Journal of Medical Science,* edited by my friend, Dr. Isaac Hays, of Philadelphia, who will be most happy to insert it, I am satisfied. It is in my opinion, and I believe in the opinion of the world in general, the best periodical on medical science published in this country.

It will afford me great pleasure to hear from you on this or any other subject.

In the meantime believe me to be, dear sir, with great respect and esteem,

Faithfully yours,

ROBLEY DUNGLISON.

To this Dr. Beaumont replied promptly, expressing the belief that the result of their work would ultimately overcome the "systematic obstinacy" of those who took issue with them.

---

[4]*Beaumont's Experiments,* No. 104, Plattsburgh edition, 1833.

[5]*Beaumont's Experiments,* page 79, Plattsburgh edition, 1833.

WASHINGTON, 10 Feby., 1833.

To Prof. Dunglison, University of Virginia.

DEAR SIR: I have the honor to acknowledge with pride, pleasure and satisfaction yrs. of the 29th ult., which would have been more promptly answered had not the delay been necessary to obtain results of the experiments you suggested. Your last surprised me in the act of complying with your request, and, having in part anticipated your last suggestion of experimenting with dilute muriatic and acetic acids, I will transmit them, imperfect as they be, and repeat the experiment according to your last suggestion soon as practicable and send you the result. The flattering manner in which you speak of my feeble efforts to promote Physiological science by diligent prosecution of my experiments, and your readiness to aid me in my prosecution, is very gratifying indeed to my feelings, and inspires the utmost confidence in the candor and correctness of your motives, fairness of your views, and liberality of your sentiments. The result of your Chemical analysis of the Bottle of Gastric juice is highly satisfactory and important, and shall be duly appreciated and applied. It will afford me great pleasure to obtain and send you for further analysis another bottle of Gastric Juice as soon as I can collect it pure. Since you have been here I have seen the necessity, doubled my diligence, of presenting my experiments on different soluble aliments with even greater minuteness, if possible, than before; though now sensible of my inability to do strict justice to the subject from having so plainly observed your preeminent qualifications, talents, and tact of easily testing and readily finding correct and important results.

I most cheerfully submit to your examination the products of the experiments you suggested in yours of the 29th ult., together with my field notes (to borrow an expression) made on the occasion. I also send, through the kind agency of your very good friend, Mr. Trist, specimens of the products of several small and minute experiments on different articles of diet, which, should they be deemed of sufficient importance, I should like to have submitted to chemical analysis and to be informed of the results. I have just obtained the last no. of the *Am. Medl. and Physical Journal,* and have but hastily perused the review of your recent valuable Physiological work. Though the reviewer handles the subjects generally with gloved fingers, he seems inclined to pinch a little on some points, and fortunately that Stomach Digestion; and, though the suggestions advanced by you on that subject need nothing but their own correctness to support them, yet I think with your assistance we shall be able to afford the weapons not only to pary their criticisms, but to make them recoil with severity, if you choose, upon the authors themselves, rendering them effectually disinfecting agents of the systematic obstinacy of their selfish minds and Pseudo Physiological Theories on this subject. That [the views and theories?] of the Philadelphia Physiologists, as some others, on the subject of Stomach digestion and the Gastric juice are radically wrong, I have not the least doubt, but demonstrated and multiplied facts and experiments to prove, of which more hereafter.

In accordance with your first suggestion, I took āā pts. of fresh secreted saliva and dilute Muriatic acid, reduced with water, as nearly as I could judge by the sense of taste, to the natural acid flavor of the gastric juice, and submitted to it a proportional qty. of the following vegetable substances—an entire square piece—10 grs. of boiled carrot, and also immersed the same qties. and kinds of vegetables in saliva and acetic, and diluted to about the same Strength and flavor with the muriatic; and treated them with my other experiments on the bath of 100 Fahr. for 48 hours, and then took out and weighed the respective pieces. The carrot in the muriatic mixture had lost nothing, but remained whole and entire. The parsnip had lost 2 grs. in the acetous menstrum. The fluid of both remained unaltered in their soluble qualities and appearance. After continuing them on the bath, with frequent agitation, 72 hrs. longer, the Parsnip in the muriatic menstrum had lost 4 grns.—whether dissolved or its very soft and tender substance merely diffused thro this fluid, I was not able to determine. The carrot remained the same as when put in. The Parsnip acetic mixture had lost 6 grns. and the carrot 4, but appeared more to be masticated and diffused than dissolved or digested. I now mixed them all together, and continued to treat them on the bath 24 hrs. longer, at the end of which the whole remaining mass of vegetables weighed together 12 grs. only, having lost one-half in that time. The fluid appeared now little more than chymous, was rather turbid, and resembled, more than anything else I can compare it to, the famed "Philadelphia vegetable soup" made of "two potatoes and an onion, boiled to smash in ½ qt. of water." I send you the product as it is. Please examine for yourself, and, if you choose, let me know your conclusion upon this as on the other subjects submitted, observing that after the mixture the wash of the vegetables become more evident and rapid. It occurred to me to mix the two acids together first, and then observe their effect on meats. Accordingly I mixed equal parts of the diluted acids and saliva. Into 3 drachms of this mixture immersed XX grs. lean broiled beefsteak unmasticated, but chopped fine with a knife, & for a comparison put the same qty. and kind of meat into 3 drachms Gastric juice, and placed them on the bath and treated them together for 6 hours. When the meat in the gastric juice was taken out and filtered dry, weighed 2 grs. only; that in the artificial menstrum was not diminished in weight at all, but had lost its fibrous form and became a jelly-like mass, thick and tremulous, like half dissolved glue—not converted into anything like chyme, nor bore much resemblance to the contents of the gastric solution. After digesting 8 hrs. longer on the bath, the meat in the gastric juice completely dissolved and that in the acid menstrum nearly so, a trace of jelly-like particles only remaining on the filter. The fluids now even a little more similar, though not alike—that of the gastric juice was opaque, of a light brown color, and afforded a dark brownish sediment on standing; the other was also opaque, of a dark reddish gray color, depositing no sediment. 3 drachms Infu. of gall added to the gastric portion threw down a fine, loose reddish brown precipitate, and afforded a semi-transparent liquor of similar color. 3 drachms of the Infu. added to the acid mixture converted the whole into loose coagula and suspended throughout the whole fluid mass, becoming finer by agitation and very slowly subsiding towards the bottom, and affording a clear, transparent, slightly yellowish fluid. I send them also for examination. I would send more similar experiments, but perhaps they would not be of sufficient importance to excite your interest or merit your attention.

The Gastric juice shall be forthcoming as soon as possible to collect it in quantity and quality sufficient for your purpose.

I shall hasten the publication of my experiments as fast as my circumstances and abilities will permit, consistently with a view to future benefit to secure credit to myself. Shall be always happy to consult your judgment relative to it, and be ever grateful for your proffered assistance and friendly advice on the subject. Shall duly consider your suggestion of making Dr. Hays, of the *Am. Journal of Med. Science,* the medium of my publication. I expect soon to make an abstract report of some of my expts. to the Surgn. Genl., reserving the entire series for other and future arrangements, which to my limited ability seems an herculean labor. I shall attend to the experiments suggested in your last in due time, and communicate the result of my observations. I shall always feel gratified and grateful for any communications or suggestions you may be pleased to make on this or any other subject.

Please accept, Sir, the renewed assurances of my great respect and esteem, and believe me,

Yrs. truly and respectfully,

WM. BEAUMONT.

A few days later this was followed by another letter, entering into further details concerning the analyses:

WASHINGTON, Feb. 19, 1833.

To Prof. Dunglison,
University of Virginia.

DEAR SIR: Agreeably to promise, I have now the satisfaction of sending you another bottle of gastric juice for further investigation. I hope it will be safely received, and afford you an opportunity to extend the highly important analysis so happily commenced with the first.

I have also with peculiar pleasure carefully experimented upon the masticated meat and gelatin, as suggested in your last letter, and herewith send the result of my observation. I hope you will excuse the awkward and unscientific manner in which they were made and here attempted to be described. To develop valuable facts is my chief aim and ambition, and, if I communicate such to you, even in my own style, I am confident they will be duly appreciated and kindly disposed of to the best advantage; and I have the peculiar satisfaction, and not only this satisfaction, but am happy and proud to assure you, so far as I have read and understand your very valuable treatise on Physiology, that the principles and suggestions and reasoning on the subject of the Gastric functions and fluids are generally most confirmed by my experiments, so far as I have been capable of observing.

I am well pleased indeed to have it in my power to assist in verifying the correctness of your theory, and irrefutably establishing physiological facts which have for ages divided the opinions and distracted views of Physiologists, since the latter times have become fashionable sources of speculation, error or professional obstinacy. I shall continue most cheerfully to receive and faithfully attend to any suggestions you may be pleased to make while I remain here and at N. Y., if you see fit to communicate to me there. I expect to leave this for that city about the 5 or 6th of March.

The small square Phials contain the products of the gelatine treated with Gastric and Dilute muriatic and acetic acids, agreeably to your suggestion. The phials containing the similar products of the masticated meat accidentally got broken and spilt. I send the result of my observations, which I hope you will sufficiently comprehend. . . . Infusion of Nut galls added to the solution in the gastric juice threw down a fine reddish brown precipitate, and afforded an opaque fluid of a similar color. The same quantity of the Infusion added to the acid mixture converted the whole into loose coagula, and slowly threw down a more copious precipitate and left a clearer and thinner fluid, nearly transparent.

I took the same quantity of dry gelatine, and put them into the same kind and measures of gastric juice and dilute acids, and treated them after the same manner of the meat. In 5½ hours the gelatine in the gastric was all dissolved.

The precipitates I have not disturbed to ascertain the definite weights. If it be of importance, it may be done after you remove it. Please direct to me at this place till the last of this month—after that to N. Y.

I am, Sir, with much esteem,

Respectfully yours, etc.,

WILLIAM BEAUMONT.

In the early part of March Dr. Beaumont made the official "abstract report" of his work to the Surgeon-General, previously mentioned in his letter to Professor Dunglison. This report was prefaced by a general statement leading up to the experiments, which is so typical of the man, and which reveals so well the conscientious spirit in which he carried on his investigation, that I can not desist from reproducing it in part:

With an honest desire to contribute, if I may, a mite to the promotion of Medical Science, and in accordance with your wish and the general views of the Department under which I act, I have the honor very respectfully to submit for consideration the following experiments and observations upon the gastric fluid and functions of Digestion as made upon Alexis Saint Martin. . . .

He then describes the difficulties and vicissitudes under which he had worked during the past six years, with special reference to Alexis' inconstancy and idiosyncrasies, and proceeds:

These, with other cases not in my power to obviate and control, have prevented more satisfactory results perhaps, and delayed an earlier communication of the subject. Having him now reëngaged for a season, I have the satisfaction of presenting to your favorable consideration the following experiments and observations as the result of my faithful, though feeble abilities, but diligent efforts, to contribute to the advancement of Physiological improvement.

From the vast number of treatises, observations and experiments that have been made on the subject of Digestion, almost from time immemorial to the present day, by the most illustrious and learned of the ancients, and the wisest and ablest Professors and Physiologists of modern times, it may perhaps be thought that these experiments and observations are altogether gratuitous, unnecessary and useless, many having already been communicated to the profession in far more scientific and satisfactory forms. But following the examples and guided by the precepts of the most eminent and useful of the profession, and having so rare and singular an opportunity, I cannot, in justice to my views of the importance of the subject, omit communicating such facts and observations as have been in my power to collect, however imperfect and unimportant they may appear, or feeble my ability to do it satisfactorily.

But in offering the following I can most truly assert that no favorite theory, system, or hypothesis, preconceived opinions or partiality for popular authority have had any influence in making or recording them.

A mere tyro in science, with a mind free from every bias, I commenced them, as it were, by accident, and continued desultorily to prosecute them, without regard to any particular arrangement, or the confirmation of anything save plain and palpable truths and physiological facts, aiming singly at the more perfect development of the nature of the Gastric juice and process of Digestion in the human Stomach, subjects which neither time, nor talents, nor labor, nor learning has yet satisfactorily illustrated.

If, in any degree, I succeed in thus contributing to the cause of science, I shall be satisfied with having bestowed my time and patience upon the subject, simply even to afford the materials for the Physiologists to cultivate and improve.

In the meantime Dr. Beaumont's furlough had expired, and he was ordered to New York City—doubtless at his own suggestion, for there seemed no particular need of his services there at that time. He probably thought that he would be able to gather some additional information from New York scientists. In order to make it possible to have Alexis with him and have his salary continued, a special order was issued from the adjutant's office to the effect that "a furlough of three months from the date hereof is hereby granted to Sergeant Alexis St. Martin, of the detachment of orderlies at the seat of government.

Shortly after his arrival there a letter came from Dr. Dunglison, in which it is interesting to note that, far from being able to "fix with certainty . . . the other great ingredients," he had arrived at the conclusions "that the precise constituents of the gastric juice may never be accurately determined."

UNIVERSITY OF VIRGINIA, March 23, 1833.

To Dr. Beaumont, United States Army,
New York.

DEAR SIR: I am glad to find that your ardor in the prosecution of your interesting experiments continues unabated. Within these last few days I have received the letter and parcel you kindly sent me by Mr. Rives, which were delayed from some cause which I have not yet learned. I thank you much for your valuable present of the gastric [fluid] and the results of some of your other investigations. They prove that the fluid is not simply muriatic and acetic acids in dilution, as I knew before. It is probable, too, that the precise constituents that exist besides these Acids may never be accurately determined. I forgot to suggest to you in my last to assimilate your menstrum still more closely to the gastric fluid by adding a small quantity of Phosphate of soda, and seeing whether the soluble powers were at all modified. In the compound gastric fluid the organic constituents, I have no doubt, modifies the chemical action in a way that cannot be done by any of our artificial menstrums. Still, your experiments appear to me to tend to the result at which we previously seemed to arrive—that the main gastric action is one of solution, not of chemical conversion. I should have been glad if you had separated by the filter the precipitate thrown down by the Bichloride before I added the Infusion of galls, taking care to add no more Bichloride of mercury than is absolutely necessary to throw down the Albumen—that is, adding it only so long as a precipitate is thrown down.

On adding the infusion of galls to the gastric solution of gelatine, and making a similar experiment with gelatine dissolved simply in water to which no Bichloride has been added, and weighing the precipitate formed in the two cases, you will be able to see whether as much gelatine remained in the gastric solution as was contained in the aqueous solution, or whether a part might not have experienced conversion. If you have any Bichloride remaining in the solution after precipitating the Albumen, it would affect the Nutgalls. Perhaps you will be good enough to repeat your experiments with this view. I will do so if I have gastric juice enough. I feel anxious, however, to retain a little as a specimen liquor, if it will keep. A portion of that which I brought with me from Washington, and which was left at the bottom of a bottle, is undergoing decomposition, and I intend to see how far it retains the property of affecting the work of Digestion. Have you tried this experiment?

I am much gratified to learn that you are pleased with my work on Human Physiology. It has been pleasing to me to discover that many of your Experiments on Digestion are confirmatory of my positions, especially as opposite Doctrines prevail so much amongst those who have not reflected deeply on the matter.

My object in that work was to endeavor to sift truth from the large mass of "false facts" which we possess, and, if I have succeeded, I shall have attained the object of my Ambition.

Dr. Hays, the Editor of the *American Journal of Medical Science,* has requested me to say to you that if you determine to publish the results of your experiments in his Journal, he will place a number of copies in pamphlet form in your hands. I promised to mention the proposition to you, but I presume your facts will form a volume.

I shall be very anxious to see the results of your labors. I am now engaged in a work on the Influence of Physical Agents on Human Health, and I expect on the subject of Diet to obtain much useful information from your publication. It will afford me great pleasure to hear from you at all times.

Believe me, dear sir,

Very truly yours,

ROBLEY DUNGLISON.

After a six weeks' sojourn in New York City, Dr. Beaumont wrote the Surgeon-General on April 16th:

I expect to leave the city to visit my family at the North in a few days, and probably before I can receive an answer to this. Will you, therefore, advise the communications on this and other subjects to be addressed to me at Plattsburgh, N. Y., till further advised? Dr. Russell will attend to my official duties during my absence. I shall endeavor to prosecute my experimental and physiological duties to better advantage for a few weeks there than is practicable to do here. I doubt not this measure will meet your approbation. . . .

I have been unable to do much at accurate experiments and observations since I came here, so numerous and increasing are the calls of the curious, the social, the scientific and the professional. Have not yet been able to complete the series I had on hand when I left Washington,[6] but I am determined to do it soon, however, if I even have to shut myself up with Alexis in a convent, or retire to some seclusion in the country. My official duties are very light, and would not interfere at all with my experiments, could I avoid the vexatious social intercourse to which I am perpetually exposed in this City. It is an unfavorable place for the pursuit of physiological inquiries and experiments. . . .

The professional gentlemen of this City have quite too much personal, political and commercial business on hand to permit them to turn their attention to animal and physiological chemistry, whose high honours and rewards to them are to be the results. Their curiosity once gratified, they are silent and aloof from the subject.

Disappointed in his efforts to obtain any new light in New York City on the chemistry of the gastric juice, he endeavored to enlist the

[6]The third series of experiments, conducted largely in Washington, was completed in New York City.

coöperation of Professor Benjamin Silliman, of Yale University, and with this object in view made a special trip to New Haven. He had only recently returned from this visit, and promptly wrote the Surgeon-General:

I have recently made a visit to New Haven, Conn., and submitted a phial of gastric juice to Prof. Silliman for tentative examination and analysis. He, with all the other Professors and Fellows of that institution, seemed to highly appreciate and earnestly expressed their conviction of the importance of the subject, and evince an earnest disposition to engage with much zeal and sincerity in promoting the investigation of the subject. Professor Silliman suggested the propriety and importance of sending a bottle of gastric juice to Professor Berzelius, of Stockholm, as "the man" (to use his own expression), "of all others, best qualified to investigate the subject of such deep interest to mankind," and very kindly offered to propose the subject, being in direct correspondence with him. Has given me a letter of introduction to the Swedish Consul at this place, and sent his communication to Professor Berzelius for me to forward with the juice and a concise statement of some of the most interesting circumstances of the case. I have about 12 collected, and when I have got 4 or 6 more of the pure gastric fluid, shall hand [it] over to the Consul, to be forwarded to the Swedish professor, with the expectation of receiving most important results from his investigation and analysis in the course of three or four months. Professor Silliman will in the meantime attend to the analysis of what he has, and give me the results of his labours.

Benjamin Silliman, the elder,[7] held at this time the chair of chemistry and natural history at Yale University, and was a leading, if not

[7]Benjamin Silliman, Sr., son of General Gold Selleck Silliman, was born on August 8, 1779, in new Stratford Society, where his family had lately taken refuge on the alarm caused by the British raid along this coast. He was graduated at Yale College in 1796, and was tutor from 1799 to 1804. In the meantime he had read law with the Hon. Simeon Baldwin, of New Haven, and was admitted to the bar in 1802. His life work, however, was directed into entirely different channels by the establishment of a chair of chemistry and natural history at Yale. There being no eminent chemists in the country, the position was offered Silliman, with the understanding that he would be given time and opportunity to qualify himself for the position. Thorough in all things that he undertook, he became perhaps the most famous chemist and geologist of his day, delivering his first lecture at the age of 24. Within a short time he became so generally and favorably known as a lecturer that he received invitations from various cities throughout the United States. Through these lecture tours, which he now made almost annually, he did much to popularize science in America. He founded the *American Journal of Science,* afterward *Silliman's Journal,* in 1818, published an edition of *Bakewell's Geology* in 1829, and a two-volume textbook on chemistry in 1830. He was well known abroad, and was in regular correspondence with some of the leading scientists of Europe, such as Berzelius, Humboldt, Carl Ritter, Murchison, and others. After half a century in Yale, he resigned at 74 years of age, and was followed by his son, Benjamin Silliman, Jr. He died in 1864 in his 86th year, interested in science and national affairs to the last. (*Yale Portrait Catalogue,* 1892; *Pioneers of Science in America,* by Wm. Jay Youmans, 1896.)

Professor Benjamin Silliman, Sr.

the most prominent, chemist of his day in America. It was natural, therefore, that Beaumont should appeal to him for aid, though he knew full well that Silliman was in no sense a physiological chemist.

Soon after the visit to New Haven, Dr. Silliman, inclosing copies of two communications, one to Professor Jacob Berzelius and the other to Mr. Gahn, the Swedish consul, wrote:

You will be so good as to deliver the enclosed letters to Mr. Gahn at your convenience, having previously read my letter to Professor Berzelius, and noticed the obligations which I have taken the liberty to lay upon you.

I would recommend sending a pretty liberal supply of the fluid—if there were even a pint, so much the better. Allow me to suggest that one of Lynch & Clark's pint Congress water bottles, carefully marked, sealed and capped with strong leather and twine, and then cased in tin, with the lid soldered on, that no one may open it, will afford all necessary security.

I will attempt, as soon as my engagements will allow, some tentative experiments on the gastric fluid which you left with me, and I shall hope to hear from you again on this interesting subject.

Silliman's letter to Berzelius was for the purpose of introducing Beaumont and presenting the subject matter under discussion.

YALE COLLEGE,
NEW HAVEN, U. S. AMA., April 10, 1833.

Prof. Jacob Berzelius,
Stockholm, Sweden.

DEAR SIR: I had the honor to write to you on the 5th day of December to thank you for the first five volumes of your chemistry, and I have now the pleasure to thank you for the sixth volume, which arrived some weeks since.

My more immediate object, however, is to invite your attention to a subject which I trust you will think to possess no common interest. It is nothing less than the gastric fluid, obtained in quantity and apparently in purity from a healthy man, who more than ten years ago, by the accidental discharge of a musket, obtained an artificial orifice to his stomach, which remains open and admits of the introduction of a tube of caoutchouc, by which early in the morning the gastric fluid is extracted. A sufficient quantity of it for analysis will, at my request, be transmitted to you by Dr. Beaumont, an eminent Surgeon of the American Army, who took charge of the wounded man, effected his cure, and has ever since kept him in his military family (for ten years), and has carefully observed the phenomenon of this extraordinary case. Dr. Beaumont, now resident at No. 110 Broadway, New York, will transmit to you a more particular statement of the case, and you may implicitly rely upon his statement. Only one fact need be mentioned to prove that the fluid obtained from this man's stomach is really the long-sought gastric fluid—namely, that when animal muscle, farinaceous matter and any articles of food whatever are immersed in it, it performs the process of digestion, while this fluid is, by itself, imputrescible and undergoes no change by keeping some months and perhaps longer. There are many very curious circumstances connected with this case, and the most important I trust Dr. Beaumont will communicate when he transmits the fluid through your worthy consul, Mr. Gahn, of New York.

My motive for troubling you with this affair is that I am anxious that a subject of such deep interest to mankind should be investigated by (pardon me for saying) the man, of all others, best qualified for the task. I should be very desirous that Dr. Beaumont or myself might receive the result as soon as convenient.

In the meantime I shall make a few tentative experiments, but I can not flatter myself that I shall discover the peculiar agent which works such wonders in the animal system.

Hoping that your health is good, and wishing you every blessing, I remain, dear sir,

Most respectfully and truly yours,

B. SILLIMAN.

The note to Mr. Gahn simply requested his kind offices in the prompt and careful delivery of the bottle of gastric juice to Professor Berzelius.

Dr. Beaumont forthwith collected a quantity of fluid, which he delivered to Mr. Gahn, together with a package of books and the following letter:

UNITED STATES OF AMERICA,
CITY OF NEW YORK, 17th April, 1833.

Prof. Jacob Berzelius,
Stockholm, Sweden.

SIR: I do myself the honor, at the suggestion of your worthy and scientific friend and correspondent, Prof. Silliman, of transmitting for your consideration the accompanying papers and a quantity of pure gastric juice taken from the Stomach of a man in perfect health and vigor, and doubt not the importance of the subject and the interest it may excite will be sufficient apology for thus obtruding upon your notice. Profr. Silliman's communication will suggest to you the general nature and importance of the subject.

From the accompanying number of the *Medical Recorder,* Page 14th and sequel, you may learn the origin and early history of the case, and in the following pages of this sheet I shall endeavor to give you a concise view of what has been seen and done by way of experiments and observation on the Gastric fluid and functions of Digestion. . . . [Describes the accident and the nature of the wound.]

The Gastric juice I sent you is also extracted this way through an appropriate caoutchouc tube; not in free and abundant quantities, but by slow distillations from the very minute papillæ of the surface of the villous membrane—requiring much time, patience and address to obtain it in small quantity. Not more than 1½ or 2 oz. can be extracted after fasting for any period, and even this quantity requires 25, 30, or more minutes for its extraction. Numerous experiments and observations have been made in this case within the last two or three years upon the process of digestion and the chymefication of different kinds of aliments, both in the Stomach and out of it. The relative solubility in this Gastric fluid of many kinds of alimentary substances, both animal and vegetable, have been fairly tested, the comparative time and different facility of chymefication ascertained, and the natural and varying temperature of the system determined by accurate observations of the Thermometer, placed in the cavity of the Stomach, during abstinence and repletion. Various kinds of alimentary substances have been submitted to the action of this fluid, and it is found capable of completely dissolving them all, out of the stomach, when contained in a glass phial and placed upon a sand bath of the temperature of the Stomach—100° Fahr. Even solid bone, cartilage, tendon down to the softest textured aliments are completely chymefied and dissolved when submitted to the Gastric juice in bottles and kept agitated in a temperature equal to the natural warmth of the stomach; varying, however, in rapidity and perfection in proportion to the healthfulness of the secretion, purity of the gastric fluid, and the solidity, quality and peculiar nature of the aliment. Thus artificial chymefication of many kinds of food in this peculiar fluid has been fairly demonstrated—its powerfully solvent and antiseptic properties clearly ascertained and proven. But its accurate chemical analysis has not yet been accomplished, though several quantities have been submitted to some of the most eminent practical chemists of the United States. None have yet been able to obtain complete and satisfactory results—Profr. Robley Dunglison, of the University of Virginia, is the only one who has fairly approximated or made returns of anything useful and important on this subject. Profr. Silliman has now, for the first time, a quantity under examination, from the investigation of which highly important results are anticipated.

For the 5 or 6 months last past I have been prosecuting a series of experiments and observations on the subject of Digestion by this Gastric fluid under the patronage of the Medical Departments of the Government, the results of which are now preparing, and will probably be published in the course of the ensuing autumn or winter, or so soon as an accurate and satisfactory analysis of this fluid, if it be practicable, can be obtained.

Should the subject be worthy of your notice, and the accompanying fluids and documents merit your attention, and you be so happy as to succeed in obtaining satisfactory analysis of it, I do most earnestly and respectfully desire and shall be greatly obliged to you to communicate the results of your investigations, soon as convenient, either to Profr. Silliman, at New Haven, Ct., or to me, in the City of New York, by doing which you will doubtless confer a signal benefit on the American Medical public, greatly promote the cause of science, and confer honor and enduring obligation upon

Your most respectful and obedient servant,

WM. BEAUMONT.

P.S. The man will continue with me. Should more of the fluid be required to complete the investigation and analysis, it will be rapidly transmitted upon the earliest indication from you. Any suggestion you may please to make will be happily recd. and attended to.

1833-1834

FILLED WITH GREAT anticipations of the results of Berzelius' investigations, he now left for Plattsburgh with Alexis to visit his family for a few weeks, and at the same time to get away from the bustle of a large city, in order to arrange his data and prepare for the publication of his book. While in Plattsburgh he found that his cousin, Dr. Samuel Beaumont, was able to render him material assistance in his work, and he returned to New York much pleased with the results of his visit. In the meantime Alexis had left him and returned to Canada on account of the death of one of his children, with the promise, however, that he would come back to Plattsburgh in the near future. On Dr. Beaumont's return to New York he wrote to Dr. Lovell (June 1, 1833):

> I have a few days since returned from a visit to my family at Plattsburgh, where I have spent several weeks very pleasantly, and I believe advantageously as respects facilitating the publication of the experiments. I believe no detriment has occurred to the government nor dissatisfaction felt by the officers of the Dept. in consequence of my absence. I found much greater facilities afforded for writing out my notes and preparing the book for the press at Plattsburgh than in this city. . . .
>
> My cousin, Dr. Samuel Beaumont, having formerly served a regular apprenticeship, and, for some time previous to taking up the study and practice of medicine, followed the business of printing and publishing, is well qualified, and afforded me great and important assistance, without which I should have labored under great embarrassment from my own ignorance of such business. We have arranged a plan nearly of the one you suggested, and have now a rough MS. written out, sufficient, with the experiments, to make about 300 printed pages, 8vo. This is to be corrected and copied for the type, and will require our united industry and application to execute it properly in three or four weeks. I find it an immense job to make a book, and I heartily wish it were done and publicly approved of and well sold, and I were snugly established at the Arsenal near St. Louis in regular performance of my official duties again.

In the same letter he suggested that, since his duties in New York were "little or nothing, excepting examining recruits," which could be

Dr. Samuel Beaumont.

performed by others as well, he would like to be transferred to Plattsburgh or Burlington to act as recruiting officer. Here his book could be published as well as at New York or Philadelphia, and he could at the same time have the assistance of his cousin, which would enable him to finish the work in three or four months. Dr. Lovell replied that the plan suited him admirably if General Scott raised no objections, and added:

Mr. Livingston,[1] who will soon go to Paris, and who has always taken a great interest in the experiments, is extremely desirous that you and Alexis should go there, and told me yesterday that he had no doubt of finding plenty of people there who would gladly pay all expenses, and that he would attend to it as soon as we authorized him to do so. I proposed to send your book to him as soon as published, on which they could make their propositions, if inclined to do so, so that I would make the necessary arrangements to have Alexis if you should go. Let me hear from you and know what arrangements you propose, in order to enable you to return to Plattsburgh. If there should be no objection on the part of General Scott, I should think the best way would be for you to go, as you did before.

---

[1]Edward Livingston was about to leave for France to take up his duties as minister plenipotentiary.

Within a few days Beaumont informed Lovell that he had obtained General Scott's permission to make the suggested change, and wrote:

I applied in the latter part of May to Mr. Cooper for an extension of Alexis' furlough, but have not yet heard anything from him. Will you prompt him to let me know the result? If this furlough and enlistment can be consistently continued, it will help to relieve me from the heavy expense I am constantly incurring on his account. His family being sick, and one of his children having died while we were in Plattsburgh, and he became very anxious to see them, I yielded to his solicitude and suffered him to go and take care of the surviving ones, and he staid longer than I expected; but has now come back again and is in Plattsburgh with my family, where I hope soon to be again putting in and taking out those little bags of food, and to complete the last series of experiments. I shall engage him again for three or five years if he will agree, of which I expect there is no doubt. He always has been pleased with the idea of going to France. I feel most gratified at the expression of Mr. Livingston's desire that we should visit Paris, and shall duly consider the interest he takes in the subject, and make the best arrangements I can to meet your views and his. Will consult you again on the subject when further advanced. Shall send you a copy of the book as soon as completed. Hope you will give me your suggestions and advice freely; they will be received with gratitude and consideration. Have you any objections to my dedicating to you? Will you please to be frank on this subject?

Several weeks elapsed, but he heard nothing more concerning his request for transfer. The suggestion seemed agreeable to all concerned, but this was not sufficient—definite action was necessary on the part of the department. Alexis had returned to Plattsburgh in the meantime, and there were grave fears that he might become discontented. Therefore, on July 1st, Beaumont gave expression to his dissatisfaction in unmistakable terms at being thus ignored by his superiors:

Disappointed and almost discouraged at not receiving orders to repair to Plattsburgh or Burlington to attend the recruiting service there, I shall leave on permisson and go on suffrance tomorrow, expecting, of course, to receive orders to do that duty, as it will be unpleasurable to do it without. . . .

Alexis is at Plattsburgh with my family, and not entirely contented to remain, and fears are felt if I am not there soon that he will return to Canada and give me trouble and disappointment again. I hope you will excuse my importunities and anxiety. You know my views, and I have no doubt will rightly appreciate my feelings and wishes on the subjects suggested. I feel anxious to hear from you, officially and unofficially.

This promptly brought the department to a realization of their negligence in the matter, and within two or three days special order No. 101 was issued from the Adjutant-General's office:

> Surgeon Wm. Beaumont will repair without delay to Plattsburgh, and there report to the recruiting officer for duty.

This order was promptly followed by a personal note from the Surgeon-General himself, explaining the delay in the matter.

On returning to Plattsburgh, Dr. Beaumont found Alexis awaiting him. His furlough of three months, given so that he might go to New York with Dr. Beaumont, had elapsed, and, in order to keep him on the government rolls, it was necessary to transfer him from the detachment of orderlies at Washington to some other duty. This was done on June 10th by a communication from General Macomb's office:

> Sergeant Alexis St. Martin will hereafter be considered as on duty with you, and subject to your orders. He will consequently be dropped from the rolls of the detachment of orderlies, but will be mustered by you to the adjutant-general's office until further orders.

In recognition of Dr. Beaumont's admirable researches, the Columbian College of Washington, D. C., had conferred on him on March 6th of this year the honorary degree of Doctor of Medicine *(Doctorem in Arte Medici).* The institution requested, as a matter of form, that he present his credentials, and he sent them to Thomas Sewall, professor of anatomy and physiology, who, together with Dr. Lovell, had fathered the idea of conferring the degree. He wrote Professor Sewall:

> I enclose you such testimonials as I have with me, and hope they are sufficiently creditable to my moral and Professional character to answer the occasion, with your personal knowledge of my object in view.
>
> Though of the first respectability, their forms and the authorities from which they emanated may not be so high and classical an order as may be required by the laws of your institution, yet I respectfully request they may be candidly and favorably considered. The Resolution of the Michigan Medical Society refers to my first report of Alexis' case and the few experiments first published in the *Recorder,* a copy of which was communicated to that Society by request of some of its members while I was stationed at Mackinac. Will you have the goodness to preserve them in your own possession till I send for them after using them at discretion?

The degree was conferred by the Board of Curators and the faculty[2] on March 6th, but there was evidently some delay in delivering the diploma, for as late as June 1st Dr. Beaumont wrote to Dr. Lovell that it had not yet arrived. It came, however, in due time, but he was remiss in acknowledging its receipt, for on July 24th Dr. Sewall wrote, inquiring about the progress of the experiments, and incidentally wishing to know if the diploma had been received. In reply to these queries, Dr. Beaumont wrote:

Yrs. of the 24 Inst. I have this day recd., and hasten to reply. I am indeed justly chargeable with an inexcusable remissness in not having before made acknowledgment of the honor conferred on me by the college through [your exertions]. I recd. the Diploma through Dr. L., with the Deposit left in his hands, and verily intended to have made prompt and grateful acknowledgment to the Institution from where it emanated, but diffidence, arising from a lively consciousness of deserving in so small a degree the honor conferred, deterred and disqualified me for expressing myself in terms adequate to truth and the occasion. Alike sensible of the [high] motives, the scientific views and the liberality of the College, and my own inability and demerits, my best exertions have been and all my desires concentrated to merit by constant assiduity and close attention to the subject the honor conferred. My best efforts have been continued in prosecuting the experiments and preparing them for publication since I have left Washington. . . .

Permit me respectfully to communicate through you my sentiments of high consideration and sincere thankfulness to the Professors and Gentlemen of the College for the respect shown me by the proceedings of that institution, and the assurance of my best efforts to make proper returns for the proffered honors.

The Connecticut Medical Society also took recognition of Beaumont's efforts at this time, and made him an honorary member of the organization. The following is a highly complimentary notice of his election:

New Haven, July 10, 1833.

Wm. Beaumont, M.D.

Dear Sir: I herewith forward you the "Proceedings of President and Fellows of the Connt. Medical Society, in Convention, May, 1833," by which you will observe that you are elected an Honorary Member of the Connecticut Medical Society. In officially giving you notice of this election, I ought to state

[2]The following are the names of the members of the Board of Curators and the faculty as they appeared on his diploma: Curators—John P. Van Ness, C. K. Gardner, C. V. Towson, Enochus Reynolds, Rob. Johnstone, J. L. Skinner, Geo. Wood. Professores—Stephanus Chasim. Præses; Thos. Sewall, M.D., Anat. & Physiol.; Galeel Ruggles, Math. & Phil. Nat.; Thos. Henderson, M.D., Theor. & Prac.; N. W. Worthington, M.D., Mat. Med.; Fred. May, M.D., Obstets.; Thos. P. Jones, M.D., Chem.; J. C. Hall, M.D., Chirurg.

that the Connt. Med. Society has not been accustomed, like many other societies, to a lavish and indiscriminate distribution of its honors. This honor, it appears by referring to the records, has been conferred on only twenty-seven individuals since the year 1792. The aim has been to select candidates of distinguished merit, and in its present instance the society has acted with a consideration of the distinguished zeal, ability and success of your devotion to the improvement of medical science.

Most respectfully yr. obt. servt.,

CHAS. HOOKER,
Secy. Conn. Med. Secy.

After having prepared the manuscript of the former part of his work for the press, he conducted a fourth series of experiments from July 9th to November 1st, numbering 62 in all. These were devoted chiefly to the determination of the mean time of digestion of different articles of diet, naturally in the stomach and artificially in vials on the bath and in the axilla, recording some experiments, in which the vials were carried in the latter manner from six to twenty-four hours. Nothing especially new was brought out in this series, they being largely confirmatory of those that had gone before.

Up to the very last experiment he lived in hopes that additional revelations concerning the chemistry of the gastric juice would be forthcoming from some of his correspondents on whom he was dependent for this portion of the work. He could delay no longer, however. The results of his work must soon be given to the profession and to the public, for there was already on file in the office of the clerk of the southern district of New York this record:

Be it Remembered, That on the Twenty-ninth day of July, Anno Domini 1833, William Beaumont, M.D., of the said District, hath deposited in this Office the title of a Book, . . . the title of which is in the words following, to wit: Experiments and Observations on the Gastric Juice and the Physiology of Digestion, by William Beaumont, M.D., Surgeon in the U. S. Army. . . .

He must have been greatly disappointed at this critical moment on the receipt of a communication from Mr. Gahn, informing him of a delay of two and one-half months in starting the package containing the gastric juice on its way to Berzelius, when he was laboring under the impression that it had already reached its destination. The following note of July 30th explains the reason for the delay:

. . . The Phial and Pamphlet could not be forwarded by a safe conveyance until one offered by the American Brig Marcella, Captain Hiler, of Boston,

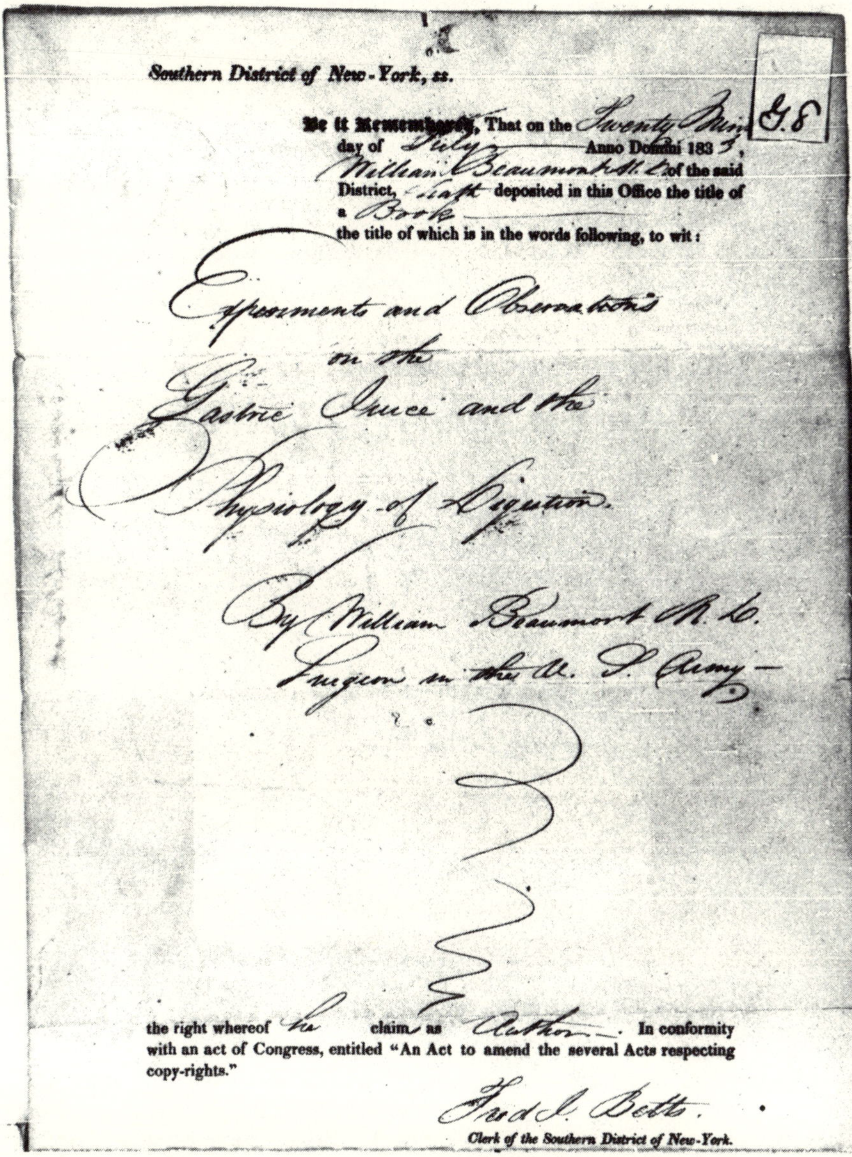

G. 8

Southern District of New-York, ss.

Be it Remembered, That on the Twenty Nine day of July Anno Domini 183 3, William Beaumont M. D. of the said District, hath deposited in this Office the title of a Book the title of which is in the words following, to wit:

Experiments and Observations on the Gastric Juice and the Physiology of Digestion

By William Beaumont M. D. Surgeon in the U. S. Army

the right whereof he claims as Author. In conformity with an act of Congress, entitled "An Act to amend the several Acts respecting copy-rights."

Fred J. Betts.

Clerk of the Southern District of New-York.

Certificate of copyright for the first edition of Beaumont's *Experiments and Ob erva-tions . . .*, dated July 29, 1833.

Washington University School of Medicine

who sailed from New York June 27th direct for Gothenburg, where he promised to hand it to the superintendent or Chief Physician of the Quarantine establishment, with necessary directions for its being forwarded to Professor Berzelius at Stockholm. Mr. Gahn will take particular care and pleasure in transmitting the earliest advice to Dr. Beaumont.

Another communication received at this time from Professor Silliman was equally disappointing. It was as follows:

I have delayed replying to your letter of June 30 because I have been hoping ere this to go into the laboratory to attempt something with the Gastric juice. I can only say that hitherto I have found it impossible, owing to unceasing and inevitable demands on my time. I write now merely that you may know that I have not forgotten you. I do not see at present anything to hinder my doing something for you within a few days, . . . and in the meantime remember what I said to you last spring—that I have no great confidence in myself on this subject. The moment I hear from Professor Berzelius I will write to you, but I think we can not expect to hear from him within several months.

A few days later Professor Silliman sent Dr. Beaumont the result of his observations, as well as an abstract of a portion of a system of chemistry by Berzelius. I present this in full, because it gives one a splendid idea of the status of our knowledge of the physiology of digestion at that time.

YALE COLLEGE LABORATORY, August 2, 1833.

To Dr. William Beaumont,

Surgeon U. S. Army.

DEAR SIR: The following citations are from the seventh volume of a system of chemistry by Professor Berzelius which I have received within a few days. It is the French translation, published the present year, and is the latest authority upon the subject of The Gastric Juice. This was a long time regarded by chemists as a kind of universal solvent for different alimentary substances—afterwards they denied it any dissolving power.

Prout, Tiedeman and Gmelin give the best notions on this subject, and explain the contradictory statements of the other authors. At one time it was said to be very fluid, clear, entirely neutral; then alkaline, then acid, and that in a high degree.

Spallanzani, 1783, after many experiments declared the gastric juice, in a state of health, to be entirely neutral; a solvent for alimentary matter within and without the body; that it did not putrefy at the ordinary temperature of the air, but preserved animal matters from putrefaction and dissolved them with the aid of heat.

Carminati, 1785, found it not acid in carnivorous animals when fasting, but quite acid in those which had eaten. This, says Brezelius, is the first ray of light which illuminated the subject.

Werner, 1800, says the mass contained in the stomach of carnivorous and herbivorous animals is acid during digestion.

Montegre, 1812, who could vomit at will, and thereby obtained the gastric juice unmixed, says it is not acid or alkaline, not a solvent, not slow to putrefy; so much like saliva that he regards it as saliva swallowed, and the traces of free acid as owing to incipient decompositions.

Prout, 1824, says it is really acid; does not contain an organic acid, but free hydrochloric or muriatic acid. (For his process see Berzelius, vol. VII, page 149.) This examination was made by taking the contents of the stomach of an animal killed soon after eating, and digesting the mass in water, filtering, and testing the fluid. His results were 39.6 parts chlorine found in a quantity—9.5 parts were combined with potassium and sodium, 7.9 with ammonium, 22.2 with hydrogen, constituting the hydrochloric acid. The acid liquid vomited by a dyspeptic gave 12.11 parts of chlorine under the saline form, 5.13 in the form of hydrochloric acid.

It was a desideratum to know the reason why the opinion that the gastric juice was neutral had been so obstinately maintained. Carminati's experiment was a guide, but those of Gmelin and Tiedeman furnished a complete solution of the enigma.

G. and T., without knowing the experiments of Prout, also established, in a different mode, the presence of free hydrochloric acid in this fluid. They discovered it was acid, and contained a deliquescent salt, not destroyed when heated to redness, and which proved to be chloride of calcium. General results—the stomach, when empty, secretes only sufficient fluid to moisten its internal face; mechanical stimulation produces a greater secretion, but never so much as the ingestion of alimentary substances.

The fluid of the empty stomach is slightly acid, sometimes neutral, and the acidity is in proportion to the quantity. Contains much mucus, which may be removed by the filter; is then clear, yellowish, and of a saline taste. Contains much water; deposits about 2 percent of solid residuum like that from the fluid of the serous membranes. The gastric juice becomes very acid when aliments have been swallowed; its free acid is chiefly hydrochloric acid. By distilling the liquid to dryness in a water bath, Gm. and T. obtained traces of acetic acid, and, in that of the horse, of butyric acid. The distilled liquid had a feeble reaction like acids; was not precipitated by nitrate of silver, as the hydrochloric acid was retained by the organic matters. Saturated with carb. of barytes and evaporated, it yielded an uncrystallizable salt, which, with sulphuric acid, gave out vapors exhaling the odors of acetic and butyric acids. The difficulty of obtaining the gastric juice pure and in large quantity, and of characterizing the substance separated by analysis, leaves the whole in much uncertainty.

Gm. and T. found no albumin in the gastric juice of the dog, and only traces of it in that of the horse. Salts in the gastric juice are principally chloride of sodium, chloride of potassium in small quantity, hydrochlorate of ammonia, a little sulphate of potassa. An alkaline carbonate or phosphate has never been found. After burning the residuum of the dried gastric juice, and taking up the soluble salts with the water, there remain lime, magnesia, traces of perox. of iron and sometimes protox of manganese, all combined with phosphoric acid, and a portion of lime with carbonic acid. Sometimes there are found sulphate of lime and chloride of sodium in the ashes. It is not improbable that the hydrofluoric acid, as well as hydrochloric acid, may exist in a free state in the gastric juice, since the fluoride of calcium is found in bones and in urine.

The secretory organs of the gastric juice are not known, nor whether the same organs secrete the unacid fluid found in the empty stomach and the acid fluid is produced during the digestion. No organ for the especial secretion of the gastric juice has yet been discovered.

Then follow the results of his investigations, on which Dr. Beaumont had laid so much hope. It will be seen that his results are very unsatisfactory and incomplete, and added practically nothing to the knowledge which Beaumont had gleaned from his own observations and the suggestions of Dunglison.

EXAMINATION OF GASTRIC FLUID, August 2, 1833.

1. The fluid, after being kept in a closely corked vial more than three months, from April to August, and most of this time in a cellar, remained unaltered, except the formation of a pellicle upon the surface, slightly discolored by red spots; a second pellicle appeared after the precipitation of the first—it was thicker and more discolored, with dark-red spots like venous blood.

2. The fluid was cloudy, like a solution of gum Arabic, but on filtering it became perfectly clear and of a light straw-yellow tinge.

3. The pellicles, which had the appearance of inspissated mucus, after being separated from the fluid became, on exposure to the air, throughout of a dark, brownish red color, resembling the inner portion of a mass of coagulated blood. This change seemed to result from a sudden oxygenation.

4. The fluid exhaled a slight odor—not disagreeable, rather aromatic, and very similar to that which it at first exhaled, but not so strong; it was then rather disagreeable.

5. Taste feebly saline, not disagreeable.

6. Test papers of litmus, athanet, and purple cabbage were decidedly reddened. Turmeric paper underwent no change, but, when previously browned by an alkali (ammonia), the gastric fluid restored the yellow color.

7. Nitrate of silver gave a dense white precipitate, which, after standing five minutes in the sun's light, turned to a dark, brownish black, thus indicating muriatic acid. Mur. and Nit. barytes gave a slight opalescence, indicating a trace of sulphuric acid; not improbably there was also some phosphoric acid.

8. Specific gravity taken in a small thin glass tube containing 201 grs. of distilled water. When filled with the gastric fluid, its weight was increased 1 grm. of the gastric liquor—therefore 202. The specific gravity is therefore about 1.005, showing but little solid matter in solution.

It is quite obvious from the above statement that the gastric fluid now considered is identical with that heretofore examined by different experimenters. It appears, however, to be more acid than the gastric fluid taken from a fasting stomach has generally been found, and I can not say how far it may have become acid by so long keeping. I trust we shall have a more full examination from Professor Berzelius, to whom (agreeably to my recommendation) you were so kind as to send an ample supply (one pint), which, from my own experience in keeping the fluid, we may hope will reach him without suffering any material change. Wishing to preserve a specimen for exhibition to my classes, I did not proceed to evaporate the small quantity (2½ ounces) of the fluid which I had left. At present we are in no condition to explain the surprising process of digestion. It is obvious that the trifling portion of chemical agents hitherto discovered in the gastric fluid (all of them mild, excepting a little free hydrochloric acid, and possibly, as Professor Berzelius suggests, a little hydrofluoric acid) can give no adequate explanation of this phenomenon, nor will it aid us materially to attribute the singular activity of the fluid to the peculiar action of the stomach, since digestion, with a temperature of about 100° F. and occasional agitation, takes place as rapidly out of the body as in the living stomach. On placing today a piece of veal in a wine glass containing some of the gastric fluid, and standing on a warm stove (stirring occasionally with a glass rod), digestion began very soon and proceeded until the fluid had spent its force, and was renewed on the addition of more fluid.

I regret that I can not contribute something important to our previous knowledge—there is much in physiology that eludes the scrutiny of chemistry. Thought may emanate from the brain, volition may cause the movement of the muscles, sentient and ever rational beings may spring from a seminal drop of very simple composition, and all kinds of aliment may dissolve with the equally mild and simple gastric fluid, but who can explain the proximate, or even the ultimate, cause in any other way than by referring it to a positive law of the Creator—often incomprehensible equally in his nature and in his works.

Yours very respectfully,

B. SILLIMAN.

PRIVATE.—I am sorry to have done so little for you, but at present it is utterly impossible for me to do more—such are my engagements in many ways. It would be very desirable that the actual publication of your work should be kept back until we can hear from Berzelius, who, with such an unprecedented quantity, will, I am persuaded, do something worth telling of.

Not satisfied to go to press before every possible source of knowledge had been exhausted, Dr. Beaumont makes one more appeal to Professor Dunglison:

I feel some delicacy in addressing you at this time, lest I may be considered unwarranted. The lively interest manifested by you in the successful prosecution of my experiments last winter induces me to believe that a few suggestions on the subject may not be unkindly considered at this time. Being about to commence the publication of all my experiments, observations, etc., I am anxious to render them as interesting as possible to the public. To do this, I wish to [include] essential and important facts which may have been developed in the course of the investigation, whether my own or the more important labors of scientific friends, so far as they may feel willing to contribute and consent to be associated with me.

The result of your analysis of the qty. of gas. ju. taken with you from Wn. is considered of much importance, to be inserted as it is. Should you have obtained any fuller and further satisfactory results from an investigation of the second parcel sent you, and will take the trouble to communicate them to me at this place soon as convenient, with your permission to incorporate them with the work, it will doubtless greatly enhance its value, and confer upon science a very important benefit. . . .

May I not anticipate your continued favorable consideration of the subject, and expect your early reply to this? I have been too laboriously occupied to be either a faithful or interesting correspondent. I hope soon to attone for any weakness I may have been thought chargeable with. . . .

At this time Major Bache, of Philadelphia, was at Plattsburgh, and Dr. Beaumont took advantage of the opportunity to send a specimen of the gastric contents to his brother, Dr. Franklin Bache,[3] with the following note:

. . . I avail myself of an opportunity of sending some gastric juice by your brother, Major Bache, and you will do me a favor if you will analyse it as soon as practicable and send me the result. I am about publishing a volume of experiments and observations, and will be glad to get your analysis before it goes to press, which will be in the course of three or four weeks. By giving your immediate attention to this subject, you will greatly oblige me and confer a benefit on the cause of science.

Dr. Bache replied (September 12, 1833):

. . . I have not the requisite convenience and apparatus to make an analysis of the juice, as you request, but I hope to be able to find time to test it, so as to determine its principle, if not all its constituents, without ascertaining the relative proportion. Animal analysis is the most difficult department of [chemistry], and there are not living now many chemists who could perform a complete analysis of the gastric juice of Alexis. Thenaud in France, Prout in

[3]Dr. Franklin Bache, an American chemist, born in Philadelphia in 1792, was professor of chemistry in the Philadelphia College of Pharmacy in 1831 and in the Jefferson Medical College in 1841. As early as 1819 he had published a system of chemistry. He was the grandson of Sarah Franklin Bache, the only daughter of Benjamin Franklin, and his father was Benjamin Franklin Bache, the "Benny" of Franklin's letters from France.

England, Berzelius in Sweden are the best analysts we have in this department, and I advise you to send specimens of the juice to them.

A few days ago I received a letter from Dr. Lovell, enclosing a prospectus of your work, requesting that I would place it in some book store. Accordingly I have placed it at the store of Mr. Judah Dobson, who has kindly agreed to receive the subscribers, and to perform any agency in the matter that may be agreeable to you. I presume you do not intend to print your work in Plattsburgh, as I am sure that it cannot be done there in that handsome style which is so essential to its success. Wishing you every success in your laudable exertion in elucidating a very important department of Physiological science, I subscribe myself. . . .

It was decided to sell the book by subscription, and, with this in view, prospectuses, or "proposals," were sent out about September 1st, presenting a synopsis of the work, at the bottom of which was space for the signatures and addresses of those subscribing to the work. These prospectuses were sent out to various friends throughout the country, some of whom personally conducted a campaign in behalf of Dr. Beaumont, others of whom deposited the prospectuses in conspicuous places such as a drugstore or bookstore. Among the names of those who acted as agents may be mentioned Dr. Chandler, his preceptor, St. Albans, Vermont; John Beaumont, his brother, Lebanon, Connecticut; Lieutenant J. W. Kingsbury, a lifelong friend, St. Louis, Missouri; Dr. Franklin Bache, Philadelphia, Pennsylvania; Mr. Ramsay Crooks, New York City; Hezekiah Howe, New Haven, Connecticut; Dr. L. Abot, Detroit, Michigan; Colonel Stearns, Malone, New York; Judge Loomis, Montpelier, Vermont; A. C. Flagg, Albany, New York; E. L. Clark's bookstore, St. Louis, Missouri; the store of Judah Dobson, Philadelphia, and many others.

After receiving several of the "proposals," Dr. Lovell wrote Dr. Beaumont on September 13th:

Your letter of the 1st inst. has been received, and I congratulate you on the prospect of finishing your book. One copy of the proposals has been sent to Dr. Franklin Bache, at Philadelphia, one to Dr. Stewart, at Baltimore, who will, I believe, give them the best direction for subscriptions whenever requested to make the result known. The two others have been retained for this place. With regard to the Paris journals, it has occurred to me that the best plan will be to transmit a copy or two of your book to Mr. Livingston through the State Department, and at the same time to suggest the plan to him for the consideration of some of their learned ones, who would then be able to form some definite opinion of the importance of the case, and be more likely to make some determinate and practical proposition, which could perhaps hard-

Washington University School of Medicine

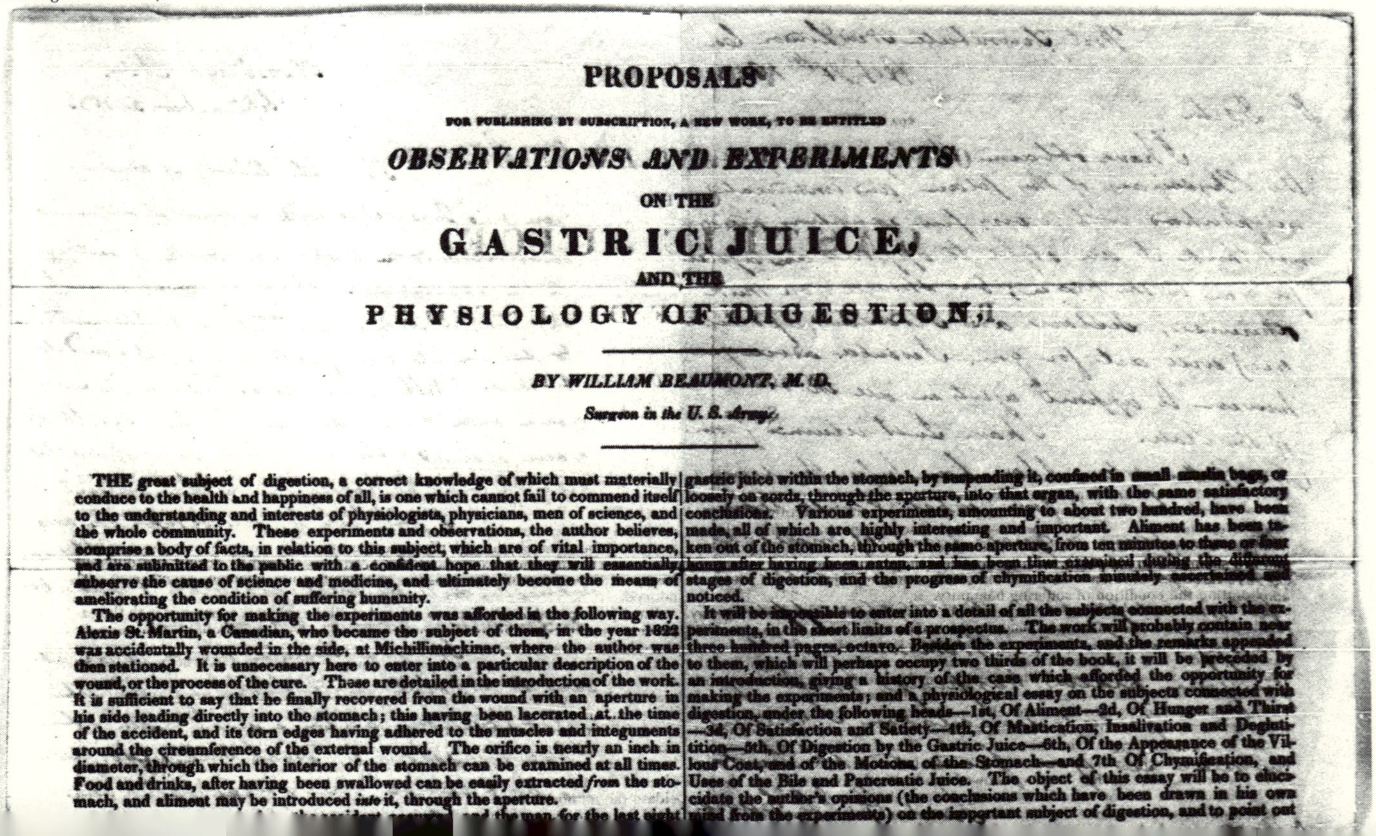

PROPOSALS

FOR PUBLISHING BY SUBSCRIPTION, A NEW WORK, TO BE ENTITLED

OBSERVATIONS AND EXPERIMENTS

ON THE

GASTRIC JUICE,

AND THE

PHYSIOLOGY OF DIGESTION,

BY *WILLIAM BEAUMONT, M.D.*

Surgeon in the U. S. Army.

THE great subject of digestion, a correct knowledge of which must materially conduce to the health and happiness of all, is one which cannot fail to commend itself to the understanding and interests of physiologists, physicians, men of science, and the whole community. These experiments and observations, the author believes, comprise a body of facts, in relation to this subject, which are of vital importance, and are submitted to the public with a confident hope that they will essentially subserve the cause of science and medicine, and ultimately become the means of ameliorating the condition of suffering humanity.

The opportunity for making the experiments was afforded in the following way. Alexis St. Martin, a Canadian, who became the subject of them, in the year 1822 was accidentally wounded in the side, at Michillimackinac, where the author was then stationed. It is unnecessary here to enter into a particular description of the wound, or the process of the cure. These are detailed in the introduction of the work. It is sufficient to say that he finally recovered from the wound with an aperture in his side leading directly into the stomach; this having been lacerated at the time of the accident, and its torn edges having adhered to the muscles and integuments around the circumference of the external wound. The orifice is nearly an inch in diameter, through which the interior of the stomach can be examined at all times. Food and drinks, after having been swallowed can be easily extracted *from* the stomach, and aliment may be introduced *into* it, through the aperture.

gastric juice within the stomach, by suspending it, confined in small muslin bags, or loosely on cords, through the aperture, into that organ, with the same satisfactory conclusions. Various experiments, amounting to about two hundred, have been made, all of which are highly interesting and important. Aliment has been taken out of the stomach, through the same aperture, from ten minutes to three or four hours after having been eaten, and has been thus examined during the different stages of digestion, and the progress of chymification minutely ascertained and noticed.

It will be impossible to enter into a detail of all the subjects connected with the experiments, in the short limits of a prospectus. The work will probably contain near three hundred pages, octavo. Besides the experiments, and the remarks appended to them, which will perhaps occupy two thirds of the book, it will be preceded by an introduction, giving a history of the case which afforded the opportunity for making the experiments; and a physiological essay on the subjects connected with digestion, under the following heads—1st, Of Aliment—2d, Of Hunger and Thirst—3d, Of Satisfaction and Satiety—4th, Of Mastication, Insalivation and Deglutition—5th, Of Digestion by the Gastric Juice—6th, Of the Appearance of the Villous Coat, and of the Motions of the Stomach—and 7th Of Chymification, and Uses of the Bile and Pancreatic Juice. The object of this essay will be to elucidate the author's opinions (the conclusions which have been drawn in his own mind from the experiments) on the important subject of digestion, and to point out

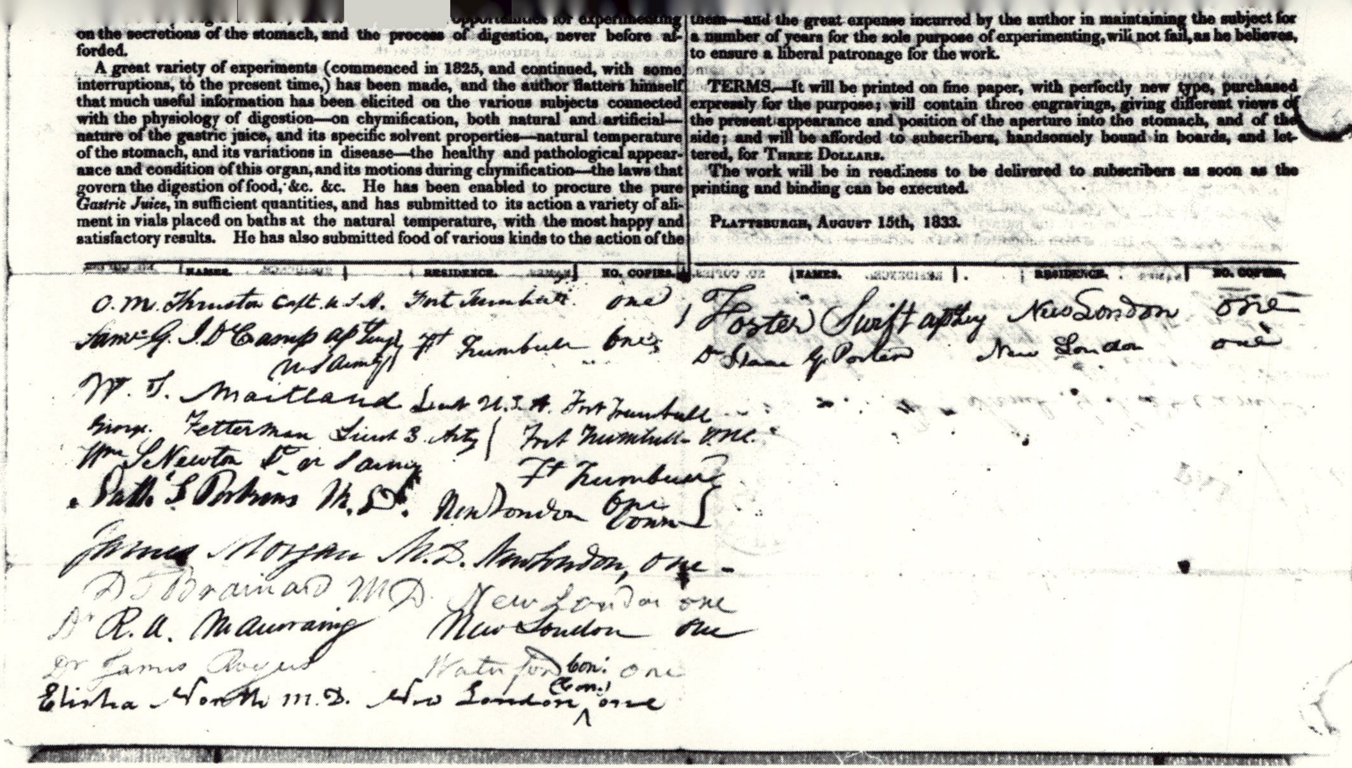

on the secretions of the stomach, and the process of digestion, never before afforded.

A great variety of experiments (commenced in 1825, and continued, with some interruptions, to the present time,) has been made, and the author flatters himself that much useful information has been elicited on the various subjects connected with the physiology of digestion—on chymification, both natural and artificial—nature of the gastric juice, and its specific solvent properties—natural temperature of the stomach, and its variations in disease—the healthy and pathological appearance and condition of this organ, and its motions during chymification—the laws that govern the digestion of food, &c. &c. He has been enabled to procure the pure *Gastric Juice*, in sufficient quantities, and has submitted to its action a variety of aliment in vials placed on baths at the natural temperature, with the most happy and satisfactory results. He has also submitted food of various kinds to the action of the them—and the great expense incurred by the author in maintaining the subject for a number of years for the sole purpose of experimenting, will not fail, as he believes, to ensure a liberal patronage for the work.

TERMS.—It will be printed on fine paper, with perfectly new type, purchased expressly for the purpose; will contain three engravings, giving different views of the present appearance and position of the aperture into the stomach, and of the side; and will be afforded to subscribers, handsomely bound in boards, and lettered, for Three Dollars.

The work will be in readiness to be delivered to subscribers as soon as the printing and binding can be executed.

Plattsburgh, August 15th, 1833.

NAMES.	RESIDENCE.	NO. COPIES.
O. M. Thurston Capt. U.S.A.	Fort Trumbull	one
Saml. G. I. DeCamp Asst. Surg. U.S. Army	Ft Trumbull	one
W. T. Maitland Lieut U.S.A.	Fort Trumbull	
George Fetterman Lieut 3. Arty	Fort Trumbull	one
Wm. S. Newton Dr. [illegible]	Ft Trumbull	
Nath. S. Perkins M.D.	New London Conn	one
James Morgan M.D.	New London	one
D. T. Brainard M.D.	New London	one
Dr R. A. Manwaring	New London	one
Dr James Rogers	Waterford Conn.	one
Elisha North M.D.	New London Conn.	one

NAMES.	RESIDENCE.	NO. COPIES.
Foster Swift Asst. Surg.	New London	one
Dr [illegible] G. Porter	New London	one

The "proposal," or combined prospectus and subscription blank, that was sent out in advance of the publication of Beaumont's *Experiments and Observations* . . .

ly be expected in a mere statement of the case. Such a letter with the books may be transmitted to Mr. L. as soon as the latter are ready. On the mode of obtaining copyrights in Europe I know nothing, but believe it to be troublesome and expensive, unless perhaps it is done through some of the bookseller's craft. . . . Although the facts on the third page of your book are somewhat apochryphal,[4] the kind feelings which invented them are not the less appreciated by, sincerely yours. . . .

When Mr. Ramsay Crooks[5] received his copy of the proposal, he replied with some excellent advice from the standpoint of the business man:

I wrote you a week ago, and by the date of this you will conclude that Sunday is not always appropriated by me to public worship.

Some medical gentlemen to whom I have spoken of your forthcoming publication, and exhibited your "proposals," encourage me to hope a respectable list of subscribers can be obtained, but they suggest that had you procured and appended to the prospectus the testimonials of some of your distinguished professional brethren as to the facts and merits of the work, it would have done you much good. If not too late, it may be well to do so yet.

I have concluded to employ a competent man to visit the principal physicians here, and subsequently those of less note. The difficulty is to get the right sort of person, but I have hopes of finding my man in a day or two.

Having no knowledge of bookmaking or bookselling, I am a poor advisor, but with your permission I think the surest road to success is to get your book out as soon as possible, and send it to such of our reviewers as will by their respectability stamp its full value upon the work, and bring it into notice at once; and I would place it in the hands of an able bookseller in each of the principal cities and towns of the Union. I am told that subscription lists do not always show the real amount of patronage, for more than one is quite willing to see his name in the company of his betters, but not so ready to pay when the collector calls.

Newspapers throughout the country were very generous and very complimentary in their reviews of Dr. Beaumont's book. The com-

---

[4]This allusion is to the dedication of the work: "To Joseph Lovell, M.D., Surgeon-General of the United States Army, whose merit justly entitles him to the rank which he holds, and whose zeal in the cause of medical science is equalled only by his ability to promote it. As a tribute of respect for his public and private virtues, and as a feeble acknowledgment for a long tried and unvarying friendship, this work is respectfully dedicated, by The Author."

[5]Mr. Crooks remained a partner in the American Fur Company practically up to the time of Mr. Astor's retirement from business, and in 1834 the charter of the company was transferred to Mr. Crooks and his associates. He was made president of the concern, with New York as his headquarters, but reverses in business necessitated an assignment in 1842, following which Mr. Crooks started a fur business for himself. He was very successful up to the time of his death in New York City, which occurred on June 6, 1859, in his seventy-third year.

ments of the lay press in this connection are both interesting and amusing. A few of the clippings that had been carefully preserved by Dr. Beaumont are reproduced in part here to give a general idea of the attitude of the press and the public toward this unique work.

Evening Post, New York, September 5, 1833:

It will be concluded readily by most men that one of the most important subjects for the study of man is his stomach. In sickness or in health, it equally claims our serious attention, and the epicure and the dyspeptic are alike engaged, though by very dissimilar modes, in answering the calls or soothing the complaints of this very essential part of the body. We all remember the ingenious apologue on this subject wherein it is shown how the rebellious members were punished for daring to despise and presuming to refuse allegiance to this craving and insatiable monarch. But with all the study that has been bestowed upon this all-important organ and the mysterious powers of digestion, we have been very much in the dark in respect to the causes of its healthful operations or of its morbid actions. Doctors have differed and patients have suffered, while their physicians disputed the reality of the gastric juice, or the effect of this or that substance introduced as food or medicine into the mysterious region from which all power flows. It seemed as if nothing short of miracle could have enabled man to view the internal operations and test the power of the juices provided by nature to carry on the animal economy; and yet it has been done, and the previous knowledge has been revealed without the aid of miracle, although by a process little short of one. . . .

Repertory, St. Albans, Vermont, September 19, 1833:

The march of knowledge is onward, its progress unceasing, its extent apparently boundless. Not a year rolls into eternity but leaves its memorial of discoveries of improvements. That which one year is "dark as midnight" is before the close of another unveiled, its hidden mysteries exposed, its properties shown and subjected to the use and benefit of mankind. All things seem to conduce to this advancement. Even pain and anguish further it. Whatever transpires adds to the already mighty mass of science. Genius stands ready to catch from passing events a mite to add to the stores of learning, and philanthropy is ever on the watch to extract therefrom wherewithal to ameliorate man's condition. Each is successful; every year witnesses new developments; each season brings with it an antidote, in a degree, to some of the "many ills that life is heir to." . . . The work is one of general utility, is calculated for general circulation, appeals to and should receive the support of every class of readers, and can not fail to deeply interest and instruct all who give it a perusal. It offers to the physician much minute instruction in his profession that can nowhere else be found; to the philosopher a new source of admiration and reflection on the complicated, yet beautiful, perfectness of the human body; to the chemist the examination of a liquid more potent than any in his laboratory; and to every man a greater knowledge of his internal construction and a fund of extremely interesting information, which in these days of indigestion is well worth one's notice to acquire, and thereby guard against the demon. The book will be published in December next, illustrated with engravings, and afforded to subscribers at three dollars.

St. Louis, September 19, 1833:

The Wonders of Nature.—Wonderful are the doings of nature! Strange and erratic is her course, setting at defiance the wisdom of man, baffling his deepest and most elaborate observation, as it were to show his impotence; and then suddenly, singularly and bounteously throwing open her inmost recesses to his view and for his benefit, placing at his disposal by accident what she denies to his most extended and most minute researches. Yes! avenues to acknowledge the most important are often opened by the merest accidents. An exemplification of this is seen in the case on which is founded the experiments on the gastric juice, etc., recently published by Dr. Beaumont. . . .

The author undoubtedly relied on the interest of the subject, the unexampled source and nature of his information, to secure for it the patronage of the community. Has he calculated wrongly? Is he to be made an example to deter others from pursuing the thankless path of costly research after truth? Or, is it his to show that an American public will ever cordially nourish the seeds for improvement which may come into their possession?

The price of the book, though high, compared with the light reading of the day, is not so when considered as to its intrinsic value. And even at the price at which it sells it will not probably remunerate the author for the expenses attendant on prosecuting the experiments, etc. I hope our citizens will give it a liberal patronage. O. P. Q.

National Gazette:

Beaumont's Experiments on Digestion.—The all-absorbing topic of the deposits of public money appears, from the silence of this long expected production of Dr. Beaumont, to have left no leisure in our reading community for an inquiry into its merits. As it, however, treats of a description of deposits in which each one must take the "responsibility on himself," even if our good president[6] should claim his right as chief executive officer to assume that duty also—which, by the way, would not startle a physiologist more than the shade of Washington must be by the novel assumption of power by his successor—a short notice of it may be useful.

Republican, St. Louis, September 24, 1833:

. . . Proposals for this work can be seen at E. L. Clark's book store, where not only the medical gentlemen are invited to patronize merit and obtain information, but the community at large, who in these times of cholera can

[6]Has reference to Andrew Jackson's financial policy, and his arbitrary and vigorous opposition to the Bank of the United States, when he vetoed the bill to recharter the institution and ordered the accumulated funds distributed among the state banks.

not have a better medical work in their hands than the "Physiology of Digestion."

Galenian, September 15, 1833:

. . . Will be read with utility and interest by all classes of every community. . . . It develops important facts relative to the digestive functions, concerning which no person of either sex should remain ignorant.

Following each of these introductions was a long synopsis, in many cases a column or more being devoted to the review.

Everything looked bright and encouraging for the reception of the book. More than eleven years had elapsed since the occurrence of the accident which brought Beaumont the opportunity which he had recognized and utilized to its fullest extent. He was, however, still dissatisfied with the results of his labors—discontented because it was incomplete, and restless because he realized that this source of physiological information had not yet yielded up to science all the riches that lay buried in it. More digging, more research, would be necessary to exhaust the rich vein of possibilities. He entered, therefore, into a new agreement with Alexis, the wording of which was practically the same as in the previous one, which had expired in October. This time it was executed, however, for a period of two years, beginning November 7, 1833, for a consideration of $400, to be paid in installments (page 126).

There being no longer occasion for his remaining in Plattsburgh, Dr. Beaumont requested and received orders, dated November 23d, to repair to Washington and report himself to the Surgeon-General. Two weeks later he informed the Adjutant-General that he had just received the order, and it would be promptly obeyed.

Just about this time the first copies of his book were being turned out. One of them was deposited on December 20th in the office of the clerk of the Southern District of New York, and others were sent complimentary to many of his notable friends in Washington. While I have no intention of presenting here a full synopsis or review of the work, for this would naturally lead to a discussion of the large field of the physiology of digestion, it will not be amiss to speak briefly of the book, and present a few characteristic abstracts from it, especially since the original may not be accessible to every reader of this volume. The book is an octavo of 280 pages, bound in pasteboard, printed on thin paper, and is entitled *Experiments and Observations on the Gastric*

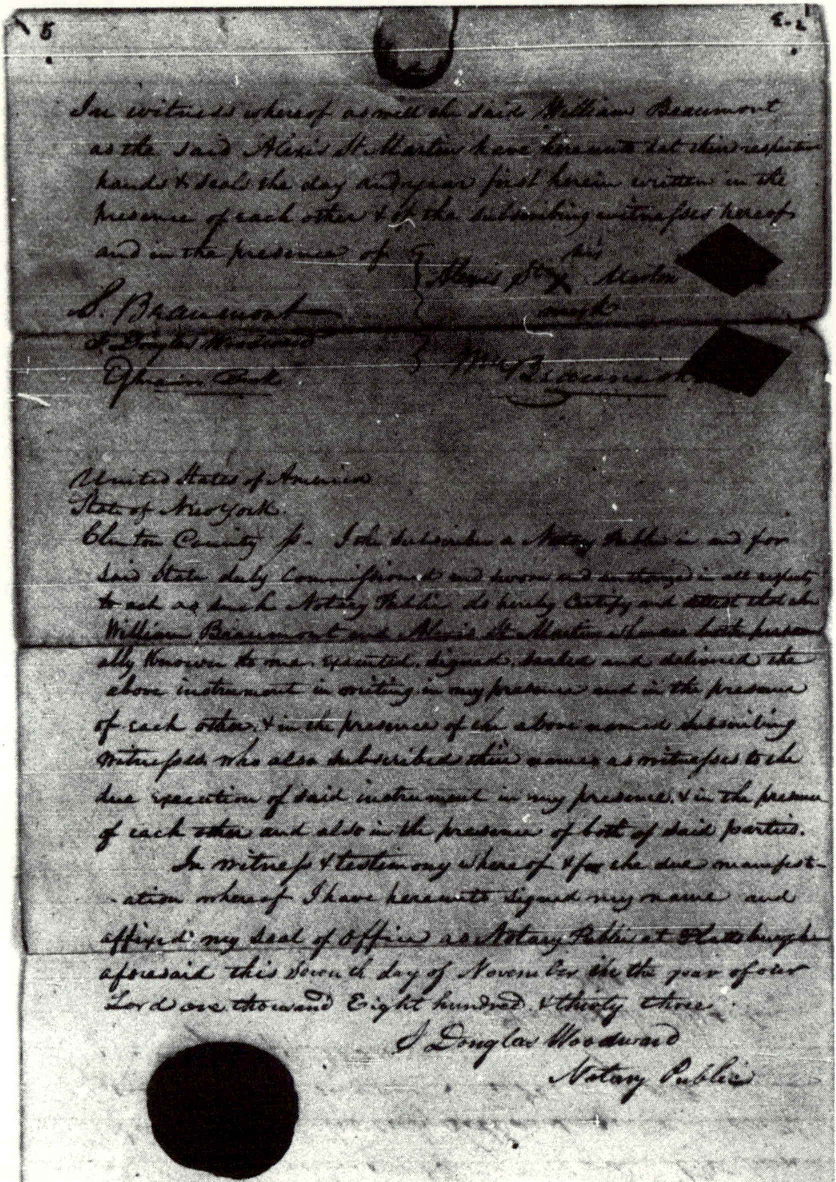

In witness whereof as well the said William Beaumont as the said Alexis St. Martin have hereunto set their respective hands & seals the day and year first herein written in the presence of each other & of the subscribing witnesses hereof and in the presence of

Alexis St. his X mark Martin

S. Beaumont
J. Douglas Woodward
Ephraim Beck

Wm Beaumont

United States of America
State of New York
Clinton County ss. I the subscriber a Notary Public in and for said State duly Commissioned and sworn and authorized in all respects to act as such Notary Public do hereby Certify and attest that the William Beaumont and Alexis St. Martin whom are both personally known to me executed, signed, sealed and delivered the above instrument in writing in my presence and in the presence of each other & in the presence of the above named subscribing witnesses who also subscribed their names as witnesses to the due execution of said instrument in my presence & in the presence of each other and also in the presence of both of said parties.

In witness & testimony whereof & for the due manifestation whereof I have hereunto signed my name and affixed my seal of Office as Notary Public at Plattsburgh aforesaid this Seventh day of November in the year of our Lord one thousand Eight hundred & thirty three.

J. Douglas Woodward
Notary Public

The last page of a contract made on November 7, 1833, between Beaumont and St. Martin for a similar purpose as the contract shown on page 81.

Juice and the Physiology of Digestion, by William Beaumont, M.D., Surgeon U. S. A., and printed by F. P. Allen,[7] Plattsburgh, 1833. The sale of the first edition was placed in the hands of Lilly, Wait & Co., of Boston.

The full import of these observations and deductions can be properly appreciated only when one takes into consideration the meagerness and uncertainty of our knowledge of digestion up to that period, and by comparison with the most recent developments of modern times. The status of our knowledge at that time is pretty well summed up in Professor Silliman's letter (page 114). The experiments of Réaumur on buzzards and Stevens on the human being during the middle and latter part of the eighteenth century had simply revealed that there was a process of solution in the stomach—at any rate, that substances contained in perforated tubes and balls were dissolved out and liberated. It remained for Spallanzani, in 1783, to prove the real solvent powers of the gastric juice within and without the body by means of fluid obtained by the swallowing of sponges attached to strings. He gave to this fluid the name of gastric juice, and established the theory of its chemical powers. Werner, Hunter, and others had shown the gastric juice to be acid in reaction, but it remained for Prout, in 1824, to determine that the acidity was due to free hydrochloric acid. Little more than this was definitely known at this time. A comparison of Beaumont's deductions with those of the most recent workers on the physiology of digestion shows that he anticipated some of their best results, notably the researches of Pavlov on the work of the digestive glands, Cannon on the mechanics of digestion, and others.

In the words of Vaughan, "he made such an exact study of the physical and chemical natures of the gastric juice that, with the exception of the discovery of pepsin, the closest research of modern times has added but litttle to the work done by him."

---

[7]Frederick P. Allen was a mere boy when Colonel Melanchton Smith, in April, 1811, purchased an old Ramage press, with two or three barrels of type, and established the Plattsburgh *Republican* in opposition to Samuel Lowell's anti-Federalist organ. A room was hired in the home of Allen's mother, and it was here the first editions were printed. At 13 years of age he became an apprentice as typesetter, and afterward served as compositor under Azariah C. Flagg's editorship. At the time of the publication of Dr. Beaumont's book in 1833, Allen was the proprietor and editor of the *Northern Intelligencer,* which afterward became the *Northern Sentinel.* The book was, therefore, published in this office.

The last word on the last page of Beaumont's book is an expression of regret at the incompleteness of his work:

I regret exceedingly that I have not been able to obtain returns from Professor Berzelius, to whom I transmitted about seven months ago a bottle of gastric juice for chemical examination. I could not, however, consistently with the expectations and wishes of my friends, further delay the publication of these experiments.

The immediate sale of the book seemed very satisfactory, though the first edition was probably a small one.[8] On December 30th the Secretary of the Navy, Levi Woodbury, wrote Dr. Beaumont:

I have to request that you will be so obliging as to furnish the department with twelve copies of your work upon the *Gastric Juice and Physiology of Digestion* for use of the naval hospitals and service, to be paid on delivery.

He was evidently considerably disappointed in the small number of copies purchased by the Government, as will be shown later in one of his communications to the Navy and War Departments. In order to show how favorably the work was received, however, by laymen as well as physicians, I present here a number of letters written by some of the famous men of the day, who were evidently close friends[9] of Dr. Beaumont. He had sent to each of these gentlemen a copy of his book, in all probability with a special binding, for he states elsewhere that he had fifty copies bound in full sheep.

PERU (N. Y.), 5 Decbr., 1833.

Dr. William Beaumont.

DEAR SIR: I thank you for your volume of *Experiments and observations on the Gastric juice and the Physiology of Digestion,* and, although I have no pretention to medical science, I have read it with great interest and satisfaction. If it had been a mere theoretical treatise on that difficult and mysterious subject, I should not presume to express an opinion on its merits, however plausible and ingenious. But the facts and demonstrations exhibited in your patient and laborious series of Experiments, the clear and discriminating inferences which you deduce, together with your modest history of the extraordinary case of Alexis St. Martin, excite my admiration and respect.

---

[8]Osler calls attention to the fact that Beaumont wrote the Surgeon-General that the first edition was to be one thousand copies, but states in the preface of the second edition that the first had been three thousand copies.

[9]Among his papers I found a large number of calling cards from these and many other men of great prominence.

The work cannot fail to correct false theories, to render a valuable accession to the common fund of useful knowledge, and to place your name on the Roll of public benefactors. With great respect and cordial esteem,

Dear Sir, yr. obedient servt.

JONAS PLATT, Judge.[10]

WASHINGTON, Dec. 31, 1933.

DEAR SIR: I thank you for the work you have presented. It is no less curious and valuable. By your skillful management of a most desperate wound you have been enabled to reveal some of the most important functions of the human system. And I sincerely trust that your useful and disinterested labors may add to your professional character and advancement.

Respectfully your obt. Svt.

LEWIS CASS.[11]

Jan. 1, 1834.

Mr. Van Buren[12] presents his compliments to Dr. Beaumont, and thanks him very kindly for a copy of his work, which Mr. V. B. was very desirous of possessing.

27 January, 1834.

R. H. Wilde[13] will not delay his acknowledgement to Dr. Beaumont until he has had an opportunity of reading the whole of the Volume the Dr. has had the politeness to send him. Though his knowledge of Physiology is very slight, he hopes it is enough to understand the value of the experiments made, partially at least, and the time, industry and expense they must have cost.

The subject is not entirely new to him, as his attention was called to it a year since by his friend, Dr. Dunglison, and a notice of the very interesting Character of the experiments, which was published in the *Intelligencer*.

An opportunity of trying such is of so very rare occurrence that the world is greatly indebted to Dr. B. for having prosecuted them so long and patient-

[10]The friendship of Jonas Platt dated back to the period of Beaumont's residence in and around Plattsburgh during the War of 1812. He had been judge of the Supreme Court, general of cavalry in the state militia, and senator. Advance in years led him to retire in 1829 to his farm in Peru, seven miles from Plattsburgh. He is said to have been "one of the most distinguished men in the state—wise, upright, and patriotic, as well as learned, eloquent, and able." His portrait now occupies a prominent place in the capitol in Albany.

[11]Secretary of War from 1831 to 1836.

[12]Martin Van Buren, at this time Vice-President of the United States.

[13]R. H. Wilde, member of Congress from Georgia, was made attorney-general of his state in 1815 and representative in 1827. He was author as well as statesman, having composed the popular song, "My Life is Like the Summer Rose," and made important contributions to the study of Dante's life.

ly, and with such success. R. H. W. desires to return his sincere thanks to Dr. Beaumont for so rich a contribution to the Stock of human knowledge on a very obscure and highly curious topic, and to assure the Dr. of his wish to aid by every means in his power in making the work as extensively known as it deserves to be.

SENATE CHAMBER, Jany. 28th, 1834.

DEAR SIR: I have the honor to acknowledge the receipt of your very acceptable book, containing *Experiments and observations on the gastric juice and the physiology of digestion,* and to express the admiration which I have felt in following the course of the curious and interesting experiments which you have made. Very respectfully, Sir,

Yr. Obedient Servant,

THOMAS H. BENTON.[14]

WASHINGTON, 4th Feby., 1834.

DEAR SIR: I have received the Copy of your interesting work, with which I had already made myself acquainted. Be pleased to accept my thanks for your attention. I deem it a work of great interest and importance; highly creditable to yourself for the manner in which the experiments are conducted, and likely to attract a great degree of attention throughout the world. I am, Dear Sir, with high respect,

Faithfully yours,

E. EVERETT.[15]

WASHINGTON, 5th February, 1834.

SIR: I have received, and looked over as far as my leisure would permit, the volume of "Experiments" you were so kind as to send me.

In whatever light the book may be viewed, it is truly a work of most surpassing interest, and I am at a loss for terms to convey my sense of the obligations under which you have placed, not your profession only, but the whole family of man. The efforts and the sacrifices required to procure and retain St. Martin, the patient perseverance bestowed for years on a course of experiments, which must have received the most delicate and unremitted attention, as well as the important and peculiar nature of your discoveries, cannot fail to be appreciated by intelligent men in all countries and in all future time, for no one can rise from a perusal of the work without an abiding

[14]Thomas Hart Benton, on Missouri's admission to the Union in 1820, was made senator, and held the office for thirty consecutive years, during which he ranked with Calhoun, Clay, and Webster, and was considered Calhoun's most formidable opponent in the Senate.

[15]Edward Everett was at this time a member of Congress, and showed much interest in the work in which Beaumont had been engaged.

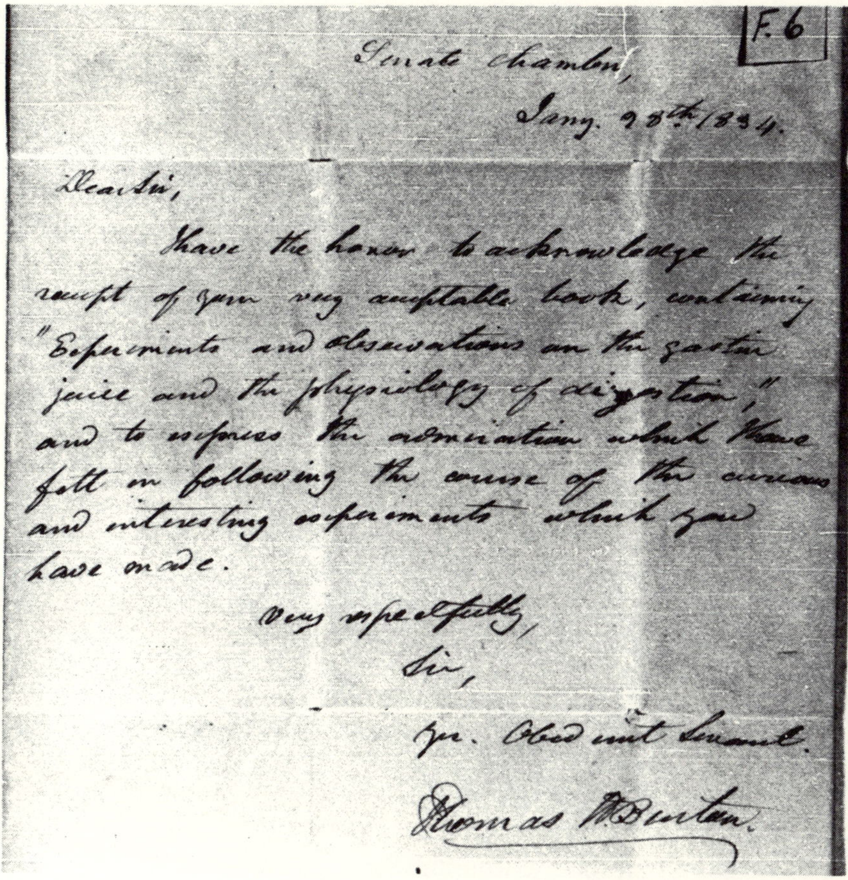

F. 6

Senate Chamber,

Jany. 28th 1834.

Dear Sir,

I have the honor to acknowledge the receipt of your very acceptable book, containing "Experiments and Observations on the gastric juice and the physiology of digestion," and to express the admiration which I have felt in following the course of the curious and interesting experiments which you have made.

Very respectfully,

Sir,

Yr. Obedient Servant.

Thomas H. Benton.

A letter from Senator Thomas Hart Benton, expressing his appreciation of a volume of Beaumont's *Experiments and Observations* . . .

sense of gratitude for the laborious and well directed researches of its author. I am very respectfully,

Yr. obliged and obt. svt.

P. C. FULLER.[16]

March 6, 1834.

Mr. Justice Story[17] begs Doctor Beaumont to accept his thanks for the valuable present of his recent work on the *Gastric Juice and the Physiology of Digestion.* The subject is full of importance and interest, and Mr. S. hopes soon

---

[16]Philo C. Fuller, Geneseo, N. Y., member of Congress 1833-36, assistant postmaster-general 1841-43, comptroller state of New York 1850-51.

to have leisure to read the whole work with care, and has received a great deal of instruction and pleasure from the portion of its contents which he has been able to peruse. Mr. S. deems it an honor to have an opportunity of placing in his library such a work from the kindness of its author.

PARIS, 16 March, 1834.

MY DEAR SIR: I am very much obliged to you for Dr. Beaumont's book of experiments, of which I have kept one copy and given the others to Orfila,[18] the Dean of the faculty of medicine, to the Academy of Science of the Institute, and to the Academy of the body of which I am a member. In the Academy of Science it has been referred to a committee to determine whether any additional experiments to those reported by Dr. Beaumont are necessary for the interest of medical science, and, if so, whether it will be expedient to send for the man here, or to request that the experiments be made in the United States. As soon as the report is made, I will send it to you.

EDWARD LIVINGSTON.[19]

To all of these panegyrics from illustrious friends were added the cheering words of his good wife Deborah:

Tomorrow I shall hear from you, and know if you have seen all the mighty puffs. . . . I need not say how much I feel gratified by the encomiums bestowed upon your work by the public. I see that one of the editors pronounced you a great scholar. That was particularly pleasing, as on that point . . . Samuel was most uneasy, as he said you nor himself were either scholars. May you see all your wishes accomplished, and be ready ere long to settle down quietly with your family, who all love you so much, is the prayer of your wife.

In spite of all these encomiums, he was, however, doomed to a certain degree of disappointment, as might be expected of one who had become so deeply engrossed in his subject that it was difficult for him to understand why every one could not feel the same interest. As mentioned in a former letter, he was desirous of having the book copyrighted and published in England. With this in view, Professor Jones, on behalf of Dr. Beaumont, communicated with a friend, evidently an author, in England, requesting him to secure a publisher in London. The friend responded with great regret and much surprise

---

[17]Joseph Story, associate justice of the United States Supreme Court.

[18]Mathieu Joseph Bonaventure Orfila was founder of the science of toxicology, professor of chemistry and toxicology, and for eighteen years dean of the medical faculty in Paris.

[19]Ambassador to France.

that he had been unable to obtain a publisher, and inclosed the following notes, which he received from two of them in response to his offer:

FLEET ST., Feb. 20th, 1834.

DEAR SIR: I have returned Beaumont's Experiments, as I do not feel inclined to make an offer for it.

Your obed. Servt.,

L. HIGHLY.

356 STRAND, Feb. 24th, 1834.

SIR: I am much obliged by the offer of Beaumont's experiments, but regret to inform you it is not a work which I should like to reprint.

I remain yours obediently,

HENRY RENSHAW.

Dr. Jones wrote Dr. Beaumont concerning the matter:

I herewith transmit to you a letter and papers received from Mr. Vail, by which you will perceive that he has not been inattentive to the request made relating to your book. It would have afforded me the most sincere pleasure to have aided you in producing a more welcome result; it would seem, however, that the deposits have been removed in England as well as with us. It is a known fact, however, that the publishing business has been much depressed for two or three years past, both in England and France. That your labors will be appreciated and recorded to your honor by the scientific physiologists, I am well convinced, and I am sure, also, that this will be very welcome remuneration; it is one, however, which will not "make the pot boil"—a thing of some importance in our mundane transactions. It is really as necessary to have something for the gastric juice to act upon as it is to know how it operates.

That Dr. Beaumont had not been neglecting the material side of life while so deeply engrossed in the fascinating problems of physiology, that he was providing his family with that "something for the gastric juice to act upon" at the same time that he was providing the world with a knowledge of "how it operates," is pretty well demonstrated by a "memo" which he jotted down just before leaving Plattsburgh on December 8, 1833, calling attention to his belongings:

I own a lot of Land at Prairie du Chien, M. T., about 140 acres, near Fort Crawford; title good; Warrantee deed, duly executed by Mr. H. Dousman, and on the records at the Clerk's office of Crawford City, Michigan Ter; a house built thereon and a small Stable; value of the property $1200, and increasing. T. P. Burnett, Esqr., of Prairie du Chien, my Agent and attorney for Renting, paying taxes, etc. Rent due from 1st of April, 1833, at about 150 dollars for the whole, conditional with circumstances and also an acct. of $20 against Mr. Cochran for collection.

A city Lot at Green Bay; title Warranted by Daniel Whitney, Esqr., of Green Bay. Deed on Record in Clerks office, Brown County, M. T.

Mortgage transferred from Bailey and Hohns, N. Y., on farms formerly belonging to the heirs of N. Smith, in the hands of Judge Haile, and also Mr. Wm. Swetland's receipt for the money, $175, paid by me through Mr. Haile; he has the receipt.

My library, Surgical Instruments, Notes of experiments, drawings, and engravings of Stomach, etc., with Dr. Samuel Beaumont. He has also Memorandums and instructions for superintending the distribution and sale of my Book of Expts., etc.

Mortgages, Deed Recpts., old accounts, Public and miscellaneous papers, correspondence with sundry persons on subject of my publications, etc., in a Black Leather Trunk. Alexis Contract, commencing December 1st, 1833, for two years at $200 per annum unless sooner discharged. He has $10 due him on old contract. He drew his pay as Sergeant in the U. S. Ay. up to Nov. 1st, 1833, of Major Kerby, Paymstr. U. S. A., of Brownville, at the rate of $12 monthly pay—$2.50 per mo. for clothing and 10 cts. per day for subsistance. I have drawn my pay of Maj. Kerby, paymaster, to the 1st of Nov., 1833. I have drawn from the government my regular allowance for fuel and quarters from the 1st day of July, 1833, to the time I leave Plattsburgh, and also my transportation from N. York to Plattsburgh on July 1st, per order. I shall be entitled to transportation from Plattsburgh to Washington when I shall have accomplished the journey.

My anatomical preparations are in Drs. Mower and Kane's office.

I am not aware that I owe anybody a cent, except Mr. Thos. Green and Mr. E. Smith when he has finished binding my Books. Will then settle to 8 cts. for all he binds in board and 28 cts. for 50 bound full in Sheep, and perhaps a few dollars for contingencies. Dr. Samuel will settle with him.

1834-1835

Some time prior to this Dr. Beaumont conceived the idea of asking aid of the Government to remunerate him for the expenses already incurred, and to enable him to continue his experiments on digestion, keeping constantly in mind the possibility of taking Alexis abroad for further investigation. Not being a politician, he took advantage of his close association with those who knew the ins and outs at Washington. Whenever a new idea occurred to him, Surgeon-General Lovell was the first one to whom he communicated it, and so in this case. Therefore, even before leaving Plattsburgh, he informed Dr. Lovell of his intentions, and received promptly from him this excellent advice:

> As to the application to Congress, the best and most efficient aid your friends can give you is to interest individual members to take up and urge the matter through. Memorials and testimonials formally presented to Congress soon find their way into the table drawer or breeches pocket of the chairman of some committee, where they sleep quietly unless the constituents make it their interest and business to look after them and call them up. Let plenty of letters to this effect be written after the session has commenced, or they will forget every word about it by the time they reach the confines of the District, with their brains addled with statistics, etc.

It is very probable that this particular matter prompted Beaumont to request a call to the capital as soon as he had finished his work at Plattsburgh. The Twenty-third Congress was now in session, and, if he wished any action on his behalf at this time, it would be well for him to be on the scene to start the ball rolling. This he did immediately on arriving at Washington, and before very long had succeeded in interesting a number of very prominent men in both houses of Congress. A letter from Azariah C. Flagg,[1] one of the most astute politicians in the state of New York, showed Beaumont to have been a very apt pupil in matters political as well as those scientific. Mr. Flagg had been secretary of state under Governor Clinton, and was now comptroller of New York and a member of the "Albany Regency," a group of politicians who controlled New York by a system of rewards for partisan service. The Regency numbered among its members such men as Martin Van Buren, Benjamin F. Butler, and Silas Wright, Democratic senator from New York. On December 19, 1833, Mr. Flagg wrote to Beaumont from Albany:

> I have written to Mr. Wright, telling him that if your application can be presented on such grounds as not to conflict with any of the vetoes, that you are in all respects deserving of his friendly aid. I suppose that the connection of your experiments with the medical service of the United States army and navy would justify an appropriation, to remunerate for past as well as prospective services in these experiments with reference to medical science solely, as connected with the public service. It is important that you present your memorial in such a shape that it will be easily *digested* by Congress.

Dr. Franklin Bache, of Philadelphia, to whom he had also applied for support, with the request that he obtain the opinions of other medical men in that city, replied:

> I have just received your letter reporting the memorial which you intend to present to Congress. It appears to me that you deserve much credit for your perseverance and public spirit in presenting your experiments, and, as their results are really a public benefit from their enlightening us on a subject which concerns every one, whether in health or disease, I think that private zeal ought not to be allowed to suffer in a pecuniary point of view. Your labors were undertaken and prosecuted with disinterested motives, so far as your original intention is concerned; and if, now, your expenses incurred in prosecuting a highly useful branch of inquiry are so heavy as to have determined you to apply to Congress, I think you ought to be indemnified, both as a proper reward to yourself and as an encouragement to future zealous citizens who may incur expenses in the successful investigation of useful branches of knowledge. In regard to the opinions of other medical men of this city, I would respectfully suggest that you had better ascertain them by direct communication with those whose views you would wish to possess. I think it would be highly important if your experiments could be prosecuted, and that at public expense.

---

[1]Azariah Culling Flagg succeeded Colonel Melanchton Smith as editor of the Plattsburgh *Republican* in December, 1811, and continued as such until 1826. It was evidently during this period that he and Dr. Beaumont became friends.

Having received encouragement, written and verbal, from practically every one to whom he applied, he memorialized Congress in the early part of January as follows:[2]

To the Honorable, The Senate and House of Representatives of the United States of America, in Congress assembled.

The memorial of William Beaumont, Surgeon in the Army of the United States, respectfully represents that a most favorable opportunity having been afforded for improving the science of medicine and physiology, and of ascertaining and testing the powers and functions of the human Stomach, which has never before occurred under such favorable circumstances; That Your memorialist has been at great pains and expense in relation to the individual whose care affords the facilities alluded to, a brief account of which, and Your Petitioner's exertions to persevere, sustain and bring him and his case before the public for the general improvement of the science of health, Your petitioner will relate:

Alexis St. Martin, the person referred to, was an indigent inhabitant of the county of Michilimackinac, Michg. Ter.—not a soldier, but dependent entirely on his daily labor for his subsistence and support, and without friend or relative to take care of or provide for him in the misfortune which befel him.

In the month of June, 1822, he being on the Island of Michilimackinac, in Lake Huron, was most dangerously wounded by the accidental discharge of a heavily loaded musket. The wound was received just under the left breast, and supposed at the time to have been mortal. A large portion of the side was blown off, the ribs fractured, and openings made into the cavities of the chest and abdomen, through which protruded portions of the Lungs and Stomach, much lacerated and burnt, exhibiting altogether an appalling and hopeless case. The Diaphragm was lacerated, and a perforation made directly into the cavity of the Stomach, through which food was escaping at the time Your memorialist was called to his relief.

His life was at first despaired of, but he very unexpectedly survived the immediate effects of the wound, and necessarily continued a long time under the constant professional care and treatment of your memorialist, and, by the blessing of God, finally recovered his health and strength.

At the end of about 10 months the wound was partially healed, but he was still an object altogether miserable and helpless. In this situation he was declared a "common pauper" by the civil authorities of the county, and it was resolved by them that they were not able nor required to provide for or support, and finally declined taking care of him, and, in pursuance of what they probably believed to be their public duty, authorized by the laws of the

[2]This is a copy from manuscript of the second memorial presented by Dr. Beaumont, in 1835, at the suggestion of Mr. Everett, and therefore does not include the paragraph requesting funds with which to continue the work. I was unable to find a copy of the first memorial.

Territory, were about to transport him in this condition to the place of his nativity in Lower Canada, a distance of more than fifteen hundred miles.

Believing the life of St. Martin must inevitably be sacrificed if such attempt to remove him should be carried into execution at that time, Your Memorialist, after earnest, repeated, but unavailing remonstrances against such a course of proceedings, resolved, as the only way to rescue St. Martin from impending misery and death, and to arrest the process of transportation and prevent the consequent suffering, by taking him into his own private family, where all the care and attentions were bestowed that his condition required.

St. Martin was at this time, as before intimated, altogether helpless and suffering under the debilitating effects of his wound—naked and destitute of everything. In this situation Your Memorialist received, kept, nursed, medically and surgically treated, and sustained him at much inconvenience and expense, for nearly two years; dressed his wounds daily, and for considerable part of the time twice a day; nursed him, fed him, clothed him, lodged him, and furnished him with such necessaries and comforts as his condition and suffering required.

At the end of these two years he had become able to walk and help himself a little, though unable to provide for his own necessities. In this situation Your Memorialist retained St. Martin in his family for the special purpose of making physiological experiments after he should have sufficiently recovered to admit of this being commenced.

The peculiar condition of St. Martin's stomach, your memorialist verily believed, afforded the fairest opportunity and greatest facilities for improving an important branch of medical science that ever had, or ever would probably be, presented, the wound in healing having left an open aperture of considerable size directly into the Stomach, through which food, drink and other substances could be passed into and from any part of its cavity, and by means of which experiments and observations of much importance to physiology have been and may continue to be made, with useful and satisfactory results.

His general health, after four or five years, being greatly improved, and consequently his necessity for the professional care and attentions of your memorialist being lessened, it has been more difficult to retain St. Martin for the purpose of pursuing those physiological researches which had been commenced. Consequently your memorialist had not only to be at the trouble and expense of frequently seeking him in Canada after an unexpected absence of several years, but was obliged to pay him high wages to induce him to return and submit to the necessary examinations and experiments upon his stomach and its fluids for the purposes herein expressed.

Your memorialist also respectfully represents that the expense incurred by him on account of St. Martin since the time he was wounded—June, 1822—for professional attendance, dressings, medicine, nursing, etc., at reasonable charges, upon the principles of justice, equity and the common usages of profession, would amount to not less than one thousand—$1000.00.

That the expenses incurred for subsisting, providing for, and retaining St. Martin, money expended and wages paid, chiefly to induce him to submit to the necessarily required examination and experiments, would exceed, in the opinion of your memorialist, the average sum of Five hundred dollars per annum for Eight out of nearly Twelve years since he received the wound in June, 1822, which would amount to Four thousand dollars—$4000.00.

Besides the ordinary expenses above stated, your memorialist, stationed for thirteen years preceding the autumn of 1832 at the extreme North and Western frontier military posts of our country, the better to secure and perpetuate the advantages of this signal opportunity for physiological experiments, has been under the necessity of incurring the additional expenses of transporting St. Martin from place to place—from Mackinac to Montreal in 1825, from thence with his family to Fort Crawford on the upper Mississippi, and back again to Berthier, L. C., in the years 1829 and '31, a total distance of more than five thousand miles—and subsequently, with a view to demonstrate and diffuse such interesting facts and useful information as may have been developed in this case, and afford favorable opportunity for more thorough and scientific research and investigation of this important subject. Your memorialist has also incurred the extra expense of travelling with St. Martin to several of the principal cities in the U. States—to Boston, New York, New Haven, Washington, etc.—and prosecuting the "experiments and observations on the Gastric juice and the physiology of Digestion" for several months at the private and individual responsibility and expense of your memorialist.

As your memorialist has not only supported St. Martin alone, but, in order to secure the benefits of his submission to the experiments, has been obliged to provide for and maintain his family, consisting of a wife and three small children, for more than three years, and has them now on his hands for protection and support, without public munificence, and with no other indemnification at present than the thanks of an indigent man for the preservation of his life, and the humble gratitude of his helpless family for their sustenance and protection.

Your memorialist also represents that he has, with much labor and expense, composed and published a volume of "experiments and observations on the Gastric juice and the physiology of Digestion," comprising nearly three hundred experiments, with a hope of permanently recording the facts and extending the benefits of his researches in this extraordinary case. But as this branch of professional science will probably be uninteresting to the general reader, the sale of the book will be so limited as not to promise an indemnity even for the expense of the mechanical part of the work.

Besides all these extra expenses, labors and exertions, your memorialist has continued constantly and faithfully to perform his public official duties as Surgeon of the United States Army without intermission, and without receiving additional indulgence or remuneration from the Government. These Experimental services were wholly extra, unofficial and voluntarily performed, and not even required of him either by common usage or the Regulation of the Army, but were executed in obedience to the dictates of humanity, a desire to benefit community, promote the blessings of health, and perpetuate the signal advantages of this extraordinary case and opportunity for improvement.

And your memorialist will further represent that the pay now allowed to Surgeons of the Army, a stipend barely sufficient to cancel the ordinarily current expenses of a military life, has been greatly insufficient for the last three or four years to defray the increased expenses he has necessarily incurred in sustaining, prosecuting and bringing forward this subject for public and professional consideration and improvement.

Conscious, therefore, of the inadequacy of his individual means to sustain, without injustice to himself, the expenses already incurred, or further to prosecute the subject unaided; humbly conceiving himself justly entitled to a fair remuneration for those extra services, and believing them useful and important to the interests of the medical profession, the science of health, and particularly to the Army and Navy, which, from the nature of their respective services, are more exposed to the ravages of disease and the deleterious effects of irregular subsistance and improper diet and drinks, your memorialist respectfully and earnestly solicits Congress to grant him an extra allowance of One dollar and twenty-five cents per diem, the compensation usually allowed to officers of the Army when performing extra duties, for the time he has actually been engaged in professionally attending and restoring St. Martin to health, and in making the said experiments and observations. . . . The first period to embrace the time between the 6th of June, 1822, and the last of May, 1825, during which he was recovering from his wounds under medical care and treatment of your memorialist, being ten hundred and fifty-nine days at $1.25—$1323.75. The Second period to include the time between the first of January, 1830, and the 1st of January, 1834, being 1460 days—$1825.00. During the last period the principal experiments were performed.

The merits of Your Memorialist's humble labors, and his motives for pursuing the object of these experiments and securing the benefits that may result therefrom, are left to the scrutinizing intelligence of the present age to discover, and to the justice of those who may come after to appreciate and apply.

Trusting to the liberal policy and enlightened patriotism of an American Congress for a just decision in the premises,

Very respectfully, Your Petitioner, etc., etc.,

WM. BEAUMONT,
Surgeon U. S. Ay.

The memorial was sent to a committee of which Hon. Asher Robbins, senator from Rhode Island, was chairman, and he proceeded at once to investigate the value of Beaumont's work and the merits of his claims. Among others, he requested Professor Sewall,

Professor Dunglison, and Surgeon-General Lovell for their opinions as to the value of the work, and it is needless to state that they were entirely favorable. Dr. Sewall wrote:

It gives me great pleasure to comply with your request in laying before you, as chairman of the Committee of the U. S. Senate, my opinion of the Experiments and observations of Dr. Beaumont upon the Gastric Juice and the Physiology of Digestion. No subject in Physiology has been more extensively investigated, or received a larger share of the attention of medical men, than the functions of digestion, and none surely is more worthy of consideration. But, unfortunately, no case in the history of our science before that of St. Martin has ever occurred affording equal facilities for observation and experiment. . . . And should the Doctor be enabled by the Munificence of the Government to avail himself of the extraordinary case of St. Martin to prosecute the investigations to its utmost limits, an opportunity which has never before and which may probably never again occur, I have no doubt he will establish the great principle of this important function of the animal economy upon a firm foundation.

Dr. Beaumont's sacrifices of time and money in the prosecution of these experiments have been great, and such as he cannot expect to have remunerated by the sale of his book. He justly merits, as I conceive, the grateful acknowledgements of the medical Profession, and I should hope he might be sustained in his laudable efforts by the fostering hand of our enlightened and liberal Government.

Professor Dunglison[3] reviewed the work, and added:

There are yet, however, many obscure points connected with the physiology of digestion, and many circumstances touching the relative digestibility of various aliments and the action of the Gastric secretions, that demand investigation, and any facilities that could be afforded to Dr. Beaumont for this purpose could not fail to result in the improvement of science and the advancement of the public good.

Based on the tenor of these reports, Mr. Robbins presented the memorial to the Senate on January 27th, and it was immediately referred to the Joint Committee on the Library of Congress. This committee, too, promptly began an investigation of the merits of the case, and sent requests to various physicians in Washington for their written opinions concerning the subject in hand. Thomas Henderson, professor of theory and practice of medicine in the Columbian College, sent the committee a full synopsis of the work, and added:

---

[3]Then professor of therapeutics, hygiene, and medical jurisprudence in the University of Maryland, Baltimore.

The facts already established here, whilst they confirm the conclusions of some of the gifted cultivators of the science, whose genius and sagacity led them on by more imperfect aids to the wonderful approximation of the truth, have forever put to flight the visionary hypotheses and fallacious theories which have been taught for ages in the schools of medicine and have exerted a pernicious influence over their practice. False hypotheses in medicine, above all the other sciences, must be productive of the worst consequences, as from false theories false practical deductions will necessarily be drawn, which may be destructive of human life; and surely any measures which will aid in the establishment of a single fact in an inquiry of such importance cannot be considered as unworthy the attention of the Philosopher or the Legislator. . . .

I feel it my duty, therefore, respectfully to recommend that every facility be afforded to Dr. Beaumont for the continuation of these interesting researches whilst nature (we may literally say) has drawn aside the veil and exposed to view her hidden operations, and which may soon again be closed upon the inquiring eye of science for ages to come.

A petition signed by Abel Huntington, John Thompson, and Robert Mitchell, members of the medical profession and of the House of Representatives as well, concluded with the appeal:

It would seem to be the part of wise legislation to encourage the continuance of these experiments, with prospective views of benefit to the interests of science and humanity.

The strongest appeal was sent to the committee by several members of the faculty of the Columbian College, among whom were Thomas B. Jones, professor of chemistry; I. C. Hall, professor of surgery, and Thomas Henderson. They said:

. . . It is a fact, which must strike every reflecting mind with great force and effect, that no such instance has ever before occurred. It is true that persons with wounds of the stomach have been seen, and experiments to very limited and unsatisfactory extent have been made. Two or three instances, perhaps, are mentioned in our medical records. But most assuredly no such case as the one here alluded to has ever been seen. In this case no limits are set to the extent and diversity of experimental investigation. The size of the orifice of the stomach, the quantity of gastric juice to be obtained, the control over the man (who will submit to no other experimenter), the devotion of Dr. Beaumont and his almost enthusiastic purpose to elicit all that can be drawn from it, the success already attending his inquiries, all make this case the only instance that we have ever had from which all may be learned that is to be learned. This appeal, now made to the liberal and enlightened minds of your honl. committee, will readily be excused when rightly appreciated. It comes from the science of medicine, it comes peculiarly from American medical science, it is directed to those whose judgment and learning concurrently acknowledge the force of the appeal. It asks not to be allowed to pass away, without yielding all its influence to medical inquiry, that which has never occurred before, and which, of all things, is least likely to ever occur again.

It is probably already known to your committee what Dr. Beaumont has accomplished. The reviews tell favorably of his labors; the newspapers teem with popular allusions to his experiments. The doctor has done this at great pecuniary sacrifice; he has done it from an unsophisticated spirit of scientific curiosity. There is but one feeling in the medical world toward him—a feeling of respect and gratitude for his excellent work. Dr. Beaumont knows and says that he has but a glimmering light which is to be shed by a further prosecution of the inquiry. He proposes to issue a circular to the distinguished physiologists of America and Europe, asking them to propose to him the various truths it is important to develop, and the surest methods of arriving at these truths. He knows nothing but disinterestedness and love of science. In no better hands could the case have fallen, to no better hands can you safely commit the patronage, pecuniary and official, which it is thought the Government, from its contingent resources, can amply bestow. . . .

To contribute to these results, to be known now and hereafter as the patrons by whose liberal decisions they have been rendered accessible, is matter of reasonable pride, as it will be of perpetual credit, in the annals of our science. Nor is it unworthy, in our humble opinions, of the consideration of your Honorable Committee that to facilitate the consummation of Dr. Beaumont's purposes will redound to the character of the medical staff of the U. S. Army in that one of its members shall have made such a contribution to science.

In spite of these eloquent appeals, and the fact that no less a personage than Hon. Edward Everett was fathering the movement in the House of Representatives, and had succeeded in appending an amendment to the appropriation bill granting Dr. Beaumont $10,000, the cause was lost. To the great discredit of the Twenty-third Congress, there is to be found spread upon its journals this record:[4]

The House then proceeded to consideration of bill 246, making appropriations for support of army of 1834. The amendments were read and severally put, concurred in by the House, except the 2d, which read, "For extra allowance to Surgeon Beaumont, of the U. S. A., at the rate of $1.25 per day, commencing Jan. 1, 1831, $1,825, etc.," and on the question it was decided in the negative, 56—129. The ayes and nays were called for and taken.

The defeat of the amendment is said to have been due to the "theoretical objection to the appropriation by Congress of money for

[4]Journal of House of Representatives, March 14, 1834, first session of Twenty-third Congress.

any scientific or philanthropic purposes whatever." While it was unusual in that day for Congress to make appropriations for such purposes, the real obstacle in this case was doubtless a proper lack of appreciation of the work by the members of Congress.

An amusing effusion, taken from the Washington *Evening Star,* probably voiced the sentiments of many of the wise legislators who opposed the measure. It is only too true that a little ridicule may completely offset in the popular mind the results of much serious thought and scientific labor.

To the Chairman of the Military Committee, Washington.

SIR: I perceive by the papers that a provision has been slipt into the military appropriation bill making a certain compensation to Dr. Beaumont. For what? It appears that the Doctor undertook to cure a man who had received a wound in his side. He did not heal the wound. An aperture thus remained, by which an inspection could be had of the stomach, and of the operation of the gastric juice on various articles of food. He noticed this operation, and has stated the result.

Does this display such science and genius as to entitle the Doctor to a national remuneration? and that, too, by a special act of Congress?

He is already paid for his medical services as a Surgeon in the Army, and the law of a copy-right secures him any further compensation which the public may think his published discoveries are entitled to.

Besides, sir, there is a fallacy in the whole business. Digestion is a chymical process, and no two things are more different than a chymical process carried on in the presence and, again, in the absence of the atmospheric air. The gastric juice performs its operations in whole bodies without the presence of atmospheric air; in the present instance atmospheric air is present in the stomach.

Gastric juice being essentially the same in human and animal stomach, far more satisfactory experiments have long since been made on the solvent power of that juice for the purpose of digestion by feeding dogs with various food, and under a variety of circumstances as regards sleep, exercise, hunger, repletion, etc., and then cutting open the stomach at different stages in the process of digestion. The results have been published to the world long and long ago.

MEDICUS.

Mr. Everett was much disappointed at the fate of his pet measure, and promptly conveyed his feelings in the matter to Dr. Beaumont, at the same time suggesting that he should not be discouraged, but that he try a "new tack."

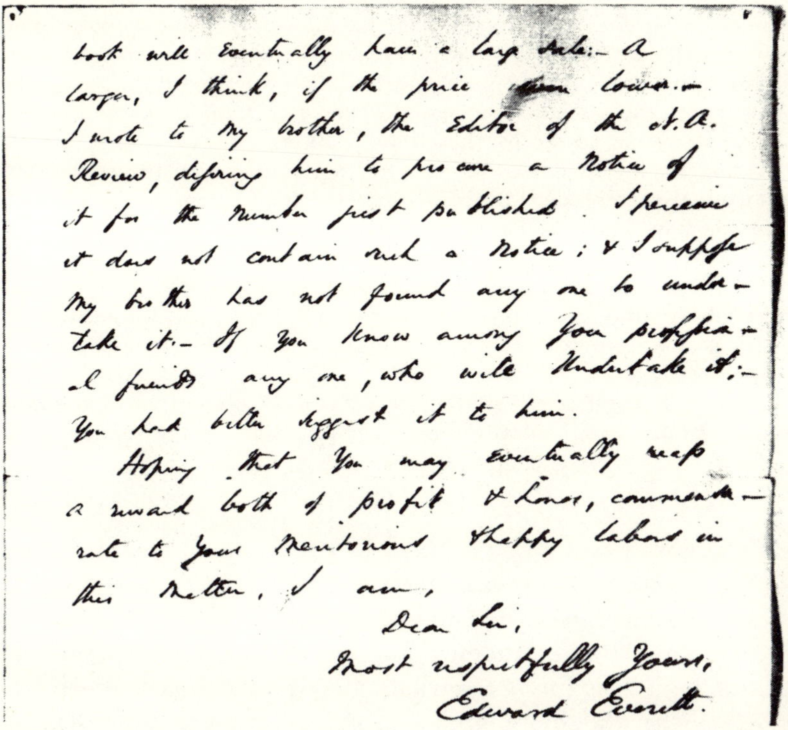
book will eventually have a large sale:— A
larger, I think, if the price were lower.—
I wrote to my brother, the Editor of the N. A.
Review, desiring him to procure a Notice of
it for the number just published. I perceive
it does not contain such a Notice; & I suppose
my brother has not found any one to under-
take it.— If you know among your profession-
al friends any one, who will undertake it:—
You had better suggest it to him.
Hoping that you may eventually reap
a reward both of profit & honor, commensu-
rate to your meritorious & happy labors in
this matter, I am,
Dear Sir,
Most respectfully Yours,
Edward Everett.

Washington University School of Medicine

A portion of the letter from Edward Everett, in which he deplores the fate of the bill asking for an appropriation by Congress for Beaumont's work.

House of Representatives, 1 April, 1834.

Dear Sir: I have your favor of the 24th March. I was much mortified at the fate of my amendment to the Appropriation bill in your favor, but not much disappointed. My only hope was that it would be carried through the House without exciting its attention. As soon as the Constitutional question was started, I feared for the result. . . .

It is true that some members were of opinion that St. Martin was a soldier. I thought so myself at first, and so stated when the amendment was first moved. The fact that he was not a soldier, to which you attach a good deal of importance in your letter, instead of strengthening, weakens the claim in the judgments of a great many of the members—not in mine. The great difficulty lies in the theoretical objection to the Appropriation by Congress of money for any scientific or philanthropic purpose whatever.

I suppose that a few members who voted against the amendment did so because they disliked to have it brought forward in that way, but I am fearful the great objection lies deeper.

Nevertheless, I would not wholly discourage you. I think, if you should present a New Memorial,[5] substantially the same as that which you presented in the Senate, setting forth what you have expended and asking some compensation, but not proposing an appropriation for the Continuance of the Experiments, there might be a chance of success. If you wish it, I will return you the Memorial, to receive these modifications.

Upon the whole, however, your reliance must be on the curiosity and intelligence of the Medical profession and the public at large. Your book will eventually have a large sale—a larger, I think, if the price were lower. I wrote to my brother, the editor of the *N. A. Review,* desiring him to procure a notice of it for the number just published. I perceive it does not contain such a notice, and I suppose my brother has not found any one to undertake it. If you know among your professional friends any one who will undertake it, you had better suggest it to him.

Hoping that you may eventually reap a reward, both of profit and honor, commensurate to your meritorious and happy labors in this matter, I am,

Dear Sir, Most respectfully Yours,

EDWARD EVERETT.

Beaumont was not one to be easily discouraged. Even if his work had not received all the recognition and encouragement at home that it deserved, it had made him an international character. His work had been presented to the Academy of Science.[6] Reviews were to be found in a number of the foreign journals, among which may be mentioned *The Athenaeum,* and, more important still, a German edition[7] had been

---

[5]This suggestion was adopted, and carried into effect at the succeeding session of Congress.

[6]*The Athenaeum of English and Foreign Literature, Science and the Fine Arts,* London, Saturday, April 19, 1834. Osler calls attention to the fact that the earlier reports of the case had been abstracted in the *Magazin der ausländischen Litteratur des gesammten eilkunde,* Hamburg, 1826, and the *Archives generales de Medicine,* Paris, 1828. The American medical publications were few in number at this period, and, while practically all of them reviewed the work, but two or three seemed to have recognized the full import of this great American scientific product. One of the most complete reviews appeared in the *American Journal of Science and Arts*, vol. XXVI, July, 1834. In the May number (1834) of the *Western Medical Gazette,* of Cincinnati, there appeared a rather long, unfavorable review. I found in manuscript a long and forceful reply to this review, which doubtless appeared in a later number of the *Gazette.* It was entitled "Observations on the Review of Beaumont on the Gastric Juice, etc.," by a reader. This reply was signed by O. P. Q., and was supposedly written by Mr. J. H. Freligh, an intimate friend of Dr. Beaumont. Judging by the style of the article and the writer's intimate knowledge of the subject in hand, one may justly conclude that this retort was either written by Beaumont himself or inspired by him.

[7]While it has been stated that the work had been translated into French, I have been unable to find any record of such a translation. The work was presented before the French Academy through the efforts of Mr. Livingston, and it is possible that this fact led to the statement that the work had been translated into French. Osler also states that he was unable to find a French edition of the work.

issued, with the title: "*Neue Versuche und Beobachtungen über den Magensaft und die Physiologie der Verdauung.* Auf eine höchst merkwürdige Weise, während einer Reihe von sieben Jahren, an einem und demselben Subjecte angestellt, von Dr. Wilhelm Beaumont, Wundarzt in der Armee der vereinigten Staaten Nordamerika's. Aus dem Englishen übersetzt von Dr. Bernhard Luden, prakt. Arzt in New York. Mit drei lithographirten Abbildungen. Leipzig, 1834. Christian Ernst Kollmann."

His determination, therefore, to continue his efforts in the further study of digestion was in no way dampened. His plans were, however, delayed by the receipt of an order on March 4th:

> Secretary of War directs that you proceed with convenient dispatch to Forts Trumbell, Walcott, Preble, Constitution, Sullivan and Hancock Barracks on a tour of inspection of the hospitals, under the head of Medical Director. Having finished this duty, your services will be required at Jefferson Barracks.

He performed this duty between April 16th and May 9th, and on May 17th sent to the Surgeon-General a full, detailed report of his investigation, which he concluded with the statement:

> In the whole course of my examinations of the several stations, Hospitals, Climates, Clothing, subsistance, exercise and employment of the Soldiery, I have not been able to discover appreciable, adequate cause for the production of any particular kinds of disease or infirmity, except that of drunkenness, dissipation or imprudence, and these are much fewer and less frequent than formerly.
>
> The moral and physical condition of the men appears to have been vastly improved by the abolition of the whiskey part of the ration and the substitution of gin, sugar, coffee, etc., in its place. Sobriety, contentment, cleanliness, and good order now generally prevail where almost constant drunkenness, riots and mutinous conduct formerly abounded in the army. Officers now derive satisfaction and pleasure from their commands, and the soldiers willingly obey and eagerly perform their various duties with alacrity and delight.
>
> The general health and moral conditions of the troops at the several posts within my observation are remarkably obvious, and the professional abilities, faithfulness and attention of the officers of the line, as well as the Medical Staff, are sufficiently evinced by the excellent policy, order, regularity and comfort in which they keep their respective Garrisons and Hospitals.

In this connection, though probably not belonging to this period, I wish to record some of his "suggestions for a plan and regulations of commercial hospitals," found in rough draft among his papers, and not dated. The ideas which he embodied here were easily a half century ahead of their time, and compare very favorably with modern ideas of industrial institutions:

That the plan be based upon the liberal, broad principles of general, just and equal beneficence and public charities set forth in the memorial and Report of the Committee.

That Ample appropriation be asked for and Secured to erect and endow suitable buildings at the points indicated in the report, and perhaps other appropriate places higher up the Mississippi and lakes, especially G. B. and P. D. C., the natural intersecting points of water communication between the great Lakes, those inland seas of our country, and the long navigable Rivers of the west.

That the institution be placed under the general supervisionary care and management of a commission appointed by the Pres. of the U. S., whose duty it should be to receive annual or semi-annual reports and returns from the subordinate branches, to devise ways and means, and make provision for the perpetuity and improvement of the institution.

That each respective Hospl. and subordinate branch have a board of commissioners or Trustees, appointed by the General commissioner, to superintend and direct and regulate the local branches; to appoint from time to time resident physicians and Surgeons, and medical and surgical Boards of consultation, purveyors or Stewards, and general superintendents and other necessary attendants for the admission of applicants, and for the regulation of imposters or unworthy invalids, etc., who may have gained admission by misrepresentation or from unavoidable suffering and immediate distress.

That the institution be founded upon a general manual labor and self-sustaining system.

That some place appropriate and, so far as practicably consistent with restoration of health and principles of humanity, useful exercise and labor be adopted and judiciously be carried into execution for the employment of such invalids and convalescents as may be capable of performing or even of learning.

That no indolent, unworthy, vicious vagrant or imposter be permitted to remain in the institution after the development of such traits of character.

That only just criterion and requisite qualifications for admission be the worthy indigent, inefficient sufferers from age, infirmities, or unavoidable misfortunes and accidental injuries.

That these tests and qualifications of applicants be ascertained and judged of, as far as practicable, by a committee of investigators appointed for that particular purpose, and that no worthy member be permitted to remain and partake of the munificence and charters of the institution after he shall have become physically able to leave it.

That the immediate and severe suffering from evident and unavoidable misfortune only can entitle an applicant of doubtful merits or of suspicious character to admittance at all, and then only till he shall have obtained relief from his immediate sufferings and inability. That the worthy old soldiers, inland sailors, invalids of the army, and craftsmen of the Western Lakes and Rivers, and transient sojourners be the special objects of the institution.

That ample, convenient and properly constructed Hospl. buildings be erected upon the most eligible salubrious and pleasant sites, and having attached to them sufficient quantity of good arable land for agricultural and horticultural purposes, with an appropriate supply of farming and gardening utensils, tools for the different kinds of mechanic arts and the various implements of domestic manufactory.

That with a view not only to facilitate convalescence, and promote health and usefulness, but the better to obviate the debilitating and pernicious effects of indolence and idleness, to detect roguery and prevent vicious habits, than which nothing is more apt to occur in such [community], every invalid and convalescent should be required to perform such labor as his particular genius, former occupation and the state of his health may admit of.

The culture of the Mulberry tree, the nurture of the worms, and the making and manufacturing of the silk would not perhaps be unworthy a trial in such an institution where the climate would admit of it.

The foregoing suggestions, hereby thrown together upon a very cursory and partial view of the important subject to which they relate, are respectfully submitted for consideration.

A few days after his return from the tour of inspection, Beaumont made a short trip with Alexis, exhibiting him to medical societies in different cities. On May 15th Milo L. North and Edward P. Terry, of New Haven, sent, on behalf of the Connecticut Medical Convention, a resolution "that the thanks of the Convention be tendered to Dr. Beaumont for his interesting exhibition of the subject of his celebrated physiological experiments." Soon after this he exhibited him in Boston, and subsequently received a most enthusiastic letter from Dr. Charles T. Jackson.[8]

No. 11 Hanover St.,
Boston, May 23, 1834.

To Dr. Beaumont.

Dear Sir: Since you left Boston I have been almost continually employed in researches on the chemical composition and the Physiological properties of the Gastric fluid which you had the goodness to furnish me with during your short stay with us.

---

[8]Charles T. Jackson was a scientist of considerable renown, who made valuable studies in chemistry, physics, and geology. He claimed to be the first, in 1832, to indicate the applicability of electricity to telegraphy, and from 1836 to 1839 he was state geologist for Massachusetts, Maine, and Rhode Island. His name is most often recalled today in connection with the acrimonious discussion concerning the discovery of ether as an anesthetic in 1842. He entered claims together with Horace Wells, Crawford W. Long, and W. T. G. Morton. Jackson ultimately relinquished his claims to the discovery in favor of Dr. Long, and in 1873 became insane, dying in an asylum.

I have become deeply interested in these researches, and regret exceedingly that you could not remain in Boston long enough to witness them. I am sorry to state also that my stock of fluid is nearly exhausted, and my researches are yet by no means finished. I have ascertained the precise quantity of Muriatic acid and of mucus in the fluid, and have also found an animal matter somewhat like gelatine, but it does not possess all its properties—perhaps it is Osmazome, but I am not yet ready to give a decisive opinion. I have obtained the acetic acid in a separate state, but have as yet not determined its precise quantity. I am now examining for Lactic acid, which I suspect exists in the fluid, but have not yet proved its presence to my own satisfaction. I am also searching for Hydrofluoric acid, which Berzelius says may possibly exist in the Gastric fluid, but the presence of which no one has yet ascertained. I have entirely satisfied myself respecting the power of the gastric fluid to digest food out of the Stomach, and have repeated all your experiments upon the chyme with bile which I took pains to obtain from the gall bladder of a child lately dead of Pneumonia acuta. The results were very like those related by yourself. I shall give you the details hereafter.

I am now going to neutralize exactly a portion of the fluid with carbonate of soda, and ascertain if, when neutralized, it still has digestive properties. I write you now especially to beg you to send me another portion of fluid—as much as you can obtain in three or four days—or, if possible, to come yourself to Boston and assist at the experiments. I think we could do something handsome in the way of Physiological researches if we had St. Martin here. I have spoken to several physicians upon the subject of your coming here, and they say that there would be no difficulty in getting up a subscription to support Alexis while experiments were making under your direction. Perhaps lectures might be given on the case, which would remunerate you for the trouble. For my own part, I am willing and should deem myself happy to perform any part of the labor you might assign to me. If you can, do endeavor to come and spend a few months with us. If you are not able to leave yourself, please send me all the Gastric juice you can obtain. It can be closely corked up in a stout green glass bottle, and sent with safety by some person coming to Boston or by the New York packets to Boston. When you send anything, please write at the same time by mail, giving me an acct. of the time at which the fluid was taken and how it was sent. Perhaps it will be less objectionable to pack up the fluid obtained at each trial in separate packages, each numbered and referred to in the letter relative to the time and under what circumstances it was taken. In this manner we may preserve separate the different specimens, and ascertain if any difference of Composition exists in the specimens. The phials might all be packed in a small box, and sent by packet or by private conveyance. Please write me soon as possible, and let me know what you intend to do. In haste, my Dear Sir,

I subscribe myself your friend and obt. servt.

CHARLES T. JACKSON.

In the meantime St. Martin had been permitted to return to Canada again to visit his family, with the promise, however, that he would meet Dr. Beaumont in Plattsburgh by June 15th, and go with him and his family to Jefferson Barracks, twelve miles south of St. Louis, the new post to which he had recently been ordered. It was Beaumont's evident intention at this time to take his family with him to Jefferson Barracks, as is indicated by the following extract from a note written to his daughter Sarah from Washington on May 29th:

. . . Be a good child; learn your lessons; improve your mind in everything useful, virtuous and becoming. Finally, become an amiable Lady in your early youth. Remember, your Father's hopes and your mother's anticipations of future joy and satisfaction are upon the right cultivation of your mind and morals. Tell your cousin Elizabeth[9] she may prepare to go with you to Jeffn. Barracks, and then continue to be your preceptress and my female protege till she can find a more agreeable protector. Improve every moment you have before you go in learning all you can at Plattsburgh. Prepare your little Library, maps, charts and all the little traveling paraphernalia you can collect, to amuse and instruct yourself and the children on our journey through the Lakes and Canals. Mr. Halsey will hand you this, and tell your Mother how I am, etc.

Kiss your Dear mother and the children affectionately for me, not forgetting Elizabeth and the rest of the family, and be assured of doting love and tenderest affections of a father, anxious for your health and happiness, and eager to be with you.

He seemed determined before leaving Washington to force the Government to take further recognition of his production, and at the same time to dispose of a large number of volumes that were left on his hands. With this in view, he sent two communications, one to the Secretary of War and another to the Secretary of the Navy, dated June 2, 1834:

To the Hon. Secretary of Navy.

Doctor Beaumont, U. S. A., having been at much labor and expense in performing and publishing a series of "experiments and observations on the Gastric juice and the physiology of Digestion," which he verily believes will be of much usefulness to the public service by conducing to promote the general health of community—the army in particular, if the work be supplied to it. And being conscious of the inadequacy of his individual means to sustain, without manifest injustice to himself and family, the expenses already incurred, or of further prosecuting the subject unaided by public patronage in

[9]Elizabeth Smith, their niece, accompanied them to St. Louis, and lived with them until she married Morgan L. Martin, of Green Bay.

some way, he very respectfully suggests the justice and propriety of, and earnestly requests, a purchase of 100 copies by the Navy Department for the use of that branch of the service.

It is conceived that they will be of much useful importance, not only to the sick and invalids at the Hospitals of our different military posts, but also to the crews, messes, and individual families in their tendency to regulate wholesome rules of dieting upon the true principles of temperance, in both eating and drinking, so essential to the preservation of good health, constitutional firmness and moral rectitude. A notice of the decision, whether adverse or in favor of this proposal, communicated soon as practically convenient, will be very desirable, as Dr. Beaumont is under orders for Jeffn. Barracks, Mo., and expects to leave the City very soon, and would wish to make arrangements to furnish the books, should a purchase be decided upon. A note dropped into the postoffice within a day or two will be received before he leaves the city.

On the following day this reply was received from Secretary Levi Woodbury:

Your communication of yesterday is received, offering to this Department 100 copies of your work upon the gastric juice, etc. This Department has already taken 12 copies of your work to supply the different Hospitals and Yards. It has not been deemed expedient heretofore to buy Medical books for the different vessels in commission. But the subject will be taken into consideration, and, should it be deemed proper to change the practice on this point, I will be happy to inform you.

He now returned to Plattsburgh to meet Alexis and to take leave of his family, for in the meantime it had evidently been determined that his stay at Jefferson Barracks would be a short one. On his arrival at Plattsburgh he wrote Major Kirby, paymaster of the army, that he expected to leave on or before July 1st, 1834, for his station "in the 'Valley of the Mississippi,' that happy abode of savages, aligators and Indian traders." "I hope," said he, "if the Deposits be not returned to Pres. Biddle's Bank,[10] that some of them may be to my empty pockets. . . . Should you have received any Book-money, and know not what better use to put it to, you might remit it if convenient. My wife wants it for 'pin money.' "

He was much disappointed to find that not only had Alexis not arrived in Plattsburgh, but that he had sent no message of any kind as to his whereabouts or his intentions. He remained here a reasonable length of time, endeavoring to get into communication with Alexis, but without success. After considerable delay he was compelled to

[10]The United States Bank: Nicholas Biddle, President.

leave without him. His feelings are best expressed in his own words in an unofficial communication to the Surgeon-General after reaching Jefferson Barracks:

A vexatious disappointment and an unexpected detention at Plattsburgh, awaiting Mons. Sergt. St. Martin's return to me at that place, prevented my arrival here several weeks sooner, and now I have the mortification to report him absent without leave. . . . I arrived at Plattsburgh on the 15th or 16th of June, sanguinely expecting to meet Alexis there in readiness to go with me and remain with me, but, instead of his having returned as he promised, I could hear nothing from him, and was left in the predicament of 1825. I wrote to him and urged his immediate return, waited 10 or 15 days longer than I otherwise should have done, and, learning nothing from him, was obliged to come on without him. This placed me in a most unpleasant and vexatious predicament. I could not feel warranted in taking time to go after him. Not to regain control of or to leave him behind seemed like abandoning an important object, and sacrificing almost all that had been done. I was compelled, however, from a sense of public duty, to proceed without him, trusting to subsequent events and circumstances to retrieve the accident. A letter[11] ford. per mail overtook me on my route containing an extract in Monsr. St. M's own language, and very expressive indeed of his assumed disposition, to wit:

"This 's just a snatch of Monsieur's ways,
Thus go's he on in *tricks* and *lies,*
And thinks to get well paid for it."

Here we have his true character. His object was to induce me to come into Canada after him, and knowing my inability there to inforce the conditions of his [agreement] to any useful purpose, intended and expected to take advantage of my anxiety to extort a much higher salary. I know well his disposition and his ugliness, and hope rightly to [defeat] them. I have taken no notice of his communication, nor shall I make any demonstrations to get him again till I return in the fall (which I hope to be permitted to do without fail), by which

---

[11]The following letter was received en route. Alexis could not write, and in his entire correspondence with Dr. Beaumont had others write his letters for him, and he would make his mark.

BERTHIER, 26th June, 1834.

DEAR SIR: My wife is not willing for me to go, for she thinks I can do a great deal better to stay at home, for on my farm she thinks will be a great deal more profit to me. I had started on the 18th of the month, and went as far as St. John's and fell sick. On account of that my wife is afraid to let me go. My wife joins me in love to you and your mistress, and all the family. Hoping this may find you all in good health, I hope you won't be angry with me, as I can do better at home. I am much obliged to you for what you have done, and, if it was in my power, I should [do] all I could for you with pleasure. You will be good enough to give my love to Mr. Green and his family. No more from

Yours,

ALEXIS ST. MARTIN.

time he will have spent all the money I advanced him to provide for his family for the year ensuing, become miserably poor and wretched, and be willing to recant his villainous obstinacy and ugliness, and then I shall be able to regain possession of him again, I have no doubt. Should it be decided by the Parisian comtte. to invite the case there, opportunity must be afforded and means used to regain controll of him immediately before it be too late. I constantly fear he may lease himself to some of the medical men in Canada, and get his case into the hands of the English Doctors. You can readily appreciate my anxiety and the deep interest I feel in the case. I hope to receive the earliest possible intelligence that may be communicated through Mr. Livingston, accompanied by your views on the subject. Should there be no call to *Europe,* I shall be extremely anxious to receive your suggested indulgence of going for my family in the fall and return to *Fort Crawford* for permanent station. My best wishes for the recovery of your health and kindest respects for Mrs. L. and family.

It is interesting to observe that the *official* report sent to the Surgeon-General upon his arrival at Jefferson Barracks read:

SIR: I have the honor to report my arrival at this post on the 29th inst. [June, 1834.] Sickness on the way prevented my arrival a few days sooner.

When Dr. Charles T. Jackson learned that Beaumont was to be stationed in the West, and that St. Martin would no longer be accessible to him for further investigation and experimentation, he was evidently desirous of forestalling the move. With this in view, a petition was circulated among the members of Congress, to which more than two hundred of their signatures were appended, and was presented by Edward Everett to Secretary Lewis Cass just about the time Beaumont started for his new post. The petition was as follows:[12]

Being informed that Doctor Charles T. Jackson, an eminent chemist of Boston, is successfully prosecuting an analysis of the gastric fluid of Alexis Saint Martin, the Canadian boy attached to Doctor Beaumont, surgeon of the United States army, and that the analysis cannot be satisfactorily accomplished without the presence of Doctor Beaumont and Saint Martin; and regarding the case as furnishing a rare and fortunate opportunity of demonstrating important principles in physiology, by which credit may be conferred on the medical science of our country and important benefits accrue to humanity; also, understanding that several national scientific bodies are anxious to draw Saint Martin from this country for the purpose of prosecuting the investigations now making by one of our countrymen, who is in every way competent to the work; and persuaded that the opportunity now afforded, if

[12]"William Beaumont and his Work," by Victor C. Vaughan. (*Physician and Surgeon,* December, 1902, page 547.)

neglected, will be lost to our country forever, we request that the Honorable Secretary of War will station Doctor Beaumont at Boston, or in the vicinity, for the term of four months, or longer if necessary for the object.

The secretary replied:

I should have considered it due to the very respectable signers of the recommendation to comply with their request in this instance, had I not ascertained that another engagement had been made by the Surgeon-General, with the consent of Doctor Beaumont, which is more acceptable to him than the proposed measure of sending him to Boston. Before I received your letter he was on his route to Saint Louis, the place of his destination, and under these circumstances I trust that you will accord with me in opinion that it is not advisable to change his station at present.

Surgeon-General Lovell wrote on the back of this paper:

Doctor Beaumont already understands that he cannot conveniently be stationed at Boston; so he has stated that this application was made without his knowledge or desire. He has gone to his station without any desire to leave it at present.

While Dr. Jackson was, no doubt, actuated by proper motives in the matter, he was evidently led away by his enthusiasm. At any rate, it seems quite certain that he had not consulted Beaumont as to his wishes on the subject.

Dr. Beaumont did not find Jefferson Barracks entirely to his liking. He was sent there, it seems, to relieve one Dr. Foot, whose position from some cause or other had become peculiarly unpleasant. Whatever these conditions were, they seemed to render Beaumont's position similarly uncongenial.

He remained at the barracks, however, throughout the summer and fall of 1834, contenting himself as well as he could, separated, as he was, from his family, the pleasant and agreeable surroundings which he experienced in the East, and from Alexis, who for the past few years had been giving him so much food for thought.

It was about this time that a letter was received by Dr. Silliman from Berzelius, to whom a quantity of gastric juice had been sent, and from whom such splendid results were expected. Of all the investigators whom Dr. Beaumont had called to his assistance, none gave a more disappointing and resultless response than did Berzelius. More than a year had elapsed since the package of "magic fluid" was forwarded to him as the one man in all the world who would be capable of extracting from it all of its hidden secrets. So sanguine had Dr.

Beaumont felt that something unusual would be forthcoming from the Swedish chemist, that he was almost tempted to delay the publication of his work until word could be received from him. The letter finally came, addressed to Professor Silliman, dated July 19, 1834:[13]

MY DEAR SIR: I had the honor to receive some time since the present which you had the goodness to make me of three bottles [vials] filled with gastric juice, drawn from a man into whose stomach there was an aperture through the abdominal integuments.

I am very grateful for the confidence you have had in me in wishing to engage me in making an analysis of it, and I regret deeply that for the following reasons I am not able to answer your expectations.

First, the gastric juice, sent in April, did not arrive at Stockholm till towards the close of the month of August. It had not become at all putrescent, but how was it possible to be assured that the animal matters dissolved in it, after a separation of almost five months from the living body, and after an exposure to the elevated temperature of the months of July and August, were still identical with those of fresh gastric juice.

But this circumstance apart, I could not make this analysis with any hope of success. I assure you that I commenced, but the difficulties immediately arrested me. On testing the gastric juice with litmus paper, I found it strongly acid. The acids are for the most part volatile. To obtain them, recourse must be had to distillation, but the operation of boiling would change the animal substance in the residuum. The quantity of gastric juice being only 266.76 grammes, I felt that I ought to sacrifice none of it, and therefore, in removing the volatile acids, I evaporated the whole in a vacuum at the temperature of the room. I had a residuum of 3.385 grams, filled with crystals of chloride of sodium.

Now, it was necessary to make a plan of the analysis, but how could I make a plan, the nature of the substance to be separated being unknown? A single mistake in the plan would destroy the whole, and I had no more of the matter to recommence.

On recalling to myself how many times I have been obliged to recommence the analysis of blood, bile, and urine, etc., because I found it necessary, time and again, to alter the plan, it was evident that I could not now attain the object with the gastric juice, of which I possessed only 3½ grammes of dry residuum. I have, therefore, put alcohol, sp. gr. 0.833, upon it and enclosed it in a vial well stopped, where it waits whatever may happen.

A great number of experiments, chemical and physiological, ought to precede the analysis. These experiments would demand almost daily to renew the supply of gastric juice—e. g., 'tis said the gastric juice dissolves the aliments swallowed, but what is this solution? Does it not consist in this, that certain parts are dissolved entirely, and that others, insoluble, but in a very

[13]Afterward published in Silliman's *Journal of Science and Arts*, vol. XXVII, No. 2, January, 1835, pages 405-407, section 28.

divided state, are diluted in the form of a thick bouilli? What are the substances dissolved, and what the part insoluble, but diluted? The fibrine of the muscle is very soluble, even out of the stomach in very diluted acids. I inquire next, is it by the free acids of the gastric juice that this solution is made in the stomach?

Would gastric juice, rendered perfectly neutral, lose the power of dissolving muscular fibre? If not, it must contain another substance, which is the true menstruum. This substance ought to be isolated and studied apart before we could have any means of determining its quantity. This we could not do without being able often to renew the experiments with fresh gastric juice. It would be necessary even to examine with the fresh gastric juice the most of the aliments which the man used, each by itself, and, by proceeding thus, we should obtain probably some sure and numerous data that would give a glimpse of what should be sought in the analysis, for that which is unsought is rarely found. You see, then, my dear Sir, how much previous knowledge I need for entering upon this analysis with hope of success.

I request you to make the proper explanation and apology to Dr. Beaumont.

Compared with the results obtained by the American chemists, this reply was most disappointing, and undeserving of the great importance that was laid upon it.

On September 4th, 1834, information came that the complications at Jefferson Barracks would be relieved by assigning Surgeon Foot to Fort Winnebago and Dr. Beaumont to Fort Crawford as soon as another surgeon could be assigned to this post. Though the transfer was promised by November, Dr. Beaumont was becoming very impatient, and immediately on the receipt of this notification sent a formal, official request for a leave of absence for the purpose of going to the North for his family, in order to return to his permanent post of duty as soon as practicable. This official request was accompanied, as usual, by one of an unofficial character. He had been living in hopes that Mr. Livingston would succeed in making arrangements to have him come to Paris with Alexis, and that by the time he had completed his temporary service he could proceed abroad with Alexis. He said:

Having received no information relative to the suggestion of going to Paris with Mons. Alexis, and, from the length of time elapsed, believing nothing more will be done about it, I enclose a formal and offl. appln. for leave of absence to go for my family in the fall, and, if no other arrangement can be made by which the objects contemplated at the time I last saw and conversed with you on the subject [can be realized], I wish to be permitted to avail myself of the indulgence soon as possible. I should prefer an order, however, to some other post. If no farther general or public measures are likely to be taken to aid in prosecuting the expts., and I am to be left to my own individual resources for further investigation of the subject, I wish to be assigned to some post more convenient than this, where I intend to take with me the subject, Mons. St. M., and such other means as I can command, and continue to prosecute the expts. and observations, etc., to the best of my limited ability, and communicate the result to the public in such manner as may be deemed most advisable. If I were not fearful of obtruding too much upon your patience, or that you would demur having anything more to do with it, I would most earnestly and sincerely desire your further advice and assistance in the proper course of proceeding. Vexed and perplexed, as I am sensible you have been more or less from the commencement, but knowing the kindness of your disposition and liberality of your views, I will presume, however, respectfully to request a frank and free comm. from you on the subject.

I have not much doubt of obtaining control of Mons. Alexis again this fall if I be permitted time and the opportunity. The pecuniary sacrifices I have already made in bringing the case before the public are all well known to you, as well as to the other Heads of the Depts. Can it not be made consistent to procure an order that would in some way aid in lightening the expenses of another trip from Missi. in the performance of a duty as much and more for the public than private benefit? Forgive the suggestion if it be improper; you know my views, and will, no doubt, rightly appreciate my motives. . . .

My preference of Posts is already before you. Ft. Crawford is still my first choice. I wish to be assigned to that station unless Ft. Columbus, N. Yk., is to become a Station for a Surgeon, in which event I would prefer that, as I have before indicated in recording my selections on the book in you office. I hope you will candidly and kindly appreciate the peculiar situation in which I am placed relative to my public and private duties, and duly consider the circumstances in which I am involved in relation to the Expts., Book, Alexis, etc., and favor my particular views and interest so far as consistently practicable without undue prejudice to the public service.

In due time the order came, and the early part of November, 1834, found him wending his way to Washington and Plattsburgh on a three months' leave of absence. Within an hour after his arrival at Washington he wrote his wife in his characteristic style a description of the country through which he had just passed.

WASHINGTON, D. C., Sat. Eve., Nov. 15, 1834.

I wrote you last, my dear Deborah, from Louisville, Ky. From there I proceeded to Cincinnati, Portsmouth and as high up the Ohio as Guyandotte, where the water was so low the boat could go no farther, and we were obliged to leave and take the stage over the Allegheny mountains through the state of Virginia. This, however, has prolonged my journey but a day or two, and has afforded me a fine opportunity of seeing an interesting and very romantic part of the United States that I have never before seen.

In taking this route, I have traversed the most celebrated parts of the State, passing almost over the very tops of the mountains, and by the White Sulphur Spring, the Hot Springs, the warm Springs and the Burning Spring, as famed in that State, and indeed they are great curiosities. I have continued in company with Genl. Ashley and family all the way. We were two whole days and part of the nights in traversing the mountains. We passed over the highest peaks, 2500 feet above the valleys, before daylight and 3 ok. on clear, serene morning. It was verily the most sublime and beautiful scene I ever witnessed. The stars shone unusually bright, and we were in the very neighborhood of them. I could almost catch the falling meteors by their tails as they dropt around us, and felt their scintillations as they passed. If I had had a good pair of ears, I am sure I should have heard Old Mars and Bellona holding their matin converse, and Venus giving lessons to a class of young cupids how to point and direct their arrows. From these heights we would suddenly descend into the deep, dark dales below, where burns the blue sulphurous gass or boils the Brimstone Springs, both hot and cold. This is no fiction—it is a fact, and, if anything on earth resembles Hell, it is these Springs. What is called the "burning Spring" is sulphurous gas or vapor, emitted from a slight excavation in a sandy soil through the pores of the earth, and, when a piece of fire or lighted torch is touched to the surface, it burns from a thousand points precisely like sulphur matches or melted brimstone, and continues till extinguished by throwing on water. The White Sulphur Spring boils up from the earth in a large fountain of perfectly clear water, with a very strong sulphur smell and taste. The Hot and warm Springs also boil up in the same way, differing in degrees of heat—one is about blood warm and the other quite hot. They are not all in the same neighborhood, but some 8 or 10 miles apart.

We also passed through Charlottesville, where is founded Mr. Jefferson's College, "The University of Virginia." I had the honor of being respectfully noticed by the medical Professors, and passed in review the celebrated dwelling place and sculpture of the Immortal Jefferson, Monticello. Passed to the banks of the Potomac, and there viewed "Mount Vernon," that sacred spot where dwelt and died and now rests the dust of the "Father of our happy country."

These are richly worth seeing, and my only regret is that I had you not with me to enhance the delights of the very interesting scenes through which I have just passed. Oh! how much you would have enjoyed it with our dear children with us!

The enclosed are leaves from a fresh rose I plucked from a bush as I passed this sacred spot, growing fresh in the open garden. I have also an unexpanded bud that I am going to try to preserve and bring you. The season is most delightful and pleasant in this latitude. Vegetation is yet fresh in the fields. A more delightful time could not have been enjoyed than we have had since we left St. Louis. Scarce an hour of foul weather have we had in the whole route till this afternoon, when it began to storm, and it is now raging furiously, but not till after our arrival here. We have had excellent roads, fine horses, elegant carriages, and good drivers, and have enjoyed fine health so far and hope to till after we meet. I arrived here an hour ago, and have not seen Dr. Lovell yet to ascertain my prospects. And tomorrow being Sunday, I fear I shall be detained till Tuesday before I can get away. I send this in advance to appease your anxiety and prepare for the reception of Your ever affectionate

WILLIAM.

Following his consultation with the Surgeon-General at Washington, he spent the greater part of his leave of absence with his family in Plattsburgh. Whatever the determining factors in their discussion of the matter, it was decided that Beaumont should return to St. Louis at the expiration of his leave of absence, to be stationed at the arsenal instead of the barracks, and that this should be his permanent station. He left Plattsburgh about January 10th or 12th, 1835, en route for St. Louis, with the understanding that he was to return in April for his family. There is evidence that he spent the 14th at Philadelphia. It goes without saying that while in Plattsburgh further efforts were made to get Alexis to join him. He could get no reply to his letters, however, before leaving, and authorized his cousin, Samuel Beaumont, to continue the correspondence. Samuel Beaumont wrote Alexis of Dr. Beaumont's great disappointment, adding:

He left word that if you answered his letter, and concluded to abide by his proposals of leaving your family there for the present and come out yourself and go with him, to write back to you and say that you might stay with your family till about the middle of April, and then meet him here, prepared to go with him. And he also said, if you refused to come without your family, or asked for more money than he had already advanced or consented to let you have when you came to him, that you might remain where you were, without expectation of any further assistance or consideration from him, enjoying the fruits of ingratitude and injustice as you could. He will not advance you any more money till you come to him, and manifest a sincere disposition faithfully to serve out your engagement to him and fulfill his expectations of you. When you have done that, he will reward you justly, amply and equal to your expectations.

To cover the distance between Plattsburgh and St. Louis at that time required a period of two weeks. Dr. Beaumont therefore returned to his post of duty about February 1st, 1835, and on March 18th notified the department at Washington that he was about to return to Plattsburgh via Washington for his family. He arrived there early in April, and began preparations for removing his family to their permanent home in St. Louis. It will be remembered that since leaving Fort Crawford in 1832 he had not been in any one location long enough to justify his having the family join him. The anticipation, therefore, of once more having his loved ones about him, with a home of his own, must have been a source of the greatest satisfaction to him. Could he but bring Alexis once more into his family group, his happiness would be complete. This visit to Plattsburgh, like the preceding ones, was characterized by efforts in this direction—this time through Mr. William Morrison,[14] with whom he had previously been in correspondence. After making apologies for trespassing again upon his time and patience, he wrote:

. . . But from the peculiar circumstances and situation in which I am placed with regard to him and his case and condition, I am induced to do so from a desire, in the first place, to ascertain facts, that I may be able to do justice to him as well as to myself; and 2dly, to avoid unnecessary disappointment either to the public or to individuals. From the suggestion in your reply to my letter on the subject last fall, and also from a commn. from Alexis himself, I was then induced to believe he would, like an honest man, have returned to me to fulfill his engagement this Spring. I have heard nothing from him since. I have now just returned from St. Louis, Mo., where I have been since Decr. last, and expect to return there again in a few weeks, permanently to remain. But before I finally abandon the idea of realizing and regaining him again into my Service, I am desirous of ascertaining Alexis' situation and true disposition, and can only expect to succeed indirectly through your kindness and candor. Long, vexatious experience has too much impaired my confidence in him to rely upon his assertions and promise. I am sorry it is so, but so it is, and it is now almost a year since, by mutual understanding and agreement, I advanced him money, considerably more than was then due him, to enable him to take his family into Canada, and leave them comfortably situated for 18 or 25 mos., to return himself in 5 or 6, to meet me at Plattsburgh to go West and serve out his engagement to me, since which I have not seen or recd. anything from him but mere pretexts for his failure to return according to agreement. I made my final proposition to him last winter, just before I started for Missouri, by which I am still willing to abide if he chooses to comply without any more cavil or delay. I expect to leave here with my family for the West on or before the 10th of May.

Will you do me a last favor, if you can endure the disagreeable condescension of seeing Alexis, to ascertain his situation, disposition and deter-

---

[14]William Morrison, an agent of the American Fur Company, residing in Canada, had been a well-known figure in the history of the company for many years. As early as 1808 he was one of the partners of the "shrewd Lisa" in the Missouri Fur Company. He was now living on a small estate in Canada.

minations, if practicable, and communicate them, with your own views and opinions on the subject, to me at this place soon as possibly convenient. I would not have troubled you again on the disagreeable subject, had not expediency as well as the necessity of the case required it.

Another matter which called for his attention prior to leaving for the West was some misunderstanding existing between him and his publishers. It seems that there had been some delay in the payment of the royalty or commission on the last five hundred copies of his work. He therefore, before leaving, gave Samuel Beaumont power of attorney, authorizing him "to ask, demand, sue for, receive from Lilly, Wait and Co., now or late Book-sellers of Boston, in the state of Massachusetts, or either of them, such sum or sums of money as are now due from them, the said Lilly, Wait and Co., or the balance due from them for or on account of five hundred copies of a book entitled *Experiments and Observations on the Gastric Juice and the Physiology of Digestion.*"

No record was found as to whether an amicable settlement, or any settlement, was made with his booksellers.

1835-1840

ALL ARRANGEMENTS MADE and his affairs satisfactorily disposed of, he started with his family through the canals and lakes on practically the same journey that he described so vividly in his "Travelling Journal from Plattsburgh to Mackinac in 1820." It was his intention to proceed overland from Green Bay to Prairie du Chien, and down the Mississippi to his destination. On June 26th, 1835, however, he reported himself sick at Green Bay, and asked for an extension of his furlough. They were not detained long, for on July 23d he reported to the Surgeon-General's office his safe arrival at St. Louis in good health, and on August 1st that they were comfortably situated at the arsenal, three miles south of St. Louis. Here he lived with his family for a short time, after which he moved into the city, and requested of the Surgeon-General the privilege of entering private practice in St. Louis and community. According to a notification which he had received in April, his duties were to be divided between St. Louis, where there were many officers and their families, and the arsenal, where he was the attending physician and the purveyor of medical supplies of the western military district. Dr. DeCamp was now stationed at the barracks, where there was at that time a small number of men under General Atkinson. He felt therefore that, since much of his time must be spent in the city, he could open an office and give medical attention to citizens as well as soldiers. He had been granted the same privilege at other posts without its having in the least degree interfered with his official functions, and the request was promptly granted on this occasion. Coming, as he did, with his well-established reputation as army surgeon, and with the prestige given him through the publication of his book, together with a strong personality and splendid presence, private practice fell into his hands very quickly. The officers and their families played an important role in the social affairs of St. Louis in these days, and his introduction was therefore prompt and complete.

In spite of the busy life he was leading and a rapidly growing income, he still had time and inclination to carry out the suggestion of Mr. Everett following the sad fate in Congress of his amendment to the appropriation bill of 1834—namely, that he present a new memorial to Congress, asking for compensation for the services that he had rendered Alexis, but not proposing an appropriation for the continuance of the experiments. When, therefore, the Twenty-fourth Congress convened in the fall, an amended memorial was presented. The following extracts from the Senate journal[1] speak for themselves:

(December 28, 1835.) Mr. Linn (senator from Missouri) presented the petition of William Beaumont, M.D., praying to be remunerated for his time and expenses in prosecuting a series of experiments on the gastric fluid on the person of Alexis St. Martin.

(December 31, 1835.) On motion of Mr. Wright (Silas Wright, Jr., senator from New York), member of Committee on Finance, ordered that the Committee on Finance be discharged from the further consideration of the petition of Dr. William Beaumont, and that it be referred to the select committee appointed on the petition of Boyd Reilly.

Boyd Reilly, the inventor of an apparatus for applying vapors to the surface of the body, was evidently endeavoring to have his appliance introduced into the army. While Beaumont was in New York conducting his experiments he received a request from the Surgeon-General to purchase one of the "patent vapor baths" of Reilly, at which time it was evidently being tested by the department. Reference to the Senate journal concerning the petition of Boyd Reilly reveals the following information:

(December 28, 1835.) Resolved, that the petition of Boyd Reilly, inventor of the apparatus for applying air, gas or vapor to the surface of the human body, which was presented to the Senate at the last session, be referred to the select committee. Ordered, that Mr. Naudein, Kent, Linn, Grundy, and Robinson be the committee.

(January 27, 1836.) Mr. Naudein made a report from the select committee on Boyd Reilly [but no mention of Beaumont's petition].

Bill to purchase apparatus sent to engrossment and third reading. (March 31, 1836.) Passed the Senate.

In other words, Beaumont's memorial was "pigeonholed," and has never been heard of from that day to this. The probabilities are that by this time Dr. Beaumont's opinion of Congress was entirely in accord with that of his cousin, Andrew Beaumont, himself a member of

[1]Senate journal, first session, Twenty-fourth Congress, 1835, pages 61, 72, 123, 250, 253.

Congress from Pennsylvania.[2] Andrew was evidently a very clever fellow, and certainly a wit if one may judge from those of his letters that it has been my pleasure to peruse. He wrote:

> I get so out of all patience with the damnable morals of men in the high station (always excepting Andrew Jackson), of the reckless knavery and gambling profligacy of a majority of the people's representatives, that I have hard work to stand it. Why, damn it, many things done here in the House of the people, under the pretense of legislating for the common welfare, would disgrace in profligacy a band of your real Mississippi Blacklegs. Oh! I forgot you lived upon the banks of the Father of Rivers. I beg your pardon. But, to leave the Mississippi Blacklegs to attend to Washington Blacklegs, these worthies rob and skin the people by millions over millions, and then tell them—and too many are fools enough to believe it—"that it is all for protection!"
>
> I hope you are happily located with the wife, when, after all your labors by land and water, you can sit down in peace and comfort, surrounded by all which makes life tolerable. It is, you know, a poor story at best, and, without those endearments of those we love and know to be our bosom friends, is cheerless indeed. I don't know how we muster courage and fortitude to endure life so long as we do, but "hope," whose ray pierces the most sullen gloom, shines always lovely, and lures us on thro many a melancholy path. You will say I am hypochondriac, but these are the reflections of Sunshine as well as clouds, and I appeal, Dear Cousin, to your own heart if there is not more truth than poetry in this half comic, half tragic reflection.

"Washington blacklegs" doubtless gave Dr. Beaumont very little concern after this, for his position was at least such that he was no longer dependent on the meager pay of an army officer, or the inconsistencies of a lot of politicians, who could squander thousands for political gain and had not a dollar to spend for the promotion of science. Dr. Beaumont's pride was probably the greatest sufferer in the matter, for he evidently did not mind so much losing the money as losing the fight. As to his cousin's "half comic and half tragic reflections," the probabilities are that he took no stock in them.

Within a year after he entered private practice he stated in a letter to the Surgeon-General:

> I have no wish or intention of resigning at present, though the professional encouragements I meet with here might seem sufficient inducement to do so, were avarice and professional fame the object of my ambition. I have a very handsome, lucrative and respectable private practice, with 6 or 8,000 dollars a year, a reputation far above my deserts, and a professional popularity more than commensurate with my best practical skill or abilities.

---

[2]From 1833 to 1837.

The doctor's professional standing among laymen is best evidenced by the fact that within a year after opening his office he had a very large and lucrative practice; his standing in the profession was shown not only by a large consultation practice, but also by the action taken about this time offering him the chair of surgery in the first medical department established west of the Mississippi. Within a year of his settlement in St. Louis a correspondence, originating in the action of the Medical Society of Missouri, was carried on between a committee of the society and the president of the St. Louis University relative to the establishment of a medical department in St. Louis under the charter of the university. Full power had been granted to the institution to establish law and medical departments, which, by the charter, were to be in operation within a given period from the date of the privilege conferred.[3] It was decided then to establish a school of medicine in St. Louis, and trustees were named.[4] This board immediately proceeded to fill the different chairs in the medical faculty, and selected six physicians: Drs. Hardage Lane, Joseph Johnson, E. H. McCabe, Henry King, C. J. Carpenter, and Dr. Beaumont, to whom was offered the chair of surgery. He was duly notified by William G. Eliot, Jr.,[5] secretary of the Board of Trustees, of his election on October 4th, 1836, but promptly replied:

Highly gratified by the purport of your note of the 4th inst., advising me of my election to the chair of surgery in the Medical Department of the St. Louis University, I regret the more that it will not be consistent with my

---

[3]"Memoir of the Late Wm. Beaumont, M.D.," delivered before the St. Louis Medical Society in 1854 by Thomas Reyburn, M.D. (St. Louis *Medical and Surgical Journal,* vol. XII, March, 1854, and Minutes of the St. Louis Medical Society.)

[4]General Wm. Ashley, Rev. William G. Eliot, Dr. B. G. Farrar, Thomas Green, Henry L. Hoffman, John W. Johnson, Judge Mary P. Leduc, Colonel John O'Fallon, William Renshaw.

[5]Rev. William Greenleaf Eliot, D.D., came to St. Louis when the city was a mere village, and ever manifested a great interest in its growth, being especially devoted to the development of the educational institutions of the city. He remained a member of the Board of Trustees of the St. Louis University for a number of years. On February 22, 1853, the Legislature of Missouri granted a charter to an educational institution, to be located in St. Louis and to bear the name of Eliot Institute in his honor. In deference to his wishes, however, the name was changed to Washington Institute, and finally to Washington University. He was president of the first Board of Directors, and at various times acting professor of ethical and political science and political economy. In 1872 he was elected chancellor, which position he held until his death in 1887. He was connected for thirty-five years with Washington University, devoting the last fifteen years exclusively to the institution.

present circumstances and official connection with the U. S. service to accept the appointment without first having consulted and obtained the consent of the head of my official department, Dr. Lovell, Surgeon-General, residing at Washington City.

Presuming that prompt action and immediate prosecution of the preliminary arrangements of this incipient institution are intended by the Trustees, and not wishing to cause any impediment to the speedy confirmation of their views, or to mar their future prosperity by any unfavorable delay in the matter, I feel in duty, as well as by conditions, bound respectfully to request the kind acceptance of my declination of the honor for the reasons above assigned. Permit me here sincerely to return thanks to the Board for the [compliment] in the proferred honor, and also express my profound respect and ardent wish for the success and prosperity of the contemplated undertaking.

Through Dr. Eliot the Board of Trustees expressed its regret that circumstances made it impossible for him to definitely accept the chair, and at the same time expressed the hope that the reasons were such as might be removed by a few weeks' delay. They were desirous that the medical department should have in its infancy the benefit of his influence and talents, and expressed, therefore, their willingness to wait for a definite answer until he could communicate with the Surgeon-General. In view of the persistence of the Board of Trustees, the Doctor sent a communication to the Surgeon-General on October 31, 1836,[6] as follows:

Having recently been elected to the Chair of Surgery in the new medical school about to be established in this place, as you will perceive by the enclosed copies of communications, I conceive it my duty to consult the Department, and obtain your views and opinion of the propriety of such a course before giving a definite answer to the proposition.

Feeling great diffidence in my own qualifications and fitness for such an appointment, I was more inclined to seek an excuse than to cherish the acceptance, and urged my official connection with the Army, the uncertainty of my continued location here, and the necessity and doubts of obtaining the sanction of the Department and your approbation of such an engagement as sufficient objections, and gave them as my reasons for declining the proferred honor in the first place. But the Board of Trustees, not inclined, it seems, to let me off so easily, proposed to wait till I should have time to consult and obtain your views and decision on the subject before they proceed to elect another person.

---

[6]This is a good example of the slowness with which news traveled in those days. The Surgeon-General had died on October 18, 1836, yet in thirteen days the information had not reached an important army post less than half way across the continent.

I now submit the case for your consideration, and very respectfully and earnestly solicit your advice on the subject, either officially or privately, as you may deem most proper, and as soon as convenient. I should prefer, in your reply, a good excuse for non-acceptance.

After waiting three and a half months for a reply from the department, he finally conveyed his provisional acceptance of the chair of surgery to the secretary of the board:

I have consented to accept for the time being the proferred "Chair of Surgery" in the Medical Department of the St. Louis University, with the privilege of withdrawing from the said professorship at any time, should the heads of the Public Departments under which I am officially placed here withhold their consent or disapprove the act, and from which Departments I have not yet received any reply to a communication made on the subject some months since, owing, I doubt not, to the extreme illness and death of the Surgeon-General about the time it reached Washington, and his successor not having yet assumed the duties of that office sufficiently to attend to it.

For some reason the plans of the university did not materialize in the autumn of 1837—in fact, not for several years thereafter (1842). It may be stated here that there is no evidence that Dr. Beaumont ever delivered any lectures after the medical department opened its doors, though he promised to deliver a single course of lectures on digestion in 1851, providing he could succeed in procuring Alexis.

He was still in active correspondence with his friends in the North with reference in St. Martin. In April of this year (1837) Mr. Morrison had written from Berthier that he had just seen Alexis, that he was living about fifteen miles from Berthier, and that he was ready and "willing to leave at a moment's warning upon condition that his family goes with him." Mr. H. L. Dousman, also an officer of the American Fur Company, wrote at this time a very friendly letter to the Doctor, in which he added:

. . . Now for your Patent Digester, Mons. Alexis. We brot up Men enough from Montreal last Spring to answer us for two years; consequently we have ordered none for next season, but the company will probably send some up for the Lake Superior traders, as they do every year. They are sent via the Grand River to Sault de St. Marie, in which case Alexis could come up with them to that place and then via Mackinac and Chicago to your place. You must write immediately to Mr. Crooks at N. York, and state to him all the circumstances and your wishes, and I know he will do all in his power for you promptly, and, if they send up men, he will order Alexis to be sent with them. Write to Mr. Morrison also, and I know he will interest himself to see that Alexis is sent off at the proper time. I shall mention to Mr. Crooks in my next letter that he will confer a very particular favor on me by immediately attending to your wishes. . . . Perhaps Mr. Crooks could find means to have Alexis brot via Lake Champlain to Albany, thence to Buffalo and Chicago. I think you had better suggest to him, in case they send men up, via the Grand River. I do not believe it will be much more expensive, and it will certainly be more expeditious, as he can reach you in about a month from Canada that way. . . .

He was spurred on to still further efforts to obtain his "patent digester" through a letter received from the American Physiological Society, making inquiries concerning Alexis and the possibility of having him come to Boston for experimentation. In replying to this communication, Dr. Beaumont wrote:

I have to express my gratification for your generous applause of my efforts to improve, perhaps, the unparalleled and only opportunity for a correct investigation of the physiology of digestion, and to say to you that St. Martin, if not on his way to join me at this place for the purpose of renewing, continuing and revising the experiments already begun, is in Lower Canada, somewhere between Montreal and Quebec. He is engaged to me, and has been under contract for two or three years past, for the express purpose suggested above, but has been prevented from surrendering himself to me, partly from the situation of his family and its affairs, but more perhaps from the natural obstinacy of his disposition and unwillingness to submit himself for public experiments, and obstinacy and unwillingness most pertinaciously persisted in from the beginning, and particularly embarrassing and vexatious to me through all my former experiments. He has recently written, however, that he is now ready, and wishes to return to me; and I have made arrangements to get him on here with his family, and hope soon to see him.

He has always refused and positively declared that he would never submit to be experimented upon by any other person or persons than myself, and has ever declined all offers and inducements from individuals and societies to engage to them for that purpose. I presume no consideration whatever would induce him to engage himself to the American Physiological Society, or any other society, for the purpose of experimentation without my personal attention and direction. He is capriciously and foolishly obstinate in that respect, and has always been. It has ever been my wish and effort to place him in some scientific physiological institution, where greater and more extensive justice might be done the subject of experimenting upon him than has ever been in my power to do, but he has ever and determinedly refused his assent to such an arrangement. My present intention is to get him with me again, commence a new series of experiments under the most favorable auspices practicable in this country, associated, as I intend to be, with some of the best physiological societies and scientific men of the West, and prosecute the experimental investigation to the best advantage practicable till circumstances will allow me to take him to some of the European Physiological Schools for the greater improvement of medical science. Duly appreciating the importance of the subject and the occasion, I perfectly coincide with you and your society in the opinion that facilities it affords for so useful physiological investigations ought not to be lost sight of nor neglected.

Not long after this a letter written in excellent style and handwriting, to which was affixed Alexis' mark, was received:

I have received a letter a few months ago from a Mr. David Campbell, corresponding secretary of the American Physiological Society, of the city of Boston, in Massachusetts, offering to engage me for a term of three or six months, to be placed under the direction of an eminent physician for the purpose of experiments on digestion similar to those you have made. I have not given an answer yet, determined that I am not to accept the offer without your approbation and consent. I have made several inquiries concerning your place of residence, but without having learnt anything satisfactory. I now venture to direct this letter to Mr. Thomas Green, at Plattsburgh, in hopes he will have the kindness to forward it to you. I would wish to know if you would have me to go to you, or if you would approve of my accepting the offer made me. Should you desire me to join you with my family, I am ready at any time to go with you on such terms as you may fix yourself.

Fearing Alexis might fall into the hands of other investigators, the Doctor determined to follow the suggestion of Mr. Dousman, and applied to Mr. Crooks again to use his influence and facilities to bring him back:

. . . I should much prefer to have him without his family, if I can; but with it, if I must. To have him again has become almost indispensably necessary to answer the reiterated demands of foreign journals and institutions for further investigations of the gastric functions, etc. He engaged in 1834 for two years to come with me to this place, but it became necessary, in order to get rid of the incumbrance of his family and leave them comfortably situated with their friends in Canada, to advance $80 or $100. . . . But he disappointed both me and the public by concluding to stay with his family in Canada, where he has remained ever since. I should like to have him fulfill his agreement first, if he will; if not, to engage him on the best terms practicable, to be brought on with the voyageurs for your trade and transferred to me when he arrives in this country, as was done in 1828. He will doubtless need some advances to enable him to start, but he must not have any before his forthcoming be secured; if he do, he may play the same trick he did last—refuse to come after all. I hereby authorize any reasonable sum necessary to the comfortable leaving of his family there, or conveniently bringing them on with him here, at your discretion, the which shall be refunded at sight. . . .

In due time Mr. Crooks replied:

He can be had by advancing him probably $150 to get him clear of Lower Canada, and pay the traveling expenses of himself and family to St. Louis. His wife will not allow him to go alone, and in that event he must come round the lakes to Cleveland or Chicago and then to St. Louis. There is some risk in this, but you must incur the hazard if you wish to have the man. I would have sent him by the Grand River in our canoe to Sault St. Marie next month, had he been alone, but it cannot be done with his family. . . .

Simultaneously a letter came from Mr. Morrison, informing him that the St. Martins lost their son, Alexis, Jr., and that there were three children remaining; that the entire family was destitute of clothing, and Alexis had fallen into dissolute and dissipated habits; that he could not be trusted with money, but would squander it promptly if given to him in advance. In view of all these discouraging reports, together with the fact that Alexis seemed determined not to go without his family, Dr. Beaumont discontinued his efforts for a period of two or three years, with the hope that an opportunity would present itself to send someone after Alexis or to go himself.

In the meantime Sir Andrew Combe,[7] an eminent English physiologist, was not content with the excellent references that he had made to Beaumont's observations in his work on dietetics,[8] but published an English edition of the work in April, 1838.[9] In his preface he presents the most impartial estimate of the value of Beaumont's contributions to science that I have yet seen. At the same time, however, it may be considered a fair summary of the opinions voiced by his impartial contemporaries in general, as well as by those who in after years have endeavored to do Beaumont justice. I feel, therefore, that I can do no better in giving an estimate of his work than to present, in part, the

[7]Sir Andrew Combe was one of those unusual characters who, in spite of the great odds against which he fought, left his mark on the pages of medical science. During twenty-seven years of his life he was simultaneously making the fight against tuberculosis and adding valuable contributions to science. He was one of seventeen children and of poor parentage, and his ailment was attributed to conditions surrounding his home—offensive pools and ditches, defective ventilation, small and overcrowded rooms, unsuitable food, and inadequate clothing. He learned medicine as an apprentice in England, afterward studying in Paris. In spite of the constant strain under which he labored, he made valuable researches in physiology and phrenology, contributing not less than eighty articles on various phrenological subjects during his lifetime. He visited his brother in America in 1847, hoping to be benefited by the trip. *(Memoir of the Life and Writings of the Late Dr. Andrew Combe.)*

[8]*The Physiology of Digestion Considered in Relation to the Principles of Dietetics,* second edition; Edinburgh, 1837.

[9]*Experiments and Observations on the Gastric Juice and the Physiology of Digestion,* by Wm. Beaumont, M.D., of the United States Army; reprinted, with notes by Andrew Combe, M.D.; 1 vol., post 8vo.; Edinburgh, 1838.

remarks of Sir Andrew Combe in his preface to the English edition of Beaumont's book:

> The reasons which have induced me to reprint the present work from the American original are, first, a strong sense of its inherent importance, and of the numerous applications which may be made of the facts and principles developed in it to the prevention and cure of disease; secondly, its comparative inaccessibility to the European physiologist from the difficulty which still exists of procuring it on this side of the Atlantic; and, lastly, an earnest desire that the author should obtain that credit which is unquestionably due to his disinterested and indefatigable labors. The value of Dr. Beaumont's experiments consists partly in the admirable opportunities for observation which he enjoyed, and partly in the candid and truth-seeking spirit in which all his inquiries seem to have been conducted.
>
> So far, then, as opportunity for observation was concerned, nothing could surpass that enjoyed by Dr. Beaumont. That Dr. Beaumont eagerly and zealously availed himself of his unusual advantages, the following pages furnish ample evidence, and it would, I think, be difficult to point out any observer who excels him in devotion to truth and freedom from the trammels of theory and prejudice. Among the disciplined physiologists of Europe a more systematic experimenter might certainly have been found, but in Dr. Beaumont's instance the absence of systematized inquiry—made too generally in support of preconceived theory, and therefore apt to mislead as well as instruct—is more than compensated by the implicit reliance which one feels can be placed on the accuracy and candor of his statements. Having no theory to support, and no favorite point to establish, Dr. Beaumont tells plainly what he saw, and leaves every one to draw his own inferences; or, where he lays down conclusions, he does so with a degree of modesty and fairness of which few perhaps in his circumstances would have been capable. . . .
>
> But it may be said, singularly favorable as Dr. Beaumont's opportunities were, he has made no original discovery in the physiology of digestion. To a certain extent this is true, for, in the proper sense of the word, he has not made and does not claim to have made any discovery, but he has done what is at least equally essential for practical purposes. By separating the truth clearly and unequivocally from the numerous errors of fact and opinion with which it was mixed up, and thus converting into certainties points of doctrine in regard to which positive proofs were previously inaccessible, he has given to what was doubtful or imperfectly known a fixed and positive value which it never had before, and which, being once obtained, goes far to furnish us with a clear, connected, and consistent view of the general process and laws of digestion. Other physiologists have attempted to effect the same end by experiments performed upon the lower animals, but these are open to so many forcible objections that we cannot always adopt their conclusions, even where they seem to be most clearly deduced. Not to mention the cruelty inseparable from the performance of such experiments, the pain which the animal suffers necessarily disturbs the regularity of the functions under examination, and in

a greater or less degree vitiates the results. And even if this were not the case, the difference between the digestive organs in man and in the lower animals is so great that it would often be unsafe to assume conclusions as applicable to the former which have been verified only in the latter.

In the present volume it is proper to bear in mind the circumstances under which the experiments were made and the account of them was written. Dr. Beaumont was far from enjoying the leisure, resources, and scientific co-operation so easily accessible to the physiologists of any of our European capitals. Stationed in a comparatively remote quarter in the exercise of his duties as an army surgeon, and previously unaccustomed to minute physiological research, he conducted his inquiry under many minor disadvantages. When he came to publish, also, his want of experience in writing prevented him from making the most of his materials, and doing that justice to himself which he might otherwise have easily accomplished. In the arrangement of his experiments, for example, Dr. Beaumont has followed the order of time, and thus mingled many things together, where a more practised inquirer would have classified them according to the subjects in illustration of which they were performed, and thus given a greater unity of purpose to each of the different series of which they are composed. But, although this defect diminishes the facility of access to the results, it by no means detracts from their intrinsic value. On the contrary, the very absence of systematized arrangement leaves a character of even greater trustworthiness attached to the individual observations than if the latter had been made under the influence of some prominent guiding principle, which might have given a bias to the mind. . . .

So much, indeed, did I consider the republication of Dr. Beaumont's work as a matter of justice to him, that, had I not expected its appearance from some other quarter long ago, I would have undertaken the task at an earlier period, and even now I can account for the omission only by supposing that very few copies of the original have reached this country. Everywhere, both in British and foreign books, we meet with reference to and often inaccurate extracts from it, but almost nowhere is it spoken of as if the work itself had been consulted. I trust, then, that in now presenting it I shall be considered as rendering an acceptable service to British and Continental physiologists, as well as a pleasing act of justice to its deserving author. . . .

One cannot help admiring the unselfish spirit which actuated Combe in the reproduction of Beaumont's work. As soon as copies were out of the press, Combe sent through his brother to Dr. Beaumont an autograph volume of the new edition, together with a volume on dietetics, with the following note:

EDIN, 1 May, 1838.

MY DEAR SIR: May I beg your acceptance of the accompanying volumes as a small expression of my respect of your character and scientific labors. I need not detain you by repeating in this note the high estimation in which I hold you. The volumes herewith sent will, I trust, convince you of the fact, and that it will not be my fault if you do not receive the credit justly due to your valuable and disinterested services. I remain, My dear Sir,

Very respectfully yours,

ANDW. COMBE.

George Combe left the books with Mr. John Turnbull, Carleton House, Broadway, New York, who promised to convey the package to Dr. Beaumont, believing that he was still in Plattsburgh. On learning that he had moved to St. Louis, Mr. Turnbull sent the package there. Mr. William Combe, then residing in Albany, fearing that the package had gone astray, in view of Dr. Beaumont's change of address, wrote him in March, 1839, making inquiry. Dr. Beaumont, living at Gov. Clark's residence, responded:

I am happy now to have an opportunity of tendering my grateful acknowledgements for the special interest you have taken in the publication of my Gastric experiments and the disinterested compliment and credit conferred on me by voluntarily sending it to yr. deservedly distinguished Brother in Edinburgh, by whom it has been reprinted, with valuable annotations, and more than merited notice taken and references made in his late invaluable work on Digestion and Dietetics, both of which I had the special honor and satisfaction of reviewing a short time since thro the agency of Mr. Turnbull, of N. Y., and by the hand of Mr. John Thorburn. Mr. Turnbull had advised me on the subject several mos. before pr. mail, but, in consequence of the unavoidable delays of transportation from low water last fall and the severe frost of the winter, the package was not recd. till the opening of the navigation this Spring, a few weeks since, all in good order, with a very kind and complimentary note from your Bro., and acknowledgements of which will probably have reached him ere you receive this. Sincere thanks and continued respectful regards are due to my friend and acquaintance, Dr. McCall, and my grateful and profound consideration to yr. Bros. whenever you may be so happy as to see or communicate with them, and believe me, though an entire stranger, etc.

The reply to Sir Andrew, in thanking him for the volumes of his work, ran something like this:

I have the satisfaction of acknowledging the recent receipt of Copies of your Work on "The Phys. of Digestion and Process of Dietetics," and Reprint of my "Expts. and Observations on the Gastric Juice, etc.," accompanied by yr. very kind and complimentary note. . . . The honor and gratification confer'd by thus tendering me the tokens of your high regard and esteem, and the favorable notice taken of my work, are far more than commensurate with its worth, . . . and surpass my ability to adequate returns.[10]

---

[10]These incomplete sentences were found in rough draft, hastily scribbled on a slip of paper.

Dr. Thomas Lawson[11] had been appointed to succeed Dr. Lovell as Surgeon-General. It would seem that when he assumed the position, it was with some preconceived prejudice and animosity against his friend of former days. One cannot help concluding from a fair study of the case that this jealousy had been aroused by the many attentions and frequent favors granted Beaumont by his predecessor, Dr. Lovell. While Beaumont was conducting his experiments on St. Martin, Lovell permitted him to take leave of absence whenever necessary to further his studies, to change his post whenever occasion demanded it—in fact, gave him many liberties that probably were not granted other surgeons in the corps.

Within a year after Lawson assumed the reins of the medical department, Beaumont began to hear rumors that he was to be transferred to Jefferson Barracks, a location which would have rendered private practice in the city practically impossible. While there was no definite basis for such fears, there was an undercurrent which caused him to feel rather restless. On February 4th, 1839, Major Ethan Allen Hitchcock[12] wrote him from Washington:

> I have just had the pleasure to read your letter of the 24th inst., and the only thing that grieves me is your account of Lucretia's arm. Captain Lee[13] told me she was hurt, but I had no idea it was a fracture. I hope no effects of it will remain, and, if so, she may thank her stars for her daddy's skill. . . . I told Lee of Lawson's inquiries after you. There seems not the smallest hint of a purpose of changing your station—I mean, of ordering it; for as to changing it, I suppose that is not so easy done as talked about.

This statement doubtless put his fears at rest for a time, coming, as it did, from one in touch with the highest officials in the army. In the spring Beaumont requested of the Surgeon-General a leave of absence for the purpose of looking after some business interests at Green Bay, in reply to which Dr. Lawson wrote:

---

[11]Dr. Thomas Lawson's military career began with the War of 1812, when he was made surgeon's mate, and ended fifty years later. He was Surgeon-General for a quarter of a century, and was recognized as a capable and fearless executive. He inaugurated many important changes in the department during his administration, and welcomed any innovation that promised improvement of the service. It is noteworthy in this connection that he strongly favored private practice by army medical officers, and ruled that "when it does not interfere with their military duties, medical officers have a right to give their professional advice to whomsoever they please." It could not, therefore, have been this factor which influenced him in his attitude toward Beaumont.

[12]In a later chapter the intimate relationship existing between General Hitchcock and Dr. Beaumont is brought out in a series of friendly communications.

[13]Robert E. Lee.

I have to inform you that, as similar applications from officers having much stronger claims to indulgences than yourself have been refused on account of the pressing demands for Medical Officers, your request can not be granted. Should your private interest require your presence at Green Bay, and you think proper to employ a suitable physician to perform your duties during your absence, free of expense to the United States, you are authorized to do so, provided it meets the approbation of the Commanding General of the Department.

In the meantime Major Hitchcock had obtained other information concerning Lawson's intentions, and had informed Beaumont of it. With the knowledge and assurance, therefore, that the Surgeon-General would welcome an opportunity and a pretext for making him feel the force of his official position, Beaumont replied in a letter that was anything but diplomatic:

I regret that you should think it necessary to attach such conditions to the leave, not so much on account of the expense it may throw upon me as the consideration that seems to have prevailed with you that I am not entitled to the indulgence I requested. If for the single reason that no medical officer was disposable for the performance of duty here during my absence you had thought proper to charge the expense to me, though I might have thought it hard, it would have appeared much less onerous.

I do not know what stronger claims others may have, but I am sure that my request was perfectly reasonable, and, if the duties upon the medical staff of the Army make so exclusive and absolute a demand upon the time of officers that so reasonable a request cannot be complied with, there is great need of additional officers in the Medical Staff of the Army, but I rather think you have in this instance underestimated my claims. Unless, however, you think proper to reconsider your decision, I may be compelled to pay the expense of services here during my absence, for a just regard to my interests will, I believe, oblige me to be absent six or eight weeks.

While writing, I will refer to a subject to which my attention has been called by a letter from Maj. Hitchcock, in which I am informed that your views of the propriety of my Station at this place do not accord with those of your predecessor, and that you appear to think Jefferson Barracks a more suitable Station.

Now, My Dear Sir, without the least desire to question the correctness of your views, or the propriety of your general arrangement of the Medical Department, permit me respectfully to ask (and I hope you will do me the favor, candidly and promptly, to answer) whether you have any such arrangement in contemplation, or even seriously think of ordering me from this to Jefferson Barracks? I presume to be thus frank and explicit from agreeable recollections of former friendly associations,[14] the confidence I feel in the

[14]They had been fellow-officers in the War of 1812.

Brevet Brigadier-General Thomas Lawson, Surgeon-General of the United States Army, 1836-61.

candor and correctness of your disposition, and firm belief in your general desire to promote the greatest good of the public service with the least prejudice and inconvenience to the individual interests of your Department. I would ask then, how can Jefferson Barracks, in its present endowments, be considered more appropriate or important a Station for me, or any other medical officer of whom the same duties would be required, than this city? The amount and importance of the official duties required at that Station bear no comparison with those to be performed here. There are Gen. Atkinson, his family, a handful of recruits occasionally and some few useless old soldiers, with nothing to do but to preserve their health, if they will, by procuring and regulating their own diet, scarce liable to accidents and more seldom to disease, spending much of the time in this city, and receiving the benefits of my professional advice and prescriptions whenever wished for or requested. But here in St. Louis [1839] are located six or seven Army officers and their families, Lt. Col. Brant, Majors Lee and Stewart, Capt. Kingsbury, Capt. Lee, Lt. Reed and Ruggles, a recruiting Rendezvous, with its party and recruits to be prescribed to, examined and reported. The Arsenal, with its command, some twenty or thirty men, besides women and children, and Capt. Symington, his family and subalternes, members of military courts, sojourning officers, under orders and transient members of the Army, which are neither few nor far between, all of whom claim and receive my free and gratuitous professional services, all these together, with the duties of Medical Purveyor for the Western Division, make the official duties required of me here constant and multifarious, and ten times more important, necessary and useful than can possibly be required at the Barracks in the present state of service.

Under this view of the subject and present exigency, it is impossible to conceive the necessity, or even the propriety, of modifying the present arrangements by ordering me to any other post or station, unless it be to devolve upon me the necessity of making great sacrifices of interest and inclination, or of being compelled to resign my place in the Army, either of which alternative would be very unpleasant, and neither of which do I hope to be forced to.

I am, and always have been, willing and ready to perform any duty required of me, am now willing to continue to attend to those incumbent on me at this Station, which are equally arduous and necessary with those of any other, and of paramount importance to most others in service.

Now, uninformed of any particular exigency, unaware of having forfeited my claims to the common indulgence or my right (after my seniors)[15] to a preference of Stations, and conscious of having faithfully discharged my duties to the Government, I cannot acknowledge just cause, or even seriously anticipate an order of removal from this place. I presume some envious or discontented junior would be glad to see my Station changed, but I know you would not yield your official sanction to gratify any of that class.

Allow me to add here, without subjecting myself to the charge of vanity or egotism, that for any professional advantages I may be supposed to enjoy here I am not indebted to the Government, but to my character and exertions. These advantages, be they more or less, are not an incident to the Station, but are fruits of toil, I am sure, if not talent, and it is by no means certain that a successful applicant for my Station as a public officer would likewise succeed to the advantages I may properly characterize as personal.

I confide in the uprightness of your disposition, and rely upon the correctness and impartiality of your judgment.

---

[15]"Senior surgeons and assistant surgeons shall respectively have choice of stations on written application through the Surgeon-General, specifying the regiment or post preferred," etc. (Brown's *History of the Medical Department of the United States.*)

He evidently received no satisfaction from the department in reply to his letter, which could not do otherwise than add fuel to the fire which was already smoldering. The following letter (1839) to his friend, Dr. King, indicated that he was well aware of the fate that was in store for him, though there was still some uncertainty as to the nature of the undeserved rebuke that would be meted out to him:

Accept my thanks for yr. kind solicitude for my desire and ambition. I am in full run of successful and lucrative professional practice in this City, to the exclusion almost of the common civilities and hospitalities due to old friends and acquaintances. I am too constantly engaged in professional duties to enjoy my friends or the comforts and pleasures of domestic and social life. Official duties, though as ardent and important here as at any other Station, constitute but a small portion of my labors, and were but very little increased by the "Military of Rank," as you term it, while they were with us.

I hear frequently from various sources that I am to be removed from this station, ordered to Florida, etc., etc. But I don't believe it. I have too much confidence in the correctness and management of the Head of our Department, and the justness of my claim to this preference, to apprehend for a moment that I shall be removed at all without some more important emergency or pressing necessity than seems now to exist. So long as the Government sees fit to continue this as a Station for a Medical Officer, I shall expect to retain it, in preference to any discontented Junior official, the murmurings and importunities (if any there be) of either Succors or juniors of the Line or Staff to the [contrary] notwithstanding. The offl. duties to be performed here are as important and arduous as at most other places. I have and shall continue to perform these as faithfully as any other officer could, with equal convenience and much more advantage to myself. I know of none with fairer claims. I confide in the justice and integrity of the Heads of the Departments, and greatly appreciate your friendly and efficient influence in my behalf, and will thank you to continue if necessary.

I here enclose you a Ten dollar U. S. Bank note as my proportion of the present assessment to the memory of our mutually lamented and respected chief.[16] It ought to have been forwarded before, but the constant press of professional business is the only excuse I have to offer. Should more be necessary to complete the work, I shall freely add my proportion to any for the contribution. I hope it will not fail to reflect honor upon the survivors as well as the deceased by its design and execution.

I have only time to add my best respects for the Surgn. Genl., and hope he has ere this perfectly recovered his health, and feels disposed to let me "quiet be" where I am while I am able to perform the duties that might be

[16]The medical corps of the army erected a monument to the memory of Surgeon-General Lovell, soon after his death, in the Congressional Cemetery at Washington, D. C.

required of a successor, for I fear I am getting too old and infirm, or too indolent and selfish, to be ordered away while there are so many juniors requiring active field duties to perfect them in their professional qualifications and enhance their official usefulness.

About a month later Dr. Beaumont's worst fears were realized in the receipt of "General Order No. 48":

ADJUTANT GENERAL'S OFFICE,
WASHINGTON, Septr. 18th, 1839.

By direction of the Secretary of War, a Medical Board, to be composed of Surgeon W. Beaumont, Surgeon P. H. Craig, Surgeon R. C. Wood, and Surgeon H. A. Stinnecke, Supernumerary member, will convene at Fort Brooks, Florida, on the 15th of November, 1839, or as soon thereafter as practicable, for the purpose of examining such Surgeons and Assistant Surgeons of the Army as may be ordered to present themselves before it. . . .
By order of Major General Macomb.

On the receipt of the order, Beaumont tendered his conditional resignation, providing the department still insisted on his going to Florida. Not content with this, the department maintained that the order must be carried out before his resignation could be considered, as shown by the answer of Dr. Lawson:

Your letter of the 3d instant and the accompanying copy of a communication to the Adjutant-General of the Army have been received. The Secretary of War[17] has decided that you shall obey the order directing you to proceed to Florida before he will take into consideration the tender of your resignation.

Hearing of Beaumont's resignation, his dear friend, Major Hitchcock, wrote him from Washington a letter full of sympathy and expressing his views of the shortsighted policy of the Surgeon-General.

I will not travel any further without sending you one line at least. I hear with regret of your resignation, though it was what I expected as soon as I heard of your orders. This thing is all wrong, but it appears beyond remedy. The policy of the department has driven the best surgeon out of the army without the slightest occasion. I am sorry to lose you from the same profession in which I serve for the double reason of personal friendship and professional pride. Nothing but a miserable and contracted view of the public service could have dictated the late orders. In the medical branch of the service the officers ought to be encouraged in their efforts to obtain private practice, for in that department experience is almost everything, and the medical officer

[17] J. R. Poinsette. It is a rule of the army that an officer may not resign while under orders.

who has the most extensive private practice is not only proved by that fact to be the best physician, but confers the highest honor on the service. The late course is wrong in other respects. It is unjust to you individually, for exceptions are made to what in your case may be asserted to be a rule. . . . But the mischief is done, and all I can say now is, God bless you in your entrance into civil life, and make you prosperous and happy as you deserve to be, for a better physician or better man is not to be found in any country. . . .

P. S. Mr. Johnson arrived last evening, and tells me he left you all well. He says Captain Lee spoke of coming East, as his work[18] has been suspended. If he is still with you, pray give him my best regards.

To this Dr. Beaumont replied:

. . . It is a matter of perfect indifference to me whether I continue in the Army or not, but any attempt to cast disgrace or censure upon my character or conduct will be manfully resisted. When they expunge my name, I will expose their wickedness and partiality for rascality. But enough of this—let them mind their own business and I will mind mine.

Soon hereafter (December 20th, 1839) he wrote Adjutant-General Jones:

Having conscientiously assigned good and sufficient reasons for not complying with Order No. 48 in my communication to you of Oct. 3rd, I had hoped they would have induced the countermanding of the order or the acceptance of my resignation. But it has been stated here in presence of several gentlemen by a lawyer recently from Washington, who assumes to be in the confidence of the Secretary, that I am to be stricken from the rolls for non-compliance with the order referred to. I cannot believe such injustice will be done me. I have served more than twenty years honestly and faithfully, and know no law, rule, regulation or precedent that should prevent my withdrawing from the service when I cannot perform the duties required of me.

---

[18]On Aug. 5, 1837, Captain Robert E. Lee was sent to St. Louis in charge of a corps of engineers, where he performed a most important service in preventing the Mississippi River from changing its course and from leaving the city "high and dry." At the same time he designed a system of river improvements. His work kept him here for several years, during which time an intimate relationship sprang up between him and Dr. Beaumont. Lee spent many leisure hours at the Beaumont home and endeared himself to every member of the family. The following note from Mrs. Lee to Sarah Beaumont gives a fair idea of the intimate relationship existing between the families: "My Dear Sarah: I went [July or August, 1840] over to the great city of Washington yesterday to look for something pretty to send you, but I suppose the members of Congress have taken everything with them, for I could find nothing worth sending. Such as they are, tho, you must wear for my sake. The ring Custis sends to Cushy and the comb to Bud. The gauze cap ribbon I send to your mother. Tell her she must wear it at her musical soirees, and I send you a neck ribbon of a very fashionable style and a pair of riding gloves. Mr. Lee has packed up my letter and is waiting for these things, and so farewell again. Yours, M. C. L."

I shall not advert to other cases or draw comparisons between faithful service and dishonor, but I do expect that the former in our government will at least meet with as much forbearance and more favor than the latter. I am, and ever have been, actuated by just and honest motives toward the public service, with no more selfishness than strict duty to myself and family and the approaching infirmities of age imperatively demand. I ask no more of the government. I will not expect less.

Captain Lee had returned to Washington, having suspended his work in St. Louis, and soon afterward wrote (sometime in January of 1840) to Miss Sarah Beaumont, who was a great favorite with both Lee and Hitchcock:

Tell the Doctor that among my first visits was one to the Surgeon-General. He was very mysterious. . . . I endeavored to find out what course he had determined on. He will agree with me that this is a difficult job where the individual has not made up his own mind.

But the individual *had* made up his mind, and, even at the time of Lee's visit to the Surgeon-General, "General Order No. 2" was on its way to Beaumont. In spite of all his remonstrances and entreaties, the inevitable crisis came, and his career as an army officer was ended.

ADJ. GENERAL'S OFFICE,
WASHINGTON, Jany. 20th, 1840.

The resignation of Surgeon Wm. Beaumont has been accepted by the President, to take effect on the 31st of December, 1839.

By order of Major General Macomb.

R. JONES, Adjt. Genl.

Though his resignation was accepted, and he had not been peremptorily dropped from the rolls, he still determined to fight to the bitter end. While he had stated to Major Hitchcock that it was a matter of "perfect indifference" to him whether he remained in the army or not, he afterward explained that when the acceptance of his resignation finally came, his "pride and indignation" were aroused to such an extent that he could not yield without making a strong fight for reinstatement.

He wrote (sometime in February of 1840) Senator Lewis F. Linn, of Missouri, who had been a fellow-surgeon:

. . . I have respectfully and earnestly to request you, as guardian of the rights and privileges of the people of this state, and of myself among the rest, to defer, if in your power, any official action of the Senate upon the nomination of a successor to the vacancy I may have occasioned till I have had an opportunity to communicate my reasons for petitioning to be restored to my commission and place again as before.

I take the liberty of making this hasty communication and request (having this day unexpectedly received the order of acceptance), lest the name of my successor be sent to the Senate for confirmation before I have time to lay my case before the president for consideration, confidently believing myself fully justified in so doing, and that you will rightly appreciate and promptly attend to it. . . .

His next step was to memorialize President Van Buren, and the memorial was presented by Senators Wright and Linn:

To His Excellency, Martin Van Buren, President of the United States.

The undersigned memorialist, Wm. Beaumont, recent Surgeon of the U. S. Army, in the 56th year of his age and the 24th year of faithful duties to government, now finding himself unjustly thrust from the public service, most earnestly and respectfully petitions your excellency to be again restored to his late official rank, commission and place for the following reasons and considerations, to wit:

Your memorialist having been in the Army, with the exception of a short voluntary suspension from 1815 to 1819, and in almost continual performance of important duties on the exposed and comfortless confines of our country from 1812 to 1840, serving through the campaigns of 1812, 13 and 14 on the northern and western frontier during the late war with England, assigned to the extreme outposts and stationed at Fort Mackinac, Howard and Crawford from 1819 to 1832, doing the duties of these several posts, with their respective complements of troops, without assistance for 12 years or more, with the various additional duties, fatigues and exposures of the several predatory Indian warfares with the Winnebagoes and other hostile tribes from 1827 to 1832, and only relieved from those active field and garrison duties at the termination of the Black Hawk War, so called, in Aug., 1832.

After this, for a few months, your memorialist was permitted by the Department merely to anticipate, but not to realize, "Leave of Absence" and supension from official duties for 12 months for the special purpose of going to Europe, with a view more perfectly to complete a series of gastric experiments and observations upon the stomach of Alexis St. Martin, whose life and extraordinary constitution had been preserved by your memorialist's individual exertions, and which was then considered by the Heads of Department, members of Congress and the scientific communities in general of paramount importance to mankind, and worthy of public patronage and physiological improvement.

Before, however, one-fifth part of the time necessary to accomplish this desirable object had elapsed, and in the midst of its execution, an emergency of service required your memorialist's return to official duty, and he was put on special service in March, 1834. In June following he was sent to Jefferson Barracks to relieve a sickly, suffering surgeon at that post, where he remained on duty till Nov. or Dec., 1834. Early in 1835 your memorialist was assigned by special order to this place for the special purpose, in addition to the ordinary official duties of the station, of affording him an opportunity for prosecuting further experiments and observations upon the gastric juice of St. Martin, without prejudicially affecting or interfering with details and duties of the Medical Department.

And your memorialist has been on constant official duties, rendering the necessary medical aid and attention to 7 or 10 resident officers and their numerous families of women and children, sojourning military men, members of martial courts and inquiries, to the command at St. Louis Arsenal, the recruiting parties at this rendezvous, examining recruits, purveying medical supplies for the western posts, making returns, reports, etc., from June, 1835, to the present time, duties more constant and multiferous than [those of] any other officer in the army. And to this place and to these duties was your memorialist specially assigned by the late Surgeon-General Lovell and War Secretary Cass, with the express understanding and expectation that it was to be as permanently enjoyed as other of the senior medical officers had been and are still privileged to enjoy their respective favorite stations, as merited rewards of long and faithful service, and a rule and precedent of justice and equity adopted in the spirit of magnanimity by those Heads of Departments, and long since become the established usage of service. Instanced, Surgeon Mower has been thus indulged in a permanent location at New York for more than 20 years continuously; Surgeon Harvey at Baton Rouge since the close of the war in 1815; Ast. Surgeon Sargent at Fort Constitution from time immemorial; Turner long lived and died unmolested at Fort Woolcot; Day permitted to enjoy any place he may choose for its light duties and pleasant location; Archie completely domesticated at Fortress Monroe; King fixed at the Capital at Washington; and Weightman forever at St. Augustine. All this is just and proper, and but merited indulgence generously awarded to age, professional worth and official faithfulness and integrity.

Your memorialist is no less fully entitled and deserving of similar and equally continued indulgence. He was assigned to this station and to these duties by the same rule and upon the same principles. He is not conscious of having in any way forfeited his claim to this preference. He knows no reason why he should be deprived of it now, conceives no justifiable cause, nor believes an adequate emergency to warrant the order for his removal.

Great injustice, therefore, having been done your memorialist by General Order No. 48 of Sept. 18, 1839, peremptorily requiring him to repair to Florida for duty, and the consequent acceptance by yr. Excellency of his constrained resignation, only to obviate the difficulties of declining an order the requirements of which I could not perform from my well-known infirmity and defect.[19] And your memorialist, believing a second order to have been issued in consequence of some erroneous impression or wrong influence to his official prejudice and personal injury, now conceiving it due to himself,

[19]Refers, no doubt, to his defective hearing.

just to the government, respectful to the Department, and consistent with the principles of subordinates to make this appeal to the President of the United States for proper redress of the alleged grievance, confidently relying upon his Excellency's well-known principles of justice, liberality and patriotism for restoration of his right to the official rank, place and appointment from which the unexpected acceptance of his resignation has taken him.

Soon after this (April 3, 1840), and in the same connection, he wrote his friend, Major Hitchcock:

. . . I cared but little, you know, how they decided at first. But on the receipt of the order of acceptance my pride, Spirit and indignation were excited at such gross and egregious injustice towards me, and I resolved to resist the influence that had induced it, and, of course, memorialized the President to be restored and continued at this place.

I submitted it through Senator Silas Wright and Dr. Linn, with my reasons and explanations. I have not heard from them since. I know not how promptly they will give their attention to the subject. It was my intention to have gone on and attended to it myself, but I cannot get away; and, if I could, I am such a poor hand at managing such things, I had rather trust it with you and my other friends there, and I rely much on your friendly interest and influence in urging the case forward. . . . Do all you can for me and not injure yourself.

He had also written Captain Robert E. Lee to use his influence, in response to which Lee wrote from Washington on May 19, 1840:

I am afraid you will think that I have neglected your letter in relation to your application to the President to be restored to your rank and place in the Army. An hundred times have I determined to write to you, but as often, when on the point of executing, have I delayed from day to day in the hope of learning something satisfactory. If Mr. Wright has ever presented your Memorial to the President, it has never reached, as far as I can learn, the War Dept. Should it be so referred, it would naturally be sent to the Surgeon General for a report, and, as well as I can ascertain his sentiments by indirect approaches, he will not recommend your reinstatement, and I think I may say will oppose it. He appears to have some feeling on the subject, the cause of which I do not know. Supposing your application would come before the Adj. Gen., I had taken occasion to impress him with your usefulness, skill, attention, etc., and the loss that the service would sustain, particularly in the West, by your withdrawal. I spoke to Cooper in the same manner, without telling them your present views, as they might have thought my object was to influence their action, and they both coincided with me in regretting your loss, as did also Dr. King, but I am almost certain that the subject has not been broached to them, unless it has been within a few days. I do not know Mr. Wright, and I have disliked to call on him purposely on the subject, as he might have thought I was interfering in what did not concern me. I have, however, thrown myself in this way, without success, in the hope of learning what he considered the prospect of a favorable decision. He is so engrossed in politics and in the affairs of Congress that I suppose he cannot find time to withdraw himself from them. At any rate, I have never been able to meet him. I showed your letter to Col. Totten, who showed every desire to aid you, and promised to let me know what he could discover, but he has not been more successful than myself. When Major Hitchcock passed through we had a long consultation on the subject, and he made a special visit to the office of the Surgeon General, which he said convinced him that nothing could be done in that quarter. Now, has Mr. Wright presented your memorial to the President, and what does he say of the probabilities of its success? For your reinstatement in the Army I should personally take great pleasure, and I believe it would conduce to the interest of the service generally. I should be particularly glad to be instrumental in it, but in the present circumstances I am at a loss how to proceed without showing an obtrusive interference, which I have always seen produce injury in similar cases. You must let me know how far matters have progressed, and whether there is anything so small a man as myself can do to shove them along.

Surgeon DeCamp, then stationed at the barracks, about twelve miles south of the city, had received orders to perform the official medical duties at St. Louis and the arsenal, and expressed his readiness and willingness to do so when called upon. There seemed, however, to be a general understanding between him, Dr. Beaumont, and the officers stationed at St. Louis that it was impracticable for him to perform these duties at such a great distance, especially in cases of emergency, since "death or remediless suffering might ensue" before relief could be obtained. With this understanding, Dr. Beaumont continued to render medical services to the families of the officers at St. Louis and the arsenal until the fall of 1840, "with the full, fair, and honest conviction and belief that he was doing in strict accordance with the rules of propriety, humanity and justice a necessary and essential public good." After the expiration of a reasonable time he presented to the department for this extra service an account which he conceded to be "irregular and informal," but at the same time "correct and just." Evidently on the receipt of this account the Surgeon-General threatened one of two procedures—either he would disregard the account entirely, or paying it, would deduct the sum from the salary of Dr. DeCamp. On hearing these rumors, Dr. Beaumont expressed himself in the most caustic style concerning "the absurd opinion, contracted views, narrow-minded, vindictive spirit and petty tyrannical disposition" of the "weak, waspish and willful head of the Medical Department," and prided himself on having the "privilege of detesting the man, the motives and mind from which such egregious folly, parsimony and injustice could emanate and be promulgated."

In order to protect Surgeon DeCamp "from such preposterous abuse and injustice and petty pawn," and himself "from the stigma of being made the means and medium of such wanton and unwarranted oppression and injustice," he gave Dr. DeCamp a written statement of the case:

> . . . The least consideration might have shown Dr. Lawson, if his capacity was not at zero, that it was utterly impracticable for you or any medical officer to render at the Arsenal and in this city the services due to the sick of the Army while you were stationed at Jefferson Barracks, 10 miles distant, and the result has fully demonstrated the folly and absurdity of his ridiculous order. To guard against the effect of that folly, which might rather be called wickedness as endangering the sick for whose attention it was his duty to provide, I have given my own attention, and, having thus acquired a just claim for compensation, the Surgeon General, it appears, in a spirit every way becoming him, but disgraceful to the head of the Corps over which he presides, has come to the determination to charge you with the amount rather than allow it to me. But he can upon no principle whatever charge the amount to you, yet, if he should so far exert himself in an act of tyranny as to produce a stoppage upon you, I will readily withdraw all claims from adjudication at the head of the Department and will look to a higher officer, where I may hope to find more intelligence, more humanity, more liberality and a better sense of justice. . . .
>
> Congratulating myself upon having escaped service under the present Chief of the Medical Dept., I sincerely commiserate my associates in the Army who remain subject to his caprice and tyranny, the natural fruits of ignorance and vanity.

The final decision on this point was summed up in the following communication from the secretary of war through the Surgeon-General's office (November 4, 1840):

> . . . In respect to the case of Dr. Beaumont, he has had no authority whatever for rendering service to any one in behalf of the U. S. after he received notice that his resignation had been accepted, nor had he any proper excuse for doing so. His connection with the government then entirely ceased, and he should promptly have turned over to the officer of the Medical Department, who was there and who was ordered and ready to perform the duties, all the public property, etc., in his hands. For compensation for services rendered to any one in public employ after that period he must look to the individual and not to the government. . . .

This put an end to any further efforts on the Doctor's part to be reinstated, and thus was terminated his brilliant military career of more than twenty-five years. It seems sad indeed that his services should have been so abruptly and unsatisfactorily brought to a close, for his achievement had brought credit, not only to himself, but to the army as well. Even today his brilliant work is pointed to with pride by the officers of the medical corps of the army.[20]

[20]Surgeon-General Torney, in a recent address before the Philadelphia Medical Club, cited this as the first great scientific achievement of a member of the medical staff.

A Reprint of the First Edition of

EXPERIMENTS AND OBSERVATIONS ON THE GASTRIC JUICE AND THE PHYSIOLOGY OF DIGESTION

By WILLIAM BEAUMONT, M.D.

with an Introduction by

Robert Ely Shank, M.D.

Daguerreotype of Beaumont at the age of 66 years.

"God grant you may live a hundred years in happiness and doing good to your race. As to immortality, you have already secured your share. Harvey and Galen will find themselves behind you in this item in the great day of final settlement of all earthly accounts. As long as men have stomachs and are compelled to eat food to sustain life, so long will your researches on the physiology of digestion be known and admired." (Letter from Dr. John McCall, 1846.)

THE CONTRIBUTIONS OF
WILLIAM BEAUMONT
TO GASTRIC PHYSIOLOGY

AS SEEN IN THE PERSPECTIVE OF A CENTURY AND A HALF OF SCIENTIFIC PROGRESS

Robert Ely Shank ST. LOUIS, January 1981

IT HAS BEEN MY PRIVILEGE to read or to return to the writings of William Beaumont on a number of occasions for quite different purposes. Each additional perusal has brought new information and admiration for the man. Certainly Beaumont was the possessor of a mind prepared and motivated for discovery. Few other scientists have transmitted so vividly the thrill of discovery and total commitment to the derivation of new knowledge through experimentation. One can only wonder how these qualities were engendered in a man whose boyhood was spent on a Connecticut farm during the latter part of the eighteenth century and whose training in medicine was limited to an apprenticeship spent with a country practitioner. There was no opportunity to work in a laboratory nor to be exposed to the experimental method. Scientific and medical journals were very few and must have been of limited availability in the small communities and the military outposts in which Beaumont served. Newspapers apparently occasionally carried items of medical interest. It was by this means that he learned of the claim for the therapeutic effectiveness of a plant, water plantain, for the treatment of rabies and attempted its use in 1819 as a topical application for a woman patient who had been bitten by a cat "supposed to be rabid." It is of interest to note that while serving in a military outpost near Green Bay in 1827 he requested that "crusts of vaccine virus" be sent to him so that an outbreak of smallpox might be avoided. This was but several decades after Jenner had described the protection afforded by vaccination and sixty years before the description by Guarnieri of the viral agent producing smallpox.

Perhaps it was Beaumont's ability to observe and his compulsion to record his findings explicitly that set the stage for the experiments which were to be done on Alexis St. Martin. He struggled for nearly two years in treating the gunshot wound of the abdomen and thorax that occurred on June 6, 1822, seeking in vain through much of this period to gain closure of the wound. Eventually, with observations of the resulting gastrostomy and the stomach, there was description of some of the essential physical characteristics of this organ. For instance, he noted that gastric mucosal surfaces would protrude or herniate through the orifice without causing pain. Gentle pressure would reduce the herniation and caused little sensation. In addition, placing a solution of silver nitrate on the mucosa caused less discomfort than when it was applied to an ulcer of the skin.

Beaumont noted that during the first two weeks following the gunshot wound, all ingested foods were lost through the abdominal opening. The patient was sustained, it was claimed, by "nutritious injections" introduced per rectum. He does not describe the content of these solutions. However, this would not be readily accomplished in present-day therapy. Current practices would include parenteral feeding by intravenous routes or feeding through a tube passed through the pylorus into the duodenum.

The experiments that Beaumont later undertook on St. Martin produced the first description of gastric motility, secretion, and digestion. He observed the acidity of gastric juice and identified hydrochloric acid as a major constituent. He noted the time required for the digestion and gastric emptying of various foods, recording the facts that after a usual meal including meat and bread, digestion is complete within three to three and a half hours; fatty foods delay emptying; vegetables require longer periods for complete digestion; and bulk is a necessary component of the diet. Especially noteworthy was Beaumont's conclusion that gastric juice contains substances other than muriatic or hydrochloric acid that participate in the process of digestion. This derived from his comparison of the digestion of meat in vials containing gastric juice or dilute hydrochloric acid. Moreover, he ascertained that digestion proceeds more rapidly in gastric juice at body rather than room temperature. With the simple equipment available to him he offered first evidence of the presence of gastric enzymes, that is, pepsin and rennin. Very shortly thereafter, on the basis of these observations, Schwann undertook experiments which in 1836 brought the original description of pepsin, a contribution that might have been Beaumont's had he boiled the gastric secretion prior to testing its digestive properties. Similarly, if he had tested the effects of saliva on starchy foods in much the same way as he studied gastric juice, he might not have concluded that it had no role in digestion.

Beaumont's observations brought new methods of approach for studies of gastric physiology that were adapted and utilized by others. Among these were investigations by Richet[1] of a patient who had a gastrostomy produced surgically to provide a route for alimentation that bypassed a stricture of the esophagus which had been caused by the ingestion of lye. He confirmed many of Beaumont's findings and also recorded the facts that instillation of cold fluids into the stomach causes pain, while alcohol induces the sensation of warmth.

Later, in an important advance, Pavlov surgically constructed in dogs an innervated blind pouch of the stomach with a fistula emptying through the abdominal wall. By this means it was possible to collect gastric juice uncontaminated by food, saliva, or secretions from the duodenum. Like Beaumont, Pavlov[2] concluded that the stomach does not produce its secretions unless there is stimulation by mechanical or chemical means. Carlson,[3] on the other hand, observed that the resting, fasting stomach does secrete acid gastric juice. His observations were made on a patient and laboratory assistant who had a gastric fistula and in whom he noted also that hunger was associated with strong contractions of the stomach and with reddening of the mucosal surface.

Pavlov, as well as Beaumont, was able to demonstrate that gastric secretion and motility can be stimulated and enhanced by environmental or psychic stimuli. These effects were examined more fully by Cannon and Washburn[4] and were evaluated by Wolf and Wolff[5] in their observations of Tom, who had a gastrostomy because of an esophageal stricture produced by drinking very hot clam chowder. It was by such means that knowledge was gained of neural control of gastric secretory activity and motility through vagal pathways.

Beaumont, through his carefully recorded observations and experiments on Alexis St. Martin, provided the foundations on which the structure of modern gastric physiology has been built. The field has attracted the interests of many gifted and ingenious investigators and has brought evidence of hormonal (i.e., gastrin) as well as neural control, histologic description of specific cell types within the gastric mucosa responsible for the secretion of acid, pepsinogen, mucin, and gastrin, and elucidation of biochemical mechanisms involved in production of gastric secretions and of digestive processes within the stomach.

The following brief synopsis is provided as background for the reader who may not be closely associated with the field with the hope that it will bring a larger measure of appreciation for the writings and observations of Beaumont as this biography and his experiments and observations are perused.

Gastric secretory activity is normally stimulated or controlled in three phases. The initial, or cephalic, phase is brought into play by the sighting, smelling, tasting, chewing, or swallowing of food. Nerve impulses flow through vagal pathways, causing the release of acetylcholine near secretory cells located in the fundus and body of the stomach, which in turn secrete hydrochloric acid, pepsinogen, and mucus. Hydrochloric acid is produced by parietal or oxyntic cells, while chief cells secrete pepsinogen and other enzymes. Vagal impulses to antral mucosal cells cause the release of the hormone, gastrin, which, carried by the bloodstream to oxyntic cells, causes release of hydrochloric acid.

The presence of food in the stomach signals the onset of the gastric phase of secretion. By this means mechanoreceptors of afferent vagal fibers stimulate secretion either through local reflex pathways within the wall of the stomach or through efferent vagal discharges. Chemoreceptors responding to polypeptide digestion products also stimulate acid secretion through local reflex arcs. In addition, distention of antral portions of the stomach causes release of gastrin if the mucosa of the pyloric area is not bathed in contents of high acidity.

Finally, the intestinal phase of gastric secretion is effected when chyme is present in the duodenum or upper jejunum. This is in response to gastrin released from mucosa of these segments of the small intestine.

There is inhibition of gastric secretory activity through nervous reflexes that have not been fully elucidated but which respond to acid or fatty digestion products and solutions of high osmotic pressure within the duodenum and jejunum. Also, the hormone, cholecystokinin, which is released from the intestinal mucosa, inhibits gastrin-stimulated secretion of acid within the stomach.

The quantity of acid produced in response to a meal varies directly with the content of protein. The rate of acid production peaks in the second half hour after the start of a meal and declines slowly over several hours. When the stomach is emptied and the stimuli for acid secretion are removed, a small quantity of acid gastric juice remains.

Beaumont observed contraction of gastric muscles both in longitudinal and circular planes. Current knowledge demonstrates that as food is swallowed, the stomach musculature relaxes so that there is little rise in intragastric pressure. Within the body of the stomach, mixing movements are slight and acid secretions flow about the periphery of the mass of food but mix only slightly. As a result, salivary ptyalin activity persists in digestion of starch. As gastrin reaches the circulation, there is increase in contraction of the esophageal sphincter to prevent regurgitation of stomach contents. Eventually, longitudinal muscular contractions propel the food into the antrum, where mixing is completed and digestion of protein begins.

The rate of emptying is determined by the capability of the duodenum to neutralize the acid in the chyme, make it isotonic, and reduce the content of fat. Reflexes and hormones arising from the duodenum inhibit gastric motility. Solids and semisolids are emptied more slowly than liquids. Peristaltic waves in the antrum cease as the stomach is emptied but begin again as hunger is experienced.

Beaumont viewed with some consternation the presence of free acid within the lumen of the stomach and the tolerance of the mucosa to this acid environment. We now know that the mucosa is only slightly permeable to the acid that it secretes. This property of the gastric mucosal barrier accounts for the capability of this organ to contain acid without injury. Electrogenic pumps permit secretion of hydrogen and chloride ions at the luminal surface of oxyntic cells. The chemical mechanism by which these cells secrete acid is not known. However, hydrogen ions within the cell derive from carbonic acid. The bicarbonate ion released enters the plasma to replace the chloride ion taken from it to accompany the hydrogen ion in the secreted juice.

The literature on gastric physiology is now very extensive and is admirable in its detail. The reader can find fuller description in a number of recent reviews.[6-9] In essence, the simple but carefully re-

Tombstone at the grave of Beaumont in Bellefontaine Cemetery, St. Louis, Mo. The wreath was laid during the annual pilgrimage to the grave site conducted by the St. Louis Medical Society on November 21, 1980.

corded experimental observations of Beaumont are seen by hindsight to have provided a foundation for a temple of new knowledge of gastric function that is intricate in detail and explicit in function. The reader who in this book comes to know Beaumont and St. Martin for the first time is to be envied. They were remarkable and interesting people.[10] It is indeed appropriate that The C. V. Mosby Company should, in celebration of its seventy-fifth anniversary and as we approach the two hundredth anniversary of Dr. Beaumont's birth, again pay tribute to the man who so richly deserves recognition as the Father of American Physiology.

REFERENCES

1. Richet, C.: Des propriétés chimiques et physiologiques du sac gastrique chez l'homme et les animaux, appendice A, J. de l'Anat. et de la Physiol. **14:**170, 1878.
2. Pavlov, I.: The work of the digestive gland, English translation from the Russian by W. H. Thompson, London, 1910, C. Griffin & Co.
3. Carlson, A. J.: Contributions to the physiology of the stomach, Am. J. Physiol. **31:**151, 1912.
4. Cannon, W. B., and Washburn, A. L.: An explanation of hunger, Am. J. Physiol. **29:**441, 1912.
5. Wolf, S. G., and Wolff, H. G.: Human gastric function. In an experimental study of a man and his stomach, ed. 2, London, 1947, Oxford University Press.
6. Davenport, H. W.: A digest of digestion, Chicago, 1975, Year Book Medical Publishers, Inc.
7. Dworken, H. J.: The alimentary tract; basic principles and case problems, Philadelphia, 1974, W. B. Saunders Co.
8. Öbrink, K. J., and Flemström, editors: Proceedings of the Symposium on Gastric Ion Transport, Uppsala, Sweden, July 24-28, 1977, Acta Physiol. Scand., special supplement, 1978.
9. Jacobson, E. D., and Shanbour, L. L., editors: Physiology series one, vol. 4, Gastrointestinal physiology, London, 1974, Butterworth & Co. (Publishers), Ltd., pp. 147-156, 227-264.
10. The prepared mind: Nutrition Today **6:**26, 1971.

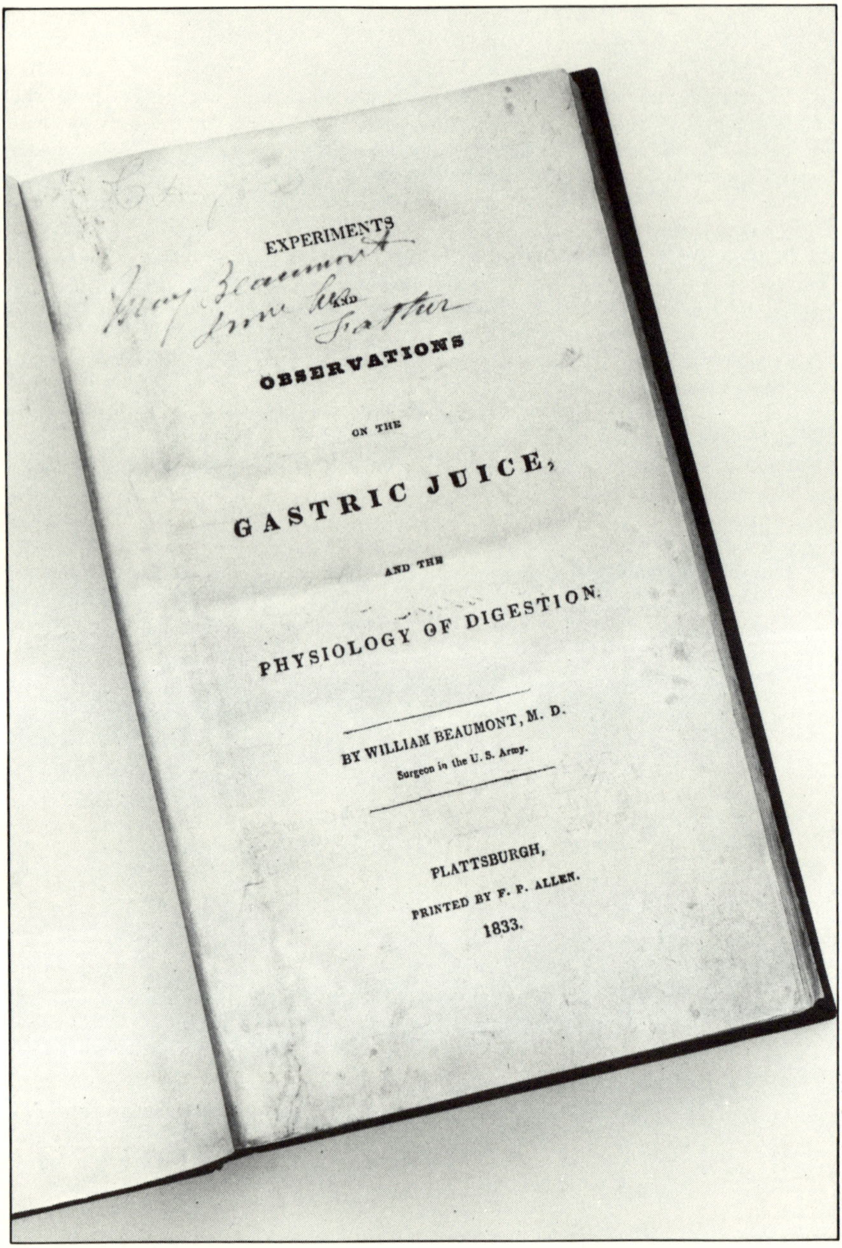

Presentation copy of the first edition of Beaumont's now famous work.

PREFACE

THE PRESENT AGE is prolific of works on physiology; therefore in offering to the public another book relative to an important branch of this science, it will perhaps be necessary to assign my motives.

They are, first, a wish to comply with the repeated and urgent solicitations of many medical men who have become partially acquainted with the facts and observations it is my intention to detail; men, in whose judgment I place confidence, and who have expressed their conviction of the deep importance of the experiments, the result of which I mean herewith to submit to the public: secondly, (and it is that which mainly influences me,) my own firm conviction that medical science will be forwarded by the publication.

I am fully aware of the importance of the subject which these experiments are intended to illustrate, as well in a pathological as in a physiological point of view; and I am therefore willing to risk the censure or neglect of critics, if I may be permitted to cast my mite into the treasury of knowledge, and to be the means, either directly or indirectly, of subserving the cause of truth, and ameliorating the condition of suffering humanity.

I make no claim to originality in my opinions, as it respects the existence and operation of the gastric juice. My experiments confirm the doctrines (with some modifications) taught by SPALLANZANI, and many of the most enlightened physiological writers. They are experiments made in the true spirit of enquiry, suggested by the very extraordinary case which gave me an opportunity of making them. I had no particular hypothesis to support; and I have therefore honestly recorded the result of each experiment exactly as it occurred.

The reader will perceive some slight seeming discrepancies, which he may find it difficult to reconcile; but he will recollect that the human machine is endowed with a vitality which modifies its movements in different states of the system, and probably produces some diversity of effects from the same causes.

I had opportunities for the examination of the interior of the stomach, and its secretions, which has never before been so fully offered to any one. This most important organ, its secretions and its operations, have been submitted to my observation in a very extraordinary manner, in a state of perfect health, and for years in succession. I have availed myself of the opportunity afforded by a concurrence of circumstances which probably can never again occur, with a zeal and perseverance proceeding from motives which my conscience approves; and I now submit the result of my experiments to an enlightened public, who I doubt not will duly appreciate the truths discovered, and the confirmation of opinions which before rested on conjecture.

I submit a body of facts which cannot be invalidated. My opinions may be doubted, denied, or approved, according as they conflict or agree with the opinions of each individual who may read them; but their worth will be best determined by the foundation on which they rest—the incontrovertible facts.

I avail myself of this opportunity to make my grateful acknowledgements to Doctor JOSEPH LOVELL, Surgeon General of the United States' Army, (to whom I am under obligations for personal kindness and official exertions in affording facilities for prosecuting the experiments;)—to Professors SILLIMAN, KNIGHT, IVES and HUBBARD, of Yale College, DUNGLISON, of the Virginia University, and SEWALL, JONES, HENDERSON and HALL, of Columbian College, for their unsolicited friendship; for the interest which they have taken in the experiments, and for the generous encouragement which they have given to the proposed publication. To Doctor SAMUEL BEAUMONT, of Plattsburgh, N. Y., I am particularly indebted for the assistance which he has rendered me in arranging and preparing my notes for the press.

INTRODUCTION

THE EXPERIMENTS which follow were commenced in 1825, and have been continued, with various interruptions, to the present time, (1833). The opportunity for making them was afforded to me in the following way.

Whilst stationed at Michilimackinac, Michigan Territory, in 1822, in the military service of the United States, the following case of surgery came under my care and treatment.

ALEXIS ST. MARTIN, who is the subject of these experiments, was a Canadian, of French descent, at the above mentioned time about eighteen years of age, of good constitution, robust and healthy. He had been engaged in the service of the American Fur Company, as a voyageur, and was accidentally wounded by the discharge of a musket, on the 6th of June, 1822.

The charge, consisting of powder and duck shot, was received in the left side of the youth, he being at a distance of not more than one yard from the muzzle of the gun. The contents entered posteriorly, and in an oblique direction, forward and inward, literally blowing off integuments and muscles of the size of a man's hand, fracturing and carrying away the anterior half of the sixth rib, fracturing the fifth, lacerating the lower portion of the left lobe of the lungs, the diaphragm, and perforating the stomach.

The whole mass of materials forced from the musket, together with fragments of clothing and pieces of fractured ribs, were driven into the muscles and cavity of the chest.

I saw him in twenty-five or thirty minutes after the accident occurred, and, on examination, found a portion of the lung, as large as a Turkey's egg, protruding through the external wound, lacerated and burnt; and immediatley below this, another protrusion, which, on further examination, proved to be a portion of the stomach, lacerated through all its coats, and pouring out the food he had taken for his breakfast, through an orifice large enough to admit the fore-finger.

In attempting to return the protruded portion of the lung, I was

St. Louis Medical Society

This crudely retouched photograph, of St. Martin at about 67 years of age, was found inserted as a frontispiece facing the title page in a copy of the original edition of Beaumont's *Experiments and Observations* . . .

prevented by a sharp point of the fractured rib, over which it had caught by its membranes; but by raising it with my finger, and clipping off the point of the rib, I was able to return it into its proper cavity, though it could not be retained there, on account of the incessant efforts to cough.

The projecting portion of the stomach was nearly as large as that of the lung. It passed through the lacerated diaphragm and external wound, mingling the food with the bloody mucus blown from the lungs.

After cleansing the wound from the charge and other extraneous matter, and replacing the stomach and lungs as far as practicable, I applied the carbonated fermenting poultice, and kept the surrounding parts constantly wet with a lotion of muriate of ammonia and vinegar; and gave internally the aq. acet. am. with camphor, in liberal quantities.

Under this treatment a strong reaction took place in about twenty-four hours, accompanied with high arterial excitement, fever, and marked symptoms of inflammation of the lining membranes of the chest and abdomen, great difficulty of breathing, and distressing cough.

He was bled to the amount of eighteen or twenty ounces, and took a cathartic. The bleeding reduced the arterial action, and gave relief. The cathartic had no effect, as it escaped from the stomach through the wound.

On the 5th day a partial sloughing of the integuments and muscles took place. Some of the protruded portions of the lung, and lacerated parts of the stomach, also sloughed, and left a perforation into the stomach, plainly to be seen, large enough to admit the whole length of my fore-finger into its cavity; and also a passage into the chest, half as large as my fist, exposing to view a part of the lung, and permitting the free escape of air and bloody mucus at every respiration.

A violent fever continued for ten days, running into a typhoid type, and the wound became very fœtid.

On the eleventh day, a more extensive sloughing took place, the febrile symptoms subsided, and the whole surface of the wound assumed a healthy and granulating appearance.

For seventeen days, all that entered his stomach by the œsophgus, soon passed out through the wound; and the only way of sustaining him was by means of nutricious injections per anus, until compresses and adhesive straps could be applied so as to retain his food. During this period no alvine evacuations could be obtained, although cathartic injections were given, and various other means were adopted to promote them.

In a few days after firm dressings were applied, and the contents of the stomach retained, the bowels became gradually excited, and, with the aid of cathartic injections, a very hard, black, fœtid stool was procured, followed by several similar ones; after which the bowels became quite regular, and continued so.

The cataplasms were continued until the sloughing was completed, and the granulating process fully established; and were afterwards occasionally resorted to, when the wound became ill conditioned. The aq. acet. am. with camphor was also continued for several weeks, in proportion to the febrile symptoms and the fœtid condition of the wound.

No sickness, nor unusual irritation of the stomach not even the slightest nausea, was manifest during the whole time; and after the fourth week, the appetite became good, digestion regular, the alvine evacuations natural, and all the functions of the system perfect and healthy.

By the adhesion of the sides of the protruded portions of the stomach to the pleura costalis and the external wound, a free exit was afforded to the contents of that organ, and effusion into the abdominal cavity was thereby prevented.

Cicatrization and contraction of the external wound commenced on the fifth week; the stomach became more firmly attached to the pleura and intercostals, by its external coats; but showed not the least disposition to close its orifice; this (the orifice) terminated as if by a natural boundary, and left the perforation, resembling, in all but a sphincter, the natural anus, with a slight prolapsus.

Whenever the wound was dressed the contents of the stomach would flow out, in proportion to the quantity recently taken. If the stomach happened to be empty, or nearly so, a partial inversion would take place, unless prevented by the application of the finger. Frequently in consequence of the derangement of the dressing, the inverted part would be found of the size of a hen's egg. No difficulty, however, was experienced in reducing it by gentle pressure with the finger, or a sponge wet with cold water, neither of which produced the least pain.

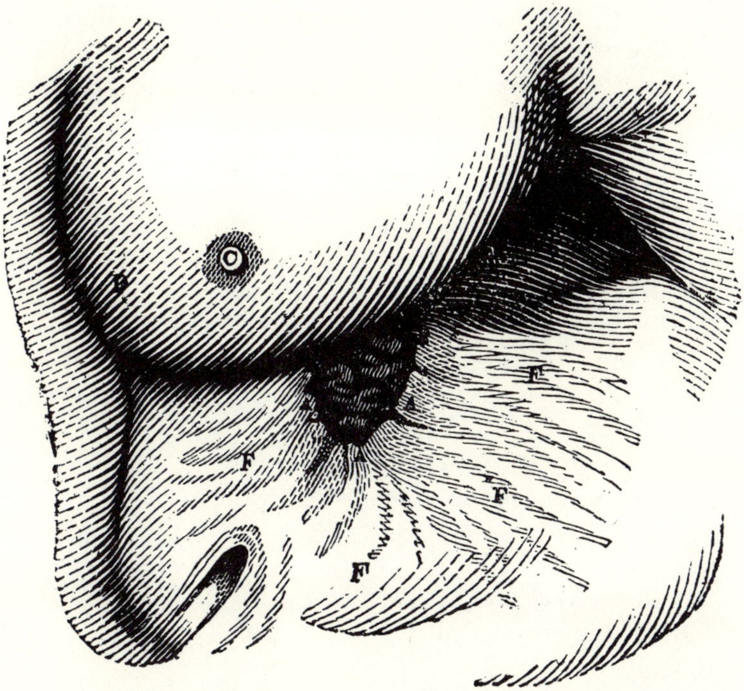

This engraving represents the ordinary appearance of the left breast and side, the aperture filled with the valve; the subject in an erect position.

A A A The circumference and edge of the aperture, within which is seen the valve.

B The attachment of the valvular portion of the stomach to the superior part of the aperture.

C The nipple.

D The anterior portion of the breast.

E The scar where the opening was made with the scalpel, and the cartilages taken out.

F F F F Cicatrice of the original wound, around the aperture.

In the seventh week, exfoliation of the ribs, and a separation of their cartilaginous ends, began to take place.

The sixth rib was denuded of its periosteum for about two inches from the fractured part, so that I was obliged to amputate it about three or four inches from its articulation with the rib. This I accomplished by dissecting back the muscles, securing the intercostal artery, and sawing off the bone with a very fine narrow saw, made for the purpose, introduced between the ribs, without injury to the neighbouring parts. Healthy granulations soon appeared, and formed soundly over the amputated end. About half the inferior edge of the fifth rib exfoliated, and separated from its cartilage.

After the removal of these pieces of bone, I attempted to contract the wound, and close the perforation of the stomach, by gradually drawing the edges together with adhesive straps, laid on in a radiated form.

The circumference of the external wound was at least twelve inches, and the orifice in the stomach nearly in the centre, two inches below the left nipple, on a line drawn from this to the point of the left ilium.

To retain his food and drinks I kept a compress and tent of lint, fitted to the shape and size of the perforation, and confined there by adhesive straps.

After trying all the means in my power for eight or ten months to close the orifice, by exciting adhesive inflammation in the lips of the wound, without the least appearance of success, I gave it up as impracticable in any other way than that of incising and bringing them together by sutures; an operation to which the patient would not submit.

By the sloughing of the injured portion of the lung, a cavity was left as large as a common sized teacup, from which continued a copious discharge of pus for three months, when it became filled with healthy granulations, firmly adhering to the pleura, and soundly cicatrized over that part of the wound.

Four months after the injury was received, an abscess formed about two inches below the wound, nearly over the cartilaginous ends of the first and second false ribs, very painful, and extremely sore, producing violent symptomatic fever. On the application of an emollient poultice it pointed externally. It was then laid open to the extent of three inches, and several shot and pieces of wad extracted. After

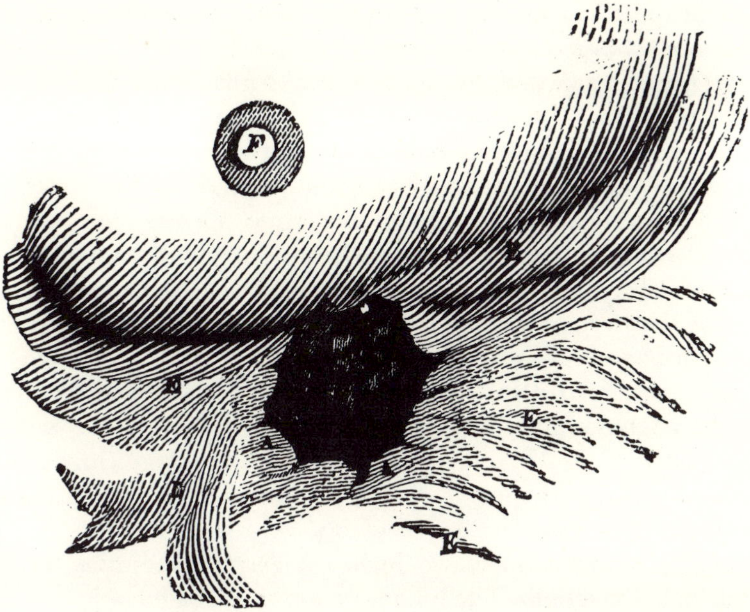

This engraving represents the appearance of the aperture with the valve depressed.

A A A Edges of the aperture through the integuments and intercostals, on the inside and around which is the union of the lacerated edges of the perforated coats of the stomach with the intercostals and skin.

B The cavity of the stomach, when the valve is depressed.

C Valve, depressed within the cavity of the stomach.

E E E E Cicatrice of the original wound.

F The nipple.

which a gum-elastic bougie could be introduced three or four inches in the longitudinal direction of the ribs, towards the spine. Great pain and soreness extended from the opening of the abscess, along the track of the cartilaginous ends of the false ribs, to the spine, with a copious discharge from the sinus.

In five or six days there came away a cartilage, one inch in length. In six or seven days more, another, an inch and a half long; and in about the same length of time, a third, two inches long, were discharged. And they continued to come away every five or six days, until *five* were discharged from the same opening, the last three inches in length. They were all entire, and evidently separated from the false ribs.

The discharge, pain and irritation, during the four or five weeks these cartilages were working out, greatly reduced the strength of the patient, produced a general febrile habit, and stopped the healing process of the original wound.

Directly after the discharge of the last cartilage, inflammation commenced over the lower end of the sternum, which by the usual applications, terminated in a few days in a large abscess, and from which, by laying it open two inches, I extracted another cartilage, three inches in length. The inflammation then abated; and in a day or two another piece came away, and the discharge subsided.

To support the patient under all these debilitating circumstances, I administered wine, with diluted muriatic acid, and thirty or forty drops of the tincture of assafœtida, three times a day; which appeared to produce the desired effect, and very much improved the condition of the wound.

On the third of January, 1823, I extracted another cartilage from the opening over the sternum, an inch and a half long; and on the fourth another, two inches and a half in length, an inch broad at one end, and narrowing to less than half an inch at the other. This must have been the ensiform cartilage of the sternum. After this the sinus closed, and there was no return of inflammation.

From the month of April, 1823, at which time he had so far recovered as to be able to walk about and do light work, enjoying his usual good appetite and digestion, he continued with me, rapidly regaining his health and strength.

By the 6th of June, 1823, one year from the time of the accident, the injured parts were all sound, and firmly cicatrized, with the excep-

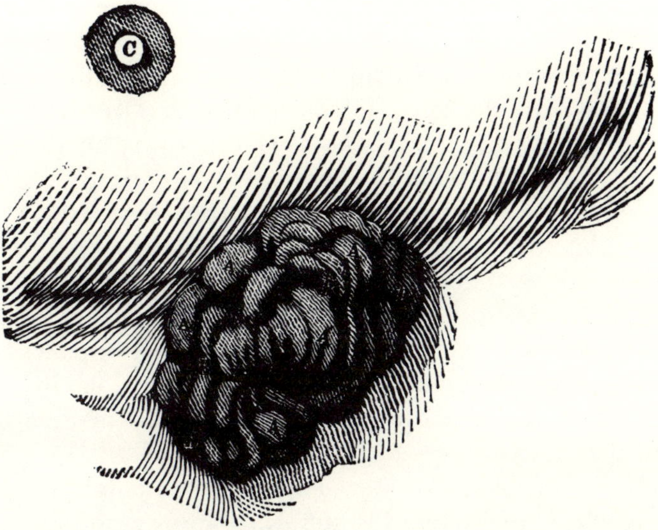

This engraving represents a portion of the stomach prolapsed through the aperture, with the inner surface inverted, and spread out over the integuments of the side.

A A A A A Folds or rugæ of the inner coats of the stomach.

B B B B Interstices, filled with mucous substance.

C The nipple.

tion of the aperture in the stomach and side. This continued much in the same situation as it was six weeks after the wound was received. The perforation was about two and a half inches in circumference, and the food and drinks constantly exuded, unless prevented by a tent, compress and bandage.

From this time he continued gradually to improve in health and strength, and the newly formed integuments over the wound became firmer and firmer. At the point where the lacerated edges of the muscular coat of the stomach and intercostal muscles met and united with the cutis vera, the *cuticle* of the external surface and the *mucous membrane* of the stomach *approached* each other very nearly. They did not unite, like those of the lips, nose, &c., but left an intermediate marginal space, of appreciable breadth, completely surrounding the aperture. This space is about a line wide; and the cutis and nervous papillæ are unprotected, as sensible and irritable as a blistered surface abraded of the cuticle. This condition of the aperture still continues, and constitutes the principal and almost only cause of pain or distress experienced from the continuance of the aperture, the introduction of instruments, &c. in the experiments, or the exudation of fluids from the gastric cavity.

Frequent dressings with soft compresses and bandages were necessarily applied, to relieve his suffering and retain his food and drinks, until the winter of 1823-4. At this time, a small fold or doubling of the coats of the stomach appeared, forming at the superior margin of the orifice, slightly protruding, and increasing till it filled the aperture, so as to supersede the necessity for the compress and bandage for retaining the contents of the stomach. This valvular formation adapted itself to the accidental orifice, so as completely to prevent the efflux of the gastric contents when the stomach was full, but was easily depressed with the finger.

In the spring of 1824 he had perfectly recovered his natural health and strength; the aperture remained; and the surrounding wound was firmly cicatrized to its edges.

In the month of May, 1825, I commenced my first series of gastric experiments with him, at Fort Mackinac, Michigan Territory. In the month of June following, I was ordered to Fort Niagara, N. Y., where, taking the man with me, I continued my experiments until August. Part of these experiments were published in 1826, in the 29th number of the Philadelphia "Medical Recorder," conducted by Doctor

Samuel Calhoun. About this time, (August 1825) I took St. Martin with me to Burlington, Vermont, and from thence to Plattsburgh, New York. From the latter place, he returned to Canada, his native place, without obtaining my consent.

Being unable to ascertain the place of his resort, I gave him up as a lost subject for physiological experiments, and returned to my post at the west again. I did not, however, remit my efforts to obtain information of his place of residence and condition.

He remained in Canada four years, during which period he married, and became the father of two children; worked hard to support his family; and enjoyed robust health and strength. In 1825, as he has informed me, he engaged with the Hudson Bay Fur Company, as a voyageur to the Indian country. He went out in 1827, and returned in 1828; and subsequently laboured hard to support his family until 1829.

Accidentally learning about this time where he was, and that he enjoyed perfect health, I made arrangements with the agents of the American Fur Company, who annually visit Canada for the purpose of procuring voyageurs, to find and engage him for my service, if practicable. After considerable difficulty, and at great expense to me, they succeeded in engaging him, and transported him from Lower Canada, with his wife and two children, to me, at Fort Crawford, Prairie du Chien, Upper Mississippi, a distance of nearly two thousand miles, in August, 1829. His stomach and side were in a similar condition as when he left me in 1825. The aperture was open, and his health good.

He now entered my service, and I commenced another series of experiments on the stomach and gastric fluids, and continued them, interruptedly, until March, 1831. During this time, in the intervals of experimenting, he performed all the duties of a common servant, chopping wood, carrying burthens, &c. with little or no suffering or inconvenience from his wound. He laboured constantly, became the father of more children, and enjoyed as good health and as much vigour as men in general. He subsisted on crude food, in abundant quantities, except when on prescribed diet, for particular experimental purposes, and under special observance.

In the spring of 1831 circumstances made it expedient for him to return with his family from Prairie du Chien to Lower Canada again. I relinquished his engagements to me for the time, on a promise that he would return when required, and gave him an outfit for himself, wife and children. They started in an open canoe, via the Mississippi, passing by St. Louis, Mo.; ascended the Ohio river; then crossed the state of Ohio, to the Lakes; and descended the Erie, Ontario, and the River St. Lawrence, to Montreal, where they arrived in June. He remained in Canada with his family until October, 1832, in good health, and at hard labour. He was in the midst of the cholera epidemic, at the time it prevailed, and passed through Canada, and withstood its ravages with impunity, while hundreds around him fell sacrifices to its fatal influence.

In November, 1832, he again engaged himself to me for twelve months, for the express purpose of submitting to another series of experiments. He joined me at Plattsburgh, N. Y., and travelled with me to the city of Washington, where, with the facilities afforded by the head of the Medical Department, the experiments were continued upon him from November, 1832, to March 1833.

During the whole of these periods, from the spring of 1824 to the present time, he has enjoyed *general* good health, and perhaps suffered much less predisposition to disease than is common to men of his age and circumstances in life. He has been active, athletic and vigorous; exercising, eating and drinking like other healthy and active people. For the last four months, he has been unusually plethoric and robust, though constantly subjected to a continued series of experiments on the interior of the stomach; allowing to be introduced or taken out at the aperture different kinds of food, drinks, elastic catheters, thermometer tubes, gastric juice, chyme, &c., almost daily, and sometimes hourly.

Such have been this man's condition and circumstances for several years past; and he now enjoys the most perfect health and constitutional soundness, with every function of the system in full force and vigour.

Mode of extracting the Gastric Juice—The usual method of extracting the gastric juice, for experiment, is by placing the subject on his right side, depressing the valve within the aperture, introducing a gum-elastic tube, of the size of a large quill, five or six inches into the stomach, and then turning him on the left side, until the orifice becomes dependent. In health, and when free from food, the stomach is *usually* entirely empty, and contracted upon itself. On introducing the tube, the fluid soon begins to flow, first by drops, then in an interrupted, and sometimes in a short continuous stream. Moving the tube about, up and down, or backwards and forwards, increases the discharge. The quantity of fluid ordinarily obtained is from four drachms to one and a half or two ounces, varying with the circumstances and condition of the stomach. Its extraction is generally attended by that peculiar sensation at the pit of the stomach, termed sinking, with some degree of faintness, which renders it necessary to stop the operation. The usual time of extracting the juice is early in the morning, before he has eaten, when the stomach is empty and clean.

On laying him horizontally on his back, pressing the hand upon the hepatic region, agitating a little, and at the same time turning him to the left side, bright yellow bile appears to flow freely through the pylorus, and passes out through the tube. Sometimes it is found mixed with the gastric juice, without this operation. This is, however, seldom the case, unless it has been excited by some other cause.

The chymous fluids are easily taken out by depressing the valve within the aperture, laying the hand over the lower part of the stomach, shaking a little, and pressing upwards. In this manner, any quantity necessary for examination and experiment can be obtained.

Valve—The valve mentioned above, is formed by a slightly inverted portion of the inner coats of the stomach, fitted exactly to fill the aperture. Its principal and most external attachment is at the upper and posterior edge of the opening. Its free portion hangs pendulous, and fills the aperture when the stomach is full, and plays up and down, simultaneously with the respiratory muscles, when empty.

On pressing down the valve when the stomach is full, the contents flow out copiously. When the stomach is nearly empty, and quiescent, the interior of the cavity may be examined to the depth of five or six inches, if kept distended by artificial means; and the food and drinks may be seen entering it, if swallowed at this time, through the ring of the œsophagus. The perforation through the walls of the stomach, is about three inches to the left of the cardia, near the left superior termination of the great curvature. When entirely empty, the stomach contracts upon itself and sometimes forces the valve through the orifice, together with an additional portion of the mucous membrane, which becomes completely inverted, and forms a tumour as large as a hen's egg. After lying on the left side, and sleeping a few hours, a still larger portion protrudes, and spreads out over the external integuments, five or six inches in circumference, fairly exhibiting the natural rugæ, villous membrane, and mucous coat, lining the gastric cavity. This appearance is almost invariably exhibited in the morning, before rising from his bed.

PLATTSBURGH, 1833.

PRELIMINARY OBSERVATIONS

I do not design, in the following remarks, to present any thing like a systematic treatise on digestion. Works of this kind, treating of the subject both physiologically and pathologically, have so multiplied of late, as to render an attempt on my part, entirely a work of supererogation, even if I believed myself qualified for the task. I consider myself but a humble inquirer after truth—a simple experimenter. And if I have been led to conclusions opposite to the opinions of many who have been considered the great luminaries of physiology, and, in some instances, from all the professors of this science, I hope the claim of sincerity will be conceded to me, when I say that such difference of opinion has been forced upon me by the convictions of experiment, and the fair deductions of reasoning.

I shall not attempt an anatomical description of the organs of digestion, for the reasons given above. In a work professedly elementary, such descriptions are essential. The medical profession are acquainted with these organisms. The general reader, if he have a wish for information of this kind, is referred to anatomical authors generally, or to the physiological writings of Richerand, Broussais, Magendie, Bostock, Fordyce, Paris, Jackson and Dunglison, the last of which, as containing the sum of what has been taught in the schools on the subject of physiology generally, arranged in a clear and distinct manner, and with the assistance of numerous plates, is well worthy his perusal.

With a view to comment on my experiments, and to elucidate my opinions on the subject of digestion, I shall divide my observations into the following heads:

Section 1st. *Of Aliment.* Section 2d. *Of Hunger and Thirst.* Section 3d. *Of Satisfaction and Satiety.* Section 4th. *Of Mastication, Insalivation and Deglutition.* Section 5th. *Of Digestion by the Gastric Juice.* Section 6th. *Of the Appearance of the Villous Coat, and of the Motions of the Stomach.* Section 7th. *Of Chylification, and Uses of the Bile and Pancreatic Juice.*

SECTION I

Of Aliment

Man is said to be an *omnivorous* animal, destined to procure his food from both the animal and vegetable kingdoms. The inhabitant of temperate climates is unquestionably so. It would be interesting to ascertain by experiment whether he could be sustained by habit, from infancy, on the productions of either of these grand divisions. If the result should be favourable to the demonstration of this proposition, though it might still more unsettle the opinions of physiologists, it would be an evidence of this truth, that man is a creature of habit and circumstance, carrying about him the effects of primeval disobedience, destined not only to earn his food by his own exertions, but to partake of such as the climate in which he resides may supply to him. Approximating to this are the habits of people of different quarters of the world—those of Asia, who live almost exclusively on vegetable and farinaceous food, and those of the northern regions of America, who derive their food principally from fish, oil and flesh.

Other substances have sometimes been used as aliment; and Professor Dunglison mentions, on the authority of Humboldt, that the Ottomaques, a tribe of Indians of South America, are in the habit of using "an unctious earth, or a species of pipe clay," as an article of diet. Whether nutriment can be supplied by such articles alone, is extremely problematical. In all countries, some persons are found who are in the constant habit of eating large quantities of clay, chalk, slate stone, &c. Such practices may be regarded as evidence, if not of a diseased, at least of a vitiated appetite; though it often happens that alkaline and absorbent substances are used medicinally with advantage, particularly where much acidity of the stomach prevails.

As it respects the inhabitants of Europe and their American descendants, as well as most other natives of temperate climates, it is well known that they derive their nourishment from both the animal and vegetable kingdoms.

The facility of digestion of different articles of diet, and the quantity of nutrient principles which they contain, have been subjects of some discrepance of opinion among physiologists. They have, however, settled down into a belief, probably as near the truth as practicable, that animal food is more readily assimilated, and affords more nutrition in a given quantity, than vegetable or farinaceous food.

Animal food has been divided into fibrine, gelatine and albumen, and a comparison drawn between their degrees of digestibility. But it will occur to every one at all acquainted with the subject, that almost every portion of animal food contains an admixture of all these principles, and it is consequently very difficult to come to a correct conclusion. The truth is, there can be no general rule on this subject. The facility of digestion is modified by so many circumstances, as health, disease, idiosyncracy, habit, and preparation of food, that a rule which would apply in one case would be incorrect in another. It depends more upon other distinctions than upon those relating to the chemical composition of the food. Albumen, (one of these chemical divisions), if taken into the stomach, either very slightly or not at all coagulated, is perhaps as rapidly chymified as any article of diet we possess. If perfectly formed into hard coagulæ, by heat or otherwise, and swallowed in large solid pieces, it experiences a very protracted digestion. The reason is obvious. In the first case the albumen becomes finely coagulated, and divided in the stomach; in the second, it is less susceptible of subdivision from its hardness. Fibrine and gelatine are affected in the same way. If tender and finely divided, they are disposed of readily; if in large and solid masses, digestion is proportionably retarded. Minuteness of division and tenderness of fibre are the two grand essentials for speedy and easy digestion. By referring to my experiments, it will be seen that those articles of diet which were submitted to the action of the gastric juice, either artificially, when out of the stomach, or in the stomach, by natural process, were dissolved in proportion to the fineness of their division or their solidity—the one rapidly, and the other slowly.

The digestion of animal and vegetable diet requires the same process, though one may afford a larger proportion of the nutrient principle than the other. Generally speaking, vegetable aliment requires more time, and probably greater powers of the gastric organs, than animal. Its digestibility is, however, dependant upon the same laws as those that govern the solution of animal food; and it is facilitated by division and tenderness.

The ultimate principles of nutriment are probably always the same, whether obtained from animal or vegetable diet. It was said by Hippocrates, that "there are many kinds of aliments, but that there is at the same time but one aliment." This opinion has been contested by most modern physiologists; but I see no reason for scepticism on this subject. Some imperfect experiments which I instituted on the operations of the hepatic and pancreatic juices, and which will be found in a subsequent part of this volume, tend to throw some light on the subject. Chyme was submitted to the action of these fluids, and they invariably produced similar effects. A fluid was separated, varying slightly in colour, but of the same apparent consistence and identity; and was increased or lessened in proportion to the quality of the food of which the chyme was formed. Whether this fluid was or was not imperfectly formed chyle, is a matter of opinion only. The circulating fluids of the system are always nearly the same, in health, and that which goes to supply and replenish them, should consequently possess the same invariable properties. Chyle, after its separation in the intestines, is probably further changed and perfected by the action of the lacteal absorbents and sanguiferous vessels, before it is completely assimilated. Chyme, from which this nutrient principle is obtained, is a compound of gastric juice and aliment. It may be regarded as a *gastrite* of whatever it is combined with, varied according to the kind of aliment used. The perfect chyle, or assimilated nutriment, probably contains the elements of all the secretions of the system; such as bone, muscle, mucus, saliva, gastric juice, &c. &c., which are separated by the action of the glands, the sanguiferous and other vessels of the system.

The action of the stomach, and its fluids, on aliment, is believed to be *sui generis,* invariably the same, in health, on all kinds. And yet it is contended by Paris, and obliquely hinted by some other modern physiologists, that as animal food "possesses a composition analogous to that of the structure it is designed to supply," it "requires little more than division and depuration," &c. It is singular that sensible men, and men of science, can allow themselves to be led to such erroneous conclusions, and will not perceive a simplicity and *uniformity* in the process of digestion, as well as in all the other operations of nature.

That the active solvent of the stomach should produce the same effect on all alimentary substances, is no more wonderful than that caloric should liquefy all kinds of matter. In both cases it only requires a longer or shorter continuance, or more or less concentrated action, of the agent, to produce the same effect. If animal food is only to be divided and depurated, blood, which is an elementary part of the body, would require no change in the stomach. But it is perfectly idle to talk in this way. The most innutritious vegetable and the most animalized substance, require the same action of the gastric solvent, as the reader will find amply demonstrated in the following experiments. It is true that one may be disposed of with ease, and the other with difficulty; but this is not always, nor indeed often, in a direct ratio to their respective proportions of nutrient principles. An innutritious diet may be disposed of as easily, the circumstances of divisibility and tenderness of fibre being equal, as a nutritious one. I do not believe that the one requires a more "complicated series of decompositions and recompositions" than the other; nor that the chyle from animal aliment is more highly "animalized" than that from the poorest diet we possess. The "digestive fever," or the excitement that follows the digestion of animal food, is the effect, not of a different *kind* of stimulus, but of the introduction of a greater *quantity* of chyle, or the nutritive principle of food, into the circulating fluids. It excites the system precisely in the same manner as ardent spirits, or other stimulus does, with the exception, that its effects are more permanent.

The quantity of nutriment required by different individuals, is as various as the individuals who partake of it. As a general rule, it may be said that persons who do not exercise much, require less nutritious diet than those who are in the habit of constant labour. What would be a natural supply in one, would be excess in another. With labouring persons, much of the excess is carried off by perspiration; and probably a great deal of nervous energy is wasted by laborious occupations, which requires to be replenished by the nutrient principles of aliment. This is a subject, however, on which we can only offer conjecture; for it is difficult to argue on a point of which we know so little. Young people who are growing, require more nutriment in proportion to their size, than those who have arrived at adult age.

The quality of nutriment is a matter of considerable importance in dietetic regulations. Bulk is, perhaps, nearly as necessary to the articles of diet as the nutrient principle. They should be so managed that one should be in proportion to the other. Too highly nutritive diet is probably as fatal to the prolongation of life and health, as that which contains an insufficient quantity of nutriment. It has been ascertained that carnivorous animals will not live on highly concentrated food alone. Dogs fed on oil or sugar, which are both converted by the digestive organs almost entirely into chyle, are found to become diseased, and die in a few weeks. The inference drawn by Paris,* that it merely "proves that an animal cannot be supported on highly concentrated aliment alone," no doubt, is a correct one; though opposed to the opinion of Magendie, the author of the experiments, who infers that death proceeds from the want of azote in these articles of diet, and that life cannot be supported on non-azotized aliment.

The following articles of the *materia alimentaria* have, in the course of these experiments, been submitted to the action of the stomach and the gastric fluids. I have attempted, in this table, to approximate towards a comparison of the digestibility of the several articles there mentioned. Precision, as to minutes, has not been attended to. When digestion has been accomplished two or three minutes either before or after a certain number of hours and quarters, I have set down the quarter to which it approached the nearest.

In a subsequent part of this volume a more particular and minute detail will be found, both of natural and artificial digestion.

Explanation of the abbreviations in the following Table—br'd, for bread—veg., for vegetables—mod., for moderate—inc., for increased—susp'd, for suspended—dur'g, for during—exp't, for experiment—h'rd bld, for hard boiled—h., for hour—m., for minute, &c.—The figures denote the time of digestion, under the circumstances mentioned at the head of the column.

---

*Paris On Diet, p. 72.

Articles of Diet	mode of cooking	meal	With br'd or veg. or both: exercise. mod.	exercise. inc.	rest	Remarks
			h. m.	h. m.	h. m.	
TRIPE, soused	fried	breakfast	1 00			
PIG'S FEET, do.	boiled	. .	1 00			
VENISON steak, fresh,	broil'd	. .	1 35			
CODFISH, dry,	boiled	dinner	2 00			
BREAD and MILK,	cold	. .	2 00			
TURKEY,	roast'd	. .	2 30			
GOOSE, wild,	roast'd	. .	2 30			
PIG, young,	. .	. .	2 30			
HASH, meat & veg.	warm	breakfast	2 30			
OYSTERS,	raw	dinner	2 45	-	-	oyster susp'd in stom. dur'g exp't.
do.	stewed	. .	3 30	-	-	
do.	raw	breakfast	3 00	-	-	Nothing but a little dry bread or cracker taken at
do.	. .	dinner	3 00	-	-	these meals.
do.	stewed	. .	3 30	-	-	
BEEF, fresh, fat & lean	roast'd	dinner	3 30			
do.	. .	. .	3 00			
do.	. .	breakfast	2 45			
do.	broil'd	. .	3 00			
do.	broil'd	breakfast			3 45	
do.	. .	. .	-	3 30	-	Exercised till fatigued.
do.	boiled	. .	4 00		-	Morbid appearance of stomach.
do.	. .	dinner	-	3 30		
do.	boiled	breakfast	3 38	-	-	Large proportion of fat.
do.	. .	supper	-	-	4 00	do.
do.	. .	breakfast	-	-	4 30	do. & in recumbent position.

do.	. .	dinner	3 30			
do.	. .	. .	-	-	4 00	
do.	. .	breakfast	-	-	4 15	
do.	. .	. .	3 30			
do.	. .	. .	-	-	4 15	
do. salted,	. .	dinner	5 30			
do. do.	. .	. .	3 30			
PORK, recently salt'd	. .	breakfast	5 15			
do. .	. .	. .	4 30			
do. .	. .	. .	5 15			Became angry during experiment
do. .	. .	. .	6 00			Unusually full meal.
do.	boiled	breakfast	4 30			
do. .	. .	. .	4 30			
do. .	. .	. .	4 30			
do. .	. .	dinner	4 30			
do. .	. .	breakfast		4 00		
do. .	. .	dinner		3 30		
do. fresh,	roast'd	dinner	6 30	-	-	Unusually full meal.
do. steak,	broil'd	. .	3 15			
do. do.	. .	breakfast	4 30			
MUTTON, fat & lean	roast'd	dinner	3 15			
do. .	broil'd	breakfast		3 00		
do. .	. .	. .	3 30			
do. .	. .	. .	4 30			Morbid appearance of stomach.
do. .	. .	dinner	4 00			
do. .	. .	breakfast	4 30	-	-	Full meal, coarsely masticated.
EGGS,	hr'd bld	. .	3 30	-	-	Bread or bread & coffee, no vegetables used with the eggs.
do.	soft bld	. .	3 00			
do.	hr'd bld	dinner	5 30	-	-	Morbid appearance of stomach.
do.	h'rd bld	breakfast	3 30	-	-	Morbid appearance of stomach.
do.	soft bld	dinner	3 00			

Continued.

Articles of Diet	mode of cooking	meal	With br'd or veg. or both — exercise. mod.	exercise. inc.	rest	Remarks
SAUSAGE, fat & lean	broil'd	breakfast	3 30	-	-	With soft boiled eggs.
do. .	. .	dinner	3 00			
do. .	fried	breakfast	4 00	-	-	Muslin bag containing same kind of diet, susp'd
do. .	. .	. .	5 00	-	-	dur'g these exp'ts—morbid appearance of stomach
do. .	broil'd	. .	3 30			also.
do. .	. .	. .		4 15		Full meal—severe exercise
FOWLS, (hens) .	boiled	. .	4 00	-	-	With bread and coffee.
do. .	. .	dinner	4 00	-	-	With bread and water.
do. .	. .	. .	4 00	-	-	do. do.
VEAL, fresh, .	broiled	breakfast	4 00	-	-	Muslin bag susp'd in stomach.
do. .	. .	dinner	4 00			
do. .	. .	breakfast	4 00			
do. .	. .	dinner	4 45	-	-	Morbid appearance of stomach.
do. .	. .	breakfast		3 45		
do. .	. .	dinner	4 30			
do. .	. .	breakfast	5 30	-	-	Morbid appearance of stomach.
SOUP, made of fresh muscular fibre of beef, and vegetables,			4 00			
do. made of the hock, with vegetables,			4 15			
BREAD, buttered, for breakfast, with coffee [morbid appearance of stomach]			4 15			
do. buttered, for breakfast, with coffee			3 45			
do. dry, for breakfast, with coffee,			4 00			
do. dry, for dinner, with dry mashed potatoes			3 45			

This Table is far from being complete. The experiments from which it has been formed, were made principally with the view of demonstrating other important principles connected with the subject of digestion. The only way of ensuring minuteness and accuracy as to the comparative digestibility of different kinds of diet, would be to try the effect of the gastric juice, in a series of experiments, first on one article of diet, and then on another, repeating and adapting them to meet all the various conditions of the stomach, and the vicissitudes and irregularities of the system, until the whole range should be completed—a Herculean task, which it would take years to accomplish. In the above table, the time is counted from the reception of the meal of various articles to the chymification of the whole: hence the conclusions are frequently indefinite, some of the articles being sooner disposed of than others. For instance, if a dinner be eaten of venison steak and fat pork, the time of digestion of the whole quantity would, in all probability, be twice as long as if venison had been used alone. Oily substances are digested with great difficulty, and the fat of all meats is converted into oil in the stomach before it is digested. Chymification is most readily effected on solid food, or rather on a soft solid, which is easily divisible into shreds or small particles. Such is particularly the character of venison, which is ascertained to be one of the most digestible of substances. The qualities of looseness of texture and susceptibility of division belong to most of those wild meats and game which are generally acknowledged to be easy of digestion. Beef and mutton, of a certain age, possess similar qualities.

The opinion advanced by Paris,* that the flesh of wild animals is more dense than the domesticated, does not correspond with the experience of those who are well acquainted with the former. Although, on making a section of wild flesh, such appearance may be indicated, yet the fibres are found to be more easily separated by mastication, or other force, and are generally tender; at least, such is the case with the flesh of those animals that are considered luxuries by the epicure. Compare, for example, the flesh of the wether and the deer, animals which have a near correspondence in their habits, and the difference will be very obvious.

The digestibility of most meats is improved by incipient putrefaction, sufficient to render the muscular fibre slightly tender.

Vegetables are generally slower of digestion than meats and fari-

*Paris On Diet, p. 72.

naceous substances, though they sometimes pass out of the stomach before them, in an undigested state. Crude vegetables, by some law of the animal economy, not well understood, are allowed, even when the stomach is in a healthy state, sometimes to pass the pyloric orifice, while other food is retained there to receive the solvent action of the gastric juice. This may depend upon their comparative indigestibility; for it is well known that cathartic medicines, various fruits, seeds, &c. which operate as laxatives, are not digested; are incapable of being retained in the stomach; and pass rapidly through the intestinal tube. When such articles are in excess, they produce considerable derangement, and sometimes fatal consequences.

Vegetable, like animal substances, are more capable of digestion in proportion to the minuteness of their division, as I have before remarked, provided they are of a soft solid; and I cannot, therefore, concur in the opinion expressed by Paris,* that potatoes are better when only boiled so as to be rendered tender, and have their shape preserved, than when boiled to a "dry, insipid powder." They may be more palatable, and contain more nutriment; but they are not so easily affected by the gastric solvent. The difference is quite obvious on submitting parcels of this vegetable, in different states of preparation, to the operation of the gastric juice, either in the stomach or out of it. Boiled, or otherwise cooked to dryness, so as to be easily mashed, potatoes very readily become reduced to a chymous state, when submitted to the action of the gastric juice. When differently prepared, and only boiled so as to be rendered barely soft, moist and tenacious, with the shape preserved, entire pieces remain long undissolved in the stomach, and very slowly yield to the action of the gastric juice in vials on the bath. Pieces of raw potato, when submitted to the operation of this fluid, in the same manner, almost entirely resist its action. Many hours elapse before the slightest appearance of digestion is observable, and this only upon the surface, where the external laminæ become a little softened, mucilaginous, and slightly farinaceous. Every physician, who has had much practice in the diseases of children, knows that partially boiled potatoes, when not sufficiently masticated, (which is always the case with children,) are frequently a source of colics and bowel complaints, and that large pieces of this vegetable pass the bowels untouched by digestion.

These remarks will apply, also, to most other vegetable aliment.

---

*Paris On Diet, p. 75.

The variety of fish, which are generally used by the citizens of this country, may be regarded as easily susceptible of digestion. The lobster, crab, and some others of the testaceous tribe, are, perhaps, exceptions.

Solid food is sooner disposed of by the stomach than fluid, and its nutritive principles are sooner carried into the circulation. It has been observed, however, that the exhaustion from abstinence is quicker removed by liquid than solid aliment. This is undoubtedly true; and it may be accounted for on the ground of a general sympathy existing between the stomach and all the other parts of the body. It is only necessary, in proof of this fact, to appeal to the experience of almost every physician. The violent spasms, contortions, &c. affecting different and remote parts of the system, that sometimes supervene on the introduction of crude or indigestible food into the stomach, are pretty clear indications of the powerful sympathy that exists between it and other organs or apparatuses.

Condiments, particularly those of the spicy kind, are non-essential to the process of digestion, in a healthy state of the system. They afford no nutrition. Though they may assist the action of a debilitated stomach for a time, their continual use never fails to produce an indirect debility of that organ. They affect it as alcohol or other stimulants do—the *present* relief afforded is at the expense of *future* suffering. Salt and vinegar are exceptions, and are not obnoxious to this charge, when used in moderation. They both assist in digestion—vinegar, by rendering muscular fibre more tender—and both together by producing a fluid having some analogy to the gastric juice.

Drinks are nearly as essential to the animal system as the more substantial food. Though not subject to digestion, they enter into the circulation, and become important agents in the ultimate changes that are undergoing in the tissues of the organism. Simple water is, perhaps, the only fluid that is called for by the wants of the economy. The artificial drinks are probably all more or less injurious; some more so than others; but none can claim exemption from the general charge. Even coffee and tea, the common beverages of all classes of people, have a tendency to debilitate the digestive organs. Let any one who is in the habit of drinking either of these articles in a weak decoction, take two or three cups made very strong, and he will soon be aware of their injurious tendency. And this is only an *addition* to the *strength* of the narcotic he is in the constant habit of using. The whole class of alcoholic liquors, whether simply fermented, or distilled, may be considered as *narcotics,* producing very little difference in their ultimate effects on the system.

The injury which a constant use of wine is known to produce on some stomachs, has been sometimes attributed to the small quantity of tartaric acid which it contains. But it is not the cream of tartar that renders wine so deleterious to many stomachs. It is the acidity produced by the acetous fermentation of the saccharine matter contained in the wine, aided, perhaps, by the alcohol which is in a state of combination with it. Beer has the same effect on the same idiosyncracies, or diseased states of the stomach. Besides, both of these fluids are in a partial stage of acetous fermentation, which is consummated by the increase of temperature in the stomach.

It would be a task of great difficulty to designate the exact kind of diet that would, if generally adopted, be the most conducive to health and longevity. A considerable variety seems to be necessary to man, in a state of civilization. This want of variety is induced by long habit, which it would probably be unsafe to break through. Whether man was originally carnivorous or granivorous, is a question which we cannot solve, and perhaps it is not worth the attempt; at present he is both, and with his present mode of existence we have to do.

The *quantity* of aliment is probably of more importance than the *quality,* to ensure health. The system requires much less than is generally supplied to it. The stomach disposes of a definite quantity. If more be taken than the actual wants of the economy require, the residue remains in the stomach, and becomes a source of irritation, and produces a consequent abberration of function, or passes into the lower bowels in an undigested state, and extends to them its deleterious influence. Dyspepsia is oftener the effect of over eating and over drinking than of any other cause.

SECTION II

Of Hunger and Thirst

Hunger is a painful sensation, referred to the region of the stomach. It is a kind provision of nature, designed to remind man, and other animated beings, of the necessity of replenishing the wastes of the system, as well as contributing to its growth. Much enquiry has been made on this subject, and many theories have been given to account for the phenomenon. It has been supposed by some, that the friction of the internal coats of the empty stomach was the cause of the sensation. This opinion is liable to several objections:—1st. A healthy stomach digests its contents in from one to three or four hours, and hunger is not usually experienced until some time after the latter period. If hunger be the effect of the friction of the parieties of the stomach, it ought to be experienced the moment that that organ has disposed of its contents. 2d. In nausea and vomiting, the stomach is brought into a situation, according to this theory, to experience the sensation of hunger; and yet we know how opposed it is to receiving any thing like food. 3d. In gastritis and fevers the sensation hardly ever occurs, though very little food shall have occupied the stomach for a long time—perhaps not for weeks. This organ, under such circumstances, is generally empty and irritable, yet the peculiar sensation in question hardly ever supervenes. Besides, hunger sometimes occurs when the stomach is partially or wholly filled. The potation of spirits or brandy and water, and some other indigestible substances of a liquid character, does not remove the sensation, although by this means the parieties of the stomach are as completely separated as by food.

It has also been suggested that the sensation of hunger is produced by the irritation of a quantity of gastric juice in the stomach, which, by its stimulus, excites the feeling. The principal objection to this doctrine is based upon the fact that the stomach contains no gastric juice, or, at any rate, but a very small quantity, in its empty state, or when aliment or other irritant is not present. Besides if it were true that it contained a quantity of the fluid, such fluid does not possess the power of producing any thing like irritation or inflammation of its coats. It is as innoxious to the stomach, as the blandest substance in nature. It exerts its influence on free aliment, but not on the living fibre.

By referring the sensation to "an energetic state of the gastric nerves, occasioned by an interval of inactivity, during which the vital powers may be supposed to accumulate,"* it appears to me that we are venturing upon unexplored grounds, of which we know but little. We are not accustomed to call those painful nervous sensations to which the system is sometimes subject, states of high nervous energy. Are they not rather states of nervous debility? or, at any rate, irregular and unhealthy motions?

That the introduction of narcotics into the stomach should destroy the appetite, proves only that they have the same effect on that organ as they have on other parts of the body; they paralyse the nerves, and render them incapable of being the media of communication to their common centre.

Many other causes have been assigned for this sensation, equally wide, probably, of the true one. It has been attributed to the "foresight of the vital principle," a phrase that means any thing, every thing, or nothing, according to the construction which each one may put upon it. Such explanations conduce nothing to the promotion of science. They are mere sounds and words, which ingeniously convey a tacit acknowledgement of their author's ignorance.

Again, the mechanical action of the liver upon the diaphragm, has been accused of producing the sensation of hunger. Some proof, more than mere assertion, is necessary to convince honest enquirers that so remote a cause should produce such effects on the stomach, the immediate seat of the feeling. Of the same nature, is the opinion of the fatigue of the contracted fibres of the stomach, or of compression of the nerves of that organ, &c. &c.

Magendie, convinced that all the theories on this subject were unsatisfactory, comes to the following comprehensive conclusion: that "Hunger is produced like all other internal sensations, by the action

*Paris On Diet, p. 55.

of the nervous system, and it has no other seat than in this system itself, and no other causes than the general laws of organization."* I cannot perceive that such explanations bring the mind to any satisfactory understanding of the subject. In such broad propositions, it is difficult to ascertain the exact meaning. If the design is to convey the impression that hunger has no "local habitation;" that it is an impression, affecting all the nerves of the system in the same manner; then the sensation would be as likely to be referred to one organ as another. It is true, that without nervous communication there would be no sensation at all. This applies as well to other parts as to the stomach. The nerves are the media of communication from the sensible parts to the centre of perceptions. They warn the encephalon not only of the injuries, but of the wants of the tissues. We are accustomed to refer local sensations and irritations to the parts *apparently* affected—desire for urination and defecation, to the bladder and rectum; for liquids, to dryness of the mouth and fauces: and we account, in like manner, for other physiological and pathological sensations. When we can arrive at the exact interpretation of an author, who says that hunger has "no other causes than the general laws of organization," it will then be time to give reasons for an assent to or dissent from the proposition.

This subject is, unquestionably, involved in considerable doubt and obscurity, and will not, it is to be apprehended, admit of a very speedy elucidation. The Author of Nature is perfect in all His works; and although we may not understand all the operations of His hands, we are compelled to acknowledge their wisdom, propriety and beauty. Man would be miserable and wretched indeed, if he depended solely on his own discretion and judgment to decide upon the quantity and quality of aliment necessary to supply the wastes, and administer to the growth, of the system. This paucity of judgment and discretion is, however, more than compensated by an irresistible sensation, which indicates the proper time for the reception of food. The immediate cause of this sensation, as we have seen, has not as yet received a very satisfactory explanation, and perhaps will not admit of one. But, although confessedly obscure, we are not denied the privilege of patient investigation, and persevering search after truth. Knowledge is progressive, as well in this as every other science; and every new discovery, and every rational hypothesis, are additions to

*Summary of Physiology, p. 196.

the general stock. Persuaded of the truth of these general propositions, and anxious mainly to elicit investigation of the subject, I submit the following *Theory of Hunger,* believing it to be as reasonable, to say the least, as any that has been propagated.

My impression is, that the sensation of hunger is produced by a *distention* of the gastric vessels, or that apparatus, whether vascular or glandular, which secretes the gastric juice; and is believed to be the effect of repletion by this fluid.

One reason, among others, for this belief, is the established fact, that the internal sensations referred to different organs, as has been previously alluded to, are caused by some modified action or condition of the parts in the tissues of the organ itself. The modification in the parts to which the sense of hunger is invariably referred, I conceive to be a distention, by the gastric juice, of a particular set of vessels or glands, constituting, in part, the erectile tissue of the villous coat of the stomach. The sensation varies according to the different degrees or states of distention, from the simplest desire to the most painful sense of hunger; and is allayed or increased in proportion to the application, or refusal, of alimentary stimulus to the excretory vessels. The greater the distention of the vessels, the more acute will be the pain: hence, the difference between a short and protracted fast. Appetite and hunger belong to the same class of sensations; they differ only in degree. In this they are like all other sensations. A little increased circulation in the vessels of the brain produces peculiarly vivid, but not absolutely unpleasant feelings, and gives force and energy to the mental volitions: carried further, it produces most painful sensations. It is unnecessary to cite further examples. Indeed, it does not need arguments to prove what is the subject of every day's observation. It is well known that the pain from acute inflammation is produced by distention of the blood vessels. Let any one, who is disposed to try the effect of vascular distention, place a ligature around the finger or arm, sufficiently tight to retard the returning blood, and the truth will be sufficiently obvious.

It is, therefore, inferred from the pain, (and no one, it is believed, will deny that *hunger* is a painful sensation, whatever may be his opinion of *appetite*) that vessels of some kind are distended; and it is demonstrated, I think, in some of the following experiments, that these are the gastric vessels. On applying aliment to the internal coat of the stomach, which, in health, is merely lubricated with mucus, innumerable minute papillæ, the orifices, undoubtedly, of the gastric vessels, immediately throw out a quantity of the fluid, which mixes with the food. This effect is too sudden, and the secretion too copious, to be accounted for on the ordinary principles and laws of secreting mucous surfaces. The quiescence and relief from the unpleasant sensation, which are experienced as soon as the vessels are emptied, are, I think, additional proofs of my opinion. It is certain, that at the introduction of every meal, or on the application of alimentary stimulus to the internal coat of the stomach, a very large secretion of a fluid, which has repeatedly been ascertained to be an alimentary solvent, immediately takes place; and that when the stomach is destitute of food or some other irritating substance, no such secretion can be found in it. And it is more than probable—it, in fact, almost amounts to demonstration, that a large quantity of this fluid must be contained in appropriate vessels, during a fast, ready to obey the call of aliment. I would not be understood to say that the whole quantity necessary for an ordinary meal is eliminated from the blood, previous to the commencement of alimentation; but that enough is contained in the gastric vessels to produce the sensation of pain or hunger.

If it be objected to this theory, that the vessels would become ruptured, or empty themselves into the cavity of the stomach, during a long fast, I reply, that this apparatus is probably constituted like many of the other organs of the system, and permits the absorption of its secretions by the lymphatic or other absorbent vessels. The male semen is constantly being secreted, and deposited in its proper seminal vessels, ready to be ejected during the venerial orgasm; and yet how many men live for years, or perhaps for a whole life, who have no intercourse with the other sex. What becomes of the semen under these circumstances? Taken up, unquestionably, by the absorbing vessels, as the gastric juice of the stomach is.

I offer this theory for consideration, persuaded that the public will allow it such weight as it may have a right to claim: more than this, I have no wish to ask.

Thirst. This sensation is felt in the mouth and fauces. Like hunger, it is a kind provision of nature, designed to remind men and animals of the necessity, not of replenishing the wasting solids of the system, but of diluting the fluids that are carrying on these processes. Although Magendie has attempted to put a stop to all inquiries on this subject, in the remark, that "Thirst is an internal sensation, an instinctive sentiment;" "the result of organization, and does not admit of any explanation;" I apprehend a remote cause of this sensation may be found in the viscidity of the blood, which requires a liquid to render it more fluid, and more susceptible of introduction into the capillaries and secreting surfaces. The proximate cause may exist in an irritation, a kind of sub-inflammation of the mucous membranes of the mouth and fauces, the effect of the viscid state of the blood, and consequently impervious state of the secretory vessels of these membranes. The sensation of dryness, or thirst, is supposed to be the effect of evaporation, the mouth and throat being constantly exposed to the atmosphere. When there is sufficient fluidity of the blood, the secretion is so much more copious than the evaporation, that a constant moisture is preserved. The sensation of thirst resides in the tissues; and it is no more "an instinctive sentiment" than any other sensation of the economy. To say that it is the "result of organization," gives no explanation, amounts to nothing, and is certainly, to say the least, a very unsatisfactory way of disposing of the question.

SECTION III

Of Satisfaction and Satiety

In the present state of civilized society, with the provocatives of the culinary art, and the incentives of high seasoned food, brandy and wines, the temptations to excess in the indulgences of the table are rather too strong to be resisted by poor human nature. It is not less the duty, however, of the watchmen on the walls to warn the city of its danger, however it may regard the premonition. Let them at least clear their own skirts from the stain of unfaithfulness, whatever may be the result.

There is no subject of dietetic economy about which people err so much, as that which relates to *quantity*. The medical profession, too, have been accessory to this error, in giving directions to dyspeptics to eat until a sense of satiety is felt. Now, this feeling, so essential to be rightly understood, never supervenes until the invalid has eaten too much, if he have an appetite, which seldom fails him. Those, even, who are not otherwise predisposed to the complaint, frequently induce a diseased state of the digestive organs by too free indulgence of the appetite. Of this fact the medical profession are, generally, not sufficiently aware. Those who lead sedentary lives and whose circumstances will permit of what is called free living, are peculiarly obnoxious to these complaints. But by paying particular attention to their sensations during the ingestion of their meals, these complaints may be avoided. There appears to be a sense of perfect intelligence conveyed from the stomach to the encephalic centre, which, in health, invariably dictates what quantity of aliment (responding to the sense of hunger, and its due satisfaction,) is naturally required for the purposes of life; and which, if noticed, and properly attended to, would prove the most salutary monitor of health, and effectual preventive of, and restorative from, disease. It is not the sense of *satiety,* for this is beyond the point of *healthful* indulgence, and is nature's earliest indication of an *abuse* and *overburthen* of her powers to replenish the system. It occurs immediately previous to this, and may be known by the pleasurable sensation of *perfect satisfaction, ease and quiescence of body and mind.* It is when the stomach says *enough,* and is distinguished from satiety by the difference of the sensations—the former feeling *enough*—the latter, *too much.* The first is produced by the timely reception into the stomach of proper aliment, in exact proportion to the requirements of nature, for the perfect digestion of which, a definite quantity of gastric juice is furnished by the proper gastric apparatus. But to effect this most agreeable of all sensations and conditions—the real Elysian satisfaction of the *reasonable* Epicure—timely attention must be paid to the preliminary processes, such as thorough mastication, and moderate or slow deglutition. These are indispensable to the due and natural supply of the stomach, at the stated periods of alimentation; for if food be swallowed too fast, and pass into the stomach imperfectly masticated, too much is received in a short time, and in too imperfect a state of preparation, to be disposed of by the gastric juice.

The quantity of gastric juice, either contained in its proper vessels, or in a state of preparation in the circulating fluids, is believed to be in exact proportion to the proper quantity of aliment required for the due supply of the system. If a more than ordinary quantity of food be taken, a part of it will be left undissolved in the stomach, and produce the usual unpleasant symptoms of indigestion. But if the ingestion of a large quantity be in proportion to the calls of nature, which sometimes happens after an unusual abstinence, it is probable that more than the usual supply of gastric juice is furnished; in which case the apparent excess is in exact ratio to the requirements of the economy; and never fails to produce a sense of quiescent gratification, and healthful enjoyment. A great deal depends upon habit, in this respect. Our western Indians, who frequently undergo long abstinences from food, eat enormous quantities, when they can procure it, with impunity.

Satiety is produced by tendering too much at once for the wants of the economy; more than the gastric juice is able to dispose of at the time; distending the muscular fibers beyond that point so admirably fixed, by the invariable and universal laws of the animal system, for agreeable sensations; disturbing the peculiarly pleasurable, undulatory motions of the rugæ of the stomach, in their operations of forming chyme; and *perhaps,* interrupting, if not diminishing, the secretion of the gastric juice. The redundant aliment, incapable of being dissolved, for want of sufficient gastric juice, remains, and becomes a source of irritation, and renders imperfect the chymification of that which would otherwise have been completed. Hence the sense of weight, and disagreeable fullness, attendant on an unusually hearty meal; the subsequent derangement of the digestive functions, and consequent acidities and vitiated contents of the prima viae, from acetic fermentation in the stomach, and imperfect formation of chyle in the intestines.

SECTION IV

Of Mastication, Insalivation and Deglutition

THESE ARE THE PRELIMINARY STEPS in the process of digestion. The comparative importance of these processes has been elevated or depressed, according to the preponderance which each of them may have received from the opinions of the different physiologists who have made them subjects of observation. As man and animals are constituted, they are all absolutely necessary to the digestion of food. But in an abstract point of view, disconnected as a mean of introducing ingestæ into the stomach, I believe I hazard nothing in saying that they may be considered as perfectly non-essential to chymification. If the *materia alimentaria* could be introduced into the stomach in a finely divided state, the operations of mastication, salivation and deglutition, would not be necessary. Aliment is as well digested and assimilated, and allays the sensation of hunger as perfectly, when introduced directly into the stomach, in a proper state of division, as when the previous steps have been taken, as may be seen by some of the following experiments. If particular importance is to be attributed to any of these previous steps, it is certainly due to mastication; though an undue importance has, of late, been given to the action of the *saliva.* Professor Jackson, of Philadelphia, who has lately published a physiological work on the "Structure and Functions of the Animal Organism," has elevated saliva to a rank in the process of digestion, seldom before claimed for it. He considers it the principal solvent, or macerating agent, of alimentary matter. He is sustained in this opinion by Montegre and others. Even Magendie is inclined to favour this belief.

It is remarked by Paris, (On Diet, p. 37,) that the introduction of saliva into the stomach is "obviously essential to a healthy digestion." That it is generally introduced into the stomach with the food is very obvious; the nature of its action is not so clear. In most of the experiments that follow, artificial digestion was performed without the admixture of saliva. Chyme formed in this way, exhibited the same sensible appearances, and was affected by re-agents in the same way, as that which was formed from food which had been previously masticated, mixed with the saliva, and swallowed. It would seem, from two or three of the experiments on artificial digestion, which were instituted for the purpose of comparison, that the mixture of saliva with the gastric juice rather retarded its solvent action. But I do not wish to deny the utility of the saliva. It is certainly important as a preliminary to digestion. Its legitimate and only use, in my opinion, is to lubricate the food, and to facilitate the passage of the bolus through the organs of deglutition. In this point of view, it is essential. Dry food cannot be swallowed until it receives an admixture of a fluid, whether it be saliva or some other liquid, is not, I conceive, a matter of much importance. Any one, disposed to try the experiment, may satisfy himself of this fact, by attempting to swallow a mouthful of dry cracker, meal or magnesia. He will find it impossible to make the organs of deglutition act till a quantity of fluid is mixed with it. Water will answer the purpose, nearly as well as saliva; though the mucous properties of this secretion may give it a slight preference.

Pathology is not, in my opinion, much indebted to Ruysch, who attributed the loss of appetite to the waste of saliva in a person who was afflicted with a fistula in one of the salivary ducts; nor to the opinion advanced by others, that the constant spitting of maniacal patients, induces loss of appetite. The truth is, that in both cases, the effects are attributed to the wrong causes. There is no difficulty in believing that a foul ulcer in the mouth would be liable to produce nausea and want of appetite; nor that maniacal patients are generally, if not always, affected with diseased organs of digestion. I have known many persons to spit freely and constantly whose appetites and digestion were perfect. Those who smoke tobacco are constantly discharging large quantities of saliva; and yet I am not aware that dyspepsia is more common with them than with others.

I entirely dissent from the opinion advanced by the author above referred to, (Paris,) that "Insalivation is as essential as mastication." The use of mastication is to separate the food into small particles, so that the solvent of the stomach may be applied to a greater extent of surface. There is no mystery about this. Every body knows that the smaller the particles of matter that are submitted to the action of a chemical agent, the more vigorously the agent will act upon them, and the sooner they will be dissolved, or decomposed.* Mastication is absolutely necessary to healthy digestion. If aliment, in large masses, be introduced into the stomach, though the gastric juice may act upon its surface, chymification will proceed so slowly, that other changes will be likely to commence in its substance before it will become completely dissolved. Besides, the stomach will not retain undigested masses for a long time, without suffering great disturbance. It is governed by certain laws with respect to aliment. After food has been retained for a certain length of time undigested, say from five to ten hours, according to the healthy or diseased state of that organ, or the quantity received into it, it is either rejected by vomiting, or is permitted to pass into the duodenum and lower bowels, where its presence almost invariably produces colic, flatulence, &c. When the stomach is unusually debilitated, food, however, is frequently retained for twenty-four hours or more, and is sometimes the cause of most distressing symptoms, producing, particularly in children, convulsions and death. I therefore consider mastication as one of the most important preliminary steps in the process of digestion.

With respect to deglutition, I shall make but a few remarks. It is important for the preservation of health, that this process should be affected slowly. If food be swallowed rapidly, more will generally be taken into the stomach before the sensation of hunger is allayed, than can be digested with ease. If due attention be paid to the previous step of thorough mastication, we shall not be so likely to err in this later one.

Swallowing very rapidly, produces irregular contractions of the muscular fibres of the œsophagus and stomach; disturbs the vermicular motions of the rugæ, and interrupts the uniform action of the gastric apparatus.

The stomach is not designed to receive more food than can be duly mixed with the gastric solvent, already in its proper vessels, or in a state of preparation in the blood vessels. Perfect harmony of action must exist throughout the whole apparatus, or derangement of healthy action will ensue.

---

*In using the word *solvent* or *solution,* in reference to the gastric juice, I wish to be understood to mean a chemical action, analogous to that of the action of mineral acids on the metals; not like the solution of sugar or salt in water.

The stomach will not admit of the introduction of food, even of a liquid kind, through the aperture, at a rapid rate. If a few spoonfuls of soup, or other liquid diet, be put in with a spoon or funnel, the rugæ gently close upon it, and gradually diffuse it through the gastric cavity, entirely excluding more during this action. When a relaxation takes place, another quantity will be received in the same manner.

If the valvular portion of the stomach be depressed, and solid food be introduced, either in entire pieces, or finely divided quantities, the same gentle contraction, or grasping motion, takes place, and continues for fifty or eighty seconds; and will not allow of the introduction of another quantity until the above time has elapsed; when the valve may again be depressed, and more food be put in. Food and drinks will be received through the aperture no faster, even when the stomach is entirely empty, than they are ordinarily received through the œsophagus.

When the subject of these experiments is so placed that the cardia can be seen, and he be allowed to swallow a mouthful of food, the same contraction of the stomach, and closing upon the bolus, is invariably observed to take place at the œsophageal ring.

SECTION V

Of Digestion by the Gastric Juice

Chymification is effected in the stomach. It is the first stage, proper, of the conversion of aliment into blood; though in the ordinary course of proceeding, as animals are constituted, some previous steps are necessary. After the aliment has been received into the stomach, it is subjected to certain evolutions, or motions, propagated by the muscular fibres of that organ; and is acted upon through the agency of some principle, which changes it from a heterogeneous mixture of the various kinds of diet, submitted to its action, to an uniform, homogeneous semi-fluid, possessing properties distinct from the elements of which it was composed. The length of time consumed in the operation is various. It depends upon the quantity or quality of the ingestæ, the healthy or diseased state of the stomach, &c. In the various experiments which I have made, the medium time may be calculated at about three and a half hours.

It has been suggested by many physiologists, and positively asserted by some, that there is considerable increase of the temperature of the stomach during the digestion of a meal. But from the result of a great number of experiments and examinations, made with a view of ascertaining the truth of this opinion, in the empty and full state of the organ, and during different stages of chymification, I am convinced that there is no alteration of temperature, unless some other circumstance should produce it. Active exercise always elevates the temperature of the stomach, whether fasting or full, about one and a half degrees.

With respect to the agent of chymification, that principle of life which converts the crude aliment into chyme, and renders it fit for the action of the hepatic and pancreatic fluids, and final assimilation and conversion into the fluids, and the various tissues of the animal organism—no part of physiology has, perhaps, so much engaged the attention of mankind, and exercised the ingenuity of physiologists. It has been a fruitful source of theoretical speculation, from the father of medicine down to the present age. It would be a waste of time to attempt to refute the doctrines of the older writers on this subject. Suffice it to say, that the theories of *Concoction, Putrefaction, Trituration, Fermentation* and *Maceration,* have been prostrated in the dust before the lights of science, and the deductions of experiment. It was reserved for SPALLANZANI to overthrow all these unfounded hypotheses, and to erect upon their ruins, a theory which will stand the test of scientific examination and experiment. He established a theory of CHEMICAL SOLUTION, and taught that chymification was owing to the solvent action of a fluid, secreted by the stomach, and operating as a true menstruum of alimentary substances. To this fluid he gave the name of GASTRIC JUICE. It does not come within the scope of this work to give a detail of the experiments and reasoning which wrought conviction in the mind of this great man. It is only necessary to say that it was the result of patient and persevering experiment and research.

The truth of SPALLANZANI's theory has been sustained, so far as relates to the most important part, the existence of a chemical solvent, by all who have made fair examinations and experiments on the subject. The experiments of TIEDEMANN and GMELIN, of LEURET and LASSAIGNE, confirm the same theory.

By far the most respectable and intelligent physiologists have now settled down in the belief that chymification is effected in the stomach, by a peculiar and specific solvent, secreted in that organ, called, after SPALLANZANI, the Gastric Juice. From the difficulty, however, of obtaining and submitting such fluid to the test of experiment, and the diversity of results in the examination of such as has been obtained, much indefiniteness is experienced on this subject. The presence of an active solvent is rather an admission on their part—a conclusion from the effect to the cause. BROUSSAIS, speaking on this subject, says: "It remains for us to know whether the portion of mucous membrane, belonging to the stomach, contains secretory organs, the office of which is to furnish a fluid, fit to produce the assimilation of nutritive substances." And, again, speaking of the gastric juice. "The question is as yet *undecided,* though, if we are to judge by analogy, we shall observe that many animals are furnished with gastric glands, supplying a digestive liquid." This author *admits* the presence of a solvent fluid in the stomach, without, however, attempting to explain its specific effects, or mode of operation; for he says, in another place, "We have expressed our opinion on this subject; but whether the gastric fluids possess an assimilating property, which, for ourselves, we admit, without pretending to demonstrate its actual presence," &c.

RICHERAND, BOSTOCK, and nearly all the authors of modern date, teach the doctrine of digestion by the gastric juice, without, however, pretending to explain its exact mode of operation. Professor DUNGLISON, whose work on "Human Physiology," taken as a whole, is, perhaps, the most comprehensive, arrives at the same conclusion. He says, "We have too many evidences in favour of the chemical action of some secretion from the stomach during digestion, to permit us to doubt for a moment of the fact." And, again—"From all these facts, then, we are justified in concluding that the food in the stomach is subjected to the action of a secretion, which alters its properties, and is the principal agent of converting it into chyme."

I have referred to these learned authors with the view of showing the exact state of the science on this subject.

Though the theory of chymification by the gastric juice, has become almost universal with physiologists, and the medical profession in general, still there are some, even of very modern date, who, with all the lights of science and experiment, from aversion to the slow and tedious processes by which truths are attained, or, perhaps, from the ambition of becoming the discoverers of some new and extraordinary process, or the projectors of some fanciful theory, deny the power of the gastric juice, or even the existence of such a fluid; and set at naught the experiments, observations and opinions of the ablest physiologists, and most experienced writers on this subject.

That chymification is effected by the *solvent* action of the gastric juice, aided by the *motions* of the stomach, and the natural *warmth* of the system, not a doubt can remain in the mind of any candid person, who has had an opportunity to observe its effects on alimentary substances, or who has the liberality to credit the opinions of those who have had such opportunities.

It has been objected to this hypothesis, that the *sensible* properties of the gastric juice contradict the opinion of its active *solvent* effect. But we should recollect that many things which make very little impression on our external senses, produce most astonishing effects in other situations. The air which we breathe, by which we are surrounded, and which, to our external senses, is almost inappreciable, is one of the most powerful and destructive agents in nature—one portion of which is capable of combining with all grades of matter, either slowly and imperceptibly, as in the gradual decay of all substances, or rapidly, as in the combustion of wood, or even the hardest metals—and which by means, inexplicable to us, sustains in life and being the whole of animated nature.

The gastric juice has been submitted to chemical examination and analysis, with various results. Perhaps in the present state of the science of chemistry it will not be practicable to ascertain its exact chemical character. The parcels heretofore submitted to analysis, have been very impure; but the result of even these partial examinations, has been to show that this fluid contains a portion of free muriatic acid, combined with the acetic, and some salts. In the winter of 1832-3, I submitted a quantity of gastric juice, with no other admixture, except a small proportion of the mucus of the stomach, to Professor DUNGLISON, for examination, who, with the assistance of the professor of chemistry of the Virginia University, effected the following analysis, and was kind enough to communicate the result to me by letter.

"UNIVERSITY OF VIRGINIA,
Feb. 6th, 1833.

"MY DEAR SIR,

"Since I last wrote you, my friend and colleague, Professor Emmett, and myself, have examined the bottle of gastric fluid which I brought with me from Washington, and we have found it to contain free *Muriatic* and *Acetic* acid, *Phosphates* and *Muriates,* with bases of *Potassa, Soda, Magnesia* and *Lime,* and an *animal matter, soluble in cold water, but insoluble in hot.* We were satisfied, you recollect, in Washington, that free muriatic acid was present, but I had no conception it existed to the amount met with in our experiments here. We distilled the gastric fluid, when the free acid passed over; the salts and animal matter remaining in the retort. The quantity of Chloride of Silver thrown down on the addition of the Nitrate of Silver, was astonishing."

I had been long convinced of the existence of free muriatic acid in the gastric fluids. Indeed, it is quite obvious to the sense of taste; and most chemists agree in this, however they may be at variance with respect to the other constituents. The analysis of Professors DUNGLISON and EMMETT is certainly as satisfactory as any that has as yet been made. It is a question, too, whether gastric juice, in so great a state of purity, has ever before been submitted to chemical analysis.

It is to be hoped that no one will be so disingenuous as to attribute to Professor DUNGLISON the design of finding the existence of certain chemical agents in the gastric juice, with the view of propping the theory of the chemical action of this fluid, which he has maintained in his work on "Human Physiology;"—or, in other words, to say, that he had determined to find certain results; and that he had accordingly found them. Those who are acquainted with him, know that his candour and fairness are above the reach of suspicion; and that he would be equally willing to retract a false opinion as to maintain a correct one. Another quantity was sent to him for further analysis; but I regret that no report has yet been received from him.

In April of the present year, (1833,) a parcel was submitted to BENJAMIN SILLIMAN, M. D. Professor of Chemistry in Yale College. Professional engagements prevented his examination of the fluid until the 2d of August, when he sent me the following result:

"*Examination of the Gastric Fluid, Aug.* 2, 1833.

"1. The Fluid, after being kept in a closely corked vial, more than three months, from April to August, and most of the time in a cellar, remained unaltered, except the formation of a pellicle upon the surface, slightly discoloured by red spots. A second pellicle appeared after the precipitation of the first. It was thicker, and more discoloured with dark red spots, like venous blood.

"2. The Fluid was cloudy, like a solution of gum arabic; but on filtering, it became perfectly clear, and of a slight, straw yellow tinge.

"3. The pellicles, which had the appearance of inspissated mucus, after being separated from the Fluid, became, after exposure to the air, throughout of a brownish red colour, resembling the inner portion of a mass of coagulated blood. This change seemed to result from a sudden oxygenation.

"4. The Fluid exhaled a slight odour—not disagreeable—rather aromatic—and very similar to that which it at first exhaled; but not so strong. It was then rather disagreeable.

"5. Taste, feebly saline—not disagreeable.

"6. Test papers of litmus, alkanet, and purple cabbage, were decidedly reddened. Turmeric paper underwent no change: but when previously browned by an alkali, (ammonia) the gastric fluid restored the yellow colour.

"7. Nitrate of Silver gave a dense white precipitate, which, after standing five minutes in the sun's light, turned to a dark, brownish black; thus indicating Muriatic Acid. Mur. and Nit. Barytes gave a slight opalescence, indicating a trace of sulphuric acid; not improbably, there was also some phosphoric acid.

"8. Specific gravity—when taken in a small, thin glass tube, containing 201 grs. of distilled water—when filled with the gastric fluid, its weight was increased 1 gr.—weight of the gastric liquor, therefore, 202 grs. The specific gravity is, therefore, about 1.005. But little solid matter in solution."

At the insistence of Professor SILLIMAN, I committed to the care of Mr. GAHN, Consul of his Swedish Majesty in New-York, a bottle, containing one pint, of gastric juice, to be transmitted by him to Professor BERZELIUS, of Stockholm, one of the most eminent chemists of the age, with a request that he would favour me with an analysis. Some unavoidable delay was experienced in forwarding the bottle; and no returns have yet been received. It is hoped, however, that they will arrive in time to be attached in an appendix to this volume.

The following results have been obtained from partial examinations and analyses of the gastric juice, or rather, in most instances, of the *mixed fluids* of the stomach.

SPALLANZANI, in 1793, after many experiments, declared the gastric juice to be entirely *neutral,* a *solvent* for alimentary matter, *within* and *without* the stomach—that it did not *putrefy* at the ordinary temperature of the stomach; but preserved animal matters from putrefaction, and dissolved them, with the aid of heat.

SCOPOLI found in the gastric juice of the rook, water, gelatine, a saponaceous matter, muriate of ammonia and phosphate of lime.

CARMINITI, in 1795, found it, in carnivorous animals, salt and bitter, and frequently acid when they had eaten, but not so when fasting.*

VIRIDET, WERNER, HUNTER and others, found the gastric juice acid.

MM. MARQUART and VAUQUELIN found albumen and free phosphoric acid in it.

TIEDEMANN and GMELIN found it to contain, on analysis, muriatic and acetic acid; mucus; very little or no albumen; salivary matter; osmazome; muriate and sulphate of soda. In the ashes, carbonate, phosphate and sulphate of lime, and chloride of calcium. Principally from carnivorous animals.

LEURET and LASSAIGNE, in a hundred parts, found water, ninety-eight, lactic acid, muriate of ammonia, muriate of soda, animal matter soluble in water, mucus, and phosphate of lime, two parts.

MONTEGRE, (1812) *who could vomit at will,*† and who analyzed the fluid so obtained, declared it not to be *acid*—not a *solvent*—not *slow to putrefy*—so much like *saliva,* that he regards it saliva swallowed.

PROUT, 1824, declares the gastric juice to be really *acid*—does not contain an organic acid, but free, hydrochloric, or muriatic acid.

These opinions are certainly discordant. The majority of evidence, however, is in favour of the existence of pretty active chemical

---

*Probably because the fluid found in the stomach when fasting, was not gastric juice.
†See remarks near the close of this section on Montegre's experiments.

agents in the gastric fluids—perhaps not sufficient, in comparison with the ordinary operations of chemistry, to account for the digestion, or solution of aliment.

The discrepance of results in the reports of those who have had opportunities of examining the process of, and have made experiments on, *artificial digestion,* by the gastric juice, as well as in the chemical examination of this fluid, has been owing more to the difficulty of obtaining it pure, in sufficient quantity, and under proper circumstances, than to any real difference in its effects. Under the circumstances in which the following experiments were made, I flatter myself that these difficulties have been obviated; and if the inferences are incorrect, the blame must be attached to the experimenter. He can only say, that the experiments were made in good faith, and with a view to elicit facts.

I think I am warranted, from the result of all the experiments, in saying, that the gastric juice, so far from being "inert as water," as some authors assert, is the most general solvent in nature, of alimentary matter—even the hardest bone cannot withstand its action. It is capable, *even out of the stomach,* of effecting perfect digestion, with the aid of due and uniform degrees of heat, (100° Fahrenheit,) and gentle agitation, as will be seen in the following experiments.

The fact that alimentary matter is *transformed,* in the stomach, into chyme, is now pretty generally conceded. The peculiar process by which the change is effected, has been, by many, considered a problem in physiology. Without pretending to explain the exact *modus operandi* of the gastric fluid, yet I am impelled by the weight of evidence, afforded by the experiments, deductions and opinions of the ablest physiologists, but more by direct experiment, to conclude that the change effected by it on aliment is *purely chemical*. We must, I think, regard this fluid as a chemical agent, and its operation as a chemical action. It is certainly every way analogous to it; and I can see no more objection to accounting for the change effected on the food, on the supposition of a chemical process, than I do in accounting for the various and diversified modifications of matter, which are operated on in the same way. The decay of the dead body is a chemical operation, separating it into its elementary principles—and why not the solution of aliment in the stomach, and its ultimate assimilation into fibrine, gelatine and albumen? Matter, in a natural sense, is indestructible. It may be differently combined; and these combinations are chemical changes. It is well known that all organic bodies are composed of very few simple principles, or substances, modified by excess or diminution of some of their constituents.

The gastric juice appears to be secreted from numberless vessels, distinct and separate from the mucous follicles. These vessels, when examined with a microscope, appear in the shape of small lucid points, or very fine papillæ, situated in the interstices of the follicles. They discharge their fluid only when solicited to do so, by the presence of aliment, or by mechanical irritation.

Pure gastric juice, when taken directly out of the stomach of a healthy adult, unmixed with any other fluid, save a portion of the mucus of the stomach, with which it is most commonly, and perhaps always combined, is a clear, transparent fluid; inodoros; a little saltish; and very percepibly acid. Its taste, when applied to the tongue, is similar to thin mucilaginous water, slightly acidulated with muriatic acid. It is readily diffusible in water, wine or spirits; slightly effervesces with alkalis; and is an effectual solvent of the *materia alimentaria*. It possesses the property of coagulating albumen, in an eminent degree; is powerfully antiseptic, checking the putrefaction of meat; and effectually restorative of healthy action, when applied to old, fœtid sores, and foul, ulcerating surfaces.

Saliva and mucus are sometimes abundantly mixed with the gastric juice. The mucus may be separated, by filtering the mixture through fine linen or muslin cambric. The gastric juice, and part of the saliva will pass through, while the mucus, and spumous or frothy part of the saliva, remains on the filter. When not separated by the filter, the mucus gives a ropiness to the fluid, that does not belong to the gastric juice, and soon falls to the bottom, in loose, white flocculi. Saliva impats to the gastric juice, an azure tinge, and frothy appearance; and, when in large proportion, renders it fœtid in a few days; whereas the *pure* gastric juice will keep for many months, without becoming fœtid.

The gastric juice does not accumulate in the cavity of the stomach, until alimentary matter be received, and excite its vessels to discharge their contents, for the immediate purpose of digestion. It then begins to exude from its proper vessels, and increases in proportion to the quantity of aliment *naturally* required, and received. A definite proportion of aliment, only, can be perfectly digested in a given quantity of the fluid. From experiments on artificial digestion, it appears that the proportion of juice to the ingestæ, is greater than is generally supposed. Its action on food is indicative of its chemical character. Like other chemical agents, it *decomposes,* or *dissolves,* and combines with, a fixed and definite quantity of matter, when its action ceases. When the juice becomes *saturated,* it refuses to dissolve more; and, if an excess of food have been taken, the residue remains in the stomach, or passes into the bowels, in a crude state, and frequently becomes a source of nervous irritation, pain and disease, for a long time; or until the *vis medicatrix naturœ* restores the vessels of this viscus to their natural and healthy actions—either with or without the aid of medicine.

Such are the appearance and properties of the gastric juice; though it is not always to be obtained pure. It varies with the changing condition of the stomach. These variations, however, depend upon the admixture of other fluids, such as saliva, water, mucus, and sometimes bile, and, perhaps, pancreatic juice. The special solvent itself—the *gastric juice*—is, probably, invariably the same substance. Derangement of the digestive organs, slight febrile excitement, fright, or any sudden affection of the passions, cause material alterations in its appearance. Overburthening the stomach produces acidity and rancidity in this organ, and retards the solvent action of the gastric juice. General febrile irritation seems entirely to suspend its secretion into the gastric cavity; and renders the villous coat dry, red and irritable. Under such circumstances, it will not respond to the call of alimentary stimulus. Fear and anger check its secretion, also:—the latter causes an influx of bile into the stomach, which impairs its solvent properties.

When food is received into the stomach, the gastric vessels are excited by its stimulus to discharge their contents, when chymification commences. It has been a favourite opinion of authors, that food, after it has been received into the stomach, should "remain there a short period before it undergoes any change;"* the common estimate is one hour. But this is an erroneous conclusion, arising from inaccuracy of observation. Why should it remain there, unchanged? It has been received into the organ which is to effect an important change upon it—the gastric juice is ready to commence its work of solution soon after the first mouthful is swallowed; and, certainly, if we admit that the gastric juice performs the office of a chemical agent, which most physiologists allow, it is contrary to all our notions of chemical

*Paris On Diet, p. 39

action, to allow it one moment to rest. It must commence its operation immediately. That it does so, is distinctly manifested by close observation of its action on food, in the healthy stomach.

But PARIS is not alone in this opinion. It appears to have been a favourite doctrine; and has been regularly handed down, from one physiologist to another, as a sort of *heir loom* to the profession. The successors in the physiological sciences seem to have been compelled to receive it with the legacy of their predecessors, without any doubt of its legitimacy; when, with a little rational examination of the subject, it would have been found a fair subject of rejection. It will be seen, by the following experiments, that it has not the slightest foundation in truth; and to them I refer the reader.

It has been said, that when one meal follows another in quick succession—or, in other words, when a subsequent meal is taken before the previous one is digested—that it *some how* disturbs the process of digestion. This is generally true; and it allows of a definite solution. It is because more is received into the stomach, in the aggregate, than the gastric juice can dissolve. And this disturbance will result, as well when too much food has been taken at once, as when too much has been received in rapid succession. But if the quantity be moderate, no ill effect will ensue. Many children are in the habit of eating as often as once an hour through the day, in small quantities, without experiencing any bad consequences. Cooks are, also, accustomed to the practice of constantly tasting of the various articles of food which they are preparing for the table; and yet I am not aware that they suffer any inconvenience from the habit. From these, and other facts, as well as from direct experiment, I think it is perfectly apparent that digestion must progress as well before as after the expiration of an hour. If, as has been suggested, the ingestion of food, in addition to the delay to itself, retards or stops the chymification of that which had been previously received, aliment, as it relates to those children who eat hourly, would be constantly accumulating; and there would remain in the stomach at night the whole quantity taken through the day: a supposition not to be credited, even by those disposed to make the most of a favourite opinion or doctrine.

Doctor WILSON PHILIP, in his "Treatise on Indigestion," says, "the layer of food lying next to the surface of the stomach, is first digested, and in proportion as this undergoes the proper change, and is moved by the muscular action of the stomach, that next in turn succeeds, to undergo the same change." That chymification commences on the surface of the food, I have no doubt; but I apprehend this to be the case as it respects each individual portion, and not the whole mass. I have frequently taken out portions from the stomach, a few minutes after they had been received into that organ, when they appeared to have received a full supply of gastric juice for perfect digestion, when submitted to the artificial mode. When a due and moderate supply of food has been received, it is probable that the whole quantity of gastric juice for its complete solution, is secreted, and mixed with it, in a short time. When an unusually full meal has been eaten, the necessary quantity for its complete solution, is not so readily supplied. If a tenacious mass of food be used, the external portion of the whole quantity is first acted on, digested, and succeeding portions presented, &c. There is no ground for the opinion inferred, that the gastric juice never leaves the parietes of the stomach, except as it chymifies food. It is a thin fluid and is governed by the same laws that other thin fluids are. From numerous examinations of the stomach, I feel warranted in saying, at least in the human subject, that there is a perfect admixture of gastric juice and food—that the particles of food are constantly changing their relations with each other—and that they are mixed with a quantity of fluid, the gastric juice, liquids that have been taken during the meal, and (as there has generally been observed a large proportion of fluid, even after a dry and solid meal,) I have been led to suspect a synthetic formation of water, from its elements. This mixture is perfectly heterogeneous at first, and is kept in constant agitation, by the *churning* motions of the stomach. If the contents of the stomach be taken out in from thirty minutes to an hour after eating, it will be found to be composed of perfectly formed chyme and particles of food, intimately mixed and blended; sometimes in larger and sometimes in smaller proportions, according to the vigorous or enfeebled state of the digestive organs, or the quantity or quality of aliment taken. Most commonly, if the meal has been moderate, the process of digestion will continue in the portion taken out, when placed on the bath at a proper temperature, and the motions of the stomach imitated.

From the circumstance that the introduction of sponge, tubes, pebbles, &c. by SPALLANZANI and others, excited the discharge of the gastric juice, and from the fact that the gum-elastic tube, in my experiments, produced the same effect, when the stomach was empty and healthy, I infer, that the first effect of aliment on the stomach, is one of *irritation* of the gastric papillæ; thus exciting the discharge of the gastric juice, and stimulating the muscular fibres of the stomach. The vermicular motions, being excited by mechanical irritation, not only carry the ingestæ into all parts of the stomach, and diffuse its mechanical influence throughout the whole inner surface of this organ; but, by this means, they uniformly mix the aliment with the gastric juice, which is constantly being secreted, in proportion to the quantity of food received into the stomach, (unless that be too much for the wants of the economy,) until chymification be completed. Some stimulus seems to be necessary to continue the motions of the stomach, after chymification is accomplished, in order to effect its complete discharge into the lower bowels. And it appears highly probable that the compound fluid of gastric juice and aliment, or chyme, by its acquired acid properties, affords this stimulus, and propagates the contractile motions of this organ, even after the mechanical irritation of the crude food ceases. This fluid acquires new chemical properties, becomes more acid and stimulating, as chymification advances, until it is completed. When it is all transferred to the duodenum, the motions of the stomach cease.

From a number of experiments on rabbits, by Doctor WILSON PHILIP,* with the view of ascertaining the process of digestion, this gentleman has brought his mind to the conclusion, that when food has been taken at different times, "the new is never mixed with the old food." With every feeling of respect for so valuable and indefatigable a contributor to physiological science, I must beg leave, however, to dissent from this opinion. In many of his experiments, the rabbits were killed soon after the introduction of a fresh quantity of food, and, generally, of a very different kind. The result was, that it was found separate from the old food, which was in an advanced stage of digestion. It was in the centre of the old food, and surrounded by it. This is precisely where a new bolus would be received, and retain its shape and consistence, in some measure, until disturbed, and broken up, by the motions of the stomach. By allowing sufficient time for the action of this organ, it is probable that the line of separation would not have been perceived. Indeed the Doctor concedes that when the second quantity of food was of the same kind as the first, and the rabbit had been left to live for some time, the line of separation was

*On Digestion.

very indistinct. It appears that he fed rabbits on *oats,* and after making them fast for *sixteen* or *seventeen* hours, he fed them as much *cabbage* as they chose to eat, "and killed them at different periods, from *one* to *eight* hours after they had eaten it;" when the line of separation between the new food and that which had been eaten from *eighteen* to *twenty-five* hours before, was, no doubt, *quite distinct.* I confess I know very little about the habits of these animals, as it respects their modes of digestion; but I should be inclined to think that if the "line of separation" between the two portions of food were not sufficiently distinct, it was not for want of *time.* In man, one fifth of the time would have been more than sufficient to have disposed of any reasonable quantity of food.

Comparative physiology, as well as comparative anatomy, is undoubtedly, very useful; but, at the same time, it will not do to make it of general application. The rabbit is a ruminating animal; and is it not probable that the "new food," found in the "small curvature," if it be in fact retained there, is detained for the purpose of regurgitation and re-mastication, before it is digested? If the circumstance be true, and there be no deception in the case, I think this must be the design of the contrivance.

Arguments from analogy may be very plausible, and are certainly very allowable, when the subject presents no other mode; but they are not conclusive. We cannot judge of the mode of digestion in the human stomach by that of animals, particularly the granivorous and ruminating animals. Carnivorous animals most resemble man in their digestive apparatus. One thing is certain, and it is capable of demonstration in the stomach of the subject of these experiments, that old and new food, if they are in the same state of comminution, are readily and speedily mixed in the stomach.

On the subject of exercise or repose, during the digestion of a meal, there has been some diversity of opinion. It has generally been conceded, however, that a state of repose is most favourable to chymification. It has been said that during the digestion of aliment, the *energies* of the system were centred on the stomach, and should not be withdrawn to any distant part; that the stomach becomes a "centre of fluxion," &c. &c. I protest, again, against the use of terms which have no definite meaning. I believe the benefits of science will be better subserved by adhering to facts, and the deductions of experiment, than by the propagation of hypotheses founded on uncertain data. From numerous trials, I am persuaded that moderate exercise conduces considerably to healthy and rapid digestion. The discovery was the result of accident, and contrary to preconceived opinions. I account for it in the following way. Gentle exercise increases the circulation of the system, and the *temperature* of the stomach. This increase of temperature is generally about one and a half degrees. Now, if the gastric juice be a solvent, its action is similar to other chemical solvents, and its rapidity is increased in proportion to the elevation of temperature. Of the reason, I leave others to judge. The effect is certain. Severe and fatiguing exercise, on the contrary, retards digestion. Two reasons present themselves for this—the debility which follows hard labour, of which the stomach partakes; and the depressed temperature of the system, consequent upon perspiration, and evaporation from the surface.

Exercise, sufficient to produce moderate perspiration, increases the secretions from the gastric cavity, and produces an accumulation of a limpid fluid, within the stomach, slightly acid, and possessing the solvent properties of the gastric juice in an inferior degree. This is probably a mixed fluid, a small proportion of which is gastric juice.

Bile is not essential to chymification. It is seldom found in the stomach, except under peculiar circumstances. I have observed that when the use of fat or oily food has been persevered in for some time, there is generally the presence of bile in the gastric fluids. Whether this be a pathological phenomenon, induced by the peculiarly indigestible nature of oily food; or whether it be a provision of nature, to assist the chymification of this particular kind of diet, I have not as yet satisfied myself. Oil is affected by the gastric juice with considerable difficulty. The alkaline properties of the bile may render it more susceptible of solution in this fluid, by altering its chemical character. Irritation of the pyloric extremity of the stomach with the end of the elastic tube, or the bulb of the thermometer, generally occasions a flow of bile into this organ. External agitation, by kneading with the hand, on the right side, over the regions of the liver and pylorus, produces the same effect. It may be laid down as a general rule, however, subject to the exceptions above mentioned, that bile is not necessary to the chymification of food in the stomach. MAGENDIE says, "I believe that, in certain morbid conditions, the bile is not introduced into this organ," (the stomach;) inferring, that in a healthy state, it is always to be found there. There can hardly be a greater mistake. With the exceptions that I have mentioned, it is never found in the gastric cavity, in a state of health; and it is only in "certain morbid conditions" that it is found there.

When bile is found with the gastric juice, the acid taste is diminished, and the flavour of the bile prevails, in proportion to the quantity in the mixture.

The resulting compound of digestion in the stomach, or *chyme,* has been described as "a homogeneous, pultaceous, greyish substance, of a sweetish, insipid taste, slightly acid," &c. In its *homogeneous* appearance, it is invariable; but not in its *colour;* that partakes very slightly of the colour of the food eaten. It is always of a lightish or greyish colour; varying in its shades and appearance, from that of cream, to a greyish, or dark coloured gruel. It is, also, more consistent at one time than at another; modified, in this respect, by the kind of diet used. This circumstance, however, does not affect its homogeneous character. A rich and consistent quantity is all alike, and of the same quality. A poorer and thinner portion is equally uniform in its appearance. Chyme from butter, fat meats, oil, &c. resembles rich cream. That from farinaceous and vegetable diet, has more the appearance of gruel. It is invariably distinctly acid.

The passage of chyme from the stomach is gradual. Portions of chyme, as they become formed, pass out, and are succeeded by other portions. In the early stages, the passage of the chyme into the duodenum, is more slowly effected than in the later stages. At first, it is more mixed with the undigested portions of aliment, and is probably separated with considerable difficulty, by the powers of the stomach. In the later stages, as the whole mass becomes more and more chymified, and fitted for the translation, the process is more rapid; and is accelerated by a peculiar contraction of the stomach, a description of which will be found in the next section. It appears to be a provision of nature, that the chyme, towards the latter stages of its formation, should become more stimulating, and operate on the pyloric extremity of the stomach, so as to produce this peculiar contraction.

After the expulsion of the last particles of chyme, the stomach becomes quiescent, and no more juice is secreted, until a fresh supply of food is presented for its action, or some other mechanical irritation is applied to its internal coat.

Water and alcohol are not affected by the gastric juice. Fluids, of all kinds, are subject to the same exemption, unless they hold in solution or suspension some animal or vegetable aliment. Fluids pass from the stomach very soon after they are received, either by absorption, or through the pylorus.

Since the general adoption of the theory of a specific, solvent fluid, others have been proposed.

M. MONTEGRE, who, it is said, had the power of vomiting at pleasure, performed a series of experiments on the fluids of the stomach, obtained in this way, which induced him to come to a different conclusion on the subject of digestion. "He conceives that what has been supposed to be the gastric juice, is, in fact, nothing but *saliva;* that it possesses no peculiar powers of acting on alimentary matter: that the principal use of the gastric juice is to dilute the food; and that the only action of the stomach consists in 'une absorption vitale et elective,' in which the absorbent vessels, in consequence of their peculiar sensibility, take up certain parts of the food, and reject others."* A complete refutation of the conclusions drawn from the experiments of MONTEGRE, will be found in the fact, which has been tested by more than two hundred examinations and experiments, made by me, on the gastric cavity, that there never exists free gastric juice in the stomach, unless excited by aliment, or other stimulants. The fluid obtained by MONTEGRE was, in all probability, a mixture of saliva (which had been unconsciously swallowed) and the mucus of the stomach. Neither of these secretions are capable of digesting aliment; nor could the peculiar products, generally obtained from the chemical analysis of the gastric juice, be found in them.

The hypotheses proposed by Professors SMITH and JACKSON, of this country, are modifications of MONTEGRE's theory.

The former of these gentlemen supposes that digestion is performed "by the *veins* of the stomach, and by the *liver*." He contends, "that the first step in the process of digestion is effected by capillary veins originating in the villi of the stomach, with absorbing extremities, and terminating in the great branches of the vena portæ;"† that this action is continued through the small intestines; that the absorbing veins take up the nutrient principles of the food, and reject, as excrementitious, the innutritious part; that these nutrient principles are mixed with the returning blood within the cavity of the abdomen,

*Note in Bostock's Physiology, vol. 2, p. 384.
†Essay on Digestion, p. 63.

and are carried into the liver, where the final processes of animalization and conversion into blood are completed.

Professor JACKSON, in a recent work, has proposed a new theory, or rather revived, in some measure, the theory of maceration. His hypothesis, as nearly as can be collected from his work, is as follows:

He supposes that digestion is performed by submitting food to the action of different fluids, each of which has "solvent powers for different principles;"* that the nutrient principles exist already formed in food, and are released from principles that are not required for nutrition, by a species of solution, or maceration. The different fluids, as saliva, mucus from the mouth, throat, stomach, intestines, the bile and pancreatic juice, are the solvents of the different innutritive principles, and separate them from nutriment. He attributes great importance to the action of the saliva; thinks it exercises a "very energetic operation on the food," &c. and denies, altogether, the existence of a specific solvent fluid.

It is unfortunate for the interests of physiological science, that it generally falls to the lot of men of vivid imaginations, and great powers of mind, to become restive under the restraints of a tedious and *routine* mode of thinking, and to strike out into bold and original hypotheses to elucidate the operations of nature, or to account for the phenomena that are constantly submitted to their inspection. The process of developing truth, by patient and persevering investigation, experiment and research, is incompatible with their notions of unrestrained genius. The drudgery of science, they leave to humbler, and more unpretending contributors. The flight of genius is, however, frequently erratic. The bold and original opinions of BROWN, for a long time unsettled the practice of medicine; and the later opinions of MONTEGRE and others, have had a like effect on the sister science of physiology. It is, however, a right, which men of genius possess, in common with others, to propose hypotheses, and to support them with such arguments and deductions as they may have in their power to bring. Great caution and circumspection ought, however, to be observed. It is dangerous to unsettle long established truths; for it is difficult to limit the extent of error. The gratification of a *morbid* desire to be distinguished as the propagator of new principles in philosophy, or as the head of a new sect, is not the only result to be

*Principles of Medicine, founded on the Structure and Functions of the Animal Organism, p. 354.

expected from such heresies. New opinions or doctrines, whether true or false, will have admirers and followers, and will lead to practical results. And the errors of one man may lead thousands into the same vortex.

These, of course, are designed as general remarks; and I have no wish to apply them, so far as bad motives are inferred, to the highly respectable gentlemen mentioned above. Honest objections, no doubt, are entertained against the doctrine of digestion by the gastric juice. That they are so entertained by these gentlemen, I have no doubt. And I cheerfully concede to them the merit of great ingenuity, talents and learning, in raising objections to the commonly received hypothesis, and ability in maintaining their peculiar opinions. But we ought not to allow ourselves to be seduced by the ingenuity of argument or the blandishments of style. Truth, like beauty, when "unadorned, is adorned the most;" and in prosecuting these experiments and inquiries, I believe I have been guided by its light. Facts are more persuasive than arguments, however ingeniously made, and by their eloquence, I hope I have able to plead for the support and maintenance of those doctrines, which have had for their advocates such men as SYDENHAM, HUNTER, SPALLANZANI, RICHERAND, ABERNETHY, BROUSSAIS, PHILIP, PARIS, BOSTOCK, the Heidleburgh and Paris Professors, DUNGLISON, and a host of other luminaries in the science of physiology.

SECTION VI

Of the Appearance of the Villous Coat, and of the Motions of the Stomach

The inner coat of the stomach, in its natural and healthy state, is of a light, or pale pink colour, varying in its hues, according to its full or empty state. It is of a soft, or velvet-like appearance, and is constantly covered with a very thin, transparent viscid mucus, lining the whole interior of the organ.

Immediately beneath the mucous coat, and apparently incorporated with the villous membrane, appear small, spheroidal, or oval shaped, glandular bodies, from which the mucous fluid appears to be secreted.

By applying aliment, or other irritants, to the internal coat of the stomach, and observing the effect through a magnifying glass, innumerable minute lucid points, and very fine nervous or vascular papillæ, can be seen arising from the villous membrane, and protruding through the mucous coat, from which distills a pure, limpid, colourless, slightly viscid fluid. This *fluid,* thus excited, is invariably distinctly acid. The *mucus* of the stomach is less fluid, more viscid or albuminous, semi-opaque, sometimes a little saltish, and does not possess the slightest character of acidity. On applying the tongue to the mucous coat of the stomach, in its empty, unirritated state, no acid taste can be perceived. When food, or other irritants, have been applied to the villous membrane, and the gastric papillæ excited, the acid taste is immediately perceptible. These papillæ, I am convinced, from observation, form a part of what is called by authors, the villi of the stomach. Other vessels, perhaps absorbing as well as secretory, compose the remainder. That some portion of the villi form the excretory ducts of the vessels, or glands, I have not the least doubt, from innumerable, ocular examinations of the process of secretion of gastric juice. The invariable effect of applying aliment to the internal, but exposed part of the gastric membrane, when in a healthy condition, has been the exudation of the solvent fluid, from the above mentioned papillæ.—Though the *apertures* of these vessels could not be seen, even with the assistance of the best microscopes that could be obtained; yet the points from which the fluid issued were clearly indicated by the gradual appearance of innumerable, very fine, lucid specks, rising through the transparent mucous coat, and seeming to burst, and discharge themselves upon the very points of the papillæ, diffusing a limpid, thin fluid over the whole interior gastric surface. This appearance is conspicuous only during alimentation, or chymification. These lucid points, I have no doubt, are the termination of the excretory ducts of the gastric vessels or glands, though the closest and most accurate observation may never be able to discern their distinct apertures.

The fluid, so discharged, is absorbed by the aliment in contact, or collects in small drops, and trickles down the sides of the stomach, to the more depending parts, and there mingles with the food, or whatever else may be contained in the gastric cavity. This fluid, the efficient cause of digestion—the true gastric juice of SPALLANZANI, I have no doubt—has generally been obtained, for experiment, by mechanical irritation of the internal coat of the stomach, produced by the introduction of a gum-elastic tube, through which it has been procured.

The gastric juice never appears to be accumulated in the cavity of the stomach while fasting; and is seldom, if ever, discharged from its proper secerning vessels, except when excited by the natural stimulus of aliment, mechanical irritation of tubes, or other excitants. When aliment is received, the juice is given out in exact proportion to its requirements for solution, except when more food has been taken than is necessary for the wants of the system.

When mechanical irritation by a non-digestible substance, as the elastic tube, stem of the thermometer, &c. has been used, the secretion is probably less than when the irritation has been produced by such substances, as are readily dissolved in the gastric juice. Alimentary stimulus, when taken into the stomach, is diffused over the whole villous surface, and excites the gastric vessels, generally, to excrete their fluids copiously; whereas the irritation of tubes, &c. is local, and produces only a partial excitement of the vessels, and a scanty flow of the gastric juice. Hence, the slowness in obtaining the clear fluid from the empty stomach, through the tube. I have never, on numerous trials, been able to obtain, at any one time, more than one and a half, or two ounces of this fluid, after the stomach had disposed of its alimentary matters, however long the period of abstinence had been. The discharge of this small quantity has generally been excited by the introduction of the tube. Ten, fifteen, or more minutes, were necessary to collect even this small quantity. Whenever fluid was obtained in larger quantity, as was sometimes the case, it invariably contained more than the usual quantity of mucus.

On viewing the interior of the stomach, the peculiar formation of the inner coats are distinctly exhibited. When empty, the rugæ appear irregularly folded upon each other, almost in a quiescent state, of a pale pink colour, with the surface merely lubricated with mucus. On the application of aliment, the action of the vessels is increased; the colour brightened; and the vermicular motions excited. The small gastric papillæ begin to discharge a clear, transparent fluid, (the alimentary solvent,) which continues abundantly to accumulate, as aliment is received for digestion.

If the mucous covering of the villous coat be wiped off, with a sponge or handkerchief, during the period of chymification, the membrane appears roughish, of a deep pink colour at first; but in a few seconds, the follicles and fine papillæ begin to pour out their respective fluids, which, being diffused over the parts abraded of mucus, restore to them their peculiar soft and velvet-like coat, and pale pink colour, corresponding with the undisturbed portions of the membrane; and the gastric juice goes on accumulating, and trickles down the sides of the stomach again.

If the membrane be wiped off when the stomach is empty, or during the period of fasting, a similar roughness, and deepened colour appear, though in a less degree; and the mucous exudation is more slowly restored. The follicles appear to swell more gradually. The fluids do not accumulate in quantity sufficient to trickle down, as during the time of chymification. The mucous coat only, appears to be restored.

The foregoing, I believe to be the natural appearances of the internal coat of the stomach, in a healthy condition of the system.

In disease, or partial derangement of the healthy function, this membrane presents various, and essentially different appearances.

In febrile diathesis, or predisposition, from whatever cause—obstructed perspiration, undue excitement by stimulating liquors, overloading the stomach with food—fear, anger, or whatever depresses or disturbs the nervous system—the villous coat becomes sometimes red and dry, at other times, pale and moist, and loses its smooth and healthy appearance; the secretions become vitiated, greatly diminished, or entirely suppressed; the mucous coat scarcely perceptible; the follicles flat and flaccid, with secretions insufficient to protect the vascular and nervous papillæ from irritation.

There are sometimes found, on the internal coat of the stomach, eruptions, or deep red pimples; not numerous, but distributed, here and there, upon the villous membrane, rising above the surface of the mucous coat. These are at first sharp pointed and red; but frequently become filled with white purulent matter. At other times, irregular, circumscribed, red patches, varying in size or extent, from half an inch to an inch and a half in circumference, are found on the internal coat. These appear to be the effect of congestion in the minute blood vessels of the stomach. There are, also, seen at times, small aphthous crusts, in connection with these red patches. Abrasions of the lining membrane, like the rolling up of the mucous coat into small shreds or strings, leaving the papillæ bare, for an indefinite space, is not an uncommon appearance.

These diseased appearances, when very slight, do not always affect, essentially, the gastric apparatus. When considerable, and, particularly, when there are corresponding symptoms of disease, as dryness of the mouth, thirst, accelerated pulse, &c. no gastric juice can be extracted, not even on the application of alimentary stimulus. Drinks received, are immediately absorbed, or otherwise disposed of; none remaining in the stomach ten minutes after being swallowed. Food, taken in this condition of the stomach, remains undigested for twenty-four or forty-eight hours, or more, increasing the derangement of the whole alimentary canal, and aggravating the general symptoms of disease.

After excessive eating or drinking, chymification is retarded; and, although the appetite be not always impaired at first, the fluids become acrid and sharp, excoriating the edges of the aperture; and almost invariably produce aphthous patches, and the other indications of a diseased state of the internal membrane, mentioned above. Vitiated bile is also found in the stomach under these circumstances; and flocculi of mucus are much more abundant than in health.

Whenever this morbid condition of the stomach occurs, with the usual accompanying symptoms of disease, there is generally a corresponding appearance of the tongue. When a healthy state of the stomach is restored, the tongue invariably becomes clean.

Motions of the Stomach. With the *anatomy* of this organ, I have, at present, nothing to do. It does not come within the limits which I have prescribed to myself. Its *motions,* as comprising a part of the process of digestion, I have endeavoured to observe as accurately as practicable, and I give the result.

The human stomach is furnished with muscular fasciculi, so arranged as to shorten its diameter in every direction. By the alternate contraction and relaxation of these bands, a great variety of motion is induced on this organ, sometimes transversely, and at other times longitudinally. These alternate contractions and relaxations, when affecting the transverse diameter, produce what are called *vermicular* or *peristaltic* motions. The effect of the contraction of the longitudinal fibres, is to approximate the splenic and pyloric extremities. When they all act together, the effect is to lessen the cavity of the stomach, and to press upon the contained aliment, if there be any in the stomach. These motions not only produce a constant disturbance, or *churning* of the contents of this organ, but they compel them, at the same time, to revolve around the interior, from point to point, and from one extremity to the other. In addition to these motions, there is a constant agitation of the stomach, produced by the respiratory muscles.

These contractions and relaxations of the muscular fasciculi, do not observe any very *exact* mode. Their motions are modified by various circumstances, such as the stimulant or non-stimulant property of the ingestæ, the healthy or unhealthy state of the internal coat of the stomach; by exercise, and by repose, &c. &c.

The ordinary course and direction of the revolutions of the food, are first, after passing the œsophageal ring, from right to left, along the small arch; thence, through the large curvature, from left to right. The bolus, as it enters the cardia, turns to the left; passes the aperture; descends into the splenic extremity; and follows the great curvature towards the pyloric end. It then returns, in the course of the smaller curvature, makes its appearance again at the aperture, in its descent into the great curvature to perform similar revolutions.

Such I have ascertained to be the revolutions of the contents of the stomach, from being able to identify particular portions of food, and from the fact, that the bulb of the thermometer, which has been frequently introduced during chymification, invariably indicates the same movements. These revolutions are completed in from one to three minutes. They are probably induced, in a great measure, by the circular or transverse muscles of the stomach, as indicated by the spiral motion of the stem of the thermometer, both in descending to the pyloric portion, and ascending to the splenic.* These motions are slower at first than after chymification has considerably advanced.

While these revolutions of the contents of the stomach are progressing, the trituration or agitation is also going on. There is a perfect admixture of the whole ingestæ, during the period of alimentation and chymification. There is nothing of the distinct lines of separation between old and new food, and peculiar central or peripheral situation of crude, as distinguished from chymified aliment, said to have been observed by PHILIP, MAGENDIE and others, in their experiments on dogs and rabbits, to be seen in the human stomach; at least in that of the subject of these experiments. The whole contents of the stomach, until chymification be nearly complete, exhibit a heterogeneous mass of solids and fluids; hard and soft; coarse and fine; crude and chymified; all intimately mixed, and circulating promiscuously through the gastric cavity, like the mixed contents of a closed vessel, gently agitated, or turned in the hand.

If a mouthful of some tenacious food be swallowed, after digestion is considerably advanced, it will be seen passing the opening, to the great curvature; and in the course of about one and a half or two minutes, it will reappear, with the general circulating contents, more or less broken to pieces, or divided into smaller pieces; and very soon loses its identity. This agitating motion has the effect, and is undoubtedly designed, to break up the bolus, as well as to separate the external and chymified portion of the particles of food, and allow the undigested portions to come in contact with the gastric juice, their proper solvent. If the motions were simply revolutionary, the central portions would retain their situation, until the outer, or chymified part, had passed into the duodenum, in successive parcels; which, it is evident, would very much retard the process of digestion.

*The terms "descending" and "ascending," are used here, as well as in many other places, relatively; because the examinations were generally made while the man was lying on his right side.

As the food becomes more and more changed from its crude to its chymified state, the acidity of the gastric fluids is considerably increased; more so in vegetable than animal diet; and the general contractile force of the muscles of the stomach is augmented in every direction; giving the contained fluids an impulse towards the pylorus.

It is probable, that from the very commencement of chymification—from the time that food is received into the stomach—until that organ becomes empty, portions of chyme are constantly passing into the duodenum, through the pyloric orifice, as the mass is presented at each successive revolution. I infer this, from the fact that the volume is constantly decreasing. This decrease of volume, however, is slow at first; but is rapidly accelerated towards the conclusion of digestion, when the whole mass becomes more or less chymified. This accelerated expulsion appears to be effected by a peculiar action of the transverse muscles, or rather of the *transverse band,* as described by SPALLANZANI, HALLER, COOPER, Sir E. HOME, and others, in their experiments on animals. This band is situated near the commencement of the more conically shaped part of the pyloric extremity, three or four inches from the smaller end. In attempting to pass a long glass thermometer tube, through the aperture, into the pyloric portion of the stomach, during the latter stages of digestion, a forcible contraction is first perceived at this point, and the bulb is stopped. In a short time, there is a gentle relaxation, when the bulb passes without difficulty, and appears to be drawn, quite forcibly, for three or four inches, towards the pyloric end. It is then released, and forced back, or suffered to rise again; at the same time, giving to the tube a circular, or rather spiral motion, and frequently revolving it completely over. These motions are distinctly indicated, and strongly felt, in holding the end of the tube between the thumb and finger; and it requires a pretty forcible grasp to prevent it from slipping from the hand, and being drawn suddenly down to the pyloric extremity. When the tube is left to its own direction, at these periods of contraction, it is drawn in, nearly its whole length, to the depth of ten inches: and when drawn back, requires considerable force, and gives to the fingers the sensation of a strong *suction* power, like drawing the piston from an exhausted tube. This ceases as soon as the relaxation occurs, and the tube rises again, of its own accord, three or four inches, when the bulb seems to be obstructed from rising further; but if pulled up an inch or two, through the stricture, it moves freely in all directions in the cardiac portions, and mostly inclines to the splenic extremity, though not disposed to make its exit at the aperture.

Above the contracting band, and towards the splenic portion of the stomach, the suction or grasping motion is not perceptible; but when the bulb is pushed down to this point, it is distinctly felt to be grasped, and confined in its movements.

These peculiar motions and contractions continue until the stomach is perfectly empty, and not a particle of food or chyme remains; when all becomes quiescent again.

If the bulb of the thermometer be suffered to be drawn down to the pyloric extremity, and detained there for a short time, or if the experiment be repeated too frequently, it causes severe distress, and a sensation like cramp, or spasm, which ceases on withdrawing the tube; but leaves a sense of soreness and tenderness at the pit of the stomach.

These peculiar contractions and relaxations, mentioned above, succeed each other, at irregular intervals, of from two to four or five minutes. Simultaneously with the contractions, there is a general shortening of the fibres of the stomach. This organ contracts upon itself in every direction; and its contents are compressed with much force. The valvular portion of the stomach is firmly thrust into the aperture; closing the orifice; preventing the egress of aliment; and obstructing the view of the interior. During the intervals of relaxation, the rugæ perform their vermicular actions, the undulatory motions of the fluids continue, and the alimentary and chymous mass appear, revolving as before, promiscuously mixed, through the splenic and cardiac portions.

All these facts, taken together, will, I think, rationally admit of the following explanation. The longitudinal muscles of the whole stomach, with the assistance of the transverse ones of the splenic and central portions, carry the contents into the pyloric extremity. The circular or transverse muscles contract progressively, from left to right. When the impulse arrives at the *transverse band,* this is excited to a more forcible contraction, and, closing upon the alimentary matter and fluids, contained in the pyloric end, prevents their regurgitation. The muscles of the pyloric end, now contracting upon the contents detained there, separate and expel some portion of the chyme. It appears that the crude food excites the contractile power of the pylorus, so as to prevent its passage into the duodenum, while the thinner, chymified portion is pressed through the valve, into the intestine. After the contractile impulse is carried to the pyloric extremity, the circular band, and all the transverse muscles, become relaxed, and a contraction commences in a reversed direction, from right to left, and carries the contents again to the splenic extremity, to undergo similar revolutions.

It would appear, then, that the discharge of the chyme from the stomach, is effected by *mechanical* impulse. But, I confess, I do not like to give an opinion. I state the circumstances as they have occurred. The idea of mechanical force, I admit, is liable to objection; but, perhaps, not more so than that of the *selecting* power of the pylorus. Whatever bias I may have in favour of the former method, has been forced upon me by the deductions of experiment and observation.

SECTION VII

Of Chylification, and Uses of the Bile and Pancreatic Juice

As food becomes chymified by the gastric juice, the contractile motions of the stomach send it into the duodenum, to receive further changes, preparatory to its assimilation to the circulating fluids of the system, by the lacteal absorbents. It is at first slowly received into this organ from the stomach; but during the later stages of chymification, its transmission becomes more accelerated. The duodenum is so constituted, that the passage of the chyme through it, is considerably retarded; and, hence, in some pathological conditions of the system, the pressure on that organ from repletion, is considerable; and frequently produces great pain and distress.

The vermicular motions of this and the other intestines, are propagated from the stomach, and are continued, after this organ has discharged all its contents, by the contained fluids, until the whole becomes assimilated. They are more or less rapid, varying at different sections of the canal; of which it is not necessary to particularize. These motions are excited by the stimulus of the chyme, and occur at intervals, on the introduction of each quantity passed through the pylorus.

The chymous mass is not changed until it arrives at, or passes the mouth of the ductus cholodochus, when the liver and pancreas are excited to discharge their respective fluids. These mix with the chyme, and produce an essential alteration in its sensible and chemical properties. At this point, the lacteal absorbents commence.

That the change from a *chymous* to a *chylous* stage is effected by the operation of the bile and pancreatic juice, there can be no doubt. Of the nature of this change, there is some diversity of opinion.—Chyle is generally described as "a white, opaque substance, considerably resembling cream in its aspect and physical properties;"* though it is *said* to vary slightly, according to the kind of aliment which had been used. It is my impression, however, that pure chyle, taken from the lacteals of a healthy subject, and produced by natural food, is invariably the same substance in the same individual. Changes that have been observed, must be reckoned as the effect of a pathological state of the system, or the absorption of a non-digesting substance. Medicines and other substances, which are not capable of digestion, are sometimes taken up by the lacteal absorbents, and may produce an alteration in the physical and chemical properties of chyle. It is possible that a small proportion of oil may escape the action of the digestive apparatus, be absorbed by the lacteals, and produce the opaque, white colour, mentioned by authors, as sometimes appearing. Countenance is given to this suggestion, by the fact, that the more opaque coloured parts of chyle are found floating on the surface; and that it is always discovered after the ingestion of oily food. At other times, it is uniform in its colour and consistence, whatever colouring matter may have been contained in the food.

I wish to be understood to say, that every species of aliment produces the same *kind* of nutrient principles. With the view of attempting an investigation of this subject, as has been previously mentioned, I instituted some imperfect experiments and examinations. For the result, see Experiments, Second Series, from 47th to 56th. By the addition of bile and dilute muriatic acid, and subsequently pancreatic juice, to chyme formed in the artificial way, as well as in the stomach, it separated into three distinct parts, a reddish brown sediment at the bottom, a whey coloured fluid in the centre, and a creamy pellicle at the top. Each repetition of the experiment produced a *similar* result; though not *exactly* alike in all. The central portion, I suspect to be imperfectly formed chyle. The sediment, from its appearance, and the coarseness of its particles, I judge, is incapable of being acted on, or taken up, by the absorbents: the creamy or oily pellicle is not only liable to the same objection, but is in too small proportion to the ingesta. The fluid part is fitted, by its fluidity, for the ready action of the absorbents; and is, moreover, in sufficient quantity for the purposes of nutrition. The change of colour and consistence is, probably, effected in the lacteal glands and vessels. The sediment and pellicle, I apprehend, are both excrementitial. The "irregular filaments," at-

*Bostock's Physiology, vol. 2d, p. 392.

tached to the valvulæ conniventes, mentioned by MAGENDIE, and which he concluded to be imperfectly formed chyle, were, undoubtedly, portions of the creamy pellicle, found in the experiments referred to.

But what is the nature of the changes effected in the duodenum? Aliment, after being introduced into the stomach, is dissolved in the gastric juice, and forms a new compound with this fluid. The constituent elements of food are various. When compounded with the gastric juice, they may, nevertheless, be said to form a *simple* compound, or a *gastrite* of *aliment.* I am indifferent about terms; and this will as well convey my meaning as any other. When this *gastrite* is introduced into the duodenum, and mixed with the hepatic and pancreatic fluids, are we not warranted, from all the facts that have been observed, in saying, that there is a general play of chemical affinities in that organ, separating the nutrient principles, and forming various new compounds from the elements of each? The chymous mass changes its colour, and loses its acidity. There is a sensible extrication of gas, as observed by MAGENDIE, and others.* In the stomach, oxygen is found mixed with a small proportion of hydrogen. In the intestines, an increased proportion of hydrogen exists, with carbonic acid, nitrogen, &c.; but no oxygen. Does not the acid of the chyme unite with the alkalis of the bile, and form new compounds? And do not other equally important changes take place? This subject, I confess, is obscure, and perhaps will not admit of a very perfect investigation.

The constant agitation which is maintained in the intestines, preserves the chyle in a state of perfect admixture with the other fluids, until absorption has taken place. By standing at rest, the separation, mentioned above, is evident and perfect.

It has been supposed that the mucus of the intestines has some agency in the formation of chyle. But I am disposed to think, with Professor DUNGLISON, and others, that the use of the mucus is to lubricate the internal coat of the intestines, and, perhaps, to dilute their contents.

It has been suggested that digestion can be perfected in the duodenum and lower bowels, when the food has not been submitted to the action of the stomach and its fluids. In two experiments by MAGENDIE, one failed, and the other was attended with partial suc-

*The escape of gas is generally observable in mixing these fluids with chyme, in my experiments.

cess. Too much reliance ought not to be placed on experiments, that require such severe and cruel vivisections, as were resorted to in these cases. It is possible, as suggested by DUNGLISON, that the presence of crude aliment in the duodenum, may excite the discharge of gastric juice in the stomach, its expulsion into the duodenum, and its consequent action on the food, before it is affected by the bile and pancreatic juice. Or, it may be that the upper part of the duodenum is furnished with vessels, which secrete a fluid similar to gastric juice.

Experiments have also been instituted with the view of ascertaining, whether chyle can be formed without the admixture of the hepatic and pancreatic fluids, with various results. BRODIE ascertained, by tying the ductus communis cholodochus in young cats, that the process of chylification was prevented, and that no chyle was found in the intestines. MAGENDIE, LEURET and LASSAIGNE, on tying this duct, discovered matter of "a rosy yellow colour," which afforded, on analysis, the same constituents of chyle, although the animals, which were the subjects of the operation, *had been kept some time without food.* There is certainly an apparent discordance in these reports. But, it is possible, they may be explained, and reconciled. It is well known that the absorbents are active during a protracted fast, (as in these last experiments) and are constantly taking up the cellular substance, for the purpose of supplying the blood vessels with these broken up solids of the system. Emaciation is the effect of absorption. The lacteals, like other absorbents, have, undoubtedly, their appropriate stimulus; but if that be withholden, they will feed on other substances, the cellular and other solid parts, within their reach. If such be the case, it will account for the rosy coloured fluid, found in the lacteals, by MAGENDIE and others.

EXPERIMENTS, &c.

FIRST SERIES

Experiment 1

August 1, 1825. At 12 o'clock, M., I introduced through the perforation, into the stomach, the following articles of diet, suspended by a silk string, and fastened at proper distances, so as to pass in without pain—viz.:—a piece of high seasoned *a la mode beef;* a piece of *raw, salted, fat pork;* a piece of *raw, salted, lean beef;* a piece of *boiled, salted beef;* a piece of *stale bread;* and a bunch of *raw, sliced cabbage;* each piece weighing about two drachms; the lad continuing his usual employment about the house.

At 1 o'clock, P. M., withdrew and examined them—found the *cabbage* and *bread* about half digested: the pieces of *meat* unchanged. Returned them into the stomach.

At 2 o'clock, P. M., withdrew them again—found the *cabbage, bread, pork,* and *boiled beef,* all cleanly digested,* and gone from the string; the other pieces of meat but very little affected. Returned them into the stomach again.

At 2 o'clock, P. M., examined again—found the *a la mode beef* partly digested: the *raw beef* was slightly macerated on the surface, but its general texture was firm and entire. The smell and taste of the fluids of the stomach were slightly rancid; and the boy complained of some pain and uneasiness at the breast. Returned them again.

The lad complaining of considerable distress and uneasiness at the stomach, general debility and lassitude, with some pain in his head, I withdrew the string, and found the remaining portions of aliment nearly in the same condition as when last examined; the fluid more rancid and sharp. The boy still complaining, I did not return them any more.

August 2. The distress at the stomach and pain in the head continuing, accompanied with costiveness, a depressed pulse, dry skin, coated tongue, and numerous white spots, or pustules, resembling coagulated lymph, spread over the inner surface of the stomach, I thought it advisable to give medicine;

*These Experiments are inserted here, as they were originally taken down in my Note Book, with very little alteration of phraseology, and none of the sense. Subsequent experiments have sometimes convinced me of errors in former ones. When this has been the case, I have generally made the corrections in the way of remarks, or observations, as in this experiment.

and, accordingly, dropped into the stomach, through the aperture, half a dozen *calomel pills,* four or five grains each; which, in about three hours, had a thorough cathartic effect, and removed all the foregoing symptoms, and the diseased appearance of the inner coat of the stomach. The effect of the medicine was the same as when administered in the usual way, by the mouth and œsophagus, except the nausea commonly occasioned by swallowing pills.

This experiment cannot be considered a fair test of the powers of the gastric juice. The cabbage, one of the articles which was, in this instance, most speedily dissolved, was cut into small, fibrous pieces, very thin, and necessarily exposed, on all its surfaces, to the action of the gastric juice. The stale bread was porous, and, of course, admitted the juice into all its interstices; and probably fell from the string as soon as softened, and before it was completely dissolved. These circumstances will account for the more rapid disappearance of these substances, than of the pieces of meat, which were in entire solid pieces when put in. To account for the disappearance of the fat pork, it is only necessary to remark, that the fat of meat is always resolved into oil, by the warmth of the stomach, before it is digested. I have generally observed that when he has fed on fat meat or butter, the whole superior portion of the contents of the stomach, if examined a short time after eating, will be found covered with an oily pellicle. This fact may account for the disappearance of the pork from the string. I think, upon the whole, and subsequent experiments have confirmed the opinion, that fat meats are less easily digested than lean, when both have received the same advantages of comminution. Generally speaking, the looser the texture, and the more tender the fibre, of animal food, the easier it is of digestion.

This experiment is important, in a pathological point of view. It confirms the opinion, that undigested portions of food in the stomach produce all the phenomena of fever; and is calculated to warn us of the danger of all excesses, where that organ is concerned. It also admonishes us of the necessity of a perfect comminution of the articles of diet.

Experiment 2

Aug. 7. At 11 o'clock, A. M., after having kept the lad fasting, for seventeen hours, I introduced the glass tube of a Thermometer (Fahrenheit's) through the perforation, into the stomach, nearly the whole length of the stem, to ascertain the natural warmth of the stomach. In fifteen minutes, or less, the mercury rose to 100°, and there remained stationary. This I determined by marking the height of the mercury on the glass, with ink, as it stood in the stomach, and then withdrawing it, and placing it on the graduated scale again.

I now introduced a gum-elastic (caoutchouc) tube, and drew off one ounce of pure gastric liquor, unmixed with any other matter, except a small portion of mucus, into a three ounce vial. I then took a solid piece of *boiled, recently salted beef,* weighing three drachms, and put it into the liquor in the vial; corked the vial tight, and placed it in a saucepan, filled with water, raised to the temperature of 100°, and kept at that point, on a nicely regulated sand bath. In *forty minutes* digestion had distinctly commenced over the surface of the meat. In *fifty minutes* the fluid had become quite opaque and cloudy; the external texture began to separate and become loose. In *sixty minutes,* chyme began to form.

At 1 o'clock, P. M., (digestion having progressed with the same regularity as in the last half hour,) the cellular texture seemed to be entirely destroyed, leaving the muscular fibres loose and unconnected, floating about in fine small shreds, very tender and soft.

At 3 o'clock, the muscular fibres had diminished one half, since last examination, at 1 o'clock.

At 5 o'clock, they were nearly all digested; a few fibres only remaining.

At 7 o'clock, the muscular texture was completely broken down; and only a few of the small fibres floating in the fluid.

At 9 o'clock, every part of the meat was completely digested.

The gastric juice, when taken from the stomach was as clear and transparent as water. The mixture in the vial was now about the colour of whey. After standing at rest a few minutes, a fine sediment, of the colour of the meat, subsided to the bottom of the vial.

Experiment 3

At the same time that I commenced the foregoing experiment, I suspended a piece of *beef,* exactly similar to that in the vial, (Ex. 2d) into the stomach, through the aperture.

At 12 o'clock, M., withdrew it, and found it about as much affected by digestion as that in the vial; there was little or no difference in their appearance. Returned it again.

At 1 o'clock, P. M., I drew out the string; but the meat was all completely digested, and gone.

The effect of the gastric juice on the piece of meat, suspended in the stomach, was exactly similar to that in the vial, only more rapid after the first half hour, and sooner completed. Digestion commenced on, and was confined to, the surface entirely, in both situations. Agitation accelerated the solution in the vial, by removing the coat that was digested on the surface; enveloping the remainder of the meat in the gastric fluid; and giving this fluid access to the undigested portions.

Experiment 4

Aug. 8. At 9 o'clock, A. M., I drew off an ounce and a half of gastric juice, into a three ounce vial; suspended two pieces of *boiled chicken,* from the breast and back, into it, and placed it in the same situation and temperature as in the second experiment; observing the same regularity and minuteness.

Digestion commenced and progressed much the same, as in the second experiment, but rather slower; the *fowl* appearing to be more difficult of digestion than the *flesh*. The texture of the *chicken* being closer than that of the *beef,* the gastric juice appeared not to insinuate itself into the interstices of the muscular fibre, so readily as into the beef; but operated entirely upon the outer surface, dissolving it as a piece of gum arabic is dissolved in the mouth, until the last particle was digested.

The colour of the fluid, after digesting the chicken, was of a greyish white, and more resembled a milky fluid than whey, which was the colour of the chyme from the beef.

The contents of both vials, kept perfectly tight, remained free from any fœtor, acidity, or offensive smell or taste, from the time of the experiments, (7th and 8th August,) to the 6th of September; at which time, that containing the solution of *boiled beef,* became very offensive and putrid; while that containing the chyme from the *boiled chicken,* was perfectly bland and sweet. Both were kept in exactly similar situations.

It is, perhaps, unnecessary to make any comments on the result of the above experiments. Each one will make up his opinion from the facts. *These* demonstrate, at least, that the stomach secretes a fluid which possesses *solvent* properties. The change in the solid substances is effected too rapidly to be accounted for on the principle of either maceration or putrefaction. I shall be able to show, in some of the following experiments, that aliment undergoes the same changes in the stomach, as is effected in the mode here adopted.

The young man, who was the subject of these experiments, left me about this time, (September, 1825) and went to Canada, the place of his former residence. The experiments were consequently suspended.

EXPERIMENTS, &c.

SECOND SERIES

FORT CRAWFORD, *Upper Mississippi,*
June 20th, 1829.

ALEXIS ST. MARTIN having returned from Canada, after an absence of nearly four years, with his stomach in the same, or very similar condition, as when he left me in September, 1825, I continued to prosecute the gastric experiments, which were commenced before he left me.

With a view to ascertain the variations of temperature, if any there were, in the interior of the stomach, under different circumstances and conditions of the system, and vicissitudes of the atmosphere, I instituted the following experiments.

Experiment 1

Dec. 6, 1829. At 9 o'clock, A. M., I introduced the glass tube of a Thermometer (Fahrenheit's) through the artificial opening into the stomach, in a healthy and empty condition, nearly the whole length of the stem. In six or eight minutes, the mercury became stationary, at 98°. Weather cloudy, damp, and almost raining—ground wet, muddy and thawing. Wind s. and mild. Thermometer, in a North exposure, 63°. Commenced raining at 11 o'clock, A. M., and continued all day, with oppressive atmosphere.

Experiment 2

Dec. 7. Introduced Thermometer at the same hour as yesterday—circumstances of stomach the same. Mercury at 98°. Weather cloudy—Atmosphere damp—Wind N.W. and light—Th: 27°.

Experiment 3

Dec. 8. Introduced Thermometer at 9 o'clock, A. M.—circumstances of stomach same as yesterday. Mercury stationary at 99°. Weather clear—Atmosphere dry—Wind S.W. and light—Th: 13°.

Experiment 4

Dec. 9. Introduced Thermometer at 9 o'clock, A. M.—circumstances similar. Mercury stationary at 99°. Weather clear—Atmosphere dry—Wind W. and light—Th: 10°.

Experiment 5

Jan. 24, 1830. Introduced Thermometer at 3 o'clock, P. M. Weather clear and cold—Th: 8° below 0.—Wind N.W. and light—stomach empty, and coats healthy. Mercury stationary at 100°.

Experiment 6

Jan. 25. Introduced Thermometer at 8 o'clock, A. M. Weather clear—Wind S.W. and light—Th: 2°—Stomach empty. Mercury stationary at 100°.

At 10 o'clock, A. M., (one hour after eating a breakfast of pork and bread,) introduced Thermometer again. Mercury stationary at 100°, as at 9 o'clock, before eating.

Experiment 7

March 17. At 10 o'clock, A. M., introduced Thermometer. Weather rainy and foggy—Wind S.W. and light—Th: 38°—Stomach empty, having eaten nothing since 7 o'clock last evening. Mercury stationary at 99°.

Experiment 8

March 18. At 8 o'clock, A. M., introduced Thermometer. Mercury stationary at 100°. Weather clear—Wind N.W.—Th: 6°.

At 9 o'clock, breakfasted on meat, biscuit and butter, with coffee. Temperature of the stomach, immediately before eating, 100°: thirty minutes after finishing breakfast the temperature had risen to 102°. Digestion rapidly advancing.

It appears, from the above experiments, that the variations of the atmosphere produce effects upon the temperature of the stomach; a dry atmosphere increasing, and a humid one diminishing it. What would be the effect of copious perspiration, in warm weather, on the temperature of the stomach? Would that of the interior of this organ be lessened by evaporation? I regret that sufficient experiments have not been made, fully to satisfy these inquires. From one or two experiments, it would seem, that the heat of the stomach was increased during the active period of digestion. This, however, was probably owing to exercise, immediately after eating, though not particularly observed and noted at the time. Subsequent experiments have not shown this result. On the contrary, the temperature has been found to be the same, in its full and empty state.

The ordinary temperature of the healthy stomach, may be fairly estimated at 100°, Fahrenheit. Some allowance ought, probably, to be made, in these experiments, for imperfect instruments. It appears, from subsequent examinations, that there is probably some difference of temperature in different regions of the stomach, it being higher at the pyloric than at the splenic end. See subsequent experiments and observations.

To ascertain whether the Gastric Juice be *accumulated* in the stomach, during periods of fasting, or even from the immediate and direct influence of hunger, I made the following experiments.

Experiment 9

Dec. 5, 1829. At 8 o'clock, A. M., after twelve hours abstinence from either food or drinks, I introduced, at the perforation, a gum-elastic tube, and drew off a drachm or two only of the gastric juice.—There was no accumulation in the stomach.

Experiment 10

Dec. 12. At 3 o'clock, P. M., introduced tube—could procure two or three drachms only—this was secreted on the irritation of the tube. Stomach contained none in a free state.

Experiment 11

Dec. 14. At 10 o'clock, P. M., after eighteen hours fasting, introduced tube, and drew off one and a half ounces of gastric juice. It was clear, and almost transparent; tasted a little saltish and acid, when applied to the tongue, similar to thin mucilage of gum arabic, slightly acidulated with muriatic acid. There was no accumulation in the stomach when the tube was introduced.

Experiment 12

March 13, 1830. At 10 o'clock, A. M.—stomach empty—introduced tube; but was unable to obtain any gastric juice. On the application of a few crumbs of bread to the inner surface of the stomach, the juice began slowly to accumulate, and flow through the tube. The crumbs of bread adhered to the mucous coat, soon became soft, and began to dissolve and digest. On viewing the villous membrane *before* applying the bread crumbs, the mucous coat and subjacent follicles only, could be observed; but immediately *afterwards,* small, sharp papillæ, and minute lucid points, situated in the interstices of, and less than, the mucous follicles, became visible; from which exuded a clear, transparent liquor. It then began to run through the tube.

Experiment 13

March 18. At 6 o'clock, P. M., after fasting from 8 o'clock, A. M., introduced tube—obtained one and a half ounces gastric juice, after having kept up the irritation, by moving the tube from point to point, for twelve or fifteen minutes. No accumulation of free juice in the stomach.

Experiment 14

Jan. 26, 1831. At 9 o'clock, A. M.—stomach empty—extracted one ounce gastric juice, slowly through the tube, with the usual admixture of mucus. Introduced food, and it began directly to flow more freely through the tube.

Experiment 15

Jan. 27. At 8 o'clock, A. M.—stomach empty—introduced elastic tube, and obtained one and a half drachms of gastric juice, by very slow distillation. Applied crumbs of bread to the villous coat, and the juice began immediately to flow freely through the tube.

Experiment 16

March 6. At 8 o'clock, A. M., extracted two ounces gastric juice, and added it to two ounces of *Madeira wine.* No visible change was produced—no coagulæ formed. They united, like pure water and wine. Heat produced no other effect.

Experiment 17

March 7. At 6 o'clock P. M.—stomach empty—extracted one and a half ounces of juice, and mixed it with the same quantity of *Jamaica spirits.* Effect same as with wine.

Experiment 18

March 8. At 8 o'clock, A. M.—stomach empty—extracted one and a half ounces of gastric juice.

Experiment 19

March 12. At 9 o'clock, A. M.—stomach empty—extracted one and a half ounces of gastric juice. Put this in a bottle.

Experiment 20

March 13. At 11 o'clock, A. M.—stomach empty—extracted two ounces of juice.

Experiment 21

March 14. At 12 o'clock, M.—stomach empty—extracted two ounces of juice.

Experiment 22

March 15. At 4 o'clock P. M.—stomach empty—extracted one and a half ounces gastric juice.

Experiment 23

March 16. At 5 o'clock P. M., introduced tube—could obtain no clear gastric juice. A little acrid fluid and frothy mucus, only, could be extracted. Villous membrane red and dry. St. Martin complained of some head ache, pain and distress about the scrobiculus cordis, lassitude and loss of appetite. Directed him to take half an ounce of *tincture of aloes and myrrh,* at 9 o'clock, P. M. This moved his bowels several times next morning. Little or no change was apparent in the appearance of the inner coat of the stomach; if any, it was a little more moist, and a shade paler, after the operation of the tincture. Gastric juice could again be obtained, but in less than usual quantity.

It would seem, from the preceding experiments, that the stomach contains *no gastric juice,* in a free state, when aliment is not present. Any digestible or irritating substance, when applied to the internal coat, excites the action of the gastric vessels. Hence, I infer that the fluid, in these experiments, was incited to discharge itself by the irritation of the tube used in extracting it.

If, as is contended for by some, a *part* of the fluid be discharged into the stomach during a fast, I see no reason why nature should withhold the other part. If we may be allowed to argue, independent of more certain data, one great objection to the opinion that the stomach contains gastric juice, in a free state, when food is withholden from it, exists in the danger of its passing out through the pyloric orifice; and thus depriving the succeeding meal of the benefit of its solvent action. It is probable that the pyloric orifice opposes no resistance to its egress; but is obedient to its summons. In this way we may account for its admitting chyme, which is an admixture, or rather, combination, of gastric juice and food, to obey the expulsive motions of the stomach, and pass out. They both appear to excite the peculiar contraction of the pyloric end of the stomach, mentioned in a former part of this work. Besides, there would be danger of the gastric juice being weakened, by the introduction of large quantities of water, or other fluids, in the intervals of eating, and thus lose its energy, and concentrated solvent properties.

The last experiment has considerable pathological importance. In febrile diathesis, very little or no gastric juice is secreted. Hence, the importance of withholding food from the stomach in febrile complaints. It can afford no nourishment; but is actually a source of irritation to that organ, and, consequently, to the whole system. No solvent can be secreted under these circumstances; and food is as insoluble in the stomach, as lead would be under ordinary circumstances.

The following, and most of the subsequent experiments of this series, were instituted with the view of ascertaining the relative difference between *natural* and *artificial* digestion; to demonstrate the performance of digestion *out of the stomach,* by the gastric juice; and, also, the *continuation* of the natural process, when *taken out* during the period of chymification.

Experiment 24

Dec. 14, 1829. At 1 o'clock, P. M., I took one and a half ounces of gastric juice, fresh from the stomach, after eighteen hours fasting, into an open mouthed vial—put into it twelve drachms *recently salted beef,* (boiled) and placed it in a basin of water, on a sand bath, and kept it at about 100° (Fahrenheit,) with frequent, gentle agitation. Digestion commenced, in a short time, on the surface of the meat, and progressed in that manner uniformly for about six hours, when its solvent action seemed to cease. The meat was at this time, nearly half dissolved; the texture of the central portion considerably loosened and tender, resembling the same kind of aliment when ejected, partly digested, from the stomach, some hours after being swallowed, as frequently seen in cases of indigestion.

The vial, continuing in the same situation, its contents varied considerable in their sensible qualities. In twenty-four hours, the digested portion separated into a reddish brown precipitate, and whey coloured fluid.

I now separated the undigested from the chymous portion, by filtration, through thin muslin. When squeezed dry, it weighed five drachms, two scruples, and eight grains, which, deducted from the twelve drachms of meat put in at first, leaves six drachms and twelve grains, digested in twelve fluid-drachms of gastric juice.

This experiment was conducted with as much precision and integrity of observation as possible, with the temperature of the digesting bath kept as near blood heat as was practicable to regulate and continue artificial warmth—the Thermometer varying, during the time, from 90° to 100°.

In this experiment, it appears, that it took *twelve* drachms of gastric juice to digest *six* drachms and *twelve* grains of aliment. No certain rule can, however, be given. Allowance must be made for the purity of the fluid, or its admixture with mucus and other secretions; for it is altogether probable that there are great variations in it, in this respect, as well as in all the other secretions of the system. It is probable, also, that different kinds of diet require different proportions of gastric juice for their solution. That its action is similar to that of other chemical agents I have no doubt. A given quantity of the fluid acts on a definite proportion of aliment, when it becomes saturated, and is inadequate to produce any further effect. There is always disturbance of the stomach when more food has been received than there is gastric juice to act upon it.

Experiment 25

Dec. 16. At 2 o'clock, P. M.—twenty minutes after having eaten an ordinary dinner of *boiled, salted beef, bread, potatoes* and *turnips,* and drank a gill only of pure *water,* I took from his stomach, through the artificial opening, a gill of the contents, into an open mouthed vial. Digestion had evidently commenced, and was perceptibly progressing, at the time. This vial and contents were immediately placed in a basin of water, on the sand bath, at 90° or 100°, and continued there for five hours.

The digestion of the contents continued to progress, until all was completely chymified.

At 7 o'clock—five hours after eating his dinner—I took out a gill of pure chyme; no particles of undigested food appearing in the mixture.

Very little difference was perceptible between this last parcel and that in the vial, digesting on the bath. The stomach had digested a little faster and more perfectly than the vial.

In this experiment, it seems, that a quantity of aliment, taken out of the stomach *twenty minutes* after having been eaten, had a sufficient admixture of gastric juice to ensure its perfect digestion when placed on the bath. An ordinary moderate meal, taken into a healthy stomach, is sooner diposed of than most physiologists are aware of; and in this case, it is probable that a sufficient quantity of gastric juice had been secreted in twenty minutes, to digest the whole quantity of aliment in the stomach. When a larger quantity has been received, though the powers of the stomach may be sufficient, ultimately to dispose of it, it would undoubtedly be found that a portion taken from the stomach a few minutes after having been eaten, would not contain a sufficient quantity of gastric juice to digest it perfectly. It is possible that the portion presented at the perforation may be in a more advanced stage of digestion, than the rest of the mass, and consequently lighter, and float on the surface of the more solid portions of the food. In ordinary cases, such would be found to be the case; but when much fat meat or oily food has been used, the oil always maintains an ascendency in the gastric cavity.

Experiment 26

Jan. 11, 1830. At 3 o'clock, P. M., dined on *bread* and *eight ounces* of *recently salted, lean beef, four ounces* of *potatoes,* and *four ounces* of *turnips,* boiled. In *fifteen minutes,* took out a portion of the contents of the stomach. The *meat* made its appearance, in an incipient stage of digestion.

At 3 o'clock, 45 minutes, took out another portion. The meat and bread only appeared, in a still more advanced stage of digestion.

The texture of the meat was, at this time, broken into small shreds, soft and pulpy, and the fluid containing it had become more opaque, and quite gruellike, or rather, glutinous, in appearance.

I put this second parcel in a vial, and placed it in water, on the sand bath, at the temperature of the stomach, (100° Fahrenheit,) as indicated by the thermometer immediately preceding its extraction, and continued it there.

At 5 o'clock, took out another quantity. Digestion had advanced in about the same ratio as from the first to the second time of extracting; and when compared with the second parcel, contained in the vial on the bath, little or no difference could be perceived in them; both were nearly in the same stage of digestion. That contained in the vial had advanced regularly and rapidly; nearly all the particles of meat had disappeared, become chymified, and changed into a reddish brown sediment, suspended in the more fluid parts, with small particles, resembling loose, white coagulæ, floating about near the surface.

On taking out the third parcel, small pieces of vegetables appeared, in a partial stage of digestion. This was also put into a vial, and placed on a bath, with the second parcel, and the same uniform temperature (100°) kept up, with frequent, gentle agitation.

At 6 o'clock, P.M., digestion had progressed equally in both. The only difference to be seen, was the particles of vegetables, in a less advanced stage than the meat.

The contents of both vials, kept on the bath, and nearly in the same temperature, until the next morning, were completely digested, except the few small particles of vegetables, which remained almost entire.

The contents of the vials, at this time, were of the consistence of thin jelly, and of a lightish brown colour; tasting peculiarly insipid, saltish and acid. After standing at rest awhile, the brownish sediment subsided towards the bottom, while small particles of whitish coloured, loose coagulæ floated about in the fluid above. The undigested particles of vegetables settled to the bottom.

In the Preliminary Observations, I have endeavoured to maintain the proposition, that chyme is *homogeneous* in its properties. It would seem from this and some other experiments, that it contains a sediment. This, however, it is believed, does not militate against its homogeneous character. Many substances, that are generally acknowledged to possess this character, deposit a sediment, on standing. The heavier parts subside, of course. It is not necessary to cite examples. It is possible, also, that mixed food contains some adventitious, indigestible substances, which are not affected by the gastric juice.

This Experiment (26th) demonstrates the comparative digestibility of *animal* and *vegetable* diet. In extracting two parcels, one in fifteen minutes and the other in three quarters of an hour, the meat only made its appearance, partially digested. In taking out a third portion, at 5 o'clock, two hours after having eaten, small particles of vegetables made their appearance. This experiment appears to confirm the opinion, pretty generally entertained by medical men, that vegetables are less easily disposed of by the gastric organs, than animal or farinaceous substances. With dyspeptics this is undoubtedly true, as experience every day teaches us; and as their stomachs have the same organization as others, are governed by the same general laws, and are only modified by debility or disease, the conclusion is natural, that they should act on aliment in the same manner, in proportion to their strength, that the healthy stomach does.

It may be inferred from this experiment, that the more perfectly chymified portions of food rise to the superior part of the stomach, as suggested in a previous observation, and are consequently exposed at the perforation, from whence parcels are taken for experiment and examination.

Experiment 27

March 17. At 12 o'clock, M.—drank half a pint of *milk.* In fifteen minutes, took a portion out of the stomach, in a fine, loosely coagulated condition, perfectly white, and suspended in a semi-transparent, whey coloured fluid. I placed this on the bath, and it continued to digest for eight hours, when the coagulæ were completely taken up. A very small proportion of light coloured sediment, settled loosely to the bottom of a cream coloured, sweetish fluid.

At the same time that he drank the milk, I put one drachm of gastric juice, warm from the stomach, into two drachms of *milk,* and placed it on the bath, at the natural temperature, (100° Fahrenheit.) In five minutes, pure, white coagulæ formed, which, in fifteen minutes, exactly resembled that taken out of the stomach. In twenty minutes, the same fine, loose coagulæ were suspended in a similar liquid. These two drachms of milk, mixed with one drachm of pure gastric juice, *out of the stomach,* gave the same result, and exhibited the same appearance, in nearly the same time, as that which was *swallowed,* and *taken from the stomach.* Scarcely a shade of difference could be perceived in four hours.

Two drachms of *milk,* coagulated by *acetous acid,* produced coagulæ very similar to the other; but the wheyey part bore no resemblance, except in mere fluidity; that formed from the gastric fluid being of an opaline, slightly yellowish cast, and the other thin, transparent and watery.

The coagulæ formed by the gastric juice, continued to digest regularly in their fluid, for about eight hours, when they were completely taken up, and converted into chyme.

The coagulæ formed by the vinegar, remained in the same condition for forty eight-hours, with no other change except mere subsidence below the watery fluid.

It is well known, and this experiment was not necessary to prove it, that milk is coagulated before it receives the *solvent* action of the gastric juice. But it has some degree of importance in demonstrating the fact, that a degree of solidity is necessary for the operation of this agent. And it is a strong argument against the doctrine of digestion by the veins of the stomach. It has been maintained by some, that the veins take up the nutritious parts of the food, immediately on their introduction into the stomach. If so, it strikes me that they should do so, as it relates to this kind of aliment, while they are in a fluid state, and more susceptible of absorption by their mouths; and not wait till they have become solidified. Wine, spirits, water and other fluids, which conduce nothing towards alimentation, are neither coagulated, nor otherwise affected by the gastric juice. These fluids are not digested; and probably enter the circulatory system without much change.

It will be seen, by succeeding experiments, that other fluid, nutritive substances, particularly the albumen of eggs, are coagulated before they receive the solvent action of the gastric juice.

Experiment 28

Jan. 25, 1831. At 1 o'clock, P. M., he ate a full dinner of *roast beef, potatoes, beets* and *bread,* and kept exercising about his usual employment, as house servant.

At 5 o'clock, 25 ms., I took out a portion of the contents of the stomach. Digestion of the different articles of food had commenced, and considerably advanced. The bread, reduced to a pultaceous condition, appeared floating about in a reddish brown fluid, of a glutinous consistence. A few small particles of the meat could also be seen in the fluid. None of the vegetables were discernible at this time. The fluids tasted slightly acid, giving the flavour peculiar to dilute muriatic acid, and very slightly bitter. A few grains of carbonate of soda, thrown into a drachm or two of this fluid, produced a slight effervescence.

At 4 o'clock, 20 mins.—took out another portion, a shade or two darker than the first. This dark colour of the chyme, I attributed to his having taken with his dinner, some of the outside, scorched pieces of the beef.

No distinct parts of the food could be seen at this time. Upon the surface of both parcels of fluids, floated a layer, of an oily or lardaceous consistence, which probably was the remains of the fat pork which he had eaten for his breakfast. The first parcel contained much more of this oily fluid than the last; which leads me to think that a considerable portion of an imperfect chyme, formed from the pork taken at about 10 o'clock, for breakfast, remained in his stomach when he ate his dinner; and then mixed with this aliment, in an imperfect state of digestion.

At 5 o'clock, 30 mins.—tried to extract another portion—could obtain nothing, except a little gastric juice. The chyme formed from his dinner appeared to have all passed from the stomach.

Experiment 29

March 6. At 9 o'clock, A. M.—breakfasted on *venison steak, cranberry jelly* and *bread,* and drank a pint of *coffee.*

Twenty minutes after eating, I took a portion from the stomach, in an incipient stage of digestion. Placed this on the bath.

At 9 o'clock, 45 ms.—took out another portion, in an advanced stage of digestion—very few small particles of food were discernible.

At 10 o'clock 10 mins.—took out another portion, completely chymified.

At 10 o'clock, 35 mins.—the stomach was entirely empty and clean—no chyme or aliment to be found in it. The breakfast, eaten at 9 o'clock, was all digested, and had passed through the pylorus, in *one hour and thirty-five minutes.*

This is an example of the great rapidity of digestion in some instances. This rapidity depends upon various circumstances—principally upon moderation in quantity, and the digestible properties of the food used. From various trials, I am confident, generally speaking, that venison is the most digestible of any diet of the fibrinous kind. In a few instances, it will be perceived, that other articles of diet were disposed of in a shorter period, than the venison was in this experiment.

Experiment 30

March 7. Mixed two drachms of *albumen* of a fresh egg, with two drachms of gastric juice, warm from the stomach, and placed it on the bath, at the natural temperature. The juice and the albumen were so much alike in their appearance, when first mixed, that the change was not perceptible; but in ten or fifteen minutes, small, white flocculi began to appear, floating about; and the mixture became of an opaque and whitish appearance. This continued slowly and uniformly to increase, for three hours, at which time, the fluid had become of a milky appearance; the small flocculi, or loose coagulæ, had mostly disappeared, and a little light coloured sediment subsided to the bottom.

At the same time of the above experiment, he swallowed the *white* of two *eggs,* unmixed with any other food. The stomach was perfectly empty at the time.

In thirty minutes, I took out and examined a portion. It exhibited a similar appearance to that mixed out of the stomach, in the vial on the bath, only more rapid in its progress.

In *one hour and thirty minutes,* I examined the cavity of the stomach, and found nothing but a little pure gastric juice. The albumen was completely digested, and disposed of.

Experiment 31

March 9. At 8 o'clock, A. M.—stomach empty—temperature 98°—took out two ounces gastric juice. Divided this into two equal parts, and put them into separate vials—to each of which, I put equal quantities of roasted beef—placed one of them on the bath, at 99°, and the other in the open air, at 34°.

I then put the same quantity of the meat into an equal quantity of clear water, and placed it with the cold gastric juice and meat in the open air, at the same temperature.

At 9 o'clock, he had finished breakfasting on the same kind of meat, with the addition of warm, light *biscuit, butter,* and a pint of *coffee.* Temperature of the stomach, immediately *before* eating, 100°. In thirty minutes *after* eating, the temperature rose to 102°.* Digestion rapidly advancing.

At 10 o'clock—took out a portion, partially digested; the biscuit the most so of any part of the breakfast. Placed this on the bath.

The meat, contained in the vial of gastric juice on the bath, was, at this time, in about the same condition as that taken from the stomach; very little difference could be perceived. The biscuit which he had eaten with his breakfast occasioned the only difference; that being reduced to a soft pulp.

The meat in the *cold* gastric juice was, at this time, much less advanced, than either that in the warm juice, or in the portion taken from the stomach.

That contained in the vial of water was merely macerated, and had no more appearance of digestion than what was effected by its being masticated, and mixed with the saliva, as were the other pieces of meat, before they were put into the gastric juice.

At 10 o'clock, 45 mins. I examined the stomach, but could find no distinct particles of food, and but very little chyme. His breakfast appeared to have been completely digested, and had left the stomach. Temperature 100°.

At 2 o'clock, P. M., the several parcels of meat placed in the gastric juice, on the bath, being about half digested, and appearing not to progress, I drew off twelve drachms of gastric juice from the empty stomach, and added four drachms to each, including the parcel taken from the stomach, at 10 o'clock, A. M., that being in about the same state of chymification with the others on the bath.

I continued the two on the bath, at 100°, and the others, (cold gastric and aqueous fluids) on the shelf, at 34°. Digestion evidently recommenced in the parcels on the bath, and again regularly progressed, after the addition of the second portions of the gastric juice; and more rapidly in the vial containing

*Probably the effect of exercise, but not noticed at the time.

the meat digesting in the gastric juice, taken out of the stomach first, than in the one containing the chymous portion, taken out at 10 o'clock, A.M., one hour after having eaten. This parcel, however, contained a solid piece of meat, which appeared to have been swallowed without being masticated; and consequently did not readily yield to the solvent action of the gastric juice. The juice was, also, in too small proportion completely to digest it.

The vials containing the cold aqueous and gastric portions, placed on the shelf, were, at 4 o'clock, P.M., but very little changed, and much alike.

These four parcels, after standing for twenty-four hours, and all suffered to get cool, exhibited the following appearances.

The portion taken from the stomach at 10 o'clock, A.M., one hour after having eaten, was the most perfectly digested, and completely converted into a thick pultaceous mass, of a reddish brown colour, with the exception of the piece of unmasticated meat, which remained entire and undigested. This emitted a sharp, rancid smell, and was slightly bitter. The vial containing the meat digesting in the gastric juice first taken out of the stomach, exhibited appearances very similar to the last, though the contents were less perfectly digested. It was not of so thick consistence; but gave the same sharp smell and bitter taste, with the addition of an empyreumatic and slightly fœtid flavour. The empyreuma, I attributed to a portion of the meat being a little dry and scorched when first put in; and the fœtor, to the temperature of the bath having been accidentally raised considerably above 100°, during the experiment.

The cold gastric and aqueous portions very nearly resembled each other; both *macerated,* but not *digested;* differing essentially from the other two, in not exhibiting any appearance of chyme. The cold gastric juice had very little, if any, more effect on the meat, than the water; and retained its peculiar taste. Its colour was darkish brown, while the latter was of a reddish grey. At 9 o'clock, A.M., of the 10th, I placed both of them on the bath, and continued them for twenty-four hours, at the natural temperature. An essential difference in the gastric liquor was produced, after being placed on the bath. Digestion evidently advanced; the colour became lighter and lighter; the meat diminished; and a thin, light, pastelike liquor formed, as in the other two portions, at first placed on the bath. The aqueous portion exhibited no other appearance than that of simple maceration in warm water. At the end of the last twenty-four hours, on the bath, appearances of incipient putrefactive fermentation began to be manifested, as the evolution of small bubbles of fœtid gas, and a change of colour from a reddish to a greenish shade.

A difference in the degrees of chymification between the several parcels, was now very evident.

The gastric portion, or that taken from the stomach, an hour after breakfast, was the most digested.

The artificial, or that portion of the gastric juice and meat, first placed on the warm bath, was next, and nearly as much digested; though a difference was observable.

The third, or portion of gastric juice and meat, first placed in a cool situation, after having been on the warm bath for six or eight hours, was the next, but considerably less digested than the second.

The fourth, or aqueous portion, exhibited no appearance of chymification.

It would seem, from this experiment, that a certain degree of heat is necessary to the action of the gastric juice. One parcel of the meat, after being exposed to the cold gastric juice for twenty-four hours, exhibited very little change; but being placed on the bath, at the end of this time, digestion commenced, and advanced, regularly, as in the other parcels. It also appears, that after the process of digestion has ceased, for want of a sufficient quantity of gastric juice, it will recommence on the addition of a fresh supply. It was necessary to add another quantity, even to that portion taken out of the stomach, to ensure its perfect digestion. This, I think, is an evidence, that the fluid is discharged into the stomach gradually and progressively, according to the requirements of the aliment. If the portion left in the stomach had received, at the time the parcel was taken out, the whole quantity it was destined to receive, it must have been imperfectly digested, and have remained in the stomach, precisely in the situation of that which was taken out, and submitted to artificial digestion; which is proved not to have received its full supply for perfect digestion. But subsequent examination demonstrated that it *was* perfectly digested, and had nearly all passed out of the stomach, in two hours. Hence, the conclusion is irresistible, that it received an additional quantity after the portion was taken from the stomach, one hour after eating.

Experiment 32

March 12. At 8 o'clock, A. M., extracted one ounce of gastric juice.

At 9 o'clock, he breakfasted on *fat pork, bread* and *potatoes.* One hour afterwards, examined contents of stomach—found a heterogeneous mixture, resembling thick porridge.

At 1 o'clock, P. M.—four hours after having eaten—took out a portion, in a complete chymous state, without any entire particles of food to be seen. It was of a milky, or rather thin, gruel-like consistence, and considerably tinged with yellow bile; a circumstance which I had but once before observed in my experiments upon him. And this I supposed to have been the effect of violent anger, which occurred about the time of taking out this parcel.

This experiment shows the effect of violent passion on the digestive apparatus. The presence of bile, in this instance, was believed to be the effect of anger. In a healthy state of the stomach, and an equable frame of mind, this substance has seldom been found in the stomach. When so found, except under peculiar circumstances of diet, it may generally be regarded as an indication of either mental or corporeal disease; and may be considered a foreign and offending substance in that organ. I believe its effect is to change the properties of chyme, (as it will be seen that it does, in subsequent experiments,) alter its homogeneous quality, and retard, or otherwise disturb its due egress into its destined receptacle, the duodenum.

Experiment 33

March 13. At 1 o'clock, P. M.—dined on *roasted beef, bread* and *potatoes.* In half an hour, examined contents of stomach—found what he had eaten reduced to a mass, resembling thick porridge.

At 2 o'clock, examined again—nearly all chymified—a few distinct particles of food, still to be seen.

At 4 o'clock, 30 mins., chymification complete.

At 6 o'clock, examined stomach—found nothing but a little gastric juice, tinged with bile.

Experiment 34

March 14. At 8 o'clock. 15 mins.—introduced two ounces of *rare, roasted beef,* suspended by a string, into the stomach; and at the same time, put one drachm of the same kind of meat into twelve drachms of gastric juice, contained in a vial, and put it into his bosom. The piece in his stomach, examined every hour, till 12 o'clock, M., exhibited an uniform, but very slow process of digestion, confined entirely to the surface of the meat. In four hours, about half of it, only, was dissolved and gone. That in the bosom, at the same time, digested still slower, owing, probably, to the circumstances, that the fluid in the vial had been taken out when the stomach was in a morbid condition, and had been permitted to get cold, even to the freezing point. This last circumstance, however, was probably, of less importance than the other. The meat in the stomach was too much confined by the string; was not permitted to move about freely in the gastric fluids by the natural motions of the stomach; and consequently did not digest so fast as it otherwise would have done. Another circumstance or two, may also, have contributed to interrupt the progress of digestion, such as anger and impatience, which were manifested by the subject, during this experiment.

This experiment shows the necessity of a perfect comminution of the articles of diet. The gastric juice acted very slowly on a large, solid piece of meat. Digestion or solution was confined entirely to the outer surface. This, in addition to the other causes, mentioned above, produced the delay in digestion.

Experiment 35

March 14. At 12 o'clock, M.—ate a *pint* of *milk,* and *four ounces of bread.* Examined stomach in thirty minutes—found the milk coagulated, and the bread reduced to a soft pulp, floating in a large proportion of fluid.

At 10 o'clock, 30 mins.—took out, and examined a portion—found it a thick pultaceous mass of bread, coagulæ and fluid, of a milky colour, slightly bitter taste, and acid smell. Placed it on the bath, where it continued to become more and more milky for an hour, when every particle seemed to be reduced to a rich fluid mass, resembling milk porridge.

The portion taken out thirty minutes after having been eaten, and kept on the bath, retained the appearance of the gastric fluid, with distinct flocculi of bread and coagulæ, floating about, and suspended in the fluid, and a little coarse precipitate at the bottom, after standing at rest a while.

At 2 o'clock—examined stomach—found it nearly empty. The bread and milk appeared to have been disposed of, and were gone from the stomach.

In this experiment, it took two hours for the digestion of a meal of bread and milk; something shorter than the usual time for the disposal of an ordinary meal. For those who have healthy and *unsophisticated* stomachs, milk appears to be one of the best articles of diet we possess. It is less stimulating than flesh, and more nutritious than vegetables. For persons who are disposed to pyrexial complaints, and who are not obliged to perform hard and exhausting labour, it is the most appropriate diet. But the stomach is a creature of habit. It can become accustomed to any kind of diet; and sudden changes are liable to derange its healthy actions. To those accustomed to what is called high living, such as strong meats, strong drinks, and high seasoned food, of all kinds, the transition to a milk diet, which contains a considerably lowered stimulation, would probably be an imprudent change. When necessary, the change should be so gradual, that the stomach should by degrees, become accommodated to it.

Experiment 36

At 2 o'clock, 30 mins.—dined on *fresh beef and vegetable soup,* and *four ounces of bread.*

At 3 o'clock, 20 ms., examined contents of stomach—found a pulpous mass, of the consistence of thick gruel, and of a semi-gelatinous appearance. The soup appeared to have had its more fluid parts absorbed; for it was, at this time, much more consistent than when eaten. It was even thicker than the contents of the stomach usually are, after eating more solid food. Placed this on the bath.

At 5 o'clock, took out another portion, of a whitish colour, and more paste-like consistence, mixed with a little thin, transparent yellowish fluid, of an acid taste. The thick part had the flavour of bile, but not the colour.

Here the uniform laws with respect to liquid diet, appeared to govern the action of the gastric juice. The soup could not be digested until it was formed into a harder mass, by the absorption of the watery part. There was a less quantity of fluid than is usual after eating more solid food. This is another striking demonstration of the laws that govern the action of the stomachic solvent. If water were permitted to remain in the stomach, it would render the soup too liquid to be acted on by the gastric juice.

Experiment 37

March 15. At 8 o'clock 30 mins. A. M.—breakfasted on *fresh sausage, light pancakes,* and a pint of *coffee.*

At 9 o'clock, 30 mins.—examined, and found the stomach full of fluids, mixed with the aliment; and a large portion of clear oil floated on the top, and presented itself at the perforation of the stomach.

At 10 o'clock, 30 ms., I took out a portion—found the cakes and particles of meat about half digested, with some oil, pure, bland and limpid, rising upon the top, untouched by digestion. Placed it on the bath.

At 12 o'clock, M., examined stomach—found no vestige of his breakfast—not a particle of oil was to be seen—nothing but pure gastric juice could be extracted, of which, I took out twelve drachms.

That portion of his breakfast, taken out at 10 o'clock and 30 minutes, was at this time, almost completely chymified, a few small particles of oil only remaining. The chymous mass of a milky colour, and thick gruel-like consistence.

Experiment 38

March 16. At 8 o'clock, 30 min. A. M.—breakfasted on *fresh meat and vegetable hash, bread,* and a pint of *coffee.*

At 10 o'clock, 30 mins.—examined—found but very few particles of his breakfast in the stomach—some oil, and a few flocculi of a brown colour, run out with a little thin fluid.

At 11 o'clock—examined again—found nothing but a little gastric juice. Breakfast was gone, and the stomach clean.

These experiments (37th and 38th,) are continued proofs of the solvent action of the gastric juice.

Experiment 39

At 2 o'clock, P. M.—same day—dined on *recently salted, lean beef, pork, potatoes, carrots, turnips* and *bread.*

At 5 o'clock—examined—found the stomach clear of food, but containing a quantity of white frothy mucus—villous coat inclined to dryness, and deeper pink colour. St. M. complained of some head ache, pain and distress at the pit of the stomach—dry skin and thirst. Directed him to take four drachms of *tincture of aloes and myrrh* at bed time. This operated two or three times next morning, and gave relief. The gastric juice, however, was not obtained in its usual quantity and quality, for twenty-four or thirty-six hours afterwards.

Experiment 40

March 18. At 9 o'clock, A.M., he breakfasted on *soused tripe* and *pig's feet, bread* and *coffee*.

At 9 o'clock, 30 mins.—took out, and examined a portion—found it in a half digested condition, tripe, pig's feet and bread all reduced to a pulp, floating in a large proportion of fluids. Placed it on the bath.

At 10 o'clock—examined stomach again—tried to extract another portion—could find little or no chyme—a very little gastric juice, with a few small, fibrous particles of tripe, and some coffee grounds. His breakfast appeared to have been digested, and had passed from the stomach, in *one hour*.

The portion first taken out, and placed on the bath, was also, at the end of one hour, reduced to nearly a complete chymous condition; a very few of the small particles of tripe and coffee grounds only left, as in the stomach.

This is an example of astonishing rapidity of gastric solution; and that, too, of articles generally regarded as rather hard of digestion. That there could be no mistake, I infer from the fact, that a portion taken out of the stomach, thirty minutes after having been received, and submitted to the artificial mode, exhibited the same result.

Experiment 41

At 1 o'clock, P.M.—same day—he ate eight ounces of *calf's foot jelly,* and nothing else.

In twenty minutes, examined stomach, and took out a portion of its contents, consisting of gastric juice, combined with the jelly, nearly all of it in a fluid form; a few particles only of entire jelly, suspended in the fluids, with a few small, yellowish white coagulæ, floating near the surface, could be perceived.

At 2 o'clock—examined again—extracted a little fluid, but found no appearance of jelly.

The operation of gastric juice on gelatine, is very difficult to be detected. Unlike albumen, it is unsusceptible of coagulation; and it is probable that the gastric juice acts upon it, in its soft-solid state. This was disposed of in a short period. It was, however, but a small quantity, and was much sooner digested than a full meal would have been.

From various trials, I am disposed to think that gelatine, if not in too concrete a state, is a very digestible article of diet.

During the examination, at this time, St. Martin swallowed part of a glass of water, and being situated in a strong light, favourable to an internal view, through the aperture, I distinctly saw the water pass into the cavity of the stomach, through the cardiac orifice—a circumstance, perhaps, never before witnessed, in a living subject. On taking repeated draughts of water, while in this position, it would gush out at the aperture, the instant it passed through the cardia. Food, swallowed in this position, could be distinctly seen to enter the stomach.

Experiment 42

April 7. At 8 o'clock, A. M.—breakfasted on *three hard boiled eggs, pancakes* and *coffee.*

At 8 o'clock, 30 mins.—examined stomach—found a heterogeneous mixture of the several articles eaten, slightly digested.

At 8 o'clock, 45 mins.—examined again—found contents reduced in quantity, and changed in quality—about half digested.

At 10 o'clock, 15 mins., no part of the breakfast remained in the stomach.

This, and the four following experiments, throw no additional light on the subject of digestion, except so far as relates to the period of chymification. This, it will be perceived, depends something upon the *quantity* eaten. The quality, however, is not to be overlooked.

Experiment 43

At 11 o'clock, 15 mins., A. M.—same day—he ate *two roasted eggs* and *three ripe apples.*

In thirty minutes, examined stomach—found a heterogeneous mixture, in an incipient stage of digestion.

At 12 o'clock, 15 mins., M.—examined again—found the stomach clear; no vestige of apples or eggs.

Experiment 44

At 2 o'clock, P. M.—same day—dined on *roasted pig* and *vegetables.*

At 3 o'clock—examined, and found it about half chymified.

At 4 o'clock, very little remained in the stomach.

At 4 o'clock, 30 minutes, nothing remained but a very little gastric juice.

Experiment 45

April 8. At 2 o'clock, P. M., he dined on *wild goose.* At 3 o'clock—stomach full of fluids, with a large portion of oil, floating on the surface; the goose flesh in small shreds, and soft; digesting rapidly.

At 4 o'clock—contents of stomach two thirds gone—that remaining, chymified.

At 4 o'clock, 30 mins., the stomach was empty and clean.

Experiment 46

April 9. At 3 o'clock, P. M., he dined on *boiled, dried codfish, potatoes, parsnips, bread,* and *drawn butter.*

At 3 o'clock, 30 mins.—examined, and took out a portion, about half digested; the potatoes the least so of any part of the dinner. The fish was broken down into small filaments; the bread and parsnips were not to be distinguished.

At 4 o'clock—examined another portion. Digestion had regularly advanced. Very few particles of fish remained entire. Some of the potatoes were distinctly to be seen.

At 4 o'clock, 30 mins.—took out, and examined another portion—all completely chymified.

At 5 o'clock—stomach empty.

The preceding Experiments, I think, plainly demonstrate the *solvent* properties of the gastric juice. When aliment is submitted to it, *out of the stomach,* its operation is rather slower than when the process of digestion is assisted by the *natural warmth* and *motions* of that organ. One reason, probably is, the difficulty of maintaining a bath at the exact, necessary temperature; and another one may present itself, in the impossibility of perfectly imitating the motions of the stomach. With all these disadvantages, however, chyme formed in this way, presents the same uniform, sensible appearance, as that, which is formed in the stomach, by natural process.

That the *cold* gastric juice should not act at all, or but very imperfectly, on aliment, is no proof, in my opinion, that it does not possess solvent powers, even on the admission that it was a *debatable* question. There are but a few chemical combinations that do not require caloric to effect their operations, and none, perhaps, that are not facilitated by it. Some, and indeed, many of them require an intense heat. I am under the impression, though I have never fairly tested the truth of it, that gastric juice would, in a sufficient length of time, act on aliment, in a cold state. But I am not anxious to contend for any extraordinary or unnecessary powers of this fluid. Nor is it necessary to prove that it will act on cold substances, or in cold situations. It is perfectly manifest, that its operation is that of a chemical agent; that it dissolves aliment out of the stomach, when the warmth and motions of that organ are imitated; and that it changes the various and heterogeneous articles, submitted to its action, to an uniform, homogeneous semi-fluid, varying, however, slightly in colour and consistence, according to the aliment used.

With a view to ascertain, if practicable, what effects were produced by the BILE and PANCREATIC JUICE, when added to CHYME, I instituted the following Experiments.

Not being able to procure human bile, in a pure state, I obtained some *beef's gall,* and for pancreatic juice, substituted *diluted muriatic acid,* (one scruple acid to six ounces water.) I was induced to use this acid, from a resemblance observed between its taste and that of the pancreatic juice, and not being able to obtain any of that fluid at the time.

These experiments are certainly very imperfect but such as they are, I submit them to the public. They may tend to pave the way to more perfect experiments on these fluids.

Experiment 47

I divided the chyme, produced in Experiment 24, Second Series, (Dec. 14th, 1829,) into two equal parts, about five drachms each. To one of which, I added one drachm of the Ox gall. Fine coagulæ were immediately produced, of a slightly yellowish green colour. To this, I then added one drachm of dilute muriatic acid; which immediately produced a white balsamic mixture. This, after standing at rest a few minutes, separated into three distinct parts; a clay coloured sediment at the bottom, a whey coloured fluid above, and a thin, oily, whitish pellicle on the top.

Experiment 48

To an ounce of the chyme, formed in Experiment 25, (Dec. 16,) I added one drachm of the Ox gall; which immediately converted it into a milky fluid, very finely coagulated. To this, I added one drachm of the diluted muriatic acid, which at first, increased the coagulæ; but immediately after, threw down a brown precipitate. This, on the addition of more bile and acid, varied in colour, according to the different proportions put in, from a light clay colour, to a dark brown, tinged with green, without any change in the colour or consistence of the fluid above.

On standing at rest, it separated into three distinct parts, a brown sediment at the bottom, a yellowish or whey coloured fluid in the middle, and a thin, milky white pellicle on the top.

Experiment 49

Having procured some fresh gall, from an Ox recently slaughtered, I added twenty drops of it to four drachms of the chyme formed in Experiment 26, (Jan. 11th, 1830.) A turbid, yellowish white fluid, or rather, very fine, cream-coloured coagulæ, immediately formed; which, after standing a few minutes, separated into bright, yellow coloured coagulæ, subsiding towards the bottom, and a turbid, milk coloured liquid above.

By adding twenty drops more of the bile to this, the coagulæ were increased, more collected together, and changed in colour, from a yellow to a greenish hue.

The addition of twenty drops more of bile, (making, in the whole, one drachm,) concentrated a deep grass green, jelly-like deposition at the bottom of the vial. The fluid above, became more milky in appearance; and the coagulæ and sediment became darker on the addition of bile.

I now added twenty drops of the dilute muriatic acid to other four drachms of the same kind of chyme, without bile. This produced no change in the colour or consistence, but increased the saline, acid taste, peculiar to the gastric and pancreatic juices, when uncombined with chyme.

By adding bile to this, the same effects and appearances were present as in the other similar experiments; viz.: a yellowish brown sediment at the bottom, a whey coloured fluid in the middle, and a white pellicle on the top.

To observe the different effects produced between a combination of bile and muriatic acid in clear water, and that of the chymous mass, I mixed equal quantities of the gall and dilute acid, one drachm each, with two ounces of water. This at first produced an effect, and exhibited an appearance, similar to that of their combination with chyme; but gradually changed to a bluish, green coloured, thin fluid, with a deep green, jelly-like deposition at the bottom, without any of the milky appearance of the chymous mixtures, or white pellicle on the top.

Experiment 50

To four drachms of gastric juice, fresh from the stomach, I added forty drops of Ox gall, which produced a turbid, yellowish green fluid, yielding no sediment.

Forty drops dilute muriatic acid, added to other four drachms of the gastric juice, effected no change in its appearance.

Equal parts of the bile and muriatic acid, mixed together, produced a fluid of exactly the same colour as the first; but was less consistent.

On mixing the two first together, and adding two drachms of chyme from the stomach, very fine coagulæ formed in a milky fluid, throwing down a brownish sediment, from a whey coloured liquor, with the same milky pellicle on the surface, as in the former experiments.

To one ounce of chyme, formed in a vial, on the bath, I added two drachms of bile. A turbid, yellowish white mixture formed, without sediment, or immediate separation of any kind.

To another ounce of the same chyme, I added two drachms of the dilute acid. No change in its appearance was perceptible.

I then mixed them together, and the appearance of both was changed. Whitish coagulæ formed, and let fall a brown sediment, leaving an opaque, whey coloured fluid above, with a pellicle or white flocculi on the surface.

Experiment 51

Bile added to the third portion of chyme, taken from the stomach one hour and ten minutes after a breakfast of venison steak, &c., Experiment 29, (March 6th, 1831,) changed it from a brownish, homogeneous paste, to a milky fluid, with small, white flocculi, floating about, or adhering to the sides of the vial: and a light brown sediment settled to the bottom.

The usual proportion of dilute muriatic acid, added to this, produced no very essential change in its appearance, causing only a little more deposition of sediment, and slightly increasing the milky colour.

Experiment 52

Bile added to the chyme formed from the eggs, digested out of the stomach, Experiment 30, (March 7th, 1831,) produced a rich, milky fluid, with a small quantity of fine, light coloured sediment, falling to the bottom.

The dilute acid, added to this, produced fine coagulæ, and formed a milk white whey, or fluid, from which, more of the light coloured sediment was precipitated.

Experiment 53

More minutely to observe the respective changes by the addition of bile and muriatic acid, in the several parcels of chyme formed in Experiment 31, (March 9, 1831,) and to note their difference, I put equal quantities of each into glasses, and added a portion of hog's gall.

In the first, (that taken from the stomach at 10 o'clock, one hour after having eaten,) fine, bright orange coloured coagulæ were immediately formed, equally diffused through a fluid of the same colour, exhibiting no perceptible sediment on standing at rest; but held the coagulæ, uniformly suspended throughout the fluid. The dilute acid, added to this, occasioned a copious sediment to fall to the bottom, and with it, all the colour of the mixture, leaving a transparent, semi-gelatinous-like fluid above, in the proportion of about three-fifths of the whole; upon the surface of which, floated a thin, white pellicle.

The second portion, (that produced on the bath) under the same treatment, exhibited nearly the same appearance, with the exception of the colour, which was a shade or two lighter. The sediment was not quite so compact; the fluid less gelatinous; and there was less of the white pellicle on the surface.

The third portion, treated like the other two, differed about as much from the second, as this did from the first. They all exhibited the same general appearance.

The fourth, or aqueous portion, under the same treatment, exhibited a wide difference. The same proportion of bile added to this, produced a similar coloured fluid, at first, with a very little coarse coagulæ—not so uniformly diffused through the liquid; but inclining more to precipitation. On adding the acid, it let fall a very small quantity of yellowish green sediment, leaving a thin, semi-transparent fluid, in more than quadruple the proportion of the other three.

Experiment 54

Bile and dilute muriatic acid, added to a portion of the bread and milk chyme, formed in Experiment 35, (March 14th,) produced their usual coagulation and precipitation, but of a lighter yellow; the sediment forming about one fourth of the mass. The small, white particles, forming the pellicle on the top, were in greater proportion than in some of the other experiments, especially those on lean meats. The fluid part was in greater portion to the sediment, and of a whey colour and consistence.

To another equal quantity of this same kind of chyme, I added bile, as in the other, and instead of muriatic acid, I used *pancreatic juice,* fresh from a recently slaughtered beef. An appearance exactly similar to that produced by the acid, was exhibited, except that the precipitate was more slowly thrown down, and in larger proportion; and the white pellicle on the surface was less. The fluid and sediment were a shade lighter, and in more equal proportions.

Experiment 55

Pancreatic juice, combined with the chyme of roast beef, formed both in and out of the stomach, increased its thin, paste-like consistence, and gave it more of a cream colour. Bile, added to this, produced fine coagulæ, suspended from the top to the bottom, without depositing any distinct sediment. Diluted muriatic acid darkened the whitish colour, a shade or two, threw down a more copious sediment, and increased the white pellicle on the top.

Experiment 56

Bile and pancreatic juice, added to the fresh meat and vegetable soup chyme, Experiment 36, (March 14th, 1832,) produced loose, cream coloured coagulæ; which, on standing, separated into three, about equal proportions; a coarse, brownish sediment, a semi-transparent, whey coloured fluid, and a thick, white pellicle at the top.

EXPERIMENTS, &c.

THIRD SERIES

WASHINGTON, D. C. 1832.

Experiment 1

Dec. 4. At 2 o'ck, 30 ms., P. M.—Weather cloudy, damp and snowing—Th: 35°—Wind N. W. and brisk—the temperature under the tongue was 99°; in the stomach, 101°. Dined, at 3 o'clock, 30 mins., on *beef soup, meat* and *bread.* 4 o'clock, 15 mins.—took out a portion—particles of beef slightly macerated, and partially digested. 5 o'clock, 15 mins.—took out another portion—digestion more advanced—meat reduced to a pulp; particles of bread and oil floating on the top. Temperature of stomach, 100°. 6 o'clock, 45 mins.—digestion not completed—contents considerably diminished. 7 o'clock, 45 mins.—stomach empty—chyme all passed out.

Experiment 2

Dec. 5. At 7 o'clock, A. M., temperature of the stomach, 100°; of the atmosphere, 30°.

At 1 o'clock, P. M.—temperature of stomach, 100°—atmosphere, 40°—he ate eleven *raw oysters,* and three *dry crackers*; and I suspended one *raw oyster* into the stomach, through the aperture, by a string. 1 o'clock, 30 mins.—examined—stomach full of fluids—digestion not much advanced. The oyster on the string appeared entire, though perhaps slightly affected on the surface. 2 o'clock—examined, and took out oyster—about one third digested, but retained its shape. 2 o'clock, 30 mins.—oyster gone from the string, except a small piece of the heart. Temperature of the stomach 101½°. Fluids less considerable. 4 o'clock, 15 mins.—stomach empty.

Experiment 3

At 3 o'ck, 45 ms., P. M., same day, he dined on *roast turkey, potatoes* and *bread.* 4 o'clock, 30 mins.—examined, and took out a portion. Turkey nearly all dissolved—vegetables half reduced. 5 o'clock, 15 mins.—took out another portion, almost completely chymified. 5 o'clock, 45 mins.—examined again—stomach nearly empty. 6 o'clock—some chyme yet remaining. 6 o'clock 15 mins.—stomach empty.

Experiment 4

Dec. 6. At 8 o'clock, 30 min., A. M., he breakfasted on *bread and butter,* and one pint of *coffee.* 9 o'clock, 45 mins.—examined—stomach full of fluids. 10 o'clock, 30 mins.—examined, and took out a portion, resembling thin gruel, in colour and consistence, with the oil of the butter floating on the top; a few small particles of the bread, and some mucus, falling to the bottom—about two thirds digested. It had a sharp, acid taste. Temperature of the stomach, 100°—atmosphere, 38°. 11 o'clock, 30 mins., stomach empty.

Experiment 5

At 4 o'clock, 30 mins., P. M., same day—he dined on *sausage* and *bread;* full meal. 5 o'clock, 30 mins.—stomach full of fluids; digestion but very little advanced. 6 o'clock, 30 mins.—digestion considerably advanced: few distinct particles of sausage and bread to be seen entire. 7 o'clock, 30 mins., stomach empty.

Experiment 6

Dec. 7. At 8 o'clock, A. M.,—examined stomach, and took out, with considerable difficulty, an ounce only, of gastric juice, and that not very pure. Some yellow bile came mixed with the latter portions. Temperature of the stomach, 99°—atmosphere 28°. He breakfasted, at 9 o'clock, on *corn and wheat bread, butter,* and *coffee.*

At 10 o'clock, 45 mins.—examined, and took out a portion—food partly digested; few small particles to be seen. Stomach full of fluids, with a thin pellicle of oil on the top. Temperature of the stomach, 100°.

At 12 o'clock, M.—stomach full of fluids—digestion not complete—particles of bread floating about in a pulpous state—oil floating on the surface.

At 12 o'clock, 30 mins., M.,—examined—contents of stomach half diminished—distinct particles of oil on the surface.

At 12 o'clock, 45 mins.—entire particles of bread, yet to be seen—quantity of fluid diminishing.

At 1 o'clock, P. M.—distinct particles of bread still floating—fluid less.

At 1 o'clock, 15 mins.—stomach empty.

Some indications of gastric derangement this morning: small aphthous patches on the mucous membrane: juice acrid and sharp, with bile mixed with it.

Experiment 7

At 3 o'clock, 30 mins., P. M., same day, he dined on *roasted mutton, bread* and *potatoes.* 4 o'clock, 45 mins.—examined—stomach full—digestion advancing. 5 o'clock, 45 mins.—contents of stomach three quarters reduced in quantity, and almost completely chymified. 6 o'clock, 30 mins.—stomach nearly empty; a little pulp of the bread only to be seen, floating in a little milky fluid. 7 o'clock—stomach empty.

Experiment 8

Dec. 8. At 5 o'clock, 30 mins., A.M.—temperature of stomach, 99°. 9 o'clock—finished breakfasting on *fried sausage, dry toast,* and a pint of *coffee.* 10 o'clock, 30 mins—stomach full of fluids—villous coat red and irritable, inclining to dryness—a thin, whitish coat on the tongue, and a similar appearance on the protruded portion of the stomach. 11 o'clock, 45 mins.—stomach full—oil floating on the top, and rancid. Temperature of stomach, 99°—atmosphere, 46°. Weather damp and cloudy.

This, and the 6th Experiment, show, that when there are indications of disease on the coats of the stomach, and on the tongue, digestion is consequently protracted; and, also, that oil is particularly hard of digestion.

Experiment 9

At 9 o'clock, A.M., same day, the vial containing the *bread and butter* aliment, taken from the stomach on the 5th inst. (Experiment 4,) at half past 10 o'clock, A.M., was placed on the bath for four hours, in the usual temperature, between 95° and 100°. Digestion commenced, and advanced regularly, partially reducing the oil to a milky fluid.

Dec. 9. At 11 o'clock, A.M.—added one ounce of gastric juice, and continued it on the bath for eight hours, when the oil became more, but not completely digested; particles of the limpid oil being still perceptible.

This affords an example of the re-commencement of digestion, after the operation had ceased, by the addition of a fresh supply of gastric juice.

Experiment 10

A 2 o'clock, 45 mins., P.M., same day, (*Dec.* 8,) I suspended a *roasted oyster,* weighing, when raw, four drachms, into the stomach, and he ate twelve of the same kind, each weighing about the same.

At 4 o'clock, 30 mins.—examined—oyster remaining on the string, not half digested—fluid in the stomach rancid. Complained of head ache, lassitude, dull pains in the left side, and across the breast—tongue furred, with a thin, yellowish coat, and inclined to dryness—eyes heavy, and countenance sallow. The villous membrane of the protruded portions of the stomach, very much resembled the appearance of the tongue, with small aphthous patches, in several places, quite irritable and tender.

I suspended observations, and dropped into the aperture at night, six grains *blue pill,* and four *aloetic pills,* (common size,) and sprinkled on the exposed surface of the stomach, five or six grains of *calomel.* Medicine operated early the next morning; relieved the symptoms of indisposition; changed the appearance of the stomach and tongue; and removed the aphthæ. On the 9th, he felt quite well; and the coats of the stomach looked healthy again.

Experiment 11

Dec. 13. At 7 o'clock, A.M.—temperature 100°—villous membrane perfectly healthy, of a pale pink colour, and uniform—mucous coat smooth and even. Extracted two ounces of gastric juice. It distilled more freely than common. More could have been obtained. I had never before seen the pure juice flow so freely. He felt in perfect health: had taken neither food or drinks since 9 o'clock, last evening.

At 9 o'clock—breakfasted on *broiled breast of mutton, bread, butter,* in usual quantity, and a pint of *coffee,* and kept exercising. Digested in three hours and a half; stomach empty and clean.

Experiment 12

At 2 o'ck, P.M., same day—stomach empty—coats clean—he dined on three *soft boiled eggs* and *bread,* and drank half a pint of *water.* 3 o'clock—digestion advancing. 4 o'clock—contents nearly gone from the stomach—yolk of eggs still visible, with a few particles of oil. 5 o'clock—very little chyme in the stomach. 5 o'clock, 15 mins.—some still remaining. Complains of slight head ache—pulse full and crowded—contents of stomach acrid—countenance rather sallow; eyes languid; tongue a little coated with a thin, yellowish fur. His bowels have not been moved since yesterday morning, at 10 o'clock; then inclined to costiveness.

N.B. After taking breakfast, he exercised moderately. About 12 o'clock, M., he walked about two miles very quick. After his return to his lodgings, he threw off his coat, and went into the open air again. Soon after which, he began to feel the pain in his head, &c.

Experiment 13

Dec. 14. At 7 o'clock, A.M.—stomach deeper colour than ordinary, and inclined to dryness—some small, aphthous patches, and spots of darker colour—mucous coat not uniform and even; some places thicker, a little elevated, and rolling up, like thin membrane, leaving a spot beneath, red and irritable. Very little juice could be extracted. I obtained a small quantity of fluid, mixed with yellow bile. It did not yield the peculiar acid taste of the gastric juice. Temperature of the stomach, 100°. St. Martin did not feel his usual appetite.

At 9 o'clock, he breakfasted on the same kind of diet as yesterday—had less appetite, and was labouring under some gastric derangement. He continued quiet, most of the time in a recumbent position. 10 o'clock—stomach full—globules of oil floating about—appearance of villous membrane, about the same; no perceptible change. 11 o'clock—stomach still full—appearances similar to those in last examination. 12 o'clock, M.—contents half diminished—particles of bread, and coat of oil on the surface. 1 o'clock, P.M.—some fluid still in the stomach, and a larger proportion of oil than at last examination. Taste of the contents, more sharp and rancid; fast leaving the stomach. At this time, I observed several small, sharp pointed, white pustules or pimples, here and there dispersed over the exposed portion of the inner coat. 1 o'clock, 30 mins.—stomach clear and clean.

Experiment 14

At 2 o'clock, P. M., same day, he dined on three *soft boiled eggs, bread and butter,* and half a pint of *water,* (same as yesterday, 2 o'clock.) Digested in three hours.

Experiment 15

Dec. 15. At 8 o'clock, A. M., I examined stomach—temperature 100°. Appearance of coats more natural and healthy than yesterday morning; less of those small, white, pointed pimples, and aphthous spots. Very little gastric juice could be obtained; not more than one ounce, and that mixed with an unusual quantity of mucus, not so clear as common. Complained, as he frequently does, during this operation, of a sense of sinking, and vertigo, after extracting this quantity. This feeling, however, subsided in a few minutes after rising.

At 8 o'clock, 30 mins. he breakfasted on *beef steak, bread* and *coffee.* At the same time, he thoroughly masticated four drachms of the steak, which I put into the gastric juice, just before taken from the stomach. To another similar quantity of gastric juice, I put the same quantity of the steak, unmasticated, and in one entire piece. I placed them both on the bath at 100°; and at the same time, I put the same quantity steak into one ounce of simple water, and treated it with the others on the bath.

At 11 o'clock, I examined the stomach, and found his breakfast nearly digested, and more than half gone from the stomach. I took out an ounce of what remained, which was almost completely chymified, a few particles of the bread, in a soft, pultaceous condition, only remaining. Compared this with the three parcels on the bath. It very nearly resembled the masticated meat in the gastric juice, but more digested, and thinner, and contained particles of oil (melted butter) and bread, which were not in the masticated food in the vial. The unmasticated meat differed considerably. It was not so thick and gelatinous-like; was of a darker colour; and the piece of meat retained its shape, and was not much diminished in size, the surface only a little wasted, softened, and covered with a cineritious coat.—The contents of the vial of masticated meat and water, suffered very little or no change since put in; no more than had been effected simply by mastication. Continued them all on the bath.

The contents of the vials, continued on the bath for twenty-four hours, exhibited the following changes. The portion taken from the stomach at 11 o'clock, remained nearly the same as when extracted, perhaps more completely chymified. That which was masticated, and put into the gastric juice, was reduced to a thick, pultaceous, semi-fluid mass, but retaining some distinct fibres of the meat, which, after standing awhile, subsided to the bottom of a yellowish, whey coloured fluid. These remaining particles of aliment, I conceived to have been left for want of a sufficient quantity of gastric juice; the quantity at first being too small to dissolve the whole of the meat put in. That portion in the vial of water had undergone no other change than that of incipient putrefaction, which was very evident. The unmasticated piece of meat had undergone an evident process of digestion. It was about half diminished, and the texture of the remaining part loose and soft. The containing fluid had become of a greyish-brown colour, opaque, with a fine, brown sediment, settling to the bottom, similar to that of the masticated meat in the gastric juice. The gastric juice, containing the unmasticated meat, when taken from the stomach, some sixty or seventy hours before, was not so pure as common; was mixed with yellow bile; and was in too small proportion to the meat. The colour and flavour of the other two portions were very similar, except that the one with the masticated meat was more sharp and acrid.

This experiment shows the necessity of mastication; and also demonstrates, that simple maceration, at the natural temperature, will not effect digestion.

Experiment 16

A dinner of *pork steak* and *bread,* taken at 1 o'clock, P.M., same day—digested in three hours, forty-five minutes.

Experiment 17

Dec. 16. At 9 o'clock, A.M., he breakfasted on *cold, pork steak, bread,* and one pint *coffee.* Digestion completed in three hours. Two hours after having eaten, a pellicle of oil was found floating on the top of the gastric contents.

On examining the stomach, an hour after the chyme had passed out, several red spots and patches, abraded of the mucous coat, tender and irritable, appeared spread over the inner surface. The tongue, too, had upon it a thin, whitish fur. Yet his appetite was rather craving.

At 2 o'clock, 30 mins., P.M., he ate a full dinner of *cold, roasted pork,* (fresh) *bread,* and a piece of *raw radish.* Digestion completed in seven hours.

Experiment 18

Dec. 17. At 8 o'clock, 30 mins., A.M., I put two drachms *fresh, fried sausage* in a fine muslin bag, and suspended it into the stomach. He immediately after breakfasted on the same kind of *sausage,* and a small piece of *broiled mutton, wheat bread,* and a pint of *coffee.* 11 o'clock, 30 mins., stomach half empty—contents of bag about half diminished. 2 o'clock, P.M., stomach empty and clean—contents of bag all gone, except fifteen grains, consisting of small pieces of cartilaginous and membranous fibres, and the spice of the sausage; which last weighed six grains; leaving only nine grains of the aliment put in. In consequence of being called out, I delayed the last examination longer than was necessary.

Experiment 19

Dec. 18. At 8 o'clock, 30 mins., A. M., I suspended two drachms masticated, *fried sausage,* confined in a muslin bag, into the stomach, and he breakfasted on the same kind of food, with *bread* and *coffee.* 11 o'clock, 30 mins., stomach half empty—contents of bag about half gone. 1 o'clock, P. M., stomach nearly empty—very little left in the bag. 1 o'clock, 30 mins., stomach clear, except the bag, which contained a little of the sausage: took this out, and it weighed one drachm, spice and all, of which there was less than yesterday. The bag, when drawn out, came from near the pylorus, and was covered with a coat of mucus and yellow bile. The contents of the stomach have been unusually acrid since yesterday morning, and he complains of unusual smarting and irritation at the edges of the aperture: countenance sallow; tongue covered with a thin, yellowish coat; and several deep red patches on the inner coat of the stomach: does not feel his usual appetite. 9 o'clock—dropped into the aperture, twelve grains *blue pill,* and five *cathartic pills*—operated early the next morning; removed the symptoms; and restored his healthy sensations and functions.

Experiment 20

Dec. 19. At 8 o'clock, 45 mins., A. M., I suspended three drachms *broiled bass,* in a muslin bag, into the stomach, and he breakfasted on the same kind of fish, with *bread,* and a small piece of *sausage,* and a pint of *coffee.* 2 o'clock, P. M.—complains of smarting at the aperture—I took out the bag—remaining contents weighed two drachms, having lost one drachm only in five hours and a quarter. Coats of the stomach did not appear healthy—deeper red than natural, with patches of still deeper colour, spread over the protruded portion. Mucous covering abraded in places, and rolled up; resembling shreds of epidermis, torn from a blistered surface.

These three last experiments, are examples of the solvent or chemical action of the gastric juice. It penetrated the muslin bags, dissolved the food, and allowed the chyme to strain out. They also indicate that irritating substances, (as, for instance, the muslin bags, in these experiments,) produce a diseased state of the stomach.

Experiment 21

Dec. 20. At 8 o'clock, 30 mins., A. M.—Coats of stomach appear healthy—considerable fluid plainly to be seen. It ran out of the aperture on turning him down; was transparent, and contained flocculi of mucus. Breakfasted on *broiled bass, toasted bread* and *coffee.* Digested in five and a half hours.

Experiment 22

At 2 o'clock, P. M., he dined on *boiled chicken,* and *wheat bread.* Digested in four and a half hours.

Experiment 23

Dec. 21. At 8 o'clock, 30 mins., A. M.—stomach not perfectly healthy—several small, deep red patches, on the exposed surface. Extracted four drachms gastric juice, tinged with yellow bile. Masticated one and a half scruples of the thigh of a *boiled chicken,* and half a scruple of bread: put them into this gastric juice, and placed the vial in the axilla. Into the same quantity of pure water, warmed to 70°, I put the same quantity and kind of aliment, and placed them in the same situation. He breakfasted at the same time, on the same kind of diet. 1 o'clock, P. M.—stomach empty. At 2 o'clock, he dined on same kind of food. 6 o'clock, 30 mins.—stomach empty.

The masticated portion put into the vial of gastric juice, placed on the bath, and frequently agitated, digested regularly and uniformly until about 2 o'clock, P. M., when the particles were all dissolved, except a few fibres. That in the vial of water, kept in the same situation, had not changed its appearance from the time it was put in.

On separating the remaining particles of food, in the gastric juice, at evening, filtering on thin muslin, and drying with paper, it weighed fifteen grains, and left four drachms and a fraction, of an opaque, milky coloured fluid.

That in the water, taken out at the same time, weighed forty grains, and left four drachms of a turbid fluid, like water, with flour stirred in it, and had a mawkish, insipid taste and smell. The first had the acid smell and taste, peculiar to the gastric contents.

Experiment 24

Dec. 22. At 8 o'clock, A. M.—examined stomach—temperature 100°. Extracted about four drachms gastric juice, pure, but not free.

At 8 o'clock, 30 mins., he breakfasted on *bread, cheese* and *coffee.* 9 o'clock, stomach full of fluids—temperature 100°. 11 o'clock—stomach full, with the cheese in a fluid form, floating on the surface; bread reduced to a pulp—temperature 100°. 12 o'clock, M.—food still in the stomach; but considerably diminished. 1 o'clock, 30 mins., P. M.—some of the cheese yet remaining—stomach nearly empty. 2 o'clock—stomach empty.

The coats of the stomach have not appeared in their usual healthy condition, for several days past—the colour darker—mucous coat unequal—some patches of a purplish colour, with aphthous edges—surface inclined to be dry—very little secretion of gastric juice—digestion slower, and less perfect than usual—bowels inactive, nothing having passed them for sixty hours.

It would seem from this experiment, that cheese was difficult of digestion. In addition to its closeness of texture, it generally contains a large proportion of oil.

Experiment 25

Dec. 23. At 6 o'clock, A. M.—temperature of stomach, 100°—pulse 65 a minute. 9 o'clock—temperature of stomach, 100°—pulse 75. Villous membrane inclined to dryness, and of a darker than natural colour; papillæ small and sharp; mucous covering scarcely perceptible; bowels costive; tongue coated with a yellowish fur, and its edges pale. I poured in, at the aperture, one ounce *Ol. Ricini,* and sprinkled over the surface of the protruded coats, five or six grains of *calomel.* He ate a light breakfast of *corn bread* and *crackers,* and drank a pint of *coffee,* immediately after.

At 2 o'clock, P. M.—stomach empty—coats look healthier. Medicine not having moved the bowels, I put in, at the aperture, twelve additional grains of *calomel,* per se.

At 5 o'clock, the stomach was in commotion—indications of the cathartic operation of the calomel: slight nausea; stomach full of a white, frothy fluid, running out at the aperture, like fermenting beer from a bottle; slight pain and motion in the bowels; and increased secretion of saliva. No motion from the bowels. Temperature of stomach, 101°. Pulse 80 beats in a minute.

At 8 o'clock, calomel had operated twice, copiously, commencing at 7. Temperature of stomach, 100°. Pulse 62, soft and mild.

Experiment 26

Dec. 25. At 8 o'clock, A. M.—weather partially cloudy—atmosphere dry, and smoky—wind E. and light—Th: 31°. Temperature of the stomach, 100° and a fraction. Pulse 55, in a recumbent position; 65, sitting erect. A few small, red spots, on the mucous surface. The gastric secretions appear as healthy as usual.

At 9 o'clock, he breakfasted on *boiled, salted, fat pork, corn bread* and *coffee.* 10 o'clock, the stomach at the same temperature as at 8 o'clock. Pulse 65 in a recumbent, and 75 in an erect position. Gastric cavity full of a heterogeneous mixture.

At 11 o'clock, 30 mins.—just returned from walking moderately, about an hour, a distance of two and a half miles; not to produce free perspiration, but gentle diaphoresis. Weather clear, calm and dry. Th: 50°. Temperature of the stomach 101°. Pulse 72, in a recumbent position; 82, sitting erect, and regular. Contents of stomach half reduced, and nearly homogeneous.

At 12 o'clock, 30 mins., M.—temperature of stomach, 100½°. Pulse 62, recumbent; 72 erect. Contents nearly gone.

At 1 o'clock, 30 mins., P. M., stomach empty.

At 9 o'clock,—weather cloudy—atmosphere dry—no wind—Th: 42°—the temperature of the stomach was 99½°. He drank half a pint of water fifteen or twenty minutes before examination. Pulse 62, recumbent; 72, erect.

This is an example of the increase of the temperature of the stomach on exercise. See also, subsequent experiments.

Experiment 27

Dec. 26. At 6 o'clock, A. M.—weather cloudy; atmosphere damp; wind N. E. and light; Th: 38°; temperature of the stomach, 99½°. Pulse 55, recumbent;

65 erect. Respirations, in a recumbent position, 15, and in a sitting position, 18 a minute.

At 8 o'clock, he returned from a walk of two miles, but not to produce perspiration. Weather damp and raining lightly. Th: 36°. Temperature of the stomach, 101°. Pulse 65 recumbent; 85, erect. Feelings of impatience here evidently accelerated his pulse, in the erect position. He was vexed at being detained a few minutes from his breakfast.

At 5 o'clock P. M.—weather rainy—wind N. E.— Th: 41°—I examined the stomach. Temperature, 99½°. Pulse 60, recumbent; 70, erect.

At 8 o'clock, the temperature of the stomach, 101°. Pulse 50, recumbent; 60, erect. Respirations, 15 a minute.

His diet through the day had been confined principally to farinaceous substances, wheat bread and crackers, in moderate quantities.

Experiment 28

Dec. 27. At 6 o'clock, A. M. Weather unpleasant. Atmosphere damp. Wind E. Th: 38°. Temperature of stomach, 99½°. Surface clean and healthy. No dark red, or aphthous patches, nor white, elevated points. Mucous coat uniform and even, of the natural colour. No excoriation or smarting at the edges of the aperture. I extracted one ounce of gastric juice, slightly tinged with yellow bile. This, I conceive to have been entirely accidental; and occasioned by the regurgitation of the bile through the pylorus, as he turned upon his back, from right to left, to favour the exit of the gastric juice. The same thing has happened several times before.

At 9 o'clock, he breakfasted on three ounces *broiled breast of mutton,* four ounces of *wheat and corn bread,* very thoroughly masticated, and a pint of *coffee.* At the same time, I put two drachms of same kind of food, equally well masticated, into the ounce of gastric juice, taken from the stomach at 6 o'clock, and the same quantity of same kind of food, masticated in the same manner, into an ounce of simple water; placed them, both together, first in the axilla, and afterwards on the bath, between 96° and 100°.

At 12 o'clock, M., stomach nearly empty. Was just able to get out one ounce for comparison, almost completely dissolved; a few small particles of bread only visible. Temperature 100°.

At 12 o'clock, 30 mins., no distinct particles of food to be seen. All chymified, and passed from the stomach. Nothing but a little frothy mucus remaining in the stomach. Coats clean; colour, pale pink. Temperature 100°.

At 2 o'clock, P. M., he dined on the same quantity and kind of food that he had taken for his breakfast, (*broiled mutton* and *bread.*) Drank nothing since morning. Temperature of stomach 100°. Th: 62°, Wind S. Weather fair, since 12 o'clock. 2 o'clock, 30 mins., stomach as full of fluids as when he drank a pint immediately after eating. No perceptible difference in appearance. 6 o'clock, stomach empty and clean. 9 o'clock, temperature of the stomach 100°. Weather same as at 2 o'clock.

The changes effected in the contents of the two vials, mentioned above, and kept in the axilla till 9 o'clock, P. M., were as follows.

In that containing the gastric juice, the food was about half dissolved, and loosely suspended towards the bottom of a reddish-grey coloured fluid.

That in the water exhibited no other appearance of digestion than what was effected by mastication, when first put in. The masticated food had subsided to the bottom of a transparent, watery fluid, as clear as when first put in.

At 8 o'clock, A.M. of the 28th, I added the two drachms of gastric juice, taken from the stomach, at that time, to the vial containing the gastric juice and the same quantity of water to the watery mixture; and placed them in the axilla again.

At 6 o'clock, P.M., examined vials—digestion had re-commenced, and advanced in the gastric juice, in proportion to the quantity added. The sediment had become more dissolved, and the fluid part increased. This sediment taken out, filtered through muslin, and pressed as dry as when put in, weighed forty-five grains only, having completely dissolved one drachm and fifteen grains; and produced a gruel-like milky coloured fluid.

That in the water, remained unchanged; and when taken out, and pressed dry, through a piece of muslin, like the other, weighed one drachm and thirty-five grains. This reduction, I suppose, was the effects of mastication, and maceration in the water for thirty-six hours.

These two parcels, kept tight corked, in a temperature between 50° and 70°, remained free from any fœtor for forty-five days. The gastric portion, at the end of this time, emitted a caseous flavour; and the aqueous portion smelt musty and sour.

This is a comparison between solution by the gastric juice, and maceration in water. These results are interesting, not only as establishing physiological principles on certain data; but they have an important practical application. They have, consequently, been frequently repeated.

The fact, that the stomach contains a quantity of fluid, soon after the ingestion of dry food, which was alluded to in the preliminary essay, is here perfectly demonstrated.

Experiment 29

Dec. 28. At 8 o'clock, A.M. Weather clear. Atmosphere dry. Wind N. Th: 34°. Temperature of stomach, 100°. Coats clean and healthy. Gastric juice scarce; extracted two drachms only, and that with considerable difficulty.

At 9 o'clock, A.M., he breakfasted on same kind of food as yesterday, in usual manner, slightly masticated, and swallowed fast, without regard to quantity. 1 o'clock, P.M., a small portion still in the stomach—nearly dissolved. 1 o'clock, 30 mins. stomach empty.

Experiment 30

Dec. 29. At 9 o'clock, A.M. Weather clear and dry. Wind N.W. and light. Th: 34°. Temperature of stomach, 100°. Coats clean and healthy. He breakfasted on *fat pork, dry toast* and *coffee*—full meal. 1 o'clock, P. M., stomach half full of a lardaceous fluid—no particle of any thing else but gastric fluids to be seen. Temperature 100°. 2 o'clock, 30 mins., stomach not empty. 3 o'clock, stomach empty and clean.

The protracted period of complete chymification in this meal, I conceive to have been principally owing to the unusual *quantity* of food taken, being disproportioned to the gastric secretions, and more than was required to replenish the natural waste of the system. The *quality* of the food had, undoubtedly, some effect.

Experiment 31

Dec. 30. At 8 o'clock, A. M. Weather clear and dry. Wind N. W. and light. Th: 26°. Stomach clean and healthy. Temperature 100°. Gastric juice pure, and distills more freely than common. Extracted one ounce, without any difficulty.

At 9 o'clock, he breakfasted on two and a half ounces of *boiled, recently salted, fat pork,* three ounces of *wheat bread,* masticated in usual manner, and one pint of *coffee.*

At the same time, I took two parcels, equal quantities, of the same kind of food, (pork and bread) half a drachm of each kind, both masticated in same manner: put one of them into the ounce of gastric juice taken from the stomach before eating; and the other, into the same quantity of simple water, of the temperature of the gastric juice; and placed them in the axilla.

At 11 o'clock, I took out of the stomach, one and a half ounces of its contents; put it into a vial, and placed it in the axilla, with the other two. The difference between this taken out of the stomach, and that in the gastric juice, was quite perceptible. The particles of aliment contained in the last, appeared more nearly dissolved, very few remaining distinct. That taken from the stomach contained a larger proportion of the entire food and floating oil. The colour of the middle portions, as well as the smell and taste, were very similar. That from the stomach was rather more rancid and sharp than that in the gastric juice in the vial. Both possessed the peculiar gastric, acid flavour.

At 1 o'clock, 30 mins., the stomach was empty and clean, and probably was so at 1 o'clock; but owing to accident, I did not examine at that time. He became intoxicated in the afternoon, and interrupted the experiments.

On the 2d of *January,* 1833, I added half an ounce of fresh gastric juice to the parcel of chyme taken from the stomach at 11 o'clock, in the above experiment, which, at this time, contained a large proportion of undigested lardaceous matter, floating on the surface. Put the vial in the axilla.

On the 3d, I added three drachms more of fresh gastric juice, to the above.

On the 6th, I added three drachms gastric juice to the above, and placed it on the bath.

On the addition of each of these portions of gastric juice, chymification recommenced, and the lardaceous portion of the aliment continued to be reduced for several hours, till the solvent power became expended, when its action would cease.

Experiment 32

Dec. 31. At 7 o'clock, A. M. Weather cloudy. Atmosphere damp and chilly. Wind S. Th: 30°. Temperature of the stomach, 100½°—colour darker red than natural, and arid. Mucous coat abraded in spots, and rolled in small shreds; more irritable than usual.

At 8 o'clock, 30 mins., breakfasted on same quantity and kind of food as yesterday, (pork, bread, &c.) At 11 o'clock, took out one and a half ounces contents from the stomach, in appearance half digested. 12 o'clock, M., took out another portion, more completely dissolved. Stomach nearly empty. 1 o'clock, stomach empty.

At 1 o'clock, 30 mins., he dined on *salted, boiled beef, potatoes, parsnips* and *bread,* full meal, without regard to quantity or mastication. 4 o'clock, 30 minutes, stomach perfectly empty.

The one and a half ounces, taken from the stomach at 11 o'clock, A. M., very nearly resembled the contents of the vial of gastric juice and masticated food of the 30th, (yesterday,) in almost every particular. That taken out at 12 o'clock, M., had more of the lardaceous, and less of the distinct fibrous particles of aliment.

The diseased appearance of the stomach at this examination, was probably the effect of intoxication the day before.

Experiment 33

Jan. 1, 1833. At 8 o'clock, A. M. Weather dark and rainy. Wind S. Th: 50°. Temperature of stomach, 100°, healthy and clean. Extracted half an ounce of gastric juice.

At 9 o'clock, I took two scruples *salted, lean beef,* (boiled,) chopped very fine, with a knife: put one scruple into the half ounce of gastric juice, and the other scruple into half an ounce of simple water; and placed them together in the axilla. At the same time, he breakfasted on two ounces of *boiled, salted, lean beef, bread,* and a pint of *coffee.*

At 12 o'clock, M., I took from the stomach one ounce of its contents, not fully digested; bread principally remaining, reduced to a pulp. Compared with the gastric juice and food in the vial, the particles of meat seemed rather more dissolved. Stomach about half empty.

At 1 o'clock, P. M., stomach empty and clean.

At 8 o'clock, 30 mins., A. M, on the 3d, I added one drachm fresh gastric juice to the vial of gastric juice and chopped beef, and one drachm of water, to the watery mixture, and placed them together in the axilla.

On the 4th, the beef in the gastric juice not being completely dissolved, I added two drachms fresh gastric juice to it; and two drachms of water to the aqueous mixture. Continued them on the bath, or in the axilla. The watery portion began now to smell quite fœtid.

At 8 o'clock, on the 5th, the meat in the gastric juice was completely dissolved, and a fine, reddish grey sediment had fallen to the bottom of an opaque, gruel-like fluid, with a pellicle of greyish white particles on the top. The aqueous portion had become more fœtid. The particles of meat were the same as when first put in, only a little macerated, and paler—the fluid transparent, but becoming darker, and a little greenish—no appearance of solution.

On the 10th, the contents of the aqueous portion were quite fœtid. The gastric portion was perfectly sweet and bland.

Experiment 34

At 1 o'clock, 30 mins., P.M. same day, he dined on *lean, salted beef* and *bread.* Digested in three and a half hours.

Experiment 35

Equal parts of *alcohol* and *gastric juice,* mixed together and agitated, produced a turbid, milky white fluid; which, after standing at rest, raised a thin, white coat of fine, loose coagulæ on the surface. When the juice and alcohol were first put together, and before agitating, the gastric juice settled to the bottom, and the alcohol remained on the top, indicating that its specific gravity was less than the fluid.

Experiment 36

Jan. 2. At 8 o'clock, A.M.—stomach empty—extracted half an ounce of gastric juice. 8 o'clock, 30 mins., he breakfasted on *dry bread* and a pint of *coffee.* 11 o'clock, stomach nearly full of a pulpous, semifluid mass, resembling thick gruel. 12 o'clock, nearly empty. 12 o'clock, 30 mins., empty and clean.

Experiment 37

At 2 o'clock, P.M., he dined on *boiled potatoes,* a small piece of *bread,* and drank a glass of water. 4 o'clock, 30 mins., stomach full of fluids, and quite acrid, of a whitish colour, with particles of potatoes floating about. 6 o'clock, stomach empty.

Experiment 38

Jan. 3. At 8 o'clock, 30 mins., A.M. Weather pleasant, smoky and clear. Th: 38°. Temperature of the stomach, 101½°, immediately after a walk of two miles, producing free perspiration, and colour in the face. Extracted half an ounce of gastric juice.

At 9 o'clock, 30 mins., he breakfasted on cold *broiled breast of veal, boiled potatoes,* and *bread.* At the same, or within fifteen minutes of the time, I suspended into the stomach, at the aperture, twenty grains of masticated lean veal, contained in a muslin bag.

At 12 o'clock, M., contents of stomach half diminished. 1 o'clock, P.M., stomach nearly empty. 1 o'clock, 30 mins., all gone from the stomach, except the muslin bag and contents. The contents appeared to be about half diminished.

At 2 o'clock, I took out the bag of veal, and pressing it as dry as I could, without forcing the remaining particles of meat through the cloth, it weighed ten grains, having lost ten grains by digestion, in four and a half hours. The veal, when first put in the bag, and suspended in the stomach, was of a clay, or greyish white colour; but when taken out and weighed, was of a palish red, or light flesh colour, and of a glutinous appearance.

Experiment 39

At 3 o'clock, P.M., same day, dined on *broiled veal* and *bread,* and drank half a pint of *water.* Digested in two hours.

Experiment 40

Jan. 4. At 8 o'clock, A.M.—Stomach healthy. Extracted two drachms gastric juice—came pure, but very slow.

At 9 o'clock, breakfasted on *broiled veal, bread* and *coffee.* 11 o'clock, stomach full—oil floating on the surface, acrid and sharp, excoriating the edges of the aperture and skin. 12 o'clock, M., chyme passing out. Stomach two thirds empty. 1 o'clock, P.M., stomach empty.

Experiment 41

At 2 o'clock, P.M., same day, he dined on *breast of broiled veal* and *bread,* and drank a tumbler of *water.* 5 o'clock, 30 mins., stomach nearly empty. 6 o'clock, examined stomach—chyme of a milky white colour. 6 o'clock, 30 mins., chyme still remaining. 7 o'clock, stomach not empty. Took out half an ounce of contents. It was a milky white fluid, with a peculiar smell, and slightly acid and bitter taste. 7 o'clock, 15 mins., stomach empty.

Experiment 42

Jan. 5. At 8 o'clock, A.M.—Stomach healthy and clean. Extracted half an ounce of gastric juice. Put it into a vial, and immersed in it fifteen grains of firm *tendon* of young beef, in a solid piece. Kept it either in the axilla, or on the bath, for twenty-four hours, when all was completely dissolved.

At 8 o'clock, 45 mins., he breakfasted on *broiled veal, bread* and *coffee,* and kept exercising. 12 o'clock, M., stomach about half empty. Took out half an ounce, completely dissolved—no distinct particles of food to be seen. 12 o'clock, 30 mins., M., all gone.

This affords an example of the digestion of tendon. Hard, solid substances require a greater quantity of gastric juice than more tender fibre, and take a longer time for their complete solution.

Experiment 43

At 1 o'clock, P.M., same day, dined on *broiled veal* and *bread,* and drank half a pint of *water.* Digestion completed in four and a half hours.

Experiment 44

Jan. 6. At 8 o'clock, A. M.—Examined stomach. Coats generally healthy—few small, erythematous patches, on mucous surface. Secretions pure. Extracted one and a half ounces clear gastric juice, containing less than the usual quantity of mucous flocculi. It ran more freely than common through the tube. More could have been obtained; but a sensation of faintness, and sinking at the pit of the stomach, being felt and complained of, I desisted. This sensation has almost uniformly occurred, whenever the gastric juice has flowed more freely than usual, and has been suffered to run out to the quantity of one and a half, or two ounces; followed by dimness of vision, and vertigo, on rising. These feelings, however, subside in a few minutes, and he feels as usual, and eats his meals with a good appetite.

At 9 o'clock, he breakfasted on *broiled veal* and *bread* again, as yesterday, and kept exercising. 1 o'clock P. M., stomach nearly empty—several small spots of dark, grumous blood, exuding from the papillæ of the inner coats, made their appearance. 2 o'clock, some appearance of the breakfast still in the stomach. 2 o'clock, 15 mins., stomach empty.

Experiment 45

At 2 o'clock, 30 mins., P. M., same day, he dined on one pint of *barley gruel*, sweetened with molasses. 4 o'clock, 30 mins., stomach empty—none of the barley gruel to be seen.

Several small, sharp pointed, white pustules made their appearance on the inner surface of the stomach, at this time; and the surface, generally, was of a paler colour, and more flaccid, than usual.

Experiment 46

Jan. 7. At 8 o'clock, A. M.—Weather cloudy, damp, and disagreeable. Th: 48°. Wind N.E. Temperature of stomach, 100°. Less of the small pustules and red patches than yesterday. Colour of the coats natural again; but little secretion of gastric juice this morning. Could obtain only a drachm or two.

At 9 o'clock, A. M.—Temperature of stomach, 100°. He breakfasted on *soft boiled eggs, soft toast,* and *coffee.* 12 o'clock, M., stomach empty.

Experiment 47

At 12 o'clock, 30 mins., M., same day, he dined on three *hard boiled eggs* and *bread.* 3 o'clock, 30 mins., stomach half empty. Remaining contents acrid. Edges of the aperture excoriated. Some pimples, and erythematous patches on the surface of the inner coats. 4 o'clock, 30 mins., stomach and contents in nearly the same condition as at last examination—very acrid and sharp—coats red. 6 o'clock, stomach empty.

These three or four last experiments demonstrate, that a diseased state of the stomach retards digestion.

Experiment 48

Jan. 8. At 8 o'clock, 30 mins., A. M.—Examined stomach. Coats healthy. None of those white pustules, and erythematous patches, observed yesterday and the day before, to be seen this morning. Colour of the lining membrane rather paler than common. Surface moist. Extracted half an ounce of gastric juice, without difficulty. A slight and momentary vertigo was felt in rising up. No faintness or sense of sinking at the scrobiculus cordis, at this extraction. I divided these four drachms of gastric juice into two, equal parts, and put them into separate vials. In a third vial, I put two drachms of simple water. To each of these three vials, I added eleven grains of the muscle of a *sheep's heart,* in an entire piece. Kept one of the vials of gastric juice and meat in the axilla, and placed the other, with the aqueous vial, in a cool place, at about 46°, agitating them alike frequently.

At 7 o'clock, P. M., the piece in the warm gastric juice was half digested; the fluid of an opaque, reddish brown colour. That in the cold gastric juice was a very little affected, the surface being covered with a thin, glutinous coat, and the fluid a little turbid. That in the water was not in the least affected. The water was perfectly transparent, as when first put in.

At 9 o'clock, A. M., of the 9th, these several pieces of muscle exhibited the following results. That in the warm gastric juice, when taken out and pressed dry, as when put in, weighed seven and a half grains. That in the cold gastric juice, treated in the same manner, weighed twelve and a half grains, having *gained,* by the absorption of gastric juice, one and a half grains. And that in the simple water, weighed eleven grains, the same as when put in, having neither lost nor gained.

The three and a half grains, that remained in the first vial, were in one entire piece, of the same shape as when first put in; but very soft and tender, hardly able to sustain sufficient pressure to be raised by the finger and thumb. It was a mere pulp.

The meat in the second vial was increased a little in size; appeared swollen, soft, slimy and tender; but had sufficient firmness of texture to resist considerable pressure, when taken up. It was not dissolved.

That in the water retained its firmness, and was unaltered in appearance, except a paleness of surface, occasioned by maceration.

At 8 o'clock, next morning, (the 10th,) the following appearances were evident.

The first piece, in the warm gastric juice, weighed one and a half grains, having lost in the last twenty-three hours, two grains only. It retained the same shape, and was of about the same consistence as yesterday. A reddish brown sediment subsided to the bottom of a rich, whey-coloured fluid.

The second piece, in the cold gastric juice, weighed nine grains and a fraction, having lost about three and a half grains.

That in the water, was unaltered, and weighed the same as when put in—eleven grains.

It may be proper to remark, that the two pieces in the cold gastric juice and water, were moved from their first position in a temperature of about

46°, and placed for the last twenty-three hours on the mantle-piece, over the fire, in my room, in a temperature of about 60°.

The loss of the two and a half grains of meat, in the cold gastric juice, was evidently the effect of digestion, occasioned, no doubt, by the increase of fourteen or fifteen degrees of temperature.

On the 10th, I added to the vial, containing the warm gastric juice and muscle, one fourth of a drachm of fresh gastric juice, warm from the stomach. Continued it in axilla, and in five hours it was dissolved to a mite, scarcely perceptible.

The piece in the cold gastric juice, kept on the mantle-piece, in a temperature between 50° and 60°, till 9 o'clock, A.M., of the 11th, weighed seven grains, retaining the same shape as yesterday, and a similar texture. The fluid had become more opaque and milky, and the sediment had increased at the bottom.

The piece in the water at this time, remained unaltered, and weighed precisely the same as at first—eleven grains.

At 9 o'clock, A.M., I placed both these in the axilla.

At 9 o'clock, P.M., the piece remaining in the second vial of gastric juice, placed in the axilla this morning, was nearly all dissolved, one grain only remaining—a soft pulp.

The piece in the water remained unaltered, and weighed the same as at first; but began to emit a strong fœtid odour, and in a few days became very putrid. This was, however, almost entirely corrected, by the addition of three drachms of fresh gastric juice on the 21st. The meat still continued its original shape and size, and no doubt, its weight, though too putrid to handle, or take out, before the addition of the gastric juice. Placed it on the bath, and it began to digest, and soon became chymified—lost its fœtid smell, and acquired a sharp acid, or rather, acrid taste.

The result of this experiment is interesting, in demonstrating the solvent properties of the gastric juice. Maceration alone will not dissolve food, nor separate its nutritious parts. It appears, also, from this experiment, that gastric juice corrects the putrid tendency of aliments; and that food is more readily dissolved after that tendency has occurred.

Experiment 49

Jan. 11. At 8 o'clock, A.M.—Weather clear and dry. Wind S.W. Th: 15°. Temperature of the stomach, 100°. Coats healthy. Extracted one ounce of gastric juice, clear and transparent—few flocculi of mucus—taste distinctly acid. Complains of the usual sense of distress at the pit of the stomach, and vertigo.

At 9 o'clock, 30 mins., he breakfasted on *pork* and *bread.* Digested in four hours and a half.

Experiment 50

At 9 o'clock, 30 mins. A. M., same day, I took three vials, and put into each two drachms pure gastric juice, fresh from the healthy stomach. To one, I added one drachm of *albumen*—white of egg—to the second, half a drachm of the *yolk*—and to the third, another drachm of *albumen*. Put the two first, in axilla, and the other on the mantle-piece.

At 9 o'clock, P. M., the albumen in the warm gastric juice, in the axilla, had become quite opaque, with loose, light coloured sediment at the bottom. The albumen in the cold gastric juice remained unaltered. That containing the yolk, exhibited the appearance of a mere mixture of fine yellow coagulæ, resembling sulphur and milk, mixed together.

On the 12th, at 8 o'clock, P. M., both vials having been continued on the bath, or in the axilla, through the day, the difference observed last evening, between the cold and warm vials of albumen, was very little increased.

The yolk was considerably altered from a loose coagulæ, generally diffused through the gastric juice, to a fine compact body of coagulæ, rising upon the top of a perfectly clear, transparent fluid, free from a particle of sediment.

Experiment 51

At 8 o'clock, 30 mins., A. M.—Stomach healthy. Extracted one ounce of gastric juice, a little tinged with yellow, whether from bile or tobacco, it was difficult to determine. He had taken some tobacco into his mouth, an hour and a half previous to the examination, and the fluid was not perceptibly bitter. There was a larger portion of frothy saliva, and flocculi of mucus, than common.

At 10 o'clock, 15 mins., he breakfasted on *boiled, salted codfish, bread* and *coffee.* Digested in two hours and a quarter.

Experiment 52

Jan. 13. At 8 o'clock, A. M.—Weather overcast, dry and smoky. Light wind. Th: 12°. Temperature of stomach, 100° and a fraction. Pulse 60, in a recumbent, and 70, in an erect position. Coats not perfectly healthy—general surface rather paler than usual—some red spots and pimples to be seen. Extracted three drachms of gastric juice, slightly acid—not so much as usual—less mucus, and more saliva than common. Neither tinge nor taste of bile.

At 9 o'clock, he breakfasted on *boiled, fat pork* and *bread.*

At 12 o'clock, M.—Stomach two thirds empty.—Temperature, 100° and a fraction.

At 2 o'clock, P. M.—Stomach nearly empty—very little pulp of bread, and lardaceous fluid to be seen. Has just returned from walking two miles or more. Temperature of stomach, 100½°.

At 1 o'clock, 30 mins., stomach empty. Temperature, 101°.

Experiment 53

Jan. 9. At 2 o'clock, P.M. same day, he dined on *boiled, fat pork, boiled cabbage* and *bread,* and drank a tumbler of *water.* Digested in five hours. 9 o'clock, temperature 100°.

Experiment 54

Jan. 14. At 8 o'clock, 40 mins., A.M.—Weather clear, dry and serene. Wind N.W. and light. Th: 28°. Stomach healthy. Coats clean. Temperature of stomach, 100°. Extracted nine drachms of pure gastric juice—distinctly acid—few flocculi of mucus, and a little appearance of frothy saliva. A slight sense of faintness and vertigo ensued, as usual, on rising, after this quantity.

At 9 o'clock, breakfasted on *boiled, fat pork* and *bread.* 12 o'clock, M., stomach about half full. Temperature, immediately after walking two and a half miles, 101½°. 1 o'clock, P.M., stomach empty and clean. Temperature 100°.

Experiment 55

At 2 o'clock, P.M., same day, he dined on *boiled, fat pork* and *bread.* Digested in three hours.

Experiment 56

Jan. 14. At 9 o'clock, A.M., I put a solid piece of *rib bone,* of an old hog, weighing ten grains, into a vial, containing three drachms of pure gastric juice, taken from the stomach this morning. Placed it in the axilla, and continued it there for twelve hours; then placed it on the shelf, in a cool place, till next morning.

—15. 9 o'clock, A.M., surface of bone evidently dissolved. Fluid quite opaque. Took out the piece; and when wiped, and dried with blotting paper, as dry as when put in, it weighed just nine grains. Immersed it again in the same juice, and placed it on the sand bath at 100°. Continued it in that temperature for twelve hours, frequently agitating it; then, as yesterday, placed it on the shelf, until next morning.

—16. 9 o'clock, A.M., appearance similar to yesterday morning. Juice a little more turbid. Bone covered with a thin, cineritious coat. Taken out and wiped, the piece weighed eight and a half grains. Immersed again in same fluid, and continued on bath twelve hours; then set on shelf again until next morning.

—17. 9 o'clock, A.M., very little alteration since yesterday. Bone taken out and wiped, weighed eight and a quarter grains. Put in again, and continued on bath fifteen hours.

—18. 12 o'clock, M., no change effected since last examination. Bone taken out and wiped, weighed precisely same as yesterday, eight and a quarter grains. Conceiving the solution of the bone had ceased from a deficiency of the gastric solvent, I now added one drachm fresh gastric juice, and continued it on the bath again, for eight hours.

—19. 12 o'clock, M., bone taken out, and wiped, as usual, weighed eight grains. Returned to bath, and continued twelve hours, it weighed seven and a half grains. Returned, and continued on bath thirty-six hours, and frequently agitated, between.

—20th and 25th, no visible change was effected. Weight same as on the 19th, seven and a half grains. The solution having ceased again, I added three drachms more of gastric juice, and continued it on the bath twenty-four hours.

—27. 10 o'clock, A.M., laminæ of bone separated, and opening on one edge. Fluid more opaque, with a little fine, brown sediment, precipitated to the bottom of the vial. Weight of bone, five and a half grains. Added two drachms of gastric juice, and continued it on the bath for eighteen hours.

—28. 10 o'clock, A.M., laminæ of bone opened. Weight, four grains. Returned, and continued on bath twelve hours.

—29. 10 o'clock, A.M., laminæ of bone entirely separated, thin as paper, and elastic as horn.—Weight, three and a quarter grains. Returned to bath twelve hours.

—30. 10 o'clock, A.M., opacity of fluid, and fine sediment, increased. Weight of bone, two and three fourths grains. Continued on bath.

—31. 10 o'clock, A.M., no change since yesterday. Weight of bone, two and three fourths grains. Added half a drachm of gastric juice, and continued it on bath twelve hours.

Feb. 1. 10 o'clock, A.M., laminæ very thin and elastic. Weight of bone two and a half grains.—Took out the pieces of bone, and put them into one drachm fresh gastric juice, in a separate vial, and continued on bath six hours.

—2. 10 o'clock, A.M., weight of bone, two and a quarter grains. Continued on bath six hours.

—3. 10 o'clock, A.M., weight of bone, two grains. Continued on bath till the 5th.

—5. 10 o'clock, A.M., no change since the 3d. Weight of bone same. Added two drachms gastric juice, and continued on bath twelve hours.

—6. 10 o'clock, A.M., bones nearly all dissolved—three fourths of a grain only remaining.

—7. Weight of bone, half a grain, very thin and transparent. The solution not being quite completed, I added two drachms more of gastric juice, and continued it on bath twelve hours.

—8. 10 o'clock, A.M., all dissolved to a mite—quarter of a grain, or less.

After the solution of the bone, the menstruum was a greyish white, opaque fluid, nearly of the colour and consistence of clear, thin gruel, with considerable fine brown sediment at the bottom of the vial, after standing at rest awhile; and had a peculiarly insipid, sweetish taste, and smell—not the least fœtor or rancidity.

It will be seen, in this experiment, that the piece of bone was dissolved in proportion to the quantity of gastric juice applied, and that the solution ceased at longer or shorter intervals, as a larger or smaller quantity was added. When the juice became saturated, as well as when the vial was removed from the bath to a low temperature, the solution ceased. It appears that it took fourteen and a half drachms of gastric juice to dissolve ten grains of solid bone.

Experiment 57

Jan. 15. At 8 o'clock, A. M.—Weather cloudy and dry. Wind N.E. and light. Th: 35°. Temperature of the stomach, 100°. At 9 o'clock, A. M., he breakfasted on *fat pork* and *bread.* 2 o'clock, P. M., stomach empty and clean. Temperature, 101°.

Experiment 58

At 2 o'clock, P. M., same day, I put fifteen grains of *raw beef steak,* divided into small pieces, into three drachms of gastric juice; and fifteen grains of *broiled beef steak,* into other three drachms of gastric juice. At the same time, I put the same quantity of *broiled steak,* divided like the others, into three drachms of saliva, fresh from the mouth. I then placed them, all together, alternately in the axilla and on the bath, and kept frequently agitating them.

At 4 o'clock, the meat in the saliva exhibited the appearance of simple maceration; the other two parcels, in the gastric juice, were considerably diminished and partially dissolved, the fluid of an opaque, whitish colour; the cooked piece, rather the most dissolved.

At 6 o'clock, the salivary portion was not much changed in appearance; the other two about half dissolved; the cooked meat in advance of the raw.

At 9 o'clock, the salivary portion began to smell slightly fœtid, and to change colour. The other two were perfectly bland, and of a sweetish flavour—the meat about three fourths dissolved, with a fine, brownish red sediment at the bottoms of the vials. Took them all off the bath, and placed them on the shelf till next morning.

At 7 o'clock, A. M., on the 16th, I placed them again on the bath till 9 o'clock, when the salivary portion had become fœtid, and was of a greenish colour. The fibres of the meat retained their shape and size; and had become pale on the surface. Light, loose coagulæ had fallen to the bottom, leaving a reddish green coloured fluid above. The gastric portions were almost completely dissolved; the cooked meat still in advance.

At 12 o'clock, M., the salivary portion was very fœtid. The remaining portions of aliment, taken from the three vials, filtered through thin muslin, and dried with blotting paper, weighed as follows:—the broiled meat, in gastric juice, one grain; the raw meat, in the same, two and a half grains; and that in the saliva, twelve grains.

This experiment demonstrates that saliva does not possess the properties of a solvent; but facilitates putrefaction. See, also, subsequent experiments. It also shows, that raw meat is susceptible of digestion by the gastric juice, though in a less degree than cooked meat.

Experiment 59

Jan. 17. At 9 o'clock, A. M.—Weather clear, and dry. Wind N.W. and light. Th: 19°. Temperature of stomach, 100°. Coats clean and healthy. Extracted ten drachms of gastric fluid; not so clear and limpid as usual; some streaks of yellow bile, and more appearance of saliva than common—acid not so perceptible as usual. I divided this into three equal parts, three and one third drachms each. To one part, I put fifteen grains firmly coagulated *albumen,* (white of egg)—to the other, fifteen grains of the soft coagulæ of the same—and to the third, fifteen grains raw *albumen*—and placed them on the bath and in axilla, alternately.

At the same time, he breakfasted on three *hard boiled eggs, bread* and *coffee.*

At 11 o'clock, examined—stomach full. Temperature, 100°. Some small red spots. Contents acrid.

At 12 o'clock, M.—just returned from walking one mile, and back again. Weather clear, dry and serene. Wind N.W. and light. Th: 23°. Temperature of stomach, 102°—nearly empty. Took out one ounce, almost completely chymified; a little pure oil floating on the surface. Put this on the bath.

At 12 o'clock, 30 mins., stomach empty.

At 9 o'clock, P. M., examined the parcels of albumen, placed in the vials of gastric juice this morning, at 9 o'clock. Of the firm coagulæ, there remained one and a quarter grains; of the soft, none; of the raw, three fourths of a grain, in loose, white coagulæ.

Experiment 60

Jan. 17. At 12 o'clock, 30 mins., M. I put twenty-five grains *lean, broiled mutton,* divided into small pieces, into five drachms of gastric juice, and same quantity into five drachms of gastric juice and fresh saliva, mixed together; and placed them on the bath.

At 9 o'clock, P. M., the meat remaining in the gastric juice, taken out and dried with paper, weighed just twelve grains; that in the mixture of gastric juice and saliva, weighed eighteen and three fourths grains. The texture of the first was considerably more dissolved and tender than the second. Returned them into their respective vials again.

At 12 o'clock, 30 mins., M. of the 18th, examined them again. The meat remaining in the gastric juice, weighed five and three fourths grains; was soft, glutinous, and of a dirty brown colour. That in the gastric juice and saliva, weighed thirteen and a quarter grains; the texture was quite firm, and retained its fibrous form, and reddish, bloody colour. Put them in the bath again.

At 4 o'clock, P. M., of the 19th, the meat in the gastric juice weighed two grains. Consistence and colour of fluids, same as yesterday. The meat in the gastric juice and saliva, weighed nine and a half grains. Fluids of a reddish brown colour, and less precipitate.

In ten days, the salivary mixture became very, putrid; but the gastric portion was perfectly sweet, and so continued for thirty days, or more.

Experiment 61

Jan. 18. With a view to ascertain the antiseptic properties of the gastric juice, I took a portion of very putrid animal matter, and added to it a quantity of gastric juice. The fœtor was at once almost completely corrected, leaving only a slight putrescent smell, with the usual flavour of the gastric juice.

Experiment 62

At 9 o'clock, A. M., same day, extracted one and a half drachms of gastric juice, and added it to two and a half drachms of *milk*. The whole was formed into loose, white coagulæ, in less than five minutes. At 1 o'clock, P. M., remaining coagulæ, after filtering through muslin, weighed thirteen grains. Returned it into the vial, and placed it on the bath again. At 9 o'clock, no coagulæ remaining—all completely dissolved.

Experiment 63

Jan. 19. At 9 o'clock, A. M., coats of stomach perfectly healthy and clean. No appearance of morbid action—tongue clean—and every indication of perfect health. There was no free fluid in the gastric cavity, until after the elastic tube was introduced, when it began slowly to distill from the end of the tube, drop by drop, perfectly transparent, and distinctly acid. I obtained about one drachm of this kind, and then gave him a mouthful of bread to eat. No sooner had he swallowed it, than the fluid commenced flowing more freely from the tube, and I obtained two drachms, less pure, however, with saliva and mucus mixed with it, and slightly tinged with yellow bile. The surface of the protruded portion of the villous coat at this time became covered with a limpid fluid, uniformly spread over its whole surface, distilling from myriads of very fine papillary points, and trickling down the sides. After letting him rise and walk about two or three minutes, I again introduced the tube, and obtained about two drachms more of very pure gastric juice, making, in the whole, five drachms.

Breakfasted on *boiled pork* and *bread*. Dined and supped on the same.

Experiment 64

Jan. 20. At 8 o'clock, 30 mins., A. M., examined stomach—appearances healthy. Extracted three drachms gastric fluids, coloured with bile, slightly acid and bitter. It ran more freely than yesterday.

At 8 o'clock, 45 mins., he swallowed four ounces of pure *gelatine*, (ichthyocolla,) prepared with boiling water, transparent, and of a tremulous consistence.

At 9 o'clock, stomach appeared nearly as full as usual after eating his ordinary meals; fluid, clear, and of the consistence of the albumen of eggs. It appeared to be the gelatine, dissolved, or diffused in the gastric juice. The juice and the liquid gelatine so much resembled each other, that I could not distinguish them apart.

At 9 o'clock, 45 mins., examined again—found the stomach almost entirely empty—was just able to obtain two drachms of fluid. It appeared to be a mixture of gelatinous chyme, gastric juice and flocculi of mucus, more opaque and ropy than the gastric juice alone, and more acid than the fluids of the stomach immediately before the gelatine was swallowed. Not the least appearance of bile or yellow colour in the gastric cavity, or fluids, after taking the gelatine: considerable vertigo followed the extraction of this last fluid. It soon passed over, however, and he ate his breakfast, (*pork* and *bread*) with his usual appetite.

The process of the solution of gelatine, is difficult to ascertain. It is not subject to coagulation; and the action of the gastric juice is not easily perceived. It is no doubt dissolved by the gastric juice, in the same manner as other aliment is. See subsequent experiments.

Experiment 65

To ascertain whether the sense of hunger would be allayed without the food being passed through the œsophagus, he fasted from breakfast time, till 4 o'clock, P. M., and became quite hungry. I then put in at the aperture, three and a half drachms of *lean, boiled beef.* The sense of hunger immediately subsided, and stopped the borborygmus, or croaking noise, caused by the motion of air in the stomach and intestines, peculiar to him since the wound, and almost always observed when the stomach is empty.

This experiment proves that the sense of hunger resides in the stomach, and is as well allayed by putting the food directly into the stomach, as when the previous steps have been gone through with. Not that I would deny the utility of the previous processes, in ordinary cases. Even the sense of taste is essential. It is placed as a sentinel, to prevent improper articles from being introduced into the stomach. See, also, subsequent experiments.

Experiment 66

Jan. 21. At 8 o'clock, A. M.—Examined stomach. Could obtain but few drop of gastric juice. Sent him to exercise in the open air for half an hour. Secretions increased—gastric juice flows pure, and more freely. Extracted three drachms.

At 8 o'clock, 30 mins., he breakfasted on *bread* and *coffee,* and a small piece of *lean pork.*

At 2 o'clock, P. M., stomach empty. Extracted two drachms of gastric juice, tinged with yellow bile; and then one drachm of pure, transparent juice, distilling, by drops, from the end of the tube.

At 2 o'clock, 30 mins., I put ten grains of *raw suet* into two drachms of gastric juice, tinged with bile, and ten grains of the same, into two drachms of pure gastric juice. Placed them both on the bath.

At 9 o'clock, the piece of suet in the juice that was tinged with bile, was considerably more dissolved than that in the clear gastric juice; and when examined with the compound microscope, the globules appeared more numerous and much smaller. This appearance was, also, clearly perceptible to the naked eye, as the mixtures stood in the vials.

At 10 o'clock, the piece in the yellow juice was all dissolved—the other, not entirely.

This, with other subsequent experiments, indicate that oily or fatty food is sooner digested when there is a small admixture of bile with the gastric juice.

Exercise, it seems, promotes the discharge of the gastric juice, as well as digestion in the stomach.

Experiment 67

Jan. 22. at 8 o'clock, 30 mins., A. M.—Stomach clean and healthy. Extracted five drachms of very clear, pure gastric juice. The first three drachms ran out quite freely; the other two drachms distilled by drops. It was not the least tinged with bile, and tasted distinctly acid. Breakfasted on *beef steak, bread* and *coffee.*

At 1 o'clock, P. M., stomach empty.

Experiment 68

At 9 o'clock, P. M., same day, St. Martin having eaten nothing since 2 o'clock, and feeling quite hungry, I put into the stomach, at the aperture, eight ounces of beef and barley soup, introduced gently through a tube, with a syringe, lukewarm. It caused no unpleasant sensation, but allayed the sense of hunger. It satisfied the *appetite;* and he said he had no desire to *eat.*

At 10 o'clock, he said he felt a little hungry again, and ate eight ounces more of the same kind of soup, which had a similar effect as the other.

Experiment 69

Jan. 23. At 9 o'clock, A. M.—Weather rainy.—Wind N.E. and light. Th: 39°. Stomach empty, clean and healthy. Temperature of stomach, 100½°.* Breakfasted on *sausage, bread* and *coffee.*

At 10 o'clock—Aspect of weather same as at 9 o'clock. Th: 40°. Stomach full of fluids—temperature 101¾°. The spirit became stationary at that point, after keeping the tube in the aperture eight or ten minutes: after which, it did not vary for ten minutes, when it was taken out.

At 12 o'clock, M., he returned from a walk of two miles. Stomach nearly empty. Temperature 101½°—stationary after being continued five minutes in the stomach.

At 12 o'clock, 30 mins., stomach empty.

---

*In this, and the subsequent experiments, I used a spirit Thermometer, taken from Pool's Barometer, which varied half a degree from those formerly used.

Experiment 70

Jan. 24. At 8 o'clock, A.M., Weather cloudy and damp. Wind N. and moderate. Th: 39°. Stomach empty, clean and healthy—temperature, 100½°. Extracted four drachms gastric juice, very little tinged with yellow.

At 9 o'clock, he returned from a short walk. Temperature of stomach, the same. Breakfasted on *bread* and *coffee.* 12 o'clock, stomach empty.

Temperature of the stomach, after walking two miles or more, 101¼°.

Experiment 71

At 1 o'clock, P.M., same day, St. Martin complaining of being quite hungry, I put into the stomach at the aperture, twelve *raw oysters,* more than middling size. The sensation was allayed, and the appetite satisfied, the same as if swallowed. He was not hungry again till half after 4 o'clock, when he ate a dozen more of the same kind of oysters, with bread.

At 10 o'clock, P.M., stomach empty and clean. Weather damp and rainy. Wind N.E. and brisk. Temperature of the stomach, 99½°. He had been covered in bed, and sleeping, for two and a half hours, from which I awoke him to introduce the Thermometer. He fell asleep again during the examination—only awoke while putting in and taking out the glass tube.

Experiment 72

Jan. 25. At 6 o'clock, A.M.—Wind Southerly, and light. Th: 36°. Examined stomach before rising from his bed. Temperature 99°. Extracted fifteen drachms gastric fluid. It flowed out unusually free; was rather more opaque, and contained less flocculi of mucus than common for the quantity. Particles of the bread eaten with his oysters at 4 o'clock, 30 mins., yesterday, were distinctly to be seen in this parcel of the juice.

At 8 o'clock, 30 mins.—Temperature of the stomach, 100½°. Coats clean and healthy. Th: 38°.

At 9 o'clock, he breakfasted on *raw oysters* and *bread.* 11 o'clock, temperature of stomach, 101°. 12 o'clock, M., he returned from a walk of two miles. Stomach empty. Temperature, 102°.

Experiment 73

Jan. 26. At 8 o'clock, A.M.—Weather clear and cold. Wind N.W. and light. Th: 30°. Stomach healthy, empty and clean. Temperature, 100½°. Extracted one drachm gastric juice, containing more than usual flocculi of mucus.

At 9 o'clock, he breakfasted on *sausage, bread* and *coffee.* 10 o'clock, Th: 34. Temperature of the stomach, 100¾°, and full of a heterogeneous fluid. 12 o'clock, M., returned from a walk. Stomach empty—temperature, 101° and a fraction. Weather clear and pleasant. Th: 39°. Wind N.W. and moderate.

From this, and other experiments, it may be clearly inferred, that in the most natural and healthy states of the stomach, there are little or no fluids, of any kind, in the gastric cavity, until excited by aliment or other irritants; and that digestion, under this condition, is the most rapidly and perfectly performed.

Experiment 74

At 2 o'clock, P. M., same day, he dined on *raw oysters* and *bread.* 5 o'clock, stomach empty.

At 6 o'clock, 40 mins., immediately after drinking a tumbler of water of the temperature of 55°, introduced Thermometer—spirit rose very slowly, and did not become stationary at the natural temperature until the tube had stood in the stomach for thirty-five minutes. 12 o'clock at night, temperature 99½°, after sleeping in bed three hours.

Experiment 75

Jan. 27. At 6 o'clock, A. M.—before rising from his bed—Weather cloudy and dry—calm—Th: 32°—Stomach empty, clean and healthy—Temperature 99½°, spirits stationary in ten minutes—he swallowed a gill of water at the temperature of 55°, which immediately diffused itself over the interior of the stomach, and discharged some at the aperture, by the side of the stem of the Thermometer, which had not been withdrawn. The spirit immediately fell to 70°; stood at that point one and a half or two minutes; and then began again very slowly to rise. Thirty minutes elapsed after taking the water, before the spirit regained the 99th degree. Before the end of that time, there was no appearance of water in the gastric cavity.

At 9 o'clock, 30 mins., he ate a full breakfast of *fresh, broiled beef,* mostly fat, *bread* and *coffee,* and continued unusually smart exercise, walking for two hours, till he became fatigued, and perspired freely.

At 11 o'clock, 30 mins.—Weather clear. Th: 43°. Just returned from walking. Stomach contained considerable chyme and oil. Aliment about two thirds gone. Temperature, 101°.

At 12 o'clock, 20 mins., M.—Stomach nearly empty; a small portion of the fluid remaining, reduced to a more perfect chymous condition, with less oil, and that in much finer globules. Appeared tinged with yellow, and tasted bitter.

At 1 o'clock, P. M., chyme gone. Very little oil remaining.

At 2 o'clock—Weather unchanged. Temperature of stomach, 101¼°. No chyme to be seen. A few particles of oil still remaining, floating on the surface of a small quantity of fluid, exhibiting considerable spumous froth and mucus.

A circumstance occurred here, not before observed in my experiments, which it may not be unimportant to mention, *i.e.*—the variations of the temperature observed in moving the Thermometer up or down in the stomach. The spirits in the tube varied proportionably to the length of the stem introduced. When the bulb sank down to the pyloric portion of the stomach, to the depth of six or eight inches, the spirit rose to 101¼°: when only immersed two or three inches, it would stand at 100½°, making a difference of three fourths of a degree. These variations were uniformly observed at every thermometrical examination.

Perhaps the difference of indication of the thermometer, may result from a more complete envelopement of the stem in the gastric cavity, at the pyloric examination, and a less one at the splenic. I give the reader possession of the fact, without pretending to account for it, with certainty.

Experiment 76

At 2 o'clock, 30 mins., P. M., same day, he dined on *raw oysters* and *bread.* 4 o'clock, 30 mins., stomach not empty. Food about half gone. Small pieces of heart of oyster, and pulp of bread, to be seen, floating in a thin, pultaceous fluid, quite acrid and sharp—no bitter taste, or yellow colour. Temperature 101½°.

A striking peculiarity in the movement of the spirit in the Thermometer was observed in this experiment. It rose from about 68° to its stationary point, 100½°, in less than five minutes after the bulb was put into the stomach. At last examination, 2 o'clock, it was fifteen minutes in making the same range. Sometimes it has been twenty-five or thirty minutes before it became stationary, and under no appreciable difference of circumstances. He had been moderately exercising, (walking) immediately before the last examination.

At 5 o'clock, he returned from walking. Temperature of stomach, 101½°. Spirit rose, and became stationary, at that point, in less than three minutes—food almost completely chymified, and half gone. Took out one ounce of thick, pultaceous, porridge-like fluid, with some small pieces of the hearts of the oysters, reduced to a jelly-like appearance—plainly acid—and slightly bitter; and had the flavour of the oysters.

At 6 o'clock, 15 mins., stomach empty and clean.

At 6 o'clock, 30 mins., he ate a full meal of *cold, boiled beef* (considerable fat) and *bread.* 10 o'clock, 30 mins., stomach empty.

Experiment 77

At 9 o'clock, A. M., of the 27th, I mixed one drachm of the clear *decoction of coffee* with three drachms of fresh gastric juice, with a view to ascertain whether it would destroy the flavour of the coffee. It had no perceptible effect. The flavour of coffee remained for ten hours, as distinct as at first. Added half a drachm of loaf sugar to the mixture, and placed it on the bath. It remained there forty-eight hours: no different effect was produced on the flavor of the coffee. It remained the same as at first.

It is probable that the decoction of coffee, like many other artificial drinks, does not admit of digestion; possesses no nutritive principles; and is carried into the circulatory system without much change.

Experiment 78

At 1 o'clock, 30 mins. P.M., of the 27th, I put fifteen grains firm beef *cartilage* into three drachms of gastric juice, and placed on bath.

At 10 o'clock, A.M., of the 28th, took out and wiped dry, it weighed six and three fourths grains.

At 10 o'clock, A.M., of the 29th, it weighed one grain.

When put in, the cartilage was cut into different sized pieces: these retained their original forms till completely dissolved—the largest piece being the last digested.

Experiment 79

Jan. 28. At 6 o'clock, 30 mins., A.M.—before rising.—Weather clear and dry. Wind S.W. and light. Th: 35°. Stomach empty, clean and healthy. Temperature 100° and a fraction—spirit stationary in five minutes. No gastric juice could be procured. Extracted about half a drachm of fluids, principally mucus.

At 8 o'clock, 45 mins.—Temperature of the stomach 100¼° when Thermometer was put three or four inches only into the splenic portion; but rose to 101° when the bulb was let down, eight or nine inches, towards the pyloric extremity.

A circumstance occurred here which I had not noticed before. On settling the stem down into the stomach, a strong contraction of the muscular fibres was indicated, when the bulb had descended near to the pyloric end, by a sudden and peculiar movement of the tube, communicated to the thumb and finger that guided it, and also felt by St. Martin himself. The stomach appeared to contract at that point forcibly, and grasp the bulb, giving it a sudden impulse downwards, so much so as to require a quick compression by the thumb and finger to prevent it from slipping suddenly into the pyloric end. This grasping sensation would continue for half a minute or more, and then appear to relax again. This action occurred every time the bulb passed this point, either up or down. When the bulb was below this point, the spirit rose three fourths of a degree; when raised above, it fell the same. Sometimes the suction motion was stronger than at others, and when the stem was released from the fingers, it would be drawn down towards the pyloric end, its whole length, ten or eleven inches, occasioning considerable distress, vertigo, and a sense of sinking at the scrobiculus cordis.

At 9 o'clock, he breakfasted as yesterday, and kept quiet, most of the time in a recumbent position, on a couch.

At 11 o'clock—Aspects of weather same as in the morning. Th: 46°. Contents of stomach about two thirds diminished. Temperature, 100¾,° at three or four inches deep, and a fraction less than 101½°, when sunk to the pyloric extremity, varying proportionably to the length of the stem introduced.

At 12 o'clock, 30 mins., M., stomach nearly empty. Temperature, 101°.

At 1 o'clock, 30 mins., stomach empty. Temperature, 100¾°, splenic end—101½°, at pyloric end.

Experiment 80

At 2 o'clock, P. M., same day, he dined the same as yesterday, on *raw oysters* and *bread.* Temperature of stomach, immediately before eating, 101½°, at pyloric extremity—100¾° at splenic end.

At 4 o'clock, 30 mins., stomach half empty. Temperature 101½° at pyloric extremity—rose quick. Took out one drachm of the chyme. Digestion nearly complete—a few particles of bread and oyster to be seen.

At 5 o'clock, 30 mins., stomach nearly empty.—Temperature 101½°, pyloric extremity—rose quick.

At 6 o'clock, stomach still contained alimentary fluids—quite acrid and sharp.

At 6 o'clock, 40 mins., stomach empty.

At 7 o'clock, he supped on *boiled beef* and *bread.*

Experiment 81

Jan. 29. At 6 o'clock, A. M.—before rising—Weather clear and dry. Wind N.E. and brisk. Th: 28°. Stomach perfectly healthy, empty and clean. Temperature 100°, at pyloric extremity, and 99° at the other. No gastric secretion. Could not extract ten drops of either gastric juice, mucus or saliva.

At 8 o'clock, 30 mins.—Stomach empty—coats perfectly healthy, and free from any appearance of aphthæ, pustules or red spots, The mucous coat was even and uniform, soft and smooth. Temperature from 100¼° to 100¾°—rose quick. Extracted three and a half drachms pure gastric juice, containing some flocculi of mucus, but no bile.

At 9 o'clock, he breakfasted on *sausage* and *bread,* and kept exercising—walking smartly for two hours. Returned from walking at 11 o'clock, 30 mins. Stomach two thirds empty. Temperature, 102° and a fraction, at pyloric end, and 101⅓° at the other—rose quick. 12 o'clock, 30 mins., M., stomach almost empty. Temperature, 101½° and 100¾°—rose moderately. 1 o'clock, P. M., stomach empty.

Experiment 82

At 1 o'clock, 30 mins., P. M., same day, he dined on *stewed oysters* and *bread,* and kept still. 5 o'clock, P. M., stomach empty. Extracted three and a half drachms pure gastric juice. At 6 o'clock, 45 mins., stomach empty. Temperature 101½° and 100¾°—rose moderately.

Experiment 83

Jan. 30. At 6 o'clock, 30 mins., A. M.—Weather cloudy and damp. Wind N.E. and light. Th: 39°. Stomach empty, clean and healthy. Temperature, immediately before rising, 99½° and 98¾°—rose slowly.

At 9 o'clock, temperature of stomach, 101½° and 100¾°—rose quick. Extracted three drachms gastric juice. It came slowly; the last mixed with yellow bile. He breakfasted on *beef steak, bread* and *coffee.*

At 11 o'clock, stomach almost empty. Temperature 101¼° and 100½°. When the bulb of the glass tube descended towards the pyloric extremity, the stomach evidently contracted upon it, and drew it forcibly down. If left free to its own motions, the tube would sink to the pylorus, the whole length of the stem, ten or eleven inches, and then rise again of its own accord. When drawn above this point of apparent contraction, into the splenic end of the stomach, towards the perforation, the motion of the bulb was reversed, in a direction towards the fundus of the stomach, not inclining, however to make its exit at the perforation; but took a sort of irregular motion, revolving the tube from right to left, so as to turn it completely around, in the space of ten or fifteen seconds. This motion was not always present, nor constantly continuous when present; but interrupted, and alternate with the appearance of contraction at the pyloric end; and distinctly evident only from about one and a half to three hours, or more, after eating, and at the time when the chyme was most rapidly leaving the gastric cavity.

At 12 o'clock, 30 mins., M., he returned from a smart walk—had been all the morning, since breakfast, hard at work, wheeling coal, an unusually severe exercise. Temperature, 102° and 100¾°—rose quick. Stomach empty.

Experiment 84

Jan. 29. At 9 o'clock, 30 mins., A. M.—To two drachms of gastric juice, I put one small, *raw oyster,* weighing one drachm; and to another two drachms of gastric juice, I added one drachm of *stewed oyster.* Set them on the bath, and agitated them frequently.

At 5 o'clock, 30 mins., P. M., the residue of the raw oyster weighed four grains—that of the stewed, weighed eight and three fourths grains. Continued them on bath.

At 9 o'clock, A. M., of the 30th, the raw oyster was completely dissolved; not a particle left, except a trace of dirty brown sediment, the excrementitious part. A grain only of the heart of the stewed oyster, was left, with a trace of the same kind of sediment, as in the raw one. The flavour of the oysters was retained to the last, and even the chymous mass partook of it.

In the article here submitted to the action of the gastric juice, cooking hardens the fibre, and renders it less susceptible of digestion than the raw. This is what we should, *a priori,* judge, from the known properties of this solvent.

Experiment 85

Jan. 30. At 10 o'clock, A. M., I put 10 grains of *boiled, lean beef,* each piece of *raw lean beef,* each piece whole and undivided, and ten grains *boiled, lean beef, chopped fine,* into three drachms fresh gastric juice, and placed them on the bath, frequently agitating, as usual.

At 12 o'clock, M., of the 31st, examined and weighed them. The raw piece weighed the same as when first put in—the lean boiled beef weighed eight grains—the chopped, three grains. Added two drachms gastric juice.

At 10 o'clock, A. M., *Feb.* 1st, balance of chopped meat weighed one grain; boiled piece, five grains; raw, ten grains.

Experiment 86

Jan. 29. At 10 o'clock, A. M., I put three equal parts of *cabbage,* one part *raw,* another *boiled,* and the third, *shaved* fine, (raw) and *macerated in vinegar,* ten grains each, into three drachms of gastric juice, and placed them on the bath.

At 5 o'clock, 30 mins., P. M., I took out and pressed dry the respective parcels. They weighed as follows: the shaved, three and three fourths grains; the raw, five and a half grains; the boiled, six and a quarter grains.

At 10 o'clock, A. M., of the 30th, took out and examined—the raw weighed five and a quarter grains; the shaved, three and a half grains; and the boiled, the same as yesterday evening, six and a quarter grains. I added two drachms gastric juice, and continued them on bath.

At 12 o'clock, M., of the 31st, the raw weighed two grains; the shaved, one and a half grains; boiled, five grains. Added one drachm of gastric juice.

Feb. 1. Raw, weighed one grain; shaved, one grain; boiled, two and a half grains.

Experiment 87

Jan. 30. At 2 o'clock, P. M., he dined on *raw oysters* and *bread,* and kept still. 5 o'clock, 30 mins., stomach empty. Temperature 101¼,° and 100½°, spirit rose moderately. 6 o'clock, 45 mins., he supped on *raw oysters* and *bread;* 10 o'clock, stomach empty.

Experiment 88

Jan. 31. At 6 o'clock, A. M.—before rising.—Weather rainy. Wind N. E., and light. Thermometer 45°. Stomach empty, clean and healthy. Temperature 100° and 98½°—rose moderately. No fluids in the gastric cavity. Could obtain but half a drachm. The peculiar contraction and relaxation or suction and pulsion motion, were evidently excited this morning by the introduction of the thermometer, but not near so strong as during chymification. When the bulb is sunk down low into the stomach, and suffered to remain there a minute or two, it gives severe pain and distress at the pyloric extremity, like the cramp, or, the sensation frequently described by persons suffering from undigested food in the stomach, and leaves a sense of soreness, if repeated a few times, as was very evident this morning.

At 9 o'clock—temperature of stomach 101½°—he breakfasted on two and a half ounces of *beef steak,* four and a half ounces *soft toast,* and *coffee.*

At 9 o'clock, 30 mins., he laid himself down on his pallet, and I set the Thermometer into his stomach, and continued faithfully and constantly to observe its motions and variations, one hour and forty minutes, until ten minutes past eleven, (two hours and ten minutes after eating.) At first the stomach was full to overflowing, of heterogeneous fluids, in much commotion, as indicated by the movement of the aliment, and of that part of the stem left out of the aperture, nearly four inches.—This commotion continued about half an hour, to ten o'clock. It then seemed to subside; the general muscular action became less, as indicated by the stem of the thermometer, and motion of the fluids, until half after ten o'clock, when a different motion appeared to commence, indicating considerable forcible contraction upon the bulb of the tube, now about six inches from the aperture towards the pylorus. An irregular turning and twisting of the stem, and a simultaneous downward movement, was succeeded by an apparent relaxation and expulsive motion. These alternate motions and appearances continued to recur every two or three minutes—not uniformly, but at irregular intervals. A sense of distress and uneasiness was felt at the point where the bulb lay, every time these contractions recurred, so as to occasion involuntary manifestations of pain, expressed in the muscular motions of his face. The Thermometer did not perceptibly vary, during all this time, from the usual standard temperature of the interior of the stomach. It was 101¼°, at the pyloric extremity, and 100½° in the splenic end, and continued so, during the whole time, ranging between these two points, according as it was moved higher or lower in the gastric cavity. At this time, 11 o'clock, 10 mins., the stomach was about half empty—and chymification rapidly advancing.

At 12 o'clock, 30 mins., M., the stomach was empty and clean. Temperature 101¼° and 100½°. Extracted two and a half drachms gastric juice.

Experiment 89

Feb. 1. At 6 o'clock, A. M.—before rising—Weather clear. Wind N. W. Th: 28°. Stomach empty, clean and healthy. Temperature 100° and 99¼°—rose moderately. No gastric juice secreted.

At 8 o'clock—Weather clear, and growing cold. Th: 26°. Temperature of stomach, immediately before going out, 101° and 100°. Returned in 30 minutes. Temperature of stomach the same. Extracted four drachms gastric juice.

At 9 o'clock, he breakfasted on *bread, sausage* and *coffee,* and kept exercising. 11 o'clock, 30 mins., stomach two thirds empty. Aspects of weather similar. Th: 29°. Temperature of stomach, 101½° and 100¾°. The same appearance of contraction and dilatation, and alternate piston motions were distinctly observed at this examination. 12 o'clock, 30 mins., stomach empty.

Experiment 90

At 2 o'clock, P. M., same day, he dined on *potatoes* and *meat.* 5 o'clock, weather clear and pleasant. Wind N. W. and light. Th: 32°. Stomach nearly empty. Temperature 102° and 101½°, after walking. 5 o'clock, 30 mins., stomach empty.

Experiment 91

Feb. 2. At 8 o'clock, 30 mins., St. Martin finished breakfasting on full meal of two and a half ounces *fried sausage,* seven and a half ounces warm *corn bread,* and a pint of *coffee.* Kept gently exercising for one hour, and then increased his exercise to severe walking, two or three miles, for two hours. Stomach full when he started, at 9 o'clock, 45 minutes.

At 12 o'clock, M., returned from walking. Stomach not entirely empty. Oil and bread perceptible. 12 o'clock, 30 mins., considerable fluid in the stomach, tinged with yellow bile. No distinct particles of food to be distinguished. 1 o'clock, P. M., stomach empty and clean. Extracted two drachms pure gastric juice.

Severe exercise, in this instance, is supposed to have retarded digestion, as well as the peculiar kind of food eaten.

Experiment 92

At 1 o'clock, 30 mins., P. M., same day, he dined on four ounces of *fresh, boiled beef,* (cold) and five ounces *bread,* and continued walking smartly, for three and a half hours, till 4 o'clock, 45 mins., P. M. Stomach nearly empty. Dinner almost completely chymified. 5 o'clock, stomach empty.

Experiment 93

Feb. 3. At 8 o'clock 45 mins., extracted four drachms gastric juice. He breakfasted on full meal, two and a half ounces *boiled beef,* seven and a half ounces *bread,* and one pint of *coffee;* and kept perfectly still. 12 o'clock, 30 mins., M., stomach not empty. 1 o'clock, P. M., stomach empty and clean. Extracted one and a half drachms gastric juice.

This indicates that a complete state of repose is unfavourable to speedy digestion.

Experiment 94

At 1 o'clock, 30 mins., P. M., same day, he dined on four ounces *fresh, boiled beef,* five ounces of *bread,* and *potatoes.* 6 o'clock, stomach not entirely empty; but none of the meat remaining. 6 o'clock 15 mins., very little of the bread and potatoes to be seen. 6 o'clock, 30 mins., stomach empty.

Experiment 95

Feb. 3. At 12 o'clock, M., I put two equal and entire pieces of *parsnip,* ten grains each, one boiled, and the other raw; the same kinds and quantities of *carrot;* and the same of *potato*—into four drachms of gastric juice, and placed them on the bath.

At 12 o'clock, M., on the 4th, the vegetables taken out and wiped or filtered as dry as when put in, showed the following result:

The piece of raw parsnip, weighed three grains; the boiled, one grain. Raw carrot, three and one fourth grains; boiled, half a grain. Raw potato, eight and a half grains; boiled, no entire particle could be distinguished—a fibrous and farinaceous residuum of six grains remained on the filter.

At 12 o'clock, M., on the 5th, the pieces of parsnip and carrot were almost entirely dissolved, a grain or two of the raw carrot, and fibrous centre of the parsnip, only remaining. About a grain of the roughish white farina of the boiled potato, remained. The raw potato was a little softened and wasted on the surface, but weighed the same as at last examination, eight and a half grains.

This is an illustration of the necessity of tenderness and susceptibility of division of the articles of diet, for speedy solution by the gastric juice. The raw potato retained nearly its weight, after the other articles were dissolved.

Experiment 96

At 3 o'clock, P. M., same day, I took two equal quantities, two drachms each, of saliva, acidulated to about the flavour of gastric juice—one with acetic, and the other with muriatic acid—and put into each, two pieces of parsnip, and two of carrot, one of each boiled, and the other raw; each weighing ten grains; and placed them on the bath.

At 3 o'clock, P. M., on the 4th, the carrot in the saliva and muriatic acid, had lost nothing—the parsnip, only two grains. In the acetous menstruum, both kinds remained the same as when put in. The fluids of both were unaltered in their sensible qualities and appearances.

After continuing them on the bath, with frequent agitation, for twenty-four hours longer, the parsnip, in the muriatic menstruum, had lost four grains—the carrot nothing. The parsnip in the acetic mixture, had lost six grains, and the carrot four grains, but appeared to have been rather macerated and diffused, than dissolved or digested.

I now mixed them all together, and continued them on the bath, for twenty-four hours longer; at the end of which time, the whole remaining mass of vegetable matter, weighed twelve grains. The fluid appeared now a little more chymous, and was rather turbid.

This is an example of a species of solution, performed by chemical agents, having some resemblance to digestion. It is not at all probable, however, that this mixture was in a state of preparation for the action of the pancreatic and hepatic fluids; but if placed in the stomach, would require the same action of the gastric juice, as other diet would.

Experiment 97

Feb. 4. At 9 o'clock, A. M., he breakfasted on two and a half ounces of *boiled beef,* six ounces of *bread,* and one pint of *coffee.* Exercised smartly for three hours. At 12 o'clock, 30 mins., M., chymification complete. Stomach empty.

Experiment 98

Feb. 5. At 9 o'clock, A. M., he breakfasted same as yesterday, and kept still. 11 o'clock, stomach nearly full. 12 o'clock, M., considerable yet in the stomach;

oil and bread plainly to be seen. 12 o' clock, 30 mins., contents of stomach not yet gone. 1 o'clock, P. M., stomach almost empty. 1 o'clock, 15 mins., stomach empty.

Experiment 99

Feb. 7. At 8 o'clock, 30 mins., A. M., I put twenty grains *boiled codfish* into three drachms gastric juice, and placed them on the bath.

At 1 o'clock, 30 mins., P. M., fish in the gastric juice, on the bath, was almost dissolved, four grains only remaining—fluid opaque, white, nearly the colour of milk. 2 o'clock, the fish in the vial, all completely dissolved.

Experiment 100

Feb. 7. At 9 o'clock, A. M., breakfasted on *boiled codfish* and *bread.* Digested in four hours and a half.

Experiment 101

Feb. 8. At 10 o'clock, 30 mins., A. M., I put two parcels, ten grains each, of *strong cheese,* one masticated, and the other an entire piece, into three drachms gastric juice. At 6 o'clock, P. M., the masticated portion was all completely digested, scarcely a trace left on the filter. The entire piece had lost four and three fourths grains—five and one fourth grains remaining undissolved, and of the same shape as when put in, having lost its superfices only. This piece continued gradually to diminish, for twenty four hours, when it was completely dissolved.

Experiment 102

Feb. 12. At 1 o'clock, 30 mins., P. M., he dined on *mutton,* and *barley soup* and *bread.* Digested in three and a quarter hours.

Experiment 103

Feb. 13. At 2 o'clock, 15 minutes P. M., he dined on *mutton* and *barley soup* and *bread.* Digested in three and a quarter hours.

Experiment 104

Feb. 14. At 9 o'clock, A. M., I took forty grains *masticated, broiled beef steak,* divided into two equal parts—put one into four drachms gastric juice, and the other, into four drachms of a mixture of dilute *muriatic* and *acetic* acids, reduced with water to the flavor of the gastric fluid, as nearly as practicable—three parts of the muriatic to one part of the acetic. Placed them together on the bath. At 6 o'clock, P. M., the meat in the gastric juice was all dissolved; that in the dilute acids, when filtered, left a residuum of nine grains, of a jelly-like consistence. The fluids, also, differed in appearance. That from the gastric juice was opaque, and of a lightish grey colour, depositing a brown sediment on standing. The other was also opaque, and of a reddish brown colour, but deposited no sediment.

This was an attempt to imitate the gastric juice. It was not satisfactory. Probably the gastric juice contains some principles inappreciable to the senses or to chemical tests, besides the acid and alkaline substances already discovered in it.

Experiment 105

At the same time of the above experiment, (104) I put the same quantities of pure dry *gelatine,* (ichthyocolla) into exactly similar quantities and kinds of fluids, and placed them all together on the bath.

At 6 o'clock, P. M., the gelatine in the gastric juice was all completely dissolved—that in the dilute acids, after being placed on the filter, left a residuum of three grains of a jelly-like substance. These two fluids differed in appearance. That from the gastric liquor was of an opaque, whitish colour, with little fine, brown sediment—that from the acid menstruum was also opaque, but of a reddish brown colour, and of a thin, mucilaginous consistence, with no sediment.

One drachm of infusion of nutgalls, added to the gastric solution, immediately afforded a rich, cream-like fluid, and slowly precipitated a fine, compact sediment. The same quantity of infusion of galls, added to the other, immediately formed the whole mass into a coarse, brown coagulum. After standing a while, it afforded a large, loose, brownish sediment, and a light coloured fluid, which, on standing, became white as milk; and the sediment became compact and remained so.

The precipitates, after the addition of the *tan* taken out and filtered, weighed as follows—that in the gastric solution, eighteen grains; the other, forty grains—the difference of weight being about equal to the quantity of gelatine put in.

Experiment 106

Feb. 15. At 9 o'clock, 45 mins., A. M., repeated the last (105th) experiment, with *gelatine,* and the gastric juice, and dilute acids, in the same proportions.

At 3 o'clock, 15 mins., P. M., the gelatine in gastric juice, all dissolved, to a mere mite—that in acid mixture, nearly so, six grains only, remaining on the filter, of a jelly-like consistence. The fluid of the gastric portion had a bluish white colour, and the other, yellowish, or about the colour of dry gelatine.

At 6 o'clock, the gelatine in the acid menstruum, all dissolved. Fluids of both, nearly similar.

One drachm infusion of nutgalls, added to each, instantly formed loose, lightish coloured coagulæ in both; threw down a compact sediment in the gastric solution, and left an opaque, milky fluid. The coarse coagulæ in the acid menstruum, continued suspended throughout the mass of fluids, for a long time, gradually subsiding. At the end of forty eight hours, it had become precipitated to the bottom into a compact mass, and exhibited distinct particles of the entire, undissolved gelatine, mixed with a dirty white coloured, curd-like substance.

Experiment 107

Feb. 15. At 1 o'clock 30 mins., P. M., he dined on *boiled codfish* and *bread.* Digested in four hours and a half.

Experiment 108

Feb. 16. At 1 o'clock, 45 mins., P. M., he dined on *mutton soup* and *bread*—6 o'clock stomach empty. Digested in four hours and a quarter.

Experiment 109

Feb. 19. At 9 o'clock, A. M., I put twenty grains of *boiled fat pork,* cut fine, into three drachms of *clear* gastric juice, and the same kind and quantity into three drachms of gastric juice, strongly tinged with yellow *bile,* with a view to ascertain whether there be any difference in their solvent effects upon fat meats. Placed both on axilla. At 1 o'clock, P. M., the pork in the gastric juice, tinged with *bile,* dissolved to less than one grain—that remaining undissolved, in the *clear* juice, weighed two grains and a half.

Experiment 110

Feb. 20. At 1 o'clock, 30 mins., P. M., I put three parcels, ten grains each, of *boiled codfish,* into three separate portions of gastric juice, one *pure,* another containing *bile,* and the third, a *clear, limpid, slightly acid fluid,* taken from the stomach after active exercise and profuse perspiration, in more abundant quantity than usual. Placed them all on the bath.

At 1 o'clock, 30 mins., P. M., of the 21st, I took out and weighed the three parcels of fish. The result was as follows: that in the pure gastric juice weighed two and a half grains; that in the yellow, three grains; and the other, six grains.

This shows that other than oily food is retarded by the admixture of bile in the gastric juice.

Experiment 111

Feb. 23. At 9 o'clock, 45 mins., P. M., I took out two parcels, one drachm each, of gastric juice, one pure, and the other of the clear, limpid fluid, extracted under the circumstances mentioned in the last experiment, and put eight grains of *lean beef,* finely cut, into each; and placed them on the bath together. After being treated alike on the bath for six or eight hours, the residuum in the pure gastric juice, weighed three grains; and that in the limpid fluid, six grains and a half.

Experiment 112

Feb. 24. At 9 o'clock, 30 mins., A. M., having extracted gastric juice, containing a large proportion of yellow bile, I put twenty grains of *strong cheese,* cut small, into two drachms of it; and the same quantity and kind of cheese, into two drachms of pure gastric juice: placed them together on the bath.

At 9 o'clock, P. M., residuum in the yellow juice, weighed five grains; that in the clear juice six grains. Returned them to the bath.

At 9 o'clock, A. M., of the 25th, the cheese in the yellow juice, all dissolved; of that in the clear juice, two grains remained.

Experiment 113

March 26. At 8 o'clock, 15 mins.—Weather clear. Stomach empty and healthy. Introduced Thermometer (Pool's glass) three fourths the length of its tube, eight or ten inches, and continued it five minutes. Spirit stationary at 100½°.

At 9 o'clock, suspended, through the aperture, into the stomach, enclosed in a muslin bag, forty grains of *broiled, fresh codfish,* previously masticated, and imbued with saliva; and he immediately afterwards breakfasted on the same kind of fish, a small quantity of *bread* and *coffee,* and kept exercising moderately.

At 11 o'clock, stomach full of fluids. 2 o'clock, P. M., chymification complete. Bag empty.

Experiment 114

March 27. At 9 o'clock, 15 mins., A. M., he breakfasted on *fresh, broiled fish (Flounder) bread* and *coffee,* and kept exercising moderately. 11 o'clock, stomach half empty—pulp of bread only appeared. 11 o'clock, 30 mins., particles of fish and bread still to be seen in the stomach. 1 o'clock, P. M., stomach entirely clear of food. Temperature 101°.

Experiment 115

I took *dilute muriatic* acid, reduced it to the strength and taste of the gastric juice, as nearly as practicable, three drachms; *dilute acetous acid,* to about the same flavour, one drachm—mixed them together, and put into this mixture, one scruple of *broiled steak,* cut fine; and the same quantity and kind of meat into four drachms of gastric juice. Placed them both on the bath. In six hours and three quarters, the meat in the gastric juice, taken out and filtered, weighed two grains only—that in the acid mixture, treated in the same way, was not dissolved; but had lost its fibrous form; and was converted into a tremulous, jelly-like mass, too tenacious to pass through the filter, and weighed more than when first put in. It did not appear like chyme, nor resemble that in the gastric juice.

After digesting eight hours longer, on the bath, the contents of the acid mixture had become nearly dissolved or diffused, and when run through the filter, left only a very little of the jelly-like mass, so abundant in the first examination. The liquid was now more like, though not exactly similar, to that of the gastric portion; this being opaque, and of a lightish grey colour, affording a dark brown sediment on standing; that from the acid menstruum, was also opaque, of reddish brown colour, but deposited no sediment.

Two drachms of the infusion of nutgalls, added to the gastric portion, threw down a fine, reddish brown precipitate, and afforded an opaque fluid, of similar colour. Two drachms of the infusion, added to the acid mixture, threw down a more copious precipitate, and left a clearer and thinner fluid, of a yellowish colour, and nearly transparent.

Experiment 116

A drachm of the concentrated, disinfecting solution of chloride of soda, prepared according to the formula of Labarraque, was added to a drachm of an extremely putrid mixture of beef, macerated in water—the putridity speedily disappeared; but not more so than when a drachm of *pure gastric juice* was added to a similar quantity of the same putrid mixture.

MICROSCOPIC EXAMINATIONS

The following Microscopic examinations, were made with Jones' compound Microscope, in presence of Professor DUNGLISON and of Capt. H. SMITH, of The Army. They afford, however, very little information on the subject of digestion, and show that no very satisfactory results are attainable from Microscopic examinations of Chyme.

I. *Pure gastric juice,* exhibited the appearance of water, except that there were perceptible, a very few minute globules.

II. The chymous product of the *gastric juice and unmasticated, lean beef,* exhibited globules of various sizes, resembling those of the blood, having a transparent centre, and opaque margin, with various very fine filaments, of apparently undigested fibrine.

III. Product of *gastric juice and albumen*—exhibited appearances resembling, considerably, those presented by the gastric juice alone—no distinct globular arrangement.

IV. Chyme from *gastric juice and tendon of veal*—exhibited numerous minute, apparently fleshy, particles—no globular appearance.

V. Chyme from *gastric juice* and *fowl* and *bread*—In the comparatively clear portion, (taken without shaking the vial) exhibited a few undissolved particles, and very few globules. A portion taken after shaking the vial, exhibited considerably more particles, and a greater number of globules.

VI. The product of the same kind of aliment, (*fowl* and *bread*) macerated in water, exhibited numerous undissolved particles, with few globules;—the globules regularly formed, as in the foregoing experiment.

VII. Product of *gastric juice,* and *soup, made from fresh beef,* exhibited globules extremely numerous, and distinctly formed, far more so, than in any of the preceding experiments—and a few particles of meat.

VIII. *Impure gastric juice,* or that with an *admixture of green bile,* when taken from the stomach, exhibited numerous amorphous particles, with few globules.

IX. Chyme, *artificially formed,* from *pork* and *bread,* exhibited numerous globules of different sizes, apparently oily.

X. *Chymous product of gastric juice* and *fat pork, formed in the stomach,* exhibited a beautiful appearance of large transparent globules, of different sizes, evidently oily.

XI. *Fat pork, macerated in pure water,* presented appearances of globules precisely similar to those in the products of digestion.

EXPERIMENTS, &c.

FOURTH SERIES

PLATTSBURGH, N. Y. 1833.

The following Gastric Experiments, and Examinations of the stomach, have been made since the manuscript of the previous part of this work was prepared for the press.

EXAMINATIONS OF THE TEMPERATURE AND APPEARANCE OF THE INTERIOR OF THE STOMACH.

I. *July* 9. 6 o'clock, A.M. Weather cloudy and damp. Wind W., light. Stomach empty and clean. Introduced glass thermometer, at the aperture, bulb nine inches down towards the pylorus—temperature 100°, Fahrenheit, before rising from his bed.

II. *July* 10. 6 o'clock, A.M. Weather clear. Wind W., brisk. Th: 63°. Stomach empty and clean. Temperature 100°, before rising. 9 o'clock, P.M. Weather clean and calm. Th: 75°. Stomach empty. Temperature 101°, after moderate exercise in open air.

III. *July* 11. 6 o'clock, A.M. Weather cloudy. Wind N.E., brisk. Th: 65°. Stomach empty and clean. Temperature 100°, before rising. 8 o'clock, 30 mins. Weather clear and dry. Wind S., brisk. Temperature of stomach 101°, after exercise. 9 o'clock 30 mins., P.M. Weather hazy. Wind S.W., light. Th: 75°. Temperature 101°.

IV. *July* 12. 6 o'clock, A.M. Weather clear. Wind W., brisk. Th: 70°. Stomach empty. Temperature 100½°, after going out into the open air. 9 o'clock, P.M. Weather clear. Wind W., light. Th: 76°. Temperature 101¾°. Stomach empty.

V. *July* 13. 5 o'clock, 30 mins., A.M. Weather clear, serene and calm. Thermometer 69°. Stomach empty, healthy and clean. Temperature 99½°, before rising from his bed. 6 o'clock, 30 mins. Weather same as at last examination. Stomach empty. Temperature 100¾°, after rising and walking out in open air, twenty or thirty minutes. 6 o'clock, 45 mins. Returned from a smart walk, exercising so as to produce gentle perspiration. Temperature 101¾°.

VI. *July* 14. 5 o'clock, 30 mins., A.M. Weather variable—heavy thunder shower, during the night. Wind S., moderate. Th: 75°. Stomach empty.—Temperature 100° on rising from bed—100¾° after walking out into the open air, and immediately back. 9 o'clock, P.M. Weather rainy—atmosphere oppressive. Th: 79°. Wind S., light. Temperature of stomach 102°. St. Martin has been in the woods all day, picking whortleberries, and has eaten no other food since 7 o'clock in the morning, till 8 at evening. Stomach full of berries and chymifying aliment, frothing and foaming like fermenting beer or cider—appears to have been drinking liquor too freely.

VII. *July* 15. 5 o'clock, 30 mins., A.M. Weather clear. Wind W., light—air damp—ground wet. Th: 74°. Stomach empty. Temperature 100°, before rising. 7 o'clock, 30 mins. Weather, wind, &c. same as at last examination. Th: 74°. Stomach empty. Temperature 102°, immediately after smart exercise. 1 o'clock, 30 mins., P.M. Weather clear and pleasant, since 8 o'clock, (till within fifteen minutes, in which interim, has fallen a light shower of rain.) Wind W., light. Th: 74°. Stomach empty. Temperature 100½°—has been at manual exercise for four hours. 9 o'clock, P.M.—Weather and wind, same. Th: 72°. Temperature 101¾°. Stomach full of chymous fluid, oil, and pulp of bread and cakes, eaten for supper, two hours previous to examination.

VIII. *July* 16. 7 o'clock, 30 mins., A.M. Weather cloudy. Wind W., light. Th: 73°. Stomach empty. Temperature 101°, after rising and before exercising. 9 o'clock, P.M. Weather cloudy, damp and chilly. Th: 70°. Temperature 101½°.

IX. *July* 28. 9 o'clock, A.M. Weather clear. Wind N.W., brisk. Th: 66°. Stomach empty—not healthy—some erythema and aphthous patches on the mucous surface. St. Martin has been drinking ardent spirits, pretty freely, for eight or ten days past—complains of no pain, nor shows symptoms of any general indisposition—says he feels well and has a good appetite.

X. *August* 1. 8 o'clock, A.M. Examined stomach before eating any thing—inner membrane morbid—considerable erythema and some aphthous patches on the exposed surface—secretions vitiated—extracted about half an ounce of gastric juice—not clear and pure as in health—quite viscid.

XI. *August* 2. 8 o'clock, A.M. Circumstances and appearances very similar to those of yesterday morning. Extracted one ounce of gastric fluids—consisting of unusual proportions of vitiated mucus, saliva, and some bile, tinged slightly with blood, appearing to exude from the surface of the erythema, and aphthous patches, which were tenderer and more irritable than usual. St. Martin complains of no sense of pain, symptoms of indisposition, or even of impaired appetite. Temperature of stomach 101°.

XII. *August* 3. 7 o'clock, A.M. Inner membrane of stomach unusually morbid—the erythematous appearance more extensive, and spots more livid than usual; from the surface of some of which, exuded small drops of grumous blood—the aphthous patches larger and more numerous—the mucous covering, thicker than common, and the gastric secretions much more vitiated. The gastric fluids extracted this morning were mixed with a large proportion of thick ropy mucus, and considerable muco-purulent matter, slightly tinged with blood, resembling the discharge from the bowels in some cases of chronic dysentery. Notwithstanding this diseased appearance of the stomach, no very essential aberration of its functions was manifested. St.

Martin complains of no symptoms indicating any general derangement of the system, except an uneasy sensation, and a tenderness at the pit of the stomach, and some vertigo, with dimness and yellowness of vision, on stooping down and rising again—has a thin, yellowish brown coat on his tongue, and his countenance is rather sallow—pulse, uniform and regular; appetite good; rests quietly, and sleeps as well as usual.

XIII. *August* 4. 8 o'clock, A. M., stomach empty; less of those aphthous patches than yesterday; erythematous appearance more extensively diffused over the inner coats, and the surface inclined to bleed; secretions vitiated. Extracted about an ounce of gastric fluids, consisting of ropy mucus, some bile, and less of the muco-purulent matter, than yesterday; flavour peculiarly fœtid and disagreeable; alkalescent and insipid; no perceptible acid; appetite good; rests well, and no indications of general disease or indisposition.

XIV. *August* 5. 8 o'clock, A. M., stomach empty; coats less morbid than yesterday; aphthous patches mostly disappeared; mucous surface more uniform, soft, and nearly of the natural, healthy colour; secretions less vitiated. Extracted two ounces gastric juice, more clear and pure, than that taken for four or five days last past, and slightly acid: but containing a larger proportion of mucus, and more opaque than usual, in a healthy condition.

XV. *August* 6. 8 o'clock, A. M., stomach empty; coats clean and healthy as usual; secretions less vitiated. Extracted two ounces gastric juice, of more natural and healthy appearance, with the usual gastric acid flavour; complains of no uneasy sensations, or the slightest symptom of indisposition; says he feels perfectly well, and has a voracious appetite; but not permitted to indulge it to satiety.—He has been restricted from full, and confined to low diet, and simple, diluent drinks, for the last few days, and has not been allowed to taste of any stimulating liquors, or to indulge in excesses of any kind.

Diseased appearances, similar to those mentioned above, have frequently presented themselves, in the course of my experiments and examinations, as the reader will have perceived. They have generally, but not always, succeeded to some appreciable cause. Improper indulgence in eating and drinking, has been the most common precursor of these diseased conditions of the coats of the stomach. The free use of ardent spirits, wine, beer, or any intoxicating liquor, when continued for some days, has invariably produced these morbid changes. Eating voraciously, or to excess; swallowing food coarsely masticated, or too fast; the introduction of solid pieces of meat, suspended by cords, into the stomach; or of muslin bags of aliment, secured in the same way; almost invariably produce similar effects, if repeated a number of times in close succession.

These morbid changes and conditions are, however, seldom indicated by any ordinary symptoms, or particular sensations described or complained of, unless when in considerable excess, or when there have been corresponding symptoms of a general affection of the system. They could not, in fact, in most cases, have been anticipated from any external symptoms; and their existence was only ascertained by actual, ocular demonstration.

It is interesting to observe to what extent the stomach, perhaps the most important organ of the *animal* system, may become diseased, without manifesting any external symptoms of such disease, or any evident signs of functional aberration. Vitiated secretions may also take place, and continue for some time, without affecting the health, in any *sensible* degree.

Extensive active or chronic disease may exist in the membranous tissues of the stomach and bowels, more frequently than has been generally believed; and it is possible that there are good grounds for the opinion advanced by a celebrated teacher of medicine, that most febrile complaints are the effects of gastric and enteric inflammations. In the case of the subject of these experiments, inflammation certainly does exist, to a considerable extent, even in an *apparent* state of health—greater than could have been believed to comport with the due operations of the gastric functions.

EXPERIMENTS, &C.

Experiment 1

Sept. 18. At 8 o'clock, 45 mins., A. M., St. Martin breakfasted on four ounces of *fresh salmon trout, fried,* three ounces of *bread,* and drank half a pint of *water.* The coats of the stomach were not perfectly healthy; some aphthous patches and dark red spots to be seen on the mucous surface; gastric juice slightly viscid; acid taste distinctly perceptible. At 10 o'clock, 15 mins., stomach entirely empty. Breakfast completely chymified and gone; nothing but a little gastric juice and flocculi of mucus, remaining in the stomach.

Experiment 2

Sept. 18. At 2 o'clock, P. M., he dined on six ounces of *boiled, fresh, salmon trout,* three ounces of *bread,* and a *potato,* and drank half a pint of *water.* Continued at work, sawing and splitting wood. He had eaten nothing from the time he took his breakfast; had been hard at work all the time; looked, and said he felt quite fatigued.

At 3 o'clock, 40 mins., stomach about half full of a nearly homogeneous semi-fluid, of a rich milk or cream colour, and about the consistence of fine corn meal gruel—a few small particles of the fish, and some of the potato, could be distinguished. 4 o'clock, 15 mins., stomach empty and clean.

Experiment 3

Sept. 20. At 1 o'clock, 15 mins., P. M., he dined on three ounces *fat pork,* and one pint of *corn* and *beans,* (green,) two ounces of *bread,* and half a pint of *water;* and kept exercising. Digested in three hours and three quarters.

Experiment 4

Sept. 21. At 8 o'clock, A. M., he breakfasted on eight ounces of *beef's liver, broiled,* two ounces of *bread,* and drank half a pint of *water.* Continued usual exercise. 9 o'clock 30 mins., stomach full of partially chymified food, considerable oil, (melted butter,) floating on the surface; black pepper mingled with it, and emitting a strong aromatic odour of the spice. 10 o'clock, 30 mins., stomach empty and clean. Extracted two drachms of gastric juice.

Experiment 5

At 1 o'clock, 30 mins., P. M., same day, St. Martin dined on one pint of rich *beef* and *vegetable soup,* made of the joint, marrow bone and muscle of the leg of an ox, three ounces of *bread,* and continued moderate exercise. 3 o'clock, 15 mins., stomach nearly full of thick, greyish white, porridge-like semi-fluid, with a thick pellicle of oil floating on the surface. 4 o'clock, P. M., stomach empty.

Experiment 6

Sept. 30. At 7 o'clock, 30 mins., A. M., he breakfasted on *bread* and *milk,* and continued his usual exercise. 8 o'clock, 30 mins., stomach nearly full of milky fluid, pulp of bread and coagulæ. 9 o'clock, contents of stomach considerably diminished since last examination—took out a portion, nearly chymified; very little fine coagulæ perceptible; bread in small particles, reduced to a greyish, soft pulp; the menstruum of a whitish, whey-colour and consistence. 9 o'clock, 30 mins., chymification complete. Stomach empty and clean.

The portion taken out of the stomach, at 9 o'clock, into a vial and continued in the axilla, till 12 o'clock, M., was almost completely chymified; small pulpous particles of bread only, discernible; the fluid of a rich whey, or gruel colour and consistence; a little loose, light coloured sediment fell to the bottom, on standing.

Experiment 7

Oct. 1. At 1 o'clock, 30 mins., P. M., St. Martin dined on *boiled, fresh, lean beef, potatoes* and *bread,* and continued his usual exercise. 4 o'clock, 15 mins., stomach empty.

Experiment 8

Oct. 2. At 1 o'clock, 30 mins., P. M., he dined on same kind of food as yesterday, *lean, boiled beef, potatoes* and *bread,* dressed with a liberal quantity of strong mustard and vinegar, and continued the same exercise. 3 o'clock, 30 mins., stomach nearly full of heterogeneous mixture. 4 o'clock, 30 mins., stomach still contains chyme and some undissolved food; fluids taste and smell quite strongly of the mustard; complains of more smarting at the edges of the aperture, than usual; some slight morbid appearance on the mucous surface. 5 o'clock, stomach empty.

These two last experiments were made under almost exactly similar conditions of the stomach, with a view to notice the effects of this kind of stimulating condiment. The result was, that it apparently retarded the process of digestion; no other appreciable cause existed, for this difference of result. The stomach presented the usual healthy appearance immediately previous to the ingestion of the meal. Nothing occurred to interfere with, or interrupt the digestive functions. The slight morbid appearance on the mucous surface, towards the close of chymification, I conceive to have been more the effect of the over excitement of the mustard than any other cause.

It would seem then, that stimulating condiments, instead of being used with impunity, are actually prejudicial to the healthy stomach. They can only be required, and taken with benefit, when the gastric apparatus is languid and relaxed, and requires stimulants to excite the tone and action of its vascular tissues.

Experiment 9

Oct. 3. At 2 o'clock, 35 mins., P. M., St. Martin ate nine ounces of *raw, ripe, sour apples.* 3 o'clock, 30 mins., stomach full of fluid and pulp of apples; quite acrid, and irritating the edges of the aperture, as is always the case when he eats acescent fruits or vegetables. 4 o'clock, stomach not empty; contents more sharp and acrid; pulp of apple still to be seen. 4 o'clock, 40 mins., stomach empty; morbid appearance of the gastric surface considerably increased.

Experiment 10

Oct. 7. At 8 o'clock, A. M., he breakfasted on *bean soup,* made with *fresh beef* and *bread.* Digested in three hours. And at 2 o'clock, P. M., he dined on the same, which was digested in three and a quarter hours.

Experiment 11

Oct. 10. At 8 o'clock, A. M. Weather fair.—Wind W., light. Th: 61°. Stomach empty and healthy. Temperature 101°, after moderate exercise. Breakfasted on *baked potatoes* and *bread.* 10 o'clock, stomach nearly empty; a little chymous fluid to be seen; quite acrid. Temperature 101½°, after usual exercise. 10 o'clock, 45 mins., stomach empty. Temperature 101½°.

Experiment 12

At 2 o'clock, P. M., same day. Weather hazy. Wind S., moderate. Th: 61°. Stomach empty and healthy. Temperature 101¾°, after exercise. Dined on *roast beef, bread, potatoes* and *boiled cabbage.* 4 o'clock. Wind S.W., brisk—raining. Th: 61°. Stomach half full of heterogeneous mass of acrid fluid, oil, beef and cabbage. Temperature 103°; had been smartly exercising for two hours. 7 o'clock, 30 mins., wind and weather same as at 4 o'clock.—Th: 63°. Stomach empty. Temperature 102°.—Exercise continued moderately till this examination.

In this experiment, the temperature of the stomach rose to 103°, one degree higher than I have ever before observed it to rise; and chymification was protracted.

Whether these two circumstances were occasioned by unusually increased exercise, and the consequent fatigue of the system, or from the nature of the aliment eaten, and the unusual fulness of the meal, I am not able positively to say; but am inclined to think, from previous observations, that they are attributable to the latter—as the usual morbid appearances, consequent on too full alimentation, followed this meal in the course of twenty four or thirty six hours—as may be seen by the two subsequent experiments.

Experiment 13

Oct. 11. 7 o'clock, 30 mins., A.M. Weather fair. Wind N.W., brisk. Th: 32°. Stomach empty.—Temperature 100½°, after moderate exercise in open air. 8 o'clock, 45 mins., wind and weather, same. Th: 38. Stomach empty. Temperature 102°—had been smartly exercising, shovelling dirt, for an hour or more, and was quite warm. Breakfasted on *stewed veal* and *bread.* 11 o'clock, stomach not empty. Temperature 102°—continues exercise. 12 o'clock, stomach contains a very little chymous fluid, and a trace of the muscular fibres of the veal. 12 o'clock, 30 mins., stomach empty.

Experiment 14

At 2 o'clock, P.M., same day, he dined on *fried veal* and *bread,* and continued moderate exercise.—6 o'clock, 30 mins., stomach empty. Temperature 101¾°. Some morbid appearance on the mucous surface.

At 8 o'clock, 30 mins., weather fair and calm. Th: 36°. Stomach empty; slightly morbid, with few aphthous spots. Temperature 101½°; had been still and quiet for three or four hours.

Experiment 15

Oct. 12. At 7 o'clock, 30 mins., A.M. Weather hazy. Wind S., light. Th: 36°. Stomach empty—coats not entirely healthy—some erythema and aphthous patches. Temperature 101°, after usual morning exercise. 8 o'clock—circumstances same as at last examination. Temperature 101°. Breakfasted on *fresh beef, fried dry,* and *bread.* 10 o'clock, stomach full of fluids; particles of beef, bread and oil, distinctly to be seen. Temperature 101°. 12 o'clock, stomach empty.

Experiment 16

Oct. 13. At 7 o'clock, A.M. Weather rainy. W., N.E., brisk. Th: 42°. Stomach empty. Temperature 101°, after morning exercise. 9 o'clock, temperature same. Breakfasted on *old, salted pork, fat* and *lean* together, *(fried)* four ounces of *bread,* and the *yolks* of *six eggs,* fried hard with the pork. 11 o'clock, contents of the stomach heterogeneous; distinct particles of lean pork, egg and oil to be seen; fluid sharp and acrid. Temperature 101°. 12 o'clock, M., oil and egg still to be seen, floating in a milky, chymous fluid; the oil, or lard on the surface, and the egg, in firm coagulæ, diffused through the fluid. Temperature 101°. 1 o'clock, 15 mins., P.M., stomach empty and clean. Temperature 101°—was quiet and inactive during this experiment.

Experiment 17

At 2 o'clock 20 mins., P.M. same day, St. Martin dined on six ounces of the *spinal marrow* of an ox, steam-cooked, and seasoned with a little *butter, vinegar, salt* and *pepper,* and three ounces of *bread.* 4 o'clock, P. M., contents of stomach a perfectly milk-white, semi-fluid pulp. Temperature 102°. 5 o'clock, 10 mins., stomach empty and clean.

Experiment 18

At 6 o'clock P. M., he ate a full meal of *boiled rice,* simply cooked in water, and seasoned with a little salt. 7 o'clock, stomach empty and clean; not a vestige of the rice to be seen.

Experiment 19

Oct. 14. At 9 o'clock, A. M., he breakfasted on the *albumen* of *six eggs, fried hard,* in pork fat. 12 o'clock 15 mins., M., all chymified—stomach empty.

Experiment 20

At 1 o'clock, P. M., same day, he dined on eight ounces, *boiled beef's brains,* seasoned with *salt,* and a small piece of *bread.* 2 o'clock, stomach full of milk-white, pulpous, or porridge-like semi-fluid; slightly acid taste, and of a bland, insipid flavour. 2 o'clock, 30 mins., stomach almost empty; scarcely any of the white, pulpous mass to be seen. Temperature 102°. 3 o'clock 15 mins., P. M., stomach empty and clean.

Experiment 21

At 3 o'clock 30 mins., P. M., same day, St. Martin ate a small head of *raw cabbage,* weighing ten ounces. 5 o'clock, 45 mins., not a particle of the cabbage in the stomach; little albuminous, or greyish, chymous fluid, only remained.

Experiment 22

At 6 o'clock, 30 mins., P. M., he ate six ounces *boiled leg of fresh mutton,* rare done, dressed with a little *melted butter* and *vinegar,* and two ounces of *bread.* 8 o'clock, stomach empty and clean.

Experiment 23

Oct. 15. At 8 o'clock, 45 mins., breakfasted on *three fresh eggs,* softly coagulated, by being broken and put raw into boiling water, and three ounces of dry *bread.* 12 o'clock, M., stomach empty.

Experiment 24

At 1 o'clock, 30 mins., P. M., he dined on *apple dumplings,* made of wheaten dough and sweet apples, boiled, one and a half pounds. 4 o'clock, all chymified, and stomach empty.

Experiment 25

Oct. 16. At 8 o'clock, 45 mins., A. M., he breakfasted on *broiled, salted pork* and *bread.* 12 o'clock, M., all chymified, and gone from the stomach.

Experiment 26

At 1 o'clock, P.M., same day, he dined on *raw, salted pork,* cut thin, and eaten with dry *bread.* Digested in three hours.

Experiment 27

At 4 o'clock, 30 mins., same day, he ate half a pound of *raw cabbage,* cut fine, and macerated in *vinegar.* 5 o'clock 45 mins., stomach entirely empty, not a vestige of cabbage to be found. Extracted four drachms of gastric juice, mixed with a very little greyish white, chymous fluid.

Experiment 28

Oct. 17. At 9 o'clock, A.M., he breakfasted on *stewed, salted pork, potatoes* and *bread.* Digested in three hours. Extracted gastric juice.

Experiment 29

At 2 o'clock, 30 mins., P.M., same day, he dined on *boiled mutton, recently salted, squash, potatoes* and *bread.* Digested in three hours.

Some morbid spots begin to make their appearance on the mucous surface again; grumous blood exuding from several small points of the membrane; tongue slightly coated; countenance rather sallow; dull pain across the forehead, and through the eyes; appetite not impaired; at bed-time, put in through the aperture four drachms of *tinct. of aloes and myrrh,* diluted with water. This had the effect of correcting the morbid appearance of the stomach, and removed the pain in the head, &c.

Experiment 30

Oct. 18. At 9 o'clock, 45 mins., A.M., he breakfasted on *boiled carrots,* and nothing else—full meal. 12 o'clock, M., examined stomach; considerable yellowish, pultaceous semi-fluid, remaining. 1 o'clock P.M., stomach empty.

Experiment 31

At 7 o'clock, P.M., he ate three large *roasted potatoes,* with a little *salt*—nothing else. 9 o'clock 30 mins., stomach empty.

Experiment 32

Oct. 19. At 9 o'clock A.M., he breakfasted on *broiled mutton* and *pancakes.* Digested in three hours and forty minutes.

Experiment 33

At 2 o'clock, 15 mins., P.M., he dined on *stewed mutton* and *pancakes.* Digested in three and a half hours.

Experiment 34

Oct. 20. At 9 o'clock, 45 mins., A.M., he breakfasted on one pint of *sago, boiled,* thick and rich, sweetened with *sugar.* 11 o'clock, 30 mins., stomach empty and clean.

There was no acrimony of the gastric contents, or smarting of the edges of the aperture, during the chymification of this meal, as is usual in most vegetable and farinaceous aliments; it seemed peculiarly grateful to the surface of the stomach; rendering the membrane soft, uniform and healthy.

Experiment 35

At 12 o'clock, M., he ate four *eggs, roasted hard,* without any thing else. 3 o'clock P. M., stomach empty; no trace of the eggs to be seen.

Experiment 36

At 4 o'clock 30 mins., P. M., he dined on *roasted duck* and *fried onions.* 8 o'clock, 30 mins., stomach not empty—distinct particles of food to be seen.—9 o'clock, stomach empty.

Experiment 37

Oct. 21. At 9 o'clock, A. M., St. Martin breakfasted on one pint of *sago, boiled* and sweetened with *sugar.* 10 o'clock, 45 mins., stomach empty and clean; no vestige of the sago remaining; no acrimony of the gastric contents, or smarting of the edges of the aperture, during the chymification of this meal.

Experiment 38

Oct. 22. At 12 o'clock, M., he ate four *fresh eggs, roasted hard.* 3 o'clock, P. M., stomach empty; no trace of the eggs to be seen.

At 4 o'clock, P. M., he dined on *roasted duck,* (domesticated,) dressed with *onions.* 8 o'clock, stomach empty.

Experiment 39

Oct. 24. At 2 o'clock, 30 mins., P. M., he ate a pint of *soft custard,* and nothing else. 5 o'clock, 15 mins., stomach empty and clean.

At 6 o'clock, he ate three ounces of *strong old cheese,* and a piece of *bread.* 9 o'clock, 30 mins., stomach empty.

Experiment 40

Oct. 26. At 9 o'clock, A. M., he breakfasted on *fricasseed chickens, bread* and *coffee.* 11 o'clock, 45 mins., stomach empty and clean.

At 12 o'clock, M., he dined on *roast chicken, bread* and *potatoes.* 4 o'clock, P. M., stomach empty.

Experiment 41

Oct. 27. At 8 o'clock, A. M., he breakfasted on *broiled chicken, bread* and *coffee.* 11 o'clock, all digested, and stomach empty and clean.

At 12 o'clock, M., he dined on *chicken soup,* with *rice.* 3 o'clock, stomach empty.

At 5 o'clock, P. M., he ate a meal of *oyster soup* and *crackers.* 8 o'clock, 30 mins., stomach empty.

Experiment 42

Oct. 26. 10 o'clock A.M., stomach empty, healthy and clean. I suspended through the aperture, into St. Martin's stomach, thirty grains precisely, of each of the following articles of diet, severally masticated and separately contained in small muslin bags, viz:—Fricasseed *breast of chicken; liver* and *gizzard* of do.; *boiled, salted salmon; boiled potato,* and *wheat bread;* and he kept moderately exercising. At 3 o'clock, P.M., took out and accurately examined the several parcels. The breast of chicken was all digested and gone from the bag, to a mere atom, less than half a grain. The liver was almost as completely dissolved as the breast, half a grain only, remaining—of the bread, about the same; less than a grain. The residuum of the gizzard, consisting principally of tendinous fascia, weighed seven and a half grains. The salmon, twelve grains, and the potato, six grains. The bags, containing these several articles, were attached to a string, at equal distances from each other, about an inch apart; and I allowed length enough for them to move freely through the stomach, and pass even to the pylorus. They were attached in the following order:—1st, the breast of chicken—2d, liver—3d, gizzard—4th bread—5th, salmon, and 6th, potato. When I withdrew them, they appeared to be retained quite forcibly at the pyloric end, requiring considerable force to start them at first, but after being drawn two or three inches, they came easily. The bags too, appeared to have been compressed, in proportion as they had been settled into the pyloric extremity, and were emptied in about the same proportion, with the exception of those containing the bread and potato, which, though above, had less remaining than that containing the gizzard. This, however, may be accounted for, from the more difficult solubility or digestibility of the tendonous parts of the gizzard. The bags seemed to have been as forcibly pressed, as if they had been firmly grasped in the hand. The four first on the string, (counting from the lower end upwards) more so, than the other two; and the fourth more than the third. These circumstances coincide with the apparent contractions of a band, or circular muscle of the stomach, indicated by the motions of the glass tube, observed in former experiments. In comparing the length of the string, and situation of the bags, with the stem and bulb of the tube, it brought the fourth bag to that point in the stomach, where the contraction upon the bulb of the thermometer has invariably been observed to take place; the third bag just below, and the fifth and sixth, above it. The sensations expressed by St. Martin, on the extraction of these bags, were also indicative of the same facts. When I first commenced pulling the string, he complained of a sense of pain and distress at the pit of the stomach, and towards the pylorus, which increased while the bags were withdrawing, and particularly at this extremity, for the first three or four inches, till they had passed the band, into the splenic end.

The effects of this experiment, upon St. Martin's feelings and appearance, were very manifest, and afford interesting and important subjects of pathological consideration. He had not eaten or drunk any thing, that morning, and felt and looked in perfect health, when the bags were introduced;

continued moderately exercising, and ate nothing but a small piece of dry bread, till they were taken out.

Soon after they were suspended in the stomach, he felt a sense of weight and distress at the scrobiculus cordis; slight vertigo and dimness of vision. These continued to increase and become quite severe, accompanied, at the latter part of the time, by slight pain in the forehead and through the eyes, and a sense of tightness or stiffness across the breast. His countenance had changed from a florid, healthy, to a sallow, sickly appearance, during the time of the experiment, and a soreness at the pit of the stomach continued after the extraction of the bags, for eight or ten hours, and had not entirely subsided the next morning.

Morbid action of the inner membranes was evident next day, with considerable erythema and aphthous appearance.

The first, second and third bags were covered with a thick mucous coat, tinged with yellow bile; the others had very little or none of this appearance.—This circumstance I conceive to have been owing to the irritation of the bag, at the pyloric extremity, inviting the bile from the duodenum to the stomach, in the latter part of this experiment. Hence the pathological indications which ensued. The same appearance and circumstances have before occurred during these experiments.

The following experiments on artificial digestion, were instituted with a view of ascertaining more particularly, the relative digestibility of many of the different kinds of aliment used in the foregoing gastric experiments, on natural chymification, and to test the correctness of the results. They are minutely detailed for the purpose of showing the manner, progress and operation of the gastric solvent, on the alimentary substances, subjected to its action. How far they may illustrate these subjects, the reader will judge for himself.

The gastric juice was taken out of the stomach in different states of purity and put into vials; when food was submitted to its action, it was placed in a temperature between 96° and 100°, Fahrenheit, and kept either in the axilla, or on a sand bath, and frequently, though not constantly, agitated.

The discrepance of results in some similar experiments will generally be found to arise from the variable degrees of purity of the gastric juice, or different circumstances of the experiments.

Experiment 43

Sept. 18. At 8 o'clock, 45 mins., A. M., I put one drachm of *fresh salmon trout, fried,* and masticated, and one drachm of *wheat bread,* into two ounces of gastric juice, taken from the stomach yesterday and this morning. The juice was not perfectly clear, but contained some viscid mucus. Placed them in the axilla and kept moving. 10 o'clock, 15 mins., residuum of aliment taken out, filtered and pressed as dry as when put in, weighed one drachm and five grains. The menstruum, after filtering, was white and opaque, about the colour and consistence of rich gruel. Mixed the residuum and fluid together again and placed the vial on the sand bath, and kept it constantly agitated for *one hour.* Taken out, filtered and dried as before, the undissolved residuum now weighed just *thirty grains.* The fluids had become thicker, and richer in colour and consistence. Put them together again into the vial, and continued them on bath and in axilla, *another hour,* though not so constantly agitated, as during the last hour. The residuum, treated in the same manner as before, now weighed *twenty four grains.* Mixed together and continued in axilla *two hours* more, the residuum weighed *twelve grains.* After continuing *three hours* longer in the axilla, the undissolved portions of aliment, consisting principally of particles of fish skin; weighed *four grains,* which became gradually diminished during is continuance *an hour* longer in the axilla.

The menstruum at this time, was of a rich gruelly colour and consistence, slightly tinged with a reddish cast, or colour of the fish. Set this aside for thirty eight or nine hours.

Sept. 20. 9 o'clock, A. M., food almost entirely reduced to chyme, of a rich, lightish coloured, gruelly appearance; some few particles of the skin of the fish remaining undissolved, with some small, apparently foreign and indigestible substances, which were probably adventitiously mixed with the food.

To observe the effect produced on this chyme, by the addition of bile, and having very opportunely obtained some, from the human stomach, by the operation of an emetic, I added *one drachm* of this pure, albuminous, orange coloured bile, to six drachms of the chyme. The first apparent change, was in the colour, which partook of the bile; then a slight effervescence was perceived, and very fine coagulæ were formed. The fluid became richer in appearance, and less opaque. The foreign or indigestible particles, were more perceptible, and small, bright particles, resembling very minute scales, or skin of fish, were also quite plain to be seen.

I now divided this into two equal parts; to one of which, I added half a drachm of dilute muriatic acid, and set it by to subside. Examined at 10 o'clock, the 21st. The vial containing the mixture of chyme, bile and muriatic acid, exhibited the following appearance: It had a thick, dense sediment, of a yellowish green colour, which occupied about one quarter, of the space. The fluid above, was of the colour of whey, and about the consistence. The vial containing the mixture of chyme and bile only, showed the following appearance: The sediment was not so dense, and its colour, as well as the supernatant liquid, was rather more yellow. Standing at rest a few days, the sediment, at the bottoms of both vials, became more compact; that in the muriatic mixture, more so than the other, and was of a deeper green colour; the fluid continued of a rich, whey colour and consistence, and a very thin pellicle, or small whitish flocculi, rose on the top, or adhered to the sides of the vial.

Experiment 44

Sept. 20. At 1 o'clock, 15 mins., P. M., I put one drachm of *boiled, green corn* and *beans,* into twelve drachms of gastric juice, and kept the vial in the axilla, or on the bath, as usual, frequently agitating it, till 7 o'clock, P. M. The residuum, at this time, taken out, weighed *twenty eight* grains, consisting wholly of the hulls or cuticular parts of the broken kernels, and one entire bean and a kernel of corn; the first of which weighed thirteen, and the other eleven grains, leaving four grains of the skins of the broken, dissolved grain. The two entire kernels, (the bean and the corn) were designedly put in whole, to test the effect of the gastric juice upon them, in the entire state. The other portion of the grain was mashed soft before put in. The pulpous portion of the broken kernels was all dissolved, and appeared completely chymified. The fluid was nearly as white as milk, and of the consistence of clear rich gruel.

The gastric juice used in this experiment, was considerably vitiated when taken from the stomach, some thirty six or forty eight hours previously, and was quite fœtid when used. This fœtor was, in a great measure, corrected after chymification of the food had commenced; the sharp, acid flavour, so peculiar to forming chyme, was increased.

Experiment 45

Sept. 21. At 8 o'clock, 15 mins., A. M., I put thirty grains of *fresh beef steak* and thirty grains of *fresh beef's liver,* (broiled, and masticated) contained loosely in separate muslin bags, into one ounce of fresh gastric juice, and kept them in axilla. At 9 o'clock, 45 mins., the two parcels of aliment, taken out and pressed as dry as when put in, weighed as follows: The *steak,* seventeen grains; the *liver,* eleven grains. Put into the vial again, and continued in the axilla, till 1 o'clock, P. M. The *steak* weighed fourteen, and the *liver* eight grains. Put into the vial again and continued in axilla for four hours; no further change was effected. They both weighed the same as at last examination. The solvent action having ceased, I added one ounce more of gastric juice, and continued in axilla, two hours and thirty minutes. The *beef* weighed five grains, and the *liver,* four; the residue of the liver consisted mostly, of membranous particles, like sections of the hepatic blood vessels, of which I conceived them to be portions.

I now mixed them both together, in one bag, and continued them in axilla, three hours, when the whole were completely dissolved and chymified, and the bag empty; with scarce a trace of aliment left on the inside. The fluid was of a greyish white, gruelly appearance. A brownish sediment was deposited on standing.

Experiment 46

Sept. 22. At 12 o'clock, 30 mins., I put thirty grains of *new cheese,* (masticated) into three drachms of gastric juice, and placed it in the axilla, eight hours and thirty minutes, when five grains of the cheese remained undissolved, or rather unchymified, as the residuum was in nearly a liquid form, consisting, principally, of oil, combined with a soft caseous substance, floating on the surface of a rich, milky fluid. A little very fine, white, compact sediment, at the bottom of the vial. At this time, it had acquired a strong acid, or peculiar acrid taste, and emitted a strong, caseous smell, even stronger than the cheese itself presented, when put in.

At 12 o'clock, M., I put one drachm of *sago,* boiled so as to leave some of the grains whole and entire, but soft and gelatinous, into three drachms of gastric juice and kept it in the axilla. When first mixed, they were so much alike, that they could only be distinguished from each other by the globular forms of the grain. But by these, however, the gastric juice could distinctly be perceived to dissolve the grains of sago, till they had all disappeared.

The fluid had now become more opaline and whitish, and in two hours and twenty minutes, no trace of the sago could be discerned. At this time the fluid had become more opaque and milky. No sediment was deposited on standing for twenty four hours. A slight acid was perceptible.

At 1 o'clock, P. M., I took three vials, the first containing two drachms of gastric juice; the second, two drachms of common vinegar; and the third, two drachms of simple water. Into each of these, I put ten grains of *raw albumen* of a fresh egg.—When first put together, they presented the following appearances: The albumen put into the gastric juice, at a temperature of about 76°, produced loose coagulæ in a few minutes, generally diffused through the juice, but soon collected into a more compact mass, and subsided towards the bottom of the vial. That put into the vinegar, produced similar coagulæ and loose mass, and fell down. That in the vial of water produced loose, light coloured flocculi, equally suspended through the water, but not inclining to collect together, like the other two.

These three parcels, kept in the axilla, and agitated for two hours, presented the following appearances: The coagulæ in the gastric juice, was half dissolved, and the menstruum of a milky appearance.

Those in the vinegar and water, remained the same, and their fluids unaltered. In five hours, that in the gastric juice was entirely dissolved, and the fluid more opaque and white; the other two remained of the same appearance as at last examination; the coagulæ in the vinegar, taken out, weighed *nine* grains—that in the water was too loose and frothy to be collected and weighed.

Experiment 47

Sept. 25. At 7 o'clock, A. M., I put twenty grains of light *sponge cake* into three drachms of gastric juice and kept it in axilla. It was all dissolved and chymified, in seven hours. The fluid was rich, yellowish white, or cream colour, and of the consistence of gruel, with a little loose, brown sediment at the bottom of the vial, after standing.

Experiment 48

At 9 o'clock, A. M., I put two *purple fox grapes,* one skinned and the other entire, into six drachms of gastric juice, and kept them in axilla, six hours, with very little alteration in their appearance; the skinned grape, weighing, when first put in, thirty four grains, weighed now, thirty grains, retaining its shape and texture. The whole grape was not affected in the least, either in shape, colour or texture. It weighed fifty four grains when put in, and the same now. Continued in axilla, twelve hours, they remained unaltered, and weighed exactly the same as at last examination. Added one ounce of fresh gastric juice, and continued them in axilla, twenty four hours, unaltered. The texture of the skinned grape, was as firm and hard as when first put in; and the fluid was unchanged in its appearance, except a slight fœtor, perceptible at the end of three or four days.

This, I think, is a fair specimen of the indigestible nature of this kind of fruit.

Experiment 49

Sept. 26. At 10 o'clock, A. M., I put thirty grains of *ripe, mellow peach,* and thirty grains of *ripe, hard apple,* into one ounce of gastric juice, and kept them in axilla, till 8 o'clock, P. M. At this time the residuum of the peach, weighed eighteen grains—the apple, twenty four grains. They were neither of them mashed or masticated, but cut into small, square pieces, strung on a string, and suspended into the juice in a vial.

At 10 o'clock, A. M., of the 27th, after having been continued in axilla, six hours longer, the peach weighed ten grains and the apple the same as at last examination, twenty four grains. The peach had now become soft and pulpous, and fallen from the string. Eight hours longer continuance in axilla, completed the digestion of the peach; but the apple remained nearly the same.

Experiment 50

Sept. 27. At 2 o'clock, P. M., I put one drachm of *albumen* of egg into four drachms of gastric juice, fresh from the healthy stomach. At first, the albumen fell to the bottom of the vial; but on being agitated, it was diffused through the juice, and in a few minutes, loose coagulæ formed, and remained suspended near the bottom of the fluid. Raised the temperature to 100°, and placed the vial in the axilla.

At the same time, I put one drachm of *albumen* into four drachms of simple water, at the same temperature, and placed it with the other in the axilla. When first put together, the albumen was diffused, in loose, light flocculi, through the water, not coagulating and collecting like that in the gastric juice, and subsiding to the bottom, but adhered to the sides of the vial, or rose to the surface.

When both vials were smartly agitated, a white, frothy mass, formed on the top of the water, filling the two ounce vial in which it was contained. The vial of albumen and gastric juice exhibited the coagulæ, broken into small particles, falling towards the bottom again. Kept in the axilla and frequently agitated, for one and a half hour, the gastric mixture had become semi-opaque and the coagulæ considerably diminished in quantity. The aqueous mixture remained unchanged; the frothy portion on top, and the fluid, perfectly limpid and clear, below. No appearance of the albumen in any shape, could be seen, except the floating froth. Indeed, the albumen seemed to have clarified the water, and rendered it clearer than at first. At 6 o'clock, P. M., the albumen in the gastric juice was completely dissolved; the fluid was white and milky, with a little very fine, dirty white precipitate falling to the bottom, on standing at rest. That in the water was strikingly different in appearance. The agitation had beaten up the albumen completely into beautiful white froth, and it lay like a snow ball or bunch of clean, raw cotton, on the surface of the water, now transparent as crystal, without the least particle of sediment to be seen.

At 7 o'clock, I added two drachms of gastric juice to the vial containing the water and albumen, and continued it in axilla. In two hours, the solvent effect of the juice, upon the frothy mass, was very evident. The white froth upon the top, was almost entirely diminished and gone. Neither could agitation re-produce it as at first; small white coagulæ, like those seen in the other vials, were now distinctly visible; the fluid had become opaque and whitish, like the other, and a little fine sediment settled to the bottom, on standing. Continued in the axilla, two hours longer, it resembled almost exactly, that in the other vial, in every particular.

Experiment 51

At 2 o'clock, P. M., I put one drachm of *yolk of egg* into four drachms of gastric juice, and another drachm into four drachms of simple *water,* and kept them, as usual, in the axilla; no difference at first could be perceived between the gastric juice and aqueous mixtures; each exhibited a yellow mixture like the egg, simply beat up with any white or watery menstruum. Six hours continuance of this treatment, produced little difference in the appearance of the two, and effected a slight modification in the gastric mixture only; this seems to have been converted into a very fine coagulæ, of a rich cream colour and consistence, and of a paler yellow than the other. In twelve hours more, a striking difference was manifest—that in the water remained the same as when first put together—a dull, yellow coloured sediment, in the proportion of about one fifth of the space occupied by the whole, had subsided to the bottom of a thin fluid, of the same colour, and now emitted a fœtid odour. That in the gastric juice had become more cream-like and lighter coloured, separating, on standing, into three distinct portions—a loose, coagulated, yellow mass, rose to the top, occupying more than half the upper space—a clear, whey-coloured fluid below, with a dirty, yellow sediment at the bottom, in about the proportion of one twelfth of the whole; not the least fœtor was perceptible.

Experiment 52

At 1 o'clock, 30 mins., P. M., I mixed one drachm of *Olive oil* with three drachms of gastric juice, and kept, frequently agitated in axilla, for eight hours.—When first put together and shaken, the mixture resembled water and oil, precisely; after continuing in the axilla, four or five hours, the oil had perceptibly diminished and chyme began to be formed, rendering the juice opaque and milky. At 10 o'clock, P. M., the oil was about one sixth diminished, the menstruum nearly the colour and consistence of milk.

Sept. 30. 8 o'clock, A. M., continued in the same manner, in the axilla, for twelve hours, the oil was proportionally diminished, and the opacity and milkiness, gradually increased.

Oct. 1. At 8 o'clock, A. M., I added one drachm of gastric juice—not clear, but considerably vitiated. Continued in axilla fourteen hours. Similar proportional decrease of the oil, and change of the colour of the fluid, were produced, and a slight fœtor was perceptible. This last circumstance, no doubt was attributable to the vitiated juice added.

Oct. 2. 10 o'clock, A. M., added three drachms of pure gastric juice, and continued in axilla, *ten* hours. This addition corrected the fœtor in a great measure. The stratum of oil was not much diminished in bulk, but considerably changed in appearance, having become quite white and frothy, exhibiting myriads of minute globules; and the colour and consistence of the fluid, were more rich and milky.

On the 3d at 10 o'clock, A. M., I divided the contents of the vial into two equal parts, and put them into two separate vials. To No. 1, I added two drachms of pure gastric juice; and to No. 2, two drachms of fresh extracted gastric juice, containing a large proportion of yellowish green bile, and continued, as usual, in axilla. The following changes were produced: The portion in No. 2 vial, which had received the yellow gastric juice, at first partook of the yellow colour of the juice added, generally diffused through the whole mass—a separation then took place; the bile seemed principally to unite with the oil, breaking it down and reducing it to very minute and almost imperceptible globules; and after remaining in the axilla, ten hours, and then standing at rest a few minutes, the under surface of the supernatant stratum of oil exhibited a milky or creamy appearance, and small, white flocculi, resembling coagulated milk or albumen; these soon became dissolved, and increased the richness of the fluid below—No sediment to be seen. The portion in No. 1 vial, to which the clear gastric juice was added, at the end of ten hours, had undergone some change. The pellicle of oil on the surface, was reduced to minute globules, of a whitish colour. The same appearance of white flocculi, or coagulæ, were exhibited upon the under surface of the supernatant stratum of oil, as in the other, but not so abundant, and the fluid was not so rich in colour and consistence.

Oct. 4. At 9 o'clock, A. M., I added two drachms more of each kind of juice, to their respective parcels, and continued them as usual, in axilla, for eleven hours. The difference between the two parcels, was now considerably increased. The fluid in No. 2 vial, was of a rich cream colour and consistence: the supernatant stratum of oil was converted into a light yellowish mass, resembling a mixture of gelatine and coagulæ; few of the globules of the oil could be distinguished; yellow flocculi adhered to the sides of the vial, above the fluid, after being agitated. When suffered to stand at rest a short time, loose yellow flocculi rose on the surface, occupying more than twice the space of the oil, before the last addition of gastric juice—no sediment subsided.

The parcel in No. 1 vial, had regularly progressed in chymification, in ratio proportional to the juice added; the supernatant, oily stratum was diminished, in thickness, nearly one third, since the last addition of gastric juice, had changed from its oily appearance, into a white, semi-gelatinous mass, intermingled with white flocculi; the fluid of the same milky appearance; a little white sediment at the bottom.

Oct. 5. At 10 o'clock, A. M., I added six drachms pure gastric juice, and six drachms of fresh extracted juice, containing about the same proportion of yellow bile as the other, to their respective vials, and put them on the bath, and kept them continually agitated for five hours. The effect was palpable and plain. The supernatant stratum, in No. 2 vial, was now completely broken down, and not a globule remained; a thin, yellow pellicle, or loose flocculi, rose upon the surface, on standing, and the fluid was of a rich cream colour and consistence, slightly tinged with bile—no sediment perceptible.

The contents of No. 1 vial, had undergone considerable change; the oily pellicle on the surface, was diminished but little in volume, but changed in appearance; had become converted into a white semi-gelatinous, or rather saponaceous consistence, and the milky richness of the fluid was increased.

This experiment is minutely and accurately detailed, with a view to demonstrate the slow, but certain digestibility of oils, and the manner they are acted upon by the gastric juice. It may be tedious, from its prolixity, but I considered its communication might be of some importance and usefulness to physiological science, the interests of which have been of paramount consideration with me, in all these experiments.

It very clearly appears, by this experiment alone, that *bile* accelerates the solution of oil, by the gastric juice; and I have no doubt, it facilitates the chymification of all fatty and oily aliments; and is required, and necessarily called into the stomach *only* for that purpose. This has been frequently indicated in the course of these experiments, by the effect which it has produced on fatty or oily aliments, when adventitiously mixed with the gastric juice.

Experiment 53

Sept. 29. At 1 o'clock, P. M., I mixed one drachm of sweet *cream,* with three drachms of clear gastric juice, and placed them in the axilla. When first put together, the juice fell to the bottom of the vial, and remained distinctly separate from the cream, till agitated, when they united, but exhibited no other immediate change of appearance. When the temperature was raised to about 80°, the whole gradually formed into very fine, creamy coagulæ. Continued in axilla twelve hours, this coagulated mass was more than half diminished, and rising to the top of an opaque white, whey-coloured liquid. Small globules of oil were now seen on the upper surface of the supernatant coagulæ—No sediment.

Oct. 1. 10 o'clock, A. M., I added one drachm of clear gastric juice, and continued in axilla, ten hours, when the creamy coagulæ were still more diminished; the globules of oil on the surface, increased, and the liquor below, resembled clear, rich gruel, occupying about one sixth of the space of the whole.

Oct. 2. 12 o'clock M., I added another drachm of gastric juice, and continued it in axilla, eight hours. The creamy coagulæ were now reduced to about one fourth, and more loose and white than at first. The globules of oil were now much increased, and formed a complete pellicle over the whole upper surface, nearly resembling soft butter, and emitted a slight rancid flavour. The richness of the chymous liquid below was proportionally increased. No sediment.

Oct. 3. 12 o'clock, M., I divided the contents of the vial into two equal parts, and put them into two separate vials. To No. 1, I added two drachms of pure gastric juice; and to No. 2, two drachms of fresh extracted gastric juice, strongly tinged with yellowish green bile, and kept them in axilla nine hours. The changes effected, after this addition, were strikingly evident, and different in the two parcels. That in No. 2, to which was added the yellowish green juice, exhibited a perfectly homogeneous, rich, gruel-like liquid, slightly tinged with the bile; the creamy coagulæ were all dissolved, and not a globule of the oil to be seen; all appeared chymified—a little dirty white sediment fell to the bottom.

The creamy coagulæ of No. 1 vial, were not completely dissolved, but reduced to a thin, loose layer, and the oily pellicle was scarcely perceptible; the globules extremely minute and whitish, and of a saponaceous consistence. The fluid below, was of a light coloured, rich, gruelly appearance. No sediment deposited. To complete the chymification of the contents of No. 1, I added two drachms more, clear gastric juice, and continued it in axilla, twelve hours longer: at the end of this time, the coagulæ were reduced to a very thin layer; the oily pellicle entirely dissolved, and the liquid of a rich gruelly colour and consistence. No sediment subsided on standing.

Experiment 54

Oct. 1. Mixed four drachms of *sweet, skimmed milk* with four drachms of gastric juice, and kept in axilla. The juice fell to the bottom, when first put together, as with the cream; but when shaken, and raised to 90° or 100° temperature, formed into loose and coarser coagulæ, than the cream, which were diffused and suspended through the milky fluid. Continued in axilla eight hours, the coagulæ were more collected, firmer, and more than half diminished.—The fluid of a light whey, or thin gruel colour and consistence, with a few loose, white flocculi, and a creamy pellicle on the top.

Oct. 2. Continued in axilla eight hours more, the coagulæ were almost completely dissolved; fluid the colour of rich strained gruel; a few light flocculi on the surface, but no creamy pellicle; a little coarse sediment, or loose, white coagulæ at the bottom.

Experiment 55

Oct. 3. Put fifteen drops of gastric juice into three drachms of *sweet milk,* at the temperature of 65°; a slight appearance of very fine coagulæ, was first exhibited, but not so as to become distinctly separated, till after the temperature was raised to 85° or 90°, when the whole mass gradually formed into a tremulous, jelly-like curd, which, after cooling, and standing at rest a few hours, separated into two about equal parts; a soft, caseous substance, and a thin, light coloured whey.

Experiment 56

Oct. 3. Put two drachms of the *soft, caseous substance,* formed in the above experiment, (55) into one ounce of gastric juice, and placed in axilla, six hours; at the end of this time, the curd, or caseous substance, was nearly all dissolved; the menstruum of a white gruel-like appearance, with a thin pellicle of loose, white coagulæ on the surface. In four hours more, it was all dissolved; the fluid richer, and perceptibly acid.

Experiment 57

Oct. 13. 9 o'clock, A. M. Into one ounce of gastric juice, I put one and a half drachms of the *medulla spinalis* of an ox, enveloped in its neurilema, boiled, and placed it on the sand bath, or in axilla, six hours. At 3 o'clock, P. M., examined—the medulla had fallen out of its envelope, and when taken out and separated from the fluid, by the filter, weighed fifteen grains; the neurilema, at the same time, weighed eighteen grains. Put these remaining portions into two drachms fresh gastric juice, and continued in axilla six hours. At 9 o'clock, P. M., the remainder of the medullary portion, weighed eight grains, and the neurilema, nine grains. Continued in axilla, three hours longer, the medullary part weighed three grains, and the neurilema, four grains. The menstruum was now a rich, milk white liquid, of nearly the consistence of cream. A loose, light sediment fell to the bottom, on standing; the fluid retained its rich, milky whiteness and creamy consistence.

Experiment 58

Oct. 14. 9 o'clock, A. M., put half a drachm of *medullary substance,* the brain of an ox, boiled, into four drachms of gastric juice, and kept it on the bath, frequently agitated, six hours, when it was all dissolved; and had produced a rich milky fluid, with a loose, light sediment.

Experiment 59

Oct. 15. Put twelve grains of solid *beef bone,* broken into small pieces, with the periosteum attached to one side, into one ounce of fresh gastric juice, and kept in axilla, twelve hours. At this time the periosteum was nearly dissolved; weight of the bone, ten grains. Added six drachms of gastric juice, considerably vitiated, and continued in axilla nine hours, and the bone weighed nine grains. The menstruum was now a whitish opaque fluid, about the consistence of clear, thin gruel, with a little light brown sediment, settling to the bottom, on standing. Added one ounce more gastric juice, and continued it in axilla, twelve hours. The weight of the bone, at the end of this time, was six grains. The opacity and richness of the fluid increased; smell, slightly fœtid. Discontinued the experiment.

The result of this, confirms the correctness of some former observations, in similar experiments, and sufficiently demonstrate the solubility of solid bone, in the gastric juice of the human stomach.

Experiment 60

Oct. 17. 1 o'clock, P. M., I put twenty grains of boiled *mutton suet,* cold, and divided into small pieces, into six drachms of gastric juice, tinged with bile, and kept it in axilla, *seven hours.* The undissolved residuum, separated by the filter, now weighed ten grains; and the fluid was as white as milk, and about the consistence of thick gruel; there was no appearance of any oily particles; it seemed to have been coagulated, and converted into chyme, like milk or albumen; the chymous part very much resembled that formed from medullary substance.—Continued in axilla, three hours longer, it was all dissolved, and the richness of the fluid considerably increased.

Experiment 61

Oct. 25. 2 o'clock, P. M., put one drachm *custard,* into one ounce of gastric juice, fresh from the stomach, and placed it in axilla. 8 o'clock, 30 mins., all dissolved and chymified; fluid, as usual, from such aliment, of colour and consistence of rich gruel.

Experiment 62

Nov. 1, 1833. To one ounce of gastric juice, *taken from the stomach in Dec.* 1832, (and which was as pure as when first extracted,) I added thirty grains of *lean, boiled mutton,* masticated. Kept in axilla, six hours, it dissolved sixteen grains. The fluid exhibited the usual appearance of chyme.

TABLE

Showing the mean time of digestion of the different Articles of Diet, naturally, in the Stomach, and artificially, in Vials, on a bath.

The proportion of gastric juice to aliment, in artificial digestion, was *generally* calculated at one ounce of the former to one drachm of the latter, the bath being kept as near as practicable at the natural temperature, 100° Fahrenheit, with frequent agitation.

Articles of Diet	*Mean time of chymification*			
	In Stomach		In Vials	
	prep.	h. m.	prep.	h. m.
Rice,	boiled	1 00		
Sago,	do.	1 45	boiled	3 15
Tapioca,	do.	2 00	do.	3 20
Barley,	do.	2 00		
Milk,	do.	2 00	do.	4 15
Do.	raw	2 15	raw	4 45
Gelatine.	boiled	2 30	boiled	4 45
Pig's feet, soused,	do.	1 00		
Tripe, do.	do.	1 00		

Articles of Diet	*Mean time of chymification*			
	In Stomach		In Vials	
	prep.	h. m.	prep.	h. m.
Brains, animal,	boiled	1 45	boiled	4 30
Venison, steak,	broiled	1 35		
Spinal marrow, animal,	boiled	2 40	do.	5 25
Turkey, domesticated,	roasted	2 30		
Do. do.	boiled	2 25		
Do. wild,	roasted	2 18		
Goose, do.	do.	2 30		
Pig, sucking	do.	2 30		
Liver, beef's, fresh,	broiled	2 00	cut fine	6 30
Lamb, fresh,	do.	2 30		
Chicken, full grown,	fricas'd	2 45		
Eggs, fresh,	h'rd bld	3 30	h'rd bld	8 00
Do. do.	soft bld	3 00	soft bld	6 30
Do. do.	fried	3 30		
Do. do.	roasted	2 15		
Do. do.	raw	2 00	raw	4 15
Do. whipped.	do.	1 30	whipped	4 00
Custard,	baked	2 45	baked	6 30
Codfish, cured dry,	boiled	2 00	boiled	5 00
Trout, salmon, fresh,	boiled	1 30	boiled	3 30
Do. do.	fried	1 30		
Bass, striped, do.	broiled	3 00		
Flounder, do.	fried	3 30		
Catfish, do.	do.	3 30		
Salmon, salted,	boiled	4 00	do.	7 45
Oysters, fresh,	raw	2 55	raw, entir	7 30
Do. do.	roasted	3 15		
Do. do.	stewed	3 30	stewed	8 25
Beef, fresh, lean, rare,	roasted	3 00	roasted	
Do. do. dry	do.	3 30	do.	7 45
Do. steak,	broiled	3 00	mastic'd.	8 15
Do. do.	do.		cut fine	8 00
Do. do.	raw		do.	8 15
Do. with salt only,	boiled	2 45		9 30
Do. with must'd, &c.	do.	3 30		
Do, fresh, lean,	do.		mastic'd.	
Do.	do.		entire p.	9 00
Do.	fried	4 00		12 30
Do. old, hard salted,	boiled	4 15		
Pork, steak,	broiled	3 15		
Pork, fat and lean,	roasted	5 15		

Continued.

Articles of Diet	*Mean time of chymification*			
	In Stomach		In Vials	
	prep.	h. m.	prep.	h. m.
Pork, recently salted,	boiled	4 30	mastic'd.	6 30
Do. do.	fried	4 15		
Do. do.	broiled	3 15		
Do. do.	raw	3 00	raw	8 30
Do. do.	stewed	3 00		
Mutton, fresh,	roasted	3 15		
Do. do.	broiled	3 00	mastic'd.	6 45
Do. do.	do.		unmas'd.	8 30
Do. do.	boiled	3 00		
Veal, fresh,	broiled	4 00		
Do. do.	fried	4 30		
Fowls, domestic,	boiled	4 00	mastic'd.	6 30
Do. do.	roasted	4 00		
Ducks, domesticated,	do.	4 00		
Do. wild.	do.	4 30		
Suet, beef, fresh,	boiled	5 30	entire p.	12 00
Suet, mutton,	boiled	4 30	divided	10 00
Butter,	melted	3 30		
Cream,			raw	25 30
Cheese, old, strong,	raw	3 30	mastic'd.	7 15
Do. do.			entire p.	18 00
Do. new, mild,			divided	8 30
Oil, Olive,			raw	60 00
Soup, beef, veg. & br'd	boiled	4 00		
Do. marrow bones,	do.	4 15		
Do. bean,	do.	3 00		
Do. barley,	do.	1 30		
Do. mutton,	do.	3 30		
Green corn & beans,	do.	3 45		
Chicken soup,	do.	3 00		
Oyster soup,	do.	3 30		
Hash, meat & veg.	warmed	2 30		
Sausage, fresh,	broiled	3 20		
Heart, animal,	fried	4 00	entire p.	13 30
Tendon,	boiled	5 30	mastic'd.	12 45
Do.			entire p.	24 00
Cartilage,	do.	4 15	mastic'd.	10 00
Do.			divided	12 00
Aponeurosis,	do.	3 00	boiled	6 30
Bone, beef's, solid,			entire p.	80 00
Do. hog's, do.			do.	80 00

Articles of Diet	*Mean time of chymification*			
	In Stomach		In Vials	
	prep.	h. m.	prep.	h. m.
Beans, pod,	boiled	2 30		
Bread, wheat, fresh,	baked	3 30	mastic'd.	4 30
Do. corn,	do.	3 15		
Cake, do.	do.	3 00		
Do. sponge,	do.	2 30	broken	6 15
Dumpling, apple,	boiled	3 00		
Apples, sour, hard,	raw	2 50	entire ps.	18 00
Do. do. mellow,	do.	2 00	mastic'd.	8 30
Do. sweet, do.	do.	1 30	do.	6 45
Parsnips,	boiled	2 30	mashed	6 45
Do.	do.		entire p.	13 15
Do.	raw		do.	18 00
Carrot, orange,	boiled	3 15	mashed	6 15
Do.			entire p.	12 30
Do.			raw, do.	17 15
Beets,	boiled	3 45		
Turnips, flat,	do.	3 30		
Potatoes, Irish,	do.	3 30	mashed	8 30
Do. do.			entire p.	14 00
Do. do.	roasted	2 30		
Do. do.	baked	2 30		
Cabbage, head,	raw	2 30	mastic'd.	12 30
Do. with vinegar,	do.	2 00	shaved	10 15
Do.	boiled	4 30	boiled	20 00
Peach, mellow,			cut small	10 00
Do. do.			mashed	6 00

The foregoing table is formed from all the experiments made upon St. Martin, since 1825, taking the average from such as were generally performed under the naturally healthy condition of the stomach, and ordinary exercise.

The mean times of artificial chymification, have been taken from such experiments as were generally made with the pure gastric juice, or such as was too slightly vitiated, to impair its solvent effect, in any essential degree.

They exhibit the average, as near as practicable, for the digestion of one drachm of alimentary matter, in one ounce of gastric juice, or in about that proportion, counting the time actually kept on the bath, or in the axilla.

Exceptions, however, must be made for the bone, oil, cream, and one or two other articles, which chymify much slower and more difficulty, than the less concentrated aliments.

Several experiments have been omitted, especially when they were of the same kinds, and produced similar results.

TABLE

Showing the temperature of the interior of the Stomach, in different conditions, taken in different seasons of the year, and at various times of the day, from 5 o'clock in the morning, till 12 o'clock at night.

	Wind and Weather		Th.	*Temperature and condition of stomach**				
				Empty		During chymin		
				Repose	Exercise	Repose	Exercise	
1829								
Dec 6	S.	Cl'dy and damp	63	98°				
7	N.W.	do do	27	98				
8	S.W.	Clear and dry	13	99				
9	W.	Clear	10	99				
1830								
Jan 21	N.W.	do and cold	0.8	100				
25	S.W.	do	2	100		100		
Mar 17	S.W.	Rainy	38	99				
18	N.W.	Clear	6	100			102	
9				98				
1832								
Dec 4	N.W.	Snowing	35		101			
5			30	100			101½	
6			38	100				
7			28	99		100		Stomach morbid.
8	-	Cl'dy and damp	46	99		99		do do
13				100				
14				100				Stomach morbid.
15				100				do do
22				100		100		
23				100	101			Stomach morbid.
25	E.	Variable	31	100		100	101	do do
26	N.E.	Cl'dy and damp	38	99½	101	99½	101	
27	E.	Foul and damp	38	99½		100		
"	S.	Clear	62	100		100		
28	N.	do	34	100				
29	N.W.	do	34	100		100		
30	"	do	26	100				
31	S.	Cl'dy and damp	30	100½				Stomach morbid.

1833 Jan 1	S.	Rainy	50	100				
3		Clear	38		101½			
7	N.E.	Cl'dy and damp	48	100				
11	S.W.	Clear	15	100				
13	C'lm	Cloudy and dry	12	100	101	100	100½	Stomach morbid.
14	N.W.	Clear	28	100			101½	
15	N.E.	Cloudy and dry	35	100	101			
17	N.W.	Clear and dry	19	100		100	102	Stomach morbid.
23	N.E.	Rainy	39	100½			101¾	
24	N.	Cl'dy and damp	39	100½	100¼			
"	N.E.	Rainy		99½				after sleeping
25	S.		36	99				before rising.
"			38	100½			102	
26	N.W.	Clear	36	100½		100¾	101	89½ aft. sl'p'g.
27	C'lm	Cloudy	32	99½			101¼	99½ bef. rising.
28	S.W.	Clear	35	101*			101½	
"	S.W.	do	46	101½		101½		
"				101½		101½		
29	N.E.	Clear	28	100¾	101½		102	100 before rising.
30	N.E.	Cl'dy and damp	39	99½	101½	101¼	102	99½ bef. rising.
31	N.E.	Rainy	45	101¼	101½	101¼		100 do do
Feb 1	N.W.	Clear	28	101			102	100 do do
Mar 26		do		100½			101	
July 9	W.	Cl'dy and damp		100				before rising.
10	W.	Clear	63	100	101			
11	N.E.	Cloudy	65	100	101			
12	W.	Clear	70	100½	101¾			
13	C'lm	Clear	69	100¾	101¾			
14	S.	Variable	75	100	102			
15	W.	Clear	74	100	102			
	W.	Clear	74	100½			101¾	
16	W.	Cloudy	73	101	101½			
28	N.W.	Clear	66	101				
Oct. 10	W.	Fair	61	100	101½		101½	
	S.	Rainy	61		101¾	102	103	
11	N.W.	Fair	32	100	102		102	
11				101¾		101¾		
12	S.	Cloudy	36	101		101		
13	N.E.	Rainy		101		101	102	

*At this, and the subsequent examinations, the bulb of the thermometer was placed three or four inches nearer the pylorus than before, and exhibited an increase of temperature, indicating a difference of three fourths of a degree, between the splenic and pyloric extremities.

Abstract of Temperature of the Stomach.

When empty, and in repose,	highest,	100¾,	Mean, 100⅓°.
do. do. do.	lowest,	98,	
do. do. exercising,	highest,	102,	Mean, 101½°.
do. do. do.	lowest,	100,	
Full, or during chym'n, in rep.	highest,	102,	Mean, 100⅕°.
do. do. do.	lowest,	90,	
do. during chym'n, in exerc.,	highest,	103,	Mean, 101½°.
do. do. do.	lowest,	100½,	

In all the observations previously to the 4th of December, 1832, the examinations were made with a Mercurial Thermometer, (Fahrenheit's) and north of latitude 43°. Subsequently, and to March, 1833, the examinations were made at Washington, D. C., in Lat. 38° 53′, with the Spirit Thermometer, from Pool's Barometer, which varied half a degree from the mercurial one. From July 9, to November, 1833, I used Pool's Glass Chemical Spirit Thermometer, (Fahrenheit's scale) at Plattsburgh, N. Y., in latitude about 44° 40′, N.

INFERENCES,

From the foregoing Experiments and Observations.

1. That *animal* and *farinaceous* aliments are more easy of digestion than *vegetable.*

2. That the susceptibility of digestion does not, however, depend altogether upon *natural* or *chemical* distinctions.

3. That digestion is facilitated by *minuteness* of *division* and *tenderness* of *fibre,* and retarded by opposite qualities.

4. That the *ultimate principles* of aliment are always the same, from whatever food they may be obtained.

5. That the action of the stomach, and its fluids are the same on *all kinds* of diet.

6. That the *digestibility* of aliment does not depend upon the *quantity* of nutrient principles that it contains.

7. That the *quantity* of food generally taken, is more than the wants of the system require; and that such excess, if persevered in, generally produces, not only functional aberration, but disease of the coats of the stomach.

8. That *bulk,* as well as *nutriment,* is necessary to the articles of diet.

9. That *oily* food is difficult of digestion, though it contains a large proportion of the nutrient principles.

10. That the *time* required for the digestion of food, is various, depending upon the quantity and quality of the food, state of the stomach, &c.; but that the time ordinarily required for the disposal of a moderate meal of the fibrous parts of meat, with bread, &c., is from three to three and a half hours.

11. That *solid* food, of a certain texture, is easier of digestion, than *fluid.*

12. That stimulating *condiments* are injurious to the healthy stomach.

13. That the use of *ardent spirits always* produces disease of the stomach, if persevered in.

14. That *hunger* is the effect of *distention* of the vessels that secrete the gastric juice.

15. That the processes of *mastication, insalivation* and *deglutition,* in an abstract point of view, do not, in any way, affect the digestion of food; or, in other words, when food is introduced directly into the stomach, in a finely divided state, without these previous steps, it is as readily and as perfectly digested as when they have been taken.

16. That *saliva* does not possess the properties of an alimentary solvent.

17. That the *first* stage of digestion is effected in the stomach.

18. That the natural *temperature* of the stomach is 100° Fahrenheit.

19. That the temperature is *not elevated* by the ingestion of food.

20. That *exercise elevates* the temperature; and that *sleep* or *rest,* in a recumbent position, *depresses* it.

21. That the *agent* of chymification is the *Gastric Juice.*

22. That it acts as a *solvent* of food, and alters its properties.

23. That its action is facilitated by the *warmth* and *motions* of the stomach.

24. That it contains free *Muriatic Acid* and some other active *chemical* principles.

25. That it is never found *free* in the gastric cavity; but is always excited to discharge itself by the introduction of *food,* or other irritants.

26. That it is secreted from vessels distinct from the mucous follicles.

27. That it is seldom obtained pure, but is generally mixed with mucus, and sometimes with saliva. When pure, it is capable of being kept for months, and perhaps for years.*

28. That it *coagulates* albumen, and afterwards *dissolves* the *coagulæ.*

29. That it *checks* the progress of putrefaction.

30. That the pure gastric juice is fluid, *clear* and *transparent;* without *odour;* a little *salt,* and perceptibly *acid.*

31. That like other chemical agents, it *commences* its action on food, as soon as it comes in *contact* with it.

32. That it is capable of *combining* with a certain and fixed *quantity*

---

*I have now (Nov. 1, 1833) in my possession, some clear gastric juice, possessing all its original properties, unchanged and undiminished, which was taken from the stomach in Dec. 1832, about eleven months ago, and has been kept tightly corked in vials.

of food, and when more aliment is presented for its action than it will dissolve, disturbance of the stomach, or "indigestion," will ensue.

33. That it becomes intimately *mixed* and *blended* with the ingestæ in the stomach, by the motions of that organ.

34. That it is *invariably* the *same substance,* modified only by *admixture* with other fluids.

35. That *gentle exercise* facilitates the digestion of food.

36. That *bile* is not ordinarily found *in the stomach,* and is *not* commonly *necessary* for the digestion of food; but

37. That, when *oily* food has been used, it assists its digestion.

38. That *chyme* is *homogeneous,* but variable in its *colour* and *consistence.*

39. That towards the *latter stages* of chymification, it becomes more *acid* and *stimulating,* and passes more rapidly from the stomach.

40. That *water, ardent spirits,* and most other *fluids* are not affected by the gastric juice, but pass from the stomach soon after they have been received.

41. That the *inner coat* of the stomach, is of a pale *pink* colour, varying in its hues, according to its full or empty state.

42. That, in health, it is constantly sheathed with a *mucous* coat.

43. That the gastric juice and mucus are *dissimilar* in their *physical* and *chemical* properties.

44. That the appearance of the interior of the stomach, *disease,* is essentially different from that of its *healthy* state.

45. That the motions of the stomach produce a constant *churning* of its contents, and *admixture* of food and gastric juice.

46. That these motions are in two directions; *transversely* and *longitudinally.*

47. That the *expulsion* of the chyme is assisted by a *transverse band,* &c.

48. That *chyle* is formed in the duodenum and small intestines, by the action of *bile* and *pancreatic juice,* on the chyme.

49. That crude *chyle* is a *semi-transparent, whey coloured* fluid.

50. That it is further changed by the action of the *lacteals* and *mesenteric glands.* This is only an *inference* from the other facts. It has not been the subject of experiment.

51. That *no other* fluid produces the same effect on food that gastric juice does; and that it is the *only solvent* of *aliment.*

I regret, exceedingly, that I have not been able to obtain returns from Professor BERZELIUS, to whom I transmitted, about seven months ago, a bottle of gastric juice for chemical examination. I could not, however, consistently with the expectations and wishes of my friends, further delay the publication of these experiments.

INDEX

C

William Beaumont: A Pioneer American Physiologist

has been typeset by Academia Typographers, Ltd. in Baskerville,
printed by Universal Printing Company on Warren's Old Style Wove,
and bound by Becktold Company in
Whitman's Skivertex-Sanigal with
Elephant Hide Paper sides.

Solar Energy Heat Pump Systems for Heating and Cooling Buildings

Coordinated and Edited by Stanley F. Gilman

The Pennsylvania State University

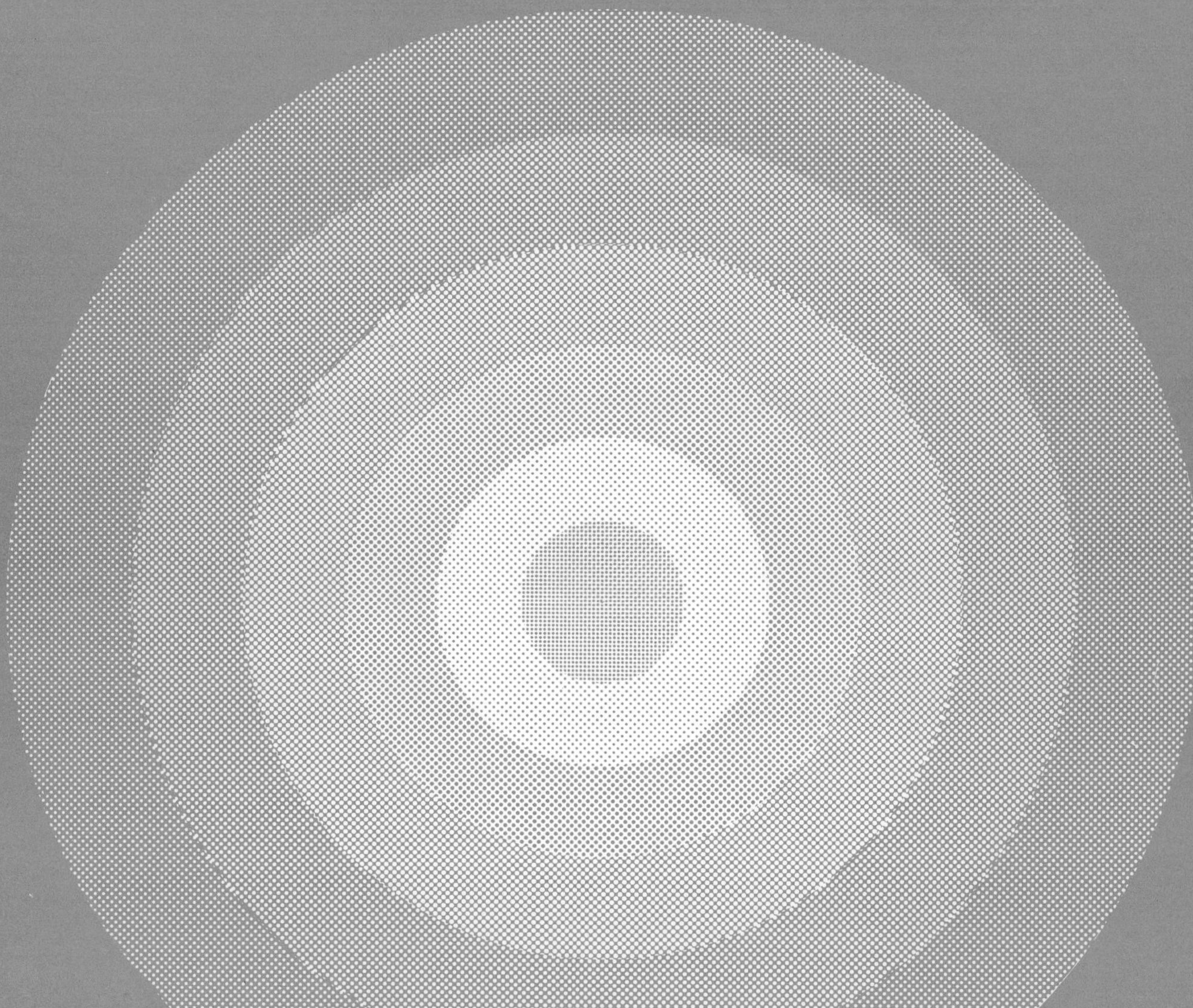

ERDA Doc. C00-2560-1
Con. 7506130

Solar Energy Heat Pump Systems for Heating and Cooling Buildings

Coordinated and Edited by
Stanley F. Gilman

Proceedings of a Workshop conducted by The Pennsylvania State University College of Engineering, June 12-14, 1975

Cosponsored by

The U.S. Energy Research and Development Administration, Division of Solar Energy

and

The Pennsylvania State University

in cooperation with

The American Society of Heating, Refrigerating and Air-Conditioning Engineers, Inc.

The Pennsylvania State University
University Park, Pennsylvania

Contents

Preface

The Workshop on Solar Energy Heat Pump Systems for Heating and Cooling Buildings was held at The Pennsylvania State University, University Park, Pennsylvania on June 12-14, 1975.

The purposes and objectives were:

To bring together researchers, consulting engineers, product development engineers, manufacturers, public utility representatives, and the interested public to exchange information and ideas on solar energy heat pump systems.

To present status reports on NSF and ERDA funded projects.

To describe and discuss recent designs and applications of solar heat pump systems.

To present equipment and systems performance characteristics, including present and future potential.

To present public utility viewpoints on the potential of heat pump systems.

To explain the present federal programs and identify future research and development needs.

A Workshop planning meeting was held in February 1975 under the direction of Mr. Harold Horowitz, then Program Manager for the National Science Foundation. Planning meeting attendees were:

H. Horowitz, National Science Foundation (Chairman)
S. Gilman, Penn State (Vice-Chairman)
R. Allen, Allied Chemical
J. Cherne, TRW
J. Comly, General Electric (Schenectady)
J. Cuba, ASHRAE
E. Drucker, Syracuse University
F. Faust, ASHRAE
R. Fischer, Battelle Memorial Institute
W. Hurlebaus, General Electric (Valley Forge)
D. Jardine, Phoenix Corporation
H. Kirschbaum, Westinghouse (Pittsburgh)
L. Loeb, General Electric (Louisville)
J. Mitchell, University of Wisconsin
A. Weinstein, Westinghouse (Baltimore)
M. Wildin, University of New Mexico

Thirty-two papers were presented to over three hundred attendees during the two and one-half day Workshop. In addition, four Working Group sessions were held on Thursday evening, June 12.

During the time the National Science Foundation was processing the Workshop Proposal, the U.S. Energy Research and Development Administration was formed; the Proposal was transferred to ERDA and subsequently approved, and financial support provided under ERDA contract No. E (11-1)-2560. The actual Workshop came under the direction of Dr. John Leech, Acting Chief, Heating and Cooling Branch, Division of Solar Energy, ERDA, Washington, D.C.

The Academic Coordinator for the University and Workshop Director was Dr. Stanley F. Gilman, Professor of Architectural Engineering; the Administrative Coordinator was F. Wally Lester of the University's Continuing Education Department.

A subcontract was let to The American Society of Heating, Refrigerating and Air-Conditioning Engineers, Inc., to provide a Proceedings Review Committee to review and edit, on a technical basis, the manuscripts submitted by the authors. Members of the Committee appointed by ASHRAE were Mr. Guy Cheek, Professional Engineer, Air-Conditioning Corporation, Greensboro, N.C. (Chairman); Mr. Harry Hale, President, Carlyle Compressor Division, Carrier Corporation, Syracuse, N.Y.; and Mr. Joseph Pietsch, Manager-Engineering, Central Air Conditioning Department, General Electric Company, Tyler, Tx.

Final review and editing was done by the Workshop Director.

The contributions of these individuals and organizations to a successful Workshop are gratefully acknowledged.

Stanley F. Gilman
Workshop Director

Introduction

Dr. N. J. Palladino

Dean, College of Engineering
Pennsylvania State University
University Park, Pa.

It is my privilege to welcome you to the Penn State Campus and to the Workshop on Solar Energy Heat Pump Systems for the Heating and Cooling of Buildings.

This conference is indeed a timely one, for energy resources and the wise use of such resources are becoming increasingly important issues.

It was my privilege not too long ago to serve as chairman of an energy subcommittee of the Governor's Science Advisory Committee here in Pennsylvania. It was our job to prepare a report for the Governor suggesting steps which the Commonwealth of Pennsylvania might take to try to help itself in providing sufficient energy for its citizens. In the process of preparing this report several general observations were made and I would like to share them with you.

It is interesting, for example, to note how various individuals, in all walks of life, perceive the energy picture. Many individuals, some of them in important positions, are saying that the energy crisis is contrived. Others blame it on the Middle East OPEC Nations. Some politicians feel that the situation is temporary and will be solved within months, or a few years at most, if particular actions are taken.

But despite such talk, it becomes quite clear from a study of the situation that the energy problems we are beginning to face are real and will reach proportions far greater than those we have experienced to date. Yet, although these problems will get worse, they can get better within the next 10 to 15 years if appropriate steps are taken now.

To improve our future energy situations we must begin now to take two important steps:

1. Modify our demands for energy.
2. Develop alternative sources of energy.

While we can take immediate steps to modify our demands by conserving energy, the development of alternative sources of energy takes time. If we are to forestall major future energy crises, we must start such work now. This is why workshops such as this are so important. Shortages will affect more than our energy needs. Oil and gas are important sources of chemicals and chemical products and should not be used much longer for heating purposes or generating electricity. Rather, they should be used chiefly for supplying materials which are not readily available from other sources, and for meeting transportation energy needs which cannot be fulfilled by other fuels.

I am concerned, however, that because recent oil shortages have been alleviated temporarily, there is a tendency to relax essential development work. Even if we proceed with such work at full speed, we will be lucky to have significant alternatives by 1985. So we must not only start now, but we must persevere.

We must take advantage of a number of expedients open to us during the next decade, including using energy more efficiently and more wisely. These expedients will bring about many readjustments in our living habits, for which we as a people must become prepared. Greater public use of rail and bus transportation, for example, must be encouraged, by improving the quality and extent of service. By such means more of our scarce fuels can be made available for transporting much needed goods and supplies.

Few people realize—or want to believe—that for our most widely used fuels—oil and gas—depletion is only several decades away. And this disbelief may be one of the big stumbling blocks in achieving long term solutions.

One reason, of course, is that energy has been so cheap that almost anyone could afford to use it. Hence it is not surprising that we rebel at paying more for energy—but pay more we will.

How did we get into our present energy predicament? As I see it, three important factors have led to, and still contribute to, our energy problems:

1. Growth in energy demands brought about by increasing per capita use of energy and by the increasing size of the population.

2. Rapid establishment of severe environmental controls on discharge of pollutants.

3. Inadequacy of facilities to produce fuels of the quality necessary to meet the new environmental standards.

The long range energy problem does not have to do with adequacy of total energy resources, for adequate resources exist for many centuries ahead. (This does not mean that some will not face eventual depletion.) But other sources exist if we learn to use them. Some of these resources, however, are in forms which must be processed to remove pollutants if they are to meet the new environmental requirements; others are in forms more costly to extract from the ground than those we are using currently.

The marketplace should be quite capable of providing incentives to convert these energy resources into available supplies to consumers. Yet the growing multiplicity of often uncoordinated government regulations has notably influenced the effectiveness of the marketplace to meet our energy needs.

The two prime candidates for increasing internal energy supplies in the United States are coal and nuclear power. But projections show that nuclear power is already growing at as good a rate as we can hope for at present, and almost solely for the generation of electricity.

Dr. Dixy Lee Ray, while Chairman of the now defunct Atomic Energy Commission, reported that as of January 1974, nuclear power was generating 5½% of the nation's electrical output. Today there are:

50 nuclear power reactors licensed to operate
58 under construction
11 on limited work authorization
100 on order
18 planned

Looking ahead, the forecast of nuclear electric generating capacity, made by Dr. Ray, is staggering:

1974—5½%	25,000,000 kw installed capacity	(42 plants)
1980—21%	132,000,000 kw installed capacity	(140 plants)
1990—44%	508,000,000 kw installed capacity	(455 plants)
2000—60%	1,200,000,000 kw installed capacity	(1000 plants)

By 1976, when nuclear energy will still be producing less than 10% of the nation's electric power, the installed capacity of our nuclear power plants will be equal in size to the entire network of electrical generating plants which existed in 1940. This is a staggering indication of the rate at which we have accelerated our demands for electrical power.

But even if nuclear power grows as much as Dr. Ray predicts, we cannot close the gap between energy demands and energy supply without increasing the energy obtained from coal and as many other sources as possible.

In any projections there are certain implicit assumptions. Perhaps the most difficult parameter to estimate is the percentage of our total energy needs that will be satisfied by electric power. Today it is about 25% in the United States, and only slightly higher in Pennsylvania. But as the pressures for a clean environment grow, this number can rise quite significantly. If such events take place, the demands for electricity—and in turn for nuclear and coal-fired power plants—may exceed the figures projected by Dr. Ray and others, if industry can keep up with the demands.

In Pennsylvania, for example, the Office of State Planning predicts that by 1985 more than 42% of the energy used in the state will be in the form of electricity; and although Pennsylvania is a leader in the application of nuclear power to the generation of electricity, because of the rapid growth in the use of electricity, the fraction generated from nuclear fuels will be only 5 to 10% of the total.

Where do solar energy systems fit into the picture? Certainly such systems are among our viable alternatives for heating and cooling of buildings. The purpose of this workshop is to help determine where such systems fit into the picture.

Your program brings together researchers, public utility representatives, and interested public representatives, to exchange information and ideas on solar energy heat pump systems.

The program will include reports on NSF and ERDA funded projects, descriptions and discussion of recent design and application of solar heat pump systems, as well as projections of future potential for such systems.

We wish to thank all participants for coming to this workshop. We wish to give particular thanks to the Energy Research and Development Administration (Division of Solar Energy) and to the American Society of Heating, Refrigerating, and Air-Conditioning Engineers for co-sponsoring this program with Penn State.

Let me again repeat my welcome and express my best wishes and those of my colleagues here at Penn State for a successful and effective program.

Part I

Chairman:
Frank H. Bridgers

President, Bridgers and Paxton Consulting Engineers, Inc.
Albuquerque, N.M.

Heat Pumps — What They Are and What They Do

Dr. Stanley F. Gilman

Professor of Architectural Engineering
Pennsylvania State University
University Park, Pa.

I will limit the scope of this discussion to conventional refrigeration compressor systems with constant speed electric motor drives. Emphasis will be on reciprocating compressor systems as they are, by far, the most widely used, but some mention will be made of centrifugal compressor systems. The other basic types—rotary and screw compressors—are rarely used in heat pump applications and will not be covered in this presentation.

Just what is a "heat pump"? All refrigeration systems are heat pumps as they pump heat from a place where it is not wanted to some other place where it should not be objectionable. Consider a household refrigerator. Heat is removed from inside the cabinet to maintain the appropriate temperatures in the food and freezer compartments. This heat is discharged into the living space outside the refrigerator and is a "heat pump," although it is commonly called a "refrigerator."

An air conditioning system for a home operates in the same way. Heat is removed from the living space and discharged to the outdoors. Conventionally, this is termed an "air conditioning system."

Now if the basic operation were reversed, i.e., if heat were removed from the outdoors and discharged to the living space, instead of the space being cooled it would be heated. This is what is commonly termed a "heat pump." It relates to the *desired effect*. When the desired effect is cooling, the system is called "air conditioning." When the desired effect is heating, the system is called "a heat pump." It is often termed a "reversed-cycle refrigeration unit," because the basic cycle is reversed to produce heating instead of its conventional function of cooling.

Because the basic components are the same, heat pumps are associated with heating and air conditioning systems; in other words, you would not install a heat pump for heating only, because all of the major components for an air conditioning system are included in the heat pump.

There are several different types of heat pumps and they are identified by the heat source and heat sink, respectively. The most common types are air-to-air, water-to-air and water-to-water.

The most widely sold is the air-to-air heat pump and its operation is illustrated in a simplified form in Fig. 1. The major components of a refrigeration system are a refrigerant compressor/drive, condenser, evaporator and throttling device (often a thermal expansion valve, TXV). For the type of heat pump being described, an additional major component is the reversing valve. Its function is to reverse the flow of refrigerant in the system so that the indoor coil acts as an evaporator on the cooling cycle (Fig. 1*a*) and as a condenser on the heating cycle (Fig. 1*b*).

Tracing the cooling and defrost cycle, refrigerant is pumped to a high pressure and temperature by the compressor and diverted by the reversing valve to the outdoor coil. A fan forces outdoor air over the coil, removing heat and condensing the refrigerant, which then flows through the TXV and into the low pressure indoor coil (evaporator). Room air forced over the indoor coil is chilled as heat is absorbed by the refrigerant in the evaporator. In absorbing heat, the liquid refrigerant boils, becomes a vapor, and flows to the compressor suction, when it is recompressed and the cycle repeated.

The indicated defrost cycle relates to the heat pump function. On the heating cycle, the functions of the indoor and outdoor coils are reversed. Since the outdoor coil acts as an evaporator the leaving air is colder than the entering air. At certain outdoor temperature conditions, frost will form on the outdoor coil, impairing the efficiency and the operation. To remove the frost, the heating cycle is reversed back to the cooling cycle for a few minutes until the hot gas has melted the frost.

There are two other common types; they use water as a heat source. One is a water-to-air heat pump. The indoor coil is a refrigerant condenser during the heating cycle and a refrigerant evaporator during the cooling cycle, similar to the Fig. 1*b* indoor coil. A water-to-refrigerant heat exchanger replaces the Fig. 1 outdoor coil. Like the air-to-air heat pump, the refrigerant flow is reversed by a four-way valve to obtain the selected heating or cooling function.

The third type is the water-to-water heat pump. Both the condenser and evaporator are water-to-refrigerant heat exchangers. Whereas the other types described above switched the refrigerant flows, this type switches the water flows. Hence, when cooling is required in a space, chilled water leaving the evaporator is directed to the space conditioning equipment. When heating is required, warm water leaving the condenser is directed to the space conditioning equipment. The refrigeration circuit is always the same. This has advantages as the condenser and evaporator designs can be optimized for peak efficiency as they serve only one function, whereas the heat exchangers of the air-to-air and water-to-air heat pumps must serve both as evaporators and condensers.

The "heart" of the system is the refrigeration compressor. A cutaway photograph is presented in Fig. 2. It is quite similar to an automobile engine: it has a crankshaft, connecting rods, pistons, valves, etc. The basic difference is that the engine has work output, whereas the compressor has work input to the refrigerant. The drive is an electric motor, shown in the upper right. The majority of compressors is "hermetic," as this one is. This means that the motor/compressor combination is in one shell. Since there are bearings and other moving parts requiring lubrication, there is oil in the system as well as refrigerant. The combination of oil, refrigerant, and the motor/compressor has been called the "uneasy trio."

The basic refrigeration/heat pump cycle can best be displayed on a P-H diagram, which is shown in a very simplified manner in Fig. 3. The vertical scale is absolute pressure and the horizontal scale is enthalpy, "total heat content" of the refrigerant. Tables and charts for many different refrigerants are available from several sources, e.g., the *ASHRAE Handbook of Fundamentals*. To the left of the saturated liquid line, the refrigerant is completely liquid. To the right of the saturated vapor line, the refrigerant is completely vapor (a gas). Between the two, the refrigerant is partially liquid and partially vapor, being mostly liquid to the left and mostly vapor to the right.

The cycle shown in Fig. 3 is a "simple" cycle for illustrative purposes only. In actuality, point A is somewhat to the right and there is some pressure drop in the refrigerant lines and heat exchangers. Nevertheless, the cycle does illustrate the basic principle.

Beginning at A, the compressor takes in the gas, which is all vapor, and compresses it to B. In the process, it puts work, W_c, into the gas and increases its enthalpy. From B to C heat is removed; i.e., rejected to the sink by the condenser. This is the heat rejection, HR, process. Since heat is removed, the vapor is condensed and is all liquid at C. From C to D it is "throttled" at constant enthalpy to the lower pressure in the evaporator. It enters the evaporator as a mixture of about 75% liquid and 25% vapor. From D to A is the evaporation process during which heat is removed (absorbed) from the "source," e.g., outdoor air. The change in enthalpy is the refrigeration effect, RE. As the refrigerant boils and absorbs heat, it vaporizes and enters the compressor as all vapor, point A, and the cycle is repeated.

The coefficient of performance (COP) is an index of the efficiency of the process. The higher the COP, the greater the efficiency. The COP is "what you get out for what you put in." As shown in Fig. 3, it is the desired effect divided by the work input. On the cooling cycle, the desired effect is the refrigeration effect, RE. On heating, the desired effect is the total heat rejection, HR. Since the compressor work input, W_c, shows up as heat (enthalpy) in the condenser, the COP for heating is always one greater than the COP for cooling. Therefore, heat pump heating has the inherent advantage of being more efficient at the same evaporator/condenser conditions.

A typical heating COP is 4.0. This means that 1 Btu of work, W_c, will provide an additional 3 Btu of heat to the space. Thus the electric power consumption, and hence the operating cost, will be 1/4.0 or 25% of that required by a straight electric resistance heating system.

Figure 4 is an example using actual values for Refrigerant 22 and a nominal value of compression efficiency. This refrigerant is used in the vast majority of reciprocating compressor applications. In the liquid-vapor mixture region, the pressure and the saturation temperature are uniquely related. As shown, condensing occurs at 110F, so the pressure in the condenser is 241 psia. The refrigerant entering the compressor at A has an enthalpy of 107 Btu/lb and a specific volume of 0.78 cu ft/lb. The work of compression, W_c, is 126 − 107 = 19 Btu/lb. The refrigeration effect, RE, is 107 − 42 = 65 Btu/lb. Hence, from Fig. 3, the cooling COP = 3.4 and the heating COP = 4.4.

If the source temperature were lower, the evaporat-

ɔuld be lower also. At an evapora- ɔf −10F, for example, the cooling ınd the heating COP to 3.1, which is n COP. Hence the greater the spread aporating and condensing tempera- ıre lift), the poorer the performance.

.or that has considerable influence on is the inherent performance charac- .e compressor. Figure 5 is a schematic of ɩı a compressor. On the downstroke of the ıı, the pressure in the cylinder must be lowered ɩ. ɘss than the valve spring and 70 psia suction gas pressures. When that pressure is reached the suction valve will open and suction gas will flow in and fill the cylinder.

On the upstroke, the gas will be compressed to a high pressure. When this pressure exceeds the total of the 241 psia discharge pressure and the valve spring pressure, the discharge valve will open and gas will flow out of the cylinder. Not all of the gas leaves the cylinder. There has to be some clearance between the cylinder at top dead center and the valve structure. Hence on the downstroke this trapped high pressure gas expands to the lowered pressure, occupies some of the cylinder volume, and limits the amount of suction gas that can be taken in. This effect is the volumetric efficiency, VE, and is primarily a function of the compression ratio, CR, the ratio of discharge to suction gas pressures. The higher the CR, the lower the VE, which results in less gas being pumped.

In addition to this effect, as the evaporating temperature is lowered, the specific volume of the suction gas increases; i.e., the density decreases. Consequently, the *pounds* of refrigerant pumped per unit of time, which directly determines the cooling or heating capacity of the system, is lowered. In Fig. 4, at 30F evaporating temperature, the specific volume entering the compressor is shown to be 0.78. For the −10F example cited, the corresponding value is 1.68, or only 46% of the density at 30F.

The principal end results of lowering the evaporating temperature, at constant condensing temperature, are:

1. Lower refrigeration effect per pound of refrigerant circulated.
2. Higher work of compression per pound of refrigerant circulated.
3. Lower COP.
4. Higher compression ratio and hence lower volumetric efficiency.
5. Lower pounds of refrigerant pumped per unit of time.
6. Lower heating and cooling capacity.
7. Lower power input.

It can be shown that if the evaporating temperature is held constant and the condensing temperature varied in the direction of increasing condensing temperature, the effects described are the same. Thus high efficiency is related to low temperature lifts, and as the lift is increased the efficiency (COP) becomes poorer.

Do we have any control over the temperature lift? Yes, but it is quite limited. The source and sink temperatures are fixed by the application. For example, for an air-source heat pump application, if it is 15F outdoors (the source) and the occupied space is to be at 70F (the sink), there is nothing that can be done about it.

However, the elevation of the condensing and evaporating temperatures of the heat pump above the sink and source temperature, respectively, is determined by the performance of the heat exchangers. The heat flow relationships are depicted in Fig. 6. The top portion relates to the condenser which is heating the room air. The 100F condensing temperature is determined by the energy balance between it and the room air being heated. This is a matter of design and cost. If considerably more heat exchange surface were added, the same amount of heat could be transferred at a lower temperature difference. If this were done, since the sink is fixed at 70F, the condenser would balance out at a lower temperature, say 90F, thus decreasing the lift. Conversely, if the efficiency of the condenser is low, e.g., it has too little heat transfer surface, then it would balance out at a higher condensing temperature, say 110F, thus increasing the lift.

The same analysis applies to the evaporator performance shown in the lower portion of Fig. 6. Increasing the amount of heat transfer surface will raise the evaporator temperature and decrease the lift. It should be noted that this will also increase the system cost.

From the above, it is clear that many factors affect the performance of a heat pump system. For high efficiency, i.e., high COP and low operating cost, the temperature lift should be a minimum. However, the designer has limited control of this. The properties of the refrigerant play a large part. A dense suction vapor is desirable for reciprocating compressors to minimize the physical size and cost. The critical temperature of Refrigerant 22 is 205F, so no condensation of vapor is possible above that temperature. Common refrigerant oils break down at 500F or so. Refrigerant stability at high temperatures is a problem; consequently, it is usual practice to design systems for maximum compressor discharge temperatures of 225F.

A heat pump system must be designed within the constraints described above. A Refrigerant 22 reciprocating compressor heat pump system simply could not operate between, e.g., a 300F source and a 500F sink.

If oil breakdown is a consideration, this may be eliminated by use of a centrifugal heat pump if a large heating capacity is required. The centrifugal compressor is like a fan, i.e., it does not have pistons, crankshaft, valves, etc., as a reciprocating compressor does. Thus no oil is necessary in the refrigeration circuit. The temperature lift (head) developed is a function of the tip speed, which is determined by the wheel diameter and the rpm (max. 3500 for two-pole electric motors). The centrifugal is a "volume gulper" and ordinarily is built for refrigeration tonnages of 200 or more. Because it is a volume gulper, a refrigerant with a high specific volume is desirable for low tonnage systems. This led to the development of Refrigerant 113, which has a specific volume at 40F saturated evaporating temperature 15 times that of Refrigerant 22.

Since the centrifugal is similar in performance characteristics to a fan, it has a similar head-capacity curve, as illustrated in Fig. 7. Since the compressor blades are, in effect, air foils, if the angle of attack is too great, they can stall just like an airplane wing. This condition is indicated by the surge line. At reduced cooling capacity (mass flow rate), if the temperature lift is above the surge line, the compressor will surge (the blades will alternately load and unload). This can be very detrimental to the machine.

The centrifugal is not a high lift machine, typical single stage duty being between 105F condensing and 40F evaporating. If high lifts (high water temperatures for heating) are required, two or more stages will be needed, with resultant increased initial cost.

In conclusion, the design and application of heat pumps must include careful consideration of many factors, such as oil and refrigerant properties, source and sink temperatures (lift), and heat exchanger performance. If designed and applied within the limits of current technology, the heat pump can be both a reliable and an energy-conserving device.

DISCUSSION

Q Why cannot the variable volume to the compressor be corrected by holding the source heat constant such as by a heat of fusion thermal reservoir at a selected temperature of, say, 70 or 80F?

A That would be an ideal situation. Unfortunately, various heat of fusion salts experimented with so far have not proved stable over prolonged cycles. As a consequence, performance degrades markedly with time. In general, you are at the mercy of the source temperature, e.g., the variation in outdoor dry-bulb temperature.

Q Do you think there will be any significant advances in compressor efficiencies or construction costs in the next five years?

A Over the years, compressor manufacturers have been forced by competitive pressures to higher speeds of reciprocating compressors, so many are now at the upper limit, i.e., two-pole speed (3500 rpm). These same pressures have reduced the amount of copper in the motor with resultant lower cost, but lower efficiency. The current emphasis on energy conservation has reversed this trend somewhat and compressor manufacturers are introducing "high efficiency" compressors, perhaps 20% more efficient than previous models. But the huge capital investment in compressor manufacturing facilities (an additional \$10–20 million) precludes any radical change. In addition, the engineering technology is such that no additional substantial increase in efficiency is plausible.

Q The COP values in the heating mode you quoted are ideal cycle values. How much are these values degraded in available equipment? What minimum degradation do you feel is possible with optimum designs?

A I apologize for not making that clear in my presentation. I used an adiabatic efficiency of 75% in my analysis, so they are realistic and not ideal cycle values. Since there is inherently a clearance volume and losses through valves, an increase to 85% adiabatic efficiency (13%) might be attainable but probably would not be economically viable.

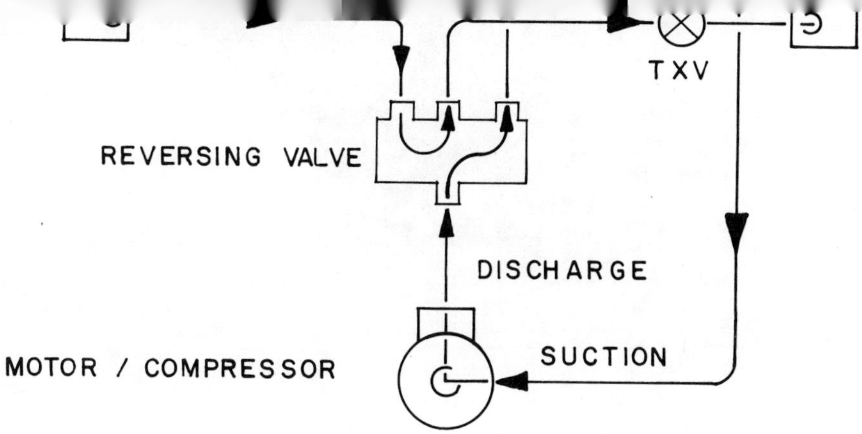

Fig. 1*a*. Basic air-to-air heat pump system, cooling and defrost cycle.

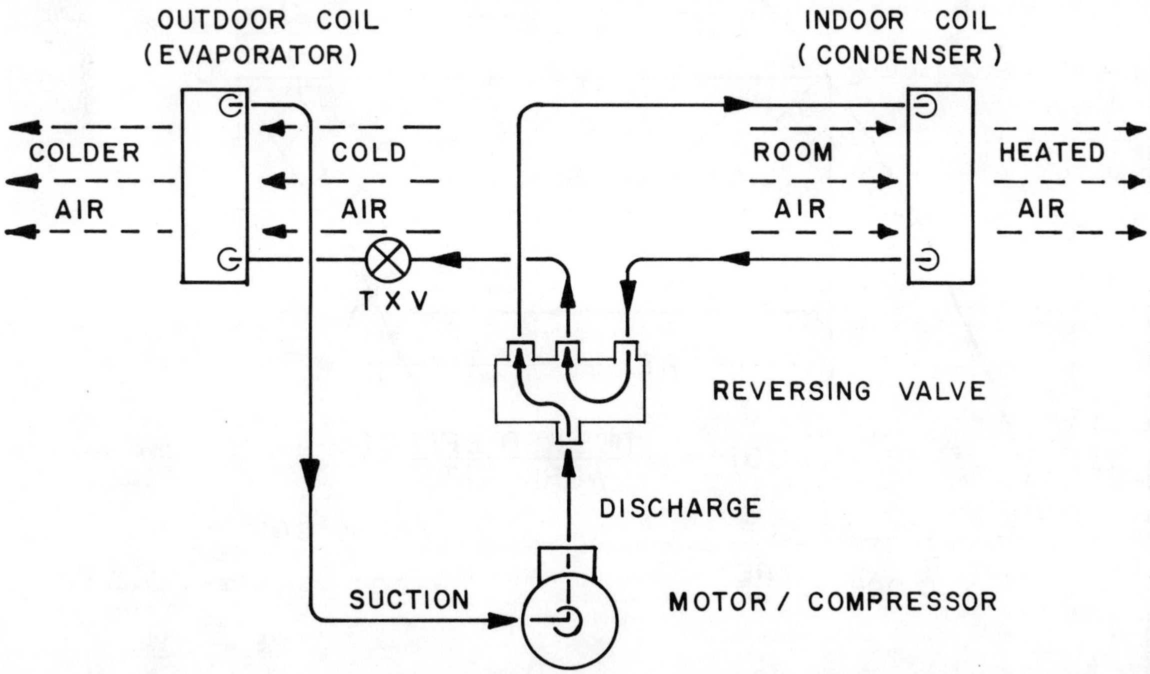

Fig. 1*b*. Basic air-to-air heat pump system, heating cycle.

Fig. 2. Hermetic heat pump motor/compressor.

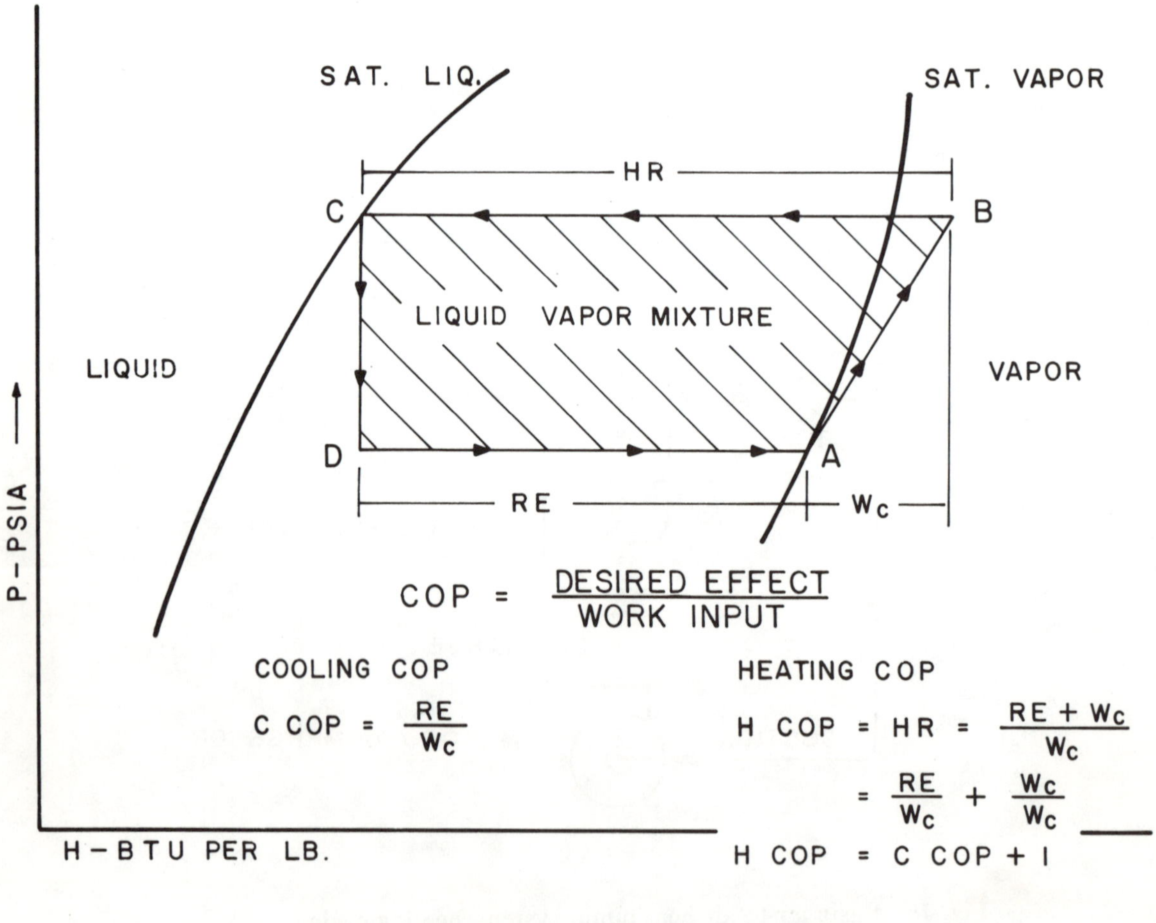

Fig. 3. Simple refrigeration/heat pump cycle.

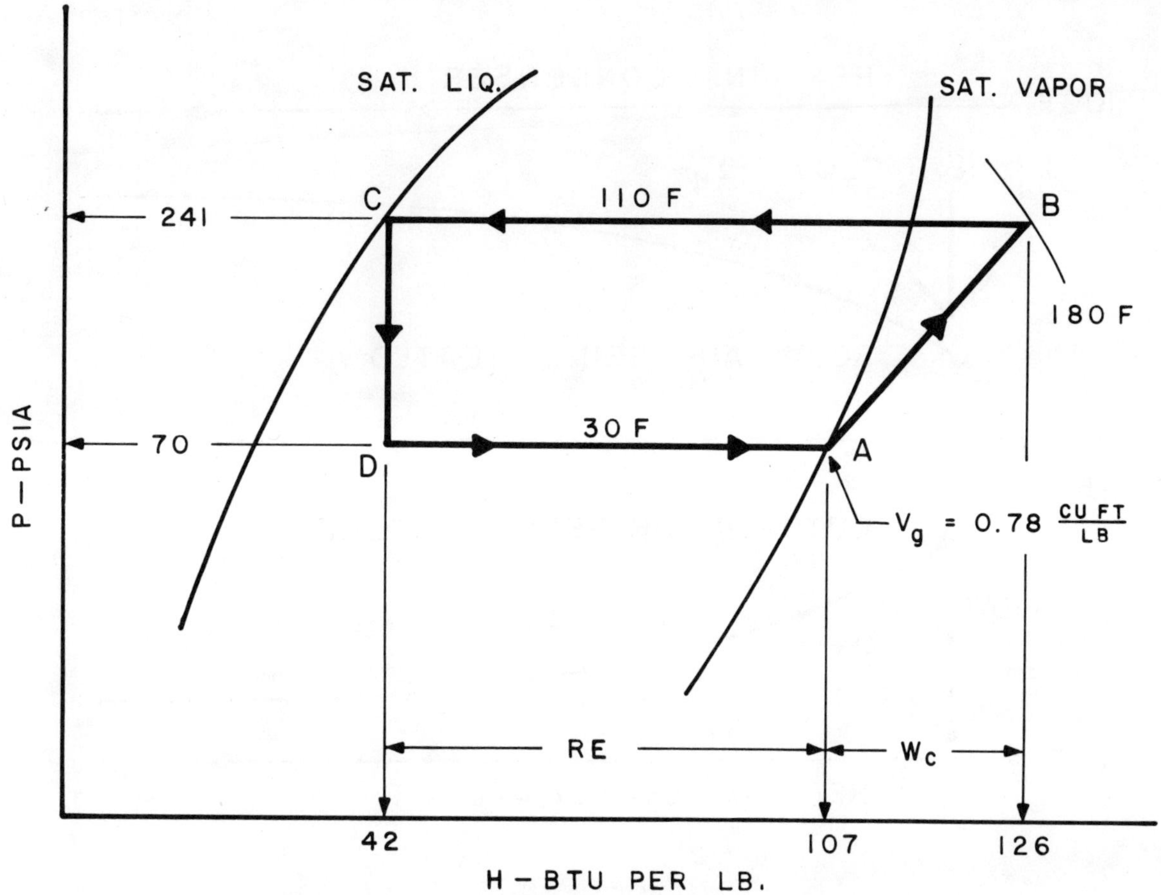

Fig. 4. Heat pump cycle.

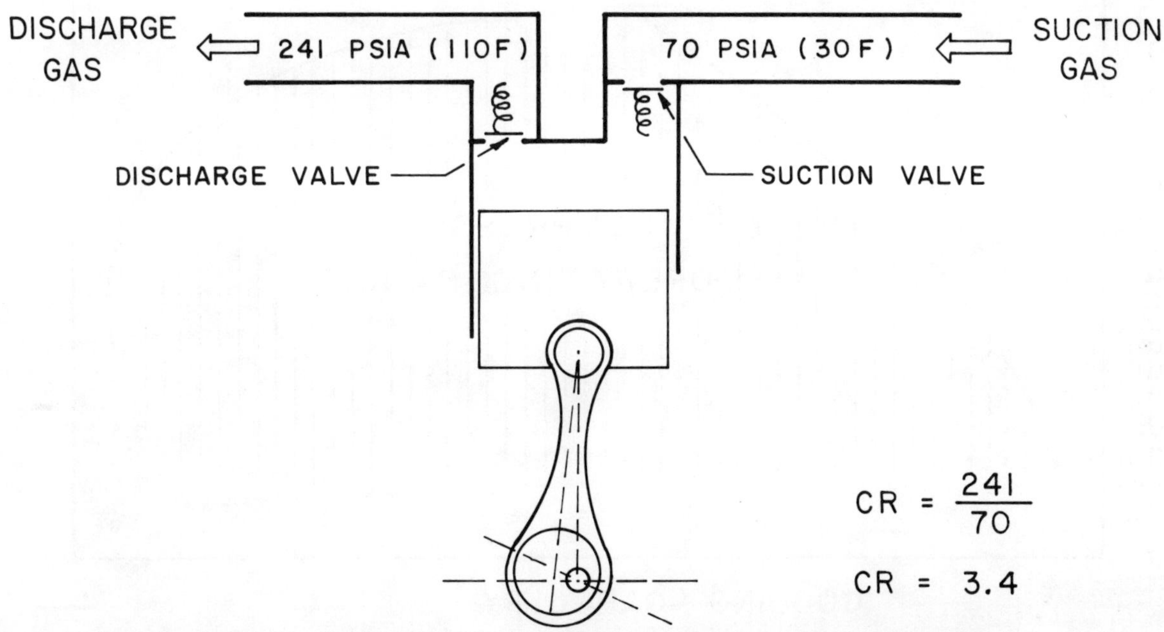

Fig. 5. Heat pump compressor schematic.

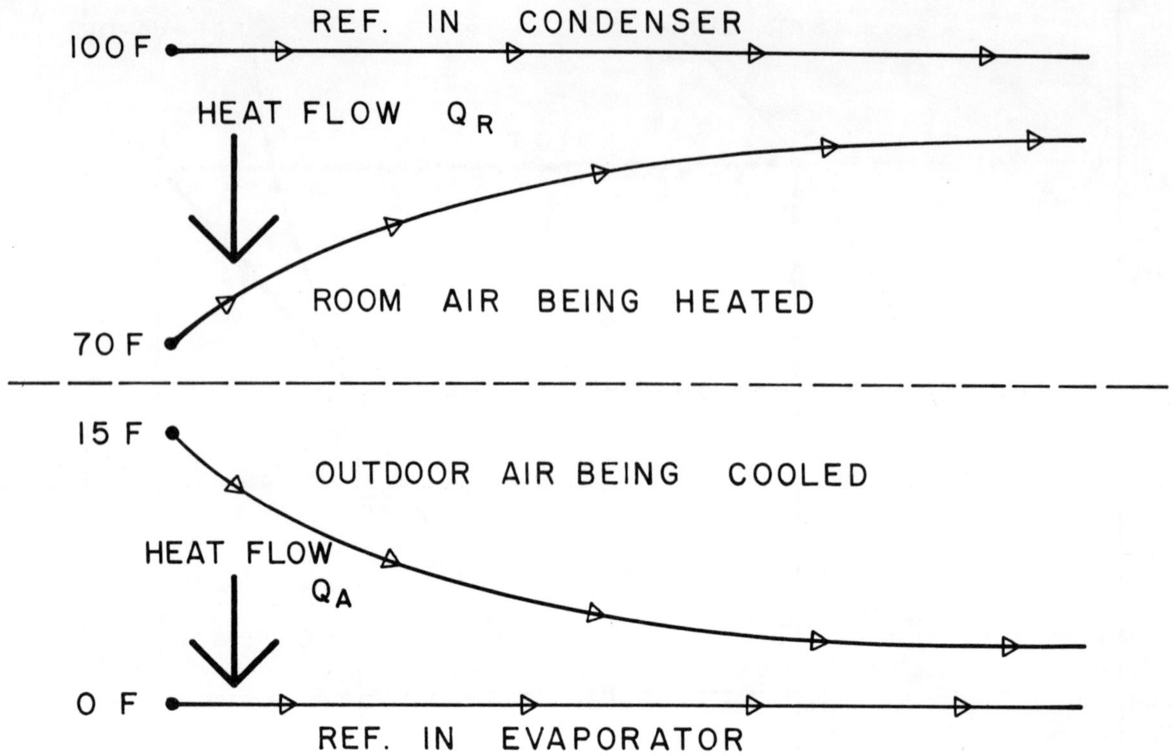

Fig. 6. Heat pump condenser and evaporator performance.

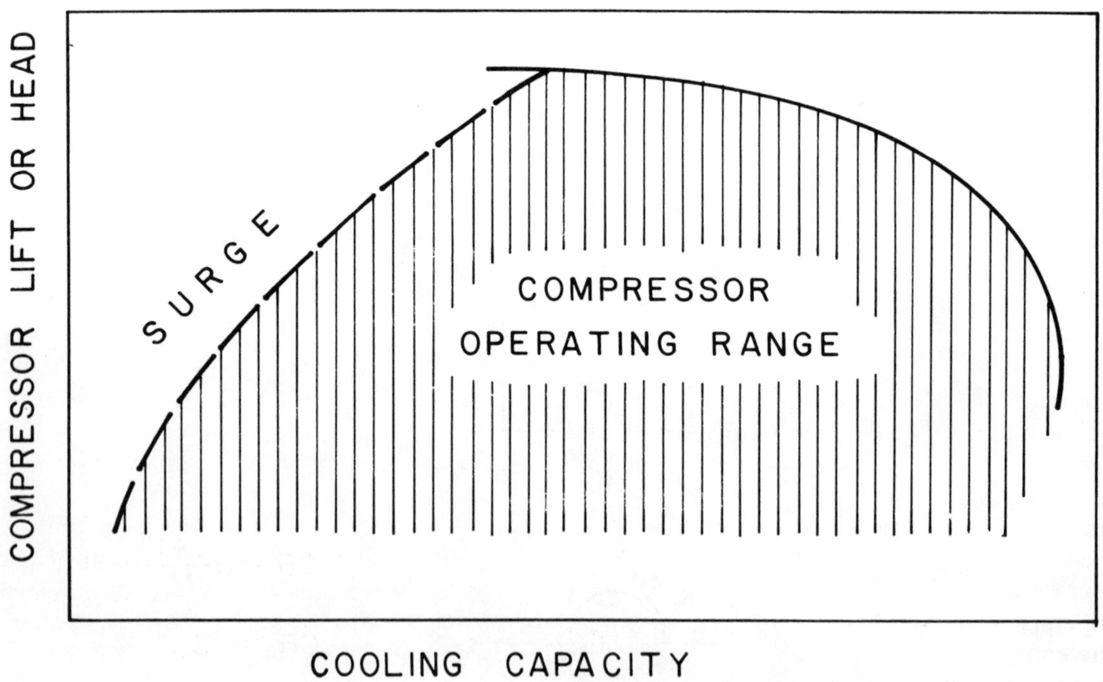

Fig. 7. Centrifugal compressor performance map.

Solar Assisted Heat Pump Evaluation

Arnold D. Cohen

Program Manager, Advanced Energy Programs
General Electric Company
King of Prussia, Pa.

This paper is based on the results of a study performed for the National Science Foundation and completed in May 1974 (Solar Heating and Cooling of Buildings—Phase 0 Feasibility and Planning Study. Final Report, NSF-RA-N74-021B. Contract NSF C-855). An analysis was performed to evaluate the technical and economic feasibility of using solar energy for heating and cooling of buildings within a time frame of 25 years—from now until AD 2000. Performance estimates were generated for nine climatic regions within the United States on a matrix of six building types. This paper addresses the solar assisted heat pump application.

By hourly computer simulation, information was derived to determine the amount of electric energy required to supplement solar energy over an entire heating and/or cooling season. By coupling this information with projected electricity costs through the year 2000 and the capital and operating expenses of the solar equipment, a life cycle economic assessment was made for selected locations and building types.

Solar Assisted Heat Pump Description

Solar energy can assist the heat pump in its heating mode by providing higher grade energy than the cold environment, which enhances the performance by lowering the electrical energy required by the heat pump.

Figure 1 is a schematic diagram of the components and circuit configuration required to utilize a heat pump for solar energy assist. This figure diagrams the heat pump operation in the heating mode, but it designates all the equipment for both heating and cooling and thus can be used for explanatory purposes.

The solar assisted heat pump system can be divided into the following subsystems:

1. A heat collection subsystem (HCS) containing a solar collector and an energy collection loop.
2. A thermal energy storage subsystem (TES) containing a thermal storage unit, an energy transfer loop, and a domestic hot water heating loop.
3. The heat pump and circuitry.
4. The air handling and distribution equipment.

Other standard components include pumps, three way valves, check valves, and fan coil units.

Heat Pump Performance Model

HEATING MODE. The performance characteristics for the heat pump model were developed from empirical correlations of existing packaged units. Performance equations are developed for the output capacity and required source supply temperature to meet the demand and the required electric power to run the heat pump at the operating conditions. The operation temperature range in the heating mode is 4.4–40.6C (40–105F). If the storage temperature rises above 40.6C, the model switches to the more efficient conventional heat exchanger operation. The heat pump model operates at the highest COP mode each hour by selecting the highest source heat between the ambient air and thermal storage supply.

Auxiliary heat, when required, is supplied by electric resistance heating.

COOLING MODE. A heat pump cooling model was derived to assess the effectiveness of a heat pump in the cooling mode. Similar performance equations for the output capacity and required power to operate the heat pump were derived. The unit is operated in an off-peak mode. The heat pump is operated continually during the evening hours to meet the cooling demands and to chill the storage tank. During the daytime peak electric load periods, the chilled water is used to provide cooling, with the heat pump operating only to meet the cooling demands and to chill the storage tank. The chilled storage operation temperature range is 4–18C (40–65F).

System Performance Analysis

SYSTEM SIMULATION. In the computation model hourly response of the system was determined from transient building load calculations derived from hourly solar radiation and weather parameters. The particular data obtained from the model are energy savings attributable to solar energy, electric energy consumption, total radiation, useful and wasted solar energy collected, and collector efficiencies.

HEAT PUMP HEATING. The overall solar system gain factor μ can also be obtained. This parameter is expressed as the fraction of the total building load D supplied by solar energy over and above that energy that would have been provided by a heat pump operated in its conventional mode (i.e., without solar assist). The value of μ may be expressed as

$$\mu = 1 - \frac{AUX}{D} (COP)_E$$

where AUX represents the energy consumption used by the heat pump system and $(COP)_E$ is the effective coefficient of performance or seasonal performance factor of the heat pump when operated in the conventional mode as an air source system.

A second significant parameter is the seasonal performance factor, SPF, which is the ratio of the total demand to the electrical energy required. This parameter may be expressed as

$$SPF = \frac{(COP)_E}{1 - \mu}$$

Some specific selection and performance comparison studies were made on heat pump system operations for evaluating cost effectiveness. Data for the two diverse climate areas of Boston and Phoenix are shown here. A plot of μ versus the ratio of collector area to roof area (A_C/A_R) is shown in Fig. 2. Similar data are presented in Fig. 3 for Phoenix.

It may be noted that significantly higher values of the solar gain factor result for Phoenix with its lower load and better insolation. Further, the curves for Phoenix tend to peak, indicating no payoff in increasing the collector area beyond a certain point.

HEAT PUMP COOLING. The initial evaluations of heat pump cooling operating in an off-peak mode were made with TES storage capacities of 50 kg/sq m, the capacity used in heating mode analyses. The off-peak operating mode assumed operation of the heat pump between the hours of 10 PM and 10 AM to satisfy the cooling loads and store excess cooling in a chilled TES. At 10 AM the heat pump was turned off and cooling loads were supplied from storage until 10 PM when the cycle was repeated. If the TES was depleted before 10 PM, the heat pump was turned on only long enough to satisfy the immediate load.

It was immediately evident that this type of off-peak operation requires larger chilled TES storage capacities if a major portion of the auxiliary energy is to be consumed during the off-peak period. This is due to the fact that the thermal capacity of a given volume TES is less for the cooling mode than it is for the heating mode. The useful TES temperature range for chilled storage is 7–21C (44.6–70F).

Figure 4 shows the influence of chilled TES weight capacity on the auxiliary energy usage for off-peak cooling of a retail store in the Charleston climate zone. Two conclusions are evident:

1. The effect on total auxiliary energy requirements is small, a decrease of only 5% for a sixfold increase in TES capacity.
2. The percentage of off-peak energy usage increases form 43.5 to 83.7% for the same sixfold increase in TES capacity.

A preliminary evaluation determined that a chilled TES capacity of 200 kg/sq m represents a good tradeoff between increased capital costs and decreased operating costs.

Life Cycle Cost Analysis

Life cycle HVAC costs were computed for selected building types. The basic elements of these costs are:

1. Capital costs of equipment, including manufacturing and distribution costs and profit. The data are predicated on high levels of production.
2. Operating costs comprised of servicing, maintenance and repair, and electricity costs over the lifetime of the system.

In addition, capitalization periods and interest profiles were established to determine the breakeven time to recoup the capital investment. Table 1 represents a typical summary of the data generated for each solar system type, building, and climate combination.

It may be noted in Table 1 that the outputs have provisions for comparison of the solar system to the

(1)	(2) INSTAL. YEAR	(3) AC/AR RATIO	(4) COLLECTOR AREA SQ. FT.	(5) LIFE CYCLE SAVINGS -- SHACOB VS. CONVENT.	(6) YEARS TO PAY OFF INITIAL INVEST.	(7) SHACOB VS. ELECTRIC RESISTANCE ENERGY SAVINGS, MBTU ELECTRIC	(8) SHACOB CAPITAL COSTS $/SQ.FT. OF AC	(9) RATIO-SHACOB LIFE CYCLE COSTS ($)/L C ENERGY SVNGS. (MBTU'S)	(10) RATIO -- SHACOB/CONV. LIFE CYCLE COSTS
BOSTON	1980	25	1300.	79857.	4.	552.0	30.03	86.93	0.61
	1990	25	1300.	79125.	4.	552.0	34.01	94.36	0.63
	2000	25	1300.	73538.	5.	552.0	38.51	101.10	0.66
MADISON	1980	50	2600.	76728.	5.	655.0	18.07	34.34	0.66
	1990	50	2600.	74598.	5.	655.0	20.46	37.23	0.69
	2000	50	2600.	67094.	6.	655.0	23.17	39.79	0.72
WASHINGTON	1980	50	2600.	83243.	5.	526.0	18.22	48.63	0.66
	1990	50	2600.	81695.	5.	526.0	20.63	52.59	0.68
	2000	50	2600.	74732.	6.	526.0	23.36	56.02	0.71
PHOENIX	1980	0	0.	9787.	4.	83.0	0.00	0.00	0.89
	1990	0	0.	9586.	4.	83.0	0.00	0.00	0.90
	2000	0	0.	8731.	5.	83.0	0.00	0.00	0.92

Table 1. Economic data for solar assisted heat pump (retail store).

competing conventional system, electrical resistance heated systems. In the study, each system combination was run for cost comparison for installation in the years 1980, 1990, and 2000 (column 2). This provides a means for evaluating the effects of the higher projected future fuel prices. Capital cost of the solar system (column 8) includes all first costs associated with the solar system collectors, storage, and the supplementary energy system equipment (HVAC). This permits a basis for comparison of first cost outlays for the solar system relative to a competing conventional HVAC system. Column 9 expresses the value of the energy saving in terms of capital costs; and, finally column 10 provides a life cycle cost effectiveness measurement parameter on the economic viability of the solar system relative to the competing conventional system. Values below 1.0 indicate the solar system to be more economical than the conventional HVAC system.

In performing the economic studies, the fuel price data were projected as a function of time and adjusted for regional variations. Both baseline and upper limit fuel costs were used. Present electric price structures closely agree with the upper limit prices.

Solar Assisted Heat Pump Results

The main parameters examined in performing the evaluations are the life cycle costs of the SHACOB (Solar Heating and Cooling of Buildings) system relative to comparable costs for a conventional system. The conventional system has been assumed as electric resistance heating. The latter parameter is one consideration in the determination of the first parameter; hence the financial aspects of energy savings are accounted for in life cycle costs. Since financial considerations will be the major factor in attracting users to solar energy systems, the ratio of life cycle costs for the SHACOB system compared to the conventional system costs was used as the prime criterion in evaluating applications for a given SHACOB system. In many instances, increasing the size of the SHACOB system, and its life cycle costs, resulted in a much greater increase in energy savings. This was a secondary criterion in the evaluations; applications were identified in which a 5% or lower increase in life cycle cost ratio produced significantly greater increases in energy savings on either a percentage or an absolute magnitude basis.

Data are presented for the heat pump operating in the heating only mode and in the heating and cooling mode in Tables 2 and 3, respectively. The tables give information for a matrix of locations and building types. For each application the performance and cost studies were performed for a range of solar collector areas. The number listed in the tables corresponds to the size with the minimum value of the ratio of SHACOB system life cycle costs to conventional system life cycle cost. A ratio less than one corresponds to a cost advantage for the SHACOB system, with decreasing values indicating greater cost savings. A ratio greater than one corresponds to a life cycle cost disadvantage for the SHACOB system, with increasing values indicating greater cost disadvantages.

For some applications, two additional numerical values, one in parentheses, are presented to the right of the minimum life cycle cost ratio. These are the life cycle cost ratios for a larger collector area and its percentage increase in annual energy savings. For

CLIMATE AREA	LOW RISE OFFICE BUILDING		EDUCATIONAL BUILDING	
	BASELINE ELEC. PRICE	UPPER LIMIT ELEC. PRICE	BASELINE ELEC. PRICE	UPPER LIMIT ELEC. PRICE
LOS ANGELES	.90	.79	1.06	.94
PHOENIX	1.05/1.10 (13%)*	.92	1.16	1.01
FORT WORTH	.87/.91 (22%)*	.78	.89	.79
CHARLESTON	.98/1.02 (24%)*	.86	.94/.99 (20%)*	.83
BOSTON	.58/.61 (7%)*	.53*	.57/.62 (8%)*	.51
MADISON	.65/.65 + (13%)*	.60*	.69/.70 (12%)*	.61*
WASH., D.C.	.56/.56 + (10%)*	.50*	.55/.59 (8%)*	.50
OMAHA	.64/.67 (7%)*	.58	.65/.69 (12%)*	.60

* - INDICATES USE OF SOLAR COLLECTORS. OTHERS ARE CONVENTIONAL HEAT PUMP OPERATION

Table 2. Life cycle cost ratios for solar assisted heat pump/electrical resistive heating, heating only.

example, in Table 2, the Fort Worth office building application has a minimum life cycle cost ratio of 0.87 when a conventional heat pump is substituted for an electrical resistance heating conventional system. But if the heat pump is changed to a solar assist heat pump configuration, 22% more energy is saved annually, for a small life cycle cost ratio increase of 0.91.

The items with an asterisk (*) in the tables indicate the use of a solar assisted heat pump, while those without the symbol represent a heat pump operated in the conventional mode.

HEATING ONLY OPERATION. The heating only operating function was investigated for the larger building applications, the industrial and education buildings. As the results on Table 2 show, the solar assisted heat pump system is cost effective when replacing electrical conventional systems. In addition, the results show that operation of the heat pump in the conventional manner is more cost effective than adding a solar system to assist the heat pump operation for the baseline electricity prices. In several cases, the solar assist mode does increase energy savings for a slight penalty in cost effectiveness. With incentives such as reduced loan rates, the use of solar energy could be encouraged.

The effect of higher electric prices indicate that cost effective applications will exist in the colder climate areas.

HEAT AND COOLING OPERATION. For this application, the heat pump is operated with solar assist for heating, and in the off-peak operating mode for cooling. Results are shown in Table 3.

When a heat pump, or in some cases a solar assisted heat pump system, is substituted for an electrical resistive system, it is seen to be cost effective for all applications except multiple family dwellings in the Los Angeles, Phoenix, and Charleston climate zones with baseline electric costs.

Operation of the heat pump in its normal mode, i.e., without solar assist for heating and without cooled storage for cooling, is generally the most cost effective application, as shown in Table 3, again for baseline prices. Those applications which show a cost advantage for solar energy use include multiple family residence in Boston and Washington; single family residence, low rise office, and retail store in Madison; multiple family residence and retail store in Washington, D.C.; single family residence and retail store in Omaha.

The use of the SHACOB system in many other applications results in increased energy savings for slight increases in life cycle costs. These applications

CLIMATE AREA	SINGLE FAMILY RESIDENCE		MULTIPLE FAMILY RESIDENCE		LOW RISE OFFICE BUILDING	
	BASELINE ELEC. PRICE	UPPER LIMIT ELEC. PRICE	BASELINE ELEC. PRICE	UPPER LIMIT ELEC. PRICE	BASELINE ELEC. PRICE	UPPER LIMIT ELEC. PRICE
LOS ANGELES	.69	.63	1.08/1.11(4%)*	1.02	.80	.79
PHOENIX	.77	.72	1.06	1.00	.85	.83
FORT WORTH	.80	.76	.93/.96(26%)*	.88	.79	.77
CHARLESTON	.74	.69	1.03	.96	.80	.76
BOSTON	.62/.66(23%)*	.59*	.71*/.76(18%)*	.65*	.61/.66(22%)*	.57
MADISON	.70*/.70+(37%)*	.64*	.76/.79(39%)*	.71*	.70*	.64*
WASH., D.C.	.62/.64(28%)*	.58*	.75*/.77(25%)*	.68*	.70/.72(56%)*	.65*
OMAHA	.75*/.77(16%)*	.71*	.88/.89(74%)*	.81*	.77/.79(68%)*	.72*

CLIMATE AREA	INDUSTRIAL BUILDING		EDUCATIONAL BUILDING		RETAIL STORE	
	BASELINE ELEC. PRICE	UPPER LIMIT ELEC. PRICE	BASELINE ELEC. PRICE	UPPER LIMIT ELEC. PRICE	BASELINE ELEC. PRICE	UPPER LIMIT ELEC. PRICE
LOS ANGELES	.80	.75	.67/.72(18%)*	.66	(1)	(1)
PHOENIX	.87/.92(22%)*	.82	.77	.74	.92	.89
FORT WORTH	.85/.88(34%)*	.80	.25/.79(25%)*	.71	.87/.90(31%)*	.83
CHARLESTON	.88/.92(34%)*	.83	.79/.82(22%)*	.75	.99	.95
BOSTON	.67/.71(14%)*	.61	.57/.61(8%)*	.53	.67/.72(36%)*	.61*
MADISON	.71/.72(18%)*	.66*	.65/.67(49%)*	.60*	.73*	.66*
WASH., D.C.	.65/.66(14%)*	.60*	.60/.64(17%)*	.56	.72*	.66*
OMAHA	.70/.75(12%)*	.65	.62/.65(12%)*	.58	.82*	.74*

Table 3. Life cycle cost ratios for solar assisted heat pump/conventional systems, heating and cooling.

are those with values to the right of the initial entry in Table 3. In general, all northern climate applications and large building applications in the warmer climates are practical and economic when the solar assisted heat pump system replaces electric conventional systems. This is important from a national energy conservation standpoint since these applications have high energy consumption rates.

Similar to the heating only application, the use of upper limit electric prices for heating and cooling indicates a pronounced shift toward economic viability for the solar assisted heat pump.

Conclusions

The results of this evaluation clearly indicate that the heat pump, either in the solar assist mode or operated in a conventional manner, is cost effective when replacing conventional electric resistance HVAC systems in most climate areas and building types. As the price of electricity rises toward the upper limit, values for the northern climate zones show a distinct cost advantage for the solar assisted heat pump.

The heat pump model used in this study was based on present conventional systems. It is expected that units specifically designed for use with solar energy and with optimized control systems will result in further improvements in cost effectiveness.

DISCUSSION

Q When you used collector area to roof area in your charts, was that considering a single or multiple story house? Wouldn't it have been more useful to use collector area to heated area of the building to eliminate this variable?

A The ratio is independent of the number of stories.

Q How sensitive is the analysis to the installed capacity of the heat pump?

A Analysis is relatively insensitive to installed capacity (See fig. 2-65 of General Electric Company Phase 0 Final Report).

Q What is the increase in solar heating fraction due to the addition of the heat pump to the system (1) in Boston? (2) in Phoenix?

A We showed the solar system gain due to *solar energy usage* in Figs. 2 and 3.

Q What kind of heat storage did you use? Can you say something about the size of the storage vessel, if you have to incorporate three days in a row without solar input?

A We used water as the heat storage medium. It is sized at about 10 lb water/sq ft collector area. It should supply two to three days of heat, depending on the location.

Q Did you extend your study of storage capacities to attempt to determine if it might be practical to eliminate the supplementary electric resistance heaters?

A Yes. It is not practical and, in many locations, not possible.

Q In your published paper in the proceedings, please provide full details on the engineering base and economics such as the insulation level expressed in Btu/degree-day/sq ft of floor space, amount of insulation, window/wall space, double glazing, etc.

A Details are published in the Phase 0 Final Report.

Q You talked about the SHACOB versus conventional life cycle costs. Could you give the assumed cost of solar heating system?

A Details are published in the Phase 0 Final Report.

Q Trend of the future is electric heating? No, thank you! It is only acceptable to us as supplemental heat to solar assisted heat pump systems.

A I agree.

Q Did I understand your statement concerning short payback periods to mean that the comparison was of electric resistance versus solar assisted heat pump systems?

A Yes.

Q In your study, what is your assumption regarding the thermal efficiency of the solar system? What is the lifetime for solar collection? What is the cost in dollars/sq ft for solar collector?

A It varies with temperature differences. Twenty years. Approximately $5.00/sq ft for the collector; installation costs are additional.

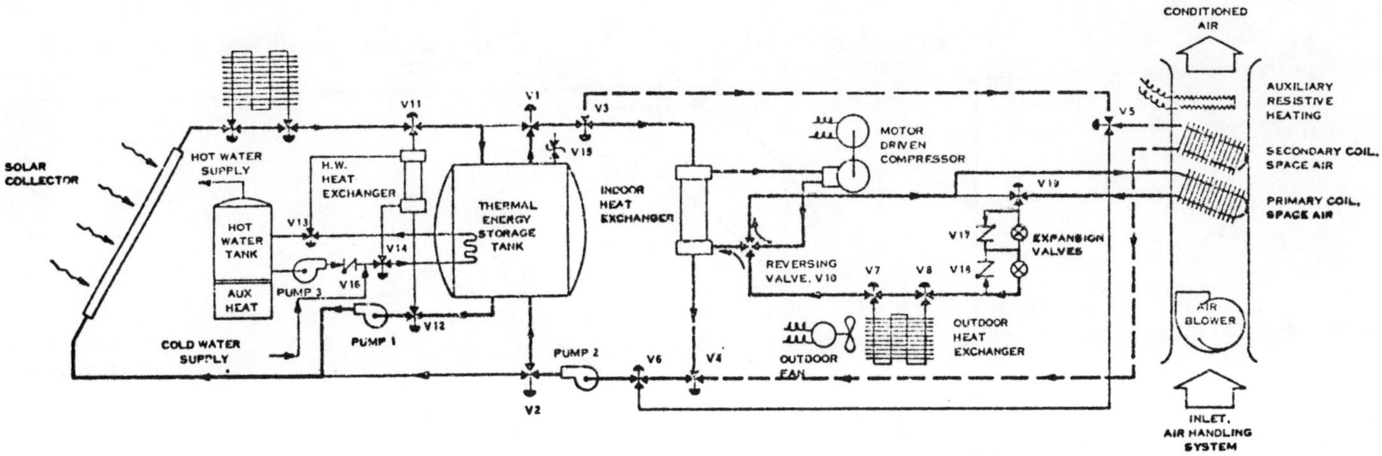

Fig. 1. Solar assisted heat pump systems heating mode.

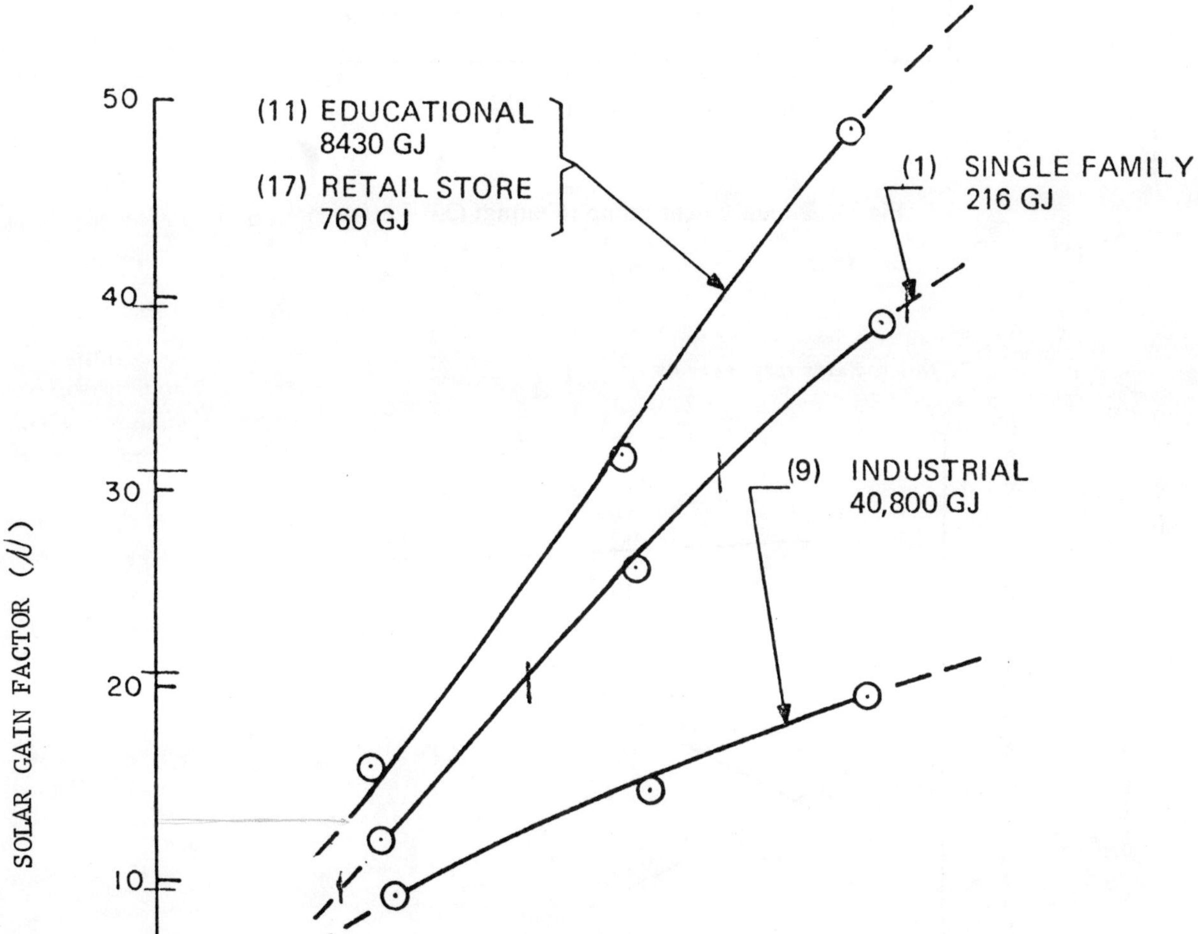

Fig. 2. Boston heat pump (heating) (42° Lat.).

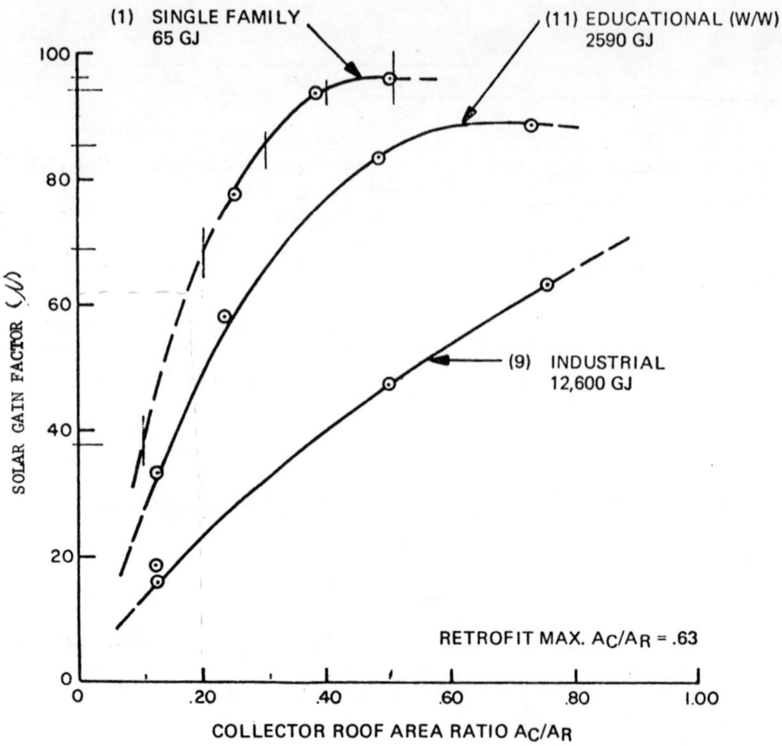

Fig. 3. Phoenix heat pump (heating) (33° Lat.).

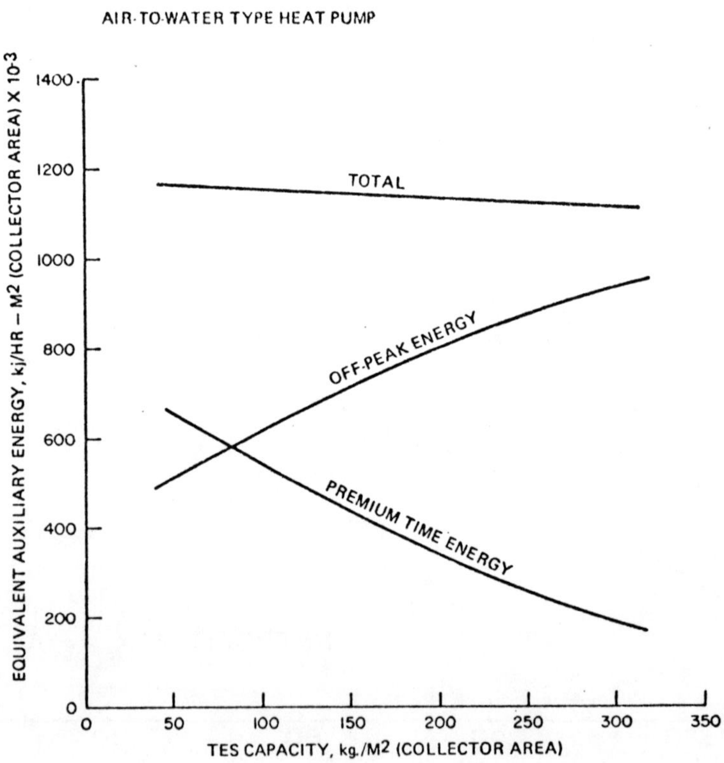

Fig. 4. Influence of TES capacity on off-peak heat pump cooling performance Charleston, Building 17.

Solar Augmented Variable Speed Heat Pump

Jack M. Cherne

Manager, Solar Energy, TRW Systems and Energy, TRW
Redondo Beach, Calif.

James C. Noe

Project Engineer
AiResearch Manufacturing Company
Torrance, Calif.

During the Phase 0 study of the use of Solar Energy for the Heating and Cooling of Buildings, TRW identified the heat pump as a component which could contribute to the utilization of solar energy and aid in the development of a solar energy industry.

The Phase 0 study was performed primarily to assess the market capture potential for solar heating and cooling application to buildings. To accomplish this, we had to examine the function and performance of all the possible systems and study the projected first costs and the operating costs of these systems.

The major conclusions of TRW's study are:

1. Heating and cooling are technically feasible.
2. Not economic today due to cheap fuel and expensive hardware.
3. Will always need a supplemental energy source.
4. Must use life cycle costs to evaluate.
5. Must involve total infrastructure.
6. Commercialization possible in:

Hot water	5 years
Heating	5–10 years
Cooling	15–20 years

These conclusions can be altered by the massive infusion of governmental support to the development of new technology and reduced costs. This effort is being taken now as a result of the three major energy bills: PL 93-409, The Solar Heating and Cooling Demonstration Act of 1974; PL 93-438, The Energy Reorganization Act of 1974; and PL 93-473, The Solar Energy Research, Development and Demonstration Act of 1974.

The National Plan for Solar Heating and Cooling developed by ERDA with the help of HUD and NASA will be the vehicle for the development of this new technology.

TRW, in conjunction with AiResearch, has developed the concept of the solar augmented variable speed heat pump. This system produces a coefficient of performance of 8.0 instead of the conventional heat pump COP of 2.3 at water source temperatures of 100F. Low water temperatures such as this are obtained with low cost solar collectors.

Preliminary data indicate that, in a typical winter season in Chicago, the solar augmented heat pump could conserve ⅔ the energy required by an oil furnace at only ⅓ the operating cost.

Heat Pumps

Heat pump technology has been in existence for many years but has had limited success as a residential heating system due to excessive mechanical problems and the high cost of operation when utilized in low temperature surroundings. The high cost of operation is due primarily to the requirement to add supplemental heat when the ambient air is at too low a temperature to supply the minimum heat source requirements of the machine. This problem is resolved when the heat pump is used in conjunction with solar collectors. Properly designed collectors can absorb solar energy even during low temperature days. Thus a higher source of energy is made available to a heat pump in comparison to conventional air source systems. We believe, therefore, that a heat pump, augmented by solar energy, is a marketable application of the equipment. Further, since the heat pump can use low temperature, high efficiency, unglazed collectors to supply the energy source, a higher return of energy per dollar invested can be obtained and may be optimum in many areas.

Another favorable factor in the use of the heat pump with a solar heated water source is that the energy can be stored during high insolation periods for use at other times.

Phase 0 Study

During the Phase 0 study performed for NSF, we concluded that the high cost of solar hardware was one of the impediments to wide commercialization of solar heating and cooling. In the analysis of the systems, we saw that the collectors were a major part of this high cost. So we looked at collector cost and plotted in Fig. 1 the Btu's collected per day per dollar cost against collector delta temperature. As you can see, it is possible to generate more Btu's per dollar of collector cost by using the least costly collector—the unglazed collector. But how do you use such low temperatures (those below a ΔT of 80F)? This is where the heat pump comes in. Conventional heat pumps are designed to operate over a wide range of source temperatures (40–100F). The design is such that the peak coefficient of performance (COP) occurs at about 70F, as shown in Fig. 2.

Most Btu's are required when the outside air temperature, and thus the source temperatures, are low. The COP's here are low and there is a practical lower limit to the source temperature (in the 40's) when pure electrical resistance heat is used. The need is to raise the source temperature to the 70's. This is a small ΔT for solar heating systems—one that is readily accomplished with low cost, unglazed collectors and small storage systems.

In addition, unglazed collectors may be used to aid the cooling function of the heat pump by night radiation to cool the stored fluid as a sink for heat rejection. Both of these uses take advantage of the ability of heat pumps to produce a high COP when the ΔT between source and sink temperatures are smallest. Today's conventional heat pumps do not do this.

Variable Speed Heat Pump

Many of the mechanical problems of heat pumps result from the use of conventional compressors designed for air conditioning applications. In heat pump applications, the refrigeration compressor is required to operate over a wide range of suction conditions (evaporator) for which it is not designed. Washing out of lubrication oil (oil trapping), operation at pressure differentials beyond the design range, and excessive discharge temperatures contribute to eventual mechanical failures. Other problems associated with the conventional heat pump are freezing of the source coil, loss of capacity at low temperature, and liquid slugging of the compressor during mode changeovers. It is anticipated that new generation heat pumps will, in most instances, be designed with mechanical unloading, compressor staging, and additional control circuits to overcome the difficulties currently experienced.

With the addition of solar energy augmentation, further requirements are imposed on the heat pump. One such requirement is the need to operate with source temperatures often exceeding the condensing temperature of the fan coil unit (condenser). Current design practices are to size the source heat exchanger and the fan coil unit for the cooling mode where suction conditions of the compressor are determined by an evaporating temperature of approximately 40F. The refrigerant mass flow rate, which affects size of the heat transfer element, is based on the suction conditions. In the heating mode, the suction conditions are related to evaporator temperatures which rise to approximately 100F. This can result in mass flow rates that are two to three times the design value. This higher flow rate causes the evaporator heat exchanger approach temperature to increase such that the conventional heat pump cannot obtain the high coefficient of performance available with high source temperatures (see Fig. 2). Also, if the compressor is allowed to operate at these high flow rates, the motor may become overloaded. One method of protecting the compressor is to add a suction pressure regulator to limit the maximum suction pressure. This would, however, affect the coefficient of performance. A higher performance design would be capable of varying the displacement of the refrigerant compressor.

Several types of compressors, including reciprocating piston with cylinder unloaders, screws, and variable speed centrifugal design, have been considered. We have concluded that a variable speed centrifugal compressor provides the highest overall efficiency for the operating range. As the source temperature rises, the speed of the compressor is decreased, maintaining a constant mass flow rate. This results in a lower power requirement, thus a higher COP. Moreover, with sufficient development, it is anticipated that a unique high speed motor and power conditioner can be used to make the compressor cost-competitive with other new generation compressors designed specifically for heat pump applications.

System Description

The solar augmented heat pump system consists of collectors designed to absorb and transfer solar heat energy to water. In the heating mode, a tank is used to store energy not immediately utilized from high insolation days to enable a series of low insolation days to be accommodated. A supplemental heater is also installed in the source fluid line to provide heat for those periods when the storage capacity has been depleted. This heater is typical of current heat pump technology. Since the output energy from the system depends on the input source energy, supplemental heating must be provided when the source energy extraction rate is less than the demand. The remainder of the system is similar to that of a conventional heat pump heating system.

In the cooling mode, the heat pump reverses the cycle and provides for cooling. This is the same as a conventional air source heat pump application. An exception occurs with solar augmentation in that a heat rejection system must be provided utilizing a water coil, whereas a conventional heat pump system rejects heat directly to the air. In the solar augmented system, however, there is a water-to-freon heat exchange. In some applications, the use of unglazed collectors provides additional cooling benefits since the system can be operated during the night to provide nocturnal rejection of heat. In this manner, the tank temperature can be reduced to near night sky conditions, providing a cool sink for operation during the following day. Rejection of heat to the surroundings during the day, however, requires a water-to-air heat exchanger cooling system.

The proposed heat pump assembly is shown schematically in Figs. 3 and 4. It consists of a variable speed centrifugal refrigerant compressor, a fan coil heat exchanger of a standard tube-fin construction, a source/sink heat exchanger constructed from copper tubing, a mode selector valve, two expansion valves, and two check valves.

In the cooling mode, refrigerant gas from the coil heat exchanger (which is acting as an evaporator) is compressed by the centrifugal compressor and delivered through the mode selector valve to the source/sink where the gas is condensed. The high pressure liquid then flows through the check valve at the outlet of the source/sink heat exchanger and expands through the expansion valve at the fan coil heat exchanger. The liquid is then evaporated in the fan coil heat exchanger and sent through the mode selector valve to the refrigerant compressor. In this manner, heat is pumped from the fan coil to the source/sink.

In the heating mode, refrigerant gas from the source/sink heat exchanger (which is acting as an evaporator) is compressed by the centrifugal compressor and delivered through the mode selector valve to the fan coil heat exchanger where the gas is condensed. The high pressure liquid flows through the check valve at the fan coil heat exchanger and expands through the expansion valve at the source/sink. The liquid is then evaporated in the source/sink heat exchanger and sent through the mode selector valve to the refrigerant compressor. In this manner, heat is pumped from the source/sink to the fan coil heat exchanger.

System Performance

The solar augmented variable speed centrifugal compressor heat pump system is expected to perform with coefficients of performance improved over those of a standard operating heat pump assembly. In analytical comparisons to similar sized standard water source heat pump assemblies, only a slight improvement in COP is obtained for the lower end of water source temperatures, but significant improvements occur with high temperature water (see Fig. 2). The true benefit of the system is that a single compressor can be designed for operating in low ambient surroundings and yet not suffer from decreased operating performance when the water source temperature is at a higher value. This is a major problem of operating a standard heat pump assembly with solar augmentation. The standard heat pump assembly will begin to operate at reduced COP when the water source temperature reaches approximately 75F and will drop off to a level of approximately 2.3 with water source temperatures at 100F. The variable speed heat pump will respond to changes in source temperature and reduce the operating speed such that the energy required to provide the heat output is correspondingly reduced. In this manner, at water source temperatures of 100F, the COP is anticipated to increase to a level of approximately 8.0. In this range, then, the variable speed heat pump will significantly outperform a conventional assembly.

A more realistic comparison of variable speed heat pump, however, is to evaluate performance against a conventional heat pump for a heating season where some percentage of supplementary heating must be supplied. Such a comparison has been made in Fig. 5; on the average, the variable speed heat pump assembly will provide approximately 79% improved COP during a period requiring 10% supplemental heating. Again referring to Fig. 2, it is seen that the overall average improved COP without supplemental heat is approximately 2 to 1. Thus for many climates, with sufficient solar insolation, the operating performance of a heat pump system can be significantly improved through use of the solar augmented variable speed heat pump system.

SYSTEM	COST PER 10^6 BTU	TOTAL OPERATING COST *	TOTAL ENERGY CONSUMED
GAS FURNACE	$1.86	$433	332 x 10^6 BTU
SOLAR AUGMENTED CONVENTIONAL HEAT PUMP	$2.50	$583	220 x 10^6 BTU **
SOLAR AUGMENTED VARIABLE SPEED HEAT PUMP	$1.39	$324	123 x 10^6 BTU **
ELECTRIC FURNACE	$8.80	$2050	776 x 10^6 BTU
OIL FURNACE	$4.40	$1024	332 x 10^6 BTU

ASSUMPTIONS:

POWER COSTS = 3¢/KWH
GAS COSTS = $1.30/1000 FT^3
OIL COSTS = $0.40/GAL.

POWER PLANT EFFICIENCY @ 30%

SEASONAL HEATING DEMAND = 233 x 10^6 BTU
GAS FURNACE EFFICIENCY = 70%
OIL " " = 70%

*EQUIPMENT COST NOT INCLUDED

**ASSUMES A 60% EFFICIENCY DOUBLE GLAZED COLLECTOR SYSTEM PROVIDING, NOMINALLY, SOLAR HEATED WATER AT 50°F ABOVE THE AMBIENT. INCLUDES AN ESTIMATED 10% SUPPLEMENTAL ELECTRIC RESISTANCE HEAT REQUIREMENT.

Table 1. Installed yearly operating costs for comparative heating systems (based on a typical winter season in Chicago).

Calculated estimates of comparative heating systems operating costs are presented in Table 1, which compares various heating systems for a typical winter season of operation in Chicago. The solar augmented conventional heat pump operating costs are approximately 28% of an electrical resistance heating system and 57% of an oil furnace. A gas furnace operating cost is approximately 35% less than the heat pump; however, the net fossil fuel energy consumption is approximately 34% more. This difference is due to the homeowner's cost of gas versus electricity. The energy consumed by a solar augmented variable speed heat pump, however, is estimated to be 56% of the energy required for a conventional heat pump and 37% of the energy required for gas and oil systems. Relative operating costs are also significant as the solar augmented variable speed heat pump is expected to operate over a heating season for approximately 75% of the cost of a gas furnace. Coupled with the dual function of providing both heating and cooling for a projected consumer package cost of $400/ton (heat pump unit), it is anticipated that an immediate market for the system is available.

In some applications, where an unglazed collector can be utilized, the heat pump can provide more efficient cooling operations during the refrigeration periods. This is due to use of nocturnal heat rejection at a rate of 10–30 BTU/hr–sq ft to cool the storage tank, resulting in a stored sink for heat rejection. Optimization of the collectors, tanks, controls, and loads will determine the value of the collectors in the cooling mode.

System Design

The system used for these analyses consisted of a sealed compressor unit with a single moving part, the rotor assembly mounted on process fluid foil bearings. The 3-in. diameter rotor turns at speeds up to 70,000 rpm as driven by a 5.3 hp three-phase induction motor. It is supplied with power from a transistorized three-phase static inverter which connects to 230-v residential power. This comparison is an outgrowth of similar centrifugal compressor assemblies AiResearch has produced for other applications. The rpm is not a problem. Typical diesel turbochargers operate at 130,000 rpm. The Refrigerant 114 is used as a lubricant for the bearings in this compressor, eliminating bearing and lubricant problems associated with conventional heat pumps. An electronic approach to the variable speed drive appears to be the most cost-competitive at this time. This approach consists of a frequency converter and high speed motor. In this manner, a two-pole motor with a frequency of approximately 1200 Hz will produce the 70,000 rpm required and, by varying the frequency over the range of 800–1200 Hz, the complete operating range can be obtained. The reason that the frequency converter offers a cost-competitive solution is that the high speed motor can be made extremely small, thereby minimizing its cost. The compressor impeller and housing is also very small and can be manufactured at much lower cost than the typical piston elements. This leaves a reasonable budget for the frequency converter. At this time, it is felt that a reasonable factory cost target of $100.00 can be obtained in large quantity production.

The cost of fully developed variable speed heat pumps will approach $400 per refrigerating ton. This cost is approximately the same for conventional heat pumps. The additional costs of a solar augmented system are the solar collectors, sized to provide the heat energy required at an estimated cost of $5.16–8.73/sq ft (unglazed versus glazed) fully installed, a storage tank sized to store several days' energy at $0.50/gal, and an outdoor air handling unit at $150/ton. The total system cost will depend on the collector/storage/environment considerations.

Summary

The potential savings in electrical energy by use of a

solar augmented heat pump can be of considerable magnitude. By narrowing the overall temperature difference between the sink/source coil and the fan coil of the heat pump, a considerable increase in COP can be achieved. In addition, by operating the solar collector at modest temperature increases above ambient, high solar collector efficiencies can be obtained. This latter operating mode can permit the use of unglazed collectors or smaller area glazed collectors. This will result in reduced system costs and enhance the market potential for solar heating and cooling systems. Finally, the improved performance achievable with a variable speed heat pump can also be used with conventional systems, thus resulting in a large electrical energy saving for the entire country and opening up a broad new market for retrofitting existing buildings presently using conventional, low performance heating and cooling systems.

DISCUSSION

Q Is your speed converter (24 v, 60 cycle, single phase) available today? Can an AiResearch compressor and frequency converter be bought today? Is the variable speed compressor available now in a 5 ton size?

A No, these are not available today.

A What is fan coil outlet temperature when source temperature is higher than fan coil inlet temperature?

A When source temperature is high enough, the heat pump is bypassed.

Q Your presentation sounded as if the function of the variable speed compressor was to compensate only for the variable boiling rate of the refrigerant with the variable temperatures of the heat source. Why not make the heat source constant with a heat of fusion thermal reservoir with a selected phase change temperature of say 85F?

A That would add to the cost of the system. The cost of the reservoir would be greater than the cost of the variable speed components.

Q Why the unequal incremental collector costs versus number of glazing layers?

A The cost of the single glazed assembly includes the delta cost of the structure to hold the glazing. The delta cost of the double glazed unit only includes the glazing.

Q How many solar houses has TRW designed and built?

A TRW has designed and built one solar heated and cooled house and one solar system for food dehydration.

Q Sol-R-Tech is now installing cost effective solar/heat-pump systems using presently available equipment. System cost for an average house is $8000 installed. Would you comment on your judgment that these systems are not yet economically feasible.

A If it takes more than 20 years to repay the difference in cost of the system with the savings in cost of the energy, then it is uneconomical.

Q Can the variable speed principle be used with reciprocating compressors as well as with centrifugal?

A Yes.

Q Your price of $8.73/sq ft for installed collector panels seems to be extremely low. Actual prices are more like $20/sq ft of installed collector; this price includes labor, piping, water storage, etc.

A I agree. However, if I used more realistic numbers, I would be harrassed even more by those who think the systems are economical today.

Q Solar hot water heating is used extensively in many parts of the world. Why won't it be economically feasible here until five years from now? What are the different factors existing in the United States which evidently do not exist in some other countries?

A Solar hot water heating was used in this country (Florida) very extensively from 1945 to 1965 but reached a peak and then declined because of availability of cheap fuel. As long as fossil fuels are available at a cost lower than the amortized cost of solar, then solar heating will have a difficult time competing.

Q Is there a noise problem at speeds of 70,000 rpm?

A No, the compressor is in a sealed can.

Q $400/ton—is this manufacturing or selling price?

A This is selling price.

Q Would you explain negative slope of COP versus source temperature above 75F?

A It is assumed that the system was optimized for a temperature of 75F.

Q How soon do you expect to have a complete variable speed solar assisted heat pump system operating?

A Commercial availability is at least five years away.

Q What must cost of fuel and solar panels be before solar energy is possible?

A Cost of fuel must double and the cost of collectors must be reduced to about one-third before solar systems will be competitive.

Q Did your TRW study consider heat pumps working alone (without solar assist)? If so, what can you say about the attractiveness of your variable speed heat pump? How does this influence your commercialization estimates?

A The variable speed system also helps the simple heat pump system.

A Rather than an unglazed collector at over $5/sq ft to supply relatively low temperature water to the heat pump evaporator, why not use an open-flow trickle-type ("Thomason design") collector? It's cheaper, and the single glazing protects the absorbing surface from weather.

A An unglazed collector is no more expensive than one of the open-flow trickle type.

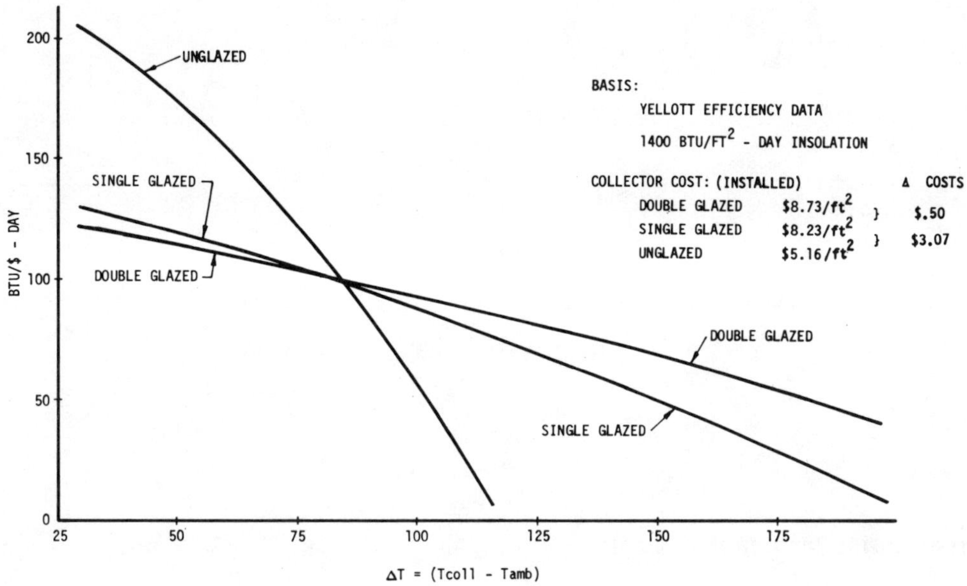

Fig. 1. Installed collector cost versus collector temperature differential.

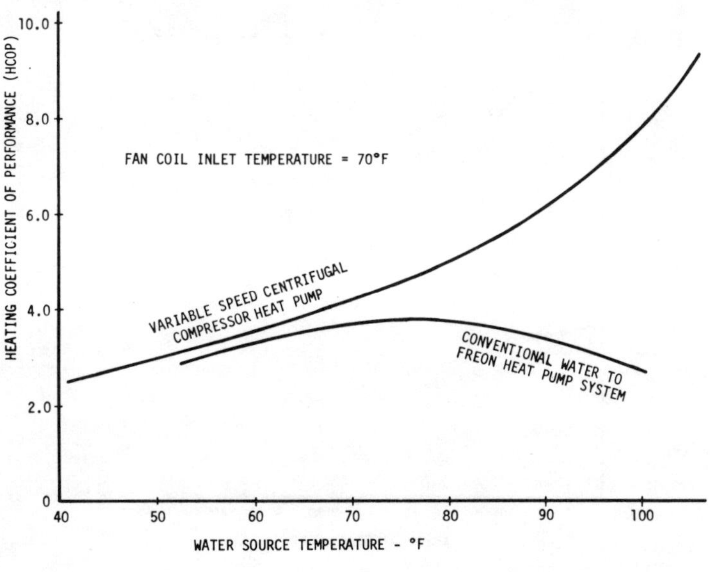

Fig. 2. Estimated performances of variable speed heat pump.

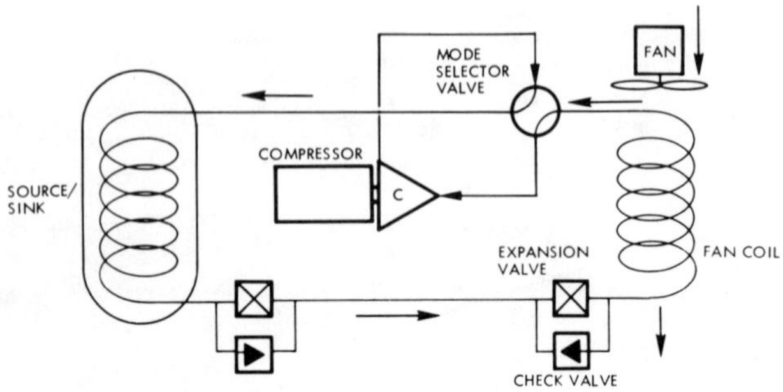

Fig. 3. Heat pump assembly, cooling mode.

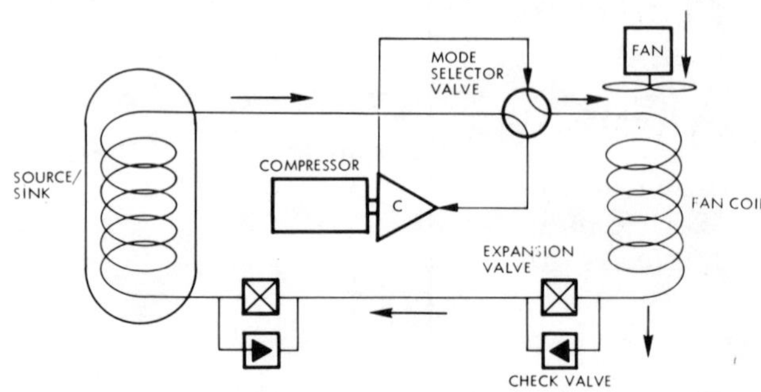

Fig. 4. Heat pump assembly, heating mode.

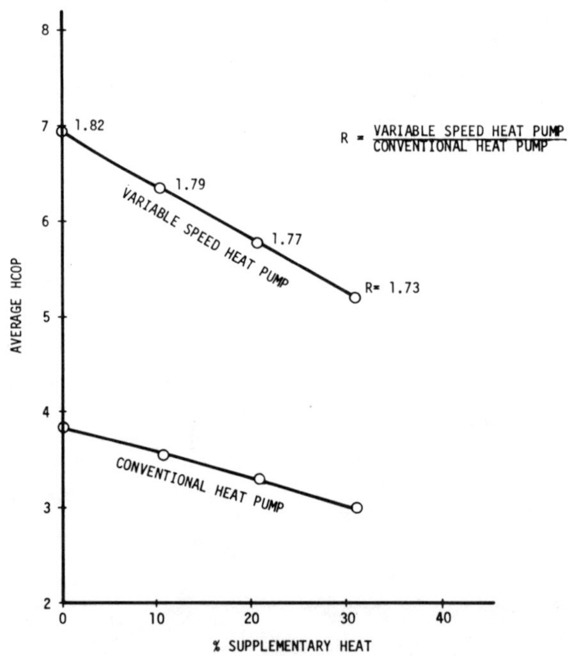

Fig. 5. COP heat pump performance data, variable speed versus conventional.

Heat Pumps Are Good — Solar Assisted Heat Pumps Are Better

Albert Weinstein

Manager, Special Energy Systems
Westinghouse Electric Corporation
Baltimore, Md.

Last year, under a National Science Foundation Contract, a comprehensive study was conducted on the "Feasibility of Solar Heating and Cooling of Buildings." The detailed results of that study are available and can be obtained from the National Technical Information Service (ref. 1). As part of that study three basic solar systems were recommended for development and application to meet the national need for heating and cooling of buildings: solar heating only; solar assisted heat pumps; and solar heating and cooling (absorption). This paper provides the rationale that underlies the recommendations for solar assisted heat pumps. The principal reasons fall within the categories: comfort requirements; performance; and energy conservation.

Comfort Cooling

Our analysis indicated that a cooling as well as heating demand exists in most of the buildings where most of the nation's population lives and works.

A demographic analysis identified the regions of the country where the population densities are greatest. Taken together with the climatic conditions that determine the type of demand for heating and cooling systems, these areas would constitute the principal potential commercial market for solar systems. The remainder of the area represents a less densely populated secondary market.

Five regions were selected which differ in terms of total heating and cooling loads and in the balance between them. The populations of these regions are shown in millions in Fig. 1 for 1969. The population distribution for the year 2000 is projected to be similar to the year 1969 except that the densities for all regions would increase.

Table 1 lists the population and the seasonal heating and cooling degree-days for the above regions and representative cities. For single-family residence requirements, they are characterized by:

1. Gulf Coast—Heating is modest and cooling requirements dominate.
2. Southeast—Heating and cooling requirements are equal.
3. Northeast—Total heating and cooling requirements are high, with heating being about three times greater than cooling.
4. Great Lakes—Heating requirement is about six times that of cooling.
5. West—Heating requirement is about the same as that for the Southeast, but the cooling need is modest.

TABLE 1. REGIONAL DATA

Regional City	Population, Millions 1969/2000 yr	Seasonal Heating (degree-days)	Seasonal Cooling (degree-days)
Gulf Coast	18/28	2000-1000	3500-2700
Mobile, Alabama		1612	3000
Southeast	28/42	4000-2000	2500-1800
Atlanta, Georgia		2826	2250
Northeast	42/64	7000-4000	2000-1000
Wilmington, Delaware		4910	1500
Great Lakes	44/69	8000-5000	2000-1000
Madison, Wisconsin		7417	1250
West	16/30	4000-2000	1500-500
Santa Maria, California		2934	500

As indicated, the above heating/cooling relationships are for single family residences. For commercial buildings which contain large numbers of people, there is almost always a cooling demand, even during the winter months. Indeed it is not unusual in these buildings for a heating and cooling demand to exist simultaneously. In those situations, the perimeter of the building has to be heated and the interior cooled.

Aspirations of people are upward. Only a few decades ago, air conditioning of homes, although technically available, was only for the privileged few and the market was very limited. Today, there are few homes built that do not have air conditioning; without it they would not sell. That is especially true for commercial buildings, particularly stores and shopping centers. It is therefore very unlikely that with the advent of solar systems, people would be willing to forgo comfort cooling.

Although there are laboratory experiments being conducted on advanced types of cooling techniques, such as Rankine Cycle Systems, only the vapor compression systems, which dominate the market, and absorption cooling systems are being manufactured on a production basis and are commercially available for widespread application to solar systems in the near term. There are advantages and limitations to each. In our Phase 0 Report we recommended that both be developed. A comparison of their characteristics would help illuminate their advantages and indicate the applications most suitable for each.

Performance (Absorption Air Conditioning Versus Heat Pumps)

A major advantage of absorption air conditioning systems is its ability to utilize directly the solar energy to operate it. The largest such system in the country is in a school in Atlanta, Georgia (Fig. 2). Ten thousand square feet of flat-plate collectors sponge up over 18 million Btu/day to heat water to temperatures of 180–200F and drive a 100-ton absorption machine. Regrettably, the coefficient of performance of such machines is low—about 0.6. Since the collectors are operating in the 200F temperature range, their efficiencies are also relatively low. The consequence of combined low collector efficiency and low COP results is final conversion of only about 20% of the solar radiation to useful cooling.

Concentrating collectors may be employed to produce higher temperatures to operate absorption systems at their full rated capacity, but these collectors throw away between 10 and 40% of solar radiation, since they use only the direct component of the sun's rays and not the diffuse. There is also the difference between heat energy and temperature. In our designs we are seeking to make maximum use of all the solar energy and employ heat pumps (in accordance with the second law of thermodynamics) to boost its temperature.

Machines in the 3-, 25-, and 100-ton range that will produce full capacity at 195F hot water input to the generator rather than 225F are being developed, which would make operation with flat-plate collectors much more attractive.

Electric-driven heat or vapor compression pumps, although capable of using solar radiation in the winter heating mode, cannot do so in the summer cooling mode. However, if the complete solar system is seen rather than the collector component, then the storage subsystem can, in fact, be seen as very useful both winter and summer. This aspect will become particularly significant when power companies can offer reduced rates for off-peak power.

In the winter, a typical heat pump would have a COP of about 1.7 at −10F outside ambient conditions and 3.0 at 55F. When the compressor output is insufficient to carry the heating load, supplementary resistance heaters, which, of course, have a COP of 1.0, are turned on. The annual efficiency of a pump system is therefore a combination of the performance of the compressor and the resistance heaters during the time they are on. Seasonal performance factor (SPF) is the ratio of a heat pump's installation (including resistance heaters) heat energy output to its electrical energy input over an entire heating season. Heat pumps manufactured in the 1970s have an SPF of about 2.0 in a "reference" house in Pittsburgh (ref. 2).

By using solar energy stored at temperatures above the ambient as the source of input to the heat pump, the compressor would operate more frequently at its higher COP and capacity and the need for supplementary resistance heaters is significantly reduced. For the higher efficiency solar-assisted heat pumps now entering the market, an SPF of 3.5 could be achievable and for future designs an SPF of 4.0 would not be unreasonable to expect.

The size of the collector array can be appreciably smaller for a solar assisted heat pump than the absorption system, as indicated in Table 2. Note that in the Great Lakes region the absorption system (HC) requires 860 sq ft of collector for a single-family residence of a nominal 1600 sq ft size, whereas the heat pump system (HP) requires but 480 sq ft—about one-half. This is due not only to the higher COP of heat pumps, but to the fact that the collectors would operate at lower temperatures, and therefore higher efficiencies. Furthermore, for the heat pump system, the collector can be tilted to maximize winter operation at the specific site, and it is not necessary to compromise for a summer and winter time tilt, as is the case for absorption systems.

TABLE 2. COLLECTOR AREA REQUIRED FOR SINGLE-FAMILY RESIDENCES (1600 sq ft)

	50% Solar Dependency		
Region	HO	HP	HC
Northeast (Wilmington)	440	400	680
Southeast (Atlanta)	240	160	560
Gulf Coast (Mobile)	120	100*	580
Great Lakes (Madison)	1,200	480	860
West (Santa Maria)	160	100*	100*

Key: Solar collector area (sq ft) required for:
HO = heating only
HP = solar-assisted heat pump
HC = heating and cooling (absorption)
* = required for domestic hot water only

Based on the above comparisons, normal system application in climatic regions requiring primarily cooling with modest amounts of heat, such as in the Gulf Coast, might best be served by solar assisted absorption air conditioning systems. In regions where heating requirements dominate and where cooling is also required, such as the Northeast and Great Lakes region, the advantage would go to solar assisted heat pump systems. In the Southeast where heating and cooling demands are about equal, in terms of tennis parlance, it may be a deuce game. As previously noted, when reduced off-peak power rates become available, the advantage could go to the solar assisted heat pump for all regions, especially for regions that have run out of gas.

Types of Solar Assisted Heat Pumps

Solar assisted heat pump systems can be of several types. They can utilize liquid in the solar collector loop, water storage, and a water-to-water heat pump as shown schematically in Fig. 3. Another system employs an air-to-air heat pump in conjunction with liquid-to-air heat exchangers and a liquid solar collector loop. Systems using air in the solar collector loop with rock storage and air-to-air heat pump are also possible.

I would also include in the heat pump category a concept of a solar assisted double-bundle condenser heat recovery chiller (ref. 3). This is shown schematically in Fig. 4.

In the summer the system would operate as a conventional central station chilled water air conditioning system; however, it would also be possible to operate at night under more favorable ambient conditions and possibly reduced off-peak power rates, and store the "coolness" in the system storage tank.

In the winter solar energy as well as heat from people and lights in the interior zone of a large building would provide the load to the evaporator coil of the compressor. Just as in the more conventional heat pump, that energy is raised to higher temperatures by the compressor, delivered to the condenser, and then distributed to the zones requiring heat. This system, suitable for large buildings, has the advantages of a heat pump without the complications associated with the reversing cycle features and provides the capability to simultaneously heat or cool different zones in the building.

Energy Conservation

The fundamental purpose of the national program is to conserve scarce forms of energy, in particular gas and oil. Although there can be disagreement over the precise dates of energy depletion, I expect that in the next two generations, the last cubic inch of natural gas will be hissing through the pilot light of someone's home furnace—and in two generations more, the last gallon of oil will be pumped into someone's automobile. But there is a large amount of coal and an increasing amount of nuclear power that can supply energy in the form of electricity to heat and cool homes for ages to come.

Dunning (ref. 2) compares energy conservation between gas, electric resistance heating, and electric heat pumps. I have extended it to include solar assisted heat pumps. As Fig. 5 indicates, to get 1 Btu of heat energy from your gas furnace, it requires 2.2 Btu of natural gas at the gas company's storage tank for an overall process efficiency of 45%. This is based on an assumption of actual efficiency to the gas furnace of 50%. It is quite possible to get efficiencies of 75–80% in the laboratory (and I have done it), but in the field the excess air admitted to the burners to assure clean burning normally reduced long term efficiencies to the 50% figure.

Referring again to Fig. 5, it is noted that the electric resistance heating, with coal or nuclear fuel at the source, has the lowest overall process efficiency, 31%, which is no surprise to homeowners in an area such as Virginia where they are paying as much as $200/month for heating. A conventional heat pump raises the overall process efficiency to 62%. Assuming future availability of higher COP solar assisted heat pumps, the overall process efficiency can be increased to 125%; i.e., to deliver one unit of useful energy to heat a home requires only 0.8 unit of coal or nuclear power energy at the power plant, thus conserving these forms of energy.

Summary

In summary, heat pumps are mass produced by many manufacturers and available for widespread use with a variety of solar systems for heating and cooling applications in most regions of the country. The solar energy will increase the seasonal performance factor of the heat pump system, improve reliability, reduce operating costs, and contribute to conservation of scarce forms of energy.

REFERENCES

1. Westinghouse Electric Corporation, ''Solar Heating and Cooling of Buildings, Phase 0,'' National Technical Information Service (PB 235-426/LK through PB 235-429/LK), May 1974.
2. Robert L. Dunning, Westinghouse Electric Corporation, ''Relative Efficiencies of Gas Furnaces and Electric Heat Pumps,'' *Nuclear Energy Digest—Volume 4*, 1973.
3. R.C. Niess, Westinghouse Electric Corporation internal memo, 26 July 1974.

DISCUSSION

Q How feasible is it to drive a heat pump compressor in winter with a solar organic Rankine cycle engine, and use the heat rejected from the engine condenser as partial input to the heat pump evaporator?

A It is technically feasible but no Rankine cycle systems are now in production. The COP is low and the economic feasibility questionable.

Q Unless supplemental resistance heat is eliminated entirely, higher COP systems could result in lower annual load factors for electric utilities in northern climates. This is not desirable.

A With a properly insulated building and solar assisted heat pump with adequate storage, the requirement for supplementary heaters could be eliminated.

Q Do your reliability data include normal maintenance calls, i.e., for oil furnace cleaning, as well as emergency calls when equipment does not function at all?

A Yes.

Q How much is the electricity bill increased to accomplish supplemental heating need with the heat pump?

A The electric bill for heating during the heating season with a heat pump could be about 50% of the cost of heating with resistance heaters.

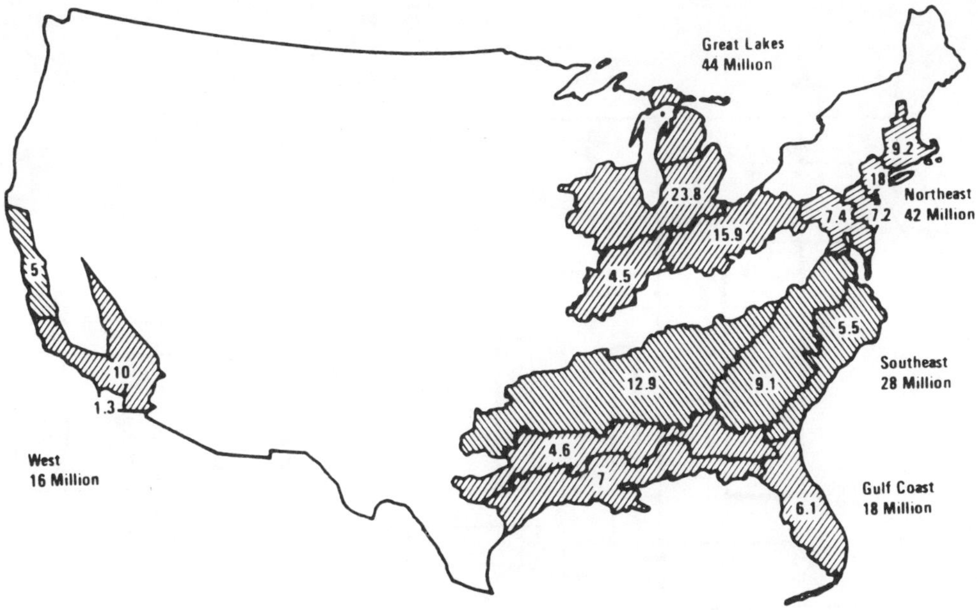

Fig. 1. 1969 demographic analysis for potential market sites for solar systems.

Fig. 2. School in Atlanta, Georgia.

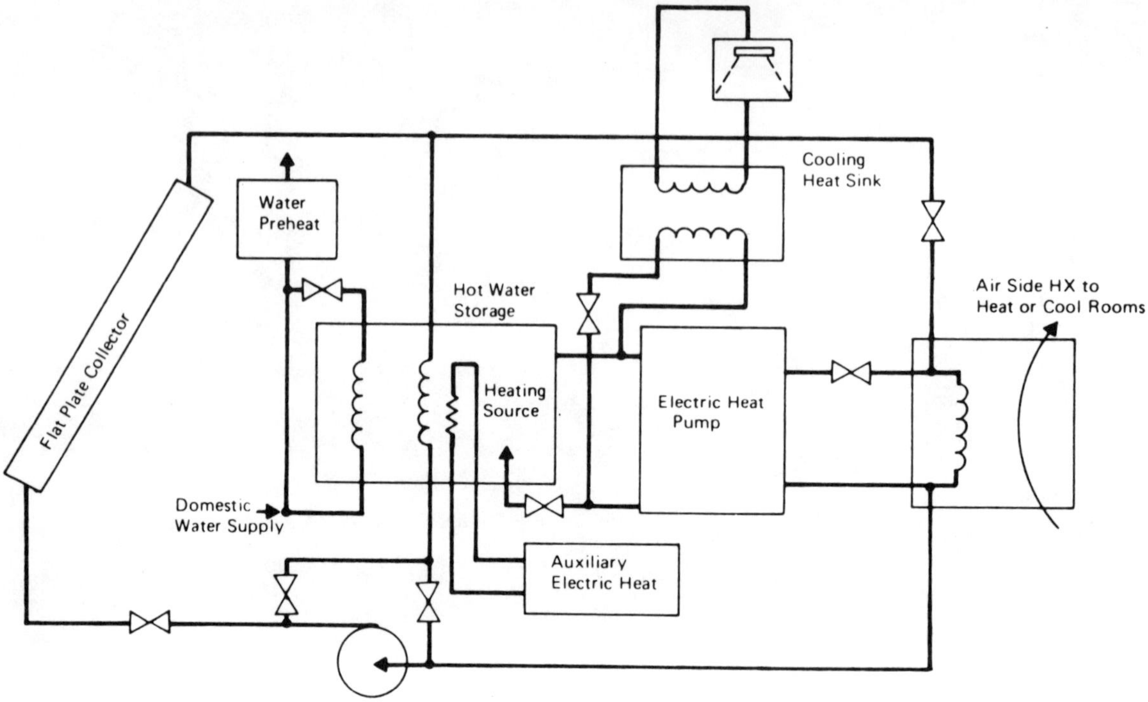

Fig. 3. Solar assisted heat pump.

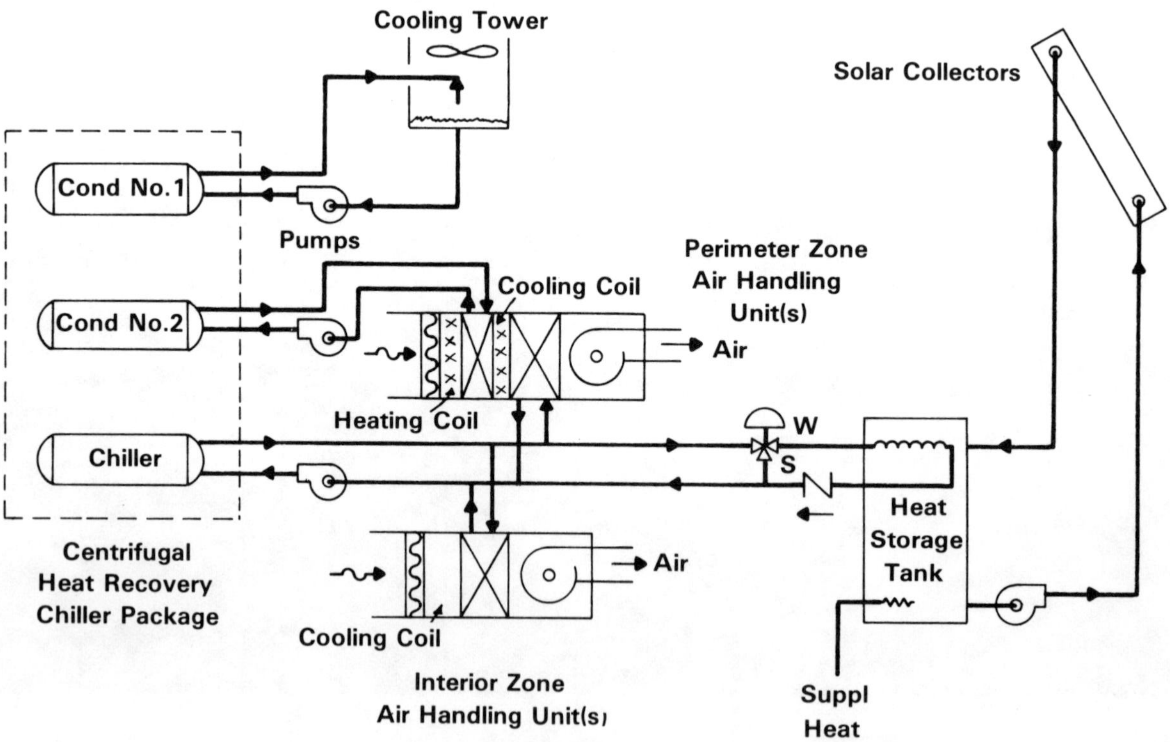

Fig. 4. Solar assisted double-bundle condenser chiller.

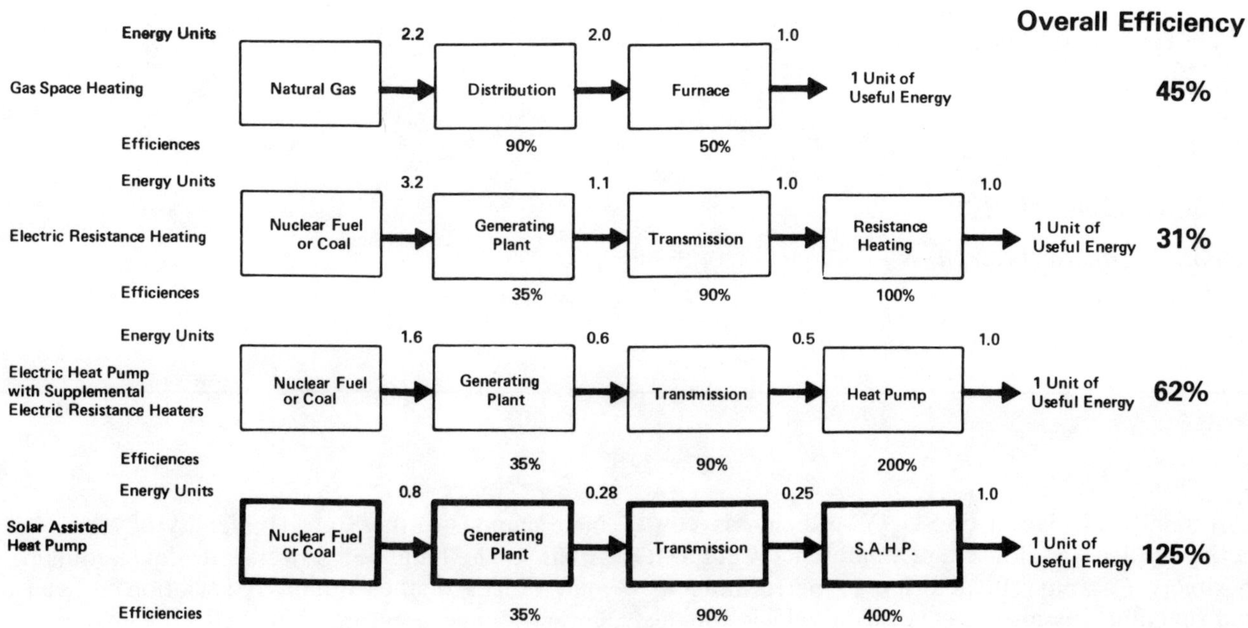

Fig. 5. Energy conservation comparison chart.

Solar Energy Heat Pump System for the Alumni House and Conference Center, State University of New York at Albany

P. Richard Rittelmann

Principal, Burt, Hill & Associates
Butler, Pa.

Richard Bodette

Principal, Rist-Frost Associates
Glens Falls, N.Y.

The Alumni Association of SUNY, Albany, is constructing an Alumni House and Conference Center at the Albany Campus. In addition to performing its stated function it is also to serve as a vehicle for the demonstration and evaluation of a solar assisted heat pump heating and cooling system. Supporting this effort is the Niagara-Mohawk Power Company and the New York State Public Service Commission. Particular efforts of the facility will be to permit a comprehensive investigation of the most effective manner in which to (*a*) combine electric heating and cooling with supplemental solar energy using thermal storage and heat pumps to form a complete system for comfort control and (*b*) attain benefits to the external electric power system in terms of peaking, load factor, etc.

The building is currently under construction and the purpose of this paper is to present some of the basic design data, schematics of this system as designed, and anticipated performance characteristics and to illustrate the capabilities of the system for evaluating a variety of operational modes.

Building Characteristics

The Alumni House Conference Center is a two-level building of approximately 7000 sq ft of floor area located on a pleasant, wooded site overlooking the main campus of the State University of New York at Albany. The building's lower entry level contains the staff offices of the Alumni Association as well as a library/lounge area and rathskeller. The upper level provides flexible meeting facilities for conferences of up to 120 people. The conference areas have been planned with relocatable walls so that meetings of different sizes can be easily accommodated. Food service facilities are located adjacent to the conference areas. From the main conference area, sliding doors open on to an outdoor terrace which blends into the surrounding wooded landscape. Figures 1 and 2 illustrate the architectural design of the building.

Prominent in the exterior design expression of the building is the solar collector array consisting of 2300 sq ft of two-glass, flat-plate solar collector mounted at 45° to the horizontal and oriented 17.5° west of south. The location of the building is at 42° 50 min north latitude.

Load Characteristics

The building was rather carefully designed to minimize energy consumption. Everything that was accomplished in construction, however, was accomplished within the vocabulary of present-day construction techniques. Typical "U" values were: walls, 0.07; roof, 0.04; glass, 0.65. The building is well sited to minimize losses due to wind exposure and has an optimum aspect ratio properly oriented with the long axis in the east-west direction. Total building losses are approximately 50,000 Btu/degree-day or approximately 7 Btu/sq ft/degree-day. Figure 3 illustrates the relationship between ambient air temperature and building load. It is an oversimplification for purposes of design calculation but is useful to graphically illustrate the relationship and relative magnitudes of the loads. Particularly significant is the variation of load with ventilation rate. If no outside air is brought into the building, it can be seen that the balance point, i.e., that point at which the internal heat gain balances the skin losses, is approximately 34F during the occupied period. If the maximum ventilation rate is used, however, the balance point is approximately 54F. Implicit in Fig. 3, however, is the fact that between 34 and 54F comfort control can theoretically be partly accomplished by varying the ventilation rate. The figure was plotted for the heating season conditions and its limitations should be recognized. This method of plotting assumes an instantaneous load application and does not recognize the dampening characteristic of time lag due to the mass of building materials.

The various balance points indicated in Fig. 3 are a function of both climatic conditions and occupancy. If we show the balance points as a function of time for a typical day in each month and also plot ambient air temperature as a function of time, we begin to get an appreciation of the effect of diurnal cycling on the building which ultimately is imposed on the mechanical system. This relationship is shown in Fig. 4. A further extension of this relationship is shown in Fig. 5 where we compare the Btu required with the Btu available internally. In conventional buildings, without an active storage capability, only a limited advantage can be taken of the internal load. We would be limited to the inherent storage characteristics of the mass of the building itself. With an active storage characteristic and heat pump capability associated with this mechanical system, it is possible to take advantage of the internal heat loads during the heating season and minimize the amount of additional energy external to the system required to achieve the design comfort conditions.

Solar Characteristics

We were fortunate in this particular project to have solar and other climatological data that would not normally be available. An important participant in the project has been the Atmospheric Sciences Research Center located on campus in close proximity to the project. Hourly data for all climatological inputs necessary for the design were available. This capability is expected to be even more important in the analysis phase of the project.

By using available insolation data, calculations were performed with procedures essentially the same as those indicated in ref. 1. The procedures were modified slightly, however, to adjust the diffuse component for greater accuracy. Reference 1 assumes the diffuse component to be evenly distributed over the hemispherical skyvault. In modifying the calculation of the diffuse, we used the technique described in ref. 2, which recommends that the diffuse portion of the sunlight be divided into two equal components, the "Circum Solar" component, originating from the vicinity of the sun, which is added to the direct solar radiation and its intensity computed accordingly, and the "Background" component, which is uniformly distributed over the skyvault. Average hourly temperatures during the daylight periods were available and those for the mid-day of each month were used for purposes of calculating the collector efficiency. The collection capability of the collector array was calculated for two different inlet temperatures. A 50F inlet was used to reflect a minimum storage tank temperature as a result of heat pump operation. An inlet of 105F was used, which was determined by the minimum useful temperature for direct solar heating. Collector production at inlet temperatures between these two limits was then available for use in predicting system performance in later calculations. Figure 6 illustrates the available insolation and the collector production at the above inlet temperatures.

The initial design of the collector array indicated the desirability of stacking five collector modules up the roof slope. To reduce installation costs it was advisable to minimize the supply and return header piping and connect the five modules in series. It was expected that at a given flow rate the efficiency of this arrangement would be lower than the efficiency of piping each module in parallel. Calculations were performed to determine the efficiency of each individual module in the series arrangement.

The combined efficiency of all five plates was approximately 43.6%, versus 50% if all plates had been piped in parallel. The reason is that the temperature rise across five plates is approximately 33F while the rise across the first plate is only 6.3F. It would be possible to increase the efficiency of series-mounted modules by increasing the mass flow rate and decreasing the temperature rise, but at a penalty in pipe sizing and pump horsepower required. It was also estimated that to limit the temperature rise across five plates to the 6.3F across one plate might result in turbulent flow in the tubes of some types of collectors, causing increased system pressure and aggravating corrosion problems.

System Characteristics

Figure 7 is the schematic flow diagram of the combined solar/heat pump system and shows the essential piping relationships as well as the important components of the system. The system uses 2300 sq ft of one-glass, flat-plate collector using inhibited water as the heat transfer fluid at a maximum pumping rate of 60 gpm with a 1½ hp pump allowing a maximum mass flow rate of 13 psf/hr. The collector array is open to the atmosphere and will drain down for freeze protection when the circulating pump stops. The collector returns will normally go through a heat exchanger supplying the domestic hot water system as a first priority load and then into the 8000-gal storage tank from which the collector pump takes suction. The heat pump selected is a 22-ton water-to-water unit. For purposes of testing and comparative analysis an air-to-water heat pump is available for optional use in the system.

An additional 8000-gal storage tank shown in the lower right of the flow diagram is used for peak shaving both in the heating and cooling mode, and is also available for storage of collected solar heat.

An electric boiler is located in the hot water supply line to furnish auxiliary energy and, as shown later in the mode descriptions, is capable of imputting auxiliary energy at the most favorable off-peak conditions. It is also possible to use the air-to-water heat exchanger for peak shaving when conditions are suitable for that function. A 275-gal insulated tank is incorporated at the suction side of the condenser for a variety of reasons:

1. The distribution system is a four-pipe method with hot water being available at all times, even during the cooling modes, for reheat purposes for humidity control. This hot water is simply the discharge from the condenser and is sent through the system before it is taken to the cooling tower to reject any heat not used. During the changeover season, the 275-gal tank allows for the balancing of water flow in the system caused by varying evaporation rates from the cooling tower and different water volume in the piping systems.

2. It is possible to manually take the collector returns directly to the 275-gal tank and blend them with the hot water return from the system going to the suction side of the condenser, to allow the early use of available solar energy. If, for instance, the 8000-gal tank had been depleted overnight, it would be necessary for the collector array to work through the inertia of that 8000 gal to provide useful heat that must be put into the system through the heat pump. It would also be a number of hours without heat pump operation before the tank would be sufficiently hot to allow heating directly. By being capable of accepting solar output directly, the utilization of available energy is increased.

3. It will be possible to investigate the amount of heat rejection possible by nocturnal radiation and evaporative cooling through the combined collector and cooling tower effects.

Operational Modes

There are far more operational modes possible with this system than would reasonably be considered for design of a commercial solar assisted heat pump system. This was purposely done to increase the project's value as an experiment. There are five basic heating modes and five basic cooling modes, each with possible variations and control strategies. For purposes of brevity a graphic technique to illustrate these modes was developed and is shown in Figs. 8 and 9. More extensive elaboration on equipment specifications, operational modes, control strategies, instrumentation, and operating results will be the subject of future papers on this project.

An operational feature, however, particularly germane to this project will be the use of weather forecasting supplied by the Atmospheric Sciences Research Center on campus to facilitate the introduction of auxiliary energy to the system under the most favorable times and conditions.

Conclusions and Recommendations

It is the intent of this paper to serve as a "progress report" as the project is still under construction. Operating data will result in far more meaningful conclusions and recommendations. Detailed computer modeling of the system has not been undertaken; however, some gross modeling of block loads, solar system operating characteristics, and heat pump system operation was done using programmable calculators to track residual heat in storage. This procedure was carried out for both average conditions and extreme conditions to check the reasonableness of major component sizing for which there is no established design methodology at present. Figures 10 through 12 illustrate the results of that modeling.

ACKNOWLEDGMENTS

The authors wish to acknowledge the assistance of John Kurtz (Richard G. Jacques Associates, Architects), Harry Gordon and Chris Johnson (Burt, Hill & Associates), and the Atmospheric Sciences

Research Center, SUNY, Albany, for their respective contributions in both the design of the project and the preparation of this report.

REFERENCES

1. *Low Temperature Engineering Applications of Solar Energy*, ASHRAE, New York, 1967.
2. Petherbridge, P., *Sunpath Diagrams and Overlays for Solar Heat Gain Calculations*, Research Series 39, Building Research Station, S.D.I.R., 1965.

DISCUSSION

Q What is the collector fluid?

A Water with inhibitors.

Q Who manufactured your collectors?

A Revere Copper and Brass.

Q What is the capital cost ratio of the heating/cooling system to total building cost for the SUNY building? How does it compare to the controlled building ratio?

A Capital cost ratio of solar HVAC system to total building cost has not yet been established. Much work is being done on a cost-plus contract. Due to the experimental nature of the system, this ratio would be very misleading unless accompanied by a complete economic analysis.

Q Is a computer model available for the dynamic problem of allowing for thermal capacity of buildings in load calculations?

A TRACE, ECUBE, WESTINGHOUSE, NBSLD, MERRIWETHER, to name only a few, are commercially available. There are numerous proprietary programs also with this capability.

Q You said a single-glazed collector was more cost-effective. Could you specify your cost assumptions on the cover plates?

A Commercially available two-glass flat-plate collectors quoted on this project varied from $6.30/sq ft to $7.80/sq ft. Using the low bid as the basis for comparison, the decrease in efficiency when collecting 70F water by reducing to one glass cover was less than 10%. This indicates that if more than $0.60/sq ft is offered for deleting one of the glazing covers, it would be more cost-effective. In our particular case, however, the low bid collector was designed specifically as a two-glass collector and the savings possible was not as much as $0.60/sq ft. Our indication in the presentation was that if a similar collector had been designed initially as a single cover collector, it is likely that more than $0.60 could be extracted from the cost.

Q What loss in insolation can be expected when the sun's rays pass through partial tree cover?

A This varies with species but we feel a 20% reduction through deciduous trees without leaves is a conservative estimate. Although we have not yet seen the data, we understand some information on this question is available in a publication of the American Society of Landscape Architects. We suggest contacting Gary Robinette at ASLA for more information.

Q Where did you get the statistical weather data? Are they available for other parts of the country? Is your preliminary work published?

A Atmospheric Sciences Research Center, SUNY, Albany, New York. No. No.

Q What is the size of the heat pump?

A 22-ton.

Q What is COP for heating, COP for cooling, and expected duty cycle for running time on/off?

A COP varies with local and other operating parameters. It is estimated at 3.89 when using 70F water as a heat source. Cooling COP should usually be in excess of 3.5, and under part load conditions can approach 4.0.

Q What are your economic conclusions?

A We have not yet attempted to draw economic conclusions. A major part of the analysis will be to operate in various modes to establish the relationship of many parameters to solar assisted heat pump system design.

Q Please clarify the 17.5° west of south orientation.

A 17.5° west of south was chosen for siting the building due to the location of existing vegetation. Slightly west of south will be a more favorable orientation to maximize solar heat production in many areas of the country. This is due to the fact that the efficiency of the collector is dependent on functions other than normal incident angle of insolation. Both the ambient air temperature and the ratio of direct radiation to total tend to be more favorable some few hours after solar noon. All factors considered, a slightly west of south orientation can often be most productive.

Q What is the ratio (width/length) of the building and how is the dimension placed?

A The aspect ratio (length to width) of the building is 1.8 to 1, with the long axis in the east-west direction.

Q What type and size computer do you intend to use?

A We do not have the answer to this question. We suggest contacting Ron Stewart at the Atmospheric Sciences Research Center, SUNY, Albany, who will be responsible for the data gathering and analysis phase of the effort.

Q Would it not be practicable to use a separate collector for domestic hot water?

A We do not feel a separate DHW collector system advisable due to the lack of diversity of being able to deploy all available collectors to any dominant load. Isolating for one function seems too limiting, and will result in duplication of system components and controls. We do, however, appreciate that some problems, such as summer operation for DHW use only, could be avoided with a separate system.

Q Can you explain fully how to limit temperature on hot summer days?

A There is no attempt to limit summertime temperatures. The collector array will drain down before boiling conditions are reached. The leaves coming out on the deciduous trees will be of some assistance in this particular project. The drained collector temperature will approach 375F in full sunlight, and this is expected to occur. The control strategy will not permit water to be pumped into the collectors when they are above 200F, to avoid both flashing into steam and the thermal shock of the collector, which would certainly create leaks.

Q What are you doing about collector water freezing, corrosion, etc.?

A Collector is copper; water will be treated with inhibitors. Final decision on inhibitors has not yet been made. Freeze protection is by drain-down.

Q Most economic calculations based on "water" storage forget that "antifreeze" is used at nearly $1/gal.

A Storage medium in this project is water, not antifreeze. Bulk ethylene glycol is currently $3.40/gal. A 60% concentration for the Albany area would have resulted in a cost of $2.04/gal for antifreeze solution. We doubt that any system with 16,000 gal total storage would consider an antifreeze solution in the entire system.

Fig. 1. SUNY Alumni House Conference Center.

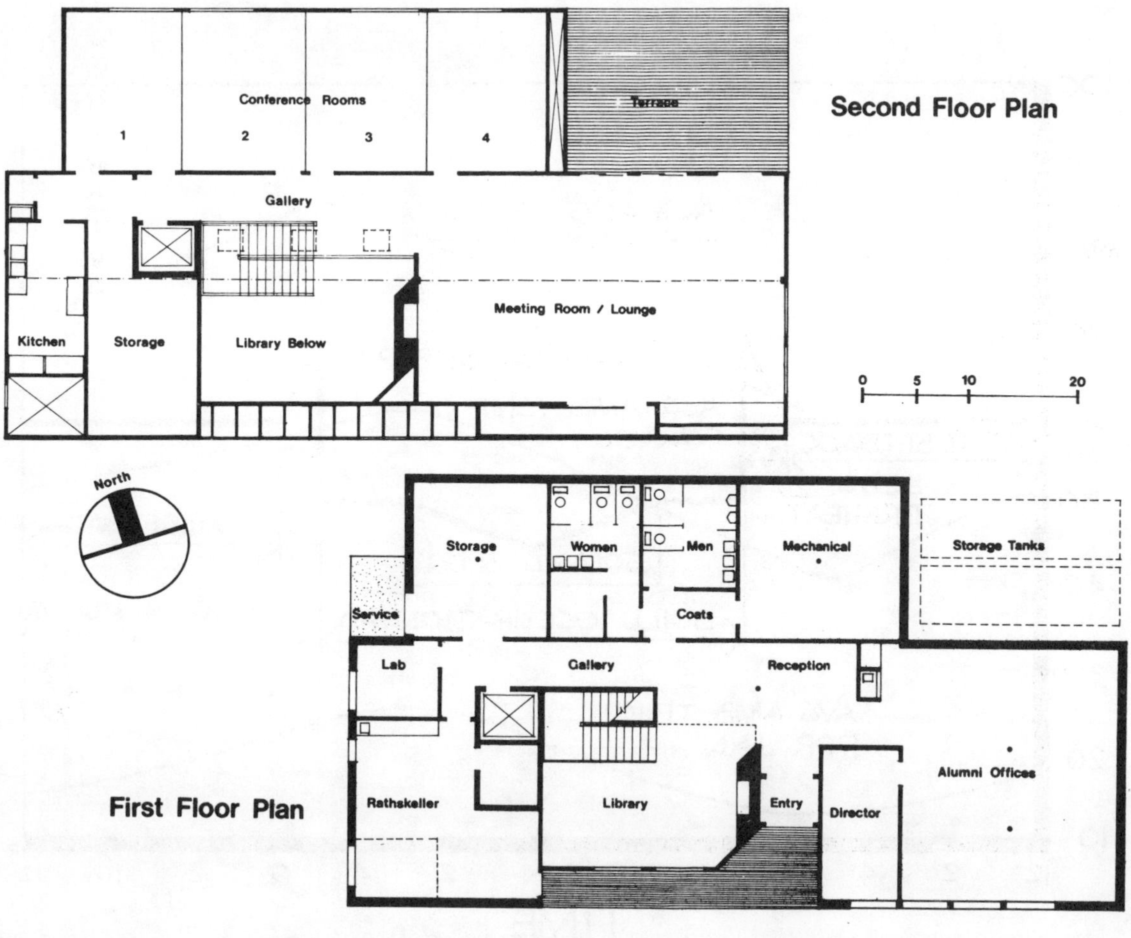

Fig. 2. Floor plans, SUNY Alumni House Conference Center.

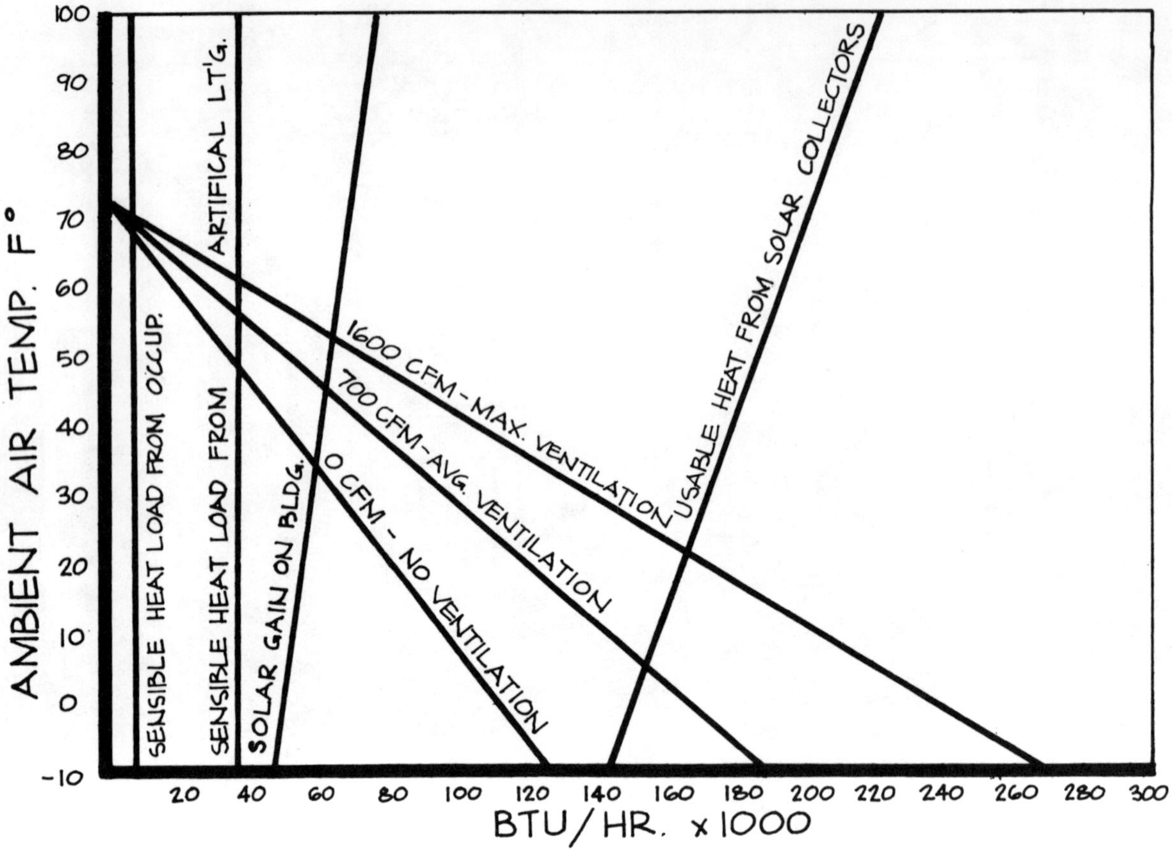

Fig. 3. Heat loads as a function of ambient air temperature.

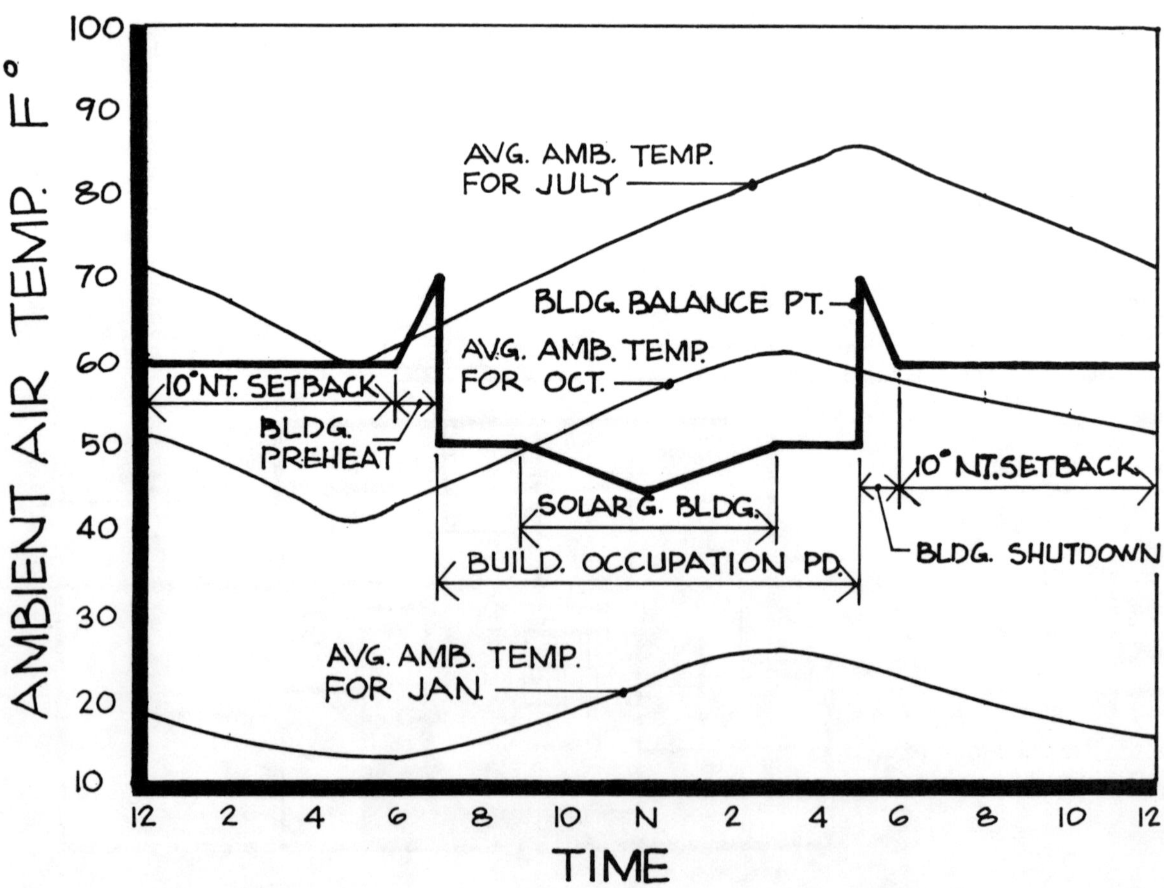

Fig. 4. Heating balance point and the schematic relation to ambient air temperature.

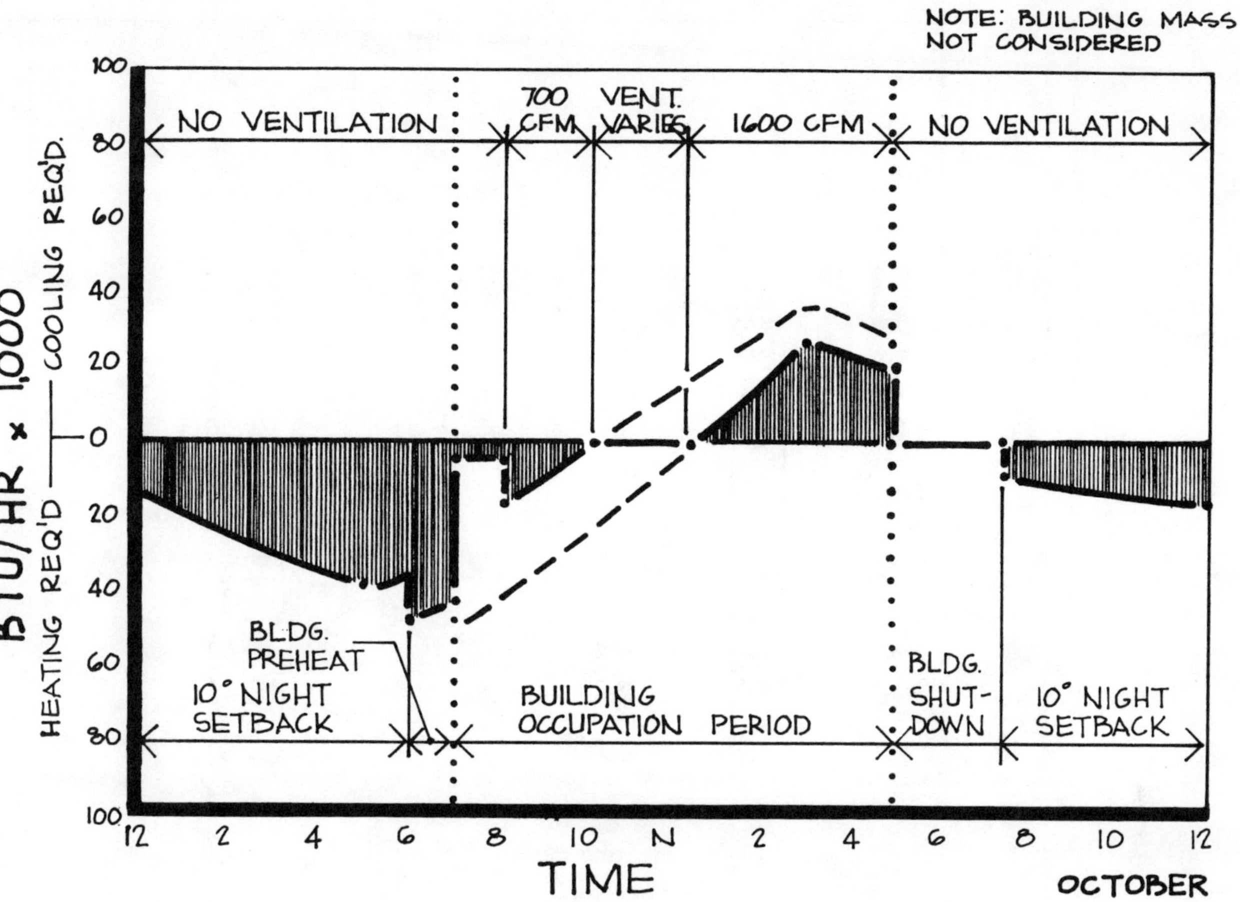

Fig. 5. Diurnal cycling of energy requirements.

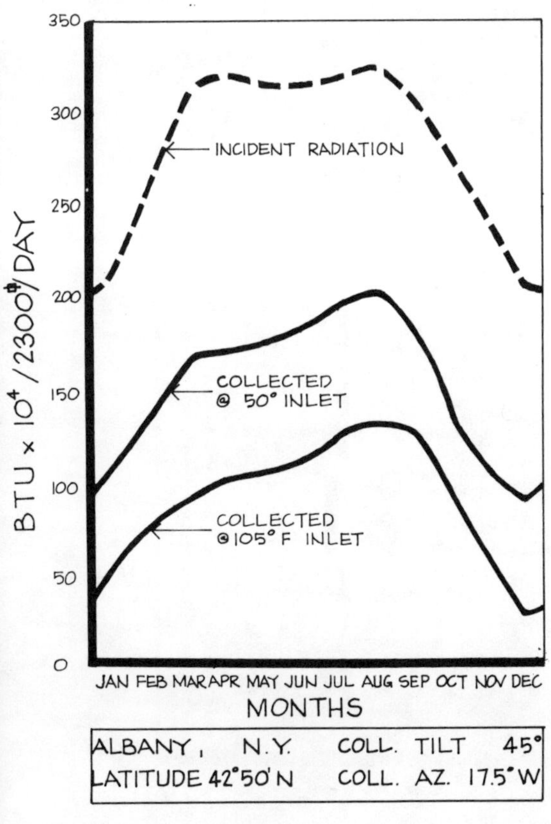

Fig. 6. Incident radiation and heat production.

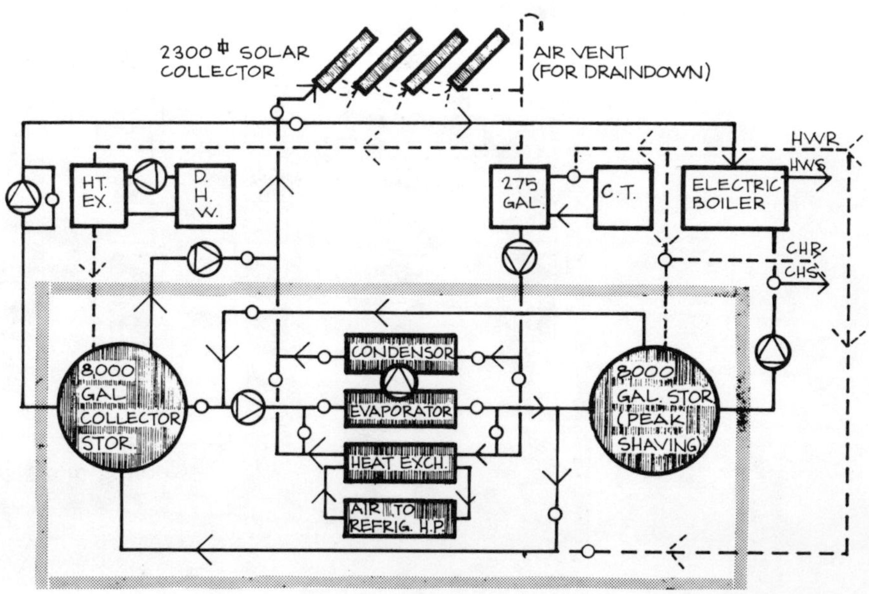

Fig. 7. Schematic system flow diagram.

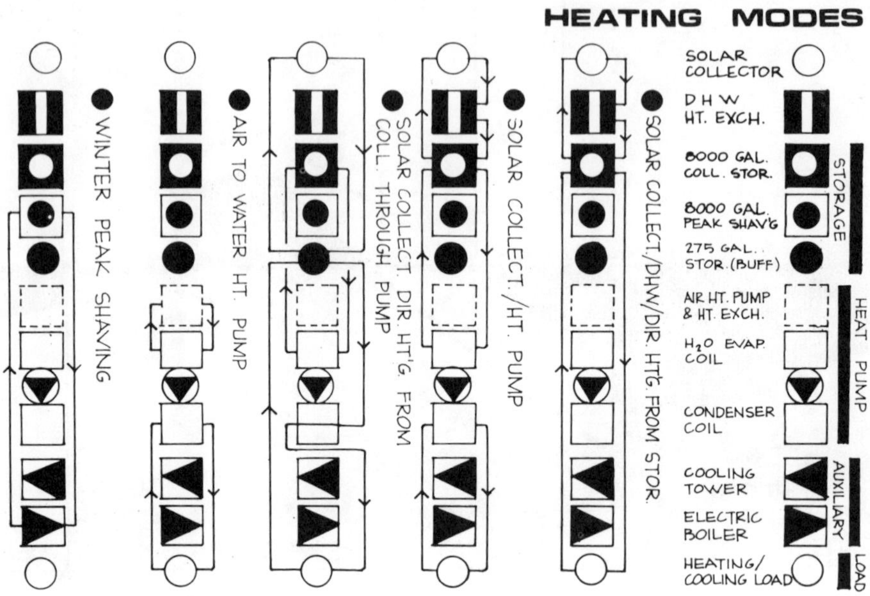

Fig. 8. Heating modes.

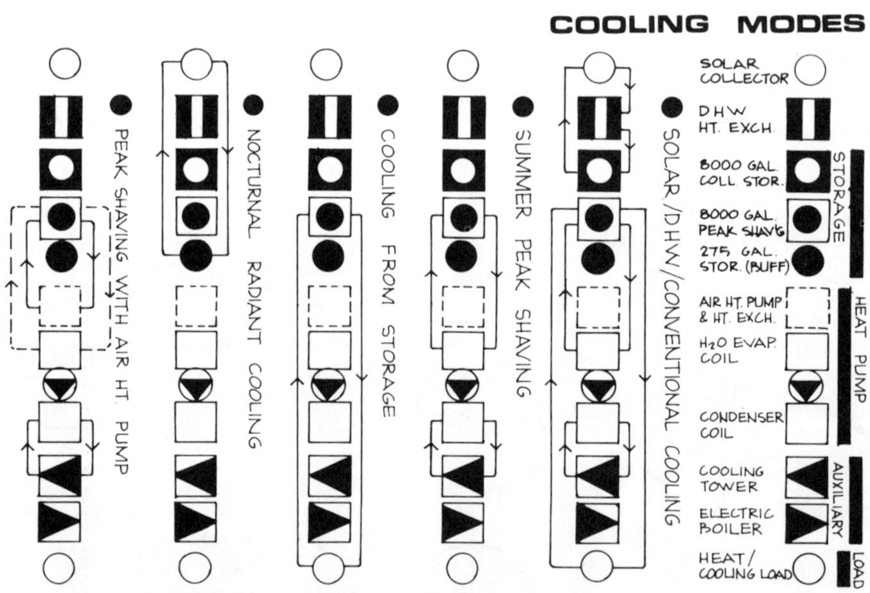

Fig. 9. Cooling modes.

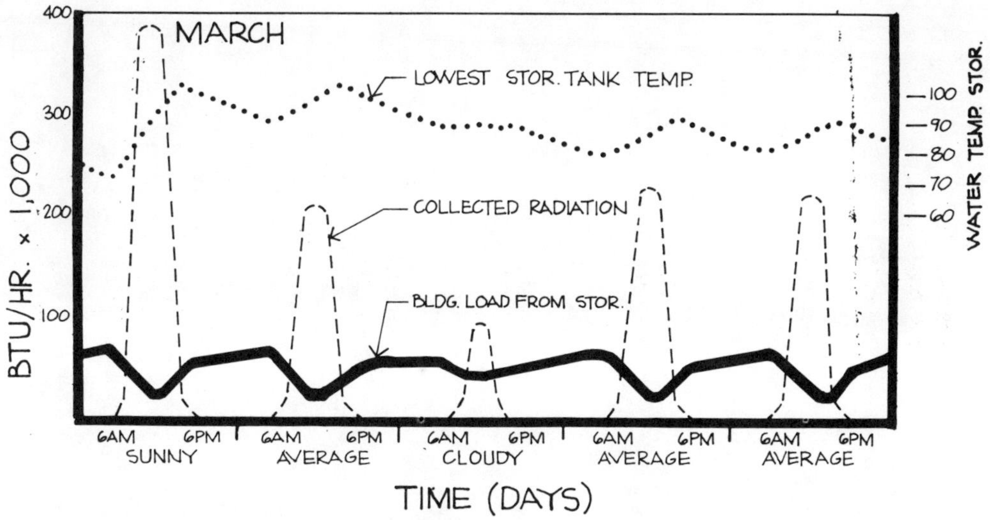

Fig. 10. System performance modeling.

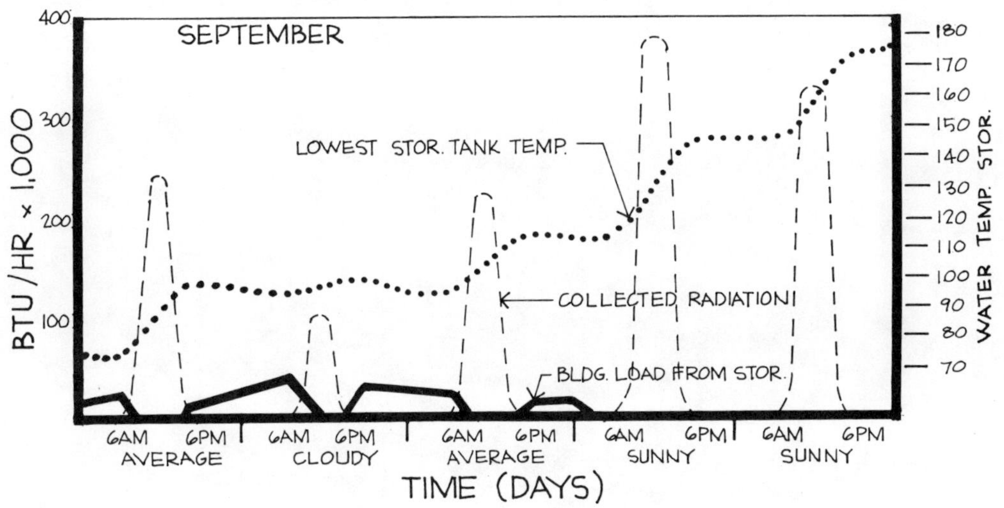

Fig. 11. System performance modeling.

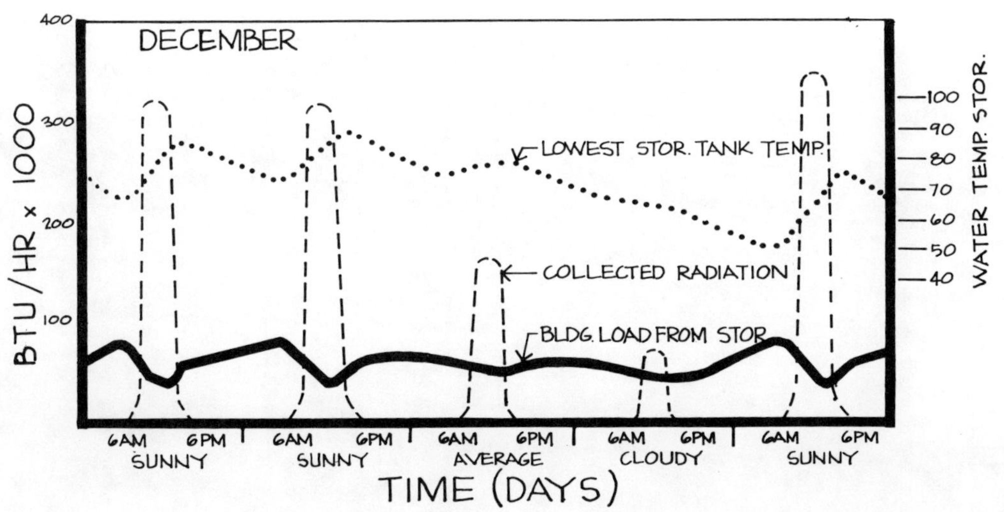

Fig. 12. System performance modeling.

Part II

Chairman:
C. Lance Charriere

Assistant Professor of Architectural Engineering
Pennsylvania State University
University Park, Pa.

Evaluation of the Alburquerque, New Mexico, Solar Building

Dr. Stanley F. Gilman

Professor of Architectural Engineering
Pennsylvania State University
University Park, Pa.

This project was funded from April 1974 through March 1975 by the National Science Foundation and is now being supported by the Energy Research and Development Administration (ERDA). Although Penn State is the prime contractor, the project is a three-team effort—Penn State, Bridgers and Paxton, Consulting Engineers, Inc. (the owners and occupiers of the building), and the University of New Mexico.

The overall objective of the program is to develop a *System Design Manual* for solar energy assisted heat pump systems that will be applicable to all geographical areas and all types of commercial or industrial buildings. To accomplish this, the major steps are:

1. Determine actual performance by analyzing field-collected data.
2. Correlate results with computer predicted performance.
3. Develop and refine a computer model of the building and system.
4. Use the computer model to generate information for the *System Design Manual*.

Figure 1 is a view of the building being evaluated. It has a solar energy assisted heat pump heating system which was originally designed by the firm for its own evaluation of the potential for solar energy. The principal elements of the system serving 4300 sq ft of single-story office space are the 750 sq ft of water-type solar collector, a 6000-gal underground water storage tank, one 7½-ton water-to-water heat pump, and 5 small tonnage water-to-air heat pumps. The collector has single glazing, is tilted 60° from the horizontal, and faces due south.

A schematic diagram of the building system is presented in Fig. 2. The solar collector (upper left) and its piping circuit contain an ethylene glycol/water solution to prevent freeze-up during cold weather. Solar energy from the collector is transferred to the pure water building system through the glycol/water heat exchanger (lower left). The water-to-water heat pump is in the center, and supplies warm or chilled water to the air handling unit (upper right) as called for by the space thermostat. The evaporative water cooler (top) controls the maximum allowable tank water temperature in the winter and is the prime heat rejector during the summer cooling operation. The five water-to-air heat pumps are all on one water-loop that is integral with the total system (lower right).

The system has several different operating modes. The solar collector circuit operates only when solar energy is to be collected. When the system water temperature is above 95F, the air handling unit is fed directly from the storage tank and the water-to-water heat pump is off. Below 95F, the heat pump operates with the storage water tank as the source and the air handling unit (building spaces) the sink. Since commercial office buildings typically have high internal loads (lights, people, etc.), during mild winter weather there is often a need for cooling, particularly in afternoons. When this situation occurs, the water-to-water heat pump shifts to a cooling mode, with the building space as a heat source and the tank water as a heat sink.

Figure 3 is an energy flow diagram for the winter heat pump heating mode. The water storage tank (top) is the central element. Energy flow from the solar collector is always positive or zero. Energy flow from the water-to-air heat pump loop is typically positive, as the principal factor is the continual cooling load caused by the general purpose and data acquisition computers in a computer room. Heat is extracted from the tank by the heat pump evaporator. This, plus the heat equivalent of the compressor power input, shows up in the condenser, from which it flows to the air handling unit coil. The fan and air duct system then transmits this energy to the building space, supplying heat thereto.

A diagram of the physical arrangement corresponding to Fig. 3 is shown in Fig. 4. One aspect of the data acquisition system is to measure the water flows through each heat exchanger, the temperature difference across them, and the power input. Hence energy flows are measured as a function of time. Check heat balances are also made. For example, the sum of the heat energy absorbed by the heat pump evaporator and the power input to the compressor should be equal to the energy flow measured across the condenser.

The "heart" of the data acquisition system is an IBM System 7 Computer. It measures some 70 points of data and sums or averages values over a basic six-minute cycle. Data collected include those from a complete weather station on the roof of the building, solar radiation on the plane of the collector, indoor conditions, and all electricity-consuming devices (compressors, pumps, lights, etc.). The frequency of measurement depends on the variable being measured. Solar radiation is measured every half second. Indoor dry-bulb temperature, which is nearly constant, is measured only every six minutes. The data are stored on disk and approximately once per day are transmitted over telephone lines to the Penn State computer facility where they are subjected to the analysis program. Results are computed on an hourly basis from the six-minute data and examined. If deeper scrutiny appears warranted, the six-minute data can be printed out or graphs can be generated by the computer, as in Fig. 5.

The computer printout in the lower left corner shows that the date is January 6, 1975, the time of start is two minutes past midnight, the plot covers a 24-hr period and 6-min data (rather than hourly data) are being plotted.

Curve A1-1 is the solar radiation incident on the plane of the collector in Btuh (sq ft). The maximum value (top line) is 350 (the scale is not shown). It is clear that it was a very sunny day in Albuquerque on January 6, 1975.

Curve A1-25 is the tank entering water temperature and shows the combined effects of heat pump heating and solar energy collection. For example, from midnight to 6 AM, the heat pump is cycling on heat pump heating, and hence removing heat from the water system. Consequently, the leaving tank water temperature (curve A1-13) is gradually decreasing. As solar energy is collected, the tank energy supply is replenished and rises to its maximum at 6 PM.

The net change over the 24-hr period was an increase of 9F in the storage tank temperature, which corresponds to 450,000 Btu of heat energy being put into storage for future use.

Any curve for any of the 70 data points can be generated for any period desired. Such visual pictures of the performance of the system or components are much more valuable and easier to examine than a mass of computed printed numerical values.

Some results for the major system components will now be presented.

In computer modeling heat pumps, one needs the relationship of COP (coefficient of performance) to heat source and sink temperatures. In the case of the water-to-air heat pumps, the sink temperature during winter heating is the occupied space temperature, which is about 70F. Hence the COP is a unique function of system water temperature (source temperature), and computer modeling is easy. COP data were obtained from Penn State laboratory tests of an actual unit. COP values range from 3.2 to 3.7 on heat pump heating and from 3.5 to 2.8 on heat pump cooling over a range of system water temperatures from 60 to 100F.

Modeling of the water-to-water heat pump is more difficult, as both source and sink temperatures may vary. Special tests have just been completed and show a COP range from 4.5 to 3.0 over a range of source and sink temperatures.

Solar collection performance on a clear day is depicted in Fig. 6 on an hourly basis. As indicated, the daily collection efficiency is 44%. This value includes the heat transfer inefficiency in the glycol/water heat exchanger as measurements of flow rate and temperature rise are made on the "pure water" side. The results indicate the collector efficiency is only fair. At the beginning of the project, it was necessary to rebond the copper tubes to the aluminum flat plate, which is a most difficult task. Collector performance calculations indicate the daily efficiency should be near 60% rather than 44%, indicating the possibility of some poor tube-to-plate bonds. This in no way affects the utilization of the results; it simply means that smaller area of modern collectors will be needed for the same applications.

In spite of the above, 721,000 Btu were collected. On this date (January 4), the 24-hr heat requirement of the structure was 990,000 Btu. Hence the collector supplied 73% of the day's requirement on a very cold day (the outdoor temperature reached a low of 7F).

On an overall basis for 100 days of test data between late December and March, the primary source of heat on 51 days was by direct heating and on 49 days heat pump heating. The average COP during the heat pump heating operation was 4.0, including approximately 10% of the total power being consumed by water pumps. Over the 100-day period, the system COP is estimated to be about 7.0, which means that the heating was accomplished with 14% of the electric power consumption of an electric resistance heating system.

Concerning the tank water storage as a system element, problems still exist in obtaining good heat balances across the tank. The piping is such that it is not possible to measure temperatures and temperature gradients in the tank. This problem is still under study.

Regarding the program ahead, during the 1975 summer the system cooling energy requirements will be measured and correlated with computer predictions. Infiltration, which is an important cooling load, will be measured by a tracer gas technique. During the summer, the heating system will be modified, primarily by installing an additional storage tank system which will be located above ground. By so doing, a "hot" tank and a "cool" tank will be available. The former will serve for direct heating and will be allowed to increase in temperature well above the present 110F maximum that is allowed. Because it is very difficult for the water-to-air heat pumps to reject heat into sinks hotter than 110F, the "cool" tank will serve as source or sink for the six heat pumps. The overall system efficiency is expected to improve markedly with the dual-tank system.

Winter heating data will again be collected and correlated with the computer model; and the development of the *System Design Manual* will be completed thereafter.

DISCUSSION

Q Why did you choose an ethylene glycol heat exchanger rather than draining the system in cold weather?

A With serpentine water coils, it is very difficult to completely drain all of the water. Trapped water will burst the tubes when it freezes. Also, the drained water must be replaced by air or some other gas. This other gas is usually nitrogen and that type of system adds substantial cost. With water, the problem is to refill the system and remove all of the air. If this is not done, air in the collector system can substantially reduce the heat transfer effectiveness.

Q What was the extent and cost of maintenance required on this 19-year-old system in order to bring it back to proper operating condition? Can you predict from this the 20-year-old life maintenance costs for this system?

A The solar collector was not in use for several years and the bond between the copper-tube and the roll-bond aluminum sheet had deteriorated substantially. This is not a typical present-day collector design. The water-to-water heat pump, water pumps, storage tank, air handling unit, and evaporative water cooler have operated without major component replacement for the entire period of 19 years. The maintenance cost has been surprisingly low.

Q During your presentation, your maximum COP on heating was 3.75 but you stated your 49 day data gave 4.0. Why the discrepancy?

A The COP of 3.75 peak was for the water-to-air heat pumps whereas the 4.0 COP was for the main water-to-water heat pump. Since the former reverses the refrigerant flow, the evaporator and condenser must provide a dual function; i.e., operate as a condenser in one mode and as an evaporator in the other mode. In contrast, with a water-to-water heat pump the water is switched; hence the condenser always operates as a condenser, and the same for the evaporator. Since the heat exchangers can be optimized, one can expect a higher COP for a water-to-water heat pump at the same source/sink temperatures.

Fig. 1. The solar building.

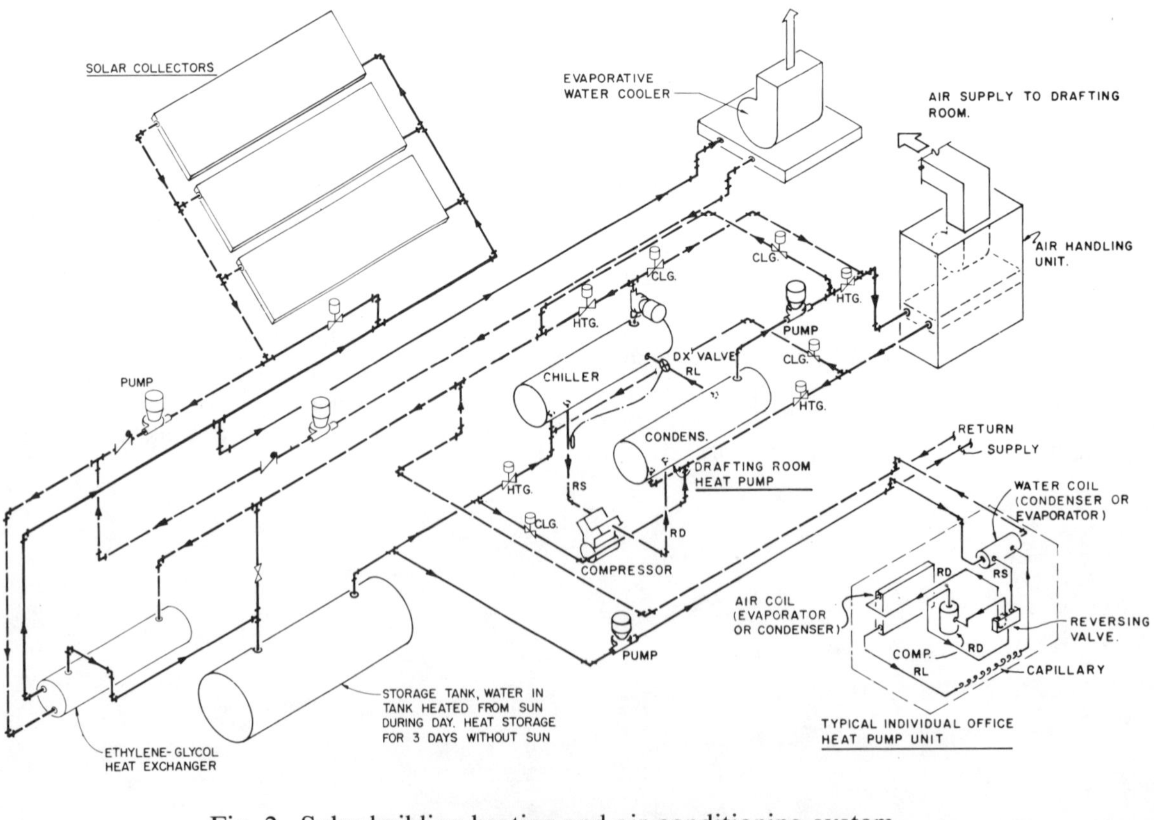

Fig. 2. Solar building heating and air conditioning system.

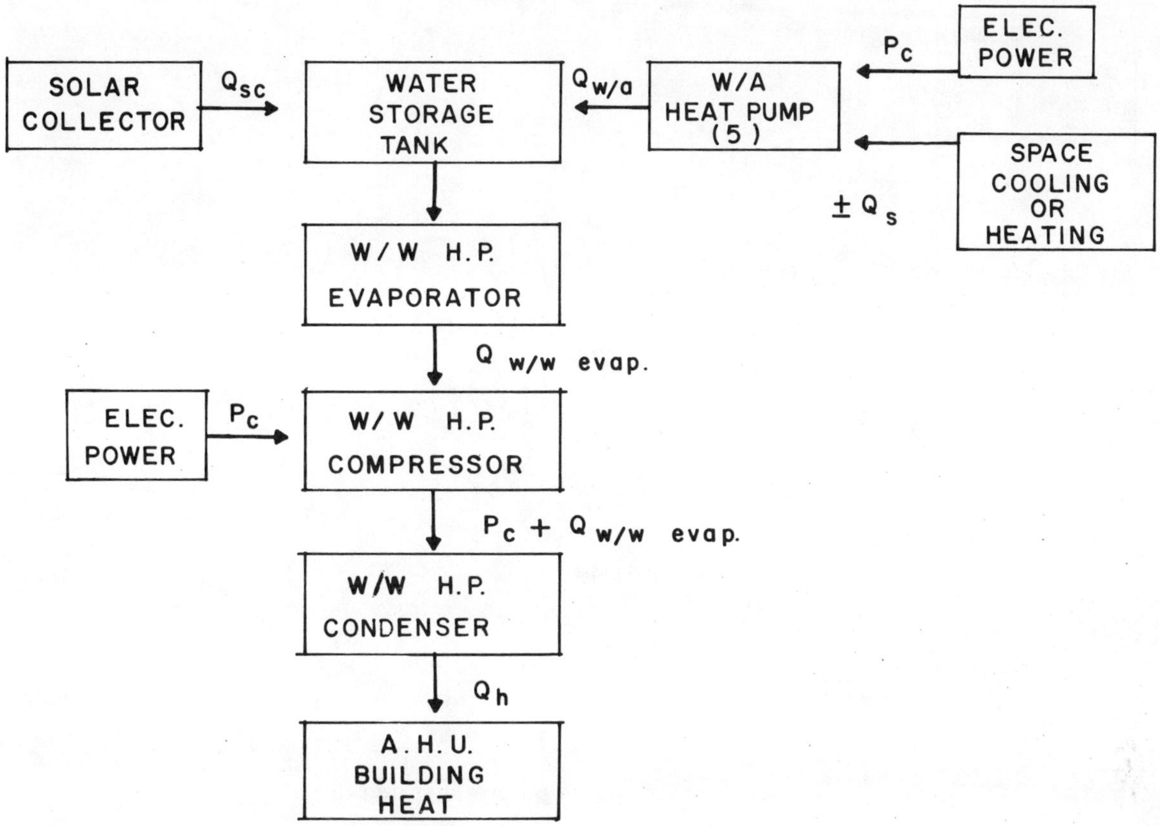

Fig. 3. Energy flows—winter heat pump heating.

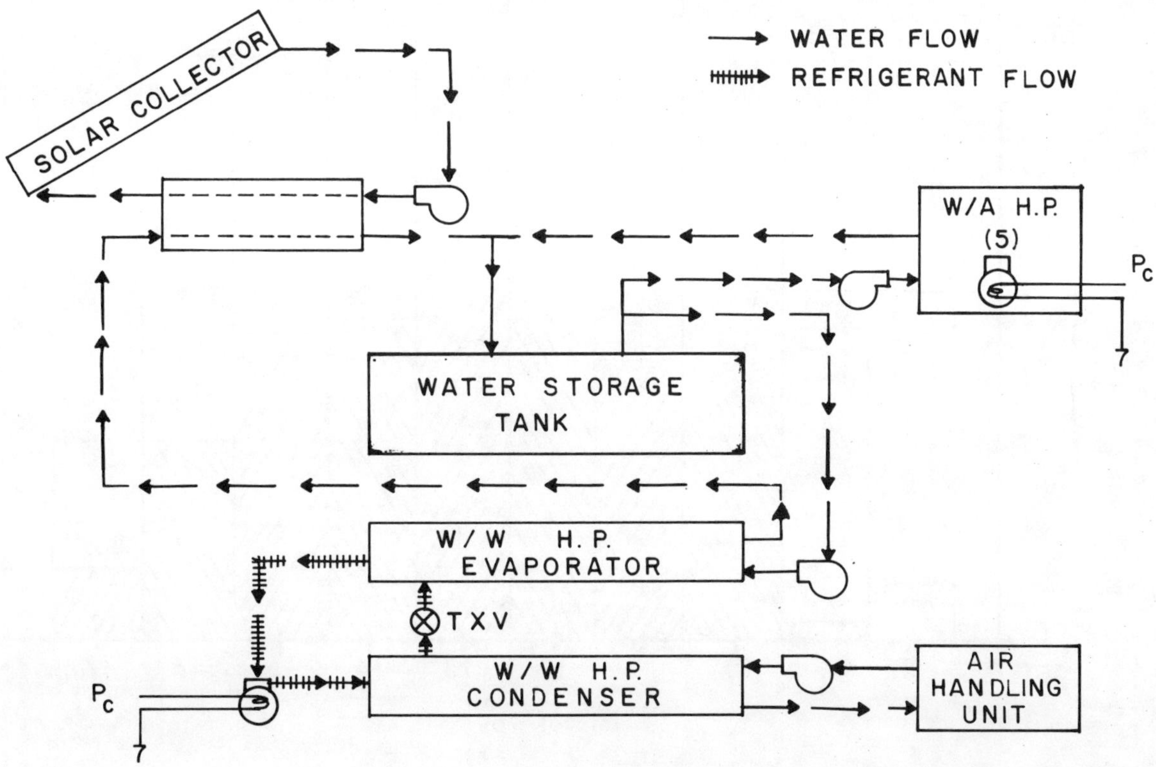

Fig. 4. System diagram—winter heat pump heating.

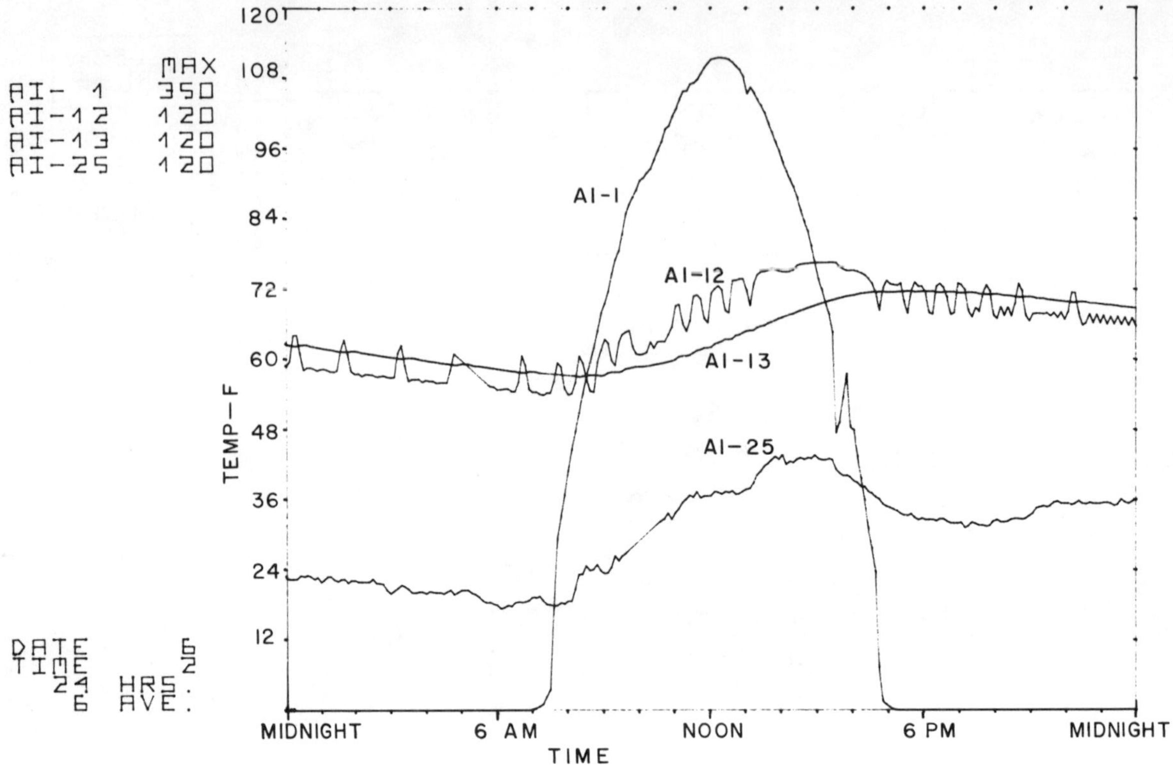

Fig. 5. Typical computer-generated data.

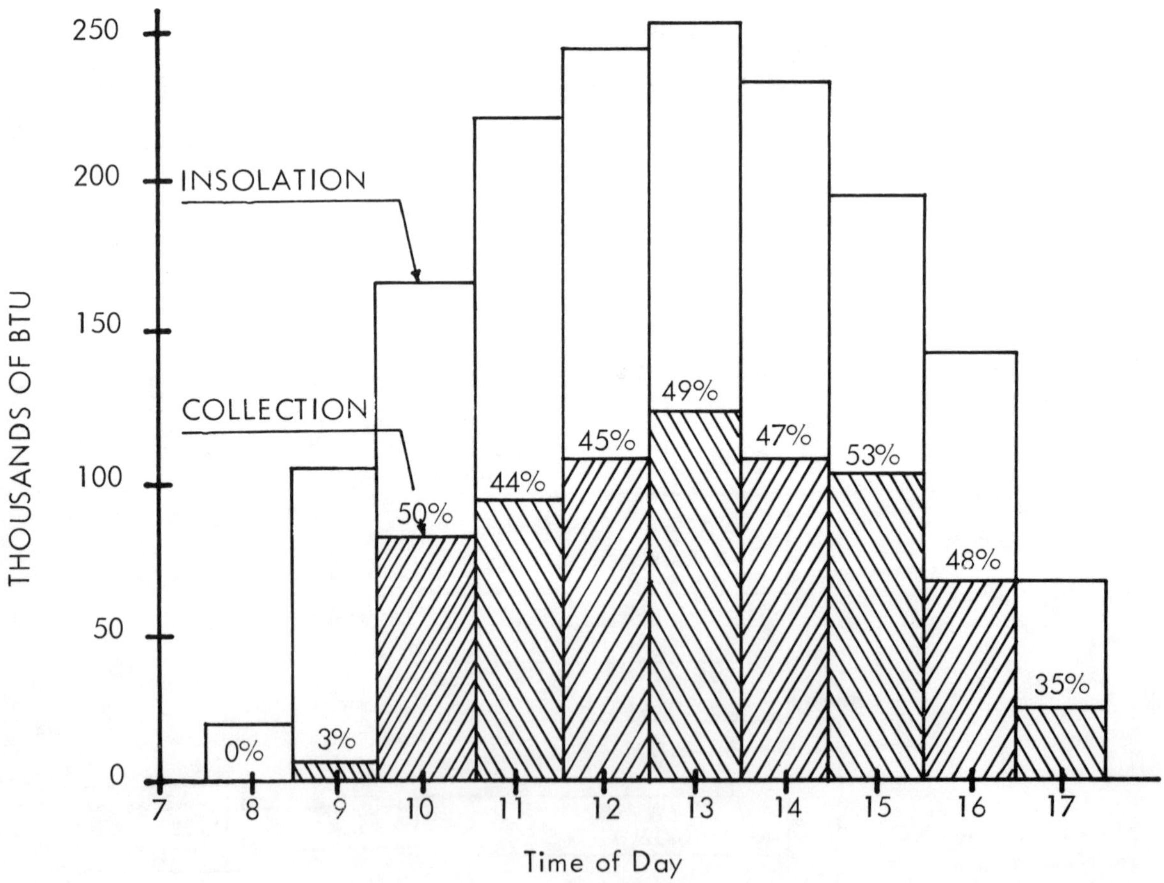

TOTAL INSOLATION - 1,637,000 BTU
TOTAL COLLECTION - 721,000 BTU DAILY EFFICIENCY - 44.0%

Fig. 6. Solar collector performance on January 4, 1975.

The Performance of a Residential Solar Assisted Heat Pump System

Douglas M. Jardine

President, Douglas Jardine, Inc.
Colorado Springs, Colo.

This paper describes the design philosophy and performance characteristics of an experimental residential solar assisted heat pump system.

The purpose of the paper is to communicate the data gained from the design and operation of the system to interested parties. Sufficient information is presented to create an understanding of the heating load characteristics of the system, and the system's response to the load. Further, the goal of the system to provide heat energy for the house with a minimum of peak demand on the utility system is pointed out. The importance of a heat storage system in optimizing the operation of a heat pump during cold weather is developed. Coverage is given to the results achieved to date in an ongoing research project. Collector performance, storage system characteristics, and heat pump performance along with demand on the utility system are the key elements of the paper.

The research results to date point to the extreme importance of the storage system for its near infinite ability to optimize the heat pump. The impact of this type of system operation on an electric utility, as revealed to date, is addressed.

Building Details and Environment

The solar house considered in this report is designed to accommodate a middle income family of five. The house utilizes a flat-plate pressurized fluid solar collector, a large variable volume uninsulated underground water storage tank, and a heat pump for space heating and cooling.

The house is a two-level, 2800 sq ft dwelling. It is designed to act as a passive energy conserver. The house is oriented due south with two banks of collectors fixed at a 55° angle. This angle puts the collector plane perpendicular to the sun November 21 and January 21, which maximizes the use of solar radiation for winter heating purposes. In addition the roof is covered with white quartz to reflect sunlight falling on it into the upper bank of collectors. There is a total of 780 sq ft of collectors on the house, evenly divided into two arrays.

The window placement and overhangs are situated in such a manner as to allow sunlight to enter the house during the winter months and exclude it in the summer. The garage is situated on the northwest side of the house. This acts as a buffer to protect the living area from the winter storms which weather statistics indicate come from the northwest. All the windows are double glazed and the house meets the insulation requirements of the new Colorado Springs City Building Codes (R11 in the walls; R15 in the ceiling).

The heating system for the house is classed as a solar-boosted heat pump system. The objectives of the system are to heat the house with solar energy to a small extent, to utilize an optimized heat pump for a backup system, and to avoid adding large loads to the local utility during times of peak demand.

The house is located in the northeastern portion of Colorado Springs, Colorado, and the elevation of the first floor is 6730 ft above sea level.

The house has a confirmed heat loss of 55,000 Btuh at an outdoor temperature of −9F with a 7 mph wind. This heat loss corresponds to a 24-hr heating load per degree-day (65F base) of 17,838 Btu/deg.

The house has a design peak air conditioning load of 42,000 Btuh.

The average monthly and yearly degree days (65F base) for Colorado Springs are as follows (ref. 1):

July	9
August	25
September	132
October	456
November	825
December	1,032
January	1,128
February	938
March	893
April	582
May	319
June	84
Total	6,423
Average winter temperature	37.3F

System Design

Figure 1 shows the components of the system, and the operation of the system when solar energy is being collected and transferred to storage as heat energy. A heat transfer fluid is utilized in the closed collector-to-storage piping system. The fluid boils at 575F and freezes at −100F, is noncorrosive, nonelectrolytic, and does not decompose. This fluid is used in order to avoid boiling, freezing, and corrosion in the system. The storage tank is located underground, is uninsulated, and contains water.

Whenever the air temperature in the collector is higher than the water temperature in storage, pump A operates and transfers heat energy from the collector to the storage system. When the storage temperature nears the boiling point, pump A cannot operate. Heat energy is stored both in the container water and the ground around the storage tank.

Figure 2 displays how the stored energy is directly utilized to heat the house, and this operation is called MODE A. When the house thermostat calls for heat, pump B circulates hot water from storage through the coil in the house air duct, while the house air fan in the heat pump circulates air through the house. This mode always operates when the storage system temperature is above 100F and the house thermostat calls for heat.

Figure 3 demonstrates the system operating mode which utilizes the heat pump for heating the house. The heat pump operates when the house thermostat calls for heat, and the storage system is below 100F. The object is to have the heat pump always operate with an ideal evaporator temperature, or to optimize the heat pump. In this system, the heat pump utilizes outdoor air as a heat sink when the outdoor temperature is above 45F. When the outdoor air temperature is below 45F, the storage system is utilized as a heat sink for the heat pump. Actually an attempt is made to simulate a fluid-to-air heat pump in this case. When the heat pump operates using the storage system as a heat sink, the three-way valve in the collector-to-storage piping system diverts the flow of heat transfer fluid, so that it is circulated between the heat exchanger in the storage system and coil B on the evaporator side of the heat pump, thus simulating a fluid-to-air heat pump connected to the storage system. Dampers are closed in the outdoor air duct during this operation.

The heat pump is equipped with a house air stream electric resistance heater, and the storage system has two electric immersion heaters in it. These resistance heaters are controlled automatically, but are placed in the control circuit manually, with disconnect switches as desired by operating test procedures.

System Operating History

The system went operational in late September 1974, and ample heat was stored to heat the house by using the operational mode shown in Fig. 2 during the fall. In early November 1974, the decision was made to utilize the resistance heaters in the storage system to accomplish a large storage buildup of heat energy with 7000 gal of water in storage. On November 29, 1974, with a storage temperature at 165F, the storage heaters and collector pump were shut off, and the stored heat was used to heat the house utilizing the operating scheme shown in Fig. 2. The storage system heated the house this way until December 15, 1974, when with the storage temperature at 94F the system no longer could maintain a 72F temperature in the house. The average outdoor air temperature during this period was 30F. This cannot be considered a valid test for several reasons. Reliance tests were then run on each operating mode until January 10, 1975, when a family was moved into the house, and the collector system was then started and automatic control was re-established in the system operation. The volume of water in storage was reduced to 1520 gal and then changed to 2000 gal in order to reduce the thermal inertia of the storage system, and it remains at that volume. Since that time the system has operated in the heat pump mode under automatic control.

The data acquisition equipment was installed and went operational at 12:34 PM on January 26, 1975, and data have been collected and analyzed continually since that time.

System Performance

The collectors have performed in a very satisfactory manner. The upper, or reflective boosted, collector array has continually collected 30% more energy than the lower collector array. Table 1 presents the actual collector performance, which accounts for collector efficiency, atmospheric conditions, and reflective input.

TABLE 1. AVERAGE DAILY HEAT ENERGY COLLECTED AND STORED (Btu/sq ft of collector/day)

Month	Year	Upper Array	Lower Array	Average Total	Upper Array Yield — % above Lower Array
January*	1975	255	193	224	32.12
February	1975	327	292	292	27.13
March	1975	296	228	262	28.82
April	1975	319	238	279	34.03
May	1975	241	192	216	25.52

*Based on six days data, commencing January 26, 1975.

The heat transfer fluid has proven very satisfactory and performed the job it was intended to do. No problem has arisen with pump seals or valve packing. Rubber flexible connectors between the collector plates and the piping headers have caused minor problems and will be replaced with solid metal connections this summer. A collector plate will be sent to the laboratory this summer for intense corrosion studies. No corrosive action is apparent in the field at this time. Thermal syphoning from the storage system to the collector is a problem, and a flow control valve is necessary to preclude this happening.

Analysis of the operation of the storage system is an interesting challenge. More complex instrumentation is needed to fully analyze the storage system. The storage system has a large thermal inertia, so it takes a long time to heat it up. The storage system was used extensively from January through May 1975 to act as a heat sink in optimizing the heat pump. Figure 4 compares the net energy of the storage system to the stored water temperature while the storage system is being used to optimize the heat pump. This figure demonstrates how the inertia of the storage system becomes greatly beneficial in optimizing the heat pump, and its long range capacity to perform this function due to the ground acting as an additional heat sink.

The passive design of the structure—the sizing, location, and overhang arrangement of the windows—has proven very satisfactory. The sun entering the windows in January has demonstrated its ability to completely heat the house from 10:00 AM to 3:30 PM with outdoor air temperatures as low as 0F.

Figure 5 shows the performance characteristics of the heat pump versus evaporator temperature, and the heat loss of the house versus outdoor air temperature. The ground temperature at the storage system remains nearly constant at 56F. Defrost cycling has been identified as a waste of energy.

This research project is conducted with the City of Colorado Springs Department of Public Utilities. As the technical research has progressed, parallel studies and research have been conducted in the area of solar rights, zoning, building codes, economics, and consumer acceptance. Particular attention has been directed toward the impact of solar heating systems on an electric utility in the area of peak demands, and the resultant capital requirements for equipment, the connection fee, and rate increase required to pay for the plant expansion.

Two important problem areas have been identified and policy is in formulation stage at this time. First, the need for collector and storage system ratings have been realized, and the Regional Building Department has adopted the policy that no building permits will be issued for collectors or storage systems that do not have performance data produced by an independent laboratory conducted in accordance with the National Bureau of Standards rating procedures.

Second, and of profound importance, it has been recognized that solar heating systems with electric powered resistance-type backup systems could have a serious effect on the demand the utility sees, and drastically reduce the utility's plant utilization factor. This condition has to be controlled, or capital demand and consumer electric rates will have to skyrocket to pay for the higher peak demand with lower plant utilization. Three options exist to control or minimize the impact of this problem:

1. Control the electric backup system so that it could not operate during times of peak load, both on a daily and a seasonal peak demand basis, or an extremely high charge could be made for electricity used for backup system operation during peak hours. This charge clearly would have to be large enough to amortize the peaking equipment costs, and be paid for by the customer having the unbalanced load. The customer should be made aware of the costs involved when he is considering installing his system.

2. Satisfy backup system power demands with individual wind-driven generators. This application is not recognized as a near term solution because the cost of wind-driven generators is in the $1500.00–3000.00/kw capacity range. The need for zoning ordinance revision has to be analyzed, and the social acceptability of residential wind generator units would have to be established.

3. Utilize the optimized heat pump with the solar collector system, providing a moderate amount of heat energy to the storage system to preheat domestic hot water and balance the heat energy required by the heat pump. This type of system has many attractive

points. A minimum collector area would be required. The sizing of the storage system is determined by heat transfer area requirements between the storage vessel and the ground, and not by sensible heat storage volume requirements. The electric utility will see an acceptable load profile, and the consumer will receive air conditioning as well as heating from his investment. The cold storage systems could be utilized to transfer air conditioning compressor demand to off-peak hours in the summer if necessary.

Table 2 shows the projected heating energy resource usage for the service area of the Electric Division of the Department of Public Utilities, The City of Colorado Springs. This table allows for the depletion of natural gas taps made available through the construction of two large propane peak shaving plants in 1974, and a utilization of all electric resistance heating and solar heating after the gas taps are depleted.

TABLE 2. PROJECTED CUSTOMER HEATING ENERGY RESOURCE USAGE, CITY OF COLORADO SPRINGS, DEPARTMENT OF PUBLIC UTILITIES (ref. 2).

	All Electric	Solar Heat	Other	Total Residential Customers	Growth (%)
1975	—	—	73,000	73,000	5.37
1976	—	—	76,300	76,300	4.52
1977	—	—	79,900	79,900	4.72
1978	3,500	—	80,100	83,600	4.63
1979	7,000	—	80,400	87,400	4.55
1980	10,400	500	80,500	91,400	4.58
1981	12,500	1,500	80,500	94,500	3.40
1982	14,200	3,000	80,500	97,700	3.40
1983	15,000	5,500	80,500	101,000	3.38
1984	15,000	9,000	80,500	104,500	3.47
1985	15,000	12,600	80,500	108,100	3.40

1966–74 growth rate 6.7%; 1975-85 projected growth rate 4.00%.

Table 3 summarizes the effect of substituting optimized heat pumps with a 3 to 1 coefficient of performance for the all electric resistance heating additions.

By utilizing the optimized heat pump in lieu of the electric resistance heating a saving in new generating plant capacity of 88,000 kw would be achieved. With an anticipated plant construction cost of $530.00/kw generating capacity, a saving of $46,640,000.00 would be realized, and with a utilization factor of 0.60 and a generation plant cycle efficiency of 36%, a saving in coal consumption of 755,352 tons would be the result during the period of 1978 through 1985.

TABLE 3. COMPARISON OF PEAK GENERATING CAPACITY REQUIRED FOR ALL OPTIMIZED HEAT PUMP HEATING LOAD ADDITIONS VERSUS ALL ELECTRIC RESISTANCE HEATING LOAD ADDITIONS FROM 1978 THROUGH 1985

Year	Kilowatts for All Resistance Heating Loads (ref. 2)	Kilowatts for All Optimized Heat Pumps*	Kilowatts Generating Capacity Saved with Optimized Heat Pumps
1978	50,000	22,000	28,000
1979	93,000	52,000	41,000
1980	140,000	80,000	60,000
1981	170,000	96,000	74,000
1982	200,000	120,000	80,000
1983	228,000	140,000	88,000
1984	254,000	170,000	84,000
1985	275,000	192,000	83,000

*Optimized heat pumps with a 3 to 1 coefficient of performance.

Conclusions

The operation of the project to date has revealed trends that will set the direction of the research effort in the future, due to the following conclusions:

1. Collector performance can be increased significantly by utilizing a commonly used building material such as white quartz gravel on roofs as a reflective material.
2. The addition of solar heating systems backup systems load can create an unfavorable condition for the utility and its customers, if the backup system operates during the utility's traditional peak load hours, days, or seasons. In effect, the utility will have to provide for meeting peak loads but will not enjoy the annual sales required to produce the revenue required to amortize plant investments without large and specific rate or connection fee relief for these systems. Long term storage development is one solution to this problem, and the development of this technology is hoped for by 1983.
3. The use of a noncorrosive, nonelectrolytic heat transfer fluid that won't freeze or boil under the temperature experienced in the system is an absolute necessity.
4. The optimized heat pump presents the most attractive alternative of all the systems being tested on this project.
5. A fluid-to-air heat pump suitable for use as an optimized heat pump, as required under the conditions identified in this research, is not commercially available. However, with suitable fluid evaporators currently available, heat pumps should perform very satisfactorily. A fluid-to-air heat pump will be installed for conducting research through next winter.

The fluid-to-air heat pump will eliminate the need for a defrost cycle with the heat pump in the heating mode.

6. An underground water storage system with a heat energy input from solar collectors has demonstrated its ability for long term support of the optimization of the heat pump. The sensible and latent heat in the ground around the storage system acts as a supplementary heat source and storage medium at the temperatures required to optimize the heat pump. Perhaps by utilizing thermal stratification in the storage system, domestic hot water could be preheated or totally heated by the thermal syphoning process through a heat exchanger in the storage system.

7. Utilizing solar energy to optimize the heat pump only, and not to heat the building directly, will dramatically reduce the first cost of the total system.

8. Employment of the optimized heat pump system in lieu of electric resistance heat as a backup to solar heating systems on a large scale will eliminate extremely high operating costs and capital requirements that will be caused by having electrical demand occur during hourly, daily, and seasonal peaks with no acceptable annual sales volume.

9. If the City of Colorado Springs should adopt the policy that it would only connect optimized heat pump systems instead of electric resistance heating systems, a cumulative market for 92,600 optimized heat pumps is projected to exist from 1978 through 1985, and the city's future electric generating plant size could be reduced by 88,000 kw as a result of the utilization of the optimized heat pump.

ACKNOWLEDGMENTS

National Science Foundation, Research Applied to National Needs Program Technical Research Grant: $125,000.00.

National Science Foundation, Office of Intergovernmental Science and Research Utilization. Social Acceptance, Solar Rights, and Economic Research Grant: $10,000.00.

The First National Bank of Colorado Springs Construction Loan: $81,500.00.

The City of Colorado Springs, Department of Public Utilities, James D. Phillips, Director, Principal Investigator.

REFERENCES

1. American Society for Heating, Refrigerating, and Air Conditioning Engineers, *Systems Handbook, 1973*, p. 43.3.
2. Stone & Webster Management Co. Study and Report for the Electric Division, Department of Public Utilities, City of Colorado Springs, May 1975. James D. Phillips, Director.

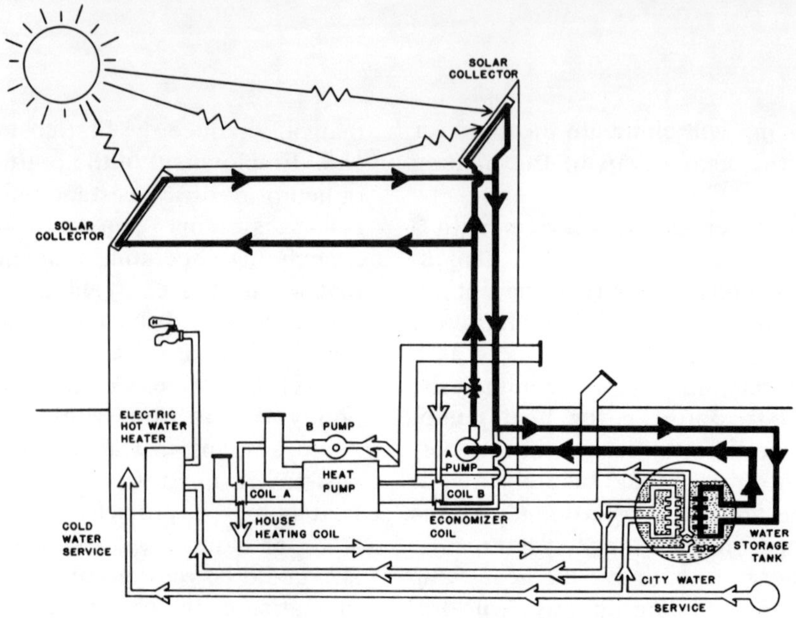

Fig. 1. Collection and storage of heat energy.

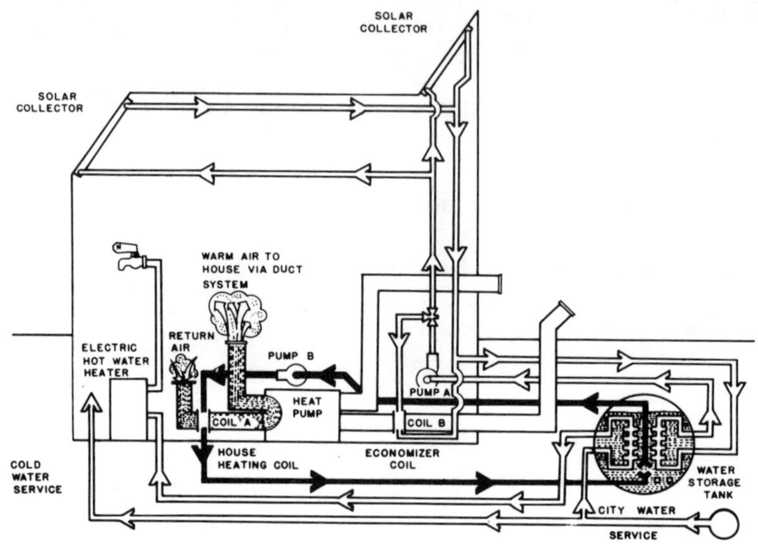

Fig. 2. Direct use of stored heat energy for space heating.

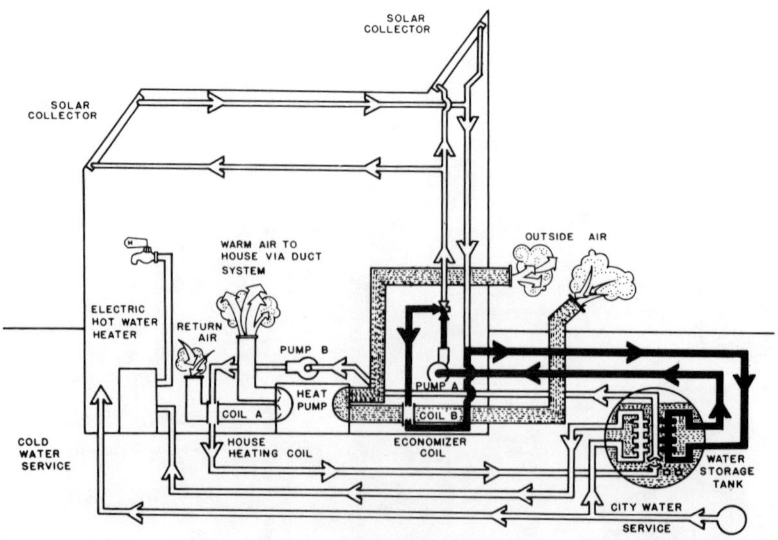

Fig. 3. Use of the optimized heat pump for space heating.

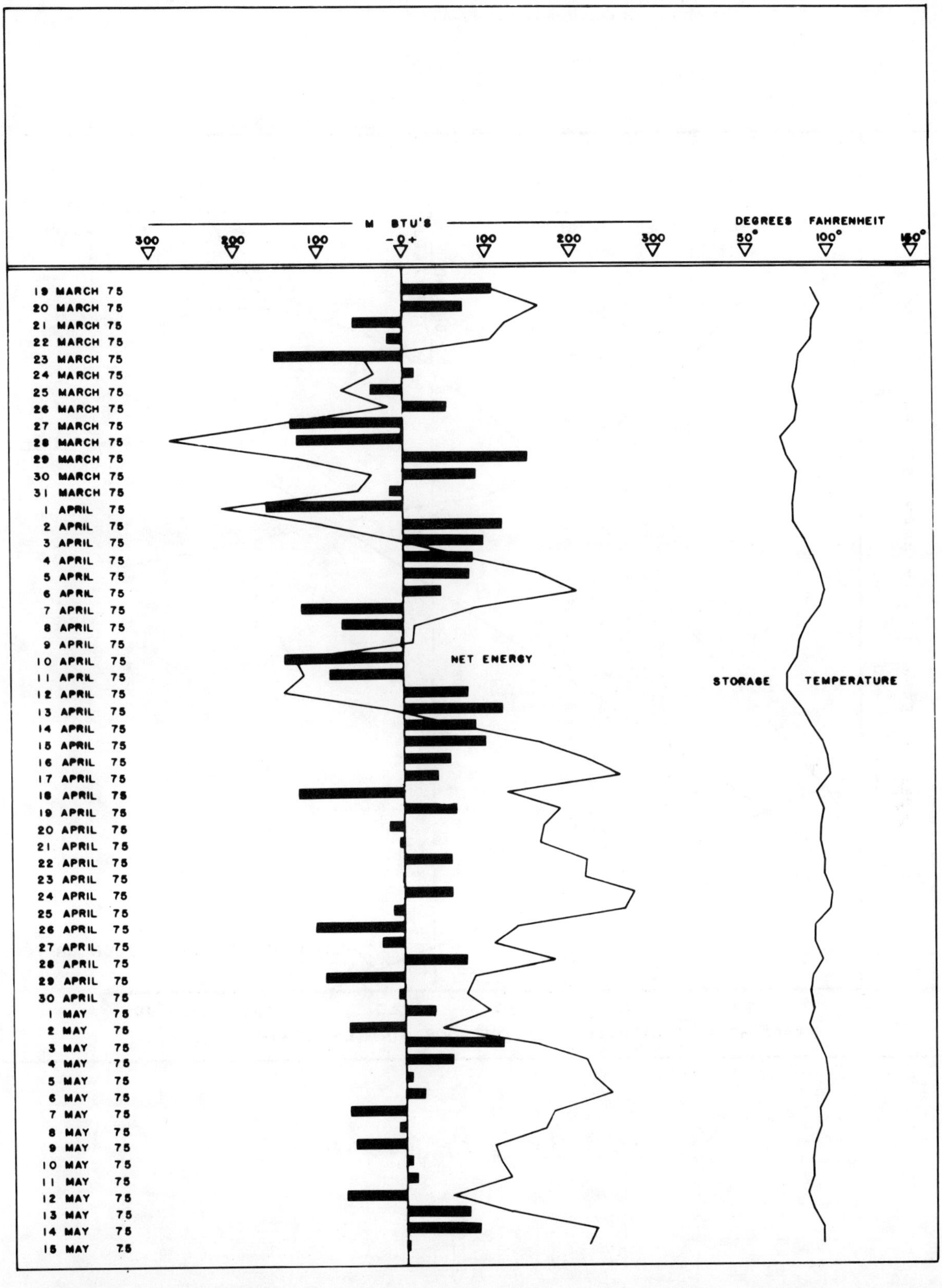

Fig. 4. Storage system net energy and temperature.

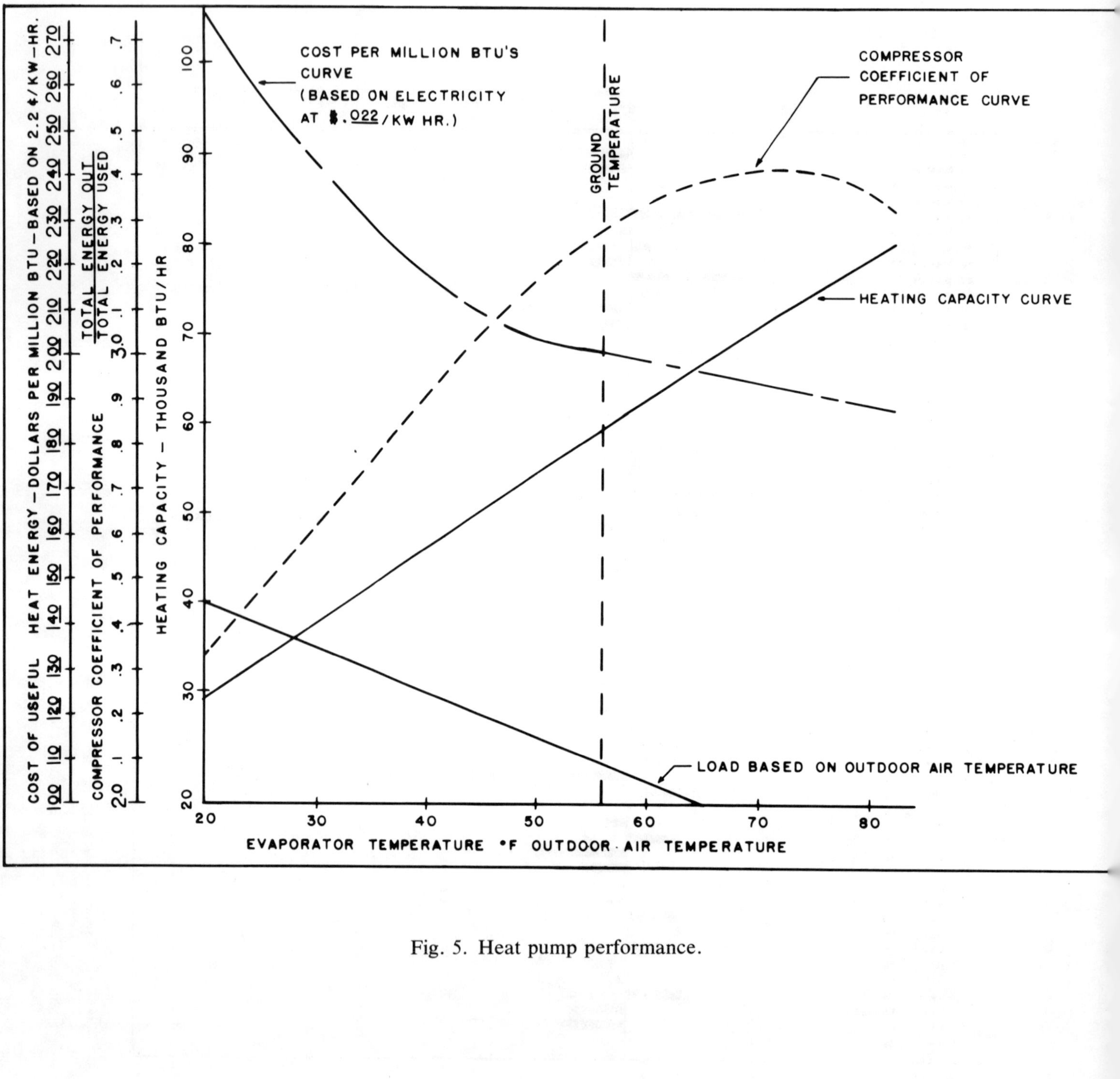

Fig. 5. Heat pump performance.

Commercial and Institutional Building Solar Heat Pump Systems

Eugene E. Drucker

Professor of Mechanical Engineering
Syracuse University
Syracuse, N.Y.

Syracuse University has conducted a detailed parametric study of solar assisted closed loop heat pump systems for heating moderate sized commercial and institutional buildings. We have operated under an NSF/RANN grant, with two subcontracting organizations: Vosbeck, Vosbeck, Kendrick and Redinger (VVKR) of Alexandria, Virginia (primarily Ved Bansal, Chief Mechanical Engineer) and the Climate Control Division of the Singer Corp.

Our simulation is based on the use of commercially available water-to-air heat pumps in the 2- to 5-ton range, manufactured by about eight different American manufacturers, and single glazed copper flat-plate collectors. Performance computations have been made for two buildings in each of three cities: Alexandria, Minneapolis, and Boston. One building is a 73,500 sq ft elementary school, modeled after the Tyler School in Alexandria, which is now equipped with a closed loop heat pump system. The second building is a five-story office building of 100,000 sq ft, also modeled after a typical existing structure.

Finally, economic analyses have been carried out on the various systems to determine the relative return on the invested capital required to build the solar components. We have studied the effects of initial cost, interest rate, energy cost escalation rate, tax rate, and maintenance cost on the overall economic picture.

Description

The schematic operation of a closed loop heat pump system is illustrated in Fig. 1. Its basic advantage with solar heat collection is that it is inherently an energy conserving system, since the energy removed from interior spaces can be used to heat the perimeter spaces of the building. Additional energy is conventionally supplied by fossil fuel or by electric resistance heating. The objective of the solar assisted system is to eliminate the need for those forms of energy for heating purposes.

There are several other advantageous features of the solar assisted heat pump system:

1. Unlike direct solar heating systems which require water temperatures of at least 120F, heat pumps can deliver heat from sources as low as 50F. Since the efficiency of solar energy collection increases as the average collector temperature decreases, high collector efficiencies can be obtained.
2. Because of this low temperature operation, single glazed collectors are adequate and less expensive than double or triple glazed collectors.
3. The energy storage capacity of a tank of water is directly proportional to the workable span of temperature. Water source heat pumps can operate with water from 50 to 120F, or a 70F span. Direct heating systems would have to work between 120 and 190F for the same storage capacity, entailing less efficient collection.
4. Domestic hot water can be produced by utilizing a water-to-water heat pump with the basic water loop as a source.
5. The solar collectors can be used as a dry cooling tower to reject excess heat in intermediate seasons, and possibly as a complete replacement for conventional cooling towers even in the summer in cooler climates.

The methodology of our research is essentially a

parametric study of a solar heat pump system, using mathematical modeling and a high speed computer. The basic model is similar for both buildings that we are studying and is conceptually simple. It is shown schematically in Fig. 2. A single liquid loop is used, corresponding to a system which would drain the collector when necessary. A copper collector is assumed, and published data for commercially available units were used.

A generalized computer program was written in FORTRAN IV. It consists of a main program and the following series of subprograms, which simulates individual components of the analysis:

1. Predict outside heat transfer coefficients.
2. Calculate SOL-AIR temperatures.
3. Predict solar radiation constants.
4. Predict solar radiation intensities.
5. Calculate steady-state heat transmission through walls.
6. Simulate the auxiliary resistance heaters.
7. Simulate heat pumps in the cooling mode.
8. Simulate heat pumps in the heating mode.
9. Simulate the thermal storage tank with water or any liquid as the storage medium.
10. Simulate solar collector behavior.
11. Calculate total heat gain through the windows.
12. Simulate domestic hot water production with water-to-water heat pump.

The basic characteristics of the program are as follows:

1. The calculational time increment is one hour. Heat loss and heat gain (load) calculations are based on steady-state heat transfer calculations.
2. Solar radiation gains by the building walls are calculated using the Sol-Air temperature concept, as outlined in the *ASHRAE Handbook of Fundamentals*.
3. Flexible schedules for ventilation, lights, and occupancy are incorporated into the program and can be adjusted for any particular building.
4. Night setback (winter) or setup (summer) is used for simulation of interior temperature. A logic routine is also used to prevent the use of cooling immediately after winter night setback to 60F.
5. Smoothed long term weather data are used, with diurnal sinusoidal variation superimposed on a best-fit annual sinusoid. Average cloud cover computed from actual weather data is also used.
6. A load calculation is performed for each zone each hour. If there is a requirement for heat, the heat pump is run for a sufficient time to provide the necessary heat quantity. A similar procedure is performed for cooling.

Results have been obtained for the school model based on the Tyler Elementary School in Alexandria, which is a two-story structure of about 75,000 sq ft total area. The school was chosen for several reasons: (*a*) it is typical of a large number of recently built, moderate sized schools, as far as construction, occupancy, glass areas, etc., are concerned; (*b*) it now contains a closed loop heat pump system, so that it is a candidate for a subsequent solar system addition; (*c*) data on electrical energy consumption in the school are available for the past several years, so that the mathematical model can be tested against actual use; and (*d*) VVKR, a member of this research team, had designed and supervised installation of the mechanical equipment of the school, and was therefore very familiar with the design and operational details of the system.

In furtherance of item *c* (above), a preliminary set of calculations was made for the school and is compared to actual electrical usage in the school for the fiscal year July 1973 through June 1974 (Table 1). The calculations do not include domestic water production, electrical energy for kitchen use, and certain other electrical usages. In some months the actual kwh usage indicated may have been determined by an estimate rather than by an actual meter reading.

TABLE 1. ELECTRICAL USAGE (kwh)

Month	Actual	Computed
1973		
July	165,000	
August	156,000	
September	185,000	101,205
October	156,000	77,697
November	149,000	97,579
December	200,000	147,175
1974		
January	168,000	168,523
February	173,000	135,740
March	132,000	103,998
April	118,000	74,735
May	140,000	
June	145,000	

Considering the various differences between actual building and model at this time, it is felt that the correspondence between actual and computed kwh usage is reasonably good, and that the computational model fairly represents the building.

Some of the results for the schools in Alexandria and Minneapolis are presented at this time. The results for Boston are generally intermediate between the characteristically warm mid-Atlantic and the cold midwestern climates. The collector area and the water storage volume were varied, and the resulting reduction in total electrical consumption (excluding lights) is shown in Fig. 3 for Alexandria. Also plotted is the amount of auxiliary electric heat used. Figure 4 is a plot of total electrical energy for the school in

Minneapolis. It can be seen that for the large collector, the incremental advantage of larger storage volumes is considerably less than it is for the small collector.

The effects of varying the collector inclination and orientation were tested. It was found that for Alexandria, the seasonal electrical consumption was minimum for a collector facing due south and inclined at 62° from the horizontal (latitude +23.5°). For other angles and directions, the variations in heat collected, auxiliary heat, energy for heat pump cooling and heat pump heating were moderate, but the total electrical consumption was almost constant. In the case of Minneapolis, the due south facing collector was best, inclined at 60° rather than latitude +23.5°.

Hot water is produced conventionally in many electrohydronic systems by direct electrical heating. The production of domestic hot water using a heat pump in Alexandria saved 10% of the total electrical energy for a 5000 sq ft, 25 psf collector-storage combination.

The effect of potential improvements in the coefficients of performance of water-to-air heat pumps was investigated. We compared the results of an arbitrary COP improvement of 1.0 to those of the current generation of heat pumps with COP from 2.5 to 3.0. In Alexandria, the kwh for heat pump heating and cooling decreased and the amounts of solar energy and auxiliary heat used increased, but the total number of kwh decreased by 9 and 11%, respectively, for 5000 and 7000 sq ft collectors. In Minneapolis, however, for a 12,000 sq ft collector, the reduction was only 3.5%. In the warm climates, electrical energy for cooling is saved in large amounts. In colder climates, larger collectors must be used in conjunction with improved heat pumps; otherwise the energy budget is balanced by using more resistance heating.

Because of the large amount of computer time required for hourly calculations with a large number of zones, most program runs were based on calculations for two days per month. To determine the effect of the number of days per month calculation, we compared a two-day and a three-day per month run for a 12,000 sq ft collector system in Minneapolis. The total electric consumption was 807,000 and 801,000 kwh, respectively. For an Alexandria-based 7000 sq ft collector system, the corresponding figures were 419,000 and 401,000, showing again that the two-day per month calculation was conservative and that longer runs give results which are more favorable for the solar system.

The results of the performance calculations for the various solar system configurations were analyzed from an economic point of view.

The objective of the economic analysis was to examine the life cycle costs of solar systems with various combinations of collector size, storage volume, building location, and other pertinent parameters which have been varied in the main performance program. The analysis includes a breakdown of cost into capital and operating components, and comparisons of total costs for solar assisted and nonsolar assisted heat pump systems.

A set of algorithms was written for digital computation of the economic parameters of solar assisted heat pump systems. The economic analysis is programmed to calculate cash flow differentials as a basis for investment decision alternatives. The analysis compares a "typical" closed loop heat pump system to a closed loop heat pump system which utilizes solar energy.

The typical closed loop heat pump system is utilized as the economic "base." The base system does not vary in construction specifications. The solar energy system varies in technological, climatic, monetary, market, and other parameters which may induce minimum capital expenditure, maximum return on investment, and minimum payback periods.

The economic analysis of the systems is being conducted for monetary investment of both taxable and nontaxable entities.

Capital costs were based on recently obtained bid prices. It was assumed that the capital cost differential between electrohydronic and solar assisted systems varied between $17.75 and $20.25/sq ft of collector, depending on the size of the collector and the storage volume. Energy escalation rates, from a base of 3.5¢/kwh, were taken to be 6%/yr, or the GSA schedule (doubling in 5 years and increasing by 5%/yr thereafter).

Figures 5 and 6 present some of the results for the 75,000 sq ft schools in Alexandria and Minneapolis. It is obvious that the smaller collectors are much more viable than the larger ones, in almost straight line proportion. It is also clear that for the smaller collectors, it is economically advantageous to install large ratios of storage volume per square foot. This economic advantage disappears for large collectors, since at all ratios, the absolute storage volume becomes large. Figures 5 and 6 are based on data which do not include domestic hot water production by heat pumping. Inclusion of this factor will increase the economic competitiveness of the solar system.

Some other parameter variations have been considered. The effect of reducing the money rate to 6% is not very dramatic. The effect of a reduced capital cost (to $15/sq ft) is a little larger. For example, in Minneapolis, with a GSA energy cost schedule, reducing the capital cost from $19 to $15/sq ft of a moderate sized collector increases the return on initial investment from 5.7 to 7.8% for taxable entities, and from 10.6 to 13.5% for nontaxable institutions. Combinations of lower money rates and lower capital cost obviously will increase rates of return even more.

Conclusions

For the configurations investigated, the solar collector system in conjunction with conventional closed loop heat pumping can substantially reduce or eliminate electric resistance heating. However, economic life cycle analysis indicates that a limited amount of collector surface with a relatively large storage volume is most cost-beneficial. For any set of variables, the economic viability of the solar addition is much greater for nontaxable than for taxable entities. In terms of location, the attractiveness of solar assisted heat pump systems for heating is greatest in the colder climates, where despite lower collector efficiencies, the total amount of solar energy actually used is maximized.

ACKNOWLEDGMENTS

The financial assistance of NSF/RANN in the conduct of this research is gratefully acknowledged.

Credit for the preparation and execution of the computer program is due to Manas Ucar. The prime responsibility for the economic analyses has been with William Fleming. The assistance of our colleagues at Singer and VVKR is also gratefully acknowledged.

DISCUSSION

Q If your economic analysis were to be applied to a single-family residence, would you see any difference between an owner-occupied residence and a residence used as rental property?

A Yes. The rental property owner could depreciate the cost of the solar system and thus have a tax advantage over the owner-occupier.

Q What temperatures do you obtain by heating domestic water with heat pump? What is the COP in heat pumping domestic hot water?

A Domestic hot water is heated to 120F by the heat pump, then raised to 140F by resistance heat. The COP for the heat pump is about 3.5.

Q How can you justify averaging the weather data? Don't you thereby lose the several-day cloud cycles which require storage and auxiliary usage?

A Averaged weather data give good results for yearly performance calculations. Several-day cloud cycles should be run to determine short-range performance.

Q When will progress and final reports be published? How can they be obtained? Will reports include copies of computer programs?

A A semiannual report dated February 1, 1975, has been issued. A final report will be issued in July or August 1975. A detailed program listing will not be included in these reports. Contact the author for more information.

Q Have you calculated a return on investment for heat pump alone for Minneapolis or Alexandria?

A No.

Q Long term weather averages are not quite equivalent to hour-by-hour computations because both heating demand and solar energy availability depend on simultaneous values of radiation, temperature, and wind. Is the program listed available?

A Program listing is not yet available.

Q The average monthly cloud cover does not agree with figures published by ASHRAE. Who is correct?

A There is agreement if it is recognized that the ASHRAE percentages of possible insolation refer to insolation at the outer edge of the earth's atmospheric envelope.

Q What percentage of collector to building area is considered a borderline economically?

A The answer depends on building size and location. For a solar assisted heat pump system, roughly 10–20% of building area should do.

Q Did your performance studies account for variable heat loss from storage as a function of storage volume?

A No, this was ignored as it was considered negligible.

Q Are all your data based on one day per month or two days per month? Does your effect of azimuth on collector performance include the effect of ambient temperature variation?

A The results are mainly based on three days per month calculations. The effects of azimuth and ambient temperature variations were considered independently.

Q In your schools, do you recover heat from exhausted air?

A The results shown did not include exhaust air heat recovery.

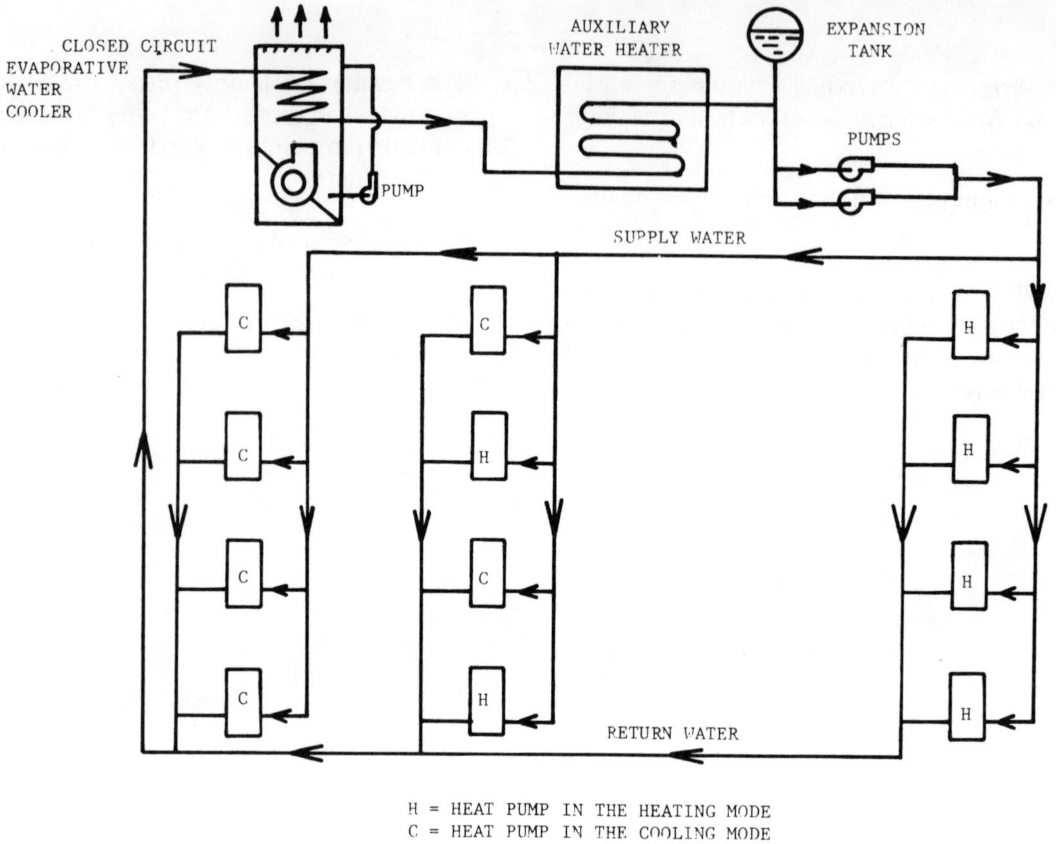

Fig. 1. Typical closed loop heat pump system.

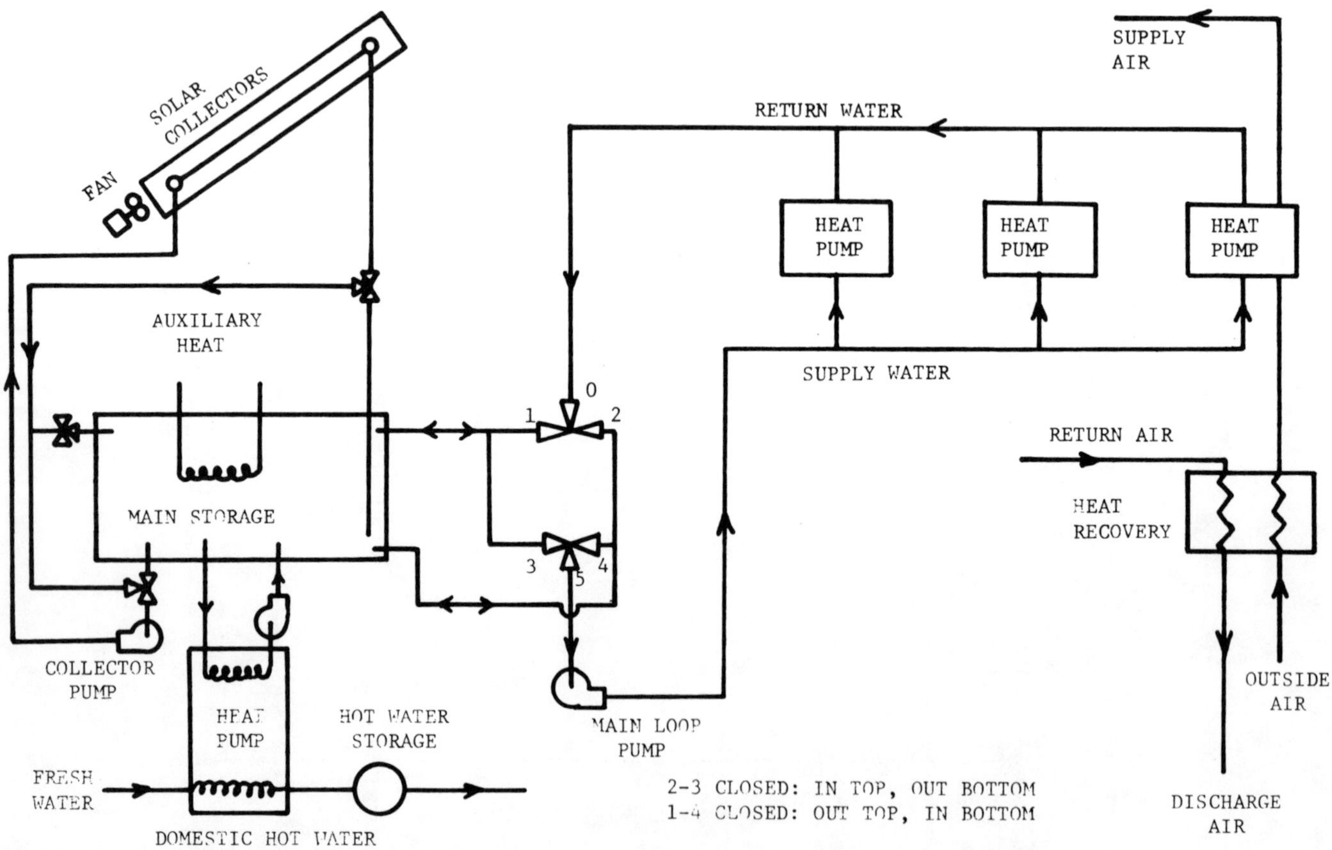

Fig. 2. Solar assisted closed loop heat pump system model.

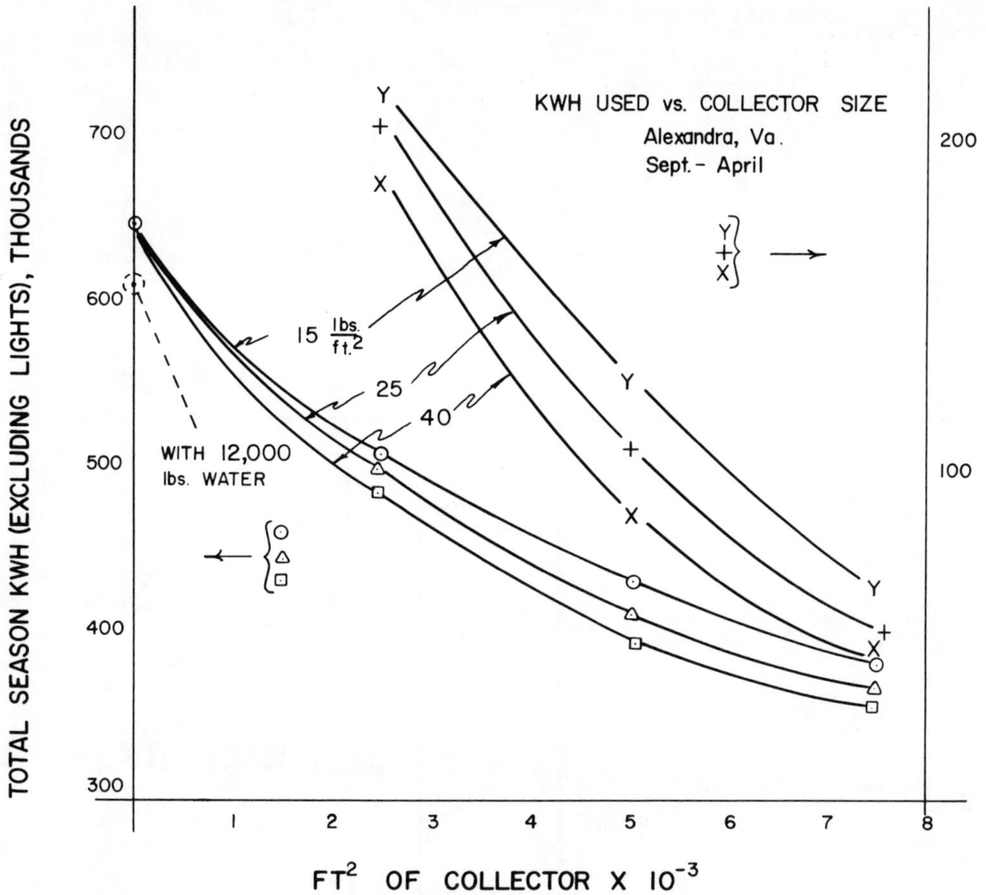

Fig. 3. System performance: Alexandria, Va., school.

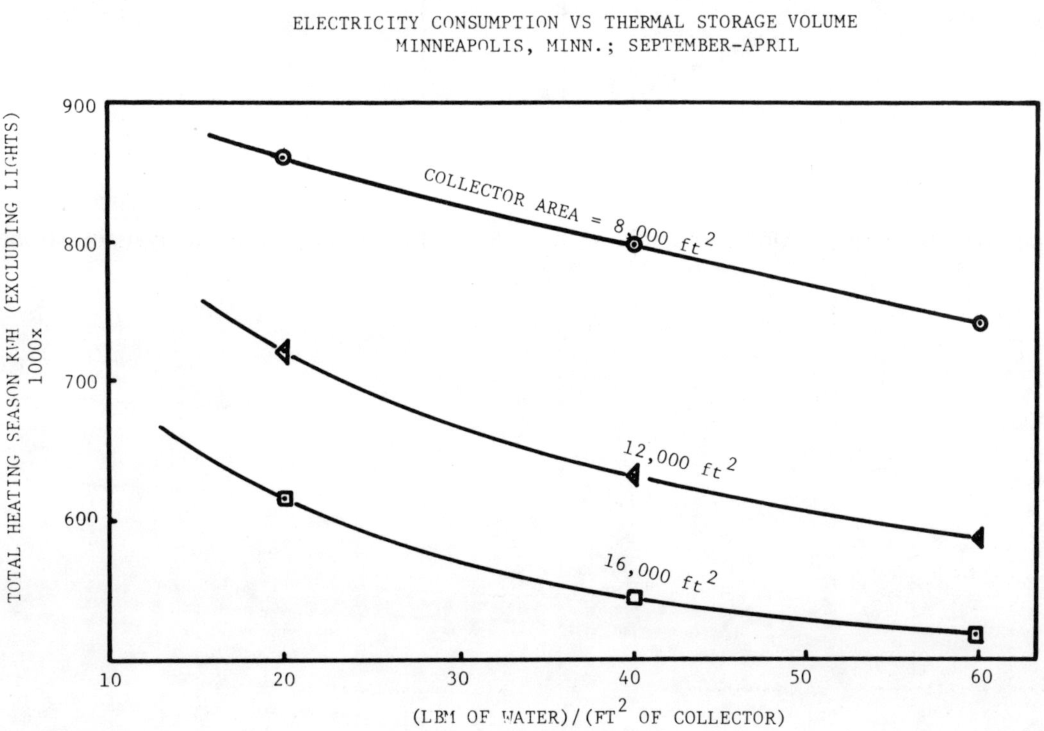

Fig. 4. System performance: Minneapolis, Minn., school.

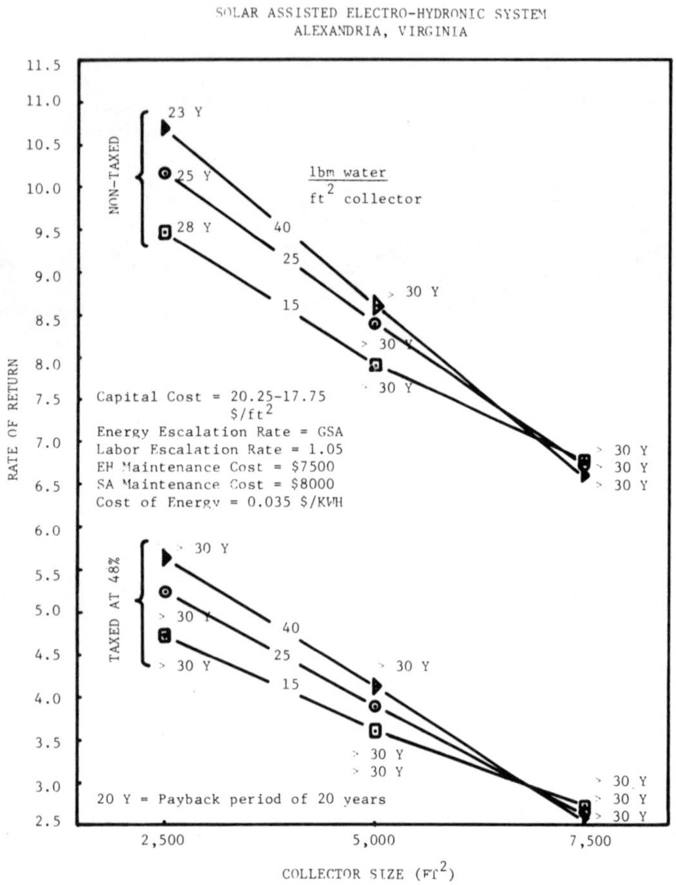

Fig. 5. Economic analysis of Alexandria school.

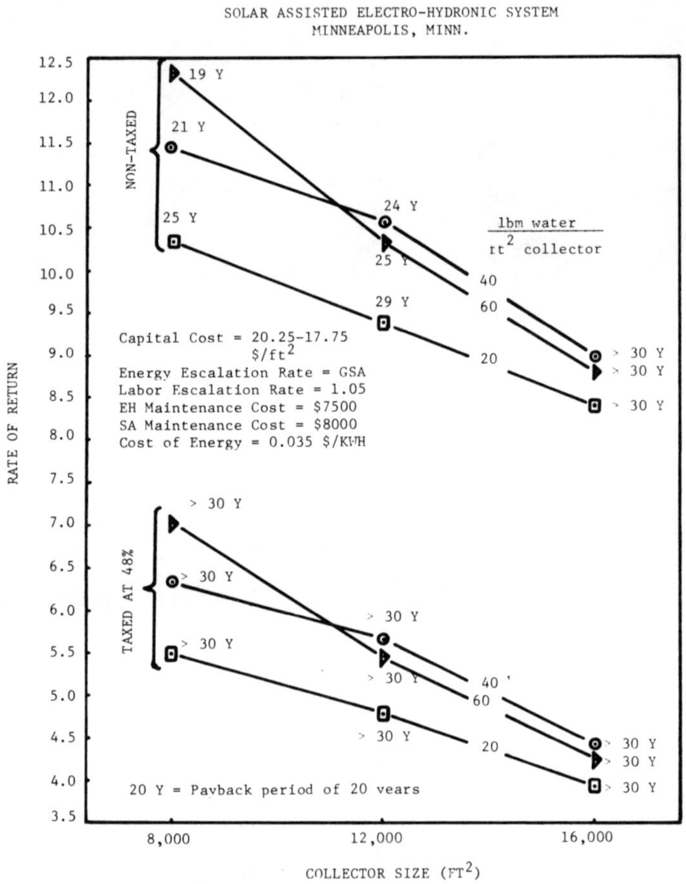

Fig. 6. Economic analysis of Minneapolis school.

Simulation of Solar Energy Heat Pump Systems

Thomas L. Freeman

Specialist, Solar Energy Laboratory
University of Wisconsin
Madison, Wisc.

A general digital computer model of a vapor compression heat pump model is described. The basic concepts involved in the models of the individual components are reviewed and the steps in the modeling procedure are discussed. The modeling strategy is to first design or "size" the four major components in the cycle to yield a user-specified design condition performance. Then the program is able to calculate off design condition operating points. A "performance map" is constructed from repeated solutions for operating points throughout the heat pump's range, which can then be interpolated in a transient simulation, for operating points for any exchange fluid inlet conditions.

A brief comparison is made of a heat pump boosted "in-line" solar system to a heat pump auxiliary "parallel" system.

Development

Most of the heat pump-solar systems described in the literature have been evaluated with a greatly simplified analytical approach to system performance. This is due partly to the lack of a general analytical model of heat pump performance. The transient nature of solar energy system performance requires that consideration be given to the transient effects, but most authors model the weather, the solar collection system, and the heat pump all on the basis of mean values. The results so obtained may be valid for steady state operation over a relatively short time, when the mean conditions prevail, but it is unlikely that these results can be interpreted as long term averages. Recent work at the University of Wisconsin Solar Energy Laboratory has been toward the development of a modular approach to digital computer simulation of solar energy systems. The heat pump model described here is an extension of that work, and makes use of the transient simulation program, TRNSYS (refs. 1, 2).

Figure 1 shows the heat pump schematic and its processes in the pressure-enthalpy plane. The heat pump considered here is a residential size, comfort heating and cooling, single stage unit. It is powered by an electric, constant speed, reciprocating compressor and employs a thermostatic expansion valve. No special hardware for liquid subcooling is assumed. The evaporator and condenser are either cross-flow gas-to-refrigerant heat exchangers or parallel-counterflow liquid-to-refrigerant heat exchangers. The model assumes a quasi-steady state, "transient" response; i.e., transient associated with momentary mass accumulations during shifts in operating points are not considered. Heat pump hardware, especially the "split" air-to-air systems, may employ long runs of refrigerant piping from the outdoor unit to the indoor unit and back again. Pressure drop and heat transfer along these are not considered in this model. No automatic mode control mechanism has been developed to change the heat pump function from heating to cooling. In place of this mode control, the heating mode is simulated independently of the cooling mode by running separate simulations for heating and cooling seasons.

Following is a discussion of the models of the vapor compression cycle components which comprise the heat pump model.

EVAPORATOR AND CONDENSER. The purpose of the evaporator and condenser models is to determine the refrigerant outlet state and the exchange fluid outlet temperature for known refrigerant and exchange fluid inlet conditions and known heat exchanger characteristics. The $\epsilon - N_{TU}$ heat exchanger equations are used to describe the evaporator and condenser performance (ref. 3). However, they are rather complex heat exchangers since the refrigerant is in at least two different states along its flowpath: a two-phase mixture and a superheated vapor. Thus, depending on the exchange fluid and its flow arrangement, the evaporator and condenser are some combination of two heat exchangers, each characterized by different heat transfer mechanisms. Functional dependence of boiling and superheated vapor heat transfer coefficients on refrigerant flowrate have been included in the models, but the condensing heat transfer coefficient is assumed to be unaffected by flowrate.

From user-specified design condition values of heat pump capacity, heat transfer coefficients, hot-to-cold side temperature differences, exchange fluid heat capacity, flowrate, and inlet temperature, and from the heat exchanger flow arrangement and hot-to-cold side area ratios, the sizing routine determines the heat transfer areas and the design condition overall heat transfer coefficients (U values). The heat exchanger equations are then solved with these fixed heat transfer areas, but with U values that vary from their design condition values in a manner prescribed by the heat transfer correlation assumptions. An optional empirical algorithm has been included in the air source evaporator model to decrease evaporator heat transfer, to account for capacity lost due to frost formation and defrost of the evaporator coil.

COMPRESSOR. The refrigerant mass flow rate through a reciprocating compressor is related to its displacement rate D, the inlet specific volume v_B, and the volumetric efficiency η_v:

$$\dot{m}_R = \frac{(\dot{D}\eta_v)}{v_B}$$

Displacement rate is simply the volume per unit time swept out by the pistons and, for the constant-speed machine, is a constant (when the heat pump is on). The inlet specific volume v_B is the evaporator outlet-state specific volume. Volumetric efficiency is a number from 0 to 1, which includes the capacity limiting effects of "clearance volumetric efficiency" and other "nonideal" effects. For a given compressor, η_v depends on the ratio of absolute pressures r_p from discharge to inlet. This function differs somewhat from one compressor design to another, but according to ref. 4, a single graphical relation (Fig. 2) fairly accurately represents a broad range of reciprocating compressors. A curve-fitting algorithm was applied to this curve to produce a functional relationship for computational determination of η_v by the model.

Since the adiabatic compression efficiency is only a weak function of the suction and discharge pressures, it has been made a user-specified constant which characterizes the speed and/or quality of the compressor. To obtain the actual electrical input to the compressor, the computer work must be divided by a motor efficiency. Compressor motor losses are not assumed to be heat gains to the refrigerant.

EXPANSION VALVE. The thermostatic expansion valve is commonly used in systems that have to operate over wide ranges of temperatures, since the valve ensures nearly optimum usage of the evaporator heat transfer surface at all operating points. This is accomplished by an adjustable orifice with "feedback" control from the evaporator outlet that ensures that just enough refrigerant is passed to maintain a uniform amount of superheating at the evaporator outlet, state B. The valve is modeled by assuming a constant enthalpy expansion and constraining state B to lie on a locus of constant superheat.

There is, however, one complication to add to this simple model. In practical heat pump systems operating at very high capacities, the evaporator can be "starved" by the valve's failure to pass enough refrigerant. This may happen when the valve is fully open but flow is restricted because the reservoir of liquid refrigerant in the condenser accumulator has been depleted. The valve no longer functions as a thermostatic valve and the assumption of constant superheat at state B gives way to a new constraint. The mass flow rate computed by the compressor equation must be equal to the mass flow computed by the hydraulic orifice equation which characterizes the valve when it is fully open.

REFRIGERANT. A means of evaluating refrigerant properties is indispensable to the component models and the solution technique. A special subroutine is used to determine the state properties (t, p, h, s, v, x) of the refrigerant when called with two independent properties. The routine performs a four-point interpolation in two dimensions with state property points previously loaded into the program. These data can be supplied for any desired working fluid and specified at any desired intervals so that the interpolation algorithm returns state properties for any refrigerant at any desired degree of accuracy. The data loaded into the program were generated from the DuPont equations for calculating refrigerant properties (ref. 5). For the studies made in this work, 400 data points in superheat and 200 data points in the two-phase region were specified.

Modeling Strategy

Consideration of the system requirements and configuration has led to a modeling strategy that consists of three basic steps: sizing, solution, and performance mapping.

SIZING. The user specifies the refrigerant type and the exchange fluid and flow configuration. He supplies a number of other parameters that characterize the hardware configuration and quality. Most important, he specifies a design condition heating or cooling capacity at a set of design condition exchange fluid inlet conditions. From this information and the system's basic design condition assumptions, the thermodynamic states of the refrigerant at each point in the cycle and the refrigerant mass flow rate may be found for the design conditions. From this information and the cycle component-defining equations, the individual components may be "sized." That is, the program calculates heat exchanger area, compressor displacement rate, and valve orifice area to yield the specified "ratings." This information is used in the next step, solutions for "off design" operating points.

SOLUTION. Once a machine is sized it must be able to respond to a set of inputs by operating in a unique manner. For a heat pump in either the heating or cooling mode, these inputs are the inlet temperatures and flow rates of the evaporator and condenser exchange fluid flowstreams. The four thermodynamic states (states A, B, C, and D in Fig. 1) and the refrigerant flow rate uniquely determine the heat pump operating point, i.e., the rate of heat addition, heat rejection, and work required.

Determination of an operating point requires an iterative solution procedure that "hunts" for refrigerant states and a flow rate that simultaneously satisfy the constraints of the compression, condensing, expansion, and evaporating processes. The solution routine employs a modified form of the two-dimensional Newton's method that repetitively predicts new condensing and evaporating pressures.

PERFORMANCE MAPPING. The solution for an operating point is a fairly lengthy calculation involving several iterations around the refrigerant cycle. It may be economical to do these calculations in a system simulation if it is not too long. However, in longer simulations, the expense becomes prohibitive. Much of the time, the heat pump is being called with inputs for which outputs have already been found. Thus a great computational savings can be realized by creating a performance map.

A performance map is a set of solutions corresponding to a set of possible combinations of input variables. Maps are constructed by repeatedly solving for the rates of heat addition and heat rejection as all the inputs are stepped through small intervals within user specified ranges. These maps are matrixes from which an interpolation routine may obtain performance data at finite intervals, and interpolate for the heat pump solution at any set of inputs within the map's range. To limit the amount of mapping required, the flow rates of the exchange fluids are assumed to be a constant (when turned on) for a particular map. This reduces the number of inputs to two (evaporator and condenser inlet temperatures). The performance maps are thus two two-dimensional matrixes of the heat addition and heat rejection for systematic combinations of evaporator and condenser exchange fluid inlet temperatures. Compressor work required and COP can easily be derived from this information.

Simulation of Systems

A limited-comparison study was undertaken in which two basically different solar heat pump system designs were simulated in Albuquerque, New Mexico, using the same meteorological data and the same heating, cooling, and service hot water loads. The systems employed similar solar collection and storage designs. Both designs supplied the service hot water load via a heat exchanger off the main storage tank.

In the "parallel" system, the heating load was met directly from the storage tank. An air-to-air heat pump operating independently of the solar system supplied space heating auxiliary when required and met all of the space cooling load. In the "in-line" system, a water-to-water heat pump was placed between the storage tank and the space heating load. A bypass was included so that storage tank water could be diverted directly to the heating load when the tank temperature rose above a preset minimum. The "in-line" heat pump rejected heat to a cooling tower when it met the space cooling load.

The simulation results indicate that the "parallel" heat pump system operates more economically than the "in-line" system, at least for the system size and location studied. As expected, a higher collector efficiency is realized by the "in-line" system because of lower collector temperatures, but this advantage was found to be offset by the inability of the collector system to provide a sufficient source for the heat pump, forcing the system to resort to auxiliary at a COP of unity (or less). It is believed that the ideal solar heat pump system must incorporate the best features of both of these designs, that is, an in-line heat pump to achieve higher collector efficiencies (and possibly cheaper initial costs) that can be "switched" to a parallel configuration with outdoor air as a source, when the available energy in storage has been depleted.

ACKNOWLEDGMENT

The research of which this study is a part is supported by NSF/RANN grant GI 34029 of the University of Wisconsin-Madison.

REFERENCES

1. Klein, S.A., et al., "A Method of Simulation of Solar Processes and Its Application," *Solar Energy*, *17*, 29-37 (1974).
2. Solar Energy Laboratory of the University of Wisconsin, "TRNSYS—A Transient Simulation Program," Engineering Experiment Station Report 38, March 1975.
3. Kays, W.M., and London, A.L., *Compact Heat Exchangers*, McGraw-Hill, New York, 1964.
4. *Trane Air Conditioning Manual*, The Trane Company, LaCrosse, Wis., 1968, chap. VI.
5. Downing, R.C., and Knight, B.W., "Computer Program for Calculating Properties of the 'Freon' Refrigerants," *'Freon' Products Bulletin, RT-52*, E.I. du Pont de Nemours and Company, 1971.

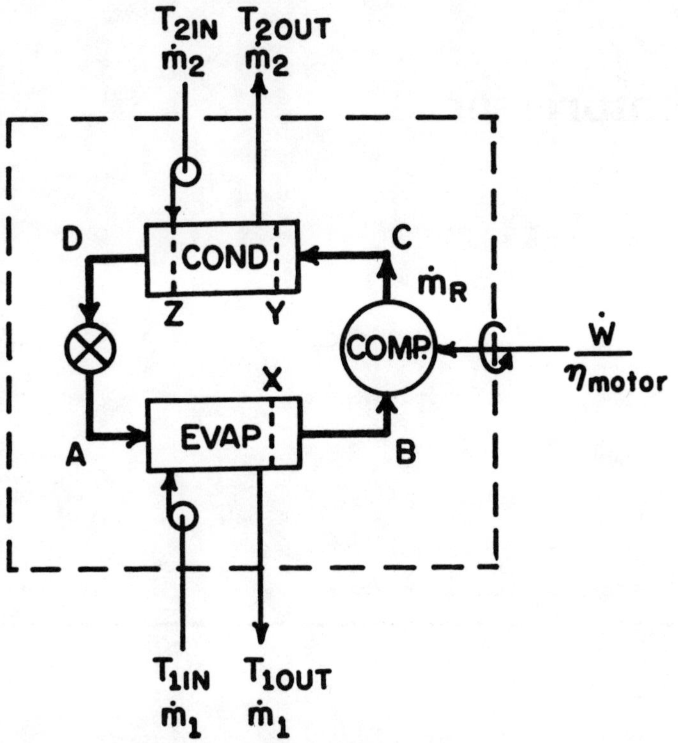

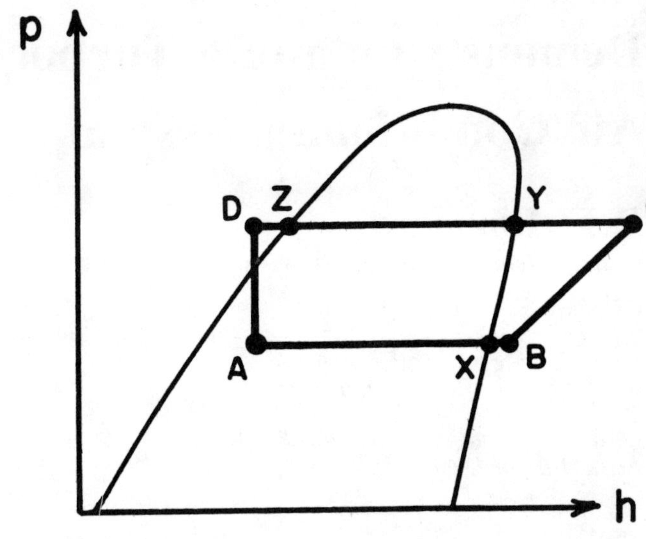

Fig. 1. Heat pump schematic and processes.

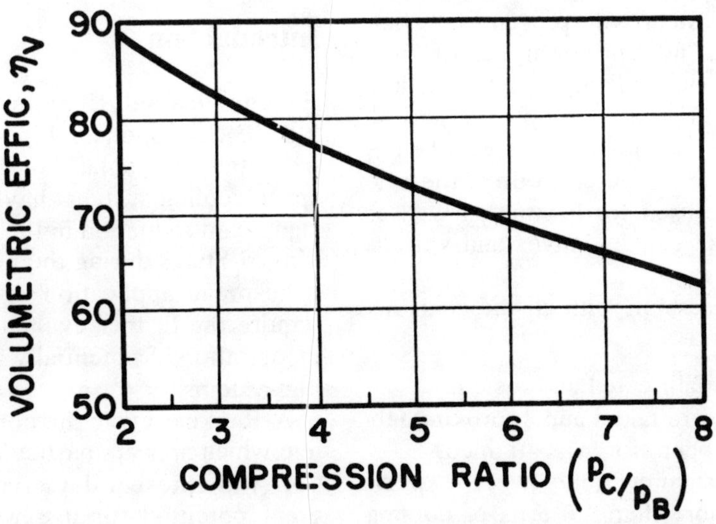

Fig. 2. Volumetric efficiency versus pressure ratio.

Demonstration of a Turbocompressor Air Conditioning System

R.F. Biancardi
Chief, Advanced Energy Conversion Systems

M.D. Meader
Analytical Research Engineer

United Technologies Research Center
East Hartford, Conn.

Summary

During the past year, under a National Science Foundation grant, a Rankine cycle turbocompressor heat pump program was conducted. The objectives of the program were to demonstrate efficient, simple, reliable operation of a laboratory system and to provide a basis for projecting ultimate system capabilities. The demonstration turbocompressor system is based on the adaptation of existing off-the-shelf key components and is capable of operating at temperature levels of 160–220F, consistent with present-day, flat-plate solar collectors. The turbocompressor heat pump is a dual-loop system, utilizing a common fluid in a power loop and a refrigeration loop. Energy collected by the solar panel (simulated by a steam heater in the demonstration system) is used to drive the system. The system can be used for heating as well as cooling by suitable use of control valves and various component arrangements.

The program was successful, with all the goals accomplished, including:

1. System design was fabricated and tested.
2. Over 70 test runs were taken and approximately 60 hours of trouble-free operation was attained.
3. All system performance goals were experimentally verified and more than 4.0 tons of cooling was produced.
4. Turbomachinery performance has been verified even though components were designed for other fluids.
5. Low temperature operation was demonstrated, with cooling produced at maximum system temperatures of 165F or less.
6. The practical, economical system capable of obtaining a COP from 0.7 to over 1.0 was shown to be feasible with further development.
7. A wide range of system cooling capacities suitable for residential and commercial applications appears feasible.

Introduction

Extensive analytical programs conducted under NSF/ERDA sponsorship during the past year have indicated that solar hot water, space heating, and space cooling systems have the potential to provide a significant reduction in the energy requirements of the United States during the next 25 years. However, the widespread application of solar energy systems will require the further evaluation, development, and demonstration of potentially attractive heating and cooling systems.

A Rankine cycle air conditioning and heating system which appears particularly well-suited for integration with present-day, flat-plate collectors and has great potential for use with medium-concentration collectors is the double-loop turbocompressor system (ref. 1). The system features a compact, efficient, yet simple hermetically sealed turbine and compressor unit based on technology developed in the past 25 years for various aerospace applications. A common nontoxic, inexpensive refrigerant such as R-11, R-113, or R-114 is utilized as the system working fluid.

Detailed analytical studies and experimental tests have shown that a thermally driven turbocompressor heat pump system is eminently practical and has the potential to be manufactured at costs lower than or comparable to those of conventional absorption air conditioning systems and heating furnaces, and can be designed for efficient operation in a wide range of sizes, from a few tons capacity to 100 tons or more (ref. 2).

Despite the potential advantages of the turbocompressor system, insufficient data were available prior to this program to prove the viability of such a solar heating and cooling system or to provide a base for projecting ultimate system capability. The work conducted during this NSF/ERDA sponsored program has provided much of the data needed to demonstrate the feasibility and potential of such a system.

Turbocompressor System Concept

The turbocompressor Rankine-cycle system concept is one of the so-called double-loop thermally driven air conditioning approaches advocated in the past decade or more (ref. 2). Shown schematically in Fig. 1 is the system concept for solar-powered heating and cooling. It utilizes a single working fluid and common condenser for both the power and cooling loops. Hot fluid from the solar collector/storage system is used to vaporize a working fluid in the vapor generator of the power loop. The vapor is expanded through a turbine which produces mechanical power to direct drive a compressor. The compressor pressurizes working fluid vapor leaving the evaporator to the level necessary for heat rejection to the atmosphere in the condenser. The exhaust of the turbine and compressor are mixed and enter the condenser, which may be air- or water-cooled.

A receiver is used at the exit of the condenser to permit full utilization of condenser heat transfer under variable load conditions. An expansion valve is used prior to the evaporator in the cooling loop to lower the working fluid pressure and temperature to the desired levels. A pump, which may be driven off the same turbocompressor shaft, circulates the working fluid in the power-conversion loop with common refrigerants as the system working fluid; materials of construction and heat exchanger components that are identical to those used in electric-powered systems can be used and thus provide the advantage of low cost manufacture. When heating is desired, the turbocompressor can be bypassed and the evaporator coil used as the heating element. Any deficiencies in solar-generated heat would be made up by a supplemental burner, or a motor-drive can be used to power the compressor and provide power in a conventional heat pump configuration.

With the same working fluid used in both the cooling and power loop, a common condenser can be utilized. Furthermore, elaborate seals are not required to separate the working fluid from each loop from mixing in the turbocompressor. Since lubrication can be provided by the working fluid itself, the turbocompressor unit can be hermetically sealed, avoiding contamination and refrigerant decomposition, a common problem in electrically driven compression systems with reciprocating expanders. Excess shaft power can also be generated, if desired, for auxiliary purposes; i.e., to drive all the system pumps and fans. Additional heat exchangers, i.e., a regenerator or precooler, can be used to increase system performance but system cost and complexity are also increased.

The attractive features of a turbocompressor system using common refrigerants as the working fluid relative to other double-loop and absorption systems are: first costs are low because a small number of simple components are required and system performance is high; a highly efficient turbocompressor unit using a single-stage turbine and compressor is possible; low-cost materials can be used readily throughout the system; mass-produced conventional design evaporator and condenser components can be utilized; the working fluid is inexpensive, nontoxic, nonflammable, readily available, with good electrical properties, and can be used for lubricating the turbomachinery bearings; no elaborate seals are required; a permanently sealed unit is possible; and high reliability can be anticipated.

Demonstration System Design and Performance Analyses

The technical approach adopted for the feasibility demonstration of the Rankine cycle turbocompressor concept is based on the adaptation of existing key components and technology. The aim is to demonstrate the capability of the system to operate safely and efficiently at temperatures and conditions consistent with present-day flat-plate solar collectors.

TEST SYSTEM SELECTION RATIONALE. The test system configuration was selected to demonstrate the key performance and operational features of the turbocompressor concept, while utilizing available or easily modified components. As such, it represents a careful balance between the utilization of available components and technology while still satisfying the overall program objectives within the specified funding and time constraints.

The turbocompressor system was assembled for this demonstration test with two separate loops, one for power generation and the other for cooling (Fig. 3). Although the same working fluid is used in each loop, separate condensers were used in the power and cooling loops. This configuration permits more convenient operation and accurate measurement of temperature and pressure conditions at the exhaust of the turbine and compressor elements and therefore provides a direct means for correlating calculated and experimental component performance. The two condensers were generally regulated to operate at the same temperature to simulate single condenser operation, although offset condenser temperature operations are possible for specific component tests.

Heat transfer at all of the system interfaces is via cold or hot water sources (see Fig. 3). This permits closer temperature control than would be feasible with air cooling. Heat input from a solar collector is simulated by a steam supply system which provides hot water to the power loop vapor generator over a range of turbine inlet temperatures between 170 and 240F. The condensers are capable of operating over a temperature range from 85 to 130F, thus simulating both water- and air-cooled operation.

Shown in Fig. 2 is the feasibility demonstration system as it appears in our laboratory installation. The overall dimensions of the system are approximately 8 ft long, 4 ft wide, and 6 ft high. Although a more compact arrangement was possible, ample room for instrumentation and any desired modifications was provided. The actual turbocompressor is less than 1 ft in diameter and 1 ft long and contains compressor and turbine wheels which are 4 in. in diameter or less.

An available turbocompressor modified to operate at low turbine inlet temperatures was used to minimize overall program cost and provide a timely demonstration of a Rankine cycle system. Several compressors originally designed for aircraft air conditioning systems were evaluated for this application. The compressor selected for the turbocompressor demonstration unit is a two-stage unit capable of producing in excess of 8 tons of cooling with R-114 at selected conditions. The turbine is a single-stage unit modified from a military aircraft environmental control system installation. The turbine and compressor elements are connected via a common shaft and operate at the same rotational speed.

WORKING FLUID SELECTION. The working fluid (R-11) used in the turbocompressor demonstration was selected on the basis of performance considerations at the desired operating conditions using the selected existing turbocompressor unit. The availability of a major system component such as the turbocompressor makes the selection process unique to this situation, since a more general approach to fluid selection would compare fluids on the basis of equipment specifically designed and optimized for the intended application. However, R-11 does provide high performance at the desired system operating conditions and would be a strong candidate as the working fluid in a fully developed solar-powered turbocompressor system, unless maximum system operating temperatures in excess of 350F become available in the future with low-cost, medium-concentration collectors and cause working fluid stability to become an important consideration. If system temperatures higher than 350F become feasible, R-114 would be recommended as the system working fluid with only a minor reduction in performance capability. A detailed discussion of the desired working fluid characteristics for a turbocompressor system is presented in refs. 2 and 3.

SELECTION OF SYSTEM OPERATING CONDITIONS. After the selection of R-11 as the demonstration system working fluid, analyses were undertaken to determine (*a*) the variation in system performance over the range of anticipated operating conditions, (*b*) the modifications required to the turbocompressor unit to operate at the desired conditions, and (*c*) specifications for the other major system components. The initial performance analyses determined the effect of the three primary system temperature levels, i.e., turbine inlet temperature, condenser temperature, and evaporator temperature, on overall system performance and capacity.

The selected system baseline operating condition is at nominal or midrange values of turbine inlet, condenser, and evaporator temperature considered: 200, 105, and 45F, respectively. The conditions are representative of turbine inlet temperatures available from a conventional flat-plate collector and condenser temperatures attainable with efficient ambient air cooling or water cooling. Changes in condenser temperature result in the greatest variation of COP, since the required compressor adiabatic head and available turbine adiabatic head are both changed. For example, lowering the condenser temperature from 105 to 90F increases the COP from 0.366 to a value of 0.55, with constant compressor and turbine efficiencies.

The performances attainable with the demonstration system turbocompressor unit over the range of operating temperature conditions considered are summarized in ref. 4, and range up to a COP approaching 0.5 and more than 4.0 tons of cooling. The results presented therein include both the effect of changing operating conditions and the effect of being constrained to specific turbine and compressor operating characteristics. The demonstration system compressor was, of course, designed for a particular application and achieves peak efficiency when operating at somewhat higher corrected head values than those encountered in this application. The compressor is therefore operating off-design, and efficiency is degraded accordingly.

Shown in Fig. 3 are the operating conditions, flow rates, and performance anticipated for the turbocompressor demonstration system at the baseline design condition of 200F turbine inlet temperature, 105F condenser, and 45F evaporator temperature.

Test Results

Over 60 performance test runs were obtained with the turbocompressor demonstration system and more than 60 hours of operation have been recorded. System operation has been achieved at turbine inlet temperatures ranging from approximately 160F to in excess of 215F, condenser temperatures from 85 to 145F, and evaporator temperatures from about 30 to 70F. The majority of the test data have been at conditions close to the nominal or baseline conditions of turbine inlet, condenser and evaporator temperatures of 200, 105, and 45F, respectively. Test results are summarized and presented in the final report (ref. 4).

The analyses indicate very close correlation between the test data and analytical predictions of system performance at condenser temperatures of 100–105F. At other condenser temperatures, wider discrepancies between the test data and analytical predictions are noted, with the test data generally higher than the analytical predictions.

A specific comparison of test data and predicted values is presented in Fig. 3: the measured test operating conditions during a selected test run near the baseline design conditions have been superimposed (in parentheses) on the demonstration system flow chart. The cooling produced is somewhat higher than predicted (4.05 tons versus 3.7 tons), but the measured COP is slightly lower than predicted. Although the turbine inlet temperature during the test run was higher than the design baseline conditions of 200F, the higher temperature is in the form of R-11 superheat, which does not significantly raise the COP of the turbocompressor system. The pressure of the R-11 vapor matches that desired for the baseline design condition; therefore this run was considered valid for comparison purposes.

While turbomachinery efficiencies are as predicted or higher, system COP at these conditions falls below that predicted, mainly because the evaporator is operating at substantially less than 45F, and the flow to the compressor has an excessive amount of superheat. At the lower pressure associated with the reduced evaporator temperature, more compressor work is required to pump the gases to the condenser pressure, and hence less flow can be circulated in the cooling loop with the available turbine power.

Other test data obtained at approximately the baseline design conditions also compare well with the predicted system performance. Data taken at a turbine inlet temperature of 200F produced 4.33 tons of cooling at an evaporator temperature of 35F with a COP of 0.34. The test data also illustrate the sensitivity of compressor performance to the desired compressor head conditions. The highest performance levels are achieved with temperature differences between the condenser and evaporator approaching 60F or more.

Although the bulk of the demonstration system testing was conducted at the nominal design conditions, operation at turbine inlet temperature other than 200F was obtained with condenser temperatures of approximately 105F. A comparison of the COP and cooling capacity data obtained from the demonstration system tests with those predicted from analyses is shown in Fig. 4 as a function of turbine inlet temperature. The predicted performance characteristics are shown in Fig. 4 by the solid line over a range of turbine inlet temperatures for a condenser temperature of 105F and a compressor inlet temperature of 50F. Since the compressor inlet temperature includes 5F of superheat, the evaporator actually operates at a temperature of 45F. The experimental data are indicated by the open circles and cover a range of condenser and evaporator temperatures close to those used for the calculated performance curves. The data indicate that the cooling produced ranged from 2.05 tons to 4.33 tons, with a measured COP ranging from 0.20 to 0.34. While substantial agreement between the experimental data and predicted performance is indicated, some differences between test and predicted performance are noted at turbine inlet temperatures of 200F and above. A portion of this difference can be attributed to the fact that during the tests the system operating conditions did not duplicate all the conditions used to analytically predict performance.

For example, the results shown in Fig. 4 are based on a comparison of calculated and experimental data for given turbine inlet temperatures. However, higher superheat conditions were generally obtained during the experimental tests than desired (e.g., about 7–8F superheat was desired, but because of the excessive pressure drop in the line between the vapor generator and turbine inlet, the superheat was sometimes 10–20F). If the experimental and calculated data are compared using turbine inlet pressure as the parameter of interest, much closer correlation is obtained.

Another reason for the differences between the test

data and analytical predictions of performance are the much lower evaporator temperatures achieved during some of the testing. A review of the data indicates that evaporator temperature was sometimes much less than 45F, with compressor inlet temperatures also less than 50F. Differences also existed between the temperature levels of the power and cooling loop condensers during some tests. During a number of runs, however, identical condenser temperatures for the power loop and the cooling loop were obtained, thus demonstrating that single condenser operation is entirely feasible and would incur no significant problems or performance penalty.

The results obtained when operating the demonstration system at condenser temperatures slightly lower than the baseline condition of 105F are presented in Fig. 5 as a function of turbine inlet temperature. Data were obtained for turbine inlet temperatures from 190 to 168F, although evaporator temperatures were generally higher than desired. During some of these tests, a COP of 0.45 was obtained with a turbine inlet temperature of 180F. The flow utilized for cooling the turbocompressor bearings was reduced during some selected runs and is believed responsible for the relatively good performance.

Test results were also obtained when operating the demonstration system at condenser temperatures of approximately 90F. While only a limited number of test runs was completed at this lower condenser temperature, experimental performance met or exceeded performance predictions, and a COP approaching 0.40 was obtained. Substantial cooling was produced with anticipated levels of COP at turbine inlet temperatures approaching 170F. Efficient operation at turbine inlet temperatures as low as 163.5F was obtained, together with evaporator temperatures of 45F or below.

Some performance characteristics were obtained for the demonstration system when operating at condenser temperatures which simulate air cooling and anticipated performance was reduced, but 2–3.0 tons of cooling were produced.

A review of all the test data over the range of turbine inlet and condenser temperature conditions investigated indicates that only approximately a 70–80F temperature difference between the vapor generator and condenser is required for the turbocompressor system to operate. With selected improvements to the system even larger temperature differences should be feasible for efficient operation.

Performance Potential

Although the present turbocompressor demonstration system has exhibited impressive performance and has generally met or exceeded the performance levels predicted by theory, much higher performance levels are achievable by incorporating specially designed, and matched, turbomachinery elements, rather than using modifications of existing equipment. Systems designed for the residential and small commercial market, i.e., 3–20 tons, can utilize turbomachinery which has component efficiencies of approximately 80%. With such machinery, the system COP can be increased by 35% above the present levels and system component sizes (including the solar collector and storage systems) reduced accordingly.

Turbocompressor systems sized for the industrial market (over 25 tons) would be more efficient because of the larger physical dimensions of the components, thus reducing some of the losses experienced in small components. These larger components should be capable of achieving individual efficiencies of 85% or more. As a result the turbocompressor product efficiency levels (i.e., compressor × turbine efficiency) achieved with the present system (50%) can be increased with development to 64% for residential systems and to 72% for industrial systems. COP levels of the turbocompressor system would correspondingly be increased to levels approaching 1.0, even when operating at 200F turbine inlet temperatures. The performance levels achievable with the turbocompressor system as a function of turbomachinery product efficiency and turbine inlet temperature are shown in Fig. 6. The improvements in system COP relative to the present system COP level are also shown.

Conclusions and Recommendations

Successful operation has been achieved with a Rankine cycle heat pump system at maximum system temperature levels consistent with present-day, flat-plate solar collectors. The system utilizes an integral turbocompressor unit and has been operated at system inlet temperatures approximately 20–30F lower than projected residential absorption air conditioning systems. As a result incorporation of the turbocompressor system into a solar-powered environmental control system could provide much higher energy utilization than presently envisioned with absorption equipment.

The demonstration turbocompressor system has been safely and routinely operated and has met (or exceeded) all of the performance and operational goals. Between 2 and 4 tons of cooling have been obtained with COP levels as high as 0.45. Water-cooled as well as simulated air-cooled condenser operation has been obtained, indicating the further potential for eliminating cooler towers in future solar-powered air conditioning installations.

The performance predicted for the turbocompressor

unit, adapted and modified from existing environmental control system elements, has been substantiated during the test program, and provides further evidence for correlating the results for fluorocarbon working fluids from designs originally developed for other working fluids. Incorporation of new turbine and compressor elements based on present design and existing component technology is predicted to permit the attainment of COP levels of 0.6–0.9 for a prototype turbocompressor system, even with present-day, flat-plate collectors that provide input heat at temperature levels of 160–200F.

Analyses of new turbocompressor systems, incorporating advanced turbomachinery as well as medium-concentration collectors so as to provide maximum system temperature levels of 250–300F, indicate that COP levels above 1.0 could be obtained even with air-cooled condenser configurations.

REFERENCES

1. Biancardi, F.R., et al., *Study of New Means of Achieving Natural-Gas Powered Air-Conditioning*, UARL Report B210313-4, November 1963.
2. Biancardi, F.R., and D.P. Darrow, *A Turbocompressor Air-Conditioning and Heating System for the Home*, ASHRAE Report No. 2104, Paper presented at ASHRAE Semiannual Meeting, Chicago, January 1969.
3. Biancardi, F.R., et al., *Feasibility Demonstration of Solar Powered Turbocompressor Air-Conditioning and Heating System*, Quarterly Progress Report No. 1, UARL Report N951923-1, September 1974.
4. Biancardi, F.R., et al., *Feasibility Demonstration of Solar Powered Turbocompressor Air-Conditioning and Heating System*, UTRC Report R75-951923-4, 1975.

DISCUSSION

Q What COP can you obtain at air-cooled condensing of about 120–125F and 205F collector temperature?

A Indicated COP at these conditions is approximately 0.4.

Q Are there turbocompressors now available as a package for commercial application of air conditioning?

A No.

Q How is your COP defined for this turbodriven heat pump?

A COP is defined as Btu cooling provided divided by Btu heat input.

Q What is the mechanical efficiency of your power loop alone?

A Power loop efficiency, defined as mechanical power delivered divided by heat input, ranged from 8 to 10%.

Q How do you feel about coupling a motor-generator to your turbocompressor?

A This is a complex problem and many tradeoffs would have to be evaluated for any specific application. In higher capacity units, incorporation of a motor-generator could be reasonably accomplished if found desirable. The approach would present greater difficulties in low capacity units.

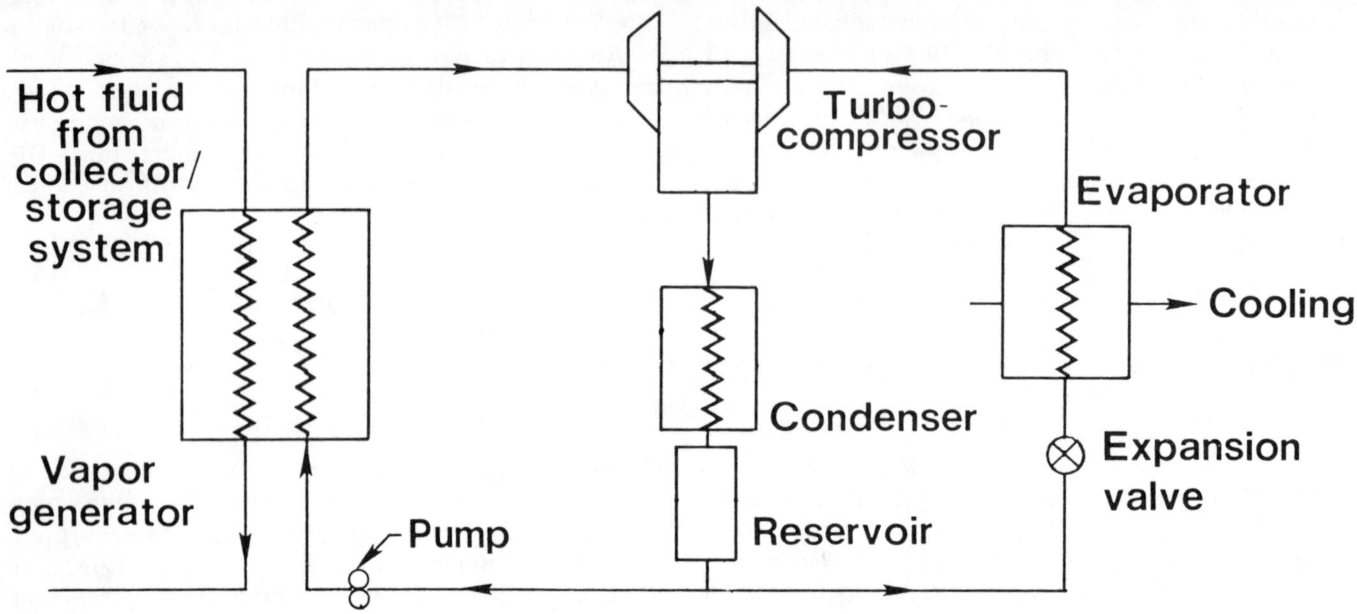

Fig. 1. Turbocompressor system.

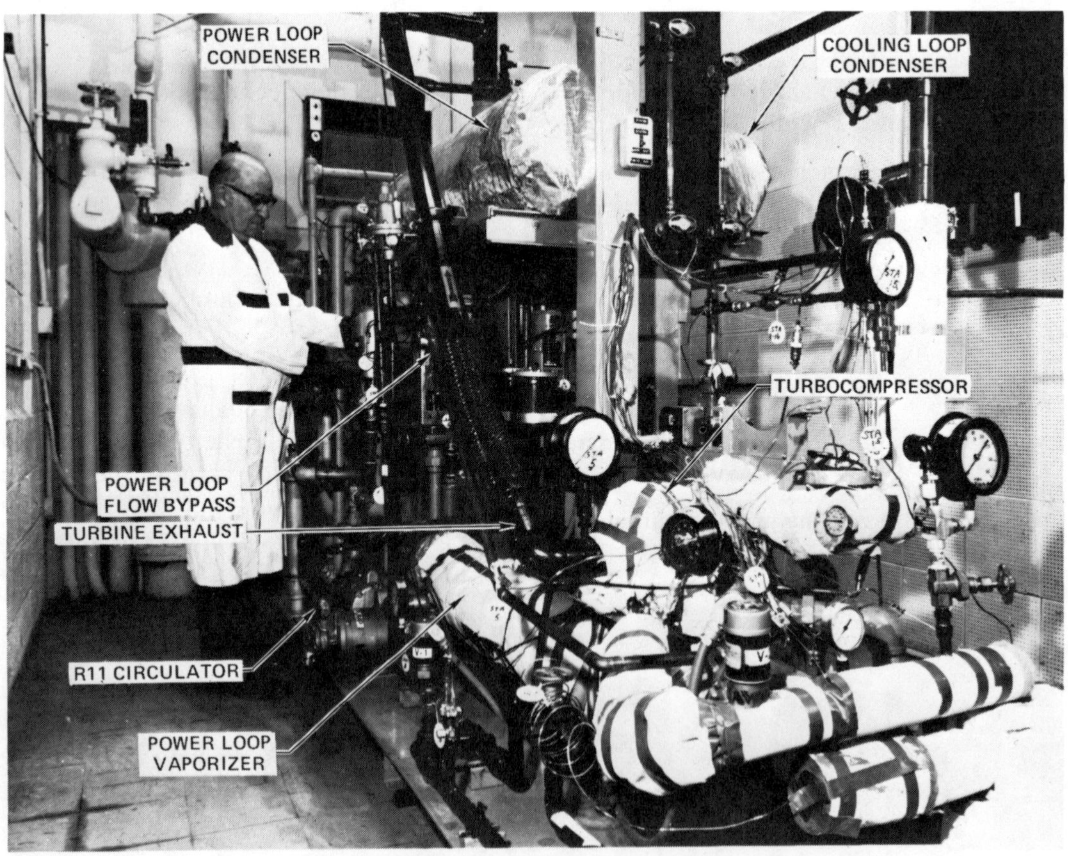

Fig. 2. Demonstration turbocompressor system.

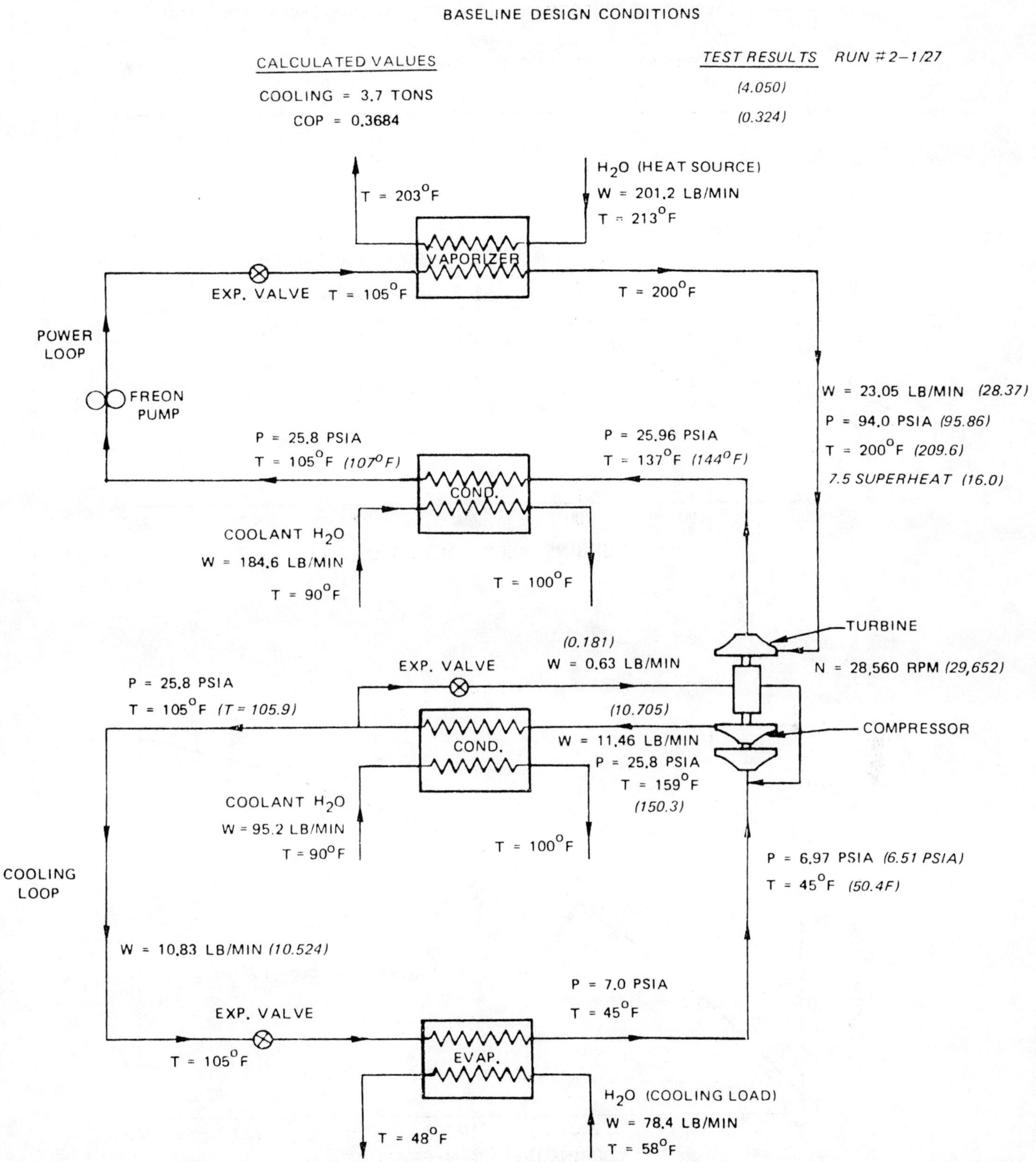

Fig. 3. Demonstration system flow chart.

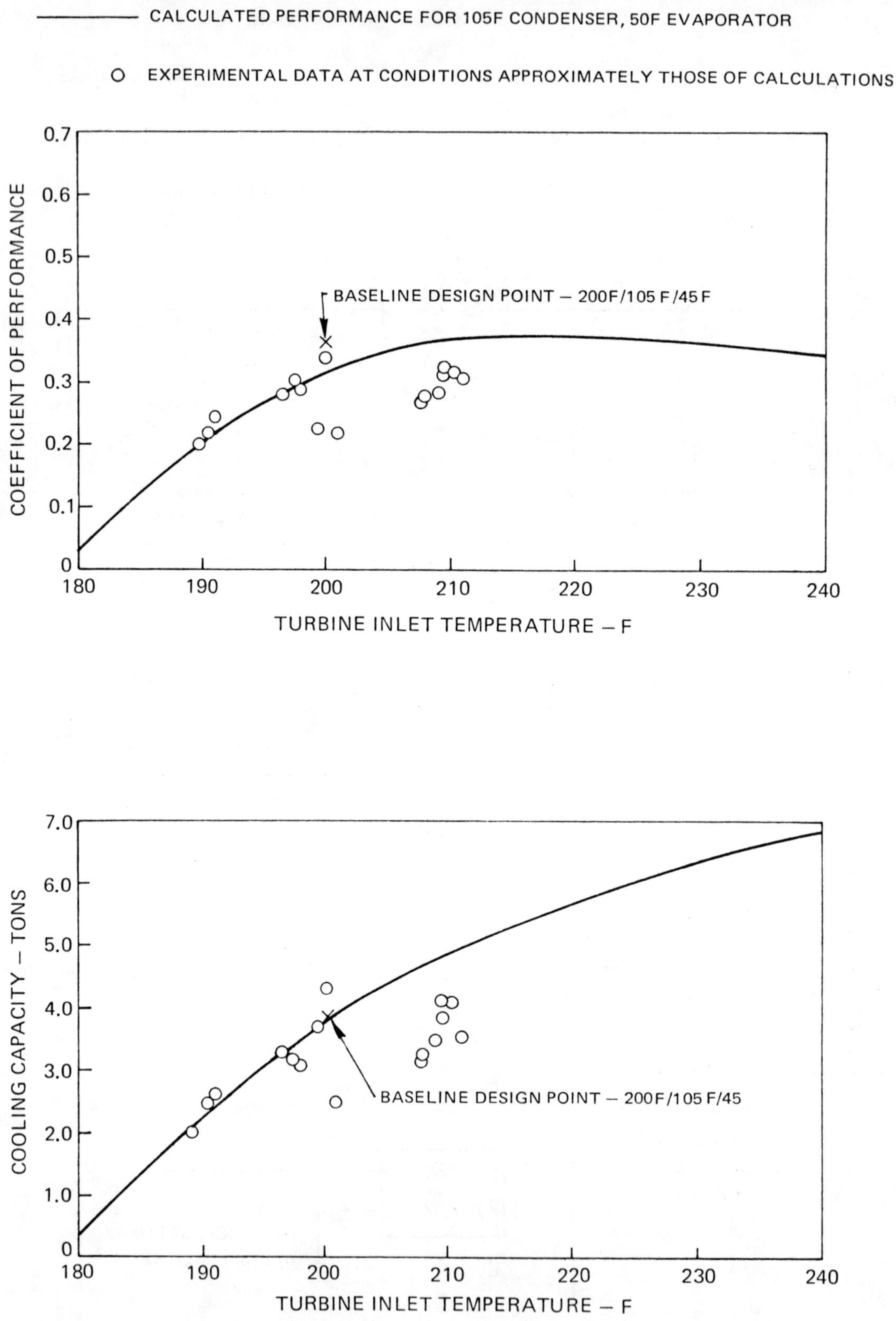

Fig. 4. Demonstration system performance characteristics.

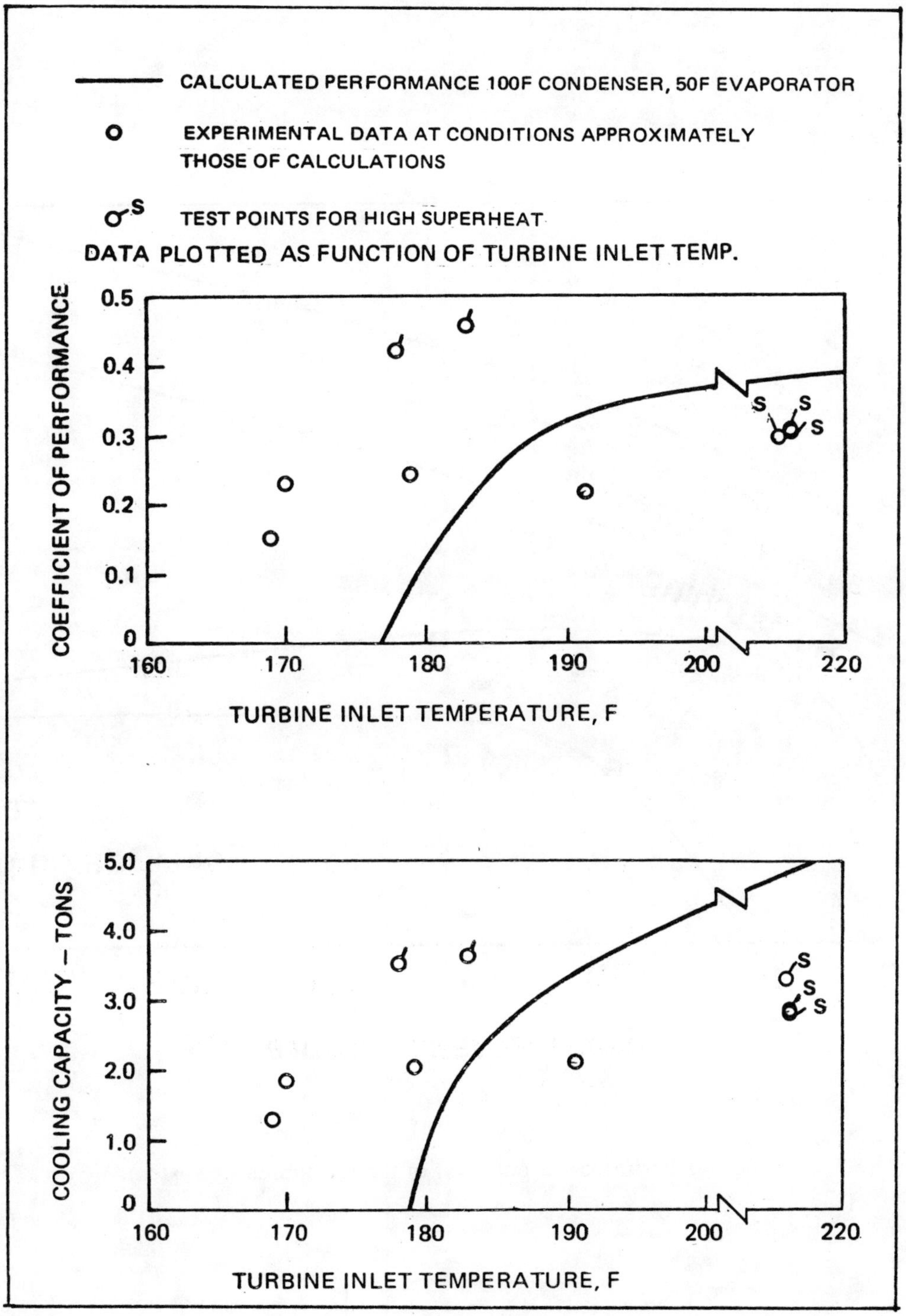

Fig. 5. Demonstration system performance characteristics.

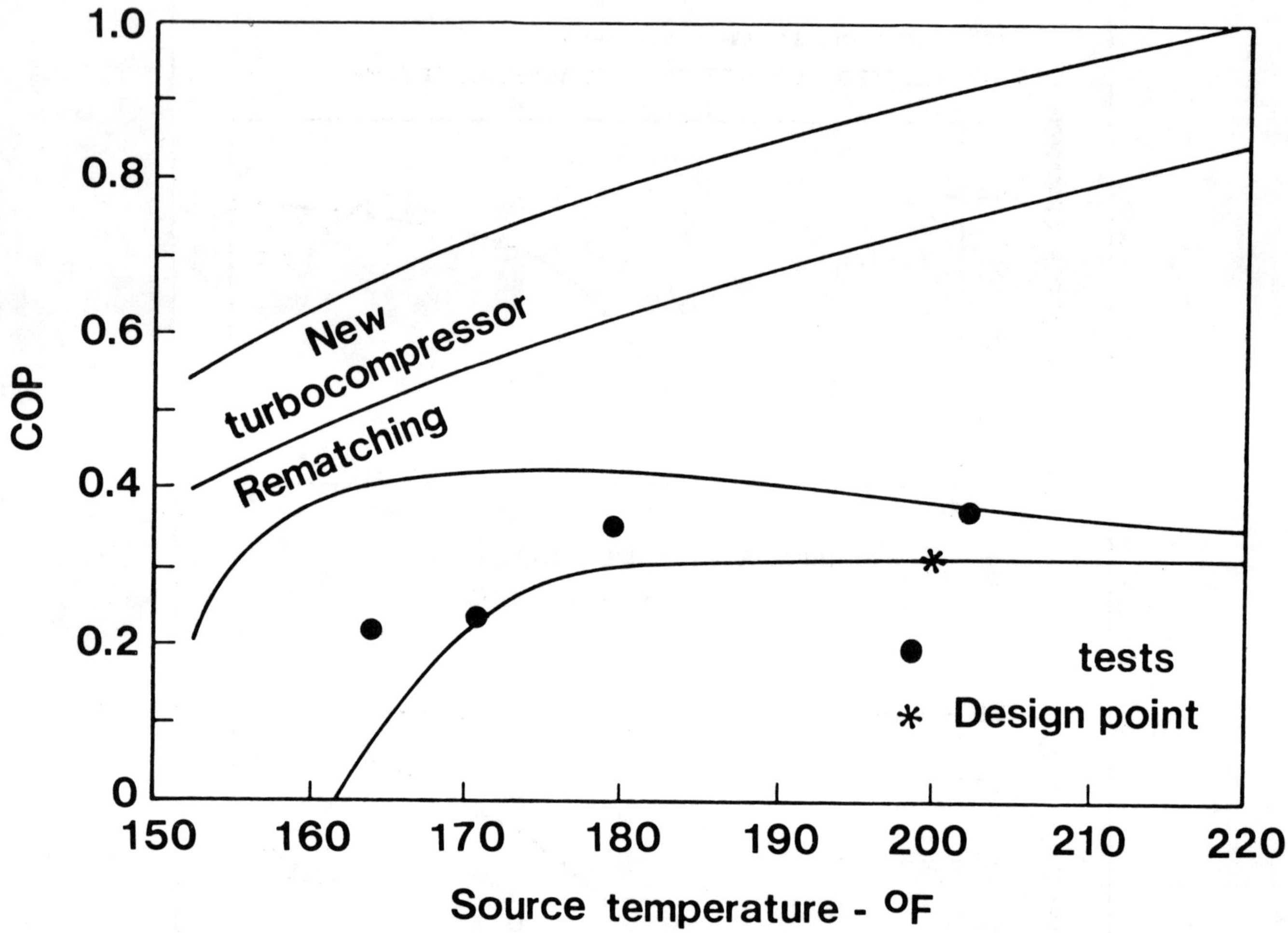

Fig. 6. Performance potential of turbocompressor system.

A Solar Heat Pump Featuring Pivoting-Tip Vanes

Robert D. Fischer
Researcher

D.H. Frieling
Associate Manager, Energy Systems

S.G. Talbert
Researcher

J.A. Eibling
Program Manager, Solar Systems

Battelle Columbus Laboratories
Columbus, Ohio

This paper reports progress in a feasibility study of a new concept of a solar-powered Rankine cycle vapor compression (RC-VC) heat pump being investigated at Battelle-Columbus.

Absorption systems and RC-VC systems are the two leading types of solar cooling systems; both types have about equal cooling performance (refs. 1–5). While both systems are in current use in the solar application, RC-VC systems are considered to be research units because of the experimental nature of some of the components in the Rankine-power loop of the candidate configurations.

An alternative to absorption solar cooling systems is desirable for residential use. RC-VC systems are a good alternative because of potential reduction in size, cost, and maintenance as well as several other attractive features. Auxiliary power input to solar-powered RC-VC systems can be supplied by an electric motor, where absorption systems require heat. This is a particular advantage where natural gas is unavailable. Still another advantage of RC-VC systems, such as described in this paper, is that the Rankine power loop can be used to drive a generator to provide some of the homeowner's electric power demand during the spring and fall, when solar energy is more than sufficient to meet heating and cooling needs.

In some climatic regions, heat pump operation is desirable to maximize the heating effect from power consumed, particularly when natural gas is unavailable. Absorption systems are not as suitable as RC-VC systems for solar heat pump use because of lower efficiency in the heating mode (ref. 6).

Design Description

Figure 1 illustrates the components of the RC-VC solar heat pump (refs. 7–9). Distinctive features of the concept include a rotary vane compressor and expander with pivoting-tip vanes, a single working fluid in the compressor and expander loops, and a common air-cooled condenser. These features are briefly discussed below.

A rationale for using the rotary-vane type of compressor and expander must include (*a*) advantages of this class of machine; i.e., compactness, low vibration, simplicity, and low cost; (*b*) ease of matching the performance characteristics of a directly coupled compressor and expander; and (*c*) an improvement in commonality of parts.

The principal reasons for selecting the pivoting-tip type of vane are reduced friction and wear and an anticipated ability to eliminate conventional lubricants from the compressor and expander loops. Power loss due to vane-tip friction is substantially reduced by using tilting-pad hydrodynamic gas bearings at the vane tips of both the compressor and expander (ref. 10); vane-tip friction is lower by an order of magnitude when pivoting-tips are used.

No liquid lubricant is required at the vane tips because the pivoting-tip vane is supported on a hydrodynamic film of gas, i.e., there is no sliding contact. Assuming leakage paths at both rotor-to-head interfaces can be sealed adequately without lubricant, there is good potential for eliminating oil from the compressor and expander loops.

A single refrigerant working fluid is used in both the compressor and expander loops to minimize performance requirements of internal shaft seals and to permit using a common condenser for both loops. A single enclosure for the rotating components is used to eliminate external shaft seals for increased reliability and hermetic sealing.

Predicted Performance

With the general features of the solar heat pump concept established, a computer simulation of an RC-VC heat pump was prepared for the following purposes:

1. To analytically determine the feasibility of the solar heat pump concept.
2. To provide fluid properties, flow rates, displacements, and shaft speed required for design of rotary-vane components and pivoting-tip gas bearings.

SIMULATION OF AN RC-VC HEAT PUMP. A schematic diagram of an RC-VC heat pump was prepared to permit visualization of flow paths and to formulate control methods. A method for computer access of refrigerant properties from ASHRAE data was formulated.

Significant features of the RC-VC heat pump simulation include:

1. The simulation is compatible with solar system simulation programs, such as SOLTRM or TRNSYS. The SOLTRM (Solar Transient Model) program uses some components from the TRNSYS program.
2. Refrigerant properties for each fluid are accessed from nine polynomial curve-fit equations.
3. Inlet conditions to the expander are assumed to be constant at any preselected value for ease of analysis using a variety of working fluids.
4. Pressure and heat loss in connecting piping is assumed negligible.
5. Rotary-vane expander and compressor are characterized by values of displacement per revolution, volumetric efficiency, and adiabatic efficiency.
6. Air-to-refrigerant heat exchange coils are characterized by overall heat transfer coefficients and a heat exchanger effectiveness of 0.75.
7. Fan power requirements are not charged to system performance.
8. Values for 31 parameters and 8 variable inputs are required.

PARAMETRIC STUDIES. Parametric studies were conducted to select design values and a working fluid for subsequent use of the heat pump in a transient simulation of a solar heating/cooling system having flat-plate collectors and a design cooling load of 36,000 Btuh.

It was assumed that compressor and expander displacements and heat exchange coil capacities would be determined primarily by requirements for cooling operation. An auxiliary optimization program based on the heat pump simulation was prepared to determine these values for a desired cooling capacity and shaft speed in a single run. The major variables explored were condenser, evaporator, expander-inlet temperatures and compressor and expander adiabatic efficiencies.

Figure 2 presents representative data from these parametric studies. Selection of design-point condensing temperature involves a tradeoff of cooling COP and capacity of the outside condenser coil. For example, cooling COP for R-12 is improved from 0.43 at 116F to 0.72 at 100F—a gain of 67%; however, this improvement requires a threefold increase in the capacity of the outside condenser coil. A compromise value of 108F was selected for design-point condenser temperature. Cooling COP is 0.56 at this value.

Selection of evaporator temperature involves a tradeoff of a cooling COP and amount of dehumidification desired. Cooling COP is improved about 23% when the evaporator temperature is increased from 45 to 55F. A value of 50F was chosen for evaporator temperature for the solar heat pump studies in order to maintain some dehumidification and to take advantage of this simple expedient of improving performance.

Selection of expander-inlet temperature involves a tradeoff between efficiency of the flat-plate collectors and heat pump performance. A value of 200F was chosen because this value can be achieved with flat-plate collectors during summer cooling operation.

The importance of maintaining high adiabatic efficiency of the compressor and expander was evident in the parametric studies. Cooling COP will drop 43% with a 20 percentage point drop in adiabatic efficiency from 85%. A constant value of 85% for adiabatic efficiency of the pivoting-tip rotary-vane components was selected even though efficiency is known to vary with pressure ratio and speed. This value includes an assumed 5-percentage-point benefit attributable to the use of pivoting-tip vanes.

Refrigerant R-12 was the preferred working fluid. Although Fig. 2 shows that the cooling COP is improved with R-11, the compressor displacement required was considered excessive. These findings are in general agreement with results from an extensive study of candidate fluids for this type of application (ref. 11).

PREFERRED SPEED OPTION. With values of key design variables established from design-point cooling operation, heat pump performance in each speed option was explored over a range of expander-inlet temperatures.

Figures 3 and 4 show cooling performance with variable and fixed speed. Figure 3 shows that rotation will cease if the expander-inlet temperature is lowered to about 180F during cooling with variable-speed operation and no electric motor power input. Cooling COP increases substantially with decreasing expander-inlet temperature because of falling condenser temperature, but capacity falls off so much that this variable-speed option is not considered practical for cooling operation.

Although design-point cooling capacity could be preserved by maintaining the expander-inlet temperature at the design-point value of 202F by electrical resistance heating of the refrigerant between the boiler and expander, Fig. 4 shows that it is better to let the expander-inlet temperature drop and hold speed constant with power input from an electric motor. With the fixed-speed option, cooling COP increased from 0.56 at 202F expander-inlet temperature to 0.84 at 160F, because of falling condenser temperature and increased energy input from the electric motor. In the limit with 100% energy input by electric motor, cooling COP for a properly designed system would increase to about 2.5.

Similar plots were prepared for heating operation with 30F ambient and 68F room temperatures. These data showed that rotation will cease when the expander-inlet temperature is lowered to about 144F during heating with variable speed and no electrical motor power input. Heating capacity with variable-speed operation at 160F expander-inlet temperature dropped to about one-fourth the value at 202F.

With fixed-speed operation, the heating capacity also dropped with decreasing expander-inlet temperature, but not nearly as much as with no electric power input. This capacity falloff with fixed-speed operation is due to lower heat rejection by the Rankine cycle at lower temperatures. COP for heating at fixed speed with 30F ambient temperature increased from 1.22 at 202F expander-inlet temperature to 1.49 at 140F.

SOLAR HEATING/COOLING WITH RANKINE HEAT PUMP. A transient simulation of a complete solar heating and cooling system was prepared to further establish the feasibility of the Battelle heat pump concept and to define the range of performance variables in an operating system. The TRNSYS computer program was obtained from the University of Wisconsin and used as a foundation for carrying out this simulation. This program contained some of the more commonly used components, such as collectors, storage tank, piping components, and temperature controls. Additional subprograms were prepared that are unique to a heat pump system, including a dual-function thermostat and building thermal-load simulation with heat gain/loss coefficients. A weather subprogram was prepared to access weather data from magnetic tape. These subprograms and those for the Rankine heat pump, boiler, and refrigerant properties, described previously, were incorporated into the systems simulation program called SOLTRM.

Input data for each component of the solar powered RC-VC heat pump system were prepared for heating and cooling during the extreme months of January and August. Bases for these studies include:

1. Single-family residence with 1600 sq ft area located in Albuquerque, New Mexico.
2. Heat gain/loss coefficients for the thermal load calculation similar to the 1974 Colorado State University house (ref. 12).
3. Hour-by-hour weather data from a 1959 Albuquerque weather tape.
4. Domestic hot water consumption schedule from ASHRAE data, which includes a total consumption of 96 gal/day at 140F (ref. 13).
5. A 450 sq ft array of flat-plate solar collectors tilted from the horizontal at an angle equal to the latitude, and constructed with two covers and a flat-black absorber plate insulated with 3 in. of fiberglass insulation.
6. Thermal energy storage capacity of about 1.6 gal of water per square foot of collector area.
7. Solar-powered RC-VC heat pump with pivoting-tip vane components designed for 3-ton cooling capacity with design point temperatures of 108F condensing, 50F evaporator, and 95F ambient; R-12 refrigerant; and compressor and expander adiabatic efficiencies of 85%.
8. Inside heat-exchange coil 1.5 times larger for heating than cooling.

9. Fixed-speed (1200 rpm) heat pump operational mode with 202F expander-inlet temperature for cooling and 140F for heating maintained by auxiliary heat input, if necessary.

10. Dual-function thermostat for on-off control of heat pump operation at set point values of 68F for heating and 75F for cooling.

With the collector area established from a brief optimization study, simulated month-long heating and cooling runs with the solar heat pump were conducted. These studies have shown that 86.9% of the total energy demand in January for space and domestic hot water heating was supplied by solar energy. About 60% (292 kwh) of the auxiliary energy for space heating was supplied by shaft power to the heat-pump compressor by the electric motor. Average heating COP for the month was about 1.5.

In August, 89.9% of the total energy demand for space cooling and domestic hot water heating is supplied by solar energy. The amount of electrical energy generated during the month was 40 kwh but would be significantly greater during the spring and fall when heating/cooling energy demand is low. Average cooling COP for the month was about 0.6, which is about the same as for an absorption cooling unit.

Another RC-VC heat pump simulation with variable-inlet conditions to the expander has been prepared for use in future studies, which will eliminate the need for electrical resistance heating.

Pivoting-Tip Performance

The conceptual sketch of the solar heat pump in Fig. 1 shows a pivoting-tip on each vane. The theory of self-acting tilting-pad bearings is well established (ref. 14), but application to tips of vanes in rotary-vane machines is not as well established. This concept has been successfully applied to high-speed fuel and oil pumps (ref. 15), and encouraging results were obtained from a preliminary design study aimed at applying the concept to vapor compressors.

The specific challenges faced in applying a tilting-pad bearing to reduce vane-tip friction in a compressor or expander are unique; e.g., load-support forces are not steady and pad area is reduced at suction or admission and discharge ports.

An analysis for computing hydrodynamic pressures in tilting-pad gas bearings has been developed at Battelle. Other bearing parameters of interest can be computed from the known pressure profile across the pad. Radial load caused by the pressure field can be determined and summed to obtain the total radial load support. Minimum film thickness in the bearing can be determined from a summation of radial forces. Stability of the bearing pad is related to the tilt angle, which can be computed from a summation of moments about the pivot.

Theoretical performance studies of representative pivoting-tip design configurations were performed to provide values for key design parameters and values of parameters which are difficult to obtain experimentally, such as pressure in the gas film, minimum film thickness, and tilt angle of the pad.

In addition to the theoretical performance studies, representative pivoting-tip vanes were evaluated in a laboratory test rig. Basic features of this apparatus include:

1. A circular cam ring which is mounted in trunnion bearings to facilitate measurement of vane-tip drag with a load cell.
2. A 6-in. diameter rotor with two pivoting-tip vanes, which is mounted concentrically within the cam ring.
3. Representative pivoting tips and rotor-head facings, which are made of carbon-graphite for operation without lubrication.
4. Provision for shaft power to be provided by an electric dynamometer for precise control of speed and measurement of overall frictional drag.
5. Provision for applying known loads on the vane tips by undervane pressure and centrifugal forces.
6. Provision for varying pressure at the tips, undervane pressure, and pressure to the gas film.

Figure 5 shows vanes and two pivoting-tip configurations used in the experiments. Configuration A is a patented design in which undervane gas is supplied to the gas film. Configuration B is identical to configuration A except that the milled slots and holes in the bearing face are eliminated. The solid carbon-graphite vane was run for comparison of friction without a pivoting tip.

Figure 6 presents representative friction and load support data for configuration A and friction data for the solid vanes during operation with nitrogen gas. Undervane and ambient pressure in the cam ring cavity were maintained at the same pressure so that vane load would be centrifugal force only. Pressure to the gas film of the pivoting tip was maintained the same as the ambient pressure at the tips. Pressure for both vanes was varied over the range of 85 to 160 psig with little noticeable effect.

Friction coefficient with the solid vanes is about an order of magnitude, or 10 times, higher than the value of 0.025 with pivoting tips. The pivoting tip supported centrifugal loads up to 1600 rpm with no noticeable increase in friction coefficient. Liftoff speed for hydrodynamic operation was below 700 rpm.

Calculated minimum film thickness (H_{min}) at a shaft speed of 1200 rpm is dependent on ambient pressure at the tip. H_{min} is 115 μin at 85 psig and 80 μin at 160 psig. Maximum calculated pressure rise in the film at this speed is about 18 psi.

Some instability was encountered during some early experiments as evidenced by jumps in friction coefficient from about 0.025 to 0.050 and chatter marks on the cam ring. Stability was improved by increasing the relative curvature of the tip with respect to the cam ring radius from the theoretically optimum value. However, a better approach may be to offset the pivot to the trailing edge of the tip.

Friction with configuration B was higher than for configuration A, with equal pressures at 85 and 120 psig ambient pressure and about the same at 160 psig.

When pressure to the gas film of configuration A was increased 24 psig above an ambient pressure of 85 psig, the friction coefficient dropped to 0.001 at 900 rpm and 0.012 at 1200 rpm. These low values indicate the potential reduction in friction which might be achieved in an operating machine during portions of rotation when the ambient pressure at the tips is lower than the maximum pressure available, assuming volumetric efficiency is not compromised excessively.

Conclusions

Major conclusions from this feasibility study include:

1. Favorable results from optimization and system performance studies of the solar heat pump concept have demonstrated that the concept continues to show much promise in application to solar heating and cooling of buildings.
2. The preferred operational mode of a solar-powered RC-VC heat pump is fixed-speed operation with auxiliary power input/output by motor/generator.
3. Favorable results from experimental studies have demonstrated the feasibility of the pivoting-tip concept. Friction coefficient is substantially reduced during operation without oil lubrication. Adequate load support is generated in the gas film.

ACKNOWLEDGMENTS

Financial assistance of the National Science Foundation-RANN and the Energy Research and Development Administration is gratefully acknowledged. The encouragement of Harold Horowitz, program manager of this study, is very much appreciated.

Thanks also go to Burke S. Looney and Richard D. Creachbaum, both of whom contributed valuable expertise to the experimental studies of pivoting-tip performance.

REFERENCES

1. Helstrom, J.C., et al., "Comparative Analysis of LiBr Absorption Dehumidification, and Solar-Rankine Cooling System Options for a 66,000 sq ft Building in Los Alamos, New Mexico," ISES Paper LA-UR-74-1062 (1974).
2. General Electric, "Solar Heating and Cooling of Buildings—Phase 0," NSF-RANN-74-012B *2* (1974), pp. 2.2–71.
3. Sargent, S.L., and Teagan, W.P., "Compression Refrigeration From a Solar-Powered Organic Rankine-Cycle Engine," ASME Paper 73-WA/Sol-8 (1973).
4. McCulloch, W.H., et al., "The Solar Community—Energy for Residential Heating, Cooling, and Electric Power," Sandia Laboratories, SLA-74-0091 (February 1974).
5. Prigmore, D.R., and Barber, R.E., "A Prototype Solar Powered, Rankine Cycle System Providing Residential Air Conditioning and Electricity," IECEC Paper 749026 (1974).
6. Balasubramaniam, M., et al., "Solar Thermal Absorption Heat Pump Break-even Coefficient of Performance," ASME Paper 74-WA/Ener-2.
7. Eibling, J.A., "A Solar Heat Pump System," *Proc. NSF/ASHRAE Conf. on Solar Cooling for Buildings*, Los Angeles, February 1974, pp. 92–96.
8. Eibling, J.A., and Frieling, D.H., "A Solar Heat Pump," ISES Paper (August 1974).
9. Fischer, R.D., et al., "A Solar-Powered Heat Pump Utilizing Pivoting-Tip Rotating Equipment—Phase I," NSF/RAN/SE/C-876/PR/75/4 (April 1975).

10. Beck, W.D., Stein, R.A., and Eibling, J.A., "Design for Minimum Friction in Rotary-Vane Refrigeration Compressors," ASHRAE Semiannual Meeting, Houston, January 1966.
11. Allen, R.A., and Stiel, L.I., "Development of New Working Fluids for Solar Rankine Heat Pumps," NSF/RANN/SE/GI42506/FR/75-2 (May 1975).
12. Lof, G.O., Progress Report on NSF Contract I040457 (1974).
13. *ASHRAE Guide and Data Handbook–Systems*, 1972, p. 37.11.
14. Gross, W.A., *Gas Film Lubrication*, John Wiley and Sons, New York, 1962.
15. Johnson, H.T., and Mitchell, R.K., "Hydrodynamic Pivoting Pad Vane Tips for High-Speed Vane Pumps," in *Gas Turbine Pumps*, published by ASME (March 1972).

DISCUSSION

Q Solar heat is collected with a lot of effort. What is the advantage of converting this energy into mechanical energy with a basically low efficiency (6%) for driving a heat pump, over a direct electrical driven heat pump? (Electricity can be generated at efficiencies up to 50% in a conventional way and we will need it anyway for lighting, etc.)

A Available components from a Rankine cooling system can be adapted for heat pump use to maximize the heating effect from solar energy and to minimize the amount of auxiliary energy consumed. Auxiliary energy can be supplied as shaft power, as in a conventional heat pump.

Q For no electric generation what is your best COP for cooling, and what are the corresponding heat source, heat sink, and refrigeration temperature?

A Cooling COP is 0.56 at design conditions (202F expander-inlet, 108F condenser, and 50F evaporator).

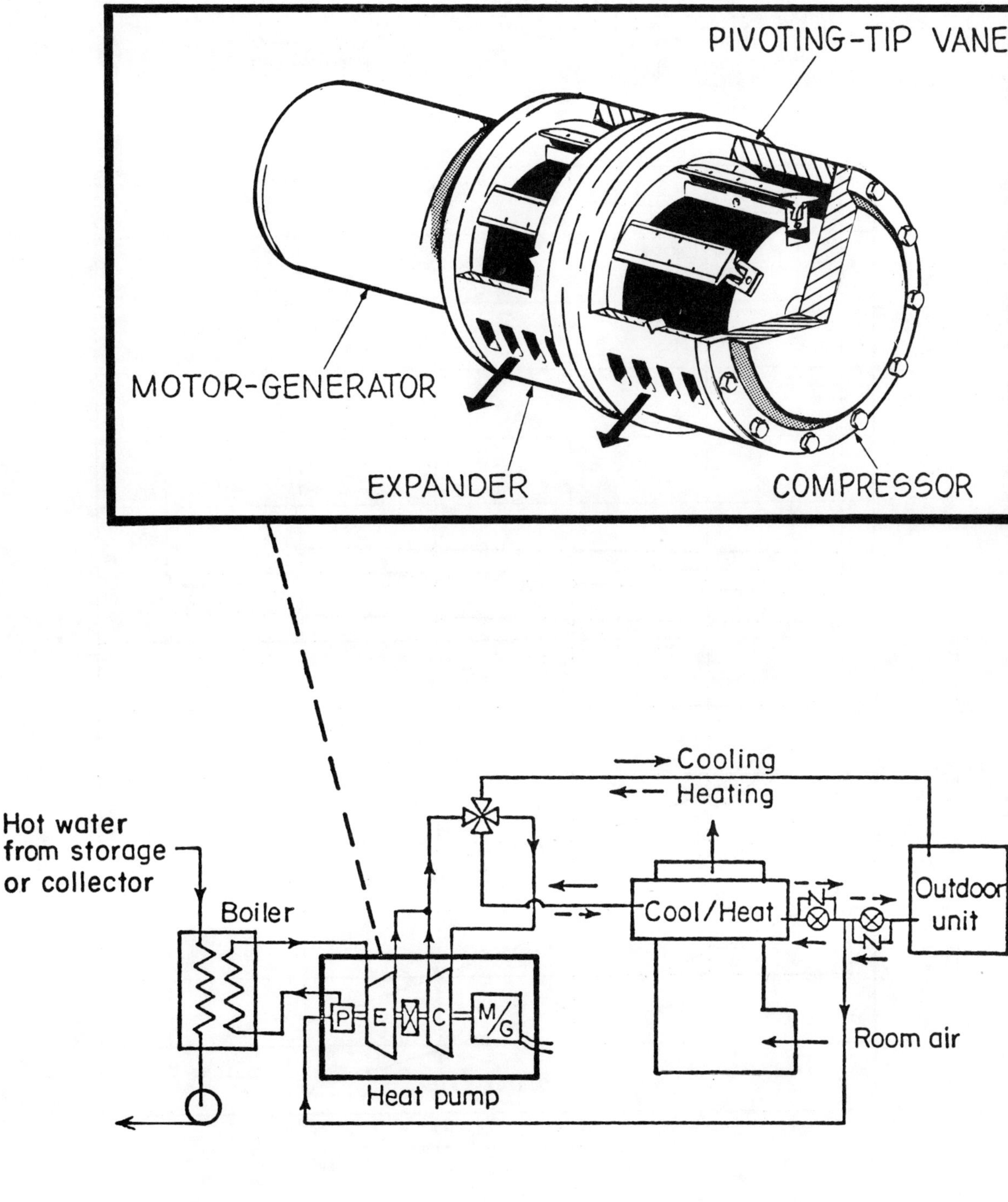

Fig. 1. Components of the Battelle solar heat pump.

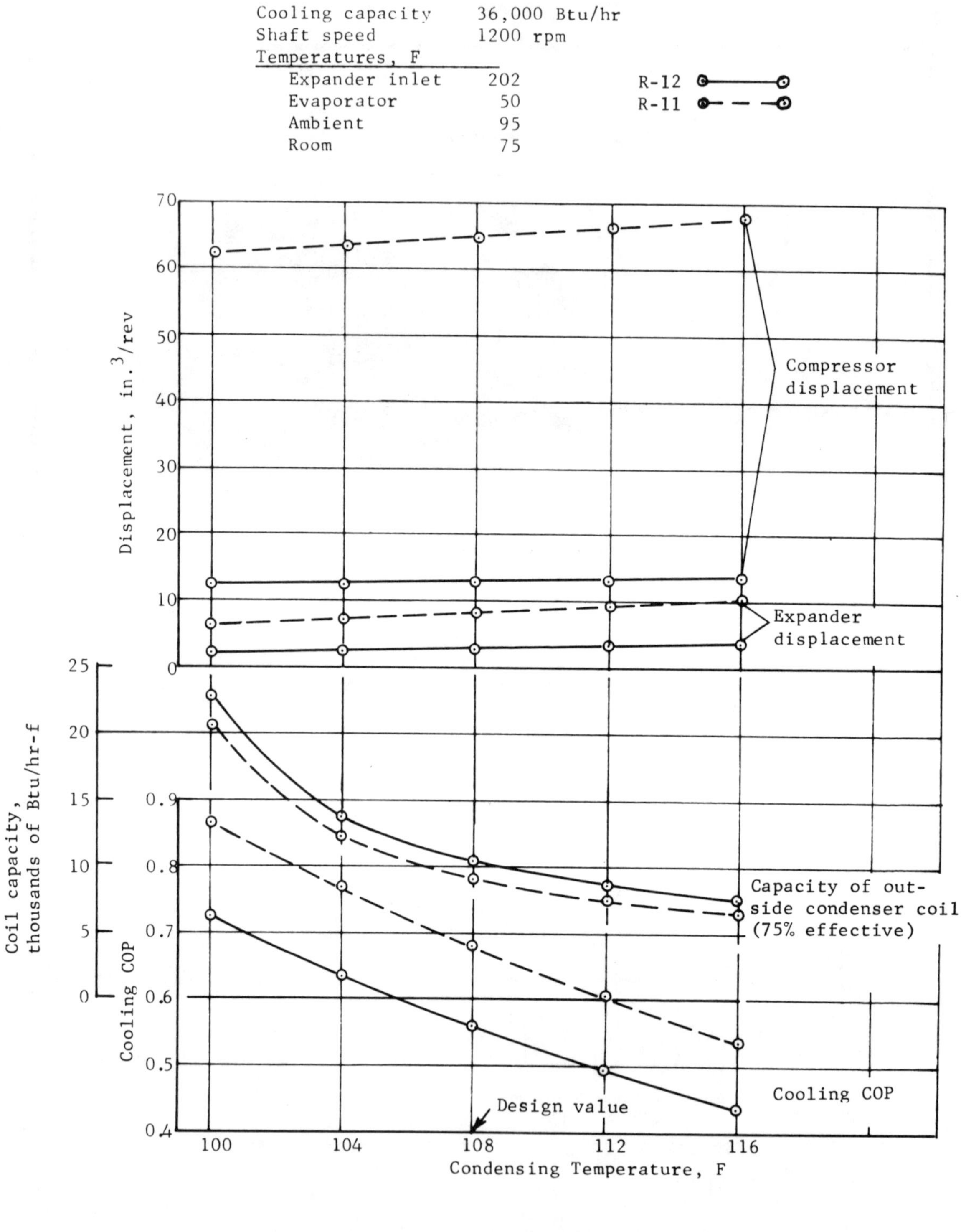

Fig. 2. Effect of condenser temperature on major heat pump design parameters.

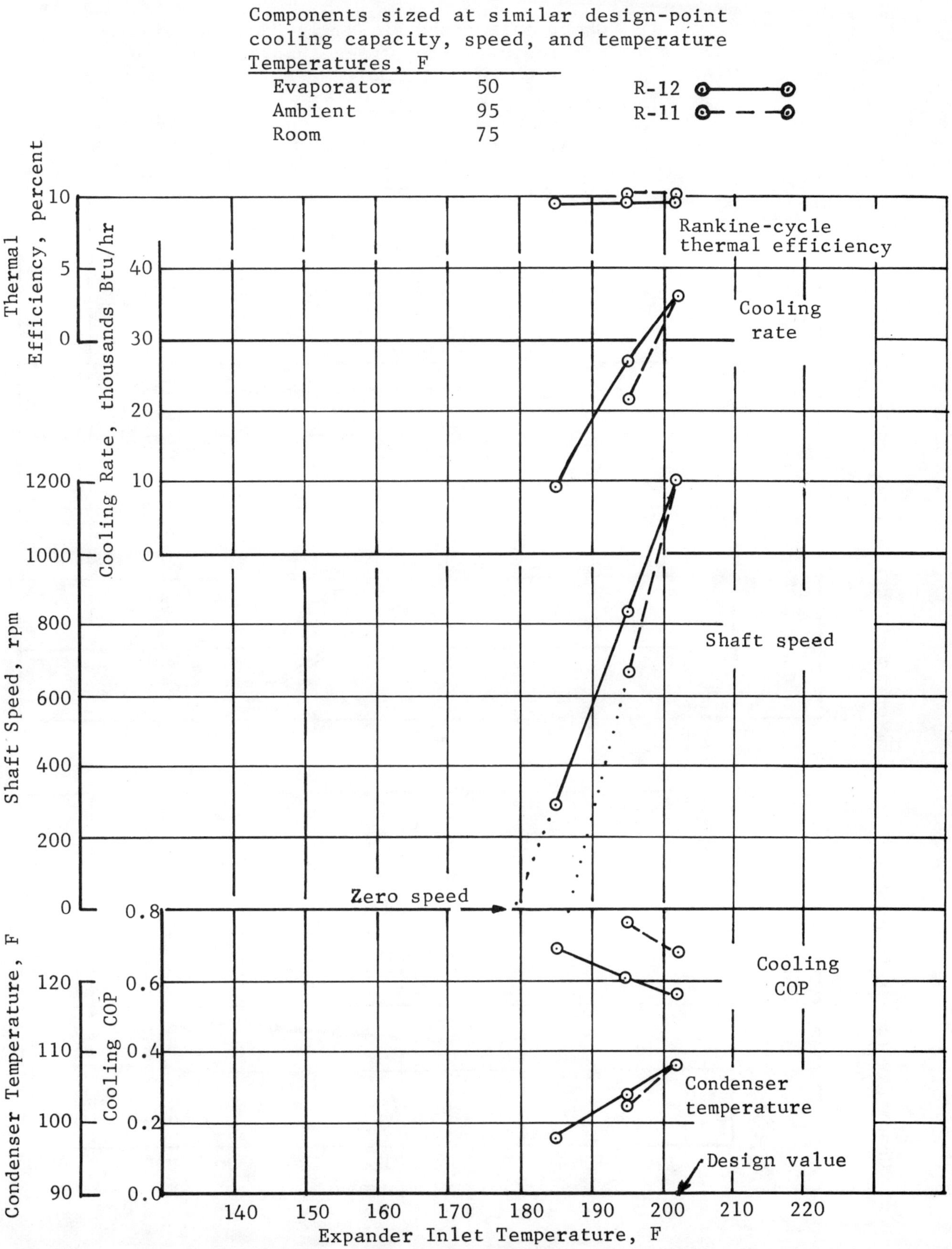

Fig. 3. Effect of expander-inlet temperature on cooling performance with variable speed.

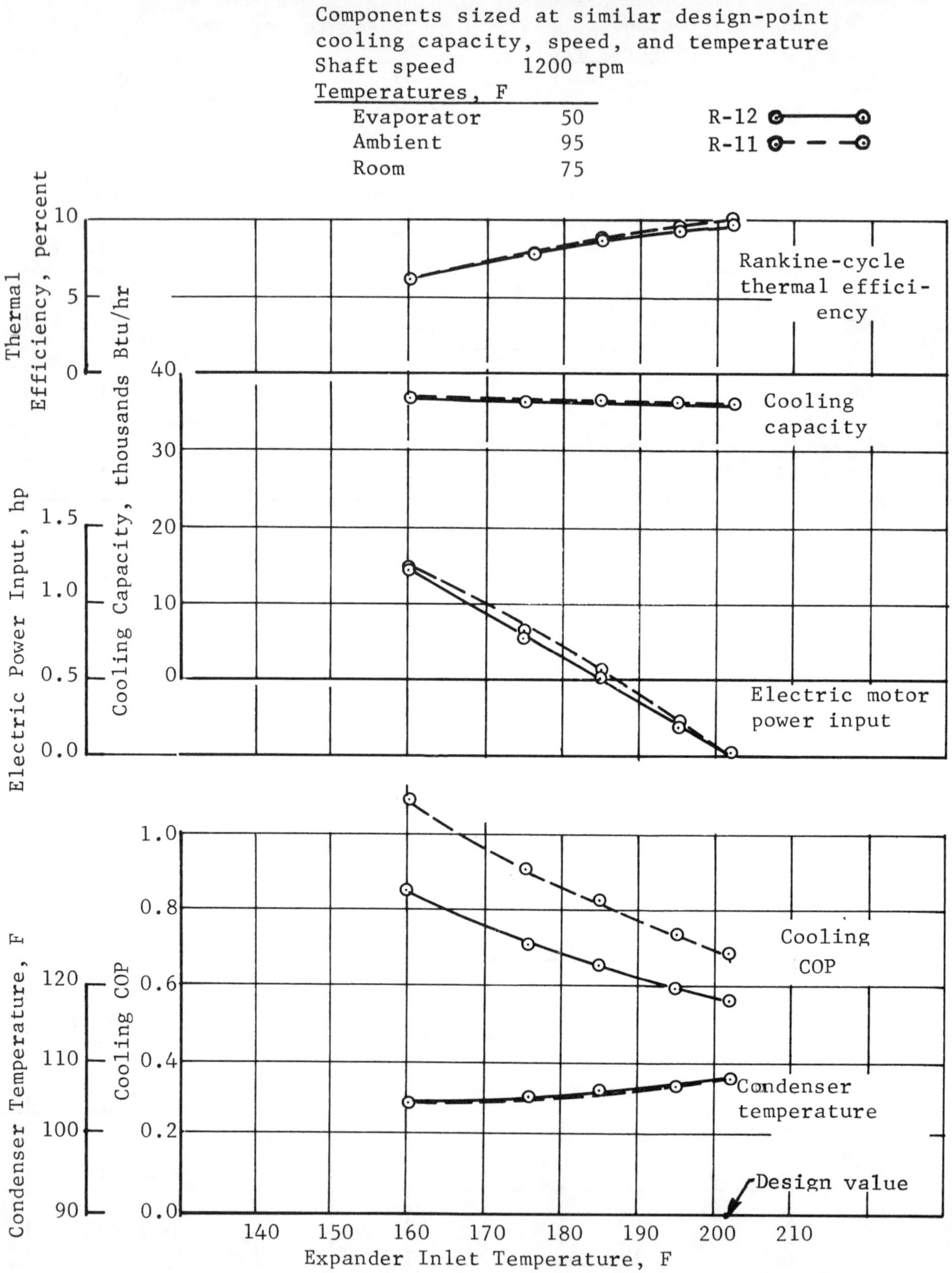

Fig. 4. Effect of expander-inlet temperature on cooling performance at fixed speed.

Fig. 5. Experimental vanes and pivoting tips.

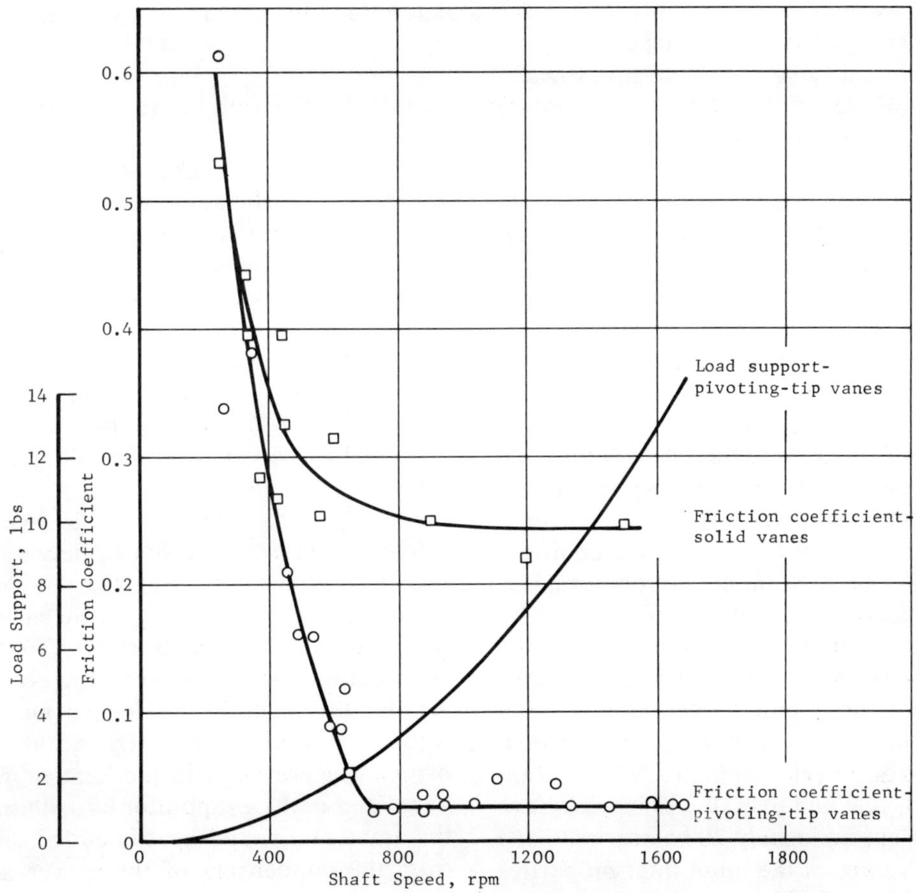

Fig. 6. Friction and load support.

Working Fluids for Solar Rankine Heat Pumps

Robert A. Allen
Research Supervisor

Dr. L.I. Stiel
Research Chemical Engineer

Research and Development Department
Specialty Chemicals Division
Allied Chemical Corporation
Buffalo, N.Y.

Considerable effort is presently being directed to the utilization of solar energy for the heating and cooling of buildings. Toward this goal, we have investigated the properties of working fluids and their potential effect on the performance of a solar powered heat pump.

Several recent studies have been conducted for dual cycle systems powered by solar energy. Prigmore and Barber (ref. 1) described a demonstration unit utilizing R-113 in the power cycle and R-12 in the compression cycle. Sargent and Teagan (ref. 2) proposed a solar powered heating and cooling unit with R-114 as the power fluid and R-22 in the compression cycle. A dual cycle system using R-11 as the working fluid in both cycles is being constructed for both cooling and heating applications (ref. 3). An economic evaluation of the dual cycle system for solar powered cooling has also been conducted (ref. 4). In addition, a solar powered turbine with monochlorobenzene as the working fluid has been described (ref. 5).

In this study, the effect of the thermodynamic properties of the working fluid(s) on the performance of a solar powered heat pump system has been analyzed. The system consists of a vapor compression cooling or heating cycle combined with a Rankine power cycle. Cycle and overall efficiencies have been calculated for approximately 20 halocarbons and other substances. Values of the input fluid properties which result in optimum efficiencies have also been determined for upper temperatures of 180 and 240F.

Analysis of Real Working Fluids

DUAL CYCLE HEAT PUMP CALCULATIONS. Figure 1 is a schematic diagram of the Rankine cycle vapor compression system. Solar heated water from storage is circulated to provide heat input Q_B to the boiler of the power cycle. The working fluid is heated in the boiler at constant pressure, and is then expanded to the condenser pressure in a work production device W_{EXP} to provide the power for the compression cycle W_{COMP}. The fluid leaving the expander is cooled and condensed, and the liquid is returned to the boiler by means of a pump. The power for the pump W_P is also provided by the expander.

The compression cycle can operate either in a cooling or heating mode. For the cooling mode, heat is removed from the building air by vaporizing the working fluid in the evaporator Q_E. The fluid leaving the evaporator is compressed to the condenser pressure, and is subsequently condensed and subcooled. The liquid is then expanded adiabatically to the evaporator pressure. In the heating mode, the fluid is vaporized in the evaporator by withdrawing heat from the ambient air, and heat is added to the building air from the condensers of the power and compression cycles.

The efficiency of the power cycle is

$$\eta = \frac{W_{EXP} - W_P}{Q_B}$$

In the cooling mode, the coefficient of performance of the compression cycle is

$$COP = \frac{Q_E}{W_{COMP}}$$

Since $W_{EXP} = W_P + W_{COMP}$, the overall coefficient of performance is

$$OCOP = \frac{Q_E}{Q_B} = \eta \cdot COP$$

In the heating mode,

$$COPH = \frac{Q_{CC}}{W_{COMP}} = 1 + COP$$

and the overall coefficient is

$$OCOPH = \frac{Q_{PC} + Q_{CC}}{Q_B} = 1 + \eta \cdot COP$$

The steps of the dual cycle system are shown on a pressure-enthalpy diagram in Fig. 2. In terms of the points on this figure, the efficiencies are given by

$$\eta = \frac{(H_3 - H_4) - (H_2 - H_1)}{(H_3 - H_2)}$$

$$COP = \frac{(H_6 - H_5)}{(H_7 - H_6)}$$

$$OCOP = \frac{(H_6 - H_5)}{(H_3 - H_2)}$$

A computer program was utilized in this study to calculate power, compression, and overall efficiencies for a number of possible working fluids for a dual-cycle heat pump.

REQUIRED FLUID INPUT PARAMETERS. The input parameters for each fluid are the normal boiling point, critical temperature, critical pressure, critical compressibility factor, molecular weight, and ideal gas heat capacities at 300 and 500K. Vapor volumes, entropies, and enthalpies are calculated by use of the Martin-Hou equation of state (ref. 6) with constants calculated with the equation of Riedel-Plank-Miller (ref. 7). Liquid enthalpies are calculated from the saturated vapor values by use of the Clapeyron equation. Saturated liquid volumes are calculated from the equation of Yen and Woods (ref. 8). The ideal gas heat capacity was assumed to vary linearly with temperature. An option of the program also enables the calculation of efficiencies for each substance with constants derived from experimental data.

REQUIRED CYCLE PARAMETERS. For computing the power cycle efficiencies, the upper temperature of the fluid before expansion T_U, condenser temperature T_{PC}, and maximum pressure limit are specified. The actual pressure in the boiler P_B is determined by equating the entropy at T_U to the entropy of the saturated vapor in the power condenser T_{PC}. If these conditions would cause the expansion to occur in the two-phase region, P_B is taken as the saturated vapor pressure at T_U, and the temperature after expansion is found by equating the entropy after expansion to that of the saturated vapor at the expander inlet. The actual work done by the expander is determined by multiplying the isentropic enthalpy drop across the expander by a specified efficiency E_{EXP}. The pump work is similarly determined by dividing the ideal enthalpy rise by an efficiency E_P. If the working fluid is superheated vapor after expansion, the efficiency can be improved by exchanging heat between the vapor and compressed liquid leaving the pump. For each value to T_U the efficiency is calculated for specified percentages of heat regeneration (Regen).

For the compression cycle, the evaporator temperature T_E, the temperature of the superheated vapor at the compressor inlet T_{SH}, the condenser temperature T_{CC}, and the temperature of the subcooled liquid T_{SC} are specified. The temperature at the compressor outlet is determined by equating the entropy at this condition to that at T_{SH} and the evaporator pressure P_E. If this condition requires compression into the two-phase region of the fluid, the outlet condition is taken as saturated vapor at the condenser pressure P_{CC}, and the inlet temperature is determined by equating the entropies at the compressor inlet and outlet. The actual work required in the compressor is found by dividing the enthalpy rise for the ideal case by the specified efficiency E_{COMP}.

For each working fluid and cycle condition, the output of the computer program supplies the heat and work requirements per ton of cooling, the efficiencies η, COP, and OCOP, the mass flow rates for the power and compression cycles (M_P and M_{COMP}), the volumetric flow rate of fluid entering the compressor (V_{COMP}), and the pressure P_E, P_{CC}, and P_B. The temperatures, pressures, volumes, enthalpies, and entropies at various points in the power and compression cycles can also be presented.

EFFICIENCIES OF KNOWN WORKING FLUIDS. The dual cycle program was utilized to determine efficiencies for a number of known refrigerants and other substances having desirable thermodynamic properties. For these evaluations the T_U was constrained between 180 and 240F, a range that might be realistically attained with flat-plate collectors and storage systems. The evaporator temperature was varied from 0 to 40F to simulate heating and cooling conditions. The condenser temperatures for both the power and compression cycles were equalized in any given analysis and were varied between 95 and 125F.

Tables 1 and 2 provide the 180F performance information for single-fluid systems operating in the cooling and heating modes, respectively. The specified cycle conditions are given at the top of each table. The fluids are listed in ascending boiling point order since this quantity helps define the type of expander and compressor one would employ.

For the Table 1 cycle conditions, the Carnot efficiency of the power cycle is 0.085, and since the power efficiency realistically would be reduced by approximately 20% because of irreversibilities of the expander and pump, the maximum value of η attainable with any fluid for this case is about 0.068. It can be seen that this maximum efficiency is approached for several of the fluids.

TABLE 2. CYCLE EFFICIENCIES—HEATING MODE

T_U = 180F, Regen = 0.5, T_{PC} = 115F, E_{EXP} = 0.8, E_P = 0.7
T_E = 0F, T_{SH} = 15F, T_{CC} = 115F, T_{SC} = 110F, E_{COMP} = 0.8

Substance	η	COPH*	OCOPH	P_E (psia)
R-22	.053	3.39	1.126	38.5
R-500	.060	3.38	1.142	28.0
R-12/31	.060	3.46	1.147	23.7
R-12	.061	3.40	1.147	22.45
R-134a	.061	3.32	1.141	21.4
R-152a	.062	3.43	1.150	19.8
R-134	.063	3.31	1.146	16.0
R-31/114	.061	3.53	1.155	11.6
iso-Butane	.070	3.37	1.166	11.5
R-142b	.070	3.46	1.173	10.1
1-Butene	.071	3.46	1.175	9.1
R-12B1	.069	3.53	1.174	8.3
R-114	.071	3.35	1.167	5.85
R-133a	.072	3.50	1.180	5.0
R-11	.071	3.61	1.186	2.4
R-12B2	.070	3.61	1.182	2.5
R-142	.071	3.62	1.187	1.2
R-133	.074	3.50	1.185	0.8

*Assumes 12,000 Btuh entering the evaporator.

TABLE 1. CYCLE EFFICIENCIES—COOLING MODE

T_U = 180F, Regen = 0.5, T_{PC} = 125F, E_{EXP} = 0.8, E_P = 0.7
T_E = 40F, T_{SH} = 65F, T_{CC} = 125F, T_{SC} = 120F, E_{COMP} = 0.8

Substance	T_B (°F)	$C°_P\|300K$ (kcal/mole °K)	η	COP	OCOP	P_E (psia)	P_{CC} (psia)	P_B (psia)	$\dot{V}_{COMP}$ (cfm/ton)	M_{COMP} (lb/min/ton)	M_P (lb/min/ton)
					T_B = −50–0F						
R-22	−41.7	13.70	.044	3.72	.163	82.1	284.3	459.4	2.2	3.0	17.7
R-500	−28.3	17.44	.050	3.76	.189	60.8	215.2	379.3	2.9	3.4	16.0
R-12/31	−21.3	15.33	.051	3.86	.195	52.0	119.2	320.1	3.2	3.3	15.3
R-12	−19.0	17.36	.052	3.81	.197	49.1	175.6	311.1	3.5	4.0	18.3
R-134a	−15.9	20.47	.051	3.73	.191	49.0	187.4	348.5	3.5	3.4	15.7
R-152a	−12.7	17.57	.052	3.84	.201	45.8	177.8	324.0	3.5	2.0	9.1
R-134	−3.7	22.54	.053	3.73	.199	39.4	166.7	330.8	4.0	3.1	13.4
					T_B = 0–50F						
R-31/114	9.6	15.25	.052	3.96	.207	28.3	119.2	214.9	5.0	2.5	10.7
iso-Butane	10.7	23.24	.060	3.84	.229	26.6	102.8	201.9	6.1	1.8	6.3
R-142b	15.4	20.26	.060	3.92	.235	24.8	103.7	210.7	6.0	2.8	10.0
1-Butene	20.4	21.14	.061	3.92	.238	22.5	91.9	184.5	6.7	3.9	5.6
R-12B1	24.6	17.89	.059	3.98	.234	20.1	82.3	159.4	7.2	4.5	16.7
R-114	38.6	27.42	.061	3.85	.233	15.0	67.4	140.5	9.7	4.6	15.7
R-133a	42.7	21.56	.062	3.97	.244	13.7	67.7	142.2	9.4	2.8	9.8
					T_B = 50–100F						
R-11	75.1	18.81	.061	4.08	0.25	6.85	35.6	76.7	17.4	3.0	10.2
R-12B2	75.9	21.84	.060	4.08	0.244	6.8	34.6	72.5	17.7	4.6	16.4
R-142	94.9	19.83	.061	4.10	0.251	4.0	26.1	61.7	25.8	1.9	6.4
					T_B > 100F						
R-113	117.4	30.44	.063	4.02	.254	2.6	16.8	40.8	42.2	3.7	11.6
CCl_4	169.5	19.97	.058	4.17	.242	0.8	6.4	15.6	112.4	2.5	8.9
Toluene	230.8	24.96	.065	4.19	.272	0.15	1.8	5.95	507.1	1.2	3.8
Chlorobenzene	268.8	23.20	.059	4.20	.248	0.1	0.8	2.6	1,193.3	1.4	4.8

The fluids listed in Table 1 are fluids having higher values of η and OCOP when compared to many other unlisted fluids with similar boiling points. It can be seen that in general the efficiencies for these fluids increase with increasing boiling point. Fluids with high boiling points such as R-11, R-113, R-142, and toluene have the largest values of η and OCOP. However, these fluids would result in very low evaporator pressures for a single fluid system, particularly for the heating mode. Several lower boiling fluids such as R-133a, R-114, and R-142b also have relatively high power and overall efficiencies and hence might prove to be more practical.

The use of the same working fluid in the power and compression cycles has several advantages for simplified equipment design. However, systems utilizing a fluid with high values of T_B and η for the power cycle and a different fluid with a lower boiling point in the compression cycle also should be considered. In Table 3, efficiencies are presented for several possible fluid pairs for operation in the cooling mode.

TABLE 3. OVERALL COEFFICIENTS OF PERFORMANCE FOR TWO-FLUID SYSTEMS—COOLING MODE CONDITIONS

$T_U = 180F$, Regen = 0.5, $T_{PC} = 125F$, $E_{EXP} = 0.8$, $E_P = 0.7$, $T_E = 40F$

	Refrigerant			
Power Fluid	R-12	R-22	R-12/31	R-31/114
R-11	.232	.227	.235	.242
R-113	.240	.234	.243	.249
R-133a	.236	.231	.239	.246
R-142	.234	.228	.237	.243
Toluene	.248	.242	.251	.257
R-114	.232	.227	.235	.242

In Table 4, power and overall efficiencies are listed for all the fluids considered for T = 240F, Regen = 0.5, and cooling mode conditions in the compression cycle. The values of P_B and M_P for these conditions are also included for each fluid.

CYCLE PARAMETERS INFLUENCING HEAT PUMP PERFORMANCE. In addition to variations in the efficiencies due to the properties of the working fluids and design parameters of the expansion and compression equipment, the cycle conditions have an important effect on the results. Values of the efficiencies were computed for all the fluids for varying values of the upper, condenser, and evaporator temperatures. The overall coefficient of performance increases strongly with increasing values of T_U, as shown in Fig. 3. The overall coefficient of performance can also be increased by reducing the condenser temperature as shown in Fig. 4. In Fig. 5, the effect of variation of the evaporator temperature is indicated. For these comparisons the fraction of the heat regeneration was found to have an insignificant effect on the overall efficiency at these solar conditions.

TABLE 4. CYCLE EFFICIENCIES FOR T_U = 240F—COOLING MODE CONDITIONS

Regen = 0.5, $T_{PC} = 125F$, $E_{EXP} = 0.8$, $E_p = 0.7$

Substance	η	OCOP*	P_B (psia)	M_p (lb/min/ton)
R-22	.076	.284	733.5	9.7
R-500	.081	.305	630.5	9.5
R-12/31	.087	.336	530.5	8.5
R-12	.085	.324	521.1	10.5
R-134a	.080	.298	603.7	9.6
R-152a	.088	.337	562.9	5.1
R-134	.081	.301	587.6	8.5
R-31/114	.094	.374	376.4	5.6
iso-Butane	.099	.381	384.5	3.6
R-142b	.100	.390	394.7	5.7
1-Butene	.104	.408	345.2	3.0
R-12B1	.103	.408	291.8	9.0
R-114	.104	.400	281.1	8.6
R-133a	.106	.419	297.6	5.4
R-11	.111	.454	157.8	5.3
R-12B2	.109	.445	145.9	8.4
R-142	.113	.464	141.0	3.2
R-113	.113	.452	90.0	6.0
CCl_4	.111	.462	37.2	4.4
Toluene	.118	.495	16.9	2.0
Chlorobenzene	.117	.493	8.9	2.3

*Assumes 12,000 Btuh entering the evaporator.

Optimum Physical Properties of Working Fluids

In order to gain insight into the effect that the physical properties of a fluid have on its thermodynamic performance, a grid-type computer program was utilized. The inputs to this program, in addition to the specified cycle conditions, are the normal boiling point, critical pressure, critical compressibility factor, and the ideal gas heat capacities at 300 and 500K. The critical temperature is generated within the program based on the relationship

$$\frac{T_B}{T_C} = \frac{(13.308 - 37.696Z_c}{(14.413 - 45.575Z_c + \log P_c)}$$

The output from this program is the efficiency, η, of the power cycle and the overall coefficient of performance, OCOP, of the dual cycle.

A range of these input values having the proper physical property interrelationships to those found among real fluids was fed into the computer. From the results obtained it was found that the most influential physical properties affecting the efficiency of the fluids are the ideal gas heat capacity and boiling point. Of somewhat less influence are the critical compressibility factor and critical pressure.

A study of Figs. 6, 7, and 8 indicates the effect of the heat capacity and boiling point of a fluid on the power cycle, vapor compression cycle, and overall dual cycle efficiencies when the maximum upper temperature is limited to 180F.

Similarly, Figs. 9 and 10 show the effect of critical compressibility factor and critical pressure on the power cycle and overall dual cycle efficiencies of single fluids.

As the permissible upper temperature is raised to 240F, the maximum overall cycle efficiency is obtained with fluids having slightly lower ideal gas heat capacities than for fluids operating at 180F. This is shown in Fig. 11. The effects of critical compressibility and critical pressure on the fluid performance at a 240F upper temperature is shown in Figs. 12 and 13.

Conclusions

Several of the real fluids considered, particularly those with high boiling points, have close to the optimum overall efficiencies possible for a solar heat pump with an upper temperature of 180F. Among the known fluids evaluated, R-11, R-113, R-142, and toluene exhibit the highest power cycle and overall efficiencies for single-fluid systems. Several lower boiling fluids such as R-133a, R-114, and R-142b also have relatively high power and overall efficiencies. A two-fluid system utilizing a substance with a high power efficiency in the power cycle and a lower boiling fluid such as R-12/31 or R-22 in the compression cycle should also give good results. The variation in the values of the overall coefficient of performance for fluids with a wide range of boiling points is less at 180F than at higher upper temperatures.

Since the halocarbon fluids should be entirely stable at upper temperatures projected for solar powered applications, they appear to be the most attractive on an overall basis. The halocarbons, in general, are much less flammable than hydrocarbon fluids under consideration.

In this study the effects of certain fluid properties on the performance of a solar heat pump also have been determined. For each input combination of boiling point, critical pressure, and critical compressibility factor, a certain value of the ideal gas heat capacity results in optimum overall efficiencies for specific cycle conditions. These optimum values of OCOP increase with increasing boiling points, critical compressibility factors, and critical pressures for the cycle conditions considered.

The actual choice of a working fluid for a solar powered heat pump will involve a number of factors in addition to the cycle efficiencies. The effect of the transport properties of the working fluid on the equipment design is an important consideration. Further, the choice of a working fluid depends on design considerations such as the size and cost parameters of the heat exchangers and compression and expansion equipment required for each system. The tabulated information on flow rates and evaporator, condenser, and boiler pressures should help to determine a suitable fluid for each heat pump application.

ACKNOWLEDGMENTS

This paper was prepared with the support of the National Science Foundation under Grant GI 42506.

NOMENCLATURE

C°_p = ideal gas heat capacity, cal/mole° K
COP = coefficient of performance for compression cycle in cooling mode
COPH = coefficient of performance for compression cycle in heating mode
E_{COMP} = efficiency of compressor in compression cycle
E_{EXP} = efficiency of expander in power cycle
E_p = efficiency of pump in power cycle
H = enthalpy of fluid, Btu/lb mole
M_{COMP} = mass flow rate in compression cycle, lb/min/ton
M_p = mass flow rate in power cycle, lb/min/ton
OCOP = overall coefficient of performance for dual loop system in cooling mode, prod-

uct of power cycle efficiency and COP

OCOPH = overall coefficient of performance for dual loop system in heating mode, 1 + OCOP

P_B = pressure in boiler, psia

P_c = critical pressure, atm

P_{CC} = pressure in compression cycle condenser, psia

P_E = pressure in evaporator, psia

Q_B = heat added to working fluid in boiler, Btuh

Q_{CC} = heat removed from fluid in compression cycle condenser, Btuh

Q_E = heat added to fluid in evaporator, Btuh

Q_{PC} = heat removed from fluid in power cycle condenser, Btuh

Q_{REGEN} = heat recovered and utilized, Btuh

Regen = fraction of maximum possible heat recovery from superheated vapor in power cycle added to fluid entering boiler

T_B = normal boiling point, °F or °K

T_{BOIL} = boiling point, at P_B, °F

T_c = critical temperature, °K

T_{CC} = temperature of fluid in compression cycle, °F

T_E = temperature of fluid in evaporator, °F

T_{PC} = temperature of fluid in power cycle condenser, °F

T_{SC} = temperature of subcooled fluid in compression cycle, °F

T_{SH} = temperature of vapor entering compressor, °F

T_U = maximum temperature of power cycle, temperature of vapor entering expander, °F

V_{COMP} = volumetric flow rate of fluid entering compressor, cu ft/min/ton

W_{COMP} = work done on fluid in compressor, Btuh

W_{EXP} = work done by fluid in expander, Btuh

W_p = work done on fluid in pump, Btuh

Z_c = critical compressibility factor

α = ratio of ideal gas heat capacity at 500K to value at 300K

η = efficiency for power cycle

REFERENCES

1. Prigmore, D.R., and Barber, R.E., 9th IECEC Conf., Paper No. 749026, San Francisco, August 1974.
2. Sargent, S.L., and Teagan, W.P., ASME Winter Annual Meeting, Paper No. 73-WA/Sol-8, Detroit, November 1973.
3. Biancardi, F.R., UARL Report No. N951923-1, September 1974.
4. Curran, H.M., Hittman Assoc. Report No. HIT-581, August 1974.
5. Bronicki, L.Y., 7th IECEC Conf., Paper No. 729057, San Diego, September 1972.
6. Reid, R.D., and Sherwood, T.K., *The Properties of Gases and Liquids*, McGraw-Hill, New York, 1966, pp. 80-82.
7. Miller, D.G., *J. Phys. Chem.*, *68*, 1399 (1966).
8. Yen, L.C., and Woods, S.S., *AIChE J.*, *12*, 95 (1966).

DISCUSSION

Q Apart from the power cycle and thermophysical properties of the fluids, their long term stability is a major problem, with toxicity, corrosiveness, etc., being also very significant. Have you considered these?

A We are aware of the variation in stability, toxicity, etc., that exists among the fluids. However, the relative differences among these fluids are highly dependent on the exact cycle condition and materials of construction a fluid would be subjected to. For example, solar systems utilizing flat-plate collectors probably would not have fluids subjected to temperature above 200F. Under these conditions, all the fluids listed in Table 1 with the possible exception of l-butene, R-12B1, and R-12B2 should be reasonably stable. As for toxicity, probably the CCl_4, toluene, and chlorobenzene are the most toxic compounds among those listed.

The thrust of this study was primarily to investigate the thermodynamic performance of the fluids and focus on the efficiency that one might expect to achieve under solar cycle conditions.

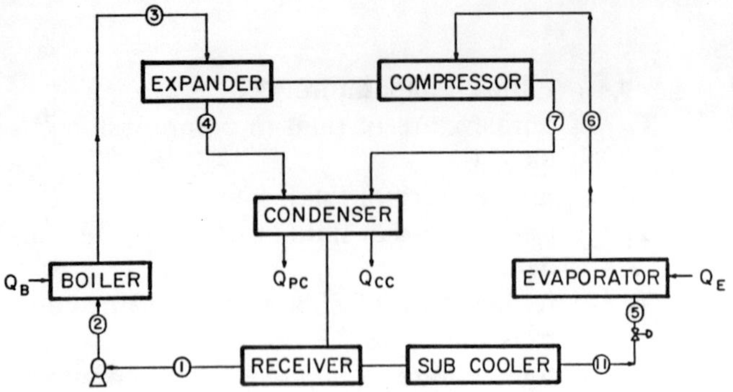

Fig. 1. Solar powered heat pump.

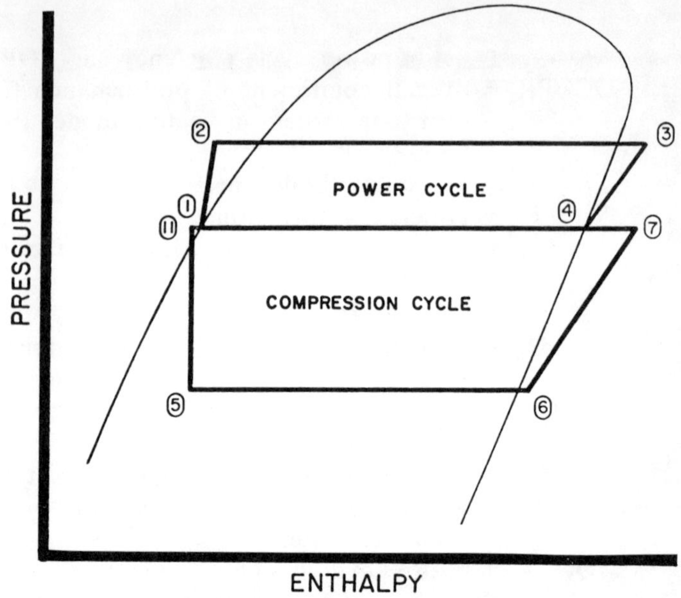

Fig. 2. Pressure-enthalpy diagram for dual cycle system.

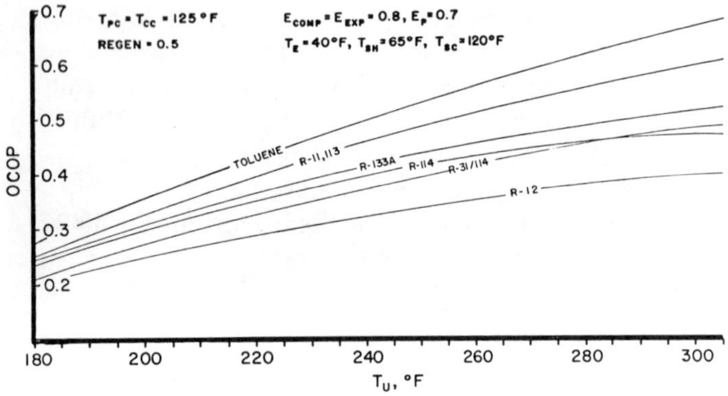

Fig. 3. Variation of OCOP with T_U.

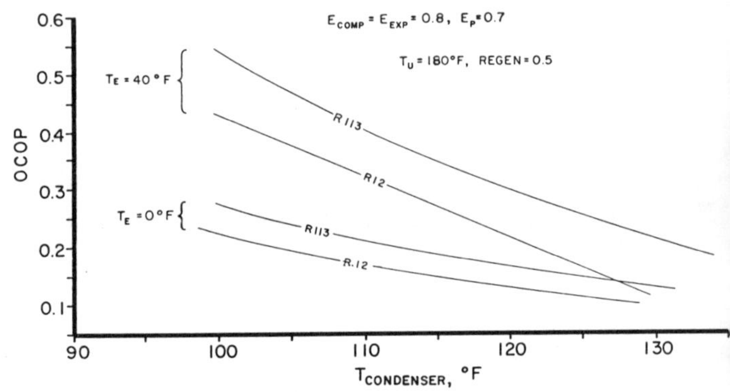

Fig. 4. Effect of condenser temperature on OCOP.

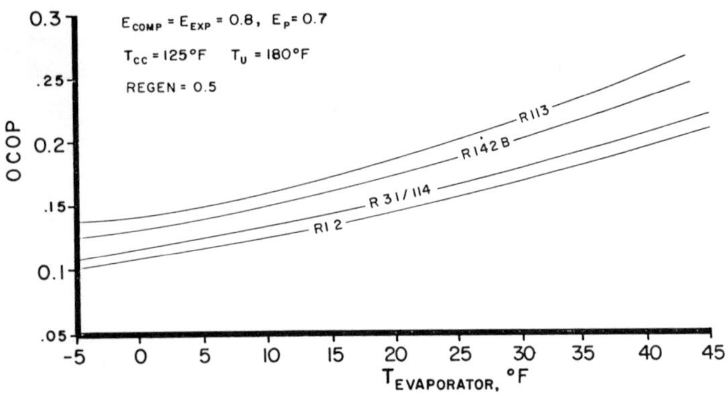

Fig. 5. Effect of T_E on OCOP.

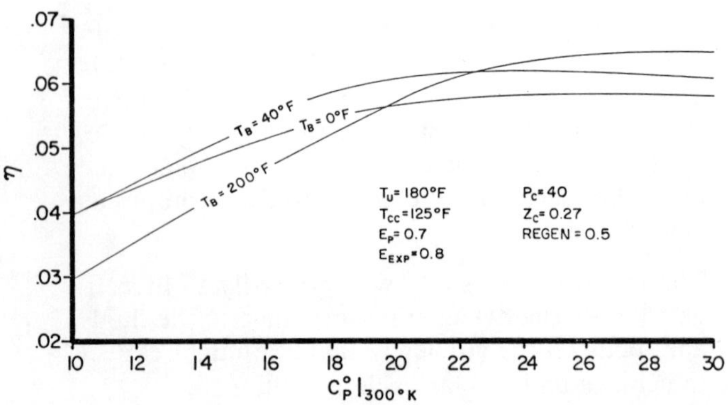

Fig. 6. Variation of η with $C^o{}_p|_{300K}$ and T_B.

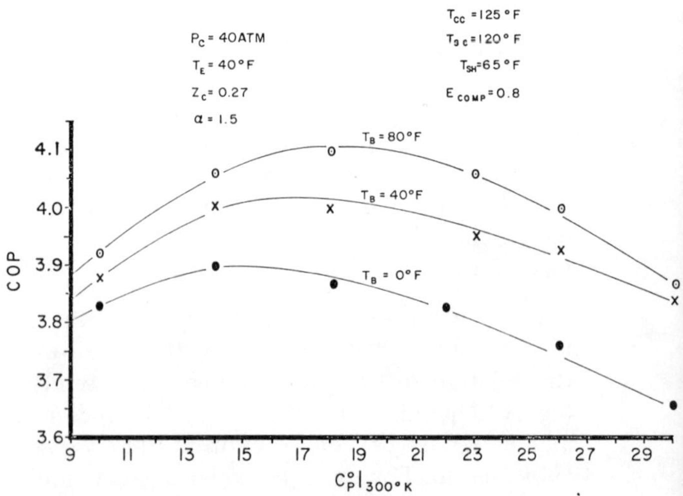

Fig. 7. Variation of COP with $C^o{}_p|_{300K}$ and T_B.

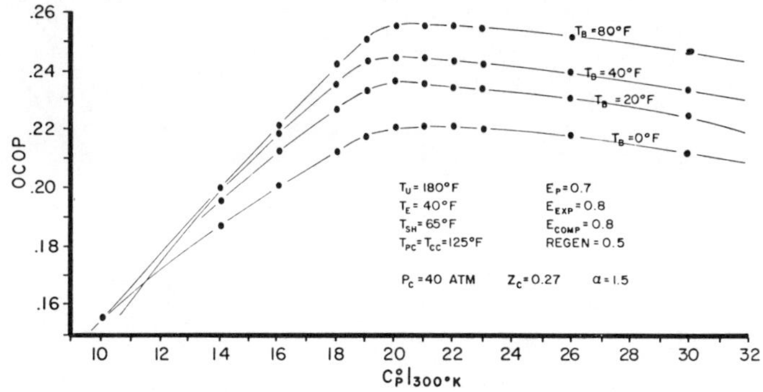

Fig. 8. Variation of OCOP with $C^o_p|_{300K}$ and T_B.

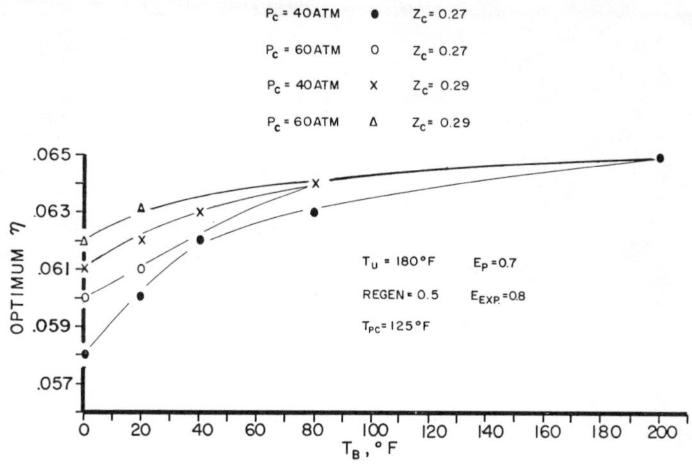

Fig. 9. Variation of optimum η with T_B, P_c, and Z_c.

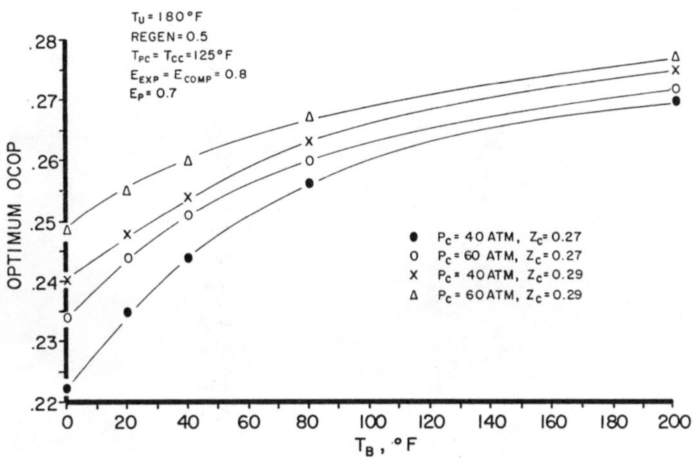

Fig. 10. Variation of optimum OCOP with T_B, P_c, and Z_c.

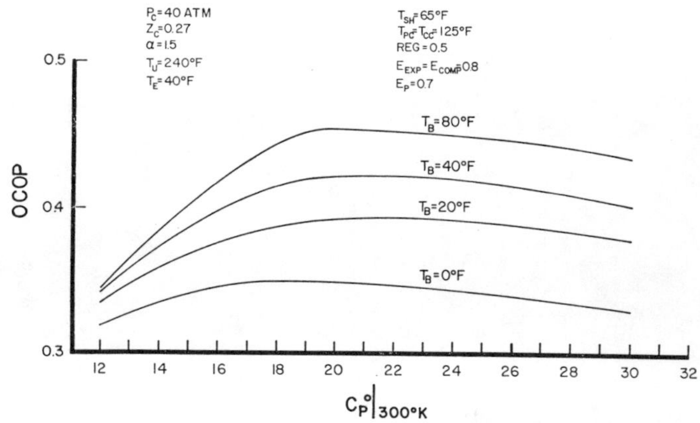

Fig. 11. Variation of OCOP with $C^o_p|_{300K}$ and T_B.

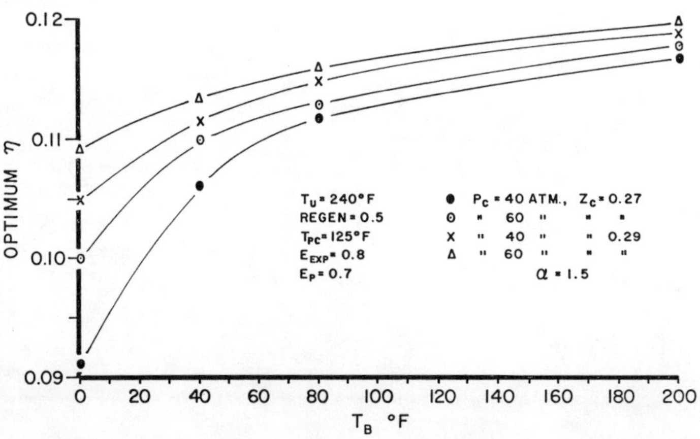

Fig. 12. Variation of optimum η with T_B, P_c, and Z_c.

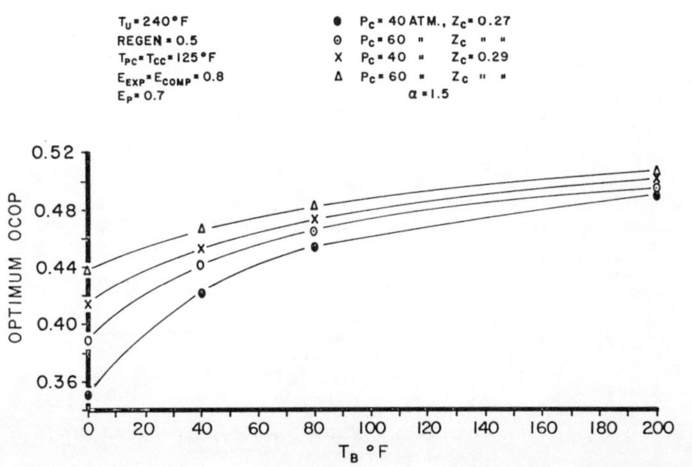

Fig. 13. Variation of optimum OCOP with T_B, P_c, and Z_c.

Part III

Chairman:
Lawrence G. Spielvogel

President, Lawrence G. Spielvogel, Inc.
Wyncote, Pa.

Solar-Boosted Super Heat Pumps

Dr. James B. Comly
Manager, Thermal Branch

Heinz Jaster
Mechanical Engineer

Dr. James Quaile
Mechanical Engineer

Corporate Research and Development
General Electric Company
Schenectady, N.Y.

The heat pump has been with us for many years, its evolution governed primarily by cost and reliability considerations. Until recently the heat pump was viewed as an air conditioning unit which could be used in heating; consequently, performance tradeoffs favored cooling performance.

Interest in residential heating has been heightened by the current economic-energy climate. Traditional heating systems employing combustion furnaces (oil and gas) are currently acceptable economically but rely solely on scarce resources. Electrical resistance heating, while requiring little initial cost, results in the highest annual heating cost. However, this type of heating requires less expenditure of scarce fuels since electrical utilities presently use these fuels for only 33% of the total energy requirements (*Federal Power Commission News*, 5 June 1975).

The heat pump, by using solar energy collected and stored in the sensible heat of the atmosphere, is able to reduce both the annual heating cost and the consumption of scarce fuels to the point where it is energy competitive with optimally sized solar systems using radiant collector panels. Advanced designs of heat pumps stressing heating performance could surpass optimum radiant solar systems; the coupling of radiant solar systems with heat pumps would result in still lower energy consumption at somewhat increased life cycle cost.

Heat Pump Efficiency

A heat pump is by definition a device which causes heat to flow from a region of low temperature to a region of higher temperature. This direction of flow is opposite to the direction required by the second law of thermodynamics so that some external energy must be added to "pump" the heat "uphill." Efficiency then can be defined as the ratio of the desired effect (heating or cooling) to the energy required, using consistent power units such as Btuh, kw, etc. When considering heating by a heat pump, the desired effect is the thermal energy arriving at the higher temperature level (house). The first law of thermodynamics tells us that this energy will be the sum of that extracted from the cold region plus the energy required by the heat pump; therefore the total energy arriving at the higher temperature level in the house will always be greater than the energy required to run the device. This means the electrical efficiency, usually called coefficient of performance (COP), will always be greater than one.

Effect of Temperature

Intuition suggests that the larger the temperature difference over which the heat must be pumped the greater the energy required and therefore the smaller the COP. The COP then varies with the temperatures in the system. The actual value of COP depends on the equipment used. There is an absolute upper limit to the COP between two given temperatures. The upper limit is the Carnot COP, given by the expression

$$COP_{max} = \frac{T_H}{T_H - T_L}$$

where

T_H = high temperature in absolute temperature

T_L = low temperature in absolute temperature

Consider a house maintained at 70F in a 30F ambient. The temperature difference is only 40F as shown in Fig. 1. Applying the above formula to these values yields a COP of 13.25. This value, however, is misleading since the actual heat pump must pump across a much wider temperature gap. In most heating systems heat is supplied at a central location and is distributed via sensible heat in the distribution fluid, e.g., via a hot air system in which house air is heated, then forced through ducts to distribute heat throughout the house. The total rate of heat distributed is proportional to the product of the mass flow rate of the air and its temperature rise. Therefore, for a given heat rate the flow rate required increases as the temperature rise decreases. In a heat pump, system efficiency is increased by minimizing this ΔT, i.e., delivering air at 100F instead of 130–160F as in combustion systems. There is a practical lower limit to this delivery temperature because of ducting and comfort considerations. Since at lower delivery temperatures more air is required, significant room air velocities are generated, which in combination with the lower temperatures may produce an unpleasant drafty effect. Good air distribution can minimize these effects, however.

A similar but smaller temperature change in the air blown across the outside heat exchanger is necessary to extract heat from it. These two temperature differences are necessary for any heat pump no matter how well it is designed, thus decreasing the maximum obtainable COP to the vicinity of 7.5. To transfer heat between the refrigerant fluid and surrounding air, additional temperature differences are necessary. For reasonably sized heat exchangers this difference is approximately 20F. The total temperature difference will amount to approximately 110F, as indicated on Fig. 1, yielding a maximum COP of 5.2. Real heat pumps operate at approximately half this value!

Efficiency of Real Heat Pumps

The COP of 5.2 could be obtained only by a unit operating on the Carnot cycle and employing no fan power to move air either inside or out. Figure 2 shows the degradation in efficiency caused by real equipment. The top line represents the previously calculated Carnot efficiency taking into account total ΔT. The effect of using the standard vapor compression cycle (all processes considered reversible except for expansion) is indicated by the second line. Using this same ideal refrigeration cycle but considering motor and compressor inefficiencies, the COP drops even lower, in fact very close to measured performance of existing devices. A true COP must include all electric power requirements, so the bottom line shows the effect of fan power.

The conclusion to be reached from this figure is that the improvements possible in the standard heat pump are limited and that maximum COP's at an outdoor temperature of 30F are on the order of 4 or 5 rather than 13, as a cursory examination may suggest.

Heat Pump Capacity

One undesirable feature of heat pumps is seen in Fig. 3. The heating capacity (curve labeled Heat Pump), the ability to deliver heat to the house, decreases with decreasing outside temperatures. This is contrary to the heat required (Demand of House), which increases at the colder temperatures. At temperatures above the intersection of these two curves (known as the theoretical balance point) the heat pump can supply all the heat required, but below this point some additional energy must be supplied. For convenience and minimum cost, the supplementary heat is supplied by resistance heaters built into the indoor air handler.

Supplementary heat is added to the house at a COP of 1, thereby reducing the total COP of the system. No matter how efficiently the heat pump operates, the greater the required makeup heat required, the lower the system efficiency. This suggests that the performance of the heating system is not only a function of heat pump COP but of the size of the heat pump used (changes in the balance point) and the seasonal temperature distribution to which the house is exposed.

Figure 4 shows the total heating demand of an 1800 sq ft house in the Boston area and indicates the result of using a 2½—ton heat pump (which results in a theoretical balance point of 30F). The area under the outermost curve is representative of the total heat required, and it also indicates the input energy needed if the house were heated with only resistance heat (COP = 1). The energy required by the heat pump system is proportional to the area under the next lower curve; it includes resistance makeup heat. The lowest curve is the energy input to the heat pump alone; therefore the difference between the two lower curves indicates supplementary heat. Totals are listed on this figure and indicate that the heat pump system required 13,900 kwh to supply 26,500 kwh worth of heat. The ratio of the total seasonal demand (26,500) to the energy required (13,900) is known as the seasonal performance factor (SPF = 1.91) and is the best measure of a heat pump system's performance.

Super Heat Pump

Figure 4 indicates that although the supplementary heat is a small fraction of the total heat supplied, it is the major fraction at lower outdoor temperatures and dilutes any pure efficiency increase in the heat pump itself in this range. If the heat pump could be made to provide more heat at low temperatures, reducing the supplementary heat required, seasonal performance would increase. Therefore, in seeking to improve the heat pump, our considerations include ideas for both efficiency and capacity improvements.

Several modifications to the standard vapor compression cycle were considered which could benefit efficiency, capacity, or both. Some examples of such cycles are shown in Fig. 5 along with estimated performance. The heating capacity has been normalized with respect to compressor size (displacement), and all cycles other than the simple cycle are assumed to have two equally sized compressors. The assumptions governing this chart are the same as those of Fig. 2, i.e., 120F condenser, evaporator 10F less than ambient, and all processes reversible except expansion.

Parallel Compression

The simplest way to increase capacity without major modification is to use parallel compressors. The two compressors, which can be of different size, can be controlled independently to accommodate differing heating or cooling loads, with a closer match between heat pump capacity and demand.

Cascade

The cascade cycle allows the use of two different refrigerants, each suited to the temperature range over which it will operate. Since each half of the cycle must pump heat over a portion of the total temperature rise, the capacity and COP of each will increase. Since the overall COP of a cascade system is given by

$$COP = \frac{(COP)_H \times (COP)_L}{(COP)_H + (COP)_L - 1}$$

where L = low side
H = high side

each heat pump must achieve very high COP to surpass a single-stage heat pump. An important advantage of this cycle is that the intermediate temperature region between the heat pumps affords an excellent area for thermal storage.

Series Compression (Compound)

The decline in heating capacity at low outdoor temperatures can be reduced by placing two compressors in series. The reduced pressure ratio across each compressor reduces clearance volume effects in reciprocating units, or vane leakage in rotary compressors.

Turbocharged

The primary cause of the low temperature decline in capacity is the decrease in refrigerant density at the compressor inlet. The turbocharged cycle increases this density by using energy extracted by an expander from the discharge vapor to compress the suction vapor in a high volumetric flow, low pressure ratio precompressor.

Two Stage (Compound)

The final cycle shown, the two stage, not only increases low temperature capacity but substantially increases the cycle COP over that of the simple cycle by a combination of reduced clearance volume effects (series compressors) and thermodynamic cycle improvements.

Modulated Heat Pump

The COP obtained in an actual piece of equipment depends not only on the cycle configuration and the efficiencies of each component, but on the correct sizing of these items to form the complete system. For instance, COP can be increased by inserting a smaller compressor in a given unit since the net effect is to oversize the heat exchangers and thus reduce the temperature drop across them. Naturally, heating capacity will decrease in this case. However, in Fig. 3, the heat pump is seen to have excess heating capacity above the balance point, so a reduction while at modest outdoor temperatures might be beneficial.

Instead of switching compressors, assume that a variable speed compressor is used which can be controlled to a desired value of heating capacity. Now by increasing or decreasing compressor rpm above or below the nominal compressor speed, the house demand can be matched both below and above the old balance point temperature (within equipment limitations). At temperatures below the old balance point temperature, system COP is increased, since less supplemental heat is required, while at temperatures above this point the heat exchangers are essentially oversized, thereby increasing COP. In addition, the ability of the heat pump to run continuously at temperatures above the balance point results in still higher COP and increased reliability, because the effects of cycling on and off are eliminated.

Solar Space Heating

Considerable excitement exists presently among members of the HVAC industry, as well as among the public, about the use of direct solar radiant energy for space heating. Most people expect solar energy to supply virtually the entire heating need of a house equipped with solar collectors. Collectors now being sold and suitable for space heating are priced between \$5 and \$15/sq ft. The cost of collectors constitutes the largest portion of the cost of a solar heating system. It is therefore necessary to optimize the size of the collector array such that large size (and therefore cost) increases do not result in relatively small increases in usable solar heat gained.

Figure 6 demonstrates the usable heat gained for heating a given house (650 Btu/hr °F) with variously sized collector arrays. The area under the curve House Heating Demand represents the seasonal heat demand of the house. The curves marked Solar Heat Collected indicate collected heat for solar collectors of the indicated size averaging 35% efficiency. The area bounded at the top by the Solar Heat Collected curves and at the sides by the House Heating Demand curve represents the usable solar energy. So, for example, a 750 sq ft array satisfies approximately 65% of the building's seasonal demand, whereas a 1925 sq ft array is required for a 100% solar heated house. This estimate does not account for cold and overcast spells, since the weather base used here is an average of a 10-year period. Various studies have shown that the optimal solar contribution is between 30 and 60%, depending on such items as house design, climatic region, and cost of alternative energy. A solar system must therefore be supplemented with a conventional heating plant, and this heating plant must supply between 40 and 70% of the building's heat demand.

Solar Boosted Heat Pumps

Figure 7 shows a schematic diagram of a solar boosted and cascaded heat pump system. The system contains a collector array with circulation pump and liquid-to-liquid heat exchanger, a large thermal storage tank filled with water, a liquid-to-air heat exchanger with circulating pump (to draw heat from thermal storage to the warm air distribution system directly when storage temperatures are above 110F), a heat pump operating between storage and indoors when storage temperatures are between 50 and 110F, a heat pump operating between outdoors and storage when storage temperatures fall below 50F, and a supplementary electrical resistance heater. A major advantage of this configuration is its ability to collect more radiant solar energy with a fixed solar array than conventional systems, since lower collection temperatures allow higher collection efficiencies. The figure also shows predicted seasonal performance for five electrical or electrically complemented heating systems. The solar nonradiant energy is heat extracted from outdoor air by the outdoor heat pump. Analyses were made for eastern New York state. The table shows performance of heat pumps operating at seasonal performance factors of 2 and 3. The higher efficiency heat pump (SPF = 3) does not represent equipment currently available. Rather, it represents the potential performance of an improved heat pump. The solar boosted heat pump is shown to be the most energy-efficient, followed by the super heat pump and a solar heating system without heat pumps.

Figure 8 shows the results of a cost analysis for these five heating systems. Costs are broken down to annual cost of amortization of the installed system, and the annual operating cost for electrical energy required to heat a house for one season. It is interesting to note that the three systems containing heat pumps are the three least expensive to own and operate.

Figure 9 expands on the information in Fig. 8. Here annual cost of ownership and operation (excluding maintenance costs) is given for these same heating systems as a function of cost of electrical energy. The solar assisted heat pump becomes less expensive than an ordinary heat pump system if the price of electricity exceeds \$0.09/kwh. Extrapolation of these curves shows a super heat pump, at the assumed SPF of 3 and the assumed cost increment of 60% over ordinary heat pumps, to be more cost effective than the solar boosted heat pump—until the price of electricity exceeds \$0.20/kwh.

Many arguments have recently been made about the inefficient use of scarce energy when homes are heated electrically. These arguments typically ignore the fact that electricity generation in the United States is based on scarce fuels (oil, gas) for only 33% of total output. Therefore even a house heated by electrical resistance heaters will, on the national average, consume less scarce fuel (oil, gas) than a house equipped with oil or gas furnaces. When one uses electrical heating in its most efficient form (heat pump heating), the use of scarce fuels is further reduced, since even current heat pumps consume approximately half as much electrical energy as resistive heaters.

Figure 10 shows the consumption of raw fuel energy for various heating systems required to supply 100 million Btu of heat to a residence. The primary fuel for the electrical systems can of course be of any type: scarce or nonscarce. Even when one does not differentiate among scarce and nonscarce fuels, the figure shows existing heat pumps to use as little fuel as combustion furnaces. Future heat pumps (SPF = 3) will require approximately 35% less fuel. The figure also shows that solar heating systems, complemented 40% by resistive heaters, require approximately as much fuel as combustion heating systems. The comparable fuel use for heat pump and solar heating systems, together with the much lower cost of heat pumps, leads to the conclusion that heat pumps are superior to solar heating systems. Only solar boosted heat pumps promise much greater conservation of scarce fuel at a reasonable cost.

DISCUSSION

Q Based on the assumptions you have made, do I understand you correctly that if the cost of electricity is less than $0.08/kwh, the conventional heat pump is better than the solar assisted heat pump insofar as heating a system economically, due to higher amortization cost of the solar assisted systems?

A Yes. The solar assisted heat pump becomes more economical than an ordinary pump only if the cost of electricity exceeds $0.08/kwh.

Q Why did your graphs and charts not show heat source temperatures at higher levels when solar energy can be simply collected with minimum effort up to 80F and more?

A Performance graphs only considered conventional heat pump applications with atmospheric air as source.

Q An early slide shows a COP less than 1 at a source temperature of less than −10F. Please comment on how this can occur. I can visualize two possibilities: compressor package is situated outside of space; fan energy is included.

A Both of your suppositions are correct. Two fans are necessary to operate an air-to-air heat pump; the fan is inside and its dissipated energy adds to the system's useful heat output; the other fan is outside; its energy is lost.

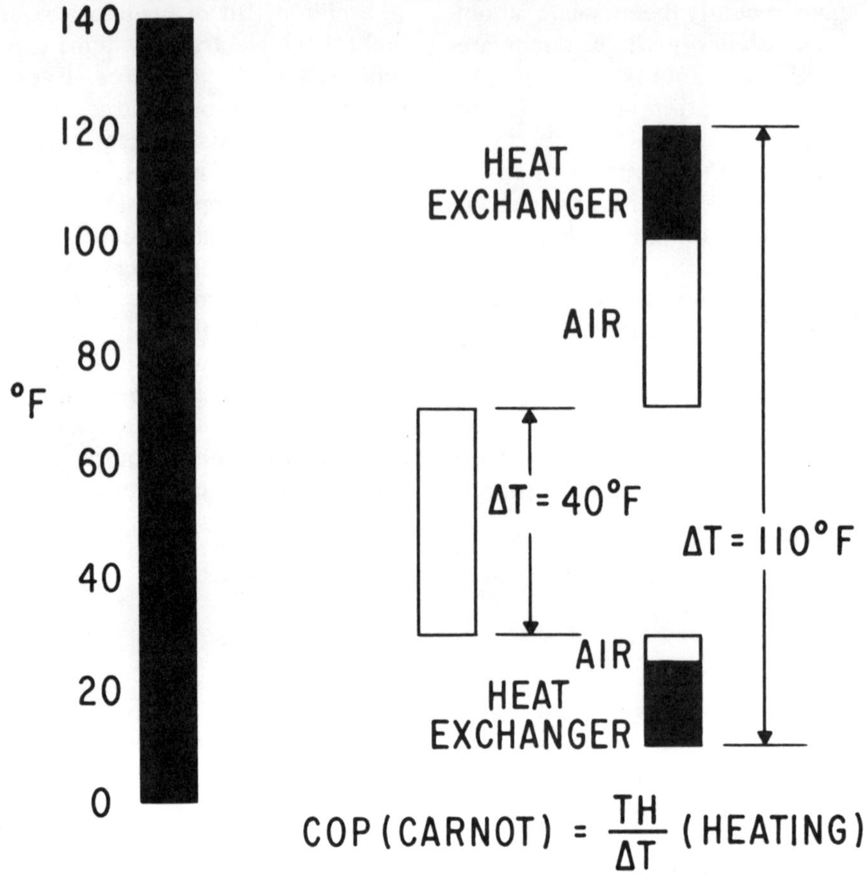

Fig. 1. Maximum heat pump efficiency calculations must be based on refrigerant temperatures rather than source and sink temperatures.

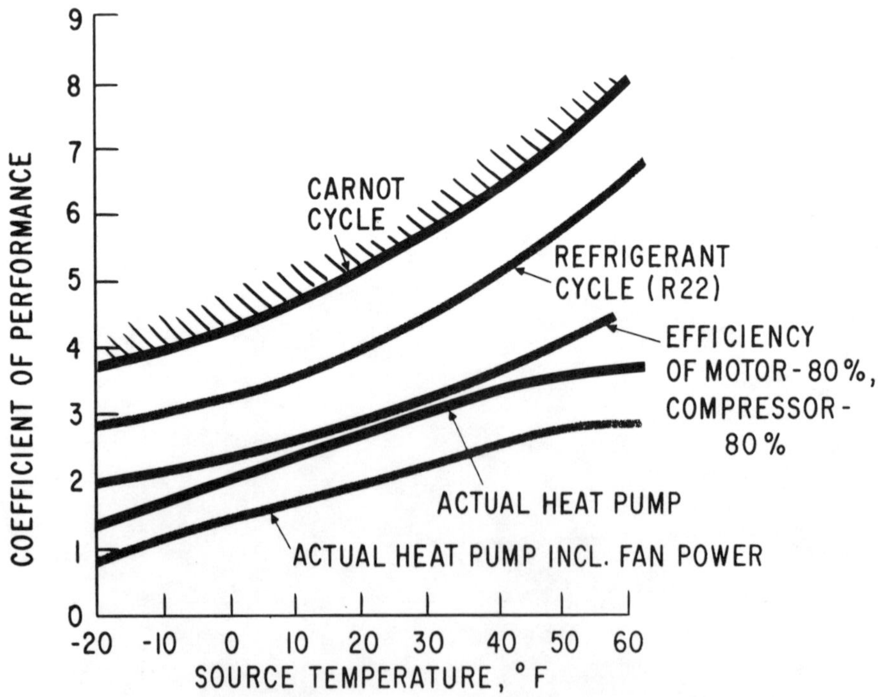

Fig. 2. Heat pump heating efficiency as it depends on source temperature for systems of different degrees of ideality.

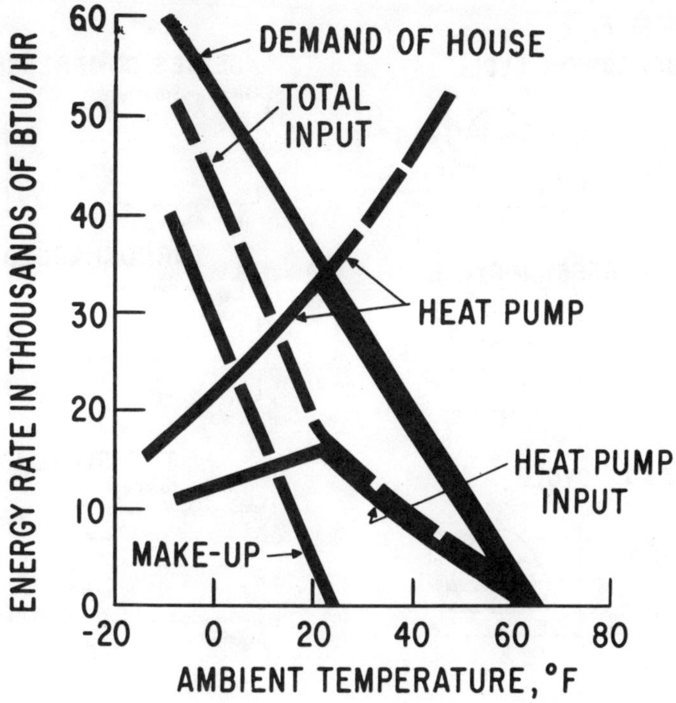

Fig. 3. Heat pump heating capacity, house demands, and supplementary heat requirements as functions of outdoor temperature.

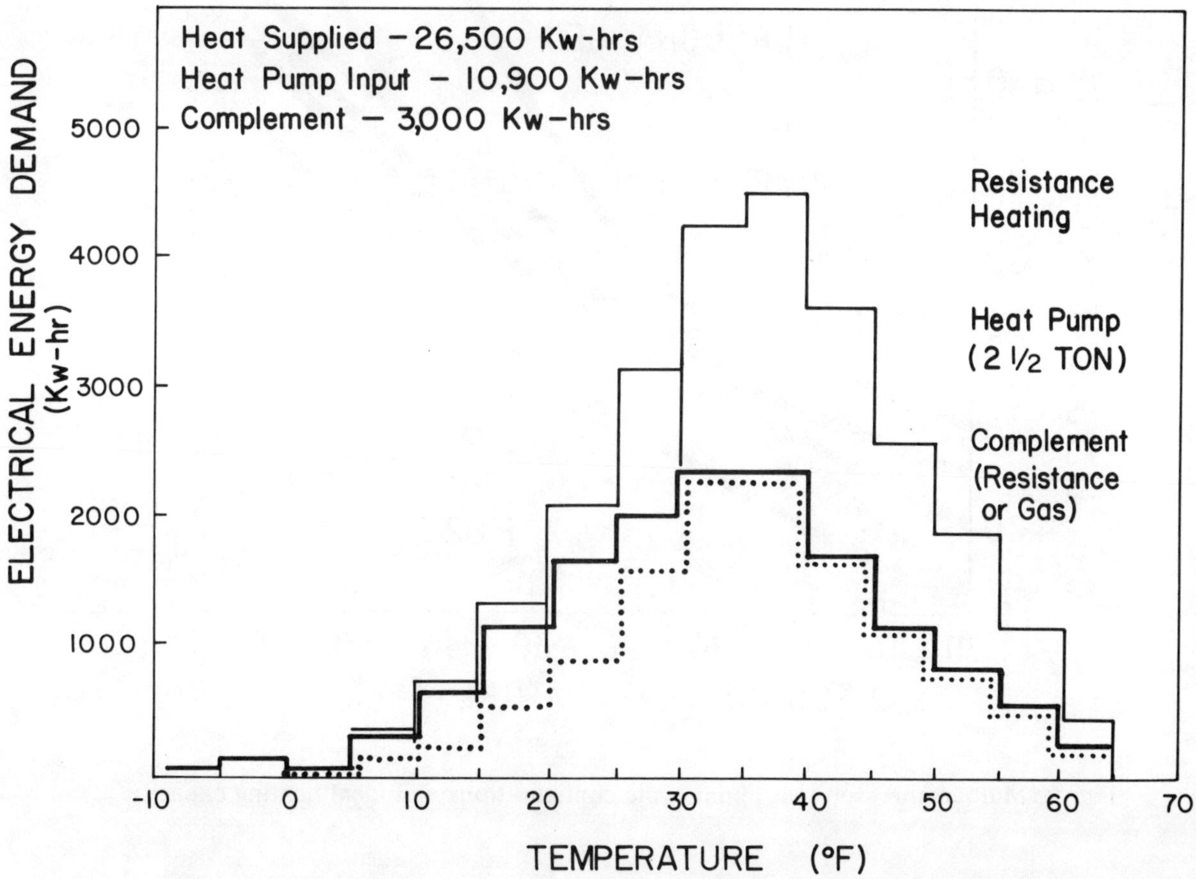

Fig. 4. Relationship between seasonal house heating demand and energy inputs for heat pump heating systems.

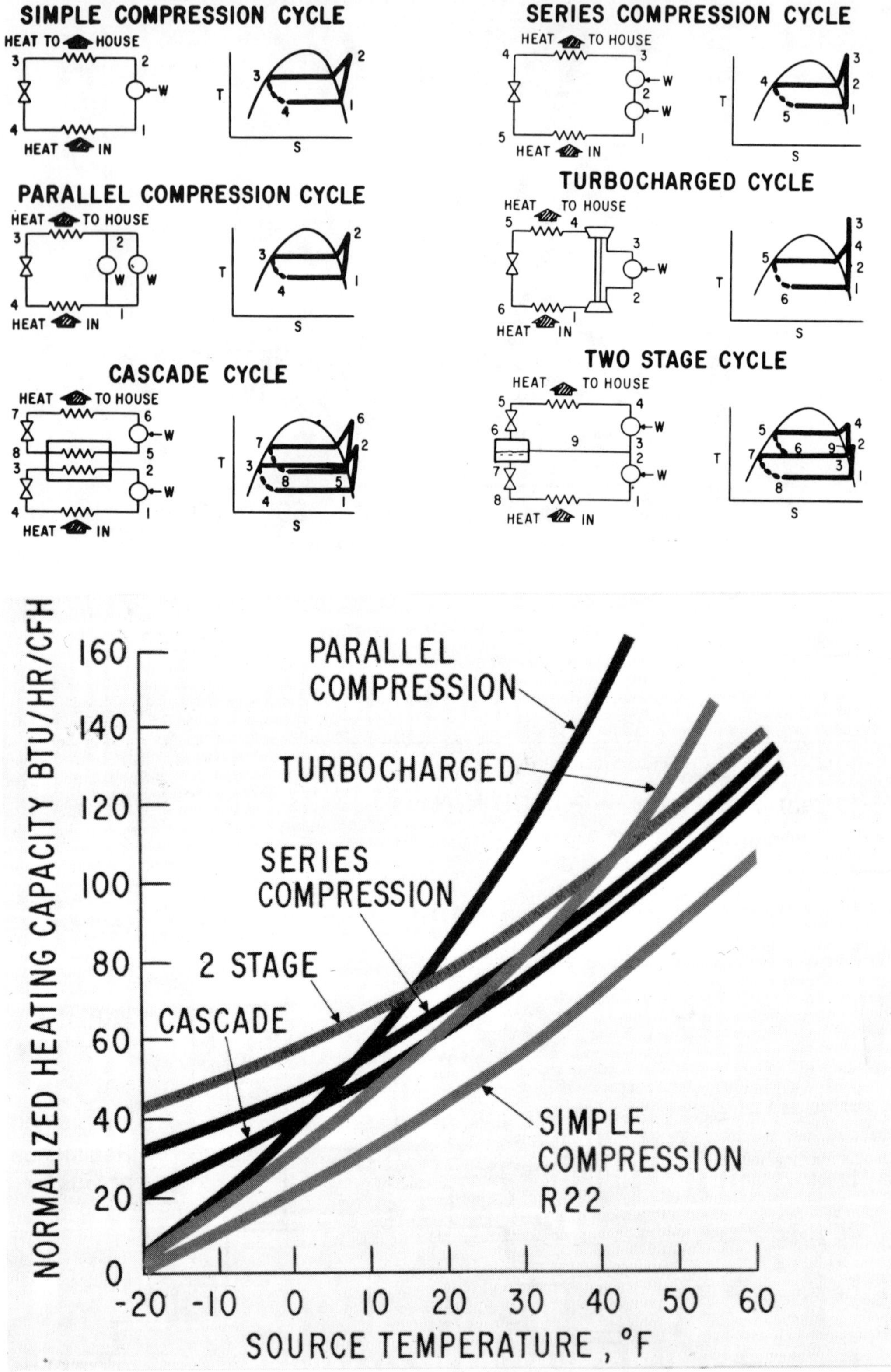

Fig. 5. Multicompressor heat pump cycle configurations and ideal heating capacity.

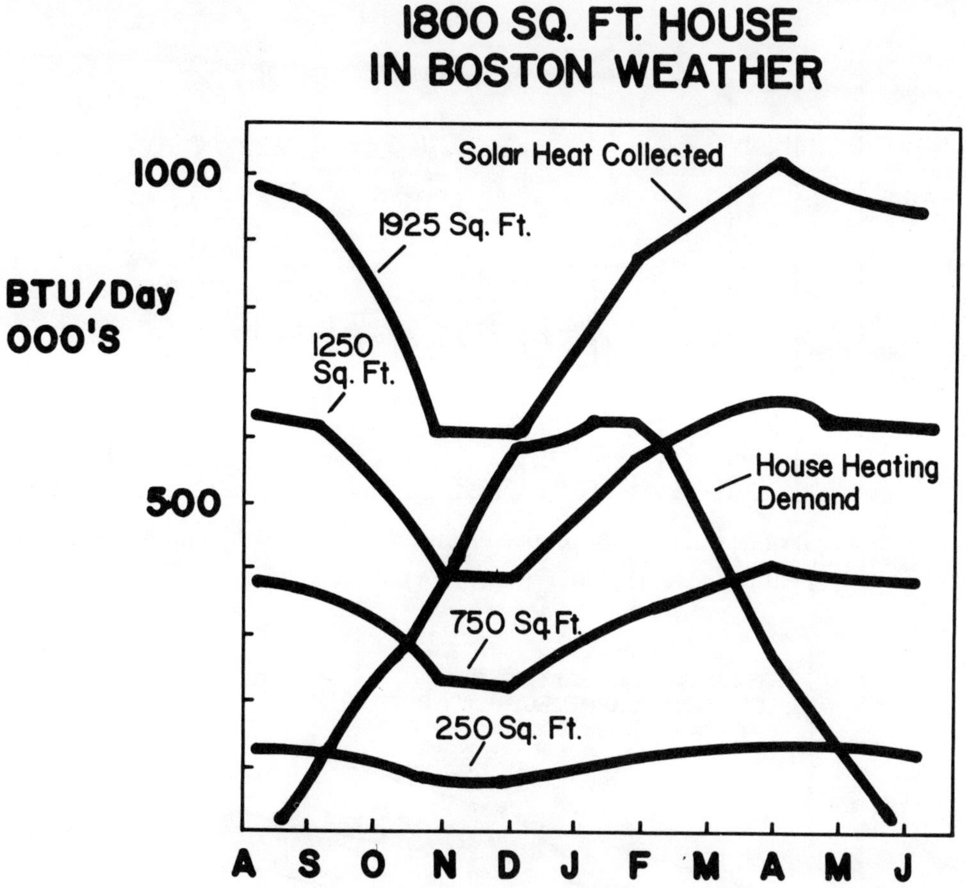

Fig. 6. Relationship between house heating demand and radiant solar energy collected by variously sized collector arrays.

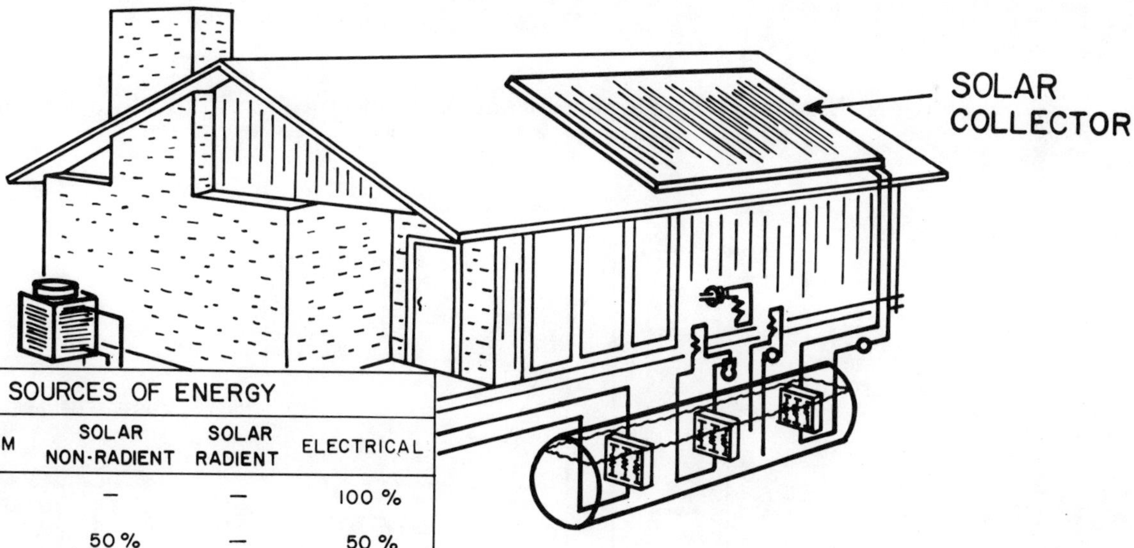

SOURCES OF ENERGY			
HEATING SYSTEM	SOLAR NON-RADIENT	SOLAR RADIENT	ELECTRICAL
RESISTIVE	—	—	100 %
CONVENTIONAL HEAT PUMP	50 %	—	50 %
60 % SOLAR AND RESISTIVE	—	60 %	40 %
SUPER HEAT PUMP	67 %	—	33 %
SOLAR BOOSTED HEAT PUMP	24 %	51 %	25 %

Fig. 7. Solar boosted cascaded heat pump system and seasonal sources of heat energy delivered to house.

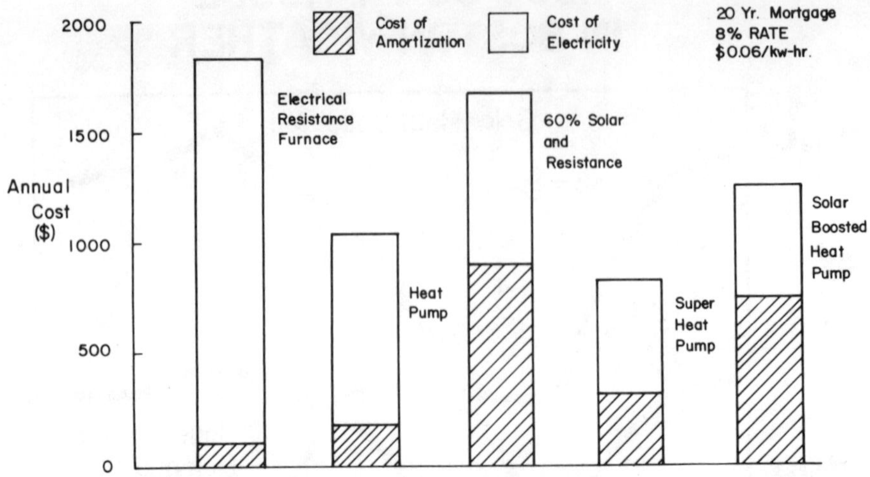

Fig. 8. Annual cost of heating for alternative electrical or electrically complemented heating systems (100 × 10^6 Btu/yr) at $0.06/kwh.

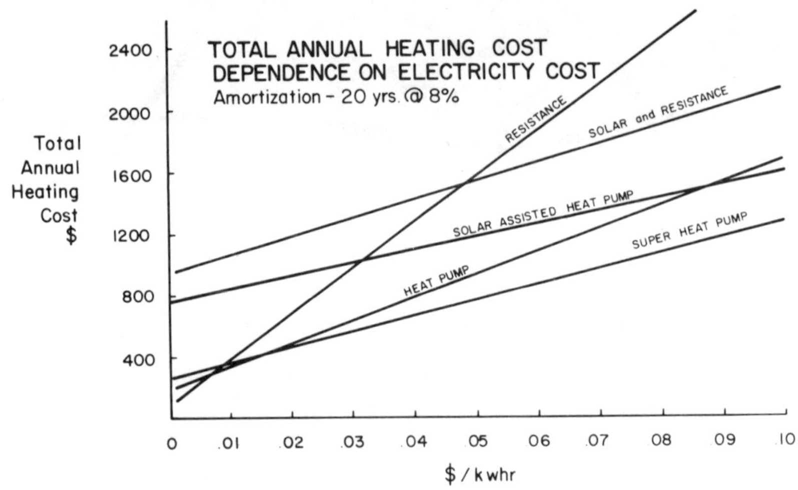

Fig. 9. Annual cost of heating for alternative electrical or electrically complemented heating systems as functions of the cost of electricity.

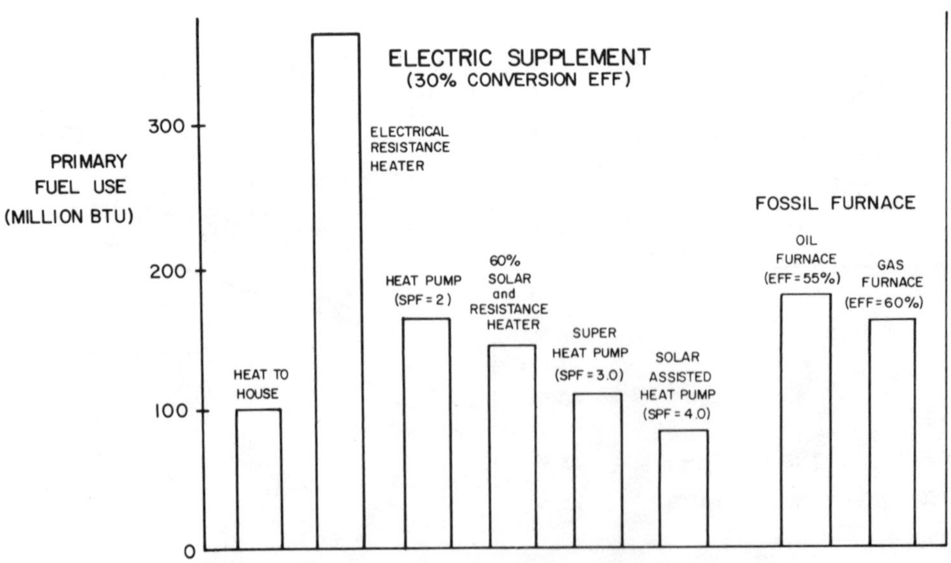

Fig. 10. Primary fuel use of alternative space heating systems.

A Solar Assisted Cascaded Heat Pump Concept

Robert Romancheck

Research and Technical Services
Pennsylvania Power and Light Company
Allentown, Pa.

One of the greatest problems faced by electric utilities today is the constant pressure being placed on them to supply the public in its greater demand for electrical power. To supply these needs, continual power plant development is required, resulting in the need for large quantities of investment capital. As a winter-peaking electric utility, the problem is even more amplified, since the addition of a new electrically heated customer results in an on-peak demand of 13–15 kw per customer, as contrasted to the typical 1–2 kw requirement of the customer using some other form of energy to provide his space heating and water heating needs, or else the typical 3–4 kw of the air conditioning customer. The easy solution would be to remove oneself from the market providing electric heat for space conditioning purposes; however, with current shortages in the supply of natural gas and fuel oil this solution becomes unacceptable. This research project is an attempt to utilize solar energy to assist a heat pump when the ambient condition is such that the heat pump does not supply sufficient heat to meet the space conditioning requirements.

Background

This project began approximately three years ago with an investigation into ways in which one could begin to evaluate the energy uses in a single-family residential unit and see whether or not these uses could be altered with minimal effect upon an average American lifestyle. We felt it was not our right to impose different lifestyles or to arbitrarily assign specific kinds of appliances for a family to use in their daily living experience.

Our design was to build a two-story unit (Fig. 1) because of the greater volume-to-surface ratios achieved in this design, and to examine some different types of construction materials in an attempt to reduce the energy loss of the structure under varying weather conditions. We believed, for example, that one of the areas that required considerable attention was the infiltration effect of outside air into a structure during the heating season.

By the use of certain building materials such as an extruded polystyrene sheathing board material and foamed-in-place urethane insulation around window and door frames, we were able to significantly reduce the building's heat loss. We also used the casement window and made use of the third piece of glass as a further energy reduction concept.

Since outside air is necessary for oxygen as well as odor dilution and the house we built was very tight, we knew we had to put into the structure a controlled ventilation feature. This system would have the capability of utilizing the waste energy from the exhausted air to preheat the incoming air, thereby imposing a smaller heating requirement upon the heating system. We have been able to test this structure under varying operating conditions and, by the use of the monitoring equipment, we have established that the heat loss reduction is approximately 45% from standard construction in our area.

The System

Since one of the unfortunate thermodynamic traits of a heat pump is decreased energy output at greater temperature differences, the common practice has been to install resistance heating to supplement the system. The unfortunate characteristic of this system design, however, is that it places upon the electric utility serving that unit an undue peaking requirement, as well as a relative low annual load factor. In past years this was much less of a consideration because of the availability and low cost of providing electrical generating equipment and service. With the increasing cost of production facilities, $80/kw in 1968 to $650/kw in 1980, it is apparent that the need to provide some other form of supplemental heating to the heat pump system is indeed a very critical concern.

Off-peak storage certainly can provide this peak shaving requirement; however, one of the traits of an electric utility; that is, winter-peaking is the nonexistence of sufficient off-peaking time periods. Additionally, the use of resistance energy to provide the heated water for this off-peaking system would inherently be of low efficiency because of system losses.

Solar energy obviously provided an alternative which not only supplies the necessary energy requirements but does so without consuming any earthbound energy sources. We installed panels in two configurations, as part of the second-floor deck fence as well as a ground level patio privacy screen. Approximately 233 sq ft of solar collection surface were installed in this vertical configuration. Whereas some may question the advisability of solar collection in a vertical mode, we have been able to achieve what we believe to be a more aesthetic and practical solar collector concept with little attendant loss of collector effectiveness.

One additional important consideration is the ability of the system design to utilize solar energy at relatively low collection temperatures, since it is supplying energy to the heat pump. The efficiency of collection is a function of the temperature difference of the absorber and the ambient temperature. Therefore, we can immediately see that much greater efficiency is obtainable through the use of this solar heat pump philosophy. Nor does this configuration restrict the individual's taste as far as architectural rendition is concerned, or impose the need to have one's house oriented in the north-south configuration, since all that is required is that the collector see the south.

Since our original concept was to investigate all of the energy uses in a single-family residential unit, and since the heat pump has the unique capability of raising low-quality energy to a higher-quality level, every attempt was made to re-integrate what is now wasted energy back into the system and make it available to the main water source heat pump. In Fig. 2 one can readily see how this low-quality energy is placed into a loop through a series of heat exchangers along with the solar energy capability of the system. The temperature varies in this warm water loop from a maximum of approximately 90F to a minimum of about 48F. In fact, the house has been satisfactorily heated from a 48F water loop during this past heating season.

Unfortunately, there are times when enough waste energy is not available from the various heat collection exchangers and it is therefore necessary to supplement this system. One of the easiest ways to do this is to install, as we have, electric resistance immersion heaters. After putting this energy into the water system we would still have to work the heat pump to extract the energy. This energy input to the heat pump is recoverable, so we maintain a coefficient of performance of 1 for the supplemental heat. The second alternative, which we also incorporated, was to use an outdoor air-to-water heat pump operating only when the water loop required energy from the outside air. This cascaded heat pump system has performed more than adequately under all types of weather conditions.

However, we find this concept does not provide the efficiency we believe is possible. We know a significant contribution could be made to this system if, in fact, the two compressors were part of one refrigerant circuit. To utilize the output of the outdoor heat pump, one must also run a circulator to get this energy to the indoor heat pump. This unfortunately produces an added energy requirement into the total system concept. Yet even in this preliminary concept we can visualize significant advantages in the use of cascaded heat pump systems, without great storage requirements, and, through proper valving, their ability to provide satisfactory cooling in the summer period.

Conclusions and Recommendations

Figure 3 shows the peak trend characteristics of the Pennsylvania Power & Light Company during the 1960–70 period. As noted previously, we are aware that a straight air-source heat pump, supplemented with electric resistance heat, can result in substantial energy savings to the customer. However, this system performs essentially as an electrically heated resistance hot-air furnace during that period of time when our system reaches its seasonal peak. The characteristics of that system are shown by the top curve of Fig. 4 and the characteristics of the solar assisted cascaded heat pump system by the lower curve. We have achieved significant reductions in demand and also believe that with some system changes it is possible to even improve this reduction. A considerable amount of waste energy in this system has been recycled, and aside from the many problems, effective use of solar energy has been made.

Our research leads us to believe that a dual-source heat pump assisted by solar, wind, waste, or other energy can adequately provide, with a high degree of reliability, all the space heating and cooling needs of a structure, significant energy savings in domestic water heating, plus, when sufficient, the direct use of solar energy for space conditioning. Figure 5 is a schematic of this system.

With modifications, we will be able to adapt our present cascaded system to this concept and test it for its efficiency. Although it is not apparent in the diagram, defrost would also be provided by the stored energy system, so as not to utilize reverse cycle defrost, thereby providing a potential for greater reliability. Practical hardware studies such as this are indeed necessary to supplement the theoretical paper analyses that are meaningless until tried.

ACKNOWLEDGMENT

Robert W. Deppen, Energy Conservation Consultant for the Pennsylvania Power & Light Company, is acknowledged as co-participant in this research effort.

DISCUSSION

Q Please comment on use of *vertical* collectors—everyone else uses a tilt!

A Energy collection maximized in December, January, and February, with vertical configuration.

Q Can too tight a structure cause suffocation?

A Unit is constantly pressurized with approximately 50 cfm of outside air.

Q What was the electric bill on the house for a monthly or bimonthly time span for the year?

A Raw data indicate a heating energy requirement of 5000–6000 kwh.

Q Have you considered enlarging your storage tank to eliminate peak demands on a conventional solar augmented heat pump for winter use?

A Yes; we don't think it is necessary.

Q What are conductance values of your house windows when open for sun, then closed with good drapes?

A If we assume low emissivity, which we believe to be true, a conductance of 0.21 with closed drapes and 0.61 with open drapes.

Q How much heat does the collector lose in a standard 7 mph wind? Can you cut the collectors out of the system when you have a net loss?

A Because of lack of an on-site weather station, collector efficiency has not been determined. Collection operation is time and temperature controlled.

Q What is the living space area of the house? What did it cost to build?

A 2000 sq ft of conditioned space. The research project cost $150,000.

Q Does your company have a demand charge for residential customers?

A No, not yet.

Q Does your utility have any special or restrictive rates concerning resistance heat supplement with heat pump installations?

A Not yet.

Q What depth of bury is necessary to provide adequate insulation for a water storage tank?

A 3–4 ft.

Q You said the storage tank would not need to be insulated if buried in the ground. Wouldn't you lose heat into ground which is about 50F?

A Ground is an excellent conductor, and under the garage floor, loss is minimized.

Q Ethylene glycol and aluminum don't mix. But why don't you blame the glycol? Aluminum works well with other transfer fluids. What corrosion inhibitor did you use with the glycol?

A The problems with aluminum were poor mechanical terminations, not necessarily corrosion.

Q What precautions were taken to prevent electrolytic corrosion? What would you have done differently on your house?

A We are changing absorbers to copper to eliminate any potential corrosion problems. As a research effort, we have done nothing.

Q Could you give a breakdown of heat provided by the various recovery devices used—refrigerator, wastewater, septic tank? What are figures in a heating season? What is done with waste heat during the cooling season?

A Breakdown of energy contribution will be released in an evaluation report probably in late 1975. Waste heat is not collected during the cooling season.

Fig. 1. Two-story experimental unit.

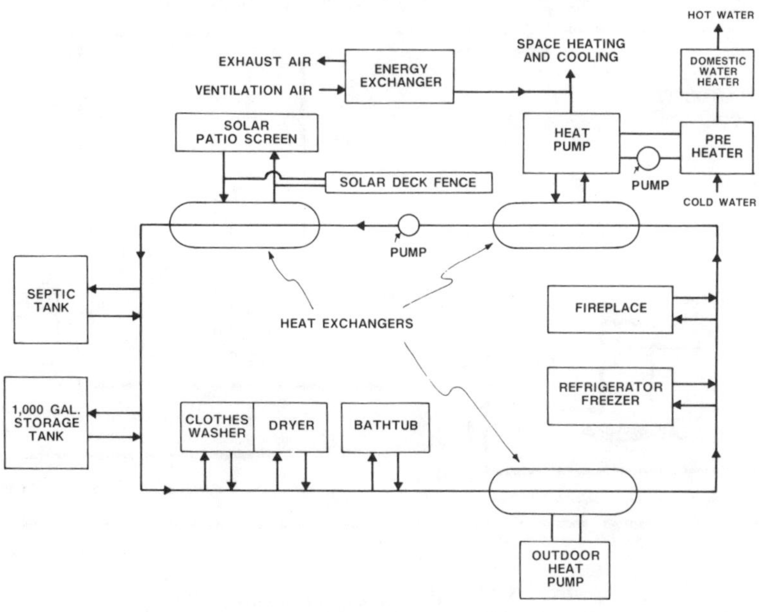

Fig. 2. Wasted energy is re-integrated into system.

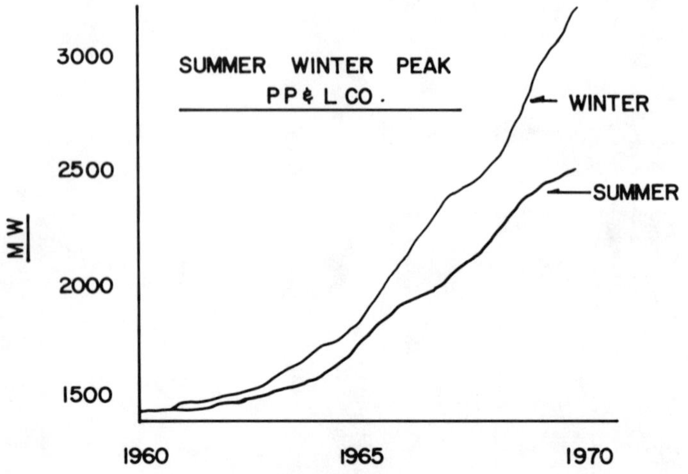

Fig. 3. Peak trend characteristics for Pennsylvania Power & Light Co., 1960–70.

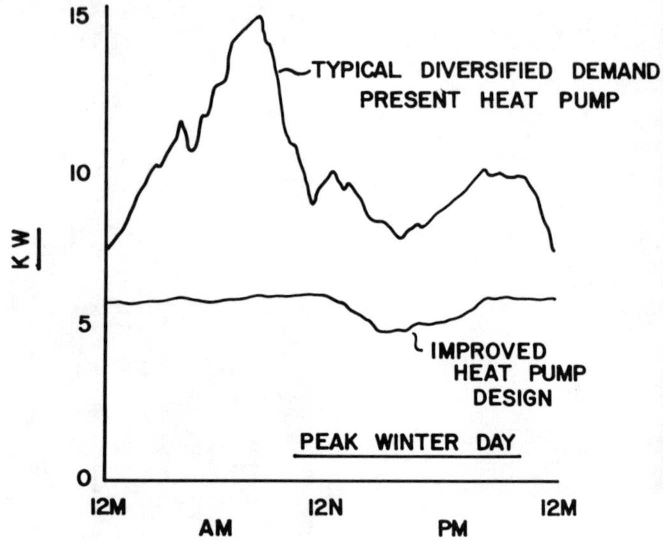

Fig. 4. System characteristics during peak winter day.

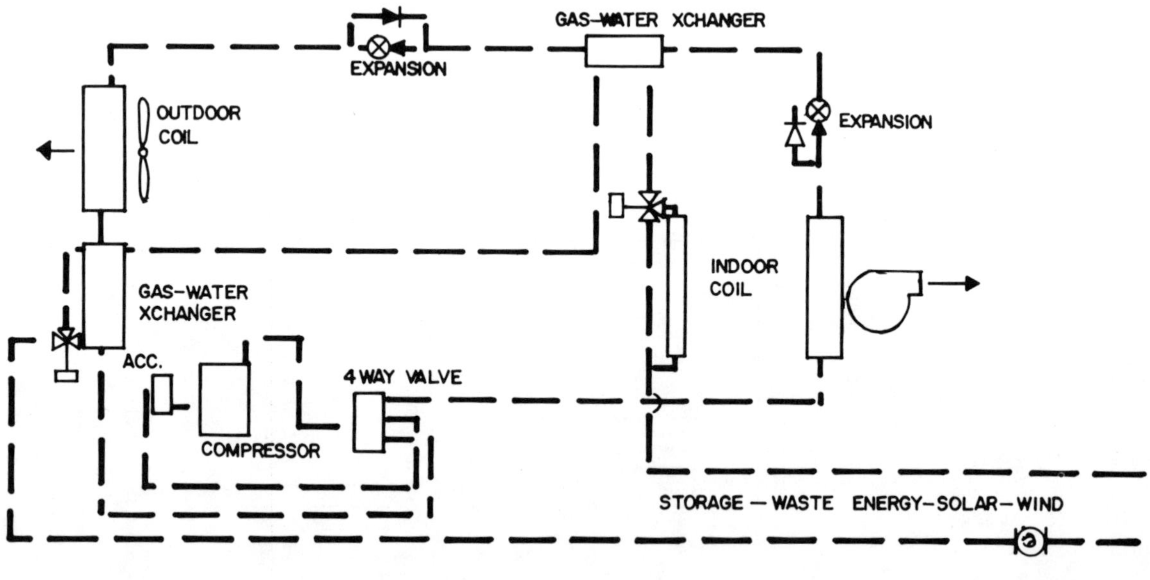

Fig. 5. Dual source heat pump.

Solar Augmented Air-to-Air Heat Pump Installation

Herman G. Barkmann
Vice President – Engineering

D. Gene Yergensen
Director – Architecture

Sunray Energy Corporation
Denver, Colo.

This paper describes briefly a residential heating installation constructed this past winter consisting of an air-to-air heat pump, a sizable rock storage system, and a solar collector air heating system. The purpose of the solar augmentation to the heat pump is to provide a source of warm air for the outdoor section of the heat pump.

The ability to store energy and provide warm air to the heat pump evaporator allows a heat pump to operate at a much higher capacity and COP during periods of low outside air temperature. This scheme helps to overcome the main drawback of the normal heat pump where the capacity curve of the heat pump and demand curve of the load have opposite slopes as a function of temperature.

A major part of the paper is devoted to a second generation system including the same components, and discussing design decisions based to some extent on lessons learned from the first system, and to a greater extent on new considerations.

We hope to indicate the advantages of such a system from the standpoint of reliability, economics, and minimal use of resource energy.

General

For the past two and a half years some of the participants of this project have been pursuing the use of solar energy for residential space heating. Several houses have been built and some operating experience is being gained. The residences, with one exception, to this time have employed air-cooled collectors along with the familiar gravel storage systems and the required blowers and ductwork. The auxiliary heating sources have been varied, including fireplaces, hot air downflow furnaces, and electric duct heaters.

The reason for following this route is our preference for air, for the following reasons:

1. The predominance of the hot air heating system with its lower first cost for residential use.
2. The temperatures encountered in the normal hot air system are considerably lower than those for the normal hot water heating system. This, then, for simple systems, allows lower collecting temperature, in turn leading to higher efficiencies of collection.
3. A probable higher reliability due to absence of problems of material compatibility and leakage.
4. Lessened need for heat exchanger.
5. Probable lower overall cost.
6. The ability for an air collector to be built either on or off the site by simple means, using quite easily available materials.

As work progressed on these projects an analysis was made of what might be the preferred backup or auxiliary fuel source. Some preference was realized for a fuel with on-site storage capability such as propane or oil. There was, however, a realization of the advantage of electricity for a backup power source because electricity generally presents the lowest installation cost and it can be made available where other fuels cannot. It should also be noted that liquified petroleum fuels have increased in price so that electricity is now competitive in cost as a fuel source.

While in the past the authors have resisted the use of electrical heating except in cases where there is no alternative, because of energy cost or use, it can be argued that where the major part of the heating load is carried by solar energy, electricity can be justified to handle the remaining part of the load. They became convinced that probably the best way to use electrical heat was by use of a heat pump where the overall consumption of electrical energy would be minimum. Studies performed by Cassel (ref. 1) have indicated that of several systems examined solar assisted heat pumps are capable of the minimum use of resource energy.

In late 1974 construction was started on a residence in Beulah, Colorado. The design of this house incorporates most of the ideas which we had been considering. Below the basement floor, gravel storage was located. Ducts were placed in the gravel so that air could be circulated through the rocks into three vertical ducts and up through collectors mounted on the roof. Air is drawn from the top of the collectors at three locations and reentered into the rocks. Ducts are also located in the rocks in such a way that air can be drawn from the rocks through the outdoor section of the heat pump, which is located on the ground floor immediately above the gravel storage, and back into the far side of the storage. It is also possible for air to be drawn from the ducts into the room air return to the heat pump, from where it is circulated through the distribution system of the house and back into the rocks through a return air grille.

By means of this fairly simple ducting arrangement we have the option of circulating air heated by the rocks either through the heat pump supply fan through the house for direct solar heating, or through the heat pump evaporator when the air temperature is too low for direct heating. Thus at low ambient air temperature, energy stored in the rocks can be used to operate the heat pump at higher evaporator temperature in order to gain a higher COP and increased heating capacity.

The house design was fairly typical for a mountain house. It is quite compact, with a steep roof, normal in areas of considerable snowfall. The south face of the roof is essentially covered with solar collectors.

This house was put into operation in late March 1975, so that no real operating data have been gained as yet.

The experience gained from the construction of this first house has led to a much more ambitious project. A group of 46 townhouses is planned to be built in Pueblo, Colorado, and all are to be heated by solar augmented heat pumps.

The system to be used is similar to the one described above, but with several improvements and added modes of operation. Nearing completion is the design of the townhouse unit with the greatest heating requirement. Let us now examine some of the design considerations for this particular unit, one with three bedrooms and located at an end. The total area including basement is 1470 sq ft, the design heat loss at −10F is 47,000 Btuh, the maximum collector area available due to the roof design is 385 sq ft, and we can readily provide a storage mass of 25 tons of gravel.

Figure 1 is a plot of the building load, the heating capacity of two different commercial heat pumps, and the COP of each of these two heat pumps, all against outside air temperature. Also shown at the top is a curve of the supply air temperature necessary to satisfy the building load at the air flow of the two heat pumps. This is a valuable curve as it indicates the temperature required for air to be drawn off the storage for direct solar heating at various outside air temperatures.

Initial calculations indicate that the 385 sq ft of collector have the capacity to supply 70% of the energy required on an annual basis, even though it supplies only 39% during the month of January. However, when one takes into account that the auxiliary heating is accomplished at COP values ranging from 2 to 4, the electrical energy that needs to be supplied for heating amounts to only some 15%, or an 85% effective heating system.

The authors are now developing a computer program whereby they can determine the percentage of auxiliary heating required as a function of collector area, and then determine least total cost system. This program differs from those available in that it uses the daily weather data available from the U.S. Weather Bureau, and can be run on a readily available computer time share service mainly used for accounting.

The authors are designing the collector systems for the maximum area available, but when the computer program is ready the authors will analyze the systems for least total cost collector area for optimum total system performance. If this is less than the maximum area available, they can adjust for that. The program will be more meaningful after construction of the first units is finished and real costs of the components are available. There are still of course the unknown future energy costs to contend with.

The authors have considered double-glazed and single-glazed collectors. An analysis by means of a modified Hottel and Whillier equation of the two collectors indicates only a 5% loss in useful energy for the single-glazed collector, while the saving in cost is 25%. This small change in energy gain between single- and double-glazed collector is due in large respect to the stratification of temperature one gets in rock storage. Experience has shown that the outlet temperature of the storage air can be expected to rise no higher than 80F. These low collector temperatures allow for high collector efficiencies with single glazing.

There are many modes of operation available to a solar assisted air-to-air heat pump system; the initial planning stage showed a possible choice of 17. It is not hard to imagine, however, that the cost of controlling some ten different dampers and four blowers can become expensive. We finally settled on the following seven modes, five automatic and two manual:

1. Heating storage from solar collectors (Fig. 2).
2. Direct heating while collecting (Fig. 3).
3. Direct heating from storage (Fig. 4).
4. Heat pump heating from warm storage or cooling from cooled storage (Fig. 5).
5. Heating house with outside air, resistance heat, or cooling with outside air (Fig. 6).
6. Cooling gravel with outside air (manual).
7. Ventilation collector (manual).

One cycle which has been omitted has considerable interest, and is currently being evaluated. This is a cycle which includes mixing of outside air and storage warmed air to maintain a fixed supply air temperature to the heat pump evaporator. This particular cycle has been omitted from our current operating plan because of the added cost of controls. Here again, however, the first cost must be evaluated with energy savings over the life of the system to determine minimum total system cost.

We are fortunate in this particular project as we will be able to evaluate and improve system design from construction of first phase to construction of the final units.

Energy Cost Comparisons for Heating

We have performed calculations to determine the energy usage and fuel costs of several heating systems for comparison with our solar assisted air-to-air heat pumps system. Our calculations of course had to be based on many assumptions. System costs are based on relatively firm costs from suppliers, and on experience gained in construction of the first house. Future energy costs are uncertain. We have estimated the cost of electricity over the next 20 years to average three times the current rate, and the cost of gas over the same period to be five times the current cost, and we have reason to believe that even these may be conservative. Our study indicates energy savings of some 80% for the solar heat pump compared to an electrical resistance heated system, and 60% compared to a gas heating system.

Costs of heating energy at expected average rate over 20 years show the following:

solar heat pump	$280 per year
gas heating	$590 per year
electrical resistance heating	$1,536 per year

Conclusions

1. We believe the solar assisted air-to-air heat pump provides an extremely flexible and useful system for heating and/or cooling or residences.

2. We find the system as designed requires a minimum of resource energy and provides a minimum lifetime total cost.

3. There are improvements to the system which could be made now or in the future. As they become available they must be evaluated from a standpoint of practicality and economics.

REFERENCE

1. Cassel, T.A.V., *Integrated Solar Energy Assisted Vapor Compression Heat Pump Systems*, National Center for Energy Management and Power, University of Pennsylvania, May 1974.

DISCUSSION

Q What maximum air temperature do you think you can operate from rock storage to heat pump evaporator? Have you considered a hot gas/water exchanger on heat pump for domestic water heating?

A We are unsure, but we expect to limit the heat pump evaporator temperature to 90F. We have not considered hot gas/water heat exchange for domestic water heating.

Q Is there any provision in your design for solar heated domestic hot water? In general, how feasible is it to supply domestic hot water with an air/heating collector?

A We have considered running domestic water preheat pipes through part of the air collector. We have not evaluated the dollar effectiveness of this, however.

Q What is approximately the cost of your storage system of 25 tons of gravel, installed (with its ducts)?

A Three hundred dollars.

Q What is the collector cost/sq ft?

A Double glazed collector—$6.00/sq ft; single glazed—$4.50/sq ft.

Q What does a cross section of the air heating solar collector look like?

A Glazing ½ in. above absorber plate of corrugated aluminum, 4-in. air duct, then insulation.

Q Do you have any quantitative information on the temperature stratification in the pebble bed? That is, what is the temperature gradient? Also, what is the pressure drop across the pebble storage bed?

A Not other than measurements of 130F air into storage of 8 ft length, 80F out. If face velocity can be maintained at 50 fpm or less, and length of bed 10 ft or less, pressure drop should not exceed ½ in. H_2O. Gradation of size is important.

Q You mentioned use of outside air cooling of the rockbed—what is done to handle the humidity? Also, in general, what consideration has been given to odor control, algae, etc., in the gravel bed for long time usage.

A We will attempt to maintain temperature always above dew point. To our knowledge odor has presented no problem in beds.

Q In your gravel storage bed, what are the following: Size of gravel used in bed? Air pressure drop through bed? Physical size of total bed for the 25 tons of gravel?

A 1½-to-3 in. diameter. Approximately ½ in. H_2O. 4 × 8 × 20 ft, approx.

Q What are the blower horsepower requirements and/or pressure drop numbers for air circulation through rock storage and collectors?

A ½ hp.

Q Please discuss in a little more detail your comment about 5% loss change when comparing single to double glass. Also, the 25% price reduction was arrived at by what method or source?

A Our experience has shown a single pane cost of about $4.50/sq ft, with a double pane cost of $6.00/sq ft, a 25% reduction. The difference in output calculated by a modified Hottel and Whillier analysis indicates a decrease in output from 950 Btu/sq ft/day to 910 Btu/sq ft/day, a loss of 4+%. You must bear in mind that with a rock bed storage, we get temperature stratification, and our collector inlet temperature very seldom rises above 80F.

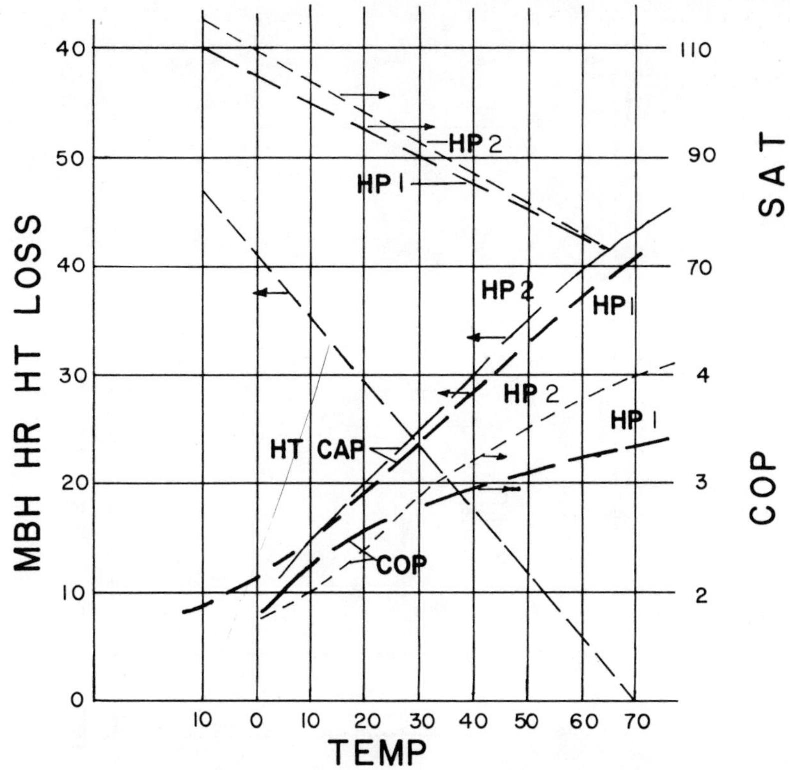

Fig. 1. Performance of two air-to-air heat pumps.

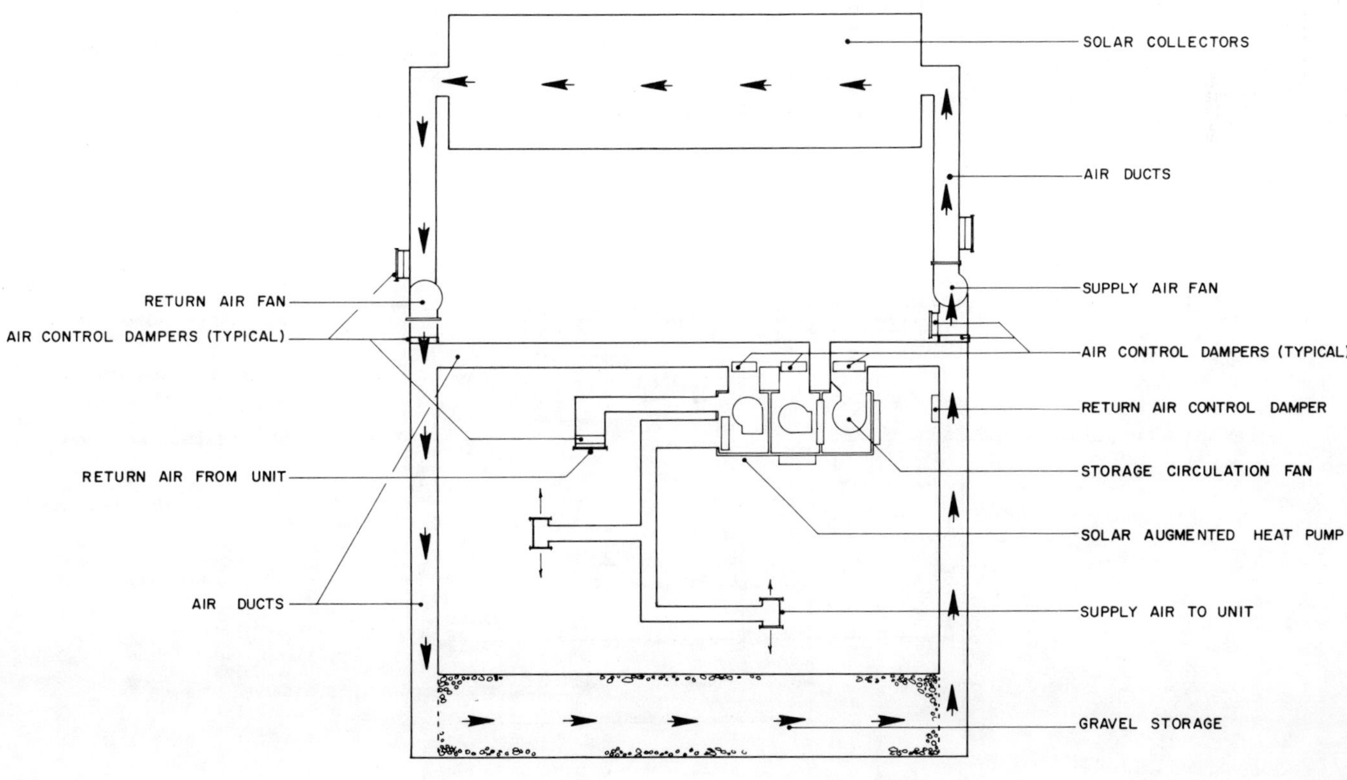

Fig. 2. Solar energy collection.

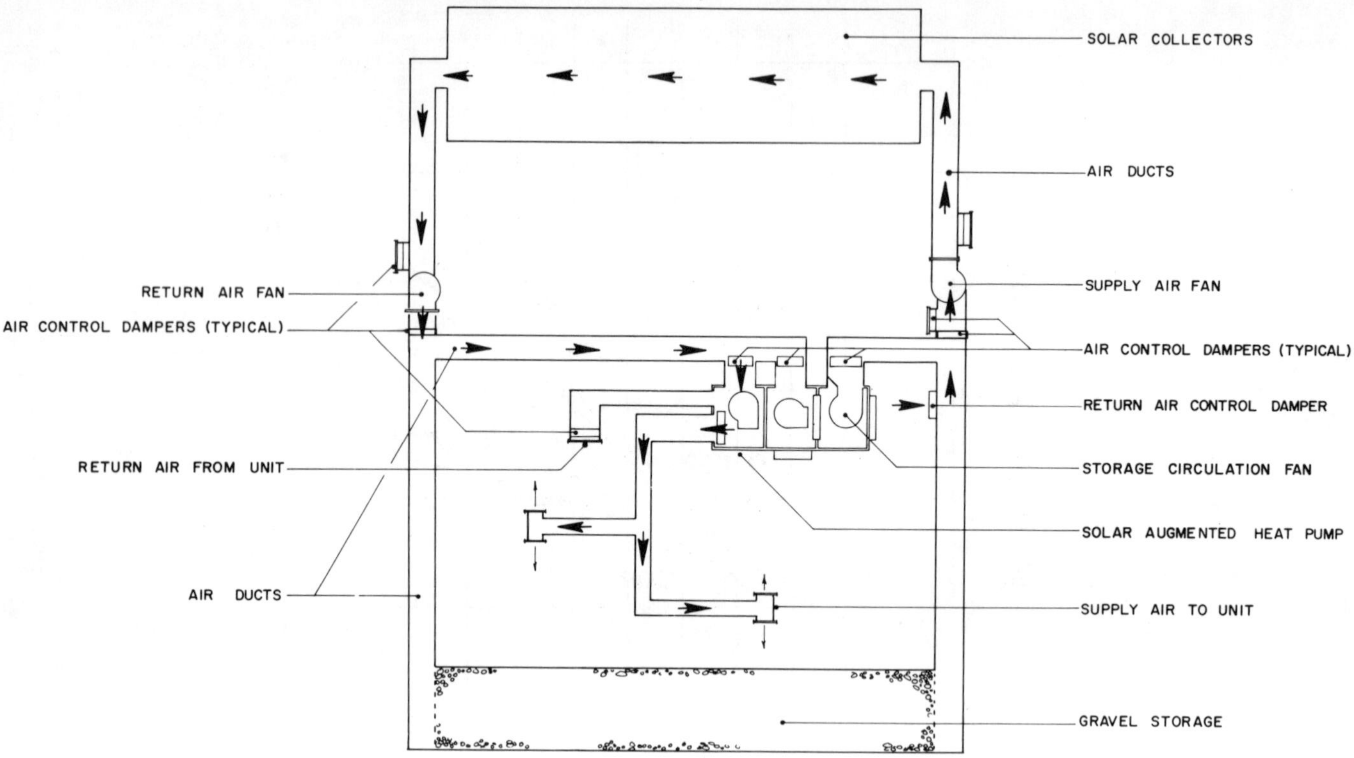

Fig. 3. Simultaneous direct heating and solar collection.

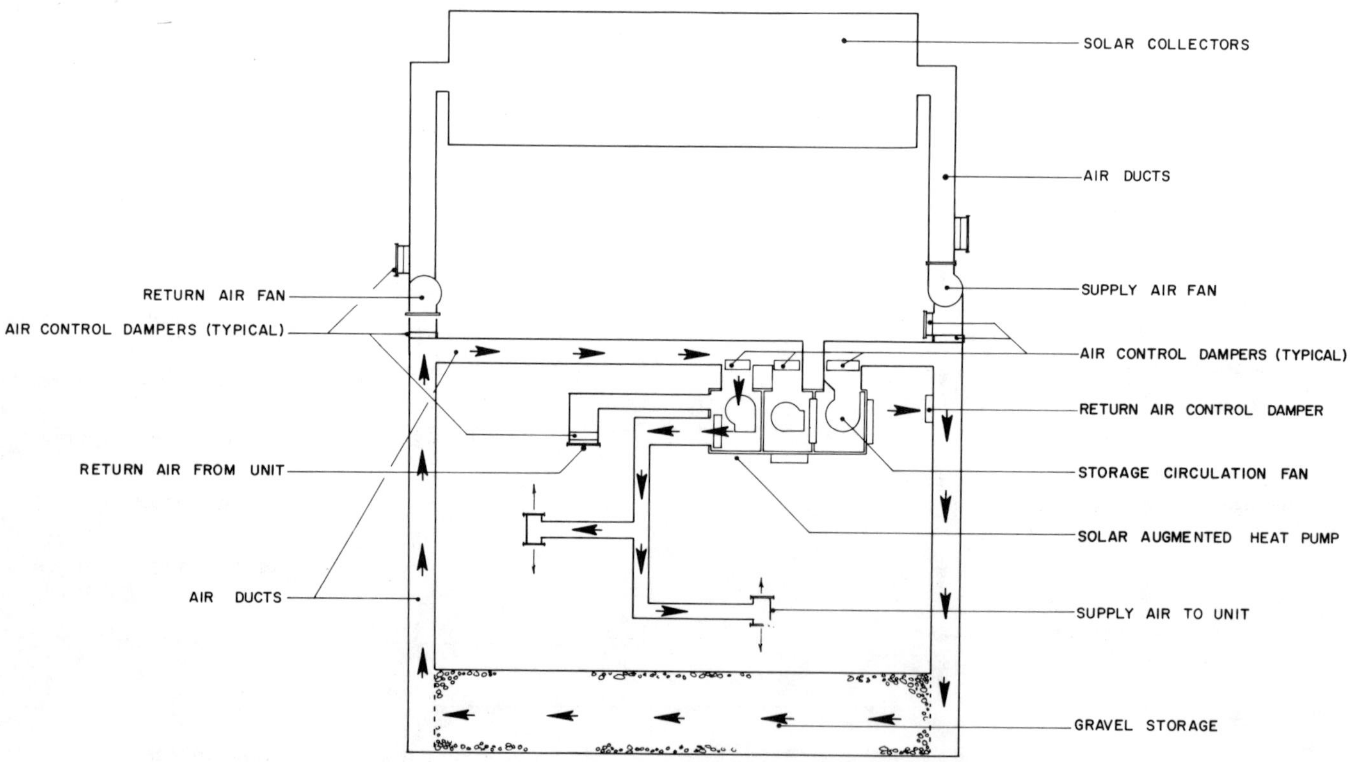

Fig. 4. Direct heating from storage.

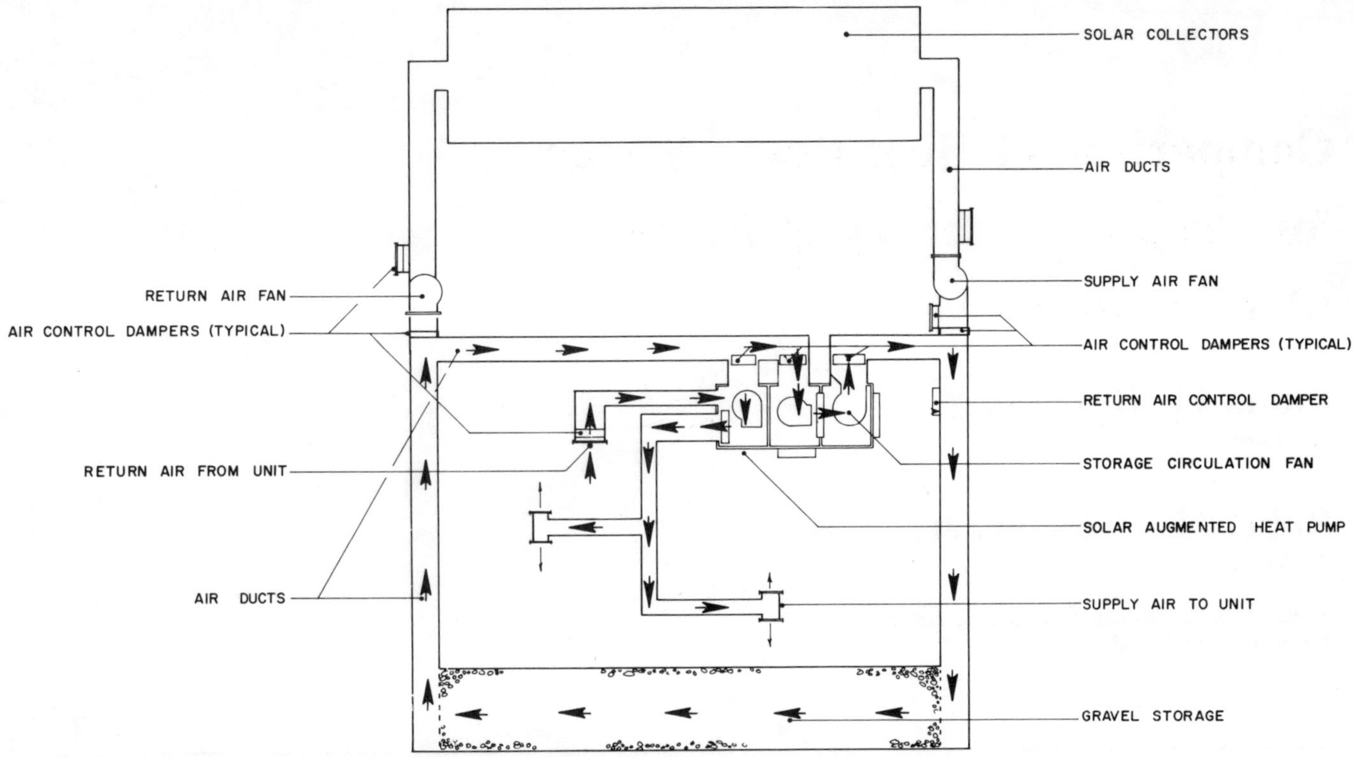

Fig. 5. Heat pump heating with warm storage air as heat source.

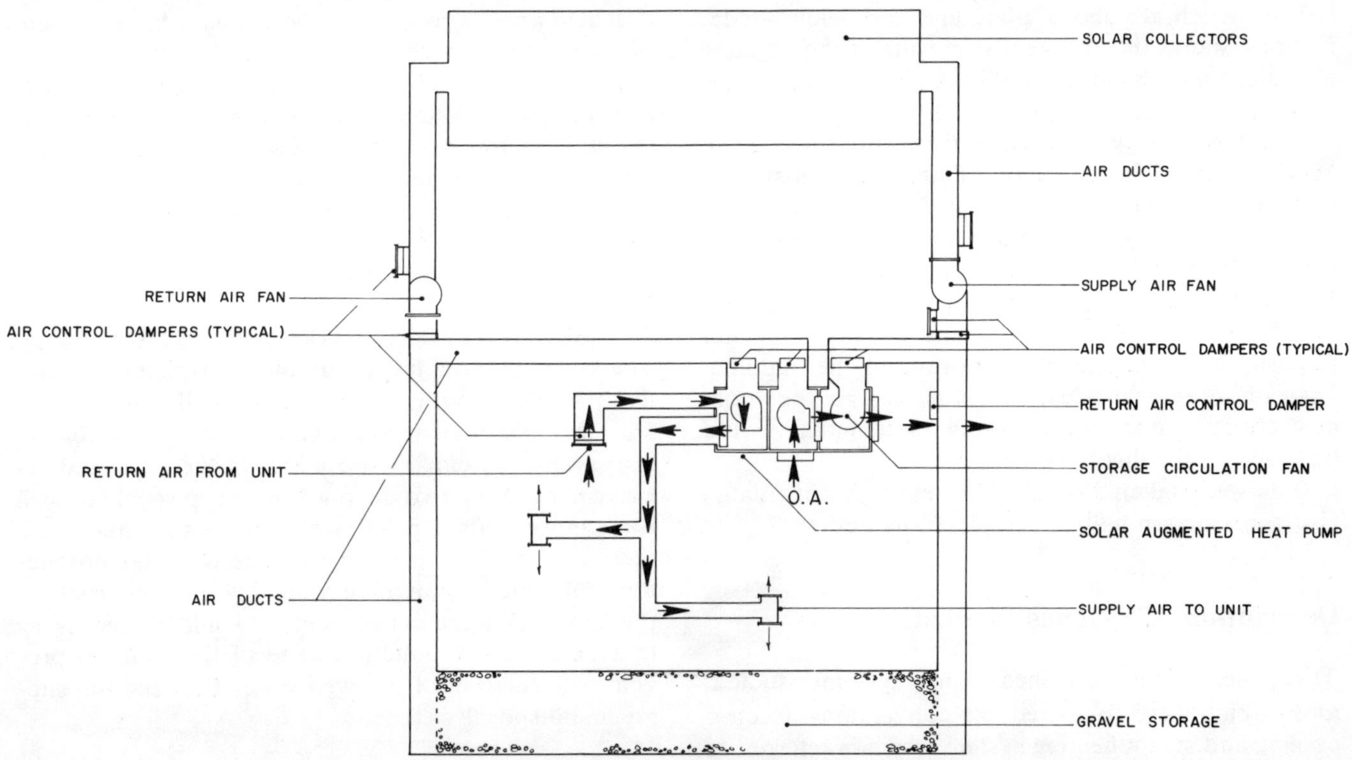

Fig. 6. Heat pump heating with outdoor air as heat source.

Comparison of Heat Pump Systems for Proposed Library Building, Idaho State University

Frank H. Bridgers
President

Robert E. Hopper, Jr.
Project Engineer

Bridgers and Paxton Consulting Engineers Inc.
Albuquerque, N.M.

The proposed Library Building at Idaho State University is a 180,000 sq ft building having five levels, four of which are above grade and one below grade. Figure 1 shows the architect's drawing of the building elevation. The building is essentially square, having nominal dimensions of 230 by 204 ft.

The University administration and the Public Works officials for the State of Idaho are most concerned that the new Library Building have an energy-conserving heating and air conditioning system. Even though the Idaho State University campus is served by a central steam heating plant, the design criteria given to the architects and engineers did not require the use of steam for heating of the building. Instead, they were instructed to design the most cost-effective energy-conserving system that would be practical for this building.

Bids were taken July 2, 1975, and it is anticipated that construction will be completed by mid-1977.

Description of Systems Studied

There were three basic heat pump systems studied, all of which were compared to a conventional electric cooling and steam heating system which is referred to as the Conventional System.

1. INTERNAL HEAT RECOVERY HEAT PUMP SYSTEM (Fig. 2). The configuration of the building made an internal energy heat recovery system a very practical application, since there was a large amount of interior space with relatively high lighting level that required year-round cooling. The internal heat recovery heat pump system recovers heat from the interior and utilizes this heat at the perimeter of the building during the winter months. At night, when the lights are off and there is no heat dissipated in the building interior, the campus steam is used as the heat source to heat the perimeter of the building.

2. WELL WATER HEAT PUMP SYSTEM (Fig.3). The Pocatello area has groundwater available at relatively shallow depths. Although the well water is only 56F, this is sufficient to provide a heat source for the heat pump system. Figure 4 shows the proposed arrangement of the two deep wells. It is possible to drill the supply and disposal wells close together since there is a shallow and a deep zone of water production, separated by an impervious layer of lava rock. If the lava rock were not present, it would be necessary to separate the production zone of the wells to prevent recirculation of the well water between the supply and disposal well.

3. SOLAR ASSISTED HEAT PUMP SYSTEM. It is possible to provide 18,000 sq ft of solar collector on the roof of the building. This amount of solar collec-

tor could provide a substantial amount of energy for heating and would greatly reduce the amount of steam required for night-time heating for System 1 (above). (Figure 5 is the schematic diagram of the solar assisted heat pump system.)

Basically, all systems would recover internal energy to the maximum extent and are supplemented by steam for System 1, well water for System 2, and solar energy for System 3. Steam also supplements the solar system since it would not be economically feasible to provide sufficient collectors to furnish all of the heating requirements for the building.

Table 1 presents the comparative coefficient of performance for a standard water chiller used as a water-to-water heat pump for the various applications. The coefficient of performance for the different systems varies because of the variation in the temperature level of the heat source. All systems are based on supplying 110F water to the building. Systems 1 and 3 are based on a parallel piping arrangement of the chillers (evaporators and condensers). System 2 is based on a reverse series flow of the two chillers and condensers. Even though the well water is available only at 56F, a well water heat exchanger is highly desirable to prevent scale or corrosion of the chiller evaporator and condenser. Because of the relatively low temperature heat source, it is necessary to design this system for 42F water leaving the evaporator. Thus the coefficient of performance is lowest for the well water heat pump system. For the internal heat recovery system, the design temperature leaving the chiller is 44F in order to recover internal heat. For the solar assisted heat pump system, it was assumed that water would be mixed to obtain a 60F temperature entering the evaporator whenever the tank water was above 60F and the building cooling load was not adequate to provide the necessary heating. The higher heat source allows a higher coefficient of performance.

TABLE 1. SUMMARY OF COMPARATIVE COP FOR WATER-TO-WATER HEAT PUMP APPLICATION (Based on Specific Model of One Manufacturer)

	Summer	Winter
System 1 (Internal Heat Recovery)	3.80	4.30
System 2 (Well Water Heat Pump, 56F well water)	3.38	3.99
System 3 (Solar Assisted Heat Pump)	3.80	4.53

The comparative energy costs in the Pocatello area used in this study are the estimated costs that will prevail when the building is completed in 1977. These costs are given in Table 2. Note that the costs of electrical energy and natural gas are extremely low compared for most areas of the United States. Campus steam energy cost includes operating and maintenance personnel at the central heating plant. Although natural gas is the most economical fuel source for heating, the availability is not reliable. The rate is based on interruptible service, making a standby fuel necessary. For the 1974–75 heating season, approximately 50% of the heating needs were supplied by coal and 50% by gas. Table 3 shows the energy cost estimates for 1997.

TABLE 2. SUMMARY OF COMPARATIVE ENERGY COSTS, 1977—IDAHO STATE UNIVERSITY, POCATELLO

Energy Source	Heating Cost Per Million Btu ($)	Cooling Cost Per Ton Hr ($)	Simultaneous Heating and Cooling, 15 MBH and 1 Ton Hr ($)
Electric resistance ($0.0135/kwh)	3.955	0.0125	0.0718
Coal ($38.00/ton)	2.262	0.0388	0.0727
Natural gas ($0.06/therm)	0.884	0.0152	0.0285
Campus steam (50% coal, 50% gas)	3.113	0.0533	0.1000
Electric heat pump, COP 4.3	0.991	0.0125	0.0138

TABLE 3. SUMMARY OF COMPARATIVE ENERGY COSTS, 1997—IDAHO STATE UNIVERSITY, POCATELLO

Energy Source	Heating Cost Per Million Btu ($)	Cooling Cost Per Ton Hr ($)	Simultaneous Heating and Cooling, 15 MBH and 1 Ton Hr ($)
Electric resistance ($0.0386/kwh)	11.311	0.0357	0.2053
Coal ($56.49/ton)	3.361	0.0577	0.1080
Natural gas ($0.192/therm)	2.837	0.0488	0.0914
Campus steam (50% coal, 50% gas)	8.041	0.1382	0.2583
Electric heat pump, COP 4.3	2.833	0.0357	0.0395

Computer Modeling of Systems

In order to obtain a comparative yearly energy consumption for the systems studied, it was necessary to create an "average" year to make the economic comparison. Fifteen-year weather records from the Pocatello weather station were used to create the "average" year. Table 4 shows the average percentage of possible sunshine for the different months of the year based on 66-yr records. The year selected for

TABLE 4. CREATED AVERAGE YEAR—SOLAR

Month	Average Value Based on 66 Years	Actual Year Selected	Actual Percentage of Possible Sunshine
January	37	1972	37
February	47	1971	49
March	58	1967	59
April	64	1973	64
May	66	1967	68
June	72	1970	71
July	82	1972	82
August	81	1966	81
September	78	1971	79
October	66	1974	69
November	48	1960	51
December	36	1966	37

each month had the closest percentage of possible sunshine to the 66-yr average. Table 5 shows the month and year selected for the temperature profile during an "average" year. The 3-hr weather observations were taken for the years indicated and used as weather data input for the energy analysis studies. Each system was analyzed on the basis of temperature data every three hours for the created "average" year. The computer printout for the conventional system and for each heat pump system provided the energy requirements for each day, together with the outdoor temperature profile, the monthly energy requirements, and the summary sheet showing the yearly energy requirements.

TABLE 5. CREATED AVERAGE YEAR—TEMPERATURE

Month	Average Number of Degree-Days	Actual Year Selected	Actual Number of Degree-Days
January	1,296	1972	1,300
February	997	1974	979
March	918	1973	895
April	591	1969	594
May	336	1972	346
June	138	1966	140
July	0	1967	0
August	20	1973	18
September	192	1974	173
October	515	1972	527
November	879	1969	878
December	1,181	1965	1,216

A preliminary analysis was made without taking into consideration the auxiliary pumping costs and cooling tower operation. The costs of pumping water (condensing water, chilled water deep well pumps) amount to as much as the energy requirements for the heat pump for System 2.

Results

Table 6 presents the comparative results of the three systems. All three systems are economically feasible based on the estimated 20-yr energy costs. The internal energy heat recovery system is the most cost effective, primarily because of the lower auxiliary pumping costs. In this study, it was assumed that the one well would be operating continuously, since it is usually not a good practice to short cycle deep well pumps because of the possibility of pumping sand. The additional pumping energy required for the solar assisted heat pump system is for the solar collector circuit and the storage tank pump.

Alternate bids will be taken on the internal energy heat recovery system (System 1) and the well water heat pump system (System 2). Energy analysis shows that System 1 is the most cost-effective from the standpoint of recovering the premium cost. This is closely followed by the well water heat pump system. If it is possible to cycle the deep well pumps, System 2 would be the most cost-effective. The solar assisted heat pump system has a payout of 12.5 years compared with a conventional system using electric driven cooling and electric resistance heating. When compared to a conventional system using electric cooling and steam heating, the payback is 19.2 years. Even though the percentage of possible sunshine for January and December is low compared to most areas of the United States (36 and 37%), the solar assisted heat pump system would be economically feasible if the energy costs were the same as the national average.

Conclusions

In making an energy cost analysis for various systems, it is important to consider the auxiliary pumping cost and cooling tower operation.

Figure 6 indicates that all of the heat pump systems are effective from an energy depletion standpoint. Systems 1 and 3 could be applied to this building if it were located anywhere in the United States. System 2 could be applied only where there is groundwater available at relatively shallow depths. The use of electric driven water chillers as water-to-water heat pumps is a proven and reliable basis for designing energy-conserving systems.

TABLE 6. SUMMARY OF SYSTEM INITIAL PREMIUM COST VERSUS OPERATING COST

Item (premium cost above conventional system)	System 1	System 2	System 3 (18,144 sq ft coll.)
Estimated premium HVAC system cost	$34,600	$34,600	$34,600
Premium cost for well water system	0	29,000	0
Solar collectors (installed)	0	0	272,160
Structure to support collectors	0	0	36,290
Storage tank (45,000 gal)	0	0	18,000
Storage tank—supplemental heat exchanger	0	0	15,000
Collector storage—heat exchanger	0	0	22,500
Total premium initial cost	$34,600	$63,600	$398,550
20-Yr Operating Cost			
Operating cost, conventional system, all electric	$1,063,620	$1,063,620	$1,063,620
Operating cost of alternate system	474,526	446,186	424,695
Savings, alternate system (20 yr)	589,094	$617,434	$638,925
Years for payback*	1.2	2.1	12.5
Operating cost, conventional system, electricity/steam	$839,560	$839,560	$839,560
Operating cost of alternate system	474,526	446,186	424,695
Savings, alternate system	$365,034	$393,374	$414,865
Years for payback*	1.9	3.2	19.2

*No interest cost included for payback period.

ACKNOWLEDGMENTS

The authors gratefully acknowledge the help of John Korbis, Physical Plant Director at Idaho State University, and Keith Sundberg of Sundberg and Associates, Architect for the project, for their assistance and cooperation in making this study.

DISCUSSION

Q Why didn't you use the well water in the solar system? Can you justify erecting the storage tank under these conditions?

A The capital investment would be too high to justify both the well water and the solar system. For solar heating, it is mandatory to have some type of storage tank and the tank could probably be justified on the basis of the cooling cycle only by shaving the electrical peak demand.

Q In drawing water from one stratum and injecting it in another stratum aren't you upsetting the underground water balance? Why use both strata?

A The small amount of water involved would not have a measurable effect. It is necessary to use both strata in order to prevent recirculation of discharge water.

Q You were so successful heating the Sims Building through windows without trying, what could you do if you tried? Did you try on subsequent buildings? Could you do without wells if necessary? Could the southside heat be suitably distributed by moving air alone and without use of heat pump?

A We probably wouldn't have done any better from the standpoint of heating if we had purposely tried to design a solar system, simply because there must be a compromise between the solar glare control and the absorption of solar energy. On subsequent buildings, we have utilized the principle of recovering heat through the sunlit side to heat the shaded side. Wells are necessary to provide a heat source when the sun is not shining. No, we could not heat the north side by moving air alone because the temperature of the air would not be high enough. It is necessary to utilize the heat pump.

Q Idaho State Library Heat Pump system did not use reverse flow through condensers but "Well Water System" did use reverse flow (both systems were series flow through chillers and condensers). Why the difference?

A System 1 and System 3 both had parallel flow through the chillers and condensers to minimize pumping costs. For the well water heat pump system, we had reverse series flow to allow extraction of heat with heat exchangers from the 56F well water. This was necessary to obtain the temperature differential for a practical heat exchanger selection.

Q Did you consider using a storage tank on your heat reclaim system for night time heating? What maximum leaving temperature on heating was used off the heating bundle?

A No, but it might have been worthwhile to do so. Leaving temperature was 110F.

Q What was the evaporator temperature in your solar assisted heat pump?

A 34F with 44F water off.

Q Do COP values include the energy input to the auxiliaries in the denominator?

A No, COP values do not include energy input of auxiliaries. The overall COP would be approximately 3.0 for System 1 and approximately 2.5 for Systems 2 and 3.

Q Doesn't reference to "*x* cents/kwh" for electric resistance heating ignore the substantial impact of electrical demand charges? With stored solar or rejected heat, the opportunity for demand limiting should be enhanced.

A The cents per kilowatt hour used is the average cost per kilowatt hour including demand charges. The storage system for the solar assisted heat pump does provide the possibility for demand limiting, which has not been considered in this study.

Fig. 1. Architect's drawing of proposed Library Building, Idaho State University.

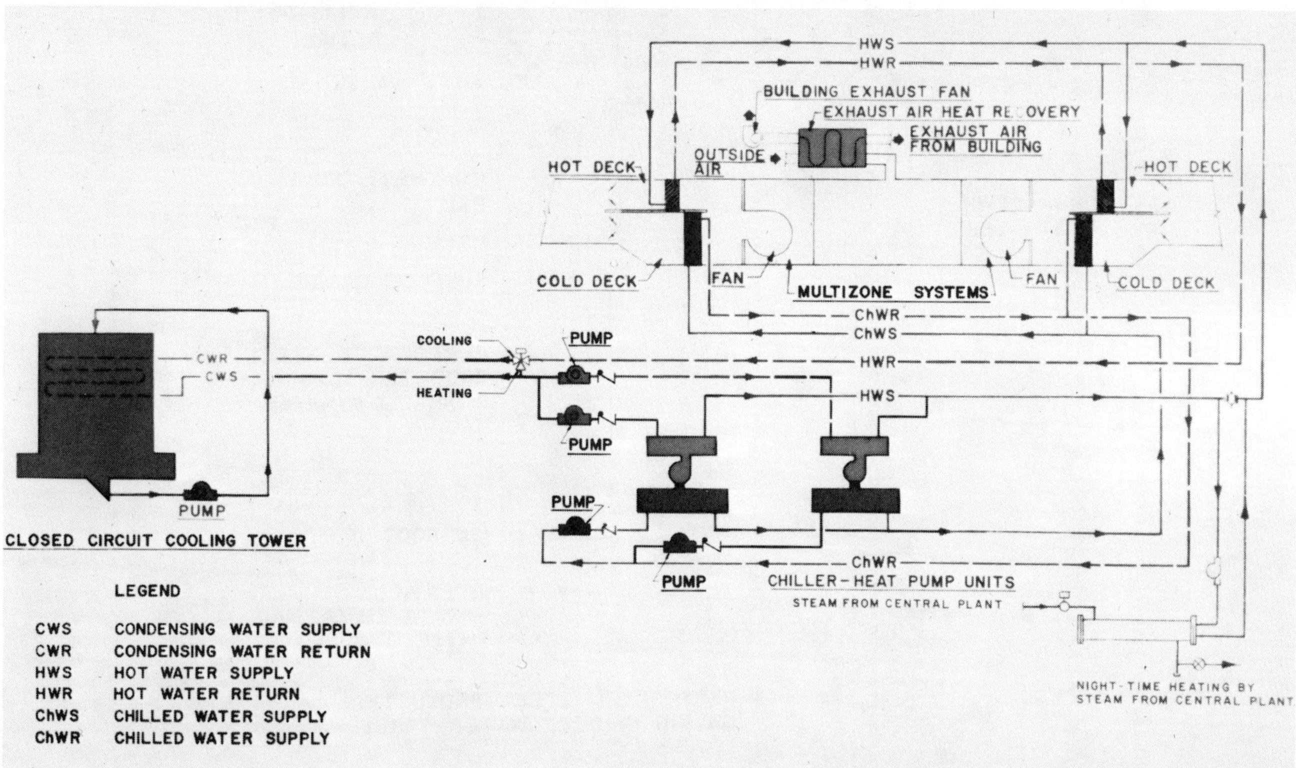

Fig. 2. Internal heat recovery heat pump system.

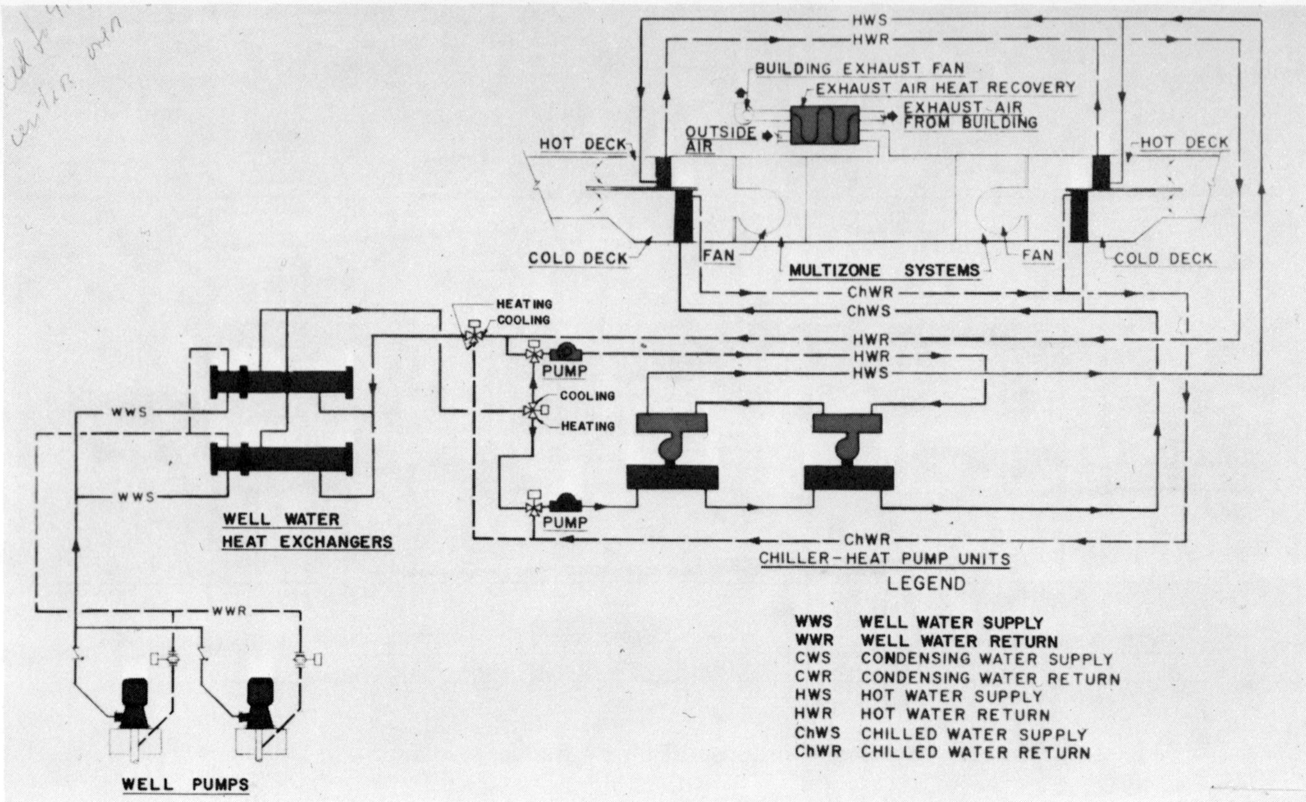

Fig. 3. Well water heat pump system.

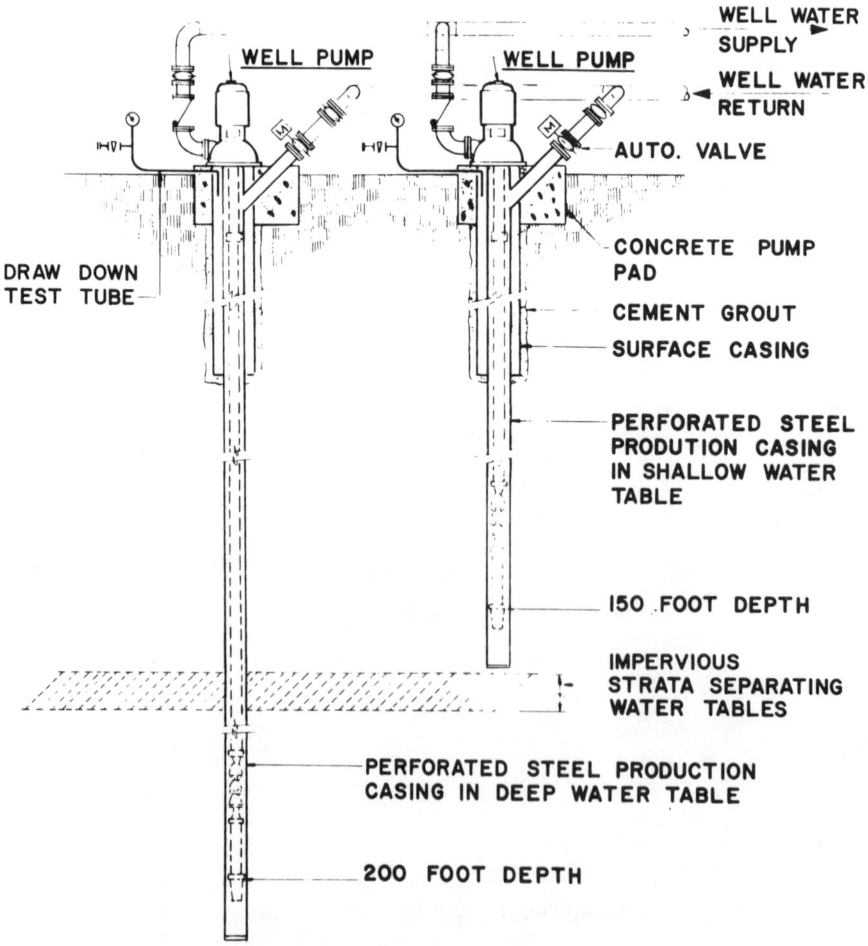

Fig. 4. Proposed arrangement of supply and disposal deep wells.

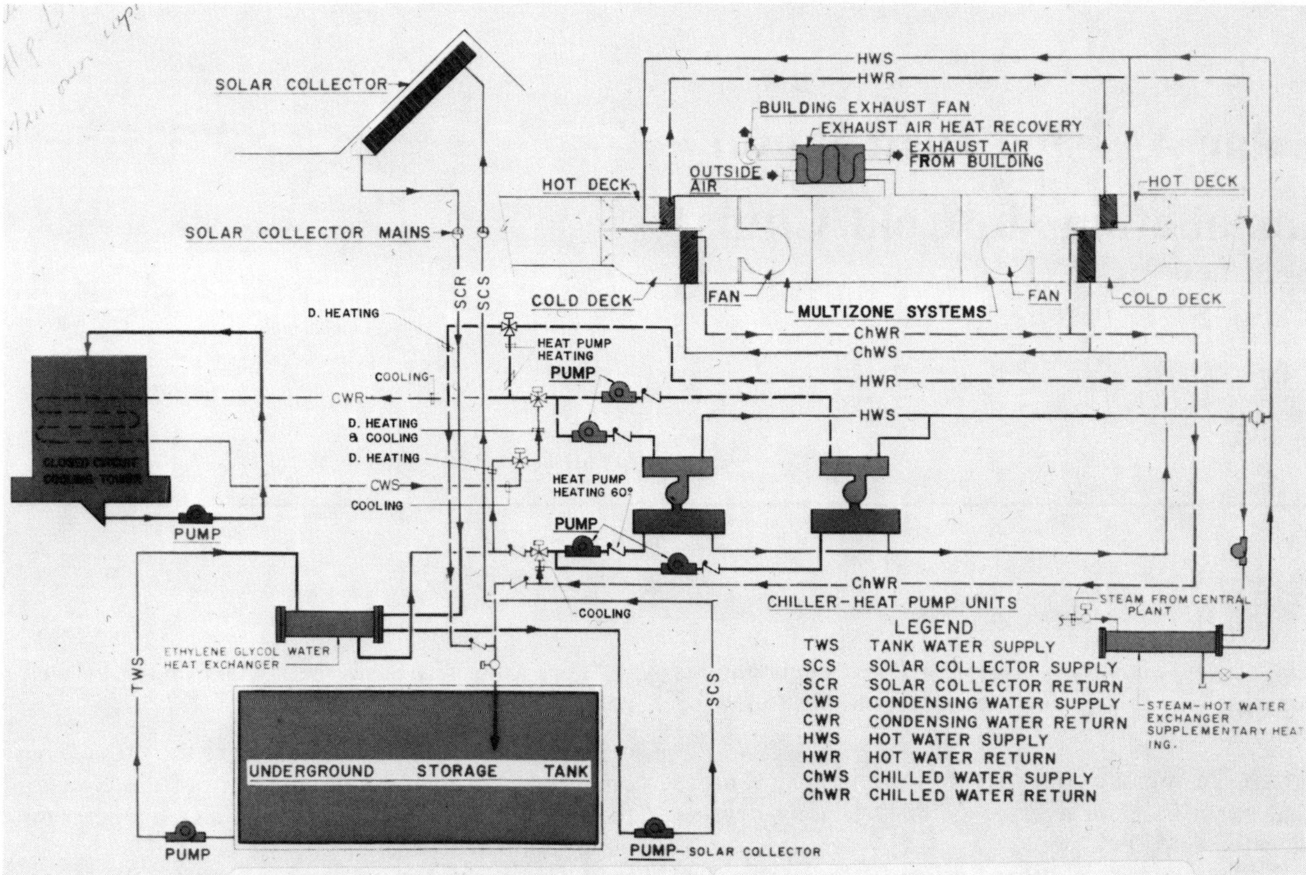

Fig. 5. Solar assisted heat pump system.

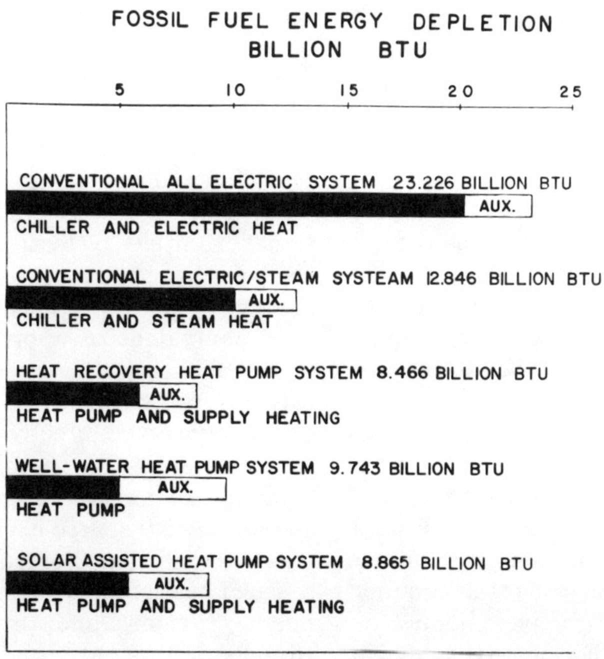

Fig. 6. Energy depletion comparison of systems studied. Note: Energy depletion includes energy used to generate electricity to drive heat pump units and related auxiliary equipment.

Solar Assisted Heat Pump Installations in Cold Climates

Fred S. Dubin

President, Dubin-Mindell-Bloome Associates
New York, N.Y.

The observations, conclusions, and recommendations presented in this paper are based on the following criteria:

1. Cold climatic zones are those geographical areas that experience an average of 4000 heating degree-days (base 65F) or more.
2. Solar energy is utilized primarily for space heating alone or also for domestic hot water.
3. Flat-plate collectors, or other nonfocusing collectors, are used to heat water or air which is used directly in heating coils or terminal heating devices when collection temperatures are adequate, and as a heat source for a heat pump to deliver warm air or warm water to the building when collection temperatures from the collector above are not high enough.
4. The heat pumps are essentially vapor-compression cycle prime-movers which are used to extract heat from water which is heated by the solar collector (by chilling it) and delivering warm water or air to the building at a temperature which is elevated above that of the warm water from the collector or available from storage.
5. Heat pumps are water-to-water, water-to-air, air-to-water, or air-to-air. The latter two types employ a water coil supplied from the solar collector or storage tank to raise the temperature of the air heat source to improve the COP of the heat pump and its available capacity for space heating.
6. When the heat pump is used in the refrigeration mode for air conditioning (without the use of solar energy), the economic feasibility of the total system is enhanced; but there are situations when the economic feasibility of the system is still attractive, although no air conditioning is required and the solar assisted heat pump operates only in the heating mode.

There are a few basic facts which might be well to review before proceeding further:

1. For any given solar collector, the efficiency increases as the average temperature of the circulating fluid in the collector decreases (inlet temperature + outlet temperature ÷ 2).
2. The coefficient of performance (COP) of a heat pump operating in the heating mode is the ratio of useful heat delivered to the building (the heat of compression plus heat extracted from the evaporator) versus the heat equivalent of the energy input to the heat pump. For a given condensing temperature, the COP increases as the temperature of the heat source increases, and the capacity of the heat pump (expressed in Btuh) also increases.
3. The total amount of electricity or fossil fuel energy required annually to heat the building and/or domestic hot water in a solar assisted heat pump system is the sum of the power required to operate the heat pump, the circulators and fans, plus any additional electricity or fossil fuel required to supplement the solar heat pump system.
4. For each kwh consumed by the heat pump, the heat equivalent of approximately 3.3 kwh is consumed at the electric generating plant. This energy is often referred to as "raw source energy," while energy consumed directly in the building is referred to as energy used "at the building boundary." From a national standpoint, we are interested in reducing the amount of raw source energy used.
5. In analyzing the economic feasibility of a solar assisted heat pump system versus either a direct solar heating system alone, or a heat pump system alone (usually both of the latter systems will require a supplementary conventional source of energy as well),

the sum of the annualized capital costs, maintenance costs, and energy costs for each system must be compared.

6. For a given solar collector, the conduction and convection losses increase as the ambient temperatures decrease, and as the wind velocity increases (except for evacuated collectors). In cold climatic zones, the losses are more severe than in warmer climates. These losses can be compensated for by increasing collector area, changing the collector configuration, or by more efficient collectors, all of which increase the capital costs of the solar collector. The collectors contribute the major costs in a solar energy system.

It should be evident now that the use of heat pumps with solar energy heating systems in the cold climatic zones can be cost-effective. With a solar assisted heat pump system, the strategy is to use the lower cost collector (rather than to use higher cost collectors to offset the conduction and convection losses which are greatest in cold climates) to provide a warm fluid heat source. In many cases, the heat storage system, mainly hot water in tanks at present, can also be smaller with the heat pump, as compared to a solar system without heat pump assistance.

As domestic gas and oil reserves diminish and imports increase in costs, we may move quite rapidly to an all-electric economy, at least until alternative energy sources are more highly developed. It behooves us then to examine our use of electricity with a very much more critical eye. The heat pump consumes only ½ to ¼ of the amount of electricity that electric resistive heating requires, and can reduce peak electric power requirements by 30–50%. Solar assisted heat pump systems which are designed so that direct solar energy supplies on the order of 50–60% of the seasonal heating requirements directly (the heat pump is inoperative during these periods) and the solar system provides a heat source for the heat pump to supply an additional 35% of the annual heat energy required (and at a relatively high COP)—leaving only about 15% or less of the annual energy use to be supplied by supplementary heat from electricity or fossil fuels—appear to be the most economically feasible for most cold climate installations. To attain that performance, it is most cost-effective to design the building and select building materials which reduce the heat loss to about 50% of that of the typical building erected in the past 10 years. The more efficient thermal building permits the use of low temperature heating media to handle the load.

The building, the solar collector and storage system, the heat pump, the heat distribution system, and the supplementary heating system must be considered synergistically. The design of any one of the components or subsystems affects the others and the total annual thermal performance and costs as well.

The upper limit of fluid temperature a single stage heat pump can deliver to a building is about 110–115F. These temperatures determine the heating coil or radiator sizes and water and air quantities, which will be greater than would be needed if designs were based on the usual design criteria of 130–160F air temperatures, or 180–200F water temperatures. Heating elements in the building, sized for the lower temperatures available from heat pumps, also permit extended use of the solar collectors directly. However, the lower fluid temperatures require larger air handling or terminal heating devices. By reducing the building heat loss, the following advantages accrue:

1. Less supplementary heat is required annually.
2. The lower media temperatures are capable of handling peak heat losses.
3. A less expensive supplementary system can be used since it will operate only a fraction of the number of hours in the heating system.
4. Reducing the heat loss lowers the quantity of air or water required, thus reducing the initial and operating costs of the circulating systems, which are already penalized by the lower media temperatures.

A variable air volume system, even in a residence, can further save energy, since peak air quantities are required less than 10% of the hours in the winter, and for a majority of the heating hours, the energy consumption for the fans is reduced. Water systems can be variable volume also.

Background

We have not found any computer programs which can provide an accurate analysis of a solar assisted heat pump system to optimize the equipment selection and building construction for each of the dozens of alternatives available. The alternatives include the number and size of storage tanks, the control cycle, the collector type and size, the specific heat pump characteristics, the building envelope, and the building HVAC systems.

In the solar assisted heat pump projects which are briefly described in the following section, we utilized a number of computer programs for building load and system analyses, and performed a large number of hand calculations before finalizing design. It remains to be seen whether the final designs are optimum or not. However, in all of the systems there are sufficient controls, valves, and piping arrangements to permit modifying the mode of operation in order to gather data for future designs. As a matter of fact, we view all solar energy projects at this time as test-bed installations to develop design procedures and analytical methods and to gather test data. The conclusions which are summarized below are based on our experiences gained during the analyses and/or design of the projects which are described below.

Summary of Conclusions

Solar collectors can produce 36–50% more usable Btu/year when coupled with heat pumps; and there is a greater savings in fossil fuels and electricity per dollar invested in collector and storage systems.

Always use solar heat fluid from the collector or storage directly to the heating coils or radiation in the building when the temperature of the fluid is high enough, before operating the heat pump.

It is always cost-effective to first reduce the building heat loss through energy conservation construction techniques and materials.

All of the buildings for which we have designed solar energy systems have energy consumption less than 50,000 Btu/sq ft/yr for all requirements—the standard developed for GSA office buildings.

Reflectors located in front of collectors to increase the concentration of radiation on the solar collectors reduce the operating hours required of the heat pump.

Each project requires a separate and detailed analysis—someone else's description of a design for their building can provide ideas, but cannot be used for any other specific project.

It is usually cost-effective to utilize solar assisted heat pumps in cold climatic zones as compared to using direct solar heat alone.

Solar assisted heat pump systems to heat domestic hot water only can be cost-effective in all climates. Cost-effective means different things to different people. For the purpose of this discussion, cost-effective means a payback period of the initial investment for solar and heat pump equipment of less than 20 years for state and federal buildings, less than 10 years for residences, and less than 3 years for speculative buildings.

When supplementary heat is required it should be introduced into the system as close to the ultimate load as possible, rather than used to boost storage temperatures.

The thermal losses and thermal inertia of the storage systems reduce the amount of effective heat gained from the supplementary source.

Use multiple storage tanks rather than a single heat storage tank, so that higher grade heat stored in the first tank will not be degraded by lower temperature fluids from the collectors.

The high temperature tank can be used as a source for direct heating and the fluid in the low temperature tank, which might otherwise have been discarded if only a single tank had been used, provides the heat source for the heat pumps.

It is generally more cost-effective to use a solar assisted heat pump during the heating mode, and the vapor-compression heat pump for cooling, rather than a system which employs solar heating directly and provides the solar cooling with an absorption refrigeration system.

It is difficult to provide a sufficient number of Btu at 180F or more for absorption cooling with standard flat-plate collectors; if solar cooling with absorption refrigeration cannot provide more than 60% of the seasonal cooling requirements, the electric- or gas engine-driven refrigeration unit will use less raw source energy for cooling.

It is essential to install efficient and compatible HVAC systems with solar energy installations. I emphasize "compatible" because a variable air volume system does save fan horsepower but must be carefully designed when used with direct expansion heat pumps.

A few particularly favorable energy conservation ideas include modular pumping; modular boilers if used for auxiliary heating in large buildings; ventilation air reduced to 5–8 cfm/person; a hot gas heat exchanger in the refrigerating system to heat domestic or service hot water; efficient lamps and luminaires to reduce the power required for lighting and hence the magnitude of the cooling load.

The heating and cooling load profiles should be controlled by the design of the building to optimize the performance of the total system when used year around.

In existing buildings which are retrofitted for solar and/or heat pump heating, use the cooling coil during the heating season, since the larger coil surface will permit the use of lower fluid temperature to attain the same thermal performance.

Current Projects

There are common characteristics in all of the projects which are briefly described below. However, each project has some unique features. It is not possible to describe each project in detail, but separate papers are available upon request which do describe each of the following buildings.

GSA ENERGY CONSERVATION DEMONSTRATION OFFICE BUILDING, MANCHESTER, NEW HAMPSHIRE (Figs. 1 and 2). This is a seven-story, 126,000 sq ft building, not including the underground garage now under construction. The solar energy system includes approximately 4700 sq ft of solar collector, mounted on the roof in four rows. The collector performance and space requirements are based on flat-plate collectors, facing south and arranged with a mechanism to provide adjustable tilt from 20 to 80°. The angle adjustment will be done seasonally to provide the amount of heat and temperature required for heating and cooling—the lower angles in the summer are best for cooling. The three storage tanks are used to store hot water from solar collectors, or waste heat from the diesel-gas engine, and chilled water in the summer. Heating coils were selected to permit operation with hot water at low temperatures to increase the collector efficiency. The solar collectors provide a heat source to supplement heat from the interior spaces for a heat pump, driven by the emergency diesel engine-generator set, and as a heat source for the closed-loop heat pump installation which is installed in three of the seven floors. There are different types of heating and lighting on each floor, so that their relative performance may be compared during operation.

In addition to the heat pump systems there are variable air volume systems; fan coil systems which can operate as two-, three-, or four-pipe systems; separate air-handling units for interior and each of the four exterior perimeter zones; blow-through air-handling units; compressors piped in series arrangements; enthalpy control with economizer cycle; recirculated toilet exhaust air through charcoal filters; air filters with low frictional resistance; electric ignition instead of gas pilots; and CO_2 monitoring system in garages. The lighting systems include photo-cell control for a natural light floor; selective lighting on all floors to include low level ballasts; furniture-mounted task lights on first floor; multilevel ballasts and polarized lighting; high output lamp sources including sodium on one floor; mercury vapor lamps in garages. Average power for lighting: 1.9 w/sq ft.

ADMINISTRATION AND RESEARCH BUILDING FOR THE CARY ARBORETUM OF THE NEW YORK BOTANICAL GARDENS, MILBROOK, NEW YORK (Fig. 3). This is a two-story, 28,000 sq ft building now under construction. It will be used as the administrative offices and research laboratories to support the arboretum program and provide classrooms for educational purposes. The major features include the following.

The lowest level is below grade and is served with light shafts to supplement electric lighting. The building is set back into a wooded area where the trees provide a windbreak on the north and sun shading on the west. The east facade is bermed up to 4 ft above the ground level. Masonry walls with insulation located on the external surface of the outside wall provide thermal mass and thermal resistance. The window area on the northern facade is less than on the west and east facades, and considerably less than the south facade. All glazing is equipped with external solar control to permit the solar rays to enter the building in the winter, especially on the south facade to reduce heating loads, and shade all windows during the summer to control the heat gain. All windows are equipped with sliding insulating shutters on the interior to reduce heat loss at night and during unoccupied periods. The U-factors of the opaque walls and roofs are less than .06. The roof is designed in a sawtooth fashion and the solar collectors are located on the south-facing slope of each peak.

The building will be equipped with approximately 6000 sq ft of solar collectors, three tanks for storage of solar heat and excess heat from an internal source for use during sunless periods. A heat pump installation, using solar heat gain as a heat source supplemented with well water from a deep well, provides winter heating with variable volume zoned air heating systems. The two 25-ton double bundle condensers and evaporators operate in a heat pump mode. Both condensers can operate with each 25-ton chiller which materially increases the COP at part-load conditions. Well water will be used directly in cooling coils for air conditioning and for condenser cooling for the small amount of refrigeration needed.

An experimental greenhouse will be fitted with thermal shades that can be closed at night to reduce heat loss and will be fitted with a combination water-air solar collector for heating; the same collectors will be used to temper makeup air for laboratory hoods. Laboratories and all cold and hot water faucets will be fitted with flow restrictors to minimize water usage. An underground cistern will collect rainwater which will be used for flushing water closets and laboratory requirements. These and other features indicate that this building will use well under 40,000 Btu/sq ft/yr for all services. The solar energy system is expected to provide virtually all of the domestic hot water and service hot water requirements and more than 70% of the yearly heating requirements. It is interesting to note that the construction budget has not been increased due to energy conservation measures, except for the cost of the solar energy system, which will add approximately $100,000 to the initial cost.

CHENEY JUNIOR HIGH SCHOOL, SPOKANE, WASHINGTON (Fig. 4). The preliminary design for this school for 900 pupils is designed to reduce heat loss and heat gain by 50% compared to conventional construction. Approximately 10,000 sq ft of solar collectors are located on the roof, providing direct heating to air-handling units in the building and as a heat source for a 100-hp double bundle condenser water-to-water heat pump. An antifreeze protection system developed for the Town Elementary School in Atlanta, Georgia, is included in the design system for the Cheney School. In order to reduce standby losses and the cost of an expensive heat exchanger and antifreeze solution, the collectors will be drained whenever the circulating pump is not circulating fluid through the collectors, and nitrogen will be introduced into the collector water passages to provide an inert atmosphere to inhibit corrosion.

GRASSY BROOK VILLAGE, NEWFANE, VERMONT. Two ten-dwelling-unit condominiums will be erected on one site. The design for the first ten units has been completed. The buildings are designed to conserve energy by the use of extra insulation, double glazing, thermal barriers which can be drawn over windows at night, and reduced infiltration by the use of special window sash. The solar energy system includes 4500 sq ft of flat-plate collector, arranged in three rows, and mounted on grade adjacent to the building. The collectors will serve all ten units. Two 10-hp water-to-water heat pumps with condensers that can be used with one or both units will utilize heated water from the solar collectors as a heat source. The sloping site permits the use of the concrete foundations as storage tanks. Insulation is applied to the exterior surfaces of each of the three 10,000-gal tanks. A separate solar assisted heat pump is used to provide domestic hot water for the ten families. In order to use concrete tanks, it was necessary to provide a heat exchanger, so that the tanks will not be subjected to high pressures due to the elevation of the solar collectors. A special oil will be used in the secondary circuit for antifreeze protection. The oil is nontoxic, has a high boiling point, low freezing point, and relative low specific weight and viscosity, making it superior to glycol, which is generally used in such systems.

SUFFOLK AND NASSAU COUNTY ENERGY STUDY, LONG ISLAND, NEW YORK. Under contract to the County of Suffolk, Department of Environmental Control, we first analyzed the Long Island Lighting Company's (LILCO) forecast for future electrical energy requirements, demographic projections, and current energy use in Long Island, a region of approximately 2,600,000 people. After adjusting LILCO's projections, a large number of energy conservation options which could be applied to existing and projected future buildings were analyzed in order to determine the effect such programs would have on peak electric demands and yearly consumption.

In the next phase of the study, we analyzed the potential reduction of peak demand and consumption that a solar energy program could accomplish. We determined that energy consumption and peak demands could be reduced by more than 40% below the 1995 forecast of LILCO. Seeking further ways of reducing consumption and peak demands, we considered the use of total energy systems for all installations in Long Island for buildings and developments with peak loads greater than 500 kw. These installations could reduce peak demand and consumption to an even greater extent.

In the next phase of this project, we worked with Professor William Heronemus of the University of Massachusetts to analyze the potential for using wind energy on Long Island. Meteorological data provided the wind patterns and velocity to determine the potential horsepower available from winds on a seasonal basis.

A number of wind generator sizes and configurations were selected for analyses, with two major modes of operation: (1) using wind energy as a solar furnace to use the electricity for heating hot water to replace all electrical resistance heating; and (2) to generate electricity for end use. Both on-shore and off-shore wind generating stations were considered. The results showed there is sufficient wind power available on Long Island which can be harnessed with conventional wind generators, although with sizes not generally in use today, to supply all of Long Island's electrical needs.

Our analyses of solar energy for an entire region indicated that the solar assisted heat pumps would not only reduce energy consumption of all fuels by a factor of 3 or more as compared to electric resistive heating, but that solar energy in combination with heat storage can also materially reduce the peak electrical demands and be instrumental in reducing the need for additional power generation facilities. The combination of solar assisted heat pumps and wind energy as a solar furnace was one option that appears to be economically competitive with nuclear power plants.

3M R & D CENTER, OAKDALE/LAKE ELMO SITE, ST. PAUL, MINNESOTA. 3M management retained a team of architects, engineers, and planners, as well as workers in other disciplines, to develop a long range master plan for research and development facilities to be erected on a new 570-acre site. The new center contemplates 10.5 million sq ft of new building for research activities to be constructed in addition to 3M's present R & D facilities, which now comprise a total of 4.5 million sq ft of space.

A major requirement of the planning team included energy conservation and minimum use of natural resources. Our energy studies recommended building construction and mechanical and electrical systems which would use no more than 60,000 Btu/sq ft/yr for office space (hours of operation would be longer than for the GSA building), and 130,000 Btu/sq ft/yr for laboratories. The study for the master plan showed that using solar assisted heat pumps in Minnesota for selected systems is feasible for systems such as tempering makeup air for laboratories (more than 8 million cfm will be required for this project), preheating heavy oil before combustion, and preheating combustion air.

Other projects for which we have completed preliminary design include two major buildings for Argonne National Laboratories at Argonne, Illinois, and a project for NASA to retrofit an existing building for solar heating and cooling, provide additional insight into the performance and economics of solar assisted heat pumps versus direct solar heating and absorption cooling, with solar energy as the heat source. Because of specific conditions peculiar to each of the individual projects, the solar assisted heat pump was not as feasible as the absorption system—again indicating the importance of making a specific analysis for all projects, even in cold climates such as Illinois. In the case of the Argonne projects, the refrigeration systems consisted of two 150-ton double bundle condensers driven by steam turbines in addition to the absorption units. The double bundle condensers are operating in a heat pump mode.

Conclusions

In order to develop solar energy and solar assisted heat pumps so that they will be commercially available in large numbers and at competitive life-cycle costs, the following is required:

1. Develop more efficient heat pumps. The theoretical maximum COP is 20, which of course could never be achieved, but heat pumps with a COP of 8 or 9 are achievable through design by a greater amount of extended heating surfaces, variable speed compressors, and different working fluids.
2. Tax incentives or subsidies are required to offset the higher capital costs, so that the nation can benefit in conservation of nonrenewable resources.
3. Manufacturers must be prepared to give a warranty with each collector for the consumer's benefit. Perhaps a government-sponsored insurance program covering the special equipment is in order.
4. Each solar project which is built should be monitored to provide data on performance, meteorological conditions, and costs as a basis for future design.
5. Schools of engineering and architecture must expand their educational programs to train design professionals who are capable of understanding and implementing intricate energy relationships and producing workable designs.
6. Educational programs are required for students in the lower grades to stimulate their interest in solar energy by early exposure to the concepts.
7. Low cost, long term mortgages must be made available for the installation of the solar assisted heat pump systems, so that cash flow does not become an impediment in taking advantage of lower life-cycle costs.
8. Manufacturers must issue a greater amount of data on part-load performance of all mechanical and electrical equipments.
9. Energy conservation measures for buildings must be pursued, with programs to develop reasonable energy budgets for all buildings, and the know-how to do so disseminated to the design professions and the public.
10. Solar energy systems with, and without, heat pumps should be considered for all new federal and state buildings. Such a program would save the taxpayer money in the long run, will not drain nonrenewable resources, and, most important, will demonstrate the feasibility of such systems to stimulate the private sector to follow suit.
11. All states should require that solar energy be used for all water heating and swimming pool heating immediately, and, in a graduated program, require that a percentage of all new construction include solar energy.

The development of other solar energy uses such as wind energy, ocean thermal differences, photovoltaic cells for direct electrical power generation, and solar collectors which can provide higher temperatures for thermal engines are all essential to prepare for the not too distant future when fossil fuels may no longer be available at a competitive price.

DISCUSSION

Q You mentioned drawbacks of panel radiant heat systems. Initial cost is one. Can you briefly list a few others?

A Disadvantages include (*a*) the lack of the capability to provide ventilation, humidification, and air conditioning, (*b*) unequal expansion of piping attached to either ceiling or floor construction caused pipe failures in the past, and (*c*) poor covering may interfere with thermal performance.

Q Does the 50% extra efficiency with heat pump assisted systems justify the price of the pump and consumption cost?

A Yes, if compared to electric heating in almost all areas of the country. Yes, if cooling is also included in the program. Not versus oil or gas if justification means a short payback period. Yes, if justification is in terms of raw sources energy conservation and hedge against curtailment. Yes, if larger payback periods can be accepted, especially as fuel costs escalate. Yes, under certain conditions, if well water, groundwater, or other heat sources are available to supplement solar heat.

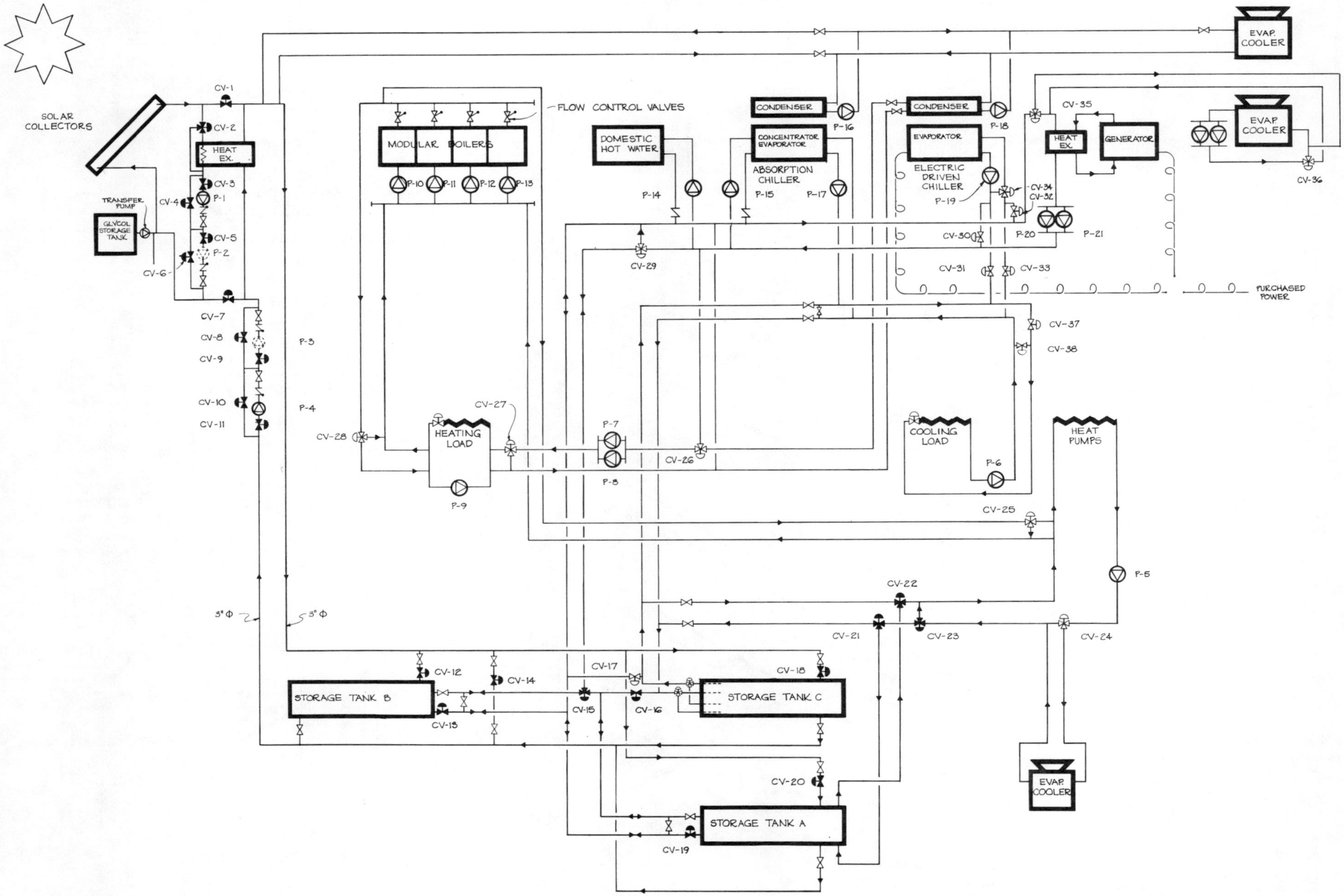

Fig. 1. Flow diagram showing interface between solar panels and environmental systems. Federal office building, Manchester, New Hampshire.

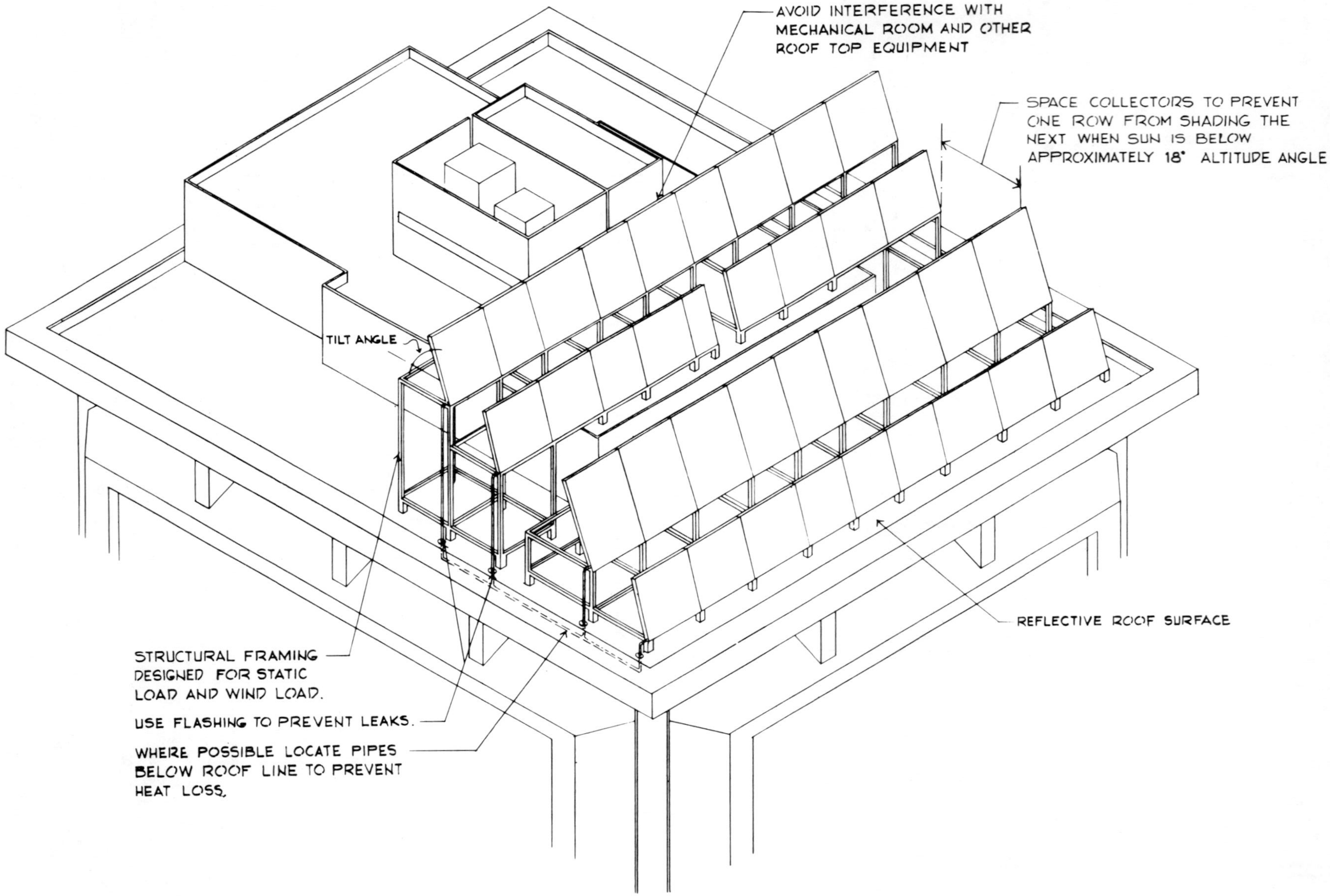

Fig. 2. Layout of solar collectors for federal office building, Manchester, New Hampshire.

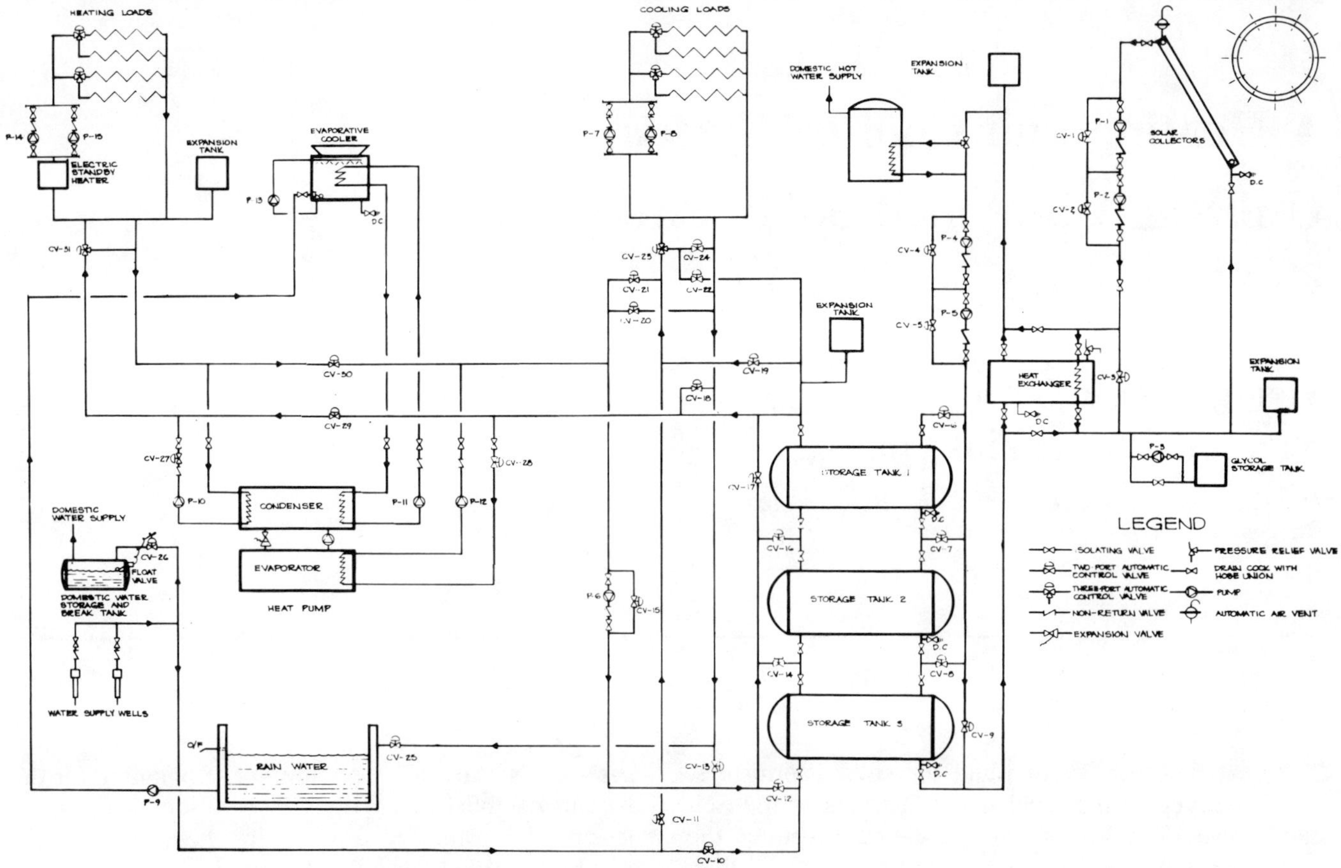

Fig. 3. Flow diagram, Cary Arboretum, Milbrook, New York.

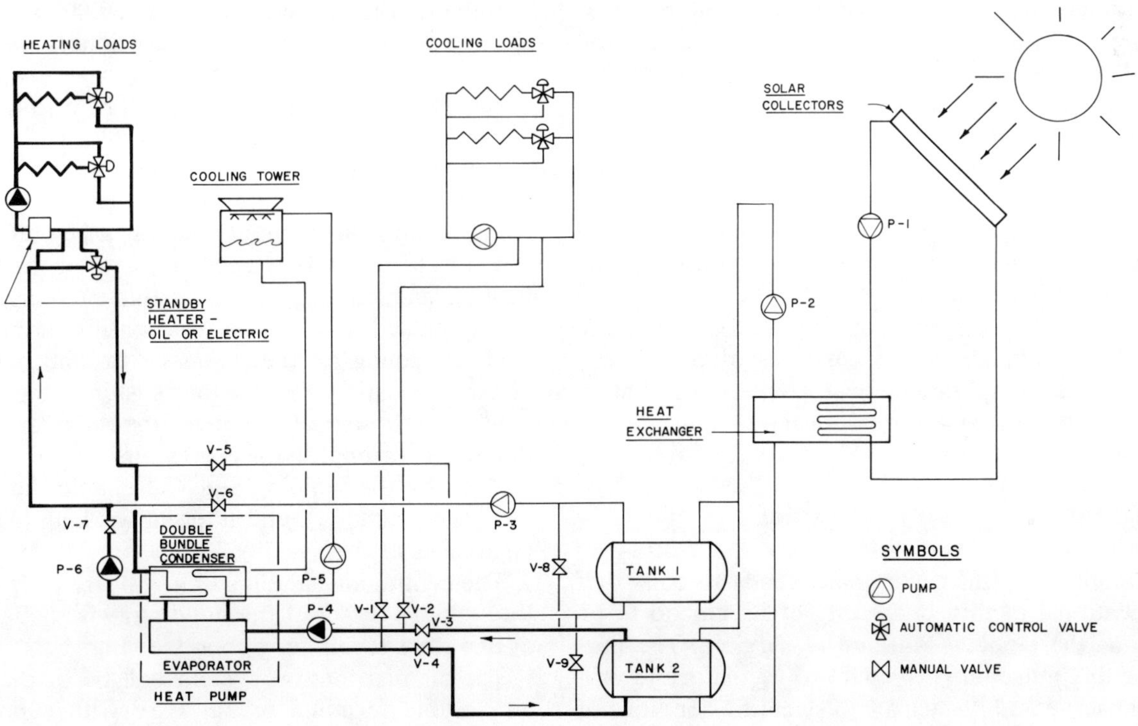

Fig. 4. Heating in the heat pump mode. Flow diagram, Cheney Junior High School, Spokane, Washington.

A Heating and Cooling System Utilizing Combined Solar Energy and Internal Heat Sources

Hsing-Chung Yu
Richard P. Hankins, Jr.

Hankins and Anderson, Inc., Consulting Engineers
Richmond, Va.

The heat recovery system and the solar energy system employed in the mechanical facilities of the Science Museum of Virginia are predicted to reduce the annual cost of thermal energy for heating, cooling, and domestic hot water from $50,000 for a conventional system to $12,000.

The heat recovery system in this building is of the low temperature, hot water type which utilizes the heat rejected by a centrifugal chiller to its condenser water for heating purposes within the building. This system has enough capacity, when recycling the internal building heat, to carry the skin load and the ventilation heating load of the building down to 20F outdoor temperature.

The solar energy system includes a 27,000 sq ft collector on the roof which is designed to provide up to 125 tons of refrigeration through the medium of an absorption refrigeration machine. It is presently the largest known such system that has been attempted for providing the energy to both heat and cool a building. This system alone can carry the domestic hot water heating load, building heat loss, and ventilation heating load down to 45F outdoor temperature.

The Science Museum of Virginia

The Commonwealth of Virginia intends to construct an educational facility in the city of Richmond to be known as the Science Museum of Virginia. The purpose of this museum is to enable Virginians and others to observe and participate in scientific activities of many kinds. An early decision in the planning for this building established energy conservation as a primary design goal, this decision having been made some time before the Arab oil embargo.

The building is to be situated in a park with the collectors on the roof. The collectors will face due south and will be tilted 30° above the horizontal to obtain maximum energy during the summer months when the demand for cooling is the greatest.

Initially, the museum will have 60,000 sq ft of floor area to be devoted to physical sciences such as astronomy, physics, chemistry, and mathematics. One of its more interesting features will be a 75-ft diameter tilted dome planetarium which will produce computer-controlled astronomical shows. The heat produced by the electronic circuitry of the planetarium is substantial and was a key factor in the decision to use a heat recovery system to serve the heating and cooling needs of this project. The museum decided it would be of scientific interest, especially to young people, to make the mechanical and electrical heating and cooling systems into a permanent exhibit, instead of concealing them, as is usually done. Accordingly, these systems will be exposed throughout the building and will enable the public to view the machinery and internal working of the principal parts of the system.

The central air-handling unit will rise 75 ft vertically through the center of the building, with a glass casing at two floor levels for public viewing.

The air distribution system will be of the variable air volume terminal reheat type with supplemental baseboard radiation, which can provide readily controllable conditions for this building with its diverse activities and schedules. The supply and return air quantities and fan horsepowers will vary with the fluctuating load in the building, and an economizer cycle will use outside air for cooling whenever possible.

The solar energy system is also on display. A bridge will take people from the mezzanine level above the main exhibit floor to an observation "bubble" projecting through the roof of the building and through the collector on the roof. This "bubble" will serve as a focal point for explaining the solar energy collector and the system to the public. These displays are reinforced by locating some of the collector piping and associated valves and equipment nearby, with thermometers and flow meters on the pipes for people to read.

The Heat Recovery System

The average power consumed in the Museum for lighting is only 2.2 w/sq ft. This imposes a cooling load of only 36.5 tons, which is less than is normally found in buildings of this kind. This low average figure is due to the use of accent lighting instead of extensive general illumination in the exhibit areas, and to the desire not to consume additional energy for lighting simply for the purpose of recycling it. However, the estimated peak equipment load averages 2.1 w/sq ft and imposes a cooling load of 32 tons, and the maximum anticipated population of 1200 people per hour provides a further 50 tons of potential internal heat.

A breakdown by percentage of the building peak design cooling load of 250 tons of refrigeration is as follows:

Transmission and solar heat gain	15%
Equipment	15
Lights	13
People	22
Ventilation air	21
Mechanical equipment	14
	100%

The peak wintertime heat loss for the building at 0F outdoor design temperature is 1.6 million Btuh of which 784,000 Btuh is transmission load and 824,000 Btuh is ventilation air load.

The minimum cooling load for the heat recovery system during normal operation will be 71 tons when necessary fan and pump cooling loads are added to the basic lighting load. This 71 tons represents 48% of the total 125 tons required of the centrifugal chiller and will require nearly 100 kw of electrical energy for operation. Thus the total heat rejection of the chiller at minimum load is 1.2 million Btuh, all of which is available for heating purposes. The coefficient of performance (COP) under this condition is calculated to be 3.5. If the equipment load is added the COP is 3.6, and if people are added the COP rises to 3.9.

It can be mathematically demonstrated that a COP ranging between 3.5 and 4.0 is justifiable in this museum when the lighting energy is available and the air-handling unit and pumps are operating at 40% capacity, even though no solar energy heat source is available. Based on an average electricity cost of \$0.03/kwh and on a cost for No. 2 fuel oil of \$0.35/gal, it is found that the electrically driven heat recovery system with a COP of 3.5 will produce 23% more useful heat than can be produced by burning equivalent input oil in a boiler.

Based on the peak heat loss of the building and the total heat rejection by the chiller at minimum load, the calculated temperature below which auxiliary energy will be required is 20F. A study of weather records for Richmond shows that only 3% of the total hours between October and May have a temperature below 20F. Therefore the theoretical requirement for an auxiliary source of heat is low.

The Solar Energy System

Despite all the advantages of the internal sources heat recovery system, the refrigeration machine must be operated in order to cool the space load and to produce the required heating capacity. It would be ideal if the building could be cooled with outside air whenever the outdoor temperature is 54F or below, and if it could be heated at the same time with solar energy. This leads to a study of the solar energy portion of the system.

The museum's Board of Directors wants to use collected solar energy to the greatest practical extent. Whatever energy is collected, but cannot be used immediately for heating, cooling, or domestic hot water, will be stored in insulated underground storage tanks for later use. Figure 1 is a simplified schematic piping diagram of the mechanical system. To obtain maximum flexibility in system flows, the piping subsystems make extensive use of primary and secondary pumping.

There are four levels of pumping. The first is the solar collector hot water heating loop. The second is the primary or high level hot water loop serving the absorption chiller generator, the domestic hot water heater, and the four 20,000-gal high level hot water storage tanks. The third is the low temperature hot water system which utilizes the heat rejected by the centrifugal chiller to its condenser water, and a 20,000-gal tank is provided to store excess heat from this system. The fourth is made up of several building heating subsystems which can draw heat from one of five sources in sequence:

1. Solar collector system.
2. Hot water drawn directly from any of the various storage tanks whenever the tank temperature is 100F or higher.
3. Heat recovery system using internal heat.
4. Heat recovery system as water-to-water heat pump using heat to artificially load the chiller from any storage tank at a temperature between 40 and 100F.
5. Standby oil-fired boiler.

The solar collector subsystem will be controlled to obtain a constant outlet water temperature by regulating the water flow through the collector using an automatic bypass valve. The outlet temperature will be set at 195F in the summer in order to provide water at a sufficiently high temperature to fire the absorption chiller, and at 140F in the winter. Once the flow reaches the pump capacity, the outlet temperature will be allowed to rise with increased solar flux.

The peak cooling load has been split into halves with both the absorption and centrifugal chillers designed to produce 125 tons of refrigeration. The two chillers are arranged in series with the absorption machine normally as the lead machine in order to make maximum use of available solar energy. Chilled water produced by the absorption chiller in excess of the instantaneous building demand will be stored in the three 20,000-gal dual temperature tanks for later use. For days when the building cooling load will approach the peak load, the centrifugal machine may be utilized to meet the morning cooling load and the entire output of the absorption chiller stored in the tanks to meet the afternoon peak load. In this way, the tanks can be used to time-phase the peak production of chilled water by the absorption chiller around solar noon to meet the peak demand for cooling several hours later. It is calculated that over 660 ton-hr of refrigeration effect can be stored in the dual temperature tanks. It is best to store excess chilled water rather than excess solar heated hot water in the summer because there is less energy loss from cold water at 50F than from hot water at 195F in buried tanks, and also because the temperature of the water entering the generator of the absorption chiller is critical.

In the heating season, the dual temperature tanks will be switched to store hot water collected by the solar collector, and thus will augment heat stored in the four high level hot water tanks. During this season, the total system will have a storage capacity of 160 hr (nearly one week) of peak transmission load calculated at 0F outdoor temperature, or 80 hr of peak transmission and ventilation load.

During the preliminary design stage of this project, an energy sources and uses chart was developed. This chart (Fig. 2) shows how much energy is estimated to be required for heating, cooling, and domestic hot water as a function of outdoor temperature.

Since solar energy is a nearly free source of energy, it will be used to the maximum extent possible whenever it is available. Accordingly, in the winter, solar energy will be used to heat virtually all of the domestic hot water consumed in the building and to provide about 60% of the building space and ventilation air heating. The heat recovery system should be able to handle the remainder of the heating, except during long cold cloudy periods when the solar energy heat will have all been consumed. Then, the boiler will have to be fired to provide auxiliary heat.

In the summer, the solar energy system will be used to do the final heating of the domestic hot water (after it has been preheated by the heat recovery system) and to fire the absorption chiller. The centrifugal chiller will be used to provide all cooling not handled by the absorption chiller.

An economic analysis of the solar energy and heat recovery systems shows that the payoff period for the solar energy system alone is 22 years, and for the combined solar energy and energy conservation system the payoff period is 15 years.

ERDA has recently awarded a contract under which a computer model will be developed to simulate the performance of the Science Museum solar energy system. The aim of the modeling is to optimize equipment sizes and to better match the solar collector and absorption chiller components. In particular, studies will be made as to the types, numbers, and sizes of the various storage tanks, through dynamic simulation of the system throughout a year of operation.

DISCUSSION

Q Did you not say that the cooling load was 71 tons? If so, why are you using 250 tons of refrigeration? If you are storing chilled water, why not peak with small capacity compressors?

A The minimum cooling load when the lights are on and the minimum number of fans and pumps are running is 71 tons. The peak cooling load for the building is 250 tons of refrigeration. Small capacity compressors could not handle the building cooling load on a cloudy humid day with a high internal heat gain when the absorption machine was unable to produce any cooling. This is stated in the paper.

Q In the title of your paper, you use the term "Internal Heat Sources." I believe this should more realistically be termed heat recovery, since the source is largely internal heat consisting of electrical (lighting) energy. As in the closed loop heat pump system, this represents a redistribution of purchased energy rather than rejecting it—hence the preferred "recovery" term.

A No disagreement.

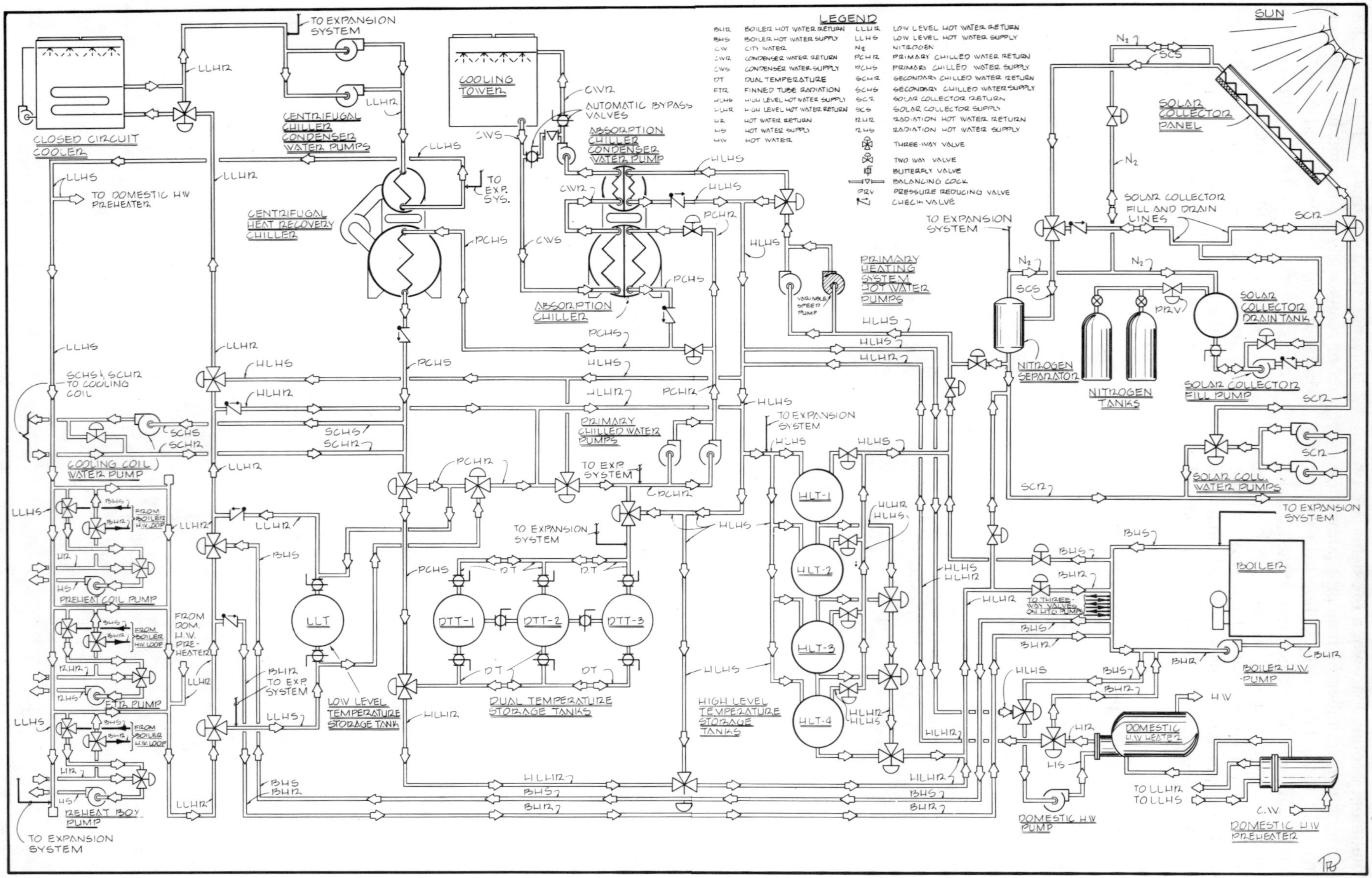

Fig. 1. Schematic piping diagram of Science Museum of Virginia.

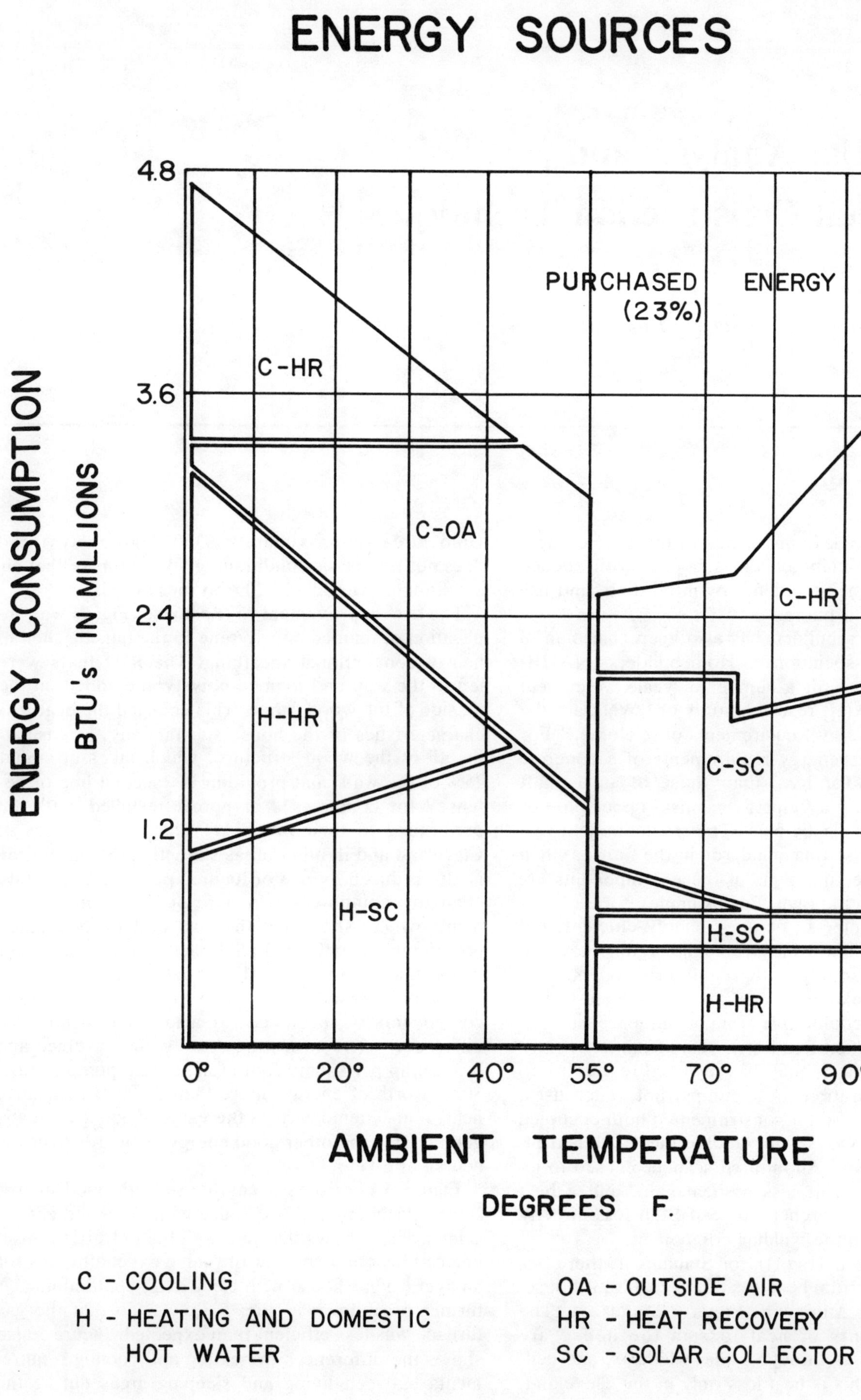

Fig. 2. Energy sources and uses chart.

Energy-Use Analysis and Residential Construction Techniques

Don J. Dirkse

Development Manager, Bob Schmitt Homes, Inc.
Strongsville, Ohio

This paper describes approaches to the development of an energy-efficient builder's house, with special emphasis on high insulation, low infiltration, and use of a nonsolar heat pump.

Bob Schmitt, a builder, has also been chairman of the National Association of Homebuilders (NAHB) Research Institute for a number of years. A current project is an NAHB research study of "What are the Comparative Energy Requirements of A Home." For example, are the energy requirements of a concrete driveway more (or less) than those of an asphalt driveway? Or of clay pipe as against copper pipe or concrete pipe or plastic pipe? The results may have a major impact in setting standards in the future, when all the energy requirements of these components are broken down and known. For example, if the results showed plastic pipe to be most energy efficient, but the code did not allow its use, a valid argument would exist when reasoning with regulatory bodies. The study is a very extensive one, and is expected to last many years. It could have a major influence on residential construction five to ten years from now.

The program at Schmitt has included research into insulation, infiltration, heat pumps, hot water uses, and solar energy, and has instrumented both occupied and unoccupied houses to validate theory and to investigate heat loss. An infrared scanner is used to locate sources of heat loss or heat gain, and a heat scope enabled researchers to read Btuh loss directly and quickly from the building component.

In the bar charts (Fig. 1), for Standard 1, there is a bar showing the total heat loss of the 2000 sq ft house, as it met FHA Minimum Property Standards. The various elements of heat loss in the house are analyzed—such as through walls, windows, and ceiling. Note that this is heat loss only in the Cleveland, Ohio, area (approximately 6350 degree-days) and does not reflect air conditioning. We estimate that air conditioning would add 20% to these totals.

The first improvement to the house (Fig. 2) was application of foamed polystyrene to the outside shell in lieu of conventional sheathing. The R-11 batts were left in the wall and foamed polystyrene added on the outside of the wood frame. This changed the heat loss characteristics of the house substantially, by insulating all of the wood structure, which takes up about 25% of the wall, and providing a "second line of defense" for compressed or poorly installed batts not performing to their full R-11 value. Test houses in Cleveland and in other areas show that the infiltration is also reduced by use of foamed polystyrene. Utility company computers which print out a budget for home owners show that the cost of 1 in. of foamed polystyrene sheathing is paid back to the homeowner in less than two years at the $0.03/kwh rate.

Figure 3 shows the "U" and "R" values of various components of the house. It also demonstrates the difference between heating costs with resistance and heat pump per square foot of area: heat pump returns $0.80 worth of energy, more than $1.00 of resistance heat. It also demonstrates the value of good windows, good doors, and other good energy standards over the course of a year.

Figure 4 compares a gas house with good energy standards and an electric house with improved energy standards. It shows that in a well-built electric house, one can heat and cool for the same expenditure as for an average gas house with electric air conditioning. A further breakdown in Fig. 5 shows too that the gas furnace was less efficient than expected. Figure 5 also shows the differences in heating and cooling requirements between living and sleeping areas during the different seasons. Note how dramatically appliances contribute to the heating and cooling requirements.

This analysis has led to erection of houses of approximately 2000 sq ft, with a complete energy bill of less than $75/month. Competitive houses in the Cleveland area, built to lesser standards, run as high as $200/month.

The normal balance point for the heat pump in the area is between 25 and 30F. With tightened design, the balance point has moved down to 15–17F, offering a further operating economy and giving at least a 1.8 SPF. Also, half the supplemental heat has been cut out, and there is only one stage of supplemental heat. That reduces original service entry costs. (Excessive supplemental heat can double normal service entry costs.)

Figure 6 shows the Sun House. (It is not a solar house. We examined solar heat two years ago, but it would cost $10,000 per house. There is no market aggregation, no single source of supply, no single source of responsibility; as a result, the costs are very high.) We took the $10,000 and put it into this house. It is U-shaped, with living on one side, sleeping on the other, and a large sky-lighted atrium in the center. There are no windows or doors on the outside of the house. The atrium is a very livable, practical place. We added about 1100 sq ft of comfortable living space to this otherwise 2500 sq ft house for less than the cost of solar heat.

Because there are no doors or windows on the outside except for the entry, the temperature difference between the outside and the atrium is about 10–20F, and there is no infiltration. Consequently, the heating bill was cut to less than half of that projected—thanks to input from the sun into the heat sink of the atrium area (Fig. 7). That does not pay for the entire atrium, but offers major living amenities. Schmitt has now built three of these houses. They are successful, though there have been problems with condensation which will be resolved.

The lowest temperatures ever recorded on our thermocouples were in the high 40's, with no supplemental heat. The fireplace will keep the entire atrium "shirt-sleeve" warm on the coldest winter evening in Cleveland. Every room in the house gets to "see" the fireplace.

The author would like to add a concluding note. Builders are concerned that the Energy Research and Development Administration (ERDA) is spending millions of dollars on solar research. This source is going to supply only a fraction of our energy needs 10 years from now. What we need to know today is how to take care of the 50 to 60 million existing homes that represent a potential for instant energy conservation.

And what about air infiltration? How many air changes are needed or required? What is the optimum of efficiency for insulation, and where should it be placed? Can we make hot water heaters more efficient? What about attic insulation and ventilation standards—should we seal up the attic and enclose that air space as a solar trap? Should we open it up even more with greater ventilation requirements and perhaps even power ventilation? What is the proper design for external combustion air for a fireplace? How can we better utilize or eliminate heat from appliances and lighting fixtures?

We need the answers to these kinds of things today; if we had them, we could save very large amounts of energy, both in new and existing construction.

DISCUSSION

Q How expensive is the sun house?

A The sun houses contain approximately 2500 sq ft of living area plus 1100 sq ft of atrium and a large garage. They sell for about $85,000 plus lot.

Q Are any energy consumption analyses available?

A Specific performance data on the sun houses for full-year operation will be available in late 1975 from Bob Schmitt Homes. Early indications show at least 30% heat savings.

Q How do you prevent excessive winter heat loss from sunlight in the atrium?

A We do not provide for controlling excessive heat loss in winter or heat gain in summer. Shuttering or shading is being considered.

Q What are the fire and toxicity hazards of foamed polystyrene?

A The fire and toxicity hazards of foamed polystyrene are equivalent to the fiberboard or plywood sheathing which it replaces. There is no difference in the fire rating or fuel contribution when used in accordance with proper installation recommendations. It is the fire, heat, and smoke from combustibles in the house which cause the greatest problem.

Q What size heat pump are you using?

A We use heat pumps averaging 2 tons. Generally, if we get up to 3-ton requirements in our large ranch homes, zoning is desirable with two or more smaller units.

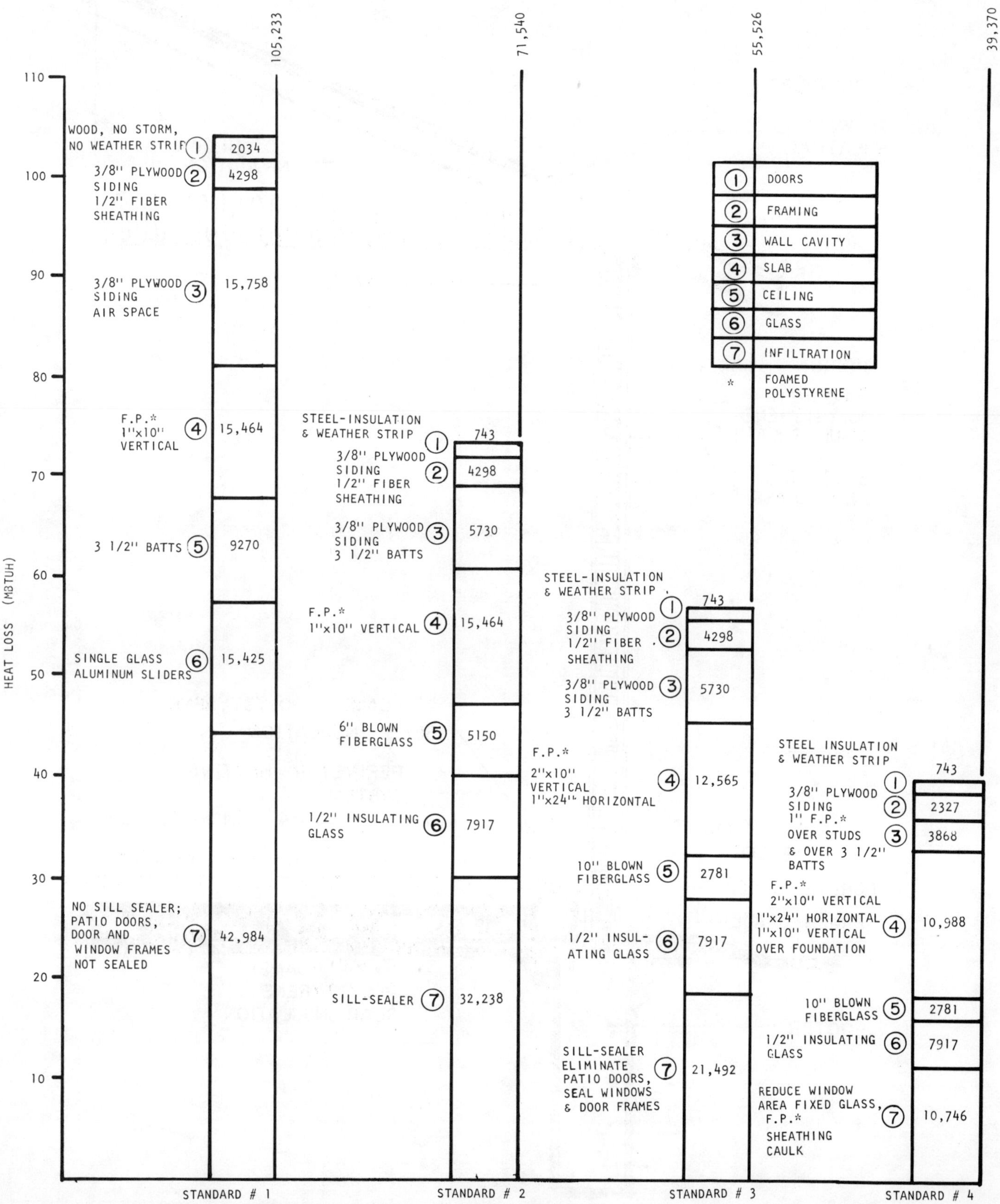

Fig. 1. Elements of heat loss in four types of construction.

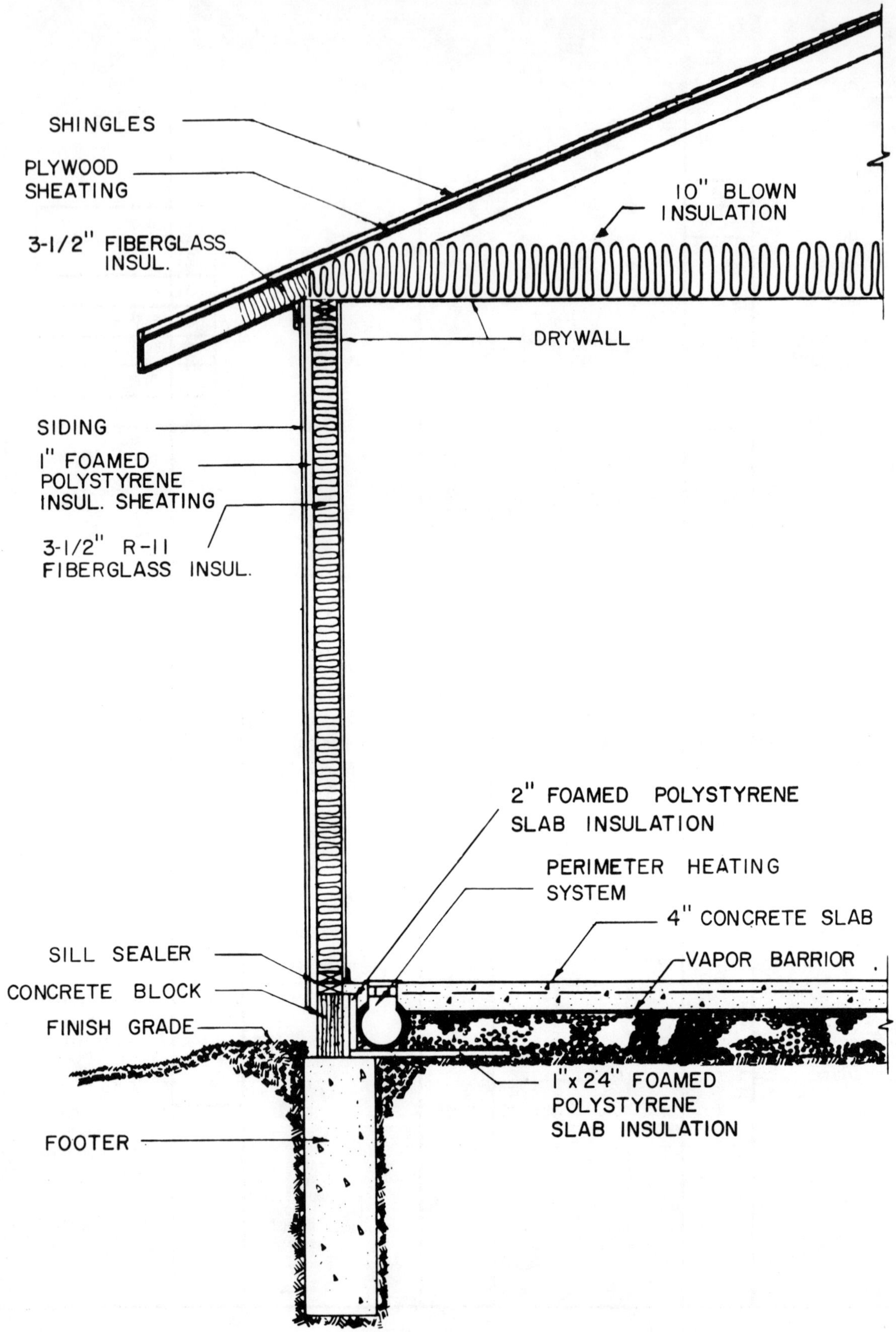

Fig. 2. Cross section of Schmitt house construction.

ENERGY LOSSES
Comparative costs per square foot of surface
for electric resistance heat and heat pump.

"U"	"R"	*Bob Schmitt Homes Specifications	Resistance Heat (COP=1)	Heat Pump (COP=1.8)
		FOUNDATION		
1.754	.57	(1) Perimeter of slab, 4" solid block, no insulation	$1.812	$1.007
.060	16.80	*(2) Perimeter of slab, 4" solid block, 3" **F. P.***	.062	.034
.510	1.96	(3) Perimeter of basement or crawl space, 8" block, no insulation	.527	.293
.136	7.37	(4) Perimeter of basement or crawl space, 8" block, 1" **F. P.***	.140	.078
		EXTERIOR WALLS (25% framing factor)		
.151	6.62	(5) 5/8" plywood siding, 25/32" fiberboard sheathing, 3 1/2" air space, 1/2" drywall	.156	.087
.084	11.90	(6) 5/8" plywood siding, 1/2" fiberboard sheathing, 3 1/2" batt; 1/2" drywall	.087	.048
.062	16.13	(7) 5/8" plywood siding, 1" **F. P.*** sheathing, 3 1/2" batt, 1/2" drywall	.064	.036
		CEILINGS		
.599	1.67	(8) No insulation (½" drywall)	.619	.344
.079	12.67	(9) Blown-in fibrous glass, 5", R-11	.082	.046
.048	20.67	(10) Blown-in fibrous glass, 8 3/4", R-19	.050	.028
.042	23.67	*(11) Blown-in fibrous glass, 10", R-22	.043	.024
.032	31.67	(12) Blown-in fibrous glass, 13 3/4", R-30	.033	.018
		WINDOWS AND DOORS		
1.13	.88	(13) Single glass	1.167	.649
.58	1.54	*(14) Double glass, ½" air space	.599	.333
.43	2.33	(15) Solid wood door, 2" nominal	.444	.247
.29	3.45	(16) Solid wood door, plus metal storm door	.300	.167
.076	13.00	*(17) Metal door, 1 3/4", urethane core, magnetic weatherstrip	.079	.044

Above "U" and "R" Values have been determined according to 1972 ASHRAE Handbook

COP = Coefficient of Performance. Resistance heat delivers $1 worth of heat for every $1 worth of electrical input. A heat pump with a COP of 1.8 delivers $1.80 worth of heat for every $1 worth of electrical input.

NOTE: The ENERGY EFFICIENCY in a Bob Schmitt home also significantly reduces cooling costs which amount to approximately 20% of the heating cost.

***F. P. = Foamed Polystyrene**

$$\frac{\text{"U"} \times 6350 \text{ Degree Days} \times \text{C18} \times 3¢/\text{KWH}}{3413} = \text{Cost Per Heating Season}$$

Fig. 3. Energy losses and values.

TOTAL ENERGY CONSUMPTION COMPARISON
TWO IDENTICAL HOUSES BOTH OCCUPIED BY FAMILY OF 3

S/L 6008: Insulated to Standard No. 3, heated and cooled with two 2 Ton Heat Pumps (Downie)
S/L 6015: Insulated to Standard No. 2, heated by 160,000 BTUH input gas furnace, cooled by Condensing Unit and Coil attached to furnace. (Wolf)

		1 2-6-73 to 3-8-73	2 3-8-73 to 4-6-73	3 4-6-73 to 5-9-73	4 5-9-73 to 6-11-73	5 6-11-73 to 7-10-73	6 7-10-73 to 8-6-73	7 8-6-73 to 9-6-73	8 9-6-73 to 10-5-73	9 9-6-73 to 11-5-73	10 11-5-73 to 12-5-73	11 12-5-73 to 1-7-74	12 1-7-74 to 2-6-74
Fuel Adjustment		.001920	.002220	.002203	.002176	.002040	.002172	.002160	.002268	.002328	.002620	.002820	.0005022
S/L 6008 (DOWNIE)													
KWH	=	4146	2560	2004	1418.	1887	1587	1816	1021	1369	2272	4414	3551
Base Cost	=	$59.07	$41.58	$33.42	$25.34	$22.90	$28.06	$31.08	$18.94	$24.54	$37.47	$58.63	$59.15
Fuel Adj. Cost	=	7.96	5.68	4.42	3.08	3.84	3.44	3.92	2.31	3.18	5.95	11.60	1.78
TOTAL	=	$67.03	$47.26	$37.84	$28.42	$26.74	$31.50	$35.00	$21.25	$27.72	$43.42	$70.23	$60.93
												TOTAL	$497.34
S/L 6015 (WOLF)													
Cu. Ft. Gas	=	23200	16000	14900	9400	6100	5600	4600	8000	10800	28000	27000	21600
Gas Cost	=	$24.73	$18.46	$17.50	$12.72	$9.84	$9.40	$8.53	$10.49	$13.92	$20.51	$28.91	$28.05
KWH	=	876	1200	1005	1320	2061	1965	1936	1399	1035	937	975	1061
Base Elec. Cost	=	$21.10	$27.82	$23.62	$30.34	$42.52	$41.17	$40.72	$32.02	$24.46	$22.36	$23.20	$32.08
Fuel Adj. Cost	=	1.68	2.66	2.21	2.87	4.20	4.26	4.18	3.17	2.40	2.46	2.74	.53
Total Elec.	=	$22.78	$30.48	$25.83	$33.21	$46.72	$45.43	$44.90	$35.19	$26.86	$24.81	$25.94	$32.61
COMB.TOTAL	=	$47.51	$48.94	$43.33	$45.93	$56.66	$54.83	$53.48	$45.68	$40.78	$45.18	$54.85	$60.66
												TOTAL	$597.73

MONTHLY ELECTRIC RATES 1973

First	20 KWH or less Per Mo.	= $1.05
Next	10 KWH Per Mo. Per KWH	= 3.4¢
Next	60 KWH Per Mo. Per KWH	= 3.1¢
Next	210 KWH Per Mo. Per KWH	= 2.7¢
Next	500 KWH Per Mo. Per KWH	= 2.1¢
Next	700 KWH Per Mo. Per KWH	= 1.5¢
For all	Excess KWH Per Mo. Per KWH	= 1.1¢

MONTHLY GAS RATES 1973

First 5000 Cu. Ft. = $8.87
For all Excess .8716/1000 Cu. Ft.

Observation:

-- The goal was to produce a house heated and cooled with a heat pump for the same energy cost as with gas heat and electric cooling. That goal has been surpassed.

-- Since the above data was assembled, the cost of energy has approximately doubled -- and is still increasing. Consequently, the high energy efficiency of construction standard #4 has become mandatory.

Fig. 4. Electric and gas house comparisons.

Electrical usage analysis on a 2,000 Sq.Ft. house heated and cooled by 2 - 2 Ton Heat Pumps:

This house is insulated to the standard of No. 3 on the attached bar chart

Period	C.E.I. Main	General Use	HEATING — LIVING AREA: Heat Pump No.1 (Living)	Aux. El.No.1	Air Handler No.1	KWH Living	SLEEPING AREA: Heat Pump No.2 (Sleeping)	Aux. El.No.2	Air Handler No.2	KWH Sleeping	H.W. Heater
Feb. 7-Mar. 7, 1973	3,796	603	791 +	211 +	170 =	1172	1,083 +	278 +	174 =	1535	486
Mar. 7-Apl. 7, 1973	2,671	580	456	69	60	585	797	102	81	980	526
Apl. 7-May 7, 1973	1,926	395	395	6	38	439	560	10	42	612	480
May 7-Jun. 7, 1973	1,397	512	161	-	13	174	175	-	13	188	523
Sep. 7-Oct. 7, 1973	1,072	557	51	-	2	53	64	-	3	67	395
Oct. 7-Nov. 7, 1973	1,704	638	216	1	18	235	340	1	30	371	460
Nov. 7-Dec. 7, 1973	2,334	713	324	4	38	366	699	13	68	780	475
Dec. 7-Jan. 7, 1974	3,964	796	930	92	101	1123	1,304	66	141	1511	534
Jan. 7-Feb. 7, 1974	3,551	635	871	82	94	1047	1,111	137	118	1366	503
Sub Totals	22,415	5,429	4,195 +	465 +	534 =	5194	6,133 +	607 +	670 =	7410	4,382
COOLING											
Jun. 7-July 7, 1973	1,706	681	199	-	29	228	212	-	18	230	567
July 7-Aug. 7, 1973	1,892	586	374	-	37	441	301	-	27	328	567
Aug. 7-Sep. 7, 1973	1,765	553	413	-	40	453	304	-	17	321	438
Sub Totals	5,363	1,820	986	+	106 =	1092	817	-	62	879	1,572
GRAND TOTAL	27,778	7,249	5,181 +	465 +	640 =	6286	6,950 +	607 +	732 =	8289	5,954

Observation:

- Living zone required less (18%) heating and more cooling(12%)than sleeping zone due to influence of appliances and people.
- Heating and cooling split approximately 80% - 20%.
- Heating used approximately 55% of the total energy during the heating season.
- Hot water required approximately 45% of the general use energy.

Fig. 5. Electric house energy breakdown.

Fig. 6. Sun House/front view.

Fig. 7. Sun House/atrium view.

Part IV

Chairman:
Joseph F. Cuba

Director of Research and Technical Services
American Society of Heating, Refrigerating and Air-Conditioning Engineers, Inc.
New York, N.Y.

Solar Assisted Heat Pumps in a Commercial Building

Dr. Jagdish N. Prasad

Senior Engineer
Syska & Hennessy, Inc., Engineers
New York, N.Y.

This paper presents an outline of design of a solar assisted heat pump system in a commercial building. At present, the building is under construction in Washington, D.C., and it is expected to be operational by June 1976. The solar system consists of 2100 sq ft of single-glazed, all-copper solar panels inclined at 63° from the horizontal, facing south, and a 4000-gal hot water storage tank. The HVAC systems consist of 245 water source heat pump units serving the perimeter areas and 19 packaged air conditioners for the interior spaces. The solar system is assisting the space heating through the heat pumps and preheating service water for the building. The cost analysis shows the net cost of the solar system is estimated at $25,000 and the payoff period about 12 years.

Description of the Building and HVAC Load

It is a five-story office building located in Washington, D.C. The gross area is about 180,000 sq ft. About 100,000 sq ft is parking area and 80,000 sq ft is conditioned for offices and rental purposes. The following are the building design data:

1. Overall heat transfer coefficient for:
   Wall—0.13 Btuh/sq ft/°F
   Roof—0.10 Btuh/sq ft/°F
   Floor—0.40 Btuh/sq ft/°F
   Window—0.56 Btuh/sq ft/°F
2. Windows are double glazed and the total glass area is about 30% of the gross wall area
3. Designed space conditions:
   Summer: Temperature—78F DB
   Relative humidity—60% maximum
   Winter: Temperature—68F DB, with no humidification, and night setback temperature of 60F DB
4. Interior load:
   Light and equipment—3 w/sq ft
   Occupancy—700 office people
   Ventilation air—8000 cfm
5. Outdoor design conditions:
   Summer—90F DB, 76F WB
   Winter—19F DB
6. HVAC design data:
   Cooling load—200 tons
   Heating load—
   (*a*) Occupied: 1200 MBH
   (*b*) Unoccupied: 600 MBH
   Supply air—100,000 cfm
7. Service hot water:
   Average consumption—1000 gal/day
   Supply water temperature—120F

HVAC System

The building is designed for office use and rental purposes and as such it calls for individual area controls. Other factors which are involved in selecting an air conditioning system for a building include installed cost, operating cost, and total owning cost. A two-pipe fan coil, a four-pipe fan coil, a variable volume, and a heat pump system were analyzed for operating and owning cost. Based on lowest operating and owning cost, water source heat pumps for perimeter systems, and packaged water-cooled air conditioning units for interior systems were selected. A two-pipe water loop system acts as a heat sink and source for the heat pumps, and as a heat sink for the interior system. Figure 1 is a schematic flow diagram of the system. This system has higher installed cost but, because of the lowest operating cost, the owning cost is minimized. Following is a brief list of advantages of the system selected:

1. Heat is redistributed from the interior spaces, which require cooling, to the perimeter areas, which require heating.
2. The evaporative cooler and the electric heater in the water loop do not operate simultaneously.
3. Excess heat from daytime operation is stored in the water loop for nighttime operation.
4. The control is simple and economical.
5. It reduces piping and ductwork. The reduction in mechanical equipment space provides additional floor area for rental.

Solar System

The solar system was not included in the project initially, but because of the energy crisis, the owner decided to include it to save some conventional fuel, irrespective of cost of the system.

The mechanical equipment is located in the penthouse on the roof of the building. The longer side of the penthouse roof faces south, ideal for solar panels. The tilt angle of the penthouse roof was set at 63° from the horizontal to maximize the energy collection for the months of December, January, and February to match the heating demand. This was done with the help of a computer program. The solar panels cover the complete south-facing penthouse roof, 2100 sq ft. The optimum area of solar panels for heating this building through heat pumps is about 8000 sq ft with a storage capacity of about 10,000 gal of water. The solar panels are specified with copper absorber plate and copper tubing to minimize corrosion problems. The solar system data are as follows:

1. Area—2100 sq ft
2. Single-glazed
3. Copper absorber plate and tubing
4. Fiberglass insulation—4 in. at back and 1 in. at sides
5. Metal casing
6. 4000 gal hot water storage tank
7. 35% ethylene glycol in the collector loop

HVAC and Solar System Operation

The following are the operation sequences:

1. Solar heat collection: Whenever the solar panel absorber plate temperature is higher than the hot water storage temperature by 10F, the collector loop and the heat exchanger loop pumps run and the heat is transferred from the collector to the storage tank.
2. Domestic water preheating: Whenever the hot water storage temperature is higher than the domestic water preheating tank by 10F, the water is circulated from the hot water storage through the heating coil of the preheat tank. The maximum temperature in the preheat tank is set at 120F.
3. Condenser water loop: (*a*) The closed-circuit loop water temperature is maintained between 60 and 90F. (*b*) The evaporative cooler operates whenever the loop water temperature exceeds 90F. (*c*) Whenever the loop water calls for heat and the hot water storage temperature is higher than the loop water temperature, water is drawn from the storage tank; otherwise the storage tank circuit is closed, and heat is supplied by the electric heater.

Heating Energy Demand and Supply

The analysis indicates the following:

1. Almost all heating energy requirements during occupied periods is supplied by the loop water which absorbs heat from the interior spaces.
2. During the months of September and May, the night heating is supplied by the stored heat in the loop from the interior spaces.
3. The night heating in October and April is supplied by the stored heat in the loop and solar hot water storage tank.
4. The night heating in the remainder of the winter months is supplied by the stored heat in the loop, solar hot water storage tank, and the electric heater in the loop.
5. Anytime heat is supplied in the space, electrical energy is consumed in the heat pumps at the rate of about 35% of the heat supplied and electrical energy consumed in the loop water heater, if any.
6. The solar system is supplying about 15% of the combined space heating and domestic hot water heating requirements.
7. The solar system is saving 265 million Btu of electrical energy per year, which is equivalent to 7500 gal of oil at the power plant.
8. Figures 2 and 3 show the monthly and yearly requirements and supply of heating energy.

Cost Analysis

The cost of the project is $5,000,000.

The following is a cost breakdown of the systems:

1. Electrical system: $300,000.

2. HVAC, including plumbing and sprinklers: $800,000.

3. Solar system: $40,000.

It has been estimated by the architect that saving in the cost of roof construction replaced by the solar panels is about $15,000. The net cost of the solar system is $25,000.

Energy savings by the solar system is 265 million Btu (77,600 kwh) of electrical energy per year.

The payoff period is about 12 years, with the following assumptions:

1. The current electric rate in Washington, D.C., is 3.5¢/kwh including demand.
2. A 15% rise in electric cost per year.
3. 10%/yr interest rate on investment.

Conclusions

1. In a commercial building with heat pump systems, the solar energy is used for unoccupied periods only. The occupied period heating is supplied by the loop water which absorbs heat from the interior systems.

2. The solar energy is very effective for domestic hot water heating.

3. Summertime energy collection is wasted, except for heating domestic water.

4. The tilt angle for solar panels should be optimized for maximum energy collection during December, January, and February to minimize wastage of energy during summer months.

DISCUSSION

Q If you are not using your solar collectors in the summer (for domestic hot water or solar cooling, etc.), how do you keep the collectors cool? Do you have any "fail-safe" emergency systems for cooling the collectors during an electrical blackout or pump failure?

A The solar collectors are used for domestic hot water heating during the summer. Anytime temperature in the collectors exceeds 200F, the circulating pumps stop. Automatic vents are provided in the collector circuit and are set at 15 psig to take care of any pressure buildup beyond that.

Q What was the construction of walls in the Washington building?

A 4-in. face brick, 1¾-in. air space, 4-in. concrete block, 1-in. rigid insulation, ½-in. dry wall.

Q Do you use the waste heat in the lighting louvers to preheat the return water?

A No

Q Do you use the solar radiation through the windows to assist building heating and cooling through the hydronic and/or heat pump systems?

A No

Q What is the cost of your collectors?

A I do not have the breakdown. Total cost, including storage and solar system controls and installation, is $40,000 for 2100 sq ft of collector area.

Q Why does supply from interior space vary each month?

A The heat generation from the interior space remains almost constant. The heating requirement varies each month, and as such the heat transferred from the interior space to the perimeter varies each month.

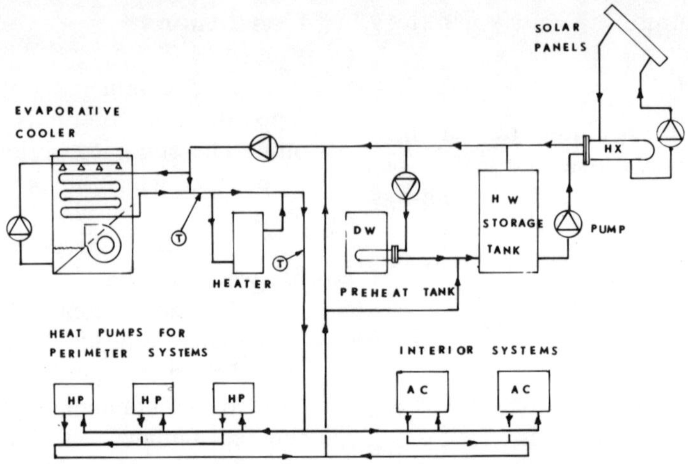

Fig. 1. Schematic flow diagram of system.

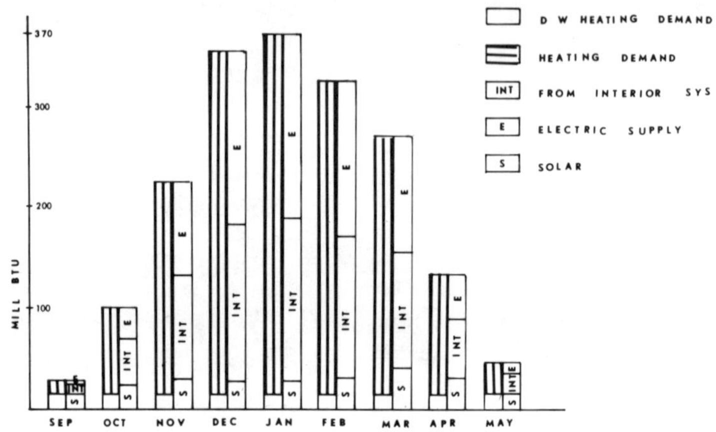

Fig. 2. Monthly heating demand and supply.

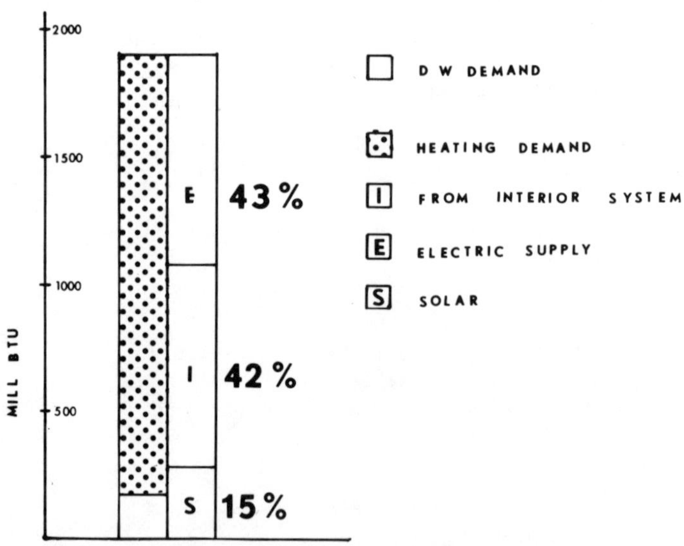

Fig. 3. Yearly heating demand and supply.

Extended Range Water Source Heat Pumps

Jay F. Miller

Application Engineer, Energy Conservation Systems
The Singer Company, Climate Control Division
Auburn, N.Y.

The application of incremental water source heat pumps to a closed water loop comprises a system which has gained wide acceptance among owners and designers of multiroom or multizone buildings in North America and Europe. Unlike most unitary heat pump systems, the closed loop system is employed to the greatest advantage in cold weather climates typified by Toronto, Minneapolis, or Syracuse. For example, metered energy consumption for Central New York State office buildings, for all electrical requirements, of all-electric office buildings, ranges between 30 and 35 kwh/sq ft/yr. School buildings in the same area have been using 24–28 kwh/sq ft/yr. The best examples of other type systems use 40–50 kwh/sq ft/yr.

As the cost of energy, initial equipment investment, field erection expense, and operating and owning costs are continually increasing, it becomes increasingly important to incorporate a heating, ventilation, and air conditioning system which is practical and efficient but also conserves available energy resources to the maximum degree possible, given the present state of the art. Whenever a combination of heating and cooling is required, and the heating and cooling loads vary from zone to zone, the closed loop heat pump system should be considered, not only for its true energy conservation characteristic, but for these other advantages as well:

1. Ultimate flexibility of zoning.
2. Maximum diversity at all times; the units only operate as required by their individual space controls.
3. Ability to heat in one zone and cool an adjacent zone.
4. Smaller mechanical rooms, because no large central refrigeration equipment is required.
5. Building volume is decreased (or usable space is increased) because ductwork is minimal, and basic energy is transmitted through electrical wiring and uninsulated pipes.
6. Less field labor is required to install than with built-up systems.
7. Simplicity of design. No complicated control valves or extensive, field erected automatic temperature control system.
8. Maximum system reliability. A failure in one unit does not affect the others.
9. No necessity to employ a licensed equipment operator, and less expensive maintenance contracts, as any unit may be removed, replaced by a spare, and the unit returned to a local repair depot for repair, and later returned to the building for use as a spare.
10. Maximum architectural design flexibility, both in basic building configuration and interior layout. Terminal units are available in under-window console models, ceiling concealed or vertical closet types, "stackable" vertical units with integral risers, large package units in capacities to 25 tons, and rooftop units for places where there is no interior place to install the equipment.
11. Minimum initial investment for speculative type office or apartment buildings, as the water loop can be designed and installed without knowing what the ultimate floor arrangement will be within the leased areas, and terminal equipment can be later purchased and installed as required.
12. Basically a constant year-round electrical demand, with any supplementary heating requirement obtainable on an off-peak basis by limiting the demand, or through the use of a load control which enables the water temperature to be increased during periods of off-peak demand.
13. Quietness of operation. Properly installed equipment can meet a sound level equivalent of NC 35 or lower.

Despite the many successful closed loop heat pump installations, there are many misconceptions about the application of the system and how it works. Additionally, criteria which once determined whether or not it would be the specified system are no longer valid, due to increasing energy costs, increases in field labor costs, and new interest in the utilization of alternate sources of energy such as geothermal or heliothermal. A brief review of what the closed loop system is, how it operates, and how it is an inherently energy-conserving system will promote greater understanding of its present operational limits, of efforts to enhance the advantages of the system through the development of new generations of system components, and of innovations in system design.

The Closed Loop Heat Pump System (Fig. 1)

A unitary cabinet-enclosed water-to-air heat pump (Fig. 2) in each room of a multiroom building supplies heating or cooling, as required. Parallel piped water-to-refrigerant heat exchangers in each unit either reject heat to, or absorb heat from, a closed water loop throughout the building. Refrigerant reversing valves in each unit permit using the water as either a heat source or a heat sink. Heat is thus conserved within the building by transferring through the loop, from rooms where there is a surplus to rooms where heating is required. In office buildings, interior core areas, which require cooling all of the time, often generate sufficient heat to "fuel" perimeter heating requirements most of the time. The closed water loop is not insulated, and is, at the present time, controlled within 70 and 90F. If a sufficient number of the heat pumps operates in the cooling mode, for a sufficiently long period of time, the loop water temperature will eventually rise to the 90F upper limit, and necessitate the rejection of surplus heat. This is most commonly accomplished with an evaporative water cooler, but heat exchange methods with groundwater, lakes, or rivers may also be used. If a sufficient number of the units operates in the heating mode for a sufficiently long period of time, the loop temperature will eventually drop to 70F and require the addition of supplementary heat to permit further operation of the heat pumps. (In many office buildings, this commonly occurs when the outside temperature drops to 10–15F—above this point, no supplementary heat is required.) Heat may be added to the loop with conventional fossil fuel boilers or with electric water heaters; or, in what is known as the "boilerless" system, the terminal units may switch from heat pump heating to electric resistance heating (at no performance penalty since system COP is 1.0 under both these conditions).

The use of solar collectors as a source of supplementary heat, and also as a means of rejecting surplus heat at night, shows considerable promise for many installations. The solar collectors function more efficiently at the low temperatures inherent to the closed loop systems than they do at the elevated temperatures required for other heating systems, and thus collector size is greatly reduced. A storage tank is generally used to offset the effect of weather persistence.

Ventilation air is normally introduced in heat exchange with exhaust air, to produce further operating economies. Large, interior zone heat pumps may be used at each floor to control final ventilation air temperature, or by passing the water leaving these large units through a reheat coil, the enthalpy of the ventilation air can be controlled with inexpensive apparatus and no additional power input.

The closed loop system operates most efficiently when the water loop is "floating" within the 70–90F range. Power consumption is limited to that required by the individual terminal units and the system loop circulating pump. There is no surplus of heat to reject, nor an additional requirement to purchase. Studies show that typical HVAC systems operate at full load less than 5% of the time, and at less than half load 70% of the time, so that further improvement of the closed loop system efficiency becomes a question of economics. For example, a storage tank which is feasible in Minneapolis may be foolish in Atlanta.

Coefficient of Performance

During the heating cycle, the net output of the unit is made up of the total of the heat of compression, the heat extracted from the closed water loop, and the heat of the air circulating fan. The coefficient of performance of heat pumps is the ratio of energy input to the energy of total output. While an individual unit might have a COP greater than 2 or 3, depending upon the temperature of the loop heat source, the entire building system may have a COP as low as 1.0; but as long as a single loop-connected unit anywhere in the building is operating in the cooling mode, the system heating COP will be greater than 1.0.

Seasonal performance factors are the best measure of system performance. Computer programs are being developed to assist the building designer in predicting energy requirements, but many used to date have questionable accuracy. Until this situation improves, actual metered experience, expressed in terms of kwh/sq ft/yr for similarly constructed buildings of similar use, remains the most reliable system comparison guide.

Annual Operating Expense

The elements which comprise the annual operating expense for the closed loop system are the sum of the energy purchased to operate the terminal units and the circulating pumps, and supplemental energy purchased to either provide supplementary heat or to reject surplus heat. The factors of building usage and weather persistence determine the amount of supplementary energy, and also determine when those requirements will occur. Night setback, outdoor reset, ventilation control, and other simple control systems can greatly reduce the basic equipment operating cost, and nearly always justify their expense. There is no need for large, expensive, and complex controls systems to maximize the advantages of the closed loop heat pump system.

Increasing the mass of the heat sink will permit the closed loop to store more energy. Storage tanks will reduce the amount of supplementary energy required, and in some cases have totally eliminated the purchase of any supplementary energy for heating, leaving only the basic equipment operating cost. Alternate methods of energy storage, such as small tanks which utilize the heat of fusion of the storage media, are being developed.

Increasing the operating range of the heat pumps has the same effect as increasing the mass of the loop, in that more energy can be stored in the loop without restricting the operability of the units.

A unit design criterion is that any terminal unit must be reversible from heating to cooling, or vice versa, at any time. This requirement is the major factor in the present definition of 70–90F range. In actual practice, ranges of 80–100F for cooling and 60–80F for heating are practical for present generation equipment at flow rates between 2 and 4 gpm/hp and should be considered by the building designer when the application permits.

Present Operating Range—A Question of Economics

The cost of control systems, storage tanks, solar collectors, etc., must be weighed against the return on the investment, and in consideration of estimates for future energy costs and energy rate structures. The economics of the marketplace also determine the configuration of the heat pump and its operational characteristics. The probability that a new generation of equipment will be developed which will have extremely high COP is slim. If one were to take any compressor manufacturer's curves, and assume an infinite condenser, the absolute maximum COP of a heat pump using those compressors can be illustrated. In actual practice, the mechanics of design and the economics of manufacturing determine the size of the air and water coils, and whether expansion valve or capillary tube refrigerant flow devices will be used. There is, however, a definite need for, and a trend toward, more efficient equipment and more innovative application of that equipment. These efforts are concentrated into three areas: (*a*) the development of more sophisticated computer programs to more accurately predict building requirements so that the optimum system and equipment can be selected; (*b*) the development of more efficient system components; and (*c*) increasing the range through which that equipment will operate.

The present operating range (70–90F) is a product of the historical development of the closed loop system. When the system was introduced at the 1967 ASHRAE exposition, approximately eight years of application and operating experience had already been gained in Florida and California. The original equipment designs, of necessity, reflected commercially available components, and the previously mentioned desire to maintain the reversibility of any unit at any time. Because these components were never stressed beyond their design limitations, water source heat pumps never experienced the bad reputation that air-to-air heat pumps earned when they were initially introduced to northern climates. Equipment as reliable as a household refrigerator was a mandatory requirement to induce responsible investors to install the system.

Extended Range Water Source Heat Pumps

The closed water loop has an upper and a lower temperature limit, both of which are in relationship to the flow rate, for unit capacity is a function of its leaving water temperature. In the heating mode, the water leaving the unit must not be lower than the dew point within the building, or the piping must be insulated, a major first cost disadvantage of significant proportions. The practical lower limit for uninsulated systems is 50F. The theoretical lower limit for insulated systems would be anything above freezing, but good design practice limits water velocities through the piping. The minimum temperature will eventually become known as units with higher COP are developed.

The upper limit has been a function of outdoor design wet bulb because of the practice of rejecting any surplus heat by evaporative means. Units can be manufactured, with available technology, to operate with a warmer heat sink, so that a dry cooler could be used.

There are advantages and disadvantages to extended range units, which may prove to be offsetting.

For example, extending the control range of the heat sink will reduce the amount of purchased energy for supplementary heating or heat rejection, but may increase the basic energy requirement for unit operation. A lower temperature limit may permit the use of a more efficient solar collector, but impose an operating penalty by causing the units to have a lower COP. We do not know the exact consequences of each possible system modification, because of the complex interrelationships involved, so we are developing some very sophisticated computer programs to assist in system designs, and we are testing experimental units to provide additional input data. These units incorporate larger and more efficient heat exchangers, a variety of refrigerant flow control devices, and "high performance" compressors. In addition, detailed studies of existing installations are being made to verify the validity of computer programs.

The closed loop heat pump system is not an overall answer for all multizone buildings, nor for the energy problems facing our industry, but it does have application far beyond the large core load office building, it does truly conserve energy, it does work with proven reliability, and it holds more promise of adaptability to alternate energy sources expected in the future than any other presently available system.

DISCUSSION

Q If the system named is used in an apartment complex where the system is all heating or all cooling, will not the addition of a storage tank with solar cells show a large saving in operating cost over the heating season versus a heating/cooling system not having thermal storage?

A An apartment building which does not have a large interior zone heat mode can have the system operating economics enhanced with the addition of a storage tank or solar collectors; however, a storage tank provides a better return on investment. Increasing the mass of the water loop reduces the amount of additional energy input required to operate either the heat rejector device or the system supplementary heater. A solar collector only reduces the amount of supplementary heat required. The solar collector may appear to be more advantageous during an operating week in January, but over the entire season a storage tank makes the larger contribution toward reducing the total system energy consumption in terms of kwh/sq ft/yr.

Q What are temperature limitations of entering air to the heat pump? What problems are encountered above the high limit or below the lower temperature limit?

A Heat pumps could theoretically be designed for any reasonable temperature limitations. In the current product, operating limits are based on the typical commercial application where it is desirable to have any unit reversible at any time. These limits are:

	Cooling	Heating
Min. Eat	52F	60F
Min. Ewt	60F	60F
Max. Eat	95/78F	85F
Max. Ewt	100F	95F

All at recommended water flow rates.

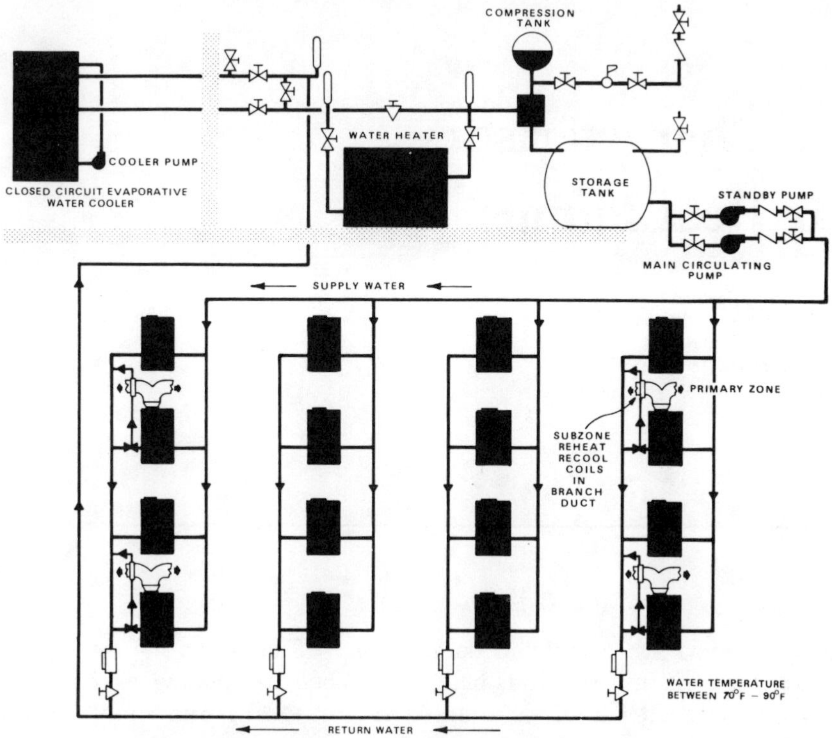

Fig. 1. Closed loop heat pump system.

With low-voltage wall thermostat calling for heating, blower, *F*, starts and reversing valve, *V*, is energized, selecting refrigerant flow path. Compressor, *P*, starts and pumps hot compressed gas to air-side coil, *A*, where the gas is condensed to a liquid by giving up heat to the air. This air, returning from the room, is warmed and delivered to the supply duct. Liquid refrigerant then passes through capillary restrictor tube, *C*, causing a pressure drop, and expands (boils) into a gas within the coaxial refrigerant-to-water exchanger, *E*. Heat is absorbed from the circulating water loop. Refrigerant gas now passes through the second path of *V* to the suction compressor, *P*, where the cycle starts over again.

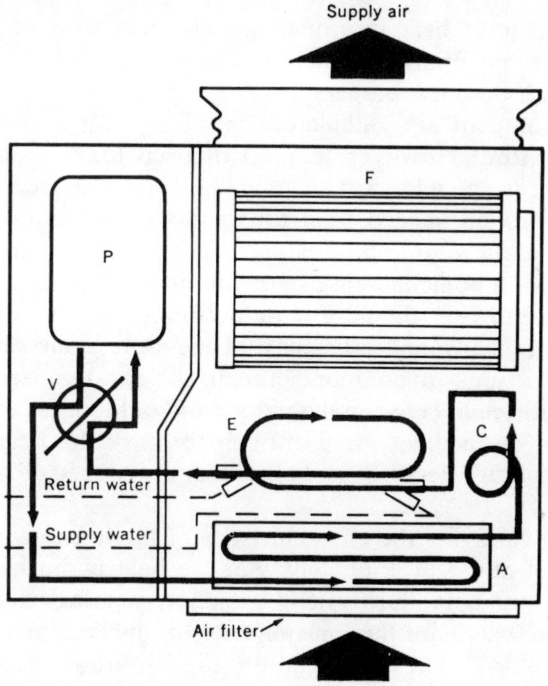

Fig. 2. Water-to-air heat pump, heating cycle.

Performance Characteristics of Air-to-Air Heat Pumps

Richard C. Barnett

Program Manager, Heat Pumps
General Electric Company
Louisville, Ky.

The data reflected in this report are calculated data intended to show relative comparisons of various heat pump configurations. The significant variations in actual field data resulting from variations in lifestyle can mask the relative performance differences. In addition, good test data for all configurations would be impractical to obtain. Calculated data are like a meter measuring relative values, and requiring field test data to calibrate the meter for accurate levels of performance.

The winter heat loss and summer heat gain of a home are generally considered to be a linear relationship with outdoor temperature (Fig. 1). The heat loss and heat gain are calculated based on 70F indoor temperature. However, due to internal loads from appliances, people, and lighting, heating is not considered to be needed until the outdoor temperature drops below 65F. The heating and cooling requirements for a home increase as the outdoor temperature varies from 70F, the desired indoor temperature.

The capacity and efficiency of the heat pump are also functions of outdoor ambient. As the temperature difference between the indoor and outdoor air increases, it becomes more difficult to move the heat; the capacities decrease and efficiency is reduced (Fig. 2).

By overlaying the characteristics of the house with the heat pump, it is obvious that the heat pump has excess capacity during mild weather, heating, and cooling. There are three major areas of application of heat pumps. These appear on Fig. 3. First, heat pumps are generally sized to handle the cooling load at design conditions. Second, heat pumps require supplemental heat when the outdoor temperature is below the heating balance point. And last, compressor operation during low temperatures affects seasonal efficiency.

When calculating a seasonal heating performance factor (SPF) one must first obtain the seasonal weather pattern in hours at various 5F-wide outdoor temperature bins (Fig. 4). Such a weather pattern is shown in the top portion of this chart for Philadelphia. Notice that the annual heating hours drop off sharply below 37F.

Shown in the bottom portion of the chart is kwh versus outdoor temperature. The heat loss of the house can be defined in terms of electrical power as well as heat flow. Simply by dividing the thermal heat loss by 3412 Btu/kwh, we can convert to equivalent electrical power or the amount of electric resistance heat required to heat the home at each outdoor temperature. The electric resistance value can be used as a basis of comparison for calculating efficiency, i.e., energy output in kw to heat the house, divided by the energy input in kw to operate the heat pump. As the outdoor temperature decreases from 62F, the heat pump power input moves considerably below the electric resistance equivalent, showing the efficiency of the heat pump. At approximately 35F, however, electric resistance heaters are required for supplementary heat and the relative efficiency decreases until at the maximum design temperature for heating (7F), the heat pump is only slightly more efficient than electric resistance heating.

The same data are shown on Fig. 5 in terms of Btuh versus outdoor temperature. The heat loss of the house is again shown as a linear function with outdoor temperature. The heat pump capacity is shown decreasing with outdoor temperature. Indicated at 62, 42, 22, and 7F is the coefficient of performance (COP), which the heat pump is able to achieve at the specific outdoor temperature. This COP differs from the seasonal performance factor (SPF) in the same way that a speedometer differs from an odometer.

To help derive the seasonal performance factor, a calculation sheet has been compiled based on ASHRAE calculation methods. The calculation procedures are indicated on the forms (Fig. 6).

Degree hours are used in calculating the equivalent electric resistance kwh which would be required to heat the house for a year. In the Philadelphia example, it is 28,101 kwh. The heat pump uses 14,450 kwh to do the heating job, thereby operating at a seasonal performance factor of 1.94, saving 13,651 kwh or 48.6% over electric resistance heating.

Diagraming this calculation procedure, we see the outdoor temperature plotted along the horizontal axis and the heat loss per year or season plotted on the vertical axis (Fig. 7). The area under the curve represents the total heat required by the house during the year. The cross-hatched area at the top of the curve represents the amount of heat being accomplished with electric supplemental heat. The remaining area is the amount of heat accomplished with the compression cycle. Only 10.3% of the total heat is accomplished with electric resistance heat. The annual heating requirement increases rapidly as the outdoor temperature decreases from 62 to 37F. The number of hours is great at these temperatures and the heat loss per hour increases with increased temperature difference between indoor and outdoor ambients. The annual heat required at lower outdoor ambients, below 37F, falls off drastically, even though the heat loss per hour continues to increase because the number of hours decreases significantly.

Supplemental heat is not required until approximately 35F outdoor temperature. The 2½-ton heat pump shown in these calculations is sized to carry the cooling load.

Increasing the unit size above the size required for cooling has a very poor economic payoff for the consumer (Fig. 8). Shown here are several size heat pumps applied to the same house in Philadelphia. Although the balance point is reduced in terms of outdoor temperature, the payoff for the larger units is considered unattractive for the consumer.

Now, using the same heat loss characteristics, Btu/°F-hr for the house in Minneapolis weather and recalculating the cooling load, we find a 2½-ton unit is apparently still required (Fig. 9). In the Minneapolis application, the peak annual heating/year occurs at 27F and the balance point is at 32F, the warm side of the heating curve. If larger units are used, the balance point moves down the curve.

The seasonal performance factor increases significantly from 1.46 to 1.60 when a 3-ton system is used (Fig. 10). In addition, the balance point coincides with the peak annual heating requirement. The payoff for the customer is approximately $100/yr in reduced operating cost and recovery of the additional first-cost within three years.

We have concluded from our studies that sizing the heat pump for the cooling load is the best economic choice for the consumer most of the time. In severe weather, like Albany, New York, and Minneapolis, it may be more economical to increase the unit size so that the balance point is at or below the temperature of maximum annual heat loss.

We are revising our sizing recommendations to include this consideration. We now recommend:

> Size the heat pump for the cooling load or so that the heating balance point is at or below the temperature yielding the maximum degree hours, whichever is larger.

The second point deals with the effects and impact of electric supplemental heat on utility peak. Today, approximately 82% of domestic utilities are summer-peaking utilities. The use of heat pumps automatically builds their cooling load. As over 50% of all new construction in 1974 was heated electrically and the effects of added installations on the winter peak are more pronounced than on the summer peak, it seems reasonable to expect that for more utilities the peaking will shift to winter in the future. Therefore, it is appropriate to spend a moment considering winter peak resulting from electric resistance supplemental heat.

The top portion of Fig. 11 shows the small effect of supplemental heat on the total heating requirement. In Philadelphia, it was 10.3%. The effect of supplemental heat on system efficiency is minor, reducing the compression cycle efficiency by 11% when compared to the total heat pump system.

The impact of the supplemental electric resistance heat is on utility peak load demand. At the heating design outdoor temperature, the heat pump system uses almost the same peak power as electric resistance heating. The utility must provide generating capacity based on peak demands. The area under each curve represents the total power used and shows that this heat pump uses 48.5% less power than electric resistance heat. The heat pump provides less return on the utilities' investment than electric resistance heating.

For the Philadelphia house and the 2½-ton heat pump, Fig. 12 shows that the average power input for the compression cycle actually levels off at 32F and declines slightly with reduced outdoor temperature. If supplemental heat could be accomplished by some other means, the utility's winter load could be very stable and even at about 3 kwh through most of the heating season.

Two basic schemes are available for modifying

supplemental heating systems. First, controls adapting a new or existing oil or gas furnace with a heat pump are available. The heat pump indoor coil is installed in the position of a normal cooling coil, which means the furnace and heat pump cannot operate simultaneously. The heat pump can be shut down at the heating balance point and all heating below the balance point is accomplished with the fossil fuel furnace. An improvement is to cycle the heat pump off with the second stage of the room thermostat; once the second stage has been satisfied, return to heat pump operation. For the Philadelphia application, this would use about 45% less fossil fuel and enable the compressor to cycle all year, keeping it warm and well lubricated.

A more flexible and possibly lower cost scheme for new construction might be to use a hydronic system, with either fossil fuel or off-peak electric heat. The stored warm water would be pumped through a water-to-air heat exchanger with the second stage of the room thermostat. The heat pump compressor would operate continuously below the balance point as normal, taking full advantage of the compression cycle.

An extension of this hydronic supplemental heat system could be a solar collector. The optimum sizing of the collector and storage tanks are being studied now. This technique for supplementing an air source heat pump with solar collectors could be the first commercially practical application of flat-plate solar collectors for space heating.

The third topic is the effect of cutting off the compressor at low outdoor ambients. Published literature indicates some products shut down the compressor when the outdoor ambient reaches 20F and do not allow the compressor to restart until the outdoor ambient is above 35F. With declining outdoor temperatures, heating is accomplished with the compression cycle and resistance supplemental heat as required down to 20F. Once the compressor shuts down at 20F, the heating is all accomplished with electric resistance heat until the outdoor temperature reaches 35F. The effects of this operation are such that in Philadelphia, the SPF decreases from 1.94 to 1.72, an 11% decrease in efficiency. In Minneapolis the SPF for the properly sized 3-ton system decreases from 1.60 to 1.31, an 18% loss in efficiency. The problem is not only higher operating cost to the consumer, but the major portion of the decrease occurs in a short period of time. More than one heat pump owner has challenged the cost savings of heat pumps because of a high monthly bill resulting from almost total electric resistance heating during a cold period.

As sales push northward into more severe climates, the reduction in efficiency is aggravated.

The heat pump, however, is an energy-efficient heating system. Figure 13 shows the relative utilization of 100 units of energy of natural gas or oil used to provide heat by direct combustion in the home or by conversion to electricity with a heat pump used in the home. The heat pump uses the electrical energy as work to move heat from the outdoor air into the home for heating. The overall utilization of the 100 units of energy is for all practical purposes equivalent.

Moving heat that is collected and stored in the outdoor air is analogous to collecting and storing heat in a solar system. In fact, the air source heat pump is a solar device. We have been told of flat-plate solar collectors heating schools, factories, and homes. If we analyze the flat-plate collector panel and water storage system, we see many parallels to the heat pump system (Fig. 14).

The sun warms the outdoor air or the collector panel. The energy is stored in the outdoor air or water tank and moved in response to a thermostat into the conditioned space to provide heat. A water pump moves the heat in a collector cell system and a refrigeration pump or compressor moves the heat in a heat pump system. Both systems use heat exchanger coils to transfer the heat to the circulating house air.

In both the solar cell and heat pump system supplemental heat is required, as economics prevent the total heating job from being accomplished solely by the solar cell storage system or heat pump refrigeration cycle.

Solar energy is available today through the air-to-air heat pump. In our zeal to develop new energy-saving ideas, we must not overlook the one product available today, which heats with the same amount of energy as a 50% solar collector/water storage system . . . the air-to-air heat pump.

DISCUSSION

Q What is the lowest ambient temperature in which an air-to-air heat pump for residential use can be used before you reach a COP of 1, compared to resistance electric heat, based on Philadelphia area?

A The outdoor temperature at which a standard air-to-air heat pump's COP reaches 1.0 is independent of climate. As a general rule, the COP reaches 1.0 between 0 and −10F and remains at 1.0 at temperatures below that point. The capacity drops off with lower outdoor ambients and the power decreases correspondingly.

Q What is the availability of equipment and technical information for heat pump mating with fossil fuel systems?

A Dual fuel systems involving fossil fuel furnaces for supplementary heating are available now from many manufacturers. The system is simply controls to switch to furnace operation on second stage of the room thermostat. Hydronic heating in place of direct combustion or electric resistance supplemental heating is also feasible now. By using a hydronic coil downstream of the heat pump refrigerant coil, the heat pump can be operated continuously below the "balance point." The hydronic supplemental heat can be provided by direct combustion of fossil fuel (limiting the fossil fuel to just what is required to make up the difference between the heat loss and heat pump capacity, and not providing the total heating required, thereby using even less fossil fuel) by off-peak electric resistance and with adequately sized storage fluid solar collector panels. Sizing of the fluid solar panels and required storage will be completed in the next nine months.

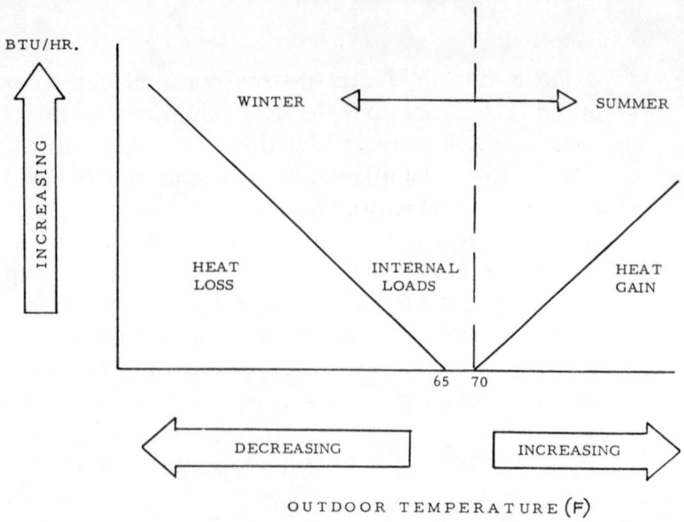

Fig. 1. House cooling and heating needs vary with outdoor temperature.

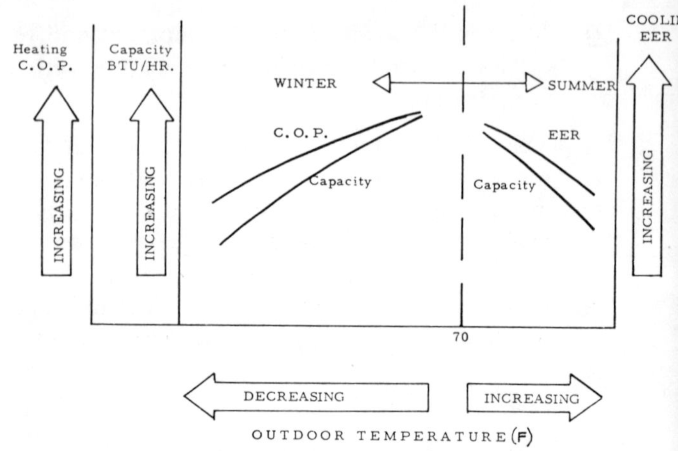

Fig. 2. Heat pump capacity and efficiency vary with outdoor temperature.

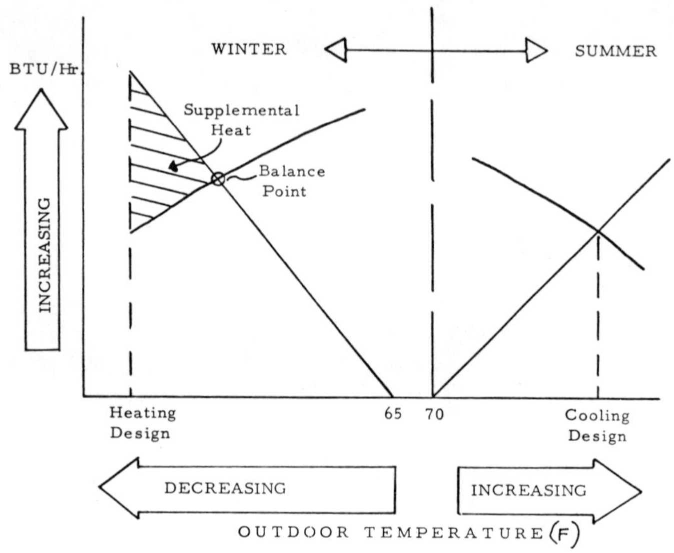

Fig. 3. Heat pump performance is inverse to house needs.

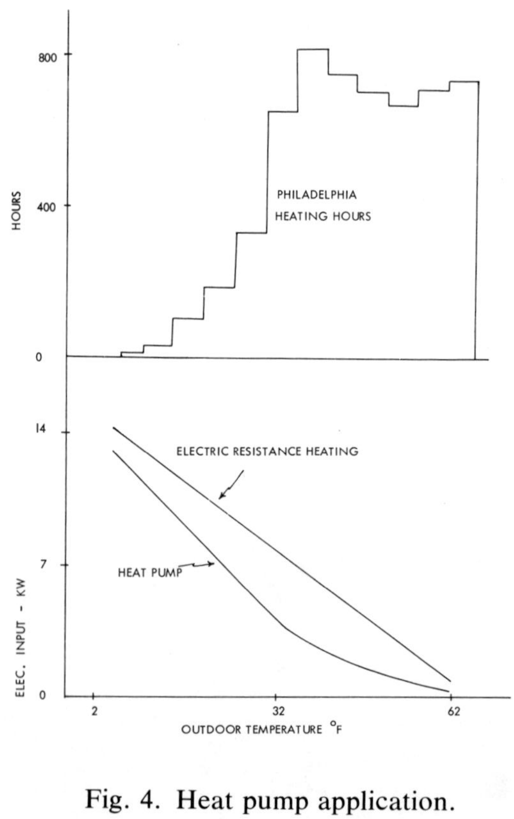

Fig. 4. Heat pump application.

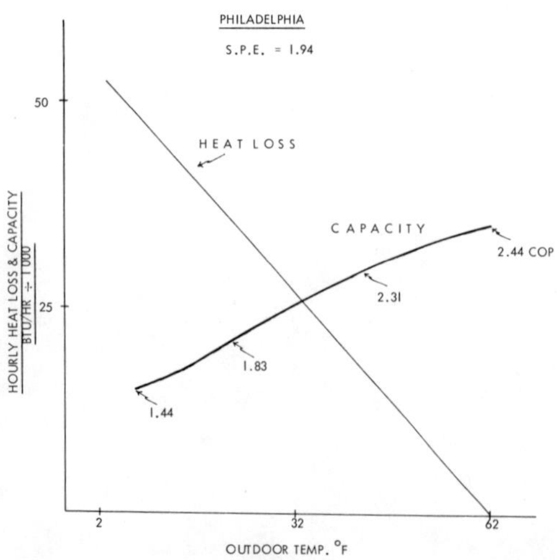

Fig. 5. Heat pump application—Philadelphia.

JOB & ADDRESS:
PHILADELPHIA / 2100 Sq. Ft. CALCULATION

by: R. C. Barnett date: 3/75

DESIGN TEMPERATURE		
I.D.	O.D.	DIFF.
70	5	65

LOAD:	29 469	52 424
	COOLING	HEATING
EQUIP.:		CFM
		1000

OUTDOOR TEMP. (5° INCREMENTS)	BTUH LOSS PER 1°F (HEAT LOSS ÷ TD)	OUTDOOR TEMP. BELOW 65°F (65 – COLUMN A)	HEAT LOSS (BTUH) B × C	HEAT PUMP HEATING CAPACITY (BTUH) MFR. DATA	HEAT PUMP RUNNING TIME (%) D ÷ E	HEAT PUMP INPUT (KW) MFR. DATA	SEASONAL HEATING HOURS U.S. W.B.	SEASONAL HEAT PUMP INPUT (KWH) F × G × H	RESISTANCE HEAT INPUT (BTUH) D minus E	RESISTANCE HEAT INPUT (KW) J ÷ 3413	SEASONAL RESISTANCE HEAT INPUT (KWH) H × K	DEGREE HOURS C × H
				← HEAT PUMP ALONE				→	← SUPPLEMENTARY HEAT		→	
A	B	C	D	E	F	G	H	I	J	K	L	M
62	806.5	3	2.42	35.0	6.9	4.2	735	213				2205
57	↑	8	6.45	34.1	18.9	4.1	710	551				5680
52		13	10.48	32.9	31.9	3.9	663	824				8619
47		18	14.52	31.4	46.2	3.8	701	1 232				12618
42		23	18.55	29.2	63.5	3.7	758	1 782				17434
37		28	22.58	27.0	83.6	3.6	818	2 463				22904
32	]	33	26.62	24.6	100	3.5	654	2 289	2.02	.59	386	21582
27		38	30.65	22.5	↓	3.4	335	1 139	8.15	2.38	800	12730
22	CONSTANT	43	34.68	20.6		3.3	189	624	14.08	4.13	780	8127
17		48	38.71	18.5		3.2	100	320	20.21	5.92	592	4800
12		53	42.75	16.6		3.1	32	99	26.15	7.66	245	1696
7		58	46.78	14.8		3.0	9	27	31.98	9.37	84	522
2		63										
−3		68										
−8		73										
−13		78										
−18 & below	↓	83										
								11 563	← TOTALS →		2 887	118917

		B		M (TOTAL)		P	
ANNUAL REQUIREMENT DUCTED RESISTANCE HEAT	=	806.5	×	118917 ÷ 3413	=	28 101	KWH
		I (TOTAL)		L (TOTAL)		N	
ANNUAL REQUIREMENT HEAT PUMP SYSTEM	=	11 563	+	2 887	=	14 450	KWH
SEASONAL PERFORMANCE FACTOR	=	(P/N)	=	28 101 / 14 450	=	1.94	S. P. F.

CAUTION: This form will provide an accurate comparison between heat pump operation and electric furnace operation. Dirty filters, condensers and evaporators as well as improper charging will adversely affect operation.

Fig. 6. Derivation of seasonal performance factor (SPF).

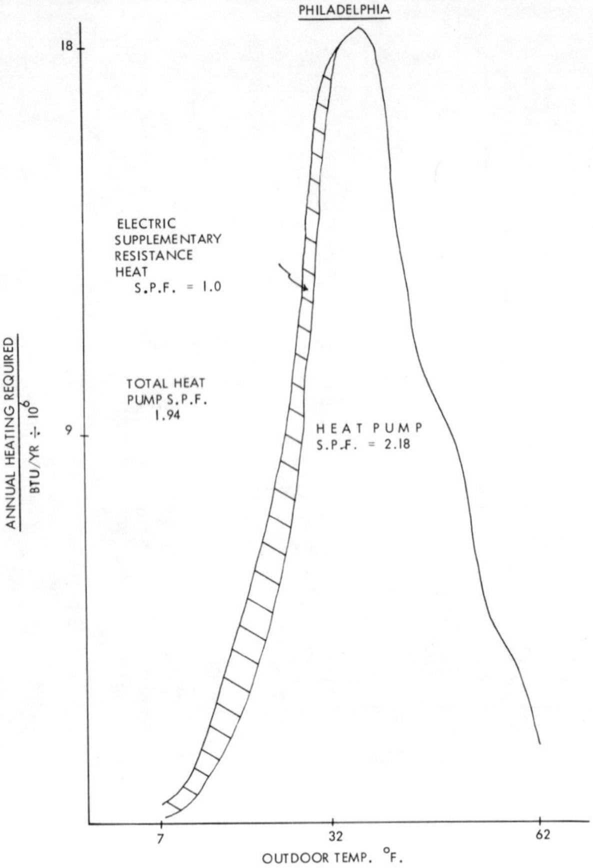

Fig. 7. Area under curve is total heat required by the Philadelphia house during year.

NOMINAL CAPACITY (TONS)	S. P. F.*	KWH SAVED	BREAKEVEN YEARS
2-1/2	1.94	----	----
3	2.19	2 000	7
4	2.13	2 000	26
5	2.18	2 000	35
10	2.55	4 000	47

* Normally, the seasonal performance factor would increase with the use of over-sized units and not hold constant for the 3 to 5 ton sizes as shown. This particular example involved climate, loading and unit size, and performance, which generated a special case and the S. P. F. remained virtually constant for the middle capacity size units shown.

Fig. 8. Heat pump sizing—Philadelphia.

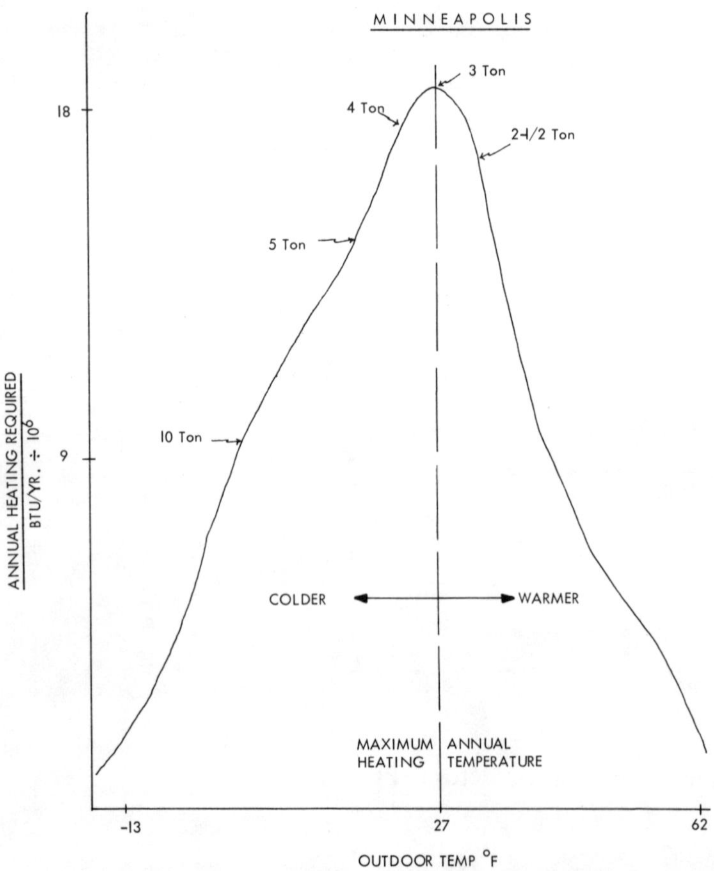

Fig. 9. Area under curve is total annual heat required in Minneapolis house.

NOMINAL CAPACITY (TONS)	S. P. F.	KWH SAVED	BREAKEVEN YEARS
2-1/2	1.46	----	---
3	1.60	3 000	3
4	1.63	3 200	10
5	1.69	4 100	13
10	2.10	9 400	17

Fig. 10. Heat pump sizing—Minneapolis.

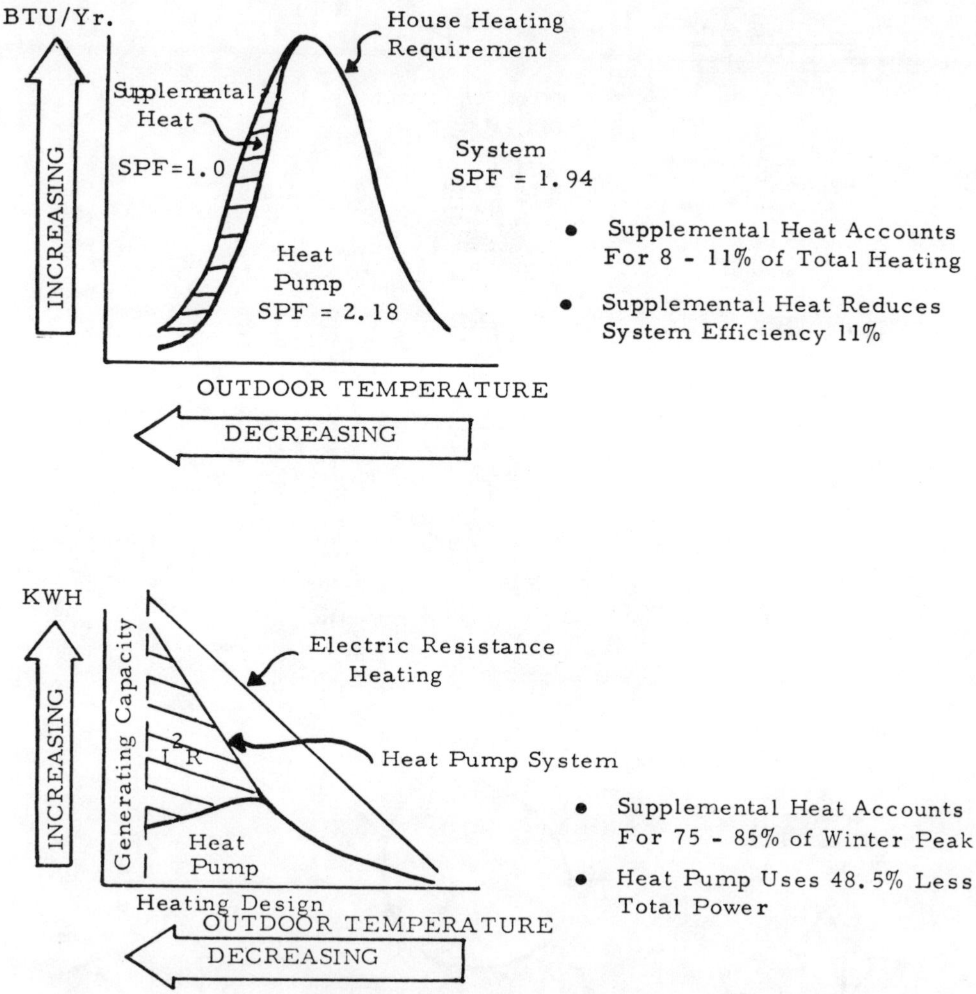

Fig. 11. Supplemental heat contributes to winter peak.

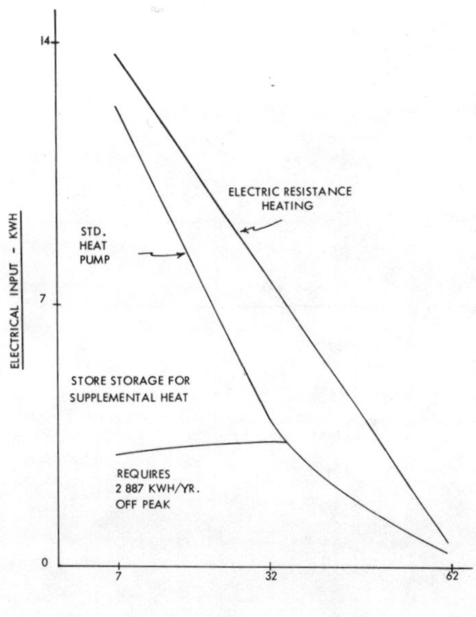

Fig. 12. Heat pump application—Philadelphia peak power.

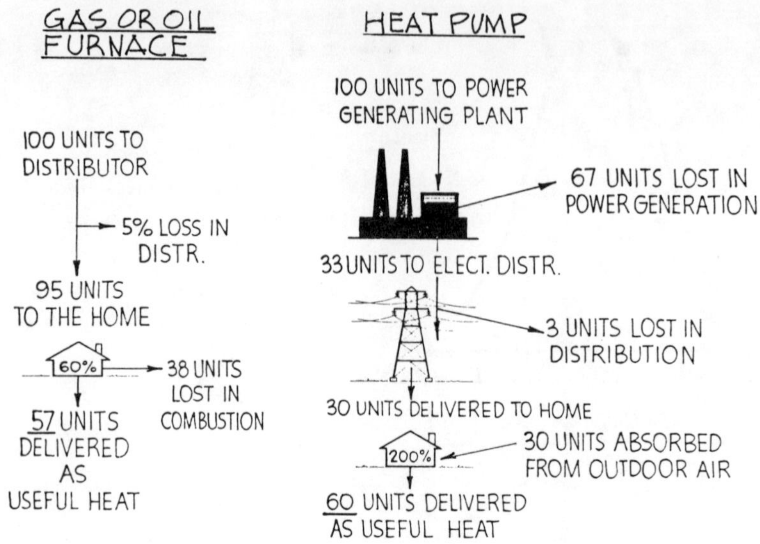

Fig. 13. The heat pump is energy-efficient.

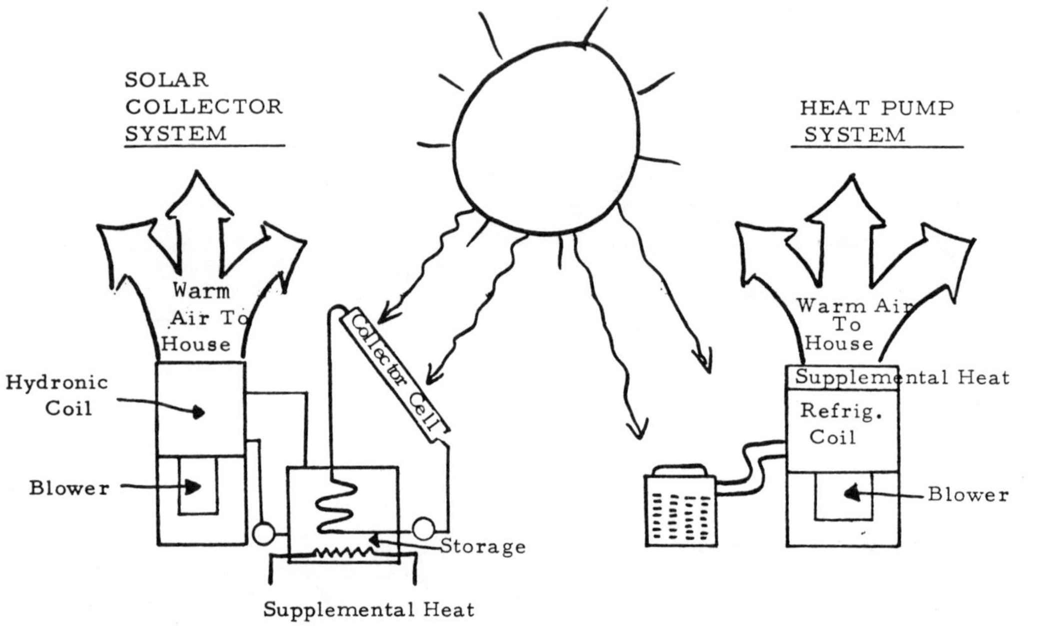

SYSTEM COMPARISON

FUNCTION	FLUID SOLAR	AIR SOLAR "HEAT PUMP"
Heat Source	Sun	Sun
Heat Collector	Fluid (Water)	Air
Heat Storage	Insulated Tank	Outside Air
Heat Mover	Water Pumps	Refrigerant Pump (Compressor)
Heat Transfer	Water-To-Air Heat Exchanger	Refrigerant-To-Air Heat Exchanger

Fig. 14. Air source heat pump is solar energy device.

Solar Heat Pumps Using Water for Cooling and Heating

Alwin B. Newton

Consultant
York, Pa.

Commercial buildings frequently use circulating water for many of their heating and cooling functions. Chilled water is circulated at temperatures typically between 40 and 45F to zone air handlers or room units which usually have local control. For heating, warm water is circulated at some temperature above 110F to similar units which may each have separate controls. Most large buildings which use water for these cooling and heating functions employ water chillers equipped with centrifugal compressors. This paper deals with the use of such equipment in conjunction with solar systems which heat and cool buildings.

Centrifugal compressors are normally driven by electric motors at constant speeds. The wheel diameter and speed of a given compressor must be chosen to achieve the maximum head required by the system and to provide the mass flow needed. Such compressors are commonly used for heat recovery, which merely means pumping heat from locations in the building which require cooling to those locations which need heating. Typically this means taking heat from the chilled water circuit after it has cooled the core areas and delivering this heat to the perimeter areas of the building. The resulting pressure ratio usually determines the wheel design requiring higher speed which can be obtained by varying gear ratio, or an increased diameter compared to the requirements for summer cooling. Hence a machine chosen for heat recovery will usually operate at a lower efficiency during the summer than a machine chosen for the optimum summer conditions. Yet the machine chosen for the optimum summer condition could not reach the required heads and flow rates for the heat recovery situation.

The higher the temperature used for perimeter heating and concurrently the lower the return water temperature from the core area, the greater the discrepancy between the optimum heat recovery design and the optimum summer cooling design. This has led to the use of lower water temperatures and higher flow rates for the perimeter heating function than might be employed with water circuits heated by fuel-burning equipment. It is also desirable to increase the temperature of the core water during winter when heat recovery is practiced, if the core can be cooled under these circumstances. Both procedures tend to reduce the head required for heat recovery, and thereby reduce the difference between the heads for summer and winter operation of the centrifugal compressor.

Solar Heating and Cooling Systems

The air-handling equipment in solar heated buildings is usually selected for full capacity at hot water temperatures of some 110F, just as in the heat recovery systems just described. The centrifugal heat pump is used to efficiently move heat around the building and thus to reduce the total solar energy needed and the size of storage. It may also be used to cool the building, or to augment a solar energized absorption system.

The principal operating modes of heat pumps include:

1. Pumping heat from a storage system at a temperature below that needed for the heated water circuit to the temperature level needed to heat the building.
2. Pumping heat from an available collector temperature lower than the storage temperature into the higher temperature storage system.

3. Pumping heat from a location in the building which needs cooling directly into the heated water circuit, as in the usual function of heat recovery.

4. Pumping heat from a location needing cooling into a partially depleted storage system.

These modes of operation impose new and unusual requirements on the centrifugal water chiller. In all four cases, the heads and capacities which are required will vary over relatively wide ranges with time.

Pumping Heat Out of Storage into the Hot Water Circuit

Figure 1 shows the types of situations which develop during prolonged periods of insufficient insolation. During this period, the building has to depend for heating on the heat that has been accumulated in storage, and we will assume the storage to be water in a tank.

Conceivably, if sufficient insolation has occurred before time zero, the storage temperature could approach 160F. If one assumes that 110F is the minimum water temperature which will adequately heat certain spaces within the building, heat can be withdrawn from the storage system over an initial time period to point F in the figure, at which time the available temperature is below 110F. At this time, the heat pump system can continue to remove heat from storage and raise its temperature level to the desired 110F as shown at B in Fig. 1. Note, however, that the length of the arrow which corresponds to the pumping head starts with a very small length close to point F and increases with time. Thus the head on the compressor also increases with time. If the load on the building remains constant, the capacity of the centrifugal system will also need to be constant at any of the heads encountered. The greatest efficiency is obtained at the lowest heads.

If the storage temperature were reduced to 40F or so, as shown in the dashed line D, the storage becomes ineffective due to problems of freeze-up in taking further heat therefrom. A centrifugal heat pump would then have to revert to an alternate source of energy E which may come from outdoor ambient conditions.

If the storage and collector system has been adequately designed, insolation should take over before the 40F storage temperature level is reached, and begin again to increase the temperature of the water in storage, as shown along line C. The heat pump will still be needed until the minimum heating temperature of 110F is reached and thereafter may be discontinued.

Pumping Heat from the Collector System into Storage

There are many hours in the operating of a solar heating system during which the insolation is insufficient to provide temperature in the collector circuit above the existing storage temperature. It is common practice to discontinue the collection of heat under these circumstances.

The quantity of heat collected by most systems varies with the temperature at which it is collected, in such a manner that more heat is received from insolation at low collector circuit temperatures than at high temperatures. Figure 2 illustrates the function of a heat pump to assist the solar system under these conditions.

The curve C in Fig. 2 illustrates a changing temperature in the collector circuit with respect to time. Initially heat can be removed from this collector circuit in considerable quantity at some low temperature such as 60F. By removing it through action of a heat pump, this heat can be delivered to storage at 120F with less auxiliary energy than would be required to directly heat the storage. Assuming that insolation gradually increases, the heat can be delivered to storage at a continually increasing temperature, up to some maximum which will probably be limited by motor characteristics of the centrifugal compressor or by inadequacies of control functions to handle increasing density of refrigerant on the suction side of the compressor. The temperature levels that can be achieved in storage by this process vary with the design of the equipment and need analysis for each individual job. At some point, line C will cross the 110–120F storage temperature, indicating that collector temperatures are now high enough to charge storage directly without intervention of the heat pump system. At this point, the heat pump would usually be discontinued.

Heat Recovery

The same heat pump which is used to extend the operating range of storage or to extend the operating range of the collector system to charge a storage system can also be used in direct heat recovery within the building. By providing sufficient piping and valve arrangements, the heat pump can accept the heat from those locations requiring cooling and deliver it to the water circuit for other locations which need heating. This is a normal application of centrifugal equipment and the same general rules and limitations apply. Since they are widely published, they are not covered here.

Heat Recovery into Storage

Since a storage system already exists in conjunction with the solar heating and cooling systems, it is logical to also use the heat pump to remove heat from core and other locations within the building which need cooling, and to deliver this heat to the storage system to raise its temperature level at times when it is partially depleted. The operating conditions and the changes in head are similar to those shown in Fig. 2. There is, however, one difference, in that the amount of heat available is limited either by the cooling requirement of those portions of the building needing cooling, or by the machine capacity itself. Hence there is likely to be a more variable load condition imposed upon the machine during this type of operation than when pumping heat from the solar collector system into the storage system.

Operating Characteristics of Centrifugal Refrigeration Equipment

The generalized performance of centrifugal compressors is shown in Fig. 3. With an electric motor drive, speed change is normally unobtainable, so a given compressor operates very closely to a single Mach number, line M. These lines pass upward to the left from a design point such as D in Fig. 3. The capacity and head requirements are determined by the existing load and the temperature differences through which the machine must pump heat, and over this the machine has no control. If the heat available to the machine decreases, the machine tends to increase the head since the speed is constant. No efficiency improvement occurs, and the compressor soon reaches a potentially unstable condition at which controls such as prerotation vanes begin to close. The locations within which this action takes place are indicated by the dashed portion of the load-head line. The limits of stability for the various vane openings are shown in Fig. 3. At some reduced load level, bypass from the discharge of the compressor to the inlet, termed hot gas bypass, is needed to prevent surging.

Both the closing of prerotation vanes and the introduction of hot gas bypass deteriorates performance in terms of the ratio of electrical kw to operate the compressor compared to thermal kw of heat pumped. Thus it is evident that stability of a compressor itself provides limits of operation and efficiency.

While heat recovery from an area in the building requiring cooling and pumping heat into storage both require high heads (and hence the design of a compressor as to speed and wheel diameter which differs from what would be used for cooling), we find an opposite situation in pumping heat in and out of storage. In Fig. 1 it is noted that one need pump heat over a relatively short temperature range when the heat pump is first called into operation. For example, the arrow shown at B in Fig. 1 starts at 82F and terminates at approximately 112F. The pumping head and pressure ratio are much smaller than if the compressor were pumping from 40 to 100F, as it frequently does in normal cooling applications. A similar situation occurs in Fig. 2 as the collector circuit temperature approaches storage temperature.

Under these conditions, several things happen. The density of the gas at the suction increases, which in turn increases the power requirements and the capacity of the machine. The head reduces and one would expect a still further increase in capacity from this factor. Both changes would cause an increase in motor horsepower which may be beyond the capability of the motor normally supplied with the equipment.

In conventional equipment now available on the market, the prerotation vanes are operated toward "closed" under this condition sufficiently to prevent motor overload. There is some deterioration of efficiency; nevertheless, the heat pump succeeds in moving the heat more economically than would be the case if fossil fuel or electrical energy were used to make the heat addition.

In essence, the constant speed centrifugal compressor tends to operate at a constant Mach number within a narrow range of capacity change and head change. Devices such as prerotation vanes and hot gas bypasses always cause a reduction in efficiency compared to the choice of another centrifugal compressor, which could operate at the new head and new capacity required.

Future Requirements

Figure 3 shows a number of Mach numbers varying from 1 to 1.6. Each of these corresponds to a different wheel speed. It is obvious that different load lines would intercept these Mach number lines at different points. Operation at these intersections would delay the need for introducing the inefficiencies of prerotation vanes or hot gas bypasses.

The need is to adjust the compressor speed to agree with the instantaneous heads imposed by the existing heat availabilities and the temperature ranges through which heat pumping must occur. Speed control can accomplish this result. For the common electric motor drives the use of an inverter to control frequency and voltage is worthy of consideration. Figure 4 shows the features of an inverter-driven centrifugal compressor. The physical size and weight of electric motors are reduced rapidly as the frequency of the electrical input increases.

For the same power, the weight of iron in the magnetic circuit reduces nearly in proportion to the frequency ratio, though the thickness of individual laminations may have to decrease with increased frequency. Torque needed for the same power is halved by doubling of the frequency. Some changes in rotors and shorting rings may be needed to absorb the stress of higher rotating speeds when the motors are used for direct drive with the objective of eliminating speed-increasing gear sets. But these appear to be the main changes needed to allow motors to meet the direct drive speeds needed by centrifugal compressors ranging from 8000 to 18,000 rpm.

It may not be generally realized that any polyphase induction motor will operate well at any combination of voltage and frequency in which the v/Hz factor is a constant. This is illustrated in Fig. 5. A 60-cycle, 2-pole, 440-v motor can operate at 30 cycles, 220 v and usually at full torque. With centrifugal compressors, the drop in torque requirements is so rapid that motors can always be used at the reduced voltage and frequency inputs with little reduction in motor efficiency. Solid state inverters are increasingly available for voltage and frequency control. The choice among inverters will depend on the characteristics of a particular unit and its load pattern. Even the gear-driven centrifugal compressors could, in most cases, be modified for operation through variable frequency, variable voltage inverters with major improvements in their efficiency.

With standard constant speed heat pump drives, it is not unusual that these low load, low head conditions require a ratio of electric to thermal kilowatts between 0.28 and 0.33 as prerotation vanes and hot gas bypass functions are added. When run at the optimum speed for the low head and low load, the ratio of electric to thermal kilowatts may drop to 0.09—a substantial saving.

While it is recognized that the inverter drives for heat pumps using water for heating and cooling are not "off the shelf" items, they are readily available on special order. As a component in life cycle costing analysis, their cost may well be offset by the fuel and energy savings which result from adapting speed to head to achieve higher overall efficiency. It should be urged that such equipment become more rapidly available as a standard available item.

Summary

It has been shown that heat pumps in systems using water for heating and cooling have a variety of functions. Heat may be pumped into storage when an excess amount of heat is available in a portion of a building, or out of storage into a heated water circuit when the storage temperature is lower than that needed to handle the heating load. Heat may be pumped into storage from a collector circuit which cannot be heated up to storage temperature. In addition, the centrifugal heat pump can provide the usual cooling function of chilled water supply and any desired portion of the resulting heat can also be placed into storage. Thus it is seen that the heat pumps operate to move heat within a building or in a system and, in doing so, reduce the requirement for energy from other types of energy sources. Since the heat pump normally operates with a coefficient of performance between 2 and 4, overall energy is saved through its use. Eventually, savings should be increased by the introduction of variable speed centrifugal equipment which may, during a large portion of its operating cycles, have a coefficient of performance in excess of 10. It is a legitimate concern of those interested in the wide and efficient usage of solar energy that such devices should be made readily available at the earliest possible date. Support of research in this direction is recommended.

DISCUSSION

Q What is the loss of COP of a conventional centrifugal chiller when used for heating at condenser temperatures of 110F? What are the normal condensing temperatures for cooling?

A The nominal maximum condensing temperature for cooling is 105F. The loss of COP in a conventional centrifugal chiller, if it were required to operate at 110F condensing temperature, depends on the location of the design point in the operating map. For example, if point D in Fig. 3 were located at the 105F condensing temperature, a small increase in head at constant speed would force point D to move parallel to the 1.6 Mach number line to a considerably lower capacity. At the same time point D would move somewhat further away from the 76.5 efficiency line with a consequent small reduction in COP. Thus both the ability to reach capacity and the change in COP must be considered.

It should also be noted that if the initial thermal difference between condensing temperature and water off the condenser were 8–10F, and if we wanted 110F water for heating, the condensing temperature would rise from 105 to at least 118F. This would obviously make a considerable difference in COP and is the reason that a different wheel design is usually required for the higher design water temperature. Figure 3 also shows that the wheel so designed would operate at a substantially lower efficiency at the reduced head normally employed for cooling design conditions.

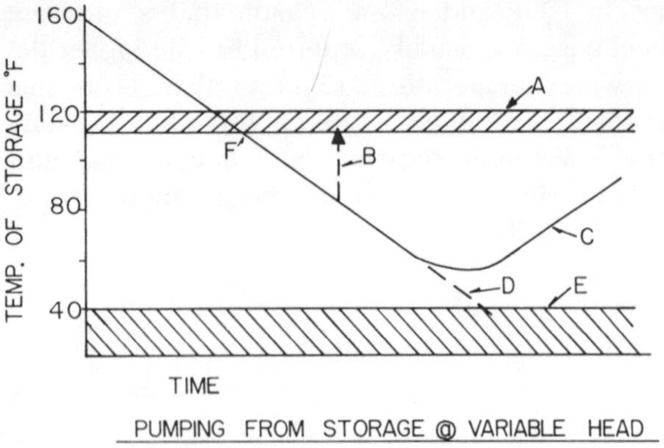

Fig. 1. Pumping heat from storage.

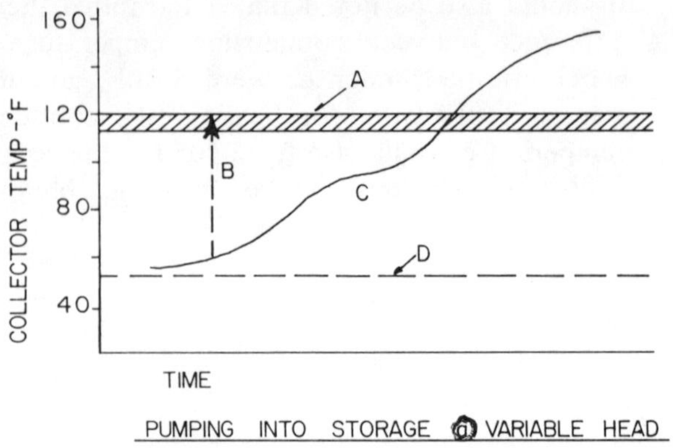

Fig. 2. Pumping heat into storage.

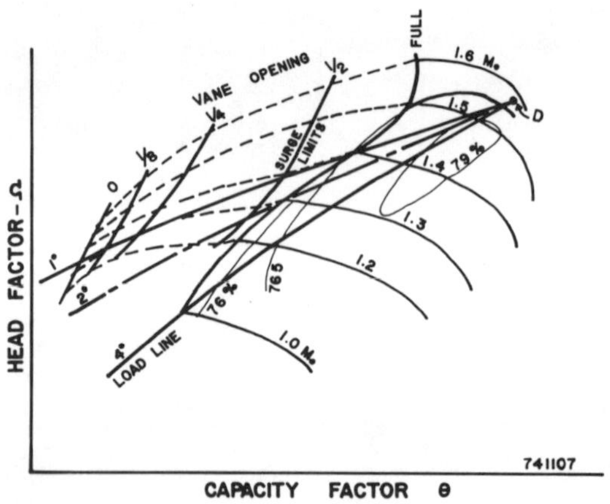

Fig. 3. Operating characteristics of centrifugal compressor.

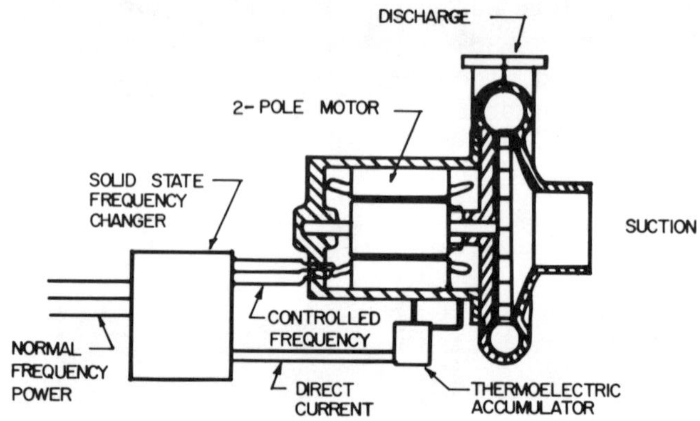

Fig. 4. Inverter driven direct drive centrifugal compressor.

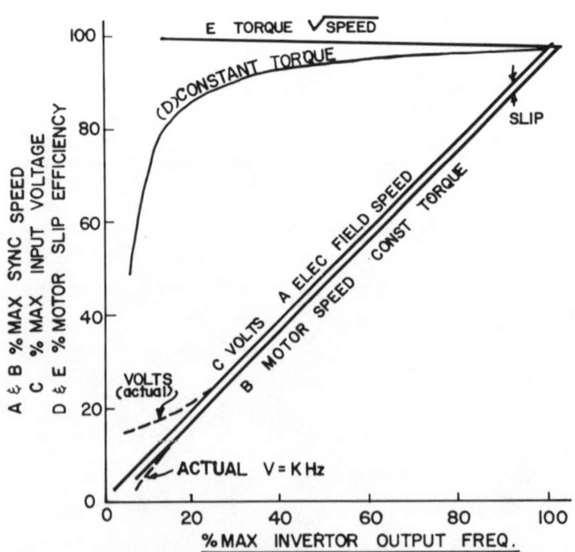

Fig. 5. Performance of typical induction motor at variable frequency.

Large Built-Up Heat Pump Systems

Robert G. Werden

Robert G. Werden & Associates, Inc., Consulting Engineers
Jenkintown, Pa.

My role on this program is to relate our experience with large built-up heat pump systems. Some of the features which have made these larger custom-designed systems successful may well be applicable to the residential unit or system.

My first topic is economics. Large commercial and industrial users only considered the following systems because of favorable economics with pay-out. Comparative life cycle cost analyses were made, favoring a greater initial first cost investment in these systems.

The trend of higher energy costs—whatever the fuel—will increasingly favor the heat pump. This is because the differential in units of energy consumed is in its favor. Consequently, it must follow that as the relative cost of all energy sources increases, this differential cost must increase. And energy costs will escalate at a more rapid rate than initial cost. All this favors the heat pump.

Figure 1 illustrates energy cost trends and provides a premise for energy cost throughout a life cycle. This figure illustrates two viewpoints from independent sources—General Services Administration and Midwest Research Institute. Though they differ in the middle years, they arrive at identical forecasts for the thirtieth year (2005).

Such escalation in energy costs, with even some reasonable increment of the escalation forecast, certainly favors consideration of air source heat pump systems. A greater investment in first cost spent with 1975 dollars will pay off handsomely in savings in year 2005 energy dollars.

Moreover, the large air source heat pump built-up systems are without the need for supplemental resistance, and have been operating for some years.

To understand this, and perhaps to spur designers of residential heat pumps to investigate the possibility, there now follows a discussion of the merit of multistage compression, in the compound and cascade configuration.

Figure 2 diagrams three compressors (or cylinders) in a conventional single-stage mode. Figure 3 shows the same displacement in a two-stage mode, with ⅔ of the displacement in the low stage and ⅓ in the high stage.

Figure 4 shows the relative heating output of these two configurations, together with the relative energy input—based on performance using three 120 cfm, R-12 machines of typical 5% clearance. It can be seen that the improvement in volumetric efficiency of the two-stage mode results in approximately 40% more capacity for no more displacement, and at a more favorable energy input.

Curve X of Fig. 4 also illustrates the added energy required in supplemental resistance, if the single-stage configuration were required to meet the capacity output of the two-stage system. It is obvious that multistage compression is beneficial.

The multistage configuration requires a greater first cost investment than single stage with supplemental resistance, but a real case can be made for investing in this increment of greater first cost—in the case of commercial and industrial owners who amortize the investment over the life cycle. Life cycle cost analysis is steadily gaining acceptance in the investment community as a valid tool in the decision making process, and is predicted to increase in acceptance as energy costs escalate.

Whether the residential market will be attuned to life cycle cost analysis in making its buying decision is yet to be determined. The ultimate outcome of consumer education, possibility of government mandates, tax incentives or penalties, rationing, escalation of energy costs—all these forces will impact, more or less, on the residential market.

In our firm's involvement in the evolution of multistage heat pumps, we have been moving away from the compound configuration to the cascade arrangement. This is because of the increased need in commercial and industrial applications for heat recovery from simultaneous cooling and heating.

In summary, we see the justification in investing in multistage rather than in single stage and supplemental resistance, because of the annual energy savings.

Naturally, we have been working with large capacity systems. But our experience in these systems and other refinements may be of interest.

First, we have available several options of refinement. Table 1 indicates the improvement in performance with subcooling, intercooling, etc. We do realize that investment in these refinements is relatively less per ton as system capacity increases. Nevertheless, Table 1 indicates the improvement in efficiency.

TABLE 1. COEFFICIENT OF PERFORMANCE*

Evaporator temperature	−40F	−30F	−20F	−10F
1 Stage without subcooling	—	1.55	1.95	2.27
1 Stage with subcooling	—	1.77	2.30	2.73
2 Stage with intercooling	2.16	2.34	2.60	2.81
2 Stage with subcooling	2.69	2.96	3.25	3.56

*Compressors only.

Next, for our larger systems, we have available the benefits of multiple compressors and/or capacity reduction. This is a great advantage in system performance. Residential unit designers should not only recognize this as a need, but should also realize that the residential buyer may well afford to pay the premium for capacity reduction with the ever rising energy costs. Unitary manufacturers should consider unloaders, since the time has now arrived when buyers will pay more for a better unit.

Third, and this relates to the value of having unloaders, is defrost. We do not experience defrost problems or long defrost time cycles. We have concluded that the period of greatest frost formation occurs at that weather condition when dry bulbs are in the range of 35F. Our large systems with unloaders provide higher suction pressures and therefore higher evaporator coil temperatures than would be available without unloaders (when the full compressor displacement rebalances at an unnecessarily low evaporator temperature with cycling). Unnecessarily low evaporator temperatures change the coil sensible heat ratio and produce greater latent capacity. This accelerates moisture dropout and frost formation. With unloaders, we materially reduce this phenomenon.

(The author then presented a series of slides showing several actual installations designed by his firm over the past 17 years. These have been operating successfully.—Ed.)

One of our firm's installations is the Moore Products Co., Springhouse, Pennsylvania, building. This system was of cascade configuration, providing simultaneous cooling and heating. Two systems, each with 300 tons cooling capacity, are installed. Each system is equipped with a centrifugal high stage, with double bundle condenser, and a reciprocating low stage serving the outdoor air coil (a component of the cooling tower). The condenser of the low stage is cooled by chilled water of the high stage, thereby transferring the heat between stages.

In conclusion, we believe that the multistage air source heat pump will become more and more feasible based on life cycle cost, as energy costs continue to escalate.

DISCUSSION

Q What is the price per ton installed for a large system?

A As I stated at the end of the discussion period following my talk, "price per ton" is a misnomer for large air source heat pump systems. This is because different buildings have different cooling/heating ratios depending upon their internal loads and building heat losses. Consequently, we have found it more desirable to relate the premium first cost to square foot of heated and cooled structure. We have found this to be $1.00/sq ft to $2.50/sq ft premium over a "conventional" system.

Q Why have you used double bundle condensers and cooling towers in lieu of single bundle condensers and evaporative coolers?

A Since large heat pumps involve centrifugal compressors, we have found that the costs of double bundle condensers and cooling towers have been less than for evaporative coolers. In addition, the annual operating cost is not penalized by having to operate the centrifugal compressor at higher heads during the cooling season. For smaller systems where reciprocating compressors have adequate displacement for the cooling and heating tonnage, evaporative condensers are used on the summer cycle. It should be remembered that a reciprocating compressor is a high head, low volume machine, whereas a centrifugal compressor is a low head, high volume machine.

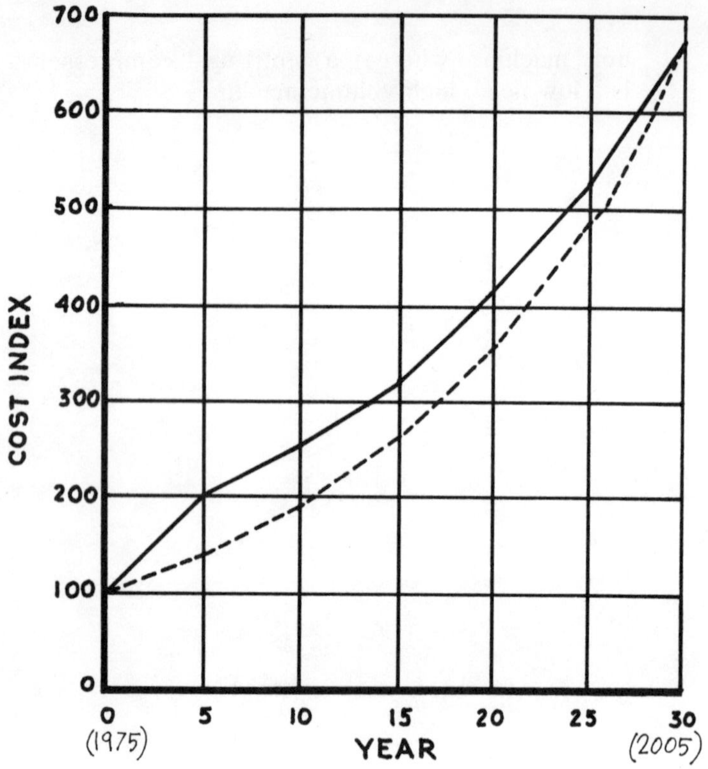

Fig. 1. Thirty-year energy cost projection. (From General Services Administration, Midwest Research Institute).

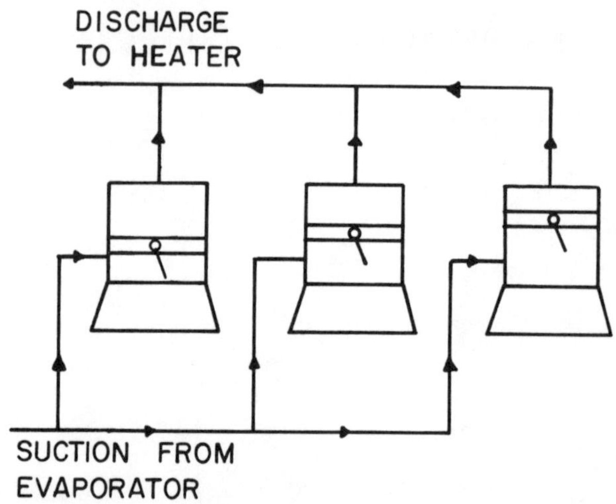

Fig. 2. Three compressors in conventional single-stage mode.

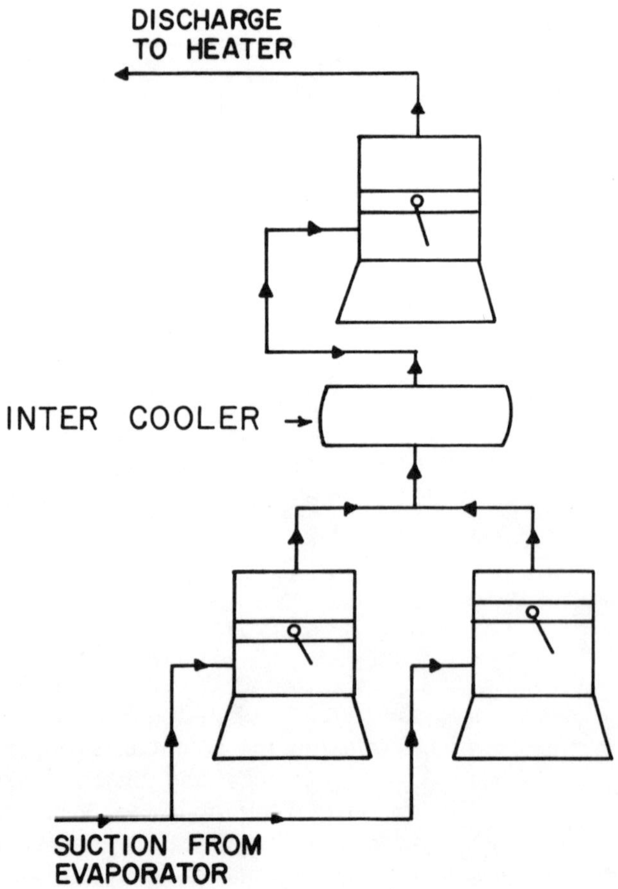

Fig. 3. Three compressors in a two-stage mode.

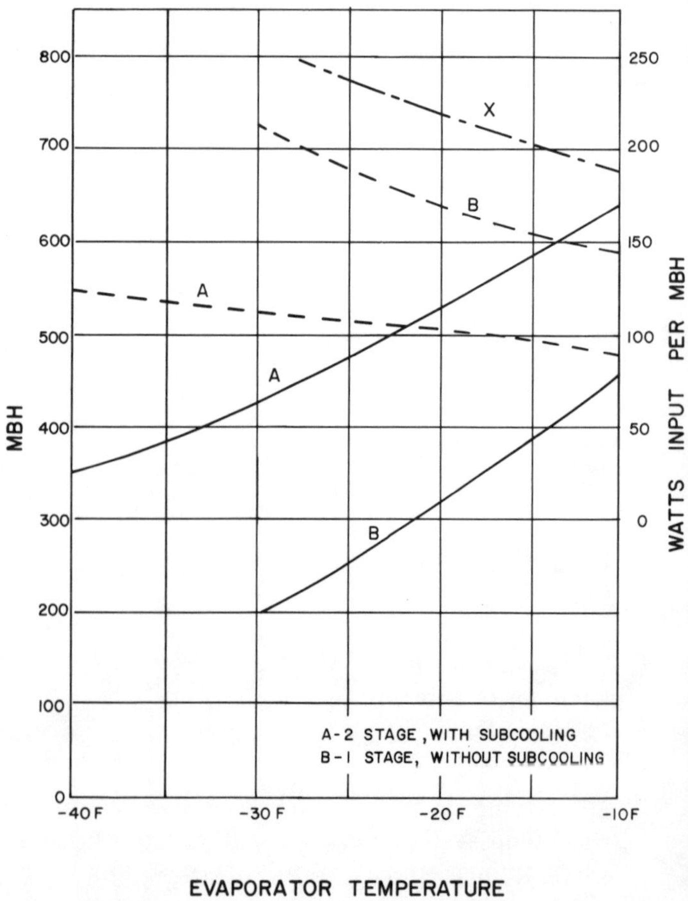

Fig. 4. Relative heating output (one- and two-stage modes).

Assessment of Selected Engine-Driven Heat Pumps

Jaroslav Wurm

Supervisor, Space Conditioning Research
Institute of Gas Technology
Chicago, Ill.

Some observers from the industrial and scientific communities are questioning the reasons being given to rouse a new wave of interest in developing engine-driven, residential heat pumps. Legitimate objections usually reflect problems of first cost, maintainability, and reliability; yet these aspects alone do not represent all possible advantages and disadvantages that must be evaluated to justify developing a particular system. Another important part of the overall and performance assessments of a heat pump system concerns the year-round energy consumption required for "delivered" comfort. Because we believe this aspect of the evaluation to be of primary importance, we have first addressed ourselves to answering the following interrelated questions: Are there any means of improving the effectiveness of known heat pumps? And are there any other refrigerators that could be used as effective heat pumps? Thus the problem is formulated in such a way that the form of energy input is not defined. Rather, the way energy will be supplied to the system (which to some extent prescribes the form of energy used) will be decided by the study.

Objectives and Methods of Design Point Performance Study

As a result of questioning the relative energy-saving potentials of applying heat pumps to heating and cooling single-family residences, two possibilities readily emerge:

1. Saving energy by designing and operating the heat pump system and its components at conditions where optimal efficiencies are achievable.

2. Saving energy by using the rejected heat of the driving (power) subsystem for heating.

For the portion of our work reported here, we have defined our basic objective as evaluating the validity of these premises.

Controlling the operating characteristics of individual components as suggested in the first premise (capacity and efficiency versus the ambient temperatures for various comfort levels) actually calls for investigating system performances corresponding to conditions for which the thermal load on a house is being satisfied by the modulated capacity of a heat pump. The energy-saving benefits of such an operation could actually reflect two aspects of capacity modulation:

1. Benefits resulting from lowering the temperature difference the heat pump must work against when milder ambient conditions occur. This is due to the decreasing load on heat exchangers of constant size and air flow, and consequently to the decreasing temperature gradient required to transfer the heat. Actually, the effect described could be studied as part of the system's overall performance at off-design operating conditions.

2. Benefits resulting from designing for and operating the refrigerator and driver at conditions where optimal efficiencies are achievable.

Because the energy saving described in the first point has been demonstrated (refs. 1–3), we have concentrated first on evaluating the potentials expressed by the second. Actually, we separated this effect by purposely analyzing system performances at design conditions (letting the heat exchangers be of variable capacity). This paper treats the results of this type of analysis for selected real systems (ref. 4).

The method used to obtain heating and cooling coefficients of performance (COP) as functions of ambient dry bulb temperatures (the most important characteristic for evaluating year-round performance of a heat pump) will become apparent from the following description. First, we used "real" components, for which the characteristics were either measured or calculated by manufacturers. The heat pump systems were then "designed" by matching these components. Only the first step of iterative matching was used. Because we were analyzing modulated systems, we had to assign them selected house-load operating characteristics. Such typical characteristics and an example of the way the heat pump performance was applied to them is presented in Fig. 1. For this configuration, the resulting COP's for capacity matching load conditions are given in Fig. 2.

Results

The most important results are presented in Fig. 3. There, for reference purposes, a band of COP's is accentuated. Its lower limit represents typical COP values for a residential, electrically driven heat pump as given by a manufacturer. These values include the average efficiency of electric power generation and transmission in the United States (29%); and for cooling, they represent the total capacity. Supplemental heat and auxiliary power are excluded. The upper limit of our reference band is a COP curve calculated for a system comparable to an electrically driven heat pump. In this case, however, the compressor capacity always equals the house load. This is accomplished by varying the compressor speed within the practical range of 1000–4000 rpm. The driving hypothetical prime mover has a constant efficiency of 30%; and, for heating, 50% of its rejected heat is recovered. The curve therefore represents the potential performance of a piston compressor with a capacity modulation. This is, however, still without the benefits resulting from decreasing temperature differences, which, as stated before, we elected to study separately.

Superimposed, one can find the performance curves of several practical systems. These COP curves were calculated by using available characteristics of refrigerators and prime movers.

Conclusions

It is recognized that a study such as ours would have to be supplemented by an analysis of costs related to manufacturing and operating a unit, especially if it were to be used as a guideline in selecting a system for preferential development. The presented analysis deals partially with two aspects of the operating characteristics of the selected systems which influence the operating-cost assessment. This represents only a first phase in evaluating the relative, year-round operating costs, yet the results are significant enough to suggest three basic conclusions:

1. The positive effect of capacity modulation on the COP of the analyzed systems derived from operating the refrigerating subsystems at optimal efficiencies is significant, but can be destroyed by improper matching with prime movers. Because the other positive effect of modulation—namely, the effect of variable temperature gradient on heat exchangers—exists, when proper attention is given to selecting and matching a driver, the systems with capacity modulation will always consume less energy as compared with unmodulated systems, and therefore should be regarded as highly desirable for increasing the energy-saving potential of a heat pump.
2. Below 30F ambient temperatures, substantial improvements in the heating COP of a residential heat pump can be achieved by using an engine to drive the refrigeration subsystem. In practice, at least 50% of an engine's rejected heat can be used directly to boost the heating COP, heating capacity, and temperature of the delivered air.
3. Capacity-modulated systems, specifically those driven by an on-site prime mover with rejected heat recovery, show a capability of operating over a wider range of ambient temperatures more efficiently.

REFERENCES

1. Cawley, R.E., and Pfarrer, D.M., "Part Load Efficiency Advantages of Two-Speed Refrigerant Compressors." Paper presented at the Conference on Improving Efficiency in HVAC Equipment and Components for Residential and Small Commercial Buildings, Ray W. Herrick Laboratories, School of Mechanical Engineering, Purdue University, 7-8 October 1974.
2. Granryd, E., "Various Heat Pumping Processes." Paper presented at the Seminar on Heat Pumps Sponsored by the Educational Services of the Finnish Society of Engineers, Helsinki, September 1974 (Swedish text).
3. Cohen, R., Hamilton, J.F., and Pearson, J.T., "Possible Energy Conservation Thru Use of Variable Compressors." Paper presented at the Con-

ference on Improving Efficiency in HVAC Equipment and Components for Residential and Small Commercial Buildings, Ray W. Herrick Laboratories, School of Mechanical Engineering, Purdue University, 7-8 October 1974.

4. Wurm, J., Rush, W.F., Weil, S.A., Dufour, R.J., and Lassota, M.J., "An Assessment of Selected Heat-Pump Systems," *Annual Report, Project HC-4-20 for American Gas Association*. Chicago, Institute of Gas Technology, March 1974.

DISCUSSION

Q What type of Stirling engine did you analyze?

A Data we have used for a Stirling engine performance were generated by the North American Philips Corporation in 1974. They relate to a four-cylinder, swash-plate-driven Stirling engine with helium as the working fluid and with a piston-swept volume of 16 cu cm. The engine, which was sized for a given load characteristic, could use any liquid or gaseous fuel.

Q What is the efficiency of the Stirling cycle used?

A The overall efficiency showed a peak value of 25.4% at 3000 rpm when the system was operated in heating mode. The efficiency was steady, but started to decrease between 2000 and 1000 rpm.

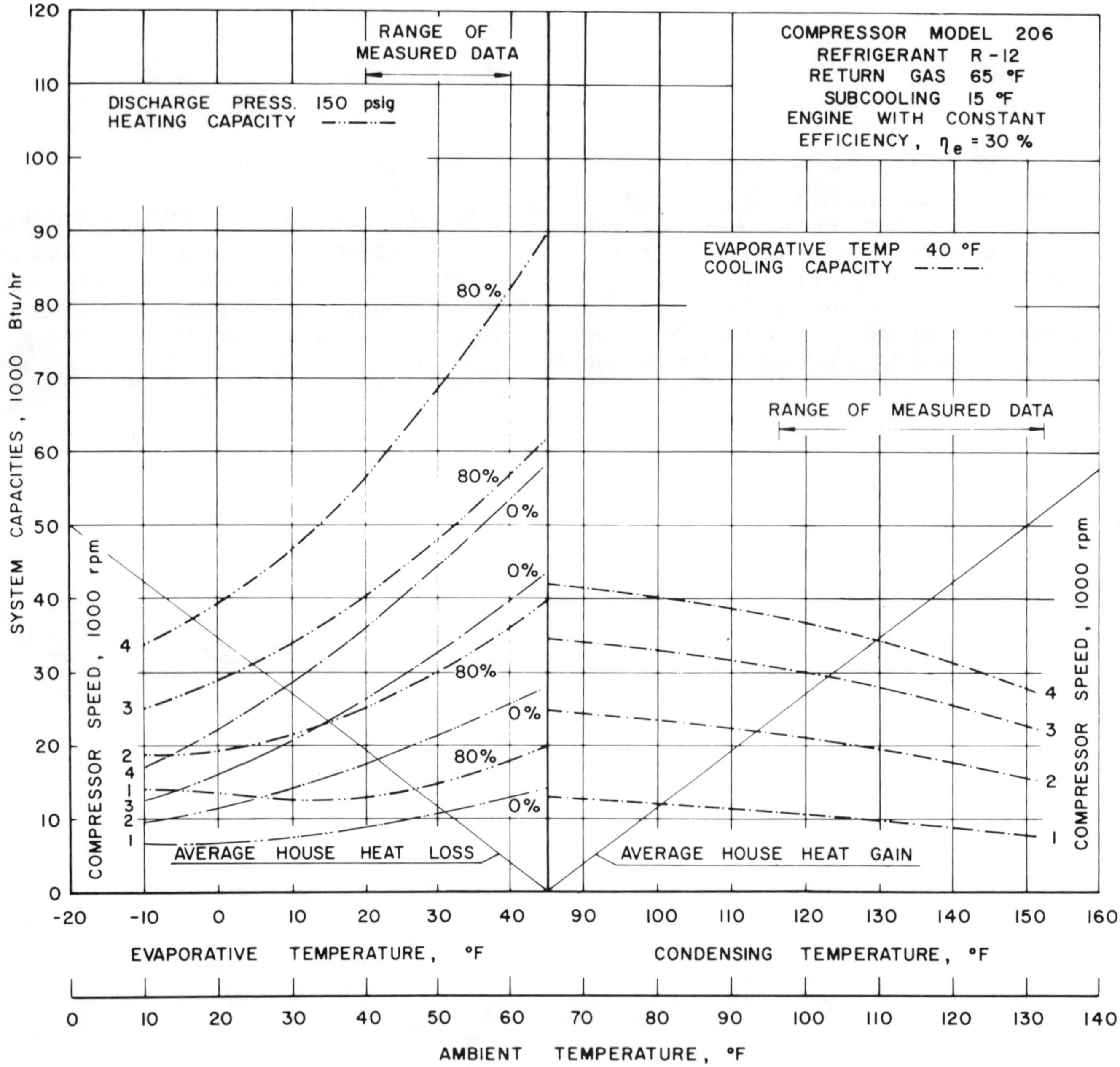

Fig. 1. Heating and cooling capacity of a Rankine piston compressor heat pump driven by a constant-efficiency prime mover (heating capacity lines correspond to 0 and 80% recovery of the engine's rejected heat).

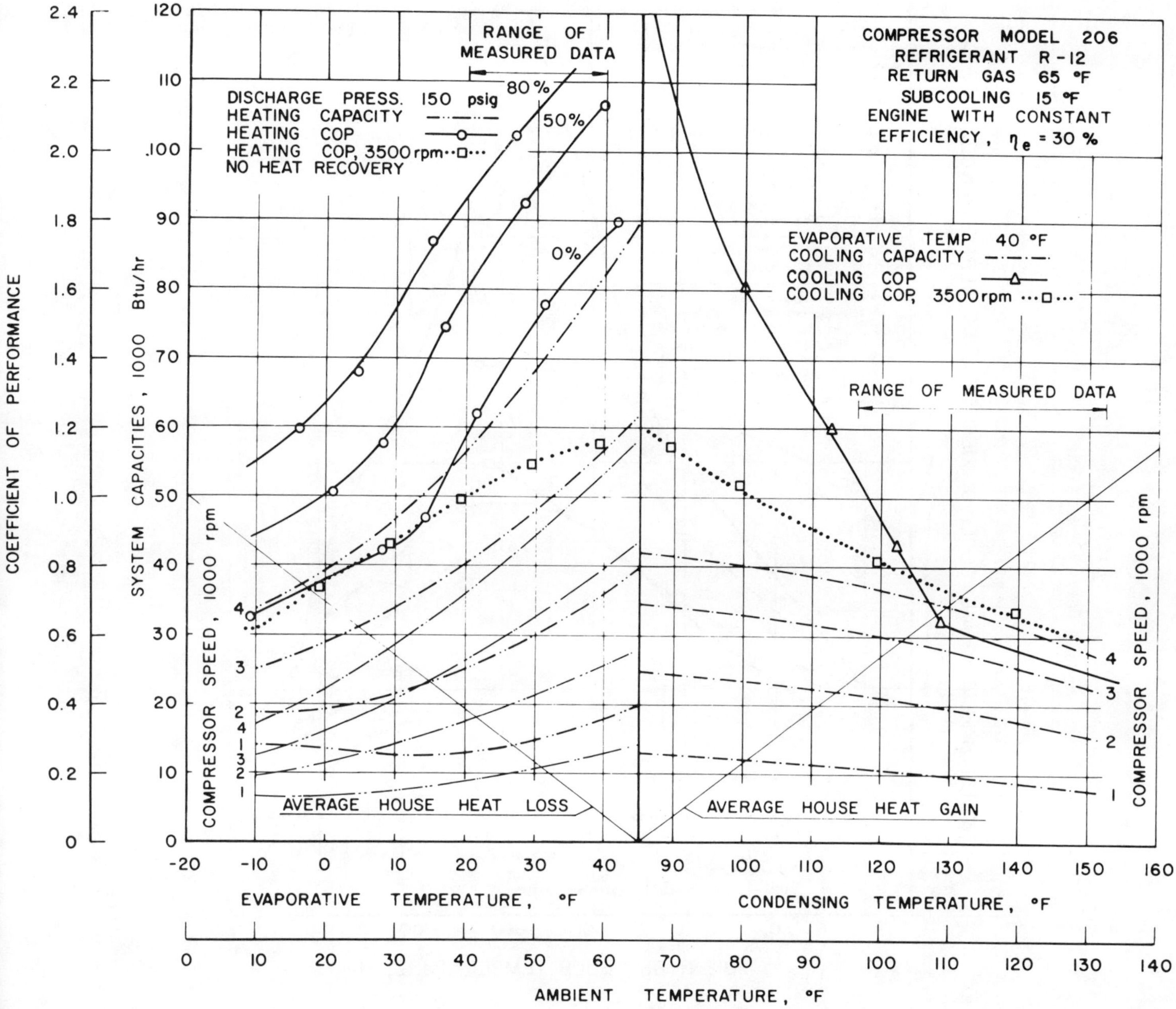

Fig. 2. Heating and cooling performance of a Rankine piston compressor heat pump driven by a constant-efficiency prime mover (heat COP lines correspond to 0, 50, and 80% recovery of the engine's rejected heat).

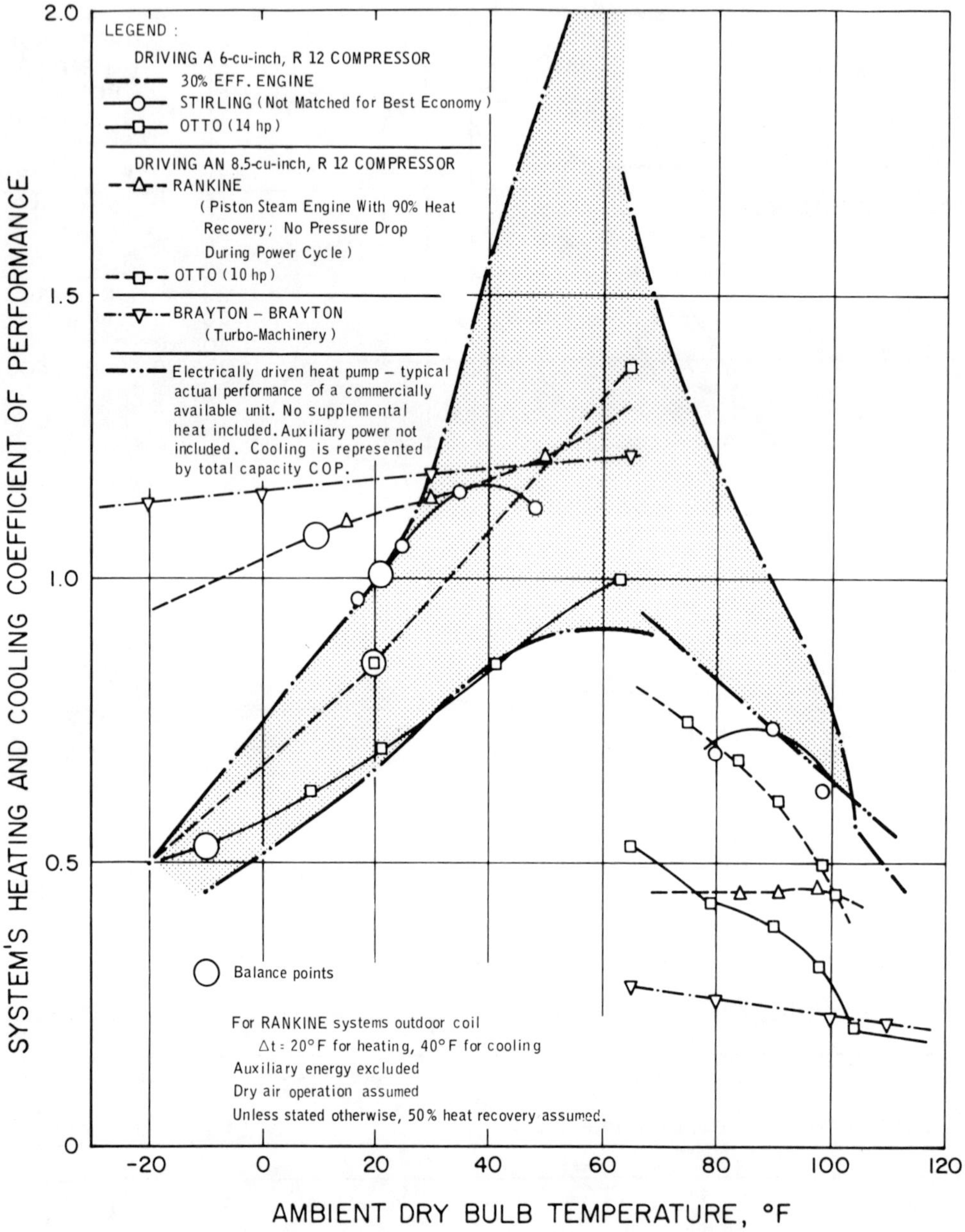

Fig. 3. Coefficient of performance of various air-to-air heat pumps (for engine-driven systems, the values do not include the benefits of decreasing temperature differences on the heat exchangers).

An Efficient Stirling Heat Engine-Heat Pump

Dr. William R. Martini

Research Professor of Chemical Engineering
Joint Center for Graduate Study
University of Washington
Richland, Wash.

The best way to use the available heat from solar collectors or from combustion is to employ a heat engine at the building site to produce energy near room temperature heat. The energy is then used in a heat pump operating between the building and a thermal reservoir. This paper will show what advantages there are to this approach to building heating and cooling if really efficient engines are used. The principle of the Stirling engine will be explained. The requirements that must be met to attain a high efficiency Stirling engine and the requirements of a solar powered heat pump will be reviewed. A simple but efficient free-displacer, free-piston heat pump will be described. Computed performances of similar but earlier heat operated heat pump designs will be presented and discussed. Key problem areas will be discussed.

Heat Engine-Heat Pump Theory

Heat engines including Stirling engines are based on a few simple facts:

1. Heated fluid expands or pressurizes and cooled fluid contracts or depressurizes.
2. Conversely, compressed fluid heats up and expanded fluid cools down.

All heat engines and heat pumps go through the same four processes (Fig. 1). The sequence used in the heat engine produces power by dropping heat from a higher to a lower level. The sequence used in the heat pump requires power to raise heat from a lower to a higher temperature level.

The maximum efficiency of a heat engine is realized when all the heat is added to the working fluid at the heat source temperature and all removed at the heat sink temperature, and when the transition from heat source to heat sink is reversible. If a heat pump satisfying analogous criteria is combined with the above heat engine, the coefficient of performance is

$$COP_C = \frac{Q_C}{Q_H} = \frac{T_C(T_H - T_W)}{T_H(T_W - T_C)} \quad (1)$$

The heating coefficient of performance is

$$COP_H = COP_C + 1 = \frac{Q_W}{Q_H} \quad (2)$$

where Q_C = heat supplied to the heat pump
Q_H = heat supplied to the heat engine
Q_W = heat removed from both heat engine and heat pump
T_C = temperature of cold part of heat pump, °R
T_H = temperature of hot part of heat engine, °R
T_W = temperature of warm part of both heat engine and heat pump, °R

As will be explained later, two optimization studies have indicated that 80% of this limiting COP is possible in a practical size machine.

Stirling Engine Theory

Figure 1 shows how the processes are accomplished in a Stirling engine operating either as a heat engine or as a heat pump. All Stirling engines have reversing flow regenerators that nearly reversibly transfer the gas between two spaces maintained at two different temperatures during the time the overall volume is nearly constant. The high efficiency of the theoretical Stirling cycle is due to postulated isothermal compression and expansion. Almost all engines so far built more nearly operate with adiabatic compression and expansion, which leads to a smaller work diagram and large irreversibilities, and therefore lower efficiency (ref. 1). In most Stirling engines the displacer and the piston are operated by two cranks 90° out of phase with each other. Also, some free-piston engines have been built (ref. 1). In either case the distinct processes indicated by the solid lines in Fig. 1 overlap so that the work diagram is smaller and more rounded (dashed line).

Requirements for High Efficiency

A relatively simple computer analysis has been conducted to determine the effect of dead volume regenerator effectiveness and live volume heat transfer on power density and efficiency (ref. 1). Figure 2 shows that the introduction of dead volume does not affect COP but reduces power density. This observation indicates that engine power can be modulated by controlling the stroke of the displacer. It also indicates that dead volume should be minimized. The larger the dead volume fraction, the more gas must be processed by the regenerator to attain a given power.

Regenerator efficiency affects COP and cooling power equally. The effect is linear for small perturbations.

In this simplified calculation there was no separate gas heater and cooler and the number of transfer units in the warm and cold spaces remains constant throughout the cycle. The number of transfer units is defined as

$$NTU_v = \frac{hA\,\Delta t}{mC_v} \qquad (3)$$

where h = heat transfer coefficient, Btu/hr sq ft °F
A = area of heat exchange surface, ft°
Δt = time for 1° of angle increment, hr (360° / revolution)
m = mass of gas in space, lb_m
C_v = heat capacity at constant volume, Btu/lb_m °F

For instance, for a conventional Stirling engine, the NTU_v is a mass average of the gas heater with a high NTU_v and the hot space usually with a low NTU_v. Figure 2 shows that lack of an adequate NTU_v is much more important for maintaining an adequate COP than it is for maintaining an adequate cooling power. The conclusion appears to be that if one wishes to approach limiting efficiency for a Stirling engine, one must have an engine with a high average NTU_v for the end spaces and a high efficiency for the regenerator.

For maximum work per cycle and, therefore, maximum efficiency, heat must be applied to the gas as it expands and removed from the gas as it is compressed. Two methods of doing this are interleaving fins and collapsible packing. Both methods divide up the two working spaces of the engine into thin layers to attain nearly perfect heat transfer. The heat transfer depends entirely on the thermal conductivity of the gas.

The collapsible packing is felt to be the more practical way of creating thin layers in the working spaces. Figure 3 shows a paper model of one type of collapsible packing design. This packing has sufficient lateral stability to act as a bearing for the regenerator. Also, the squeezing out of the last bit of gas between the foils acts as an effective snubber.

Requirements for a Solar Powered Heat Pump

Unlike a heat pump powered by fuel combustion, the solar powered heat pump must be able to operate efficiently over a wide range of input temperatures, even though the design will be optimized for the most common input temperature. It should also be capable of efficiently utilizing a range of cold heat input and warm heat output temperatures. It should be adaptable for use with low temperature flat-plate collectors, as well as high temperature focusing collectors. Heat would usually be applied to the engine from thermal storage via a hot water loop or possibly a high temperature heat pipe. The heat engine-heat pump must be inexpensive to build, have a potential for low maintenance, be easily controllable, and operate automatically.

The Free-Displacer, Free-Piston Heat Pump

The design of this type of heat pump has been evolving over the last year (refs. 1, 2). A number of computer programs have been written to evaluate this machine. These indicate that all live volumes of the machine must be isothermalized if the ultimate in efficiency is to be approached. Also, more power is obtainable from a given amount of gas passing through the regenerator if the gas expands while it is entirely in the hot space, rather than expanding from the hot space, through the regenerator, into the cold space.

One design which has a chance of attaining nearly perfect efficiency since it meets all requirements so far set down is shown in Fig. 4. Also, an engine operating on this principle has been tested; it exhibited self-starting and controllability by adjustment of the valve. The heat engine-heat pump will be explained and then the test engine will be described.

For simplicity in explanation, both heat engine and heat pump in Fig. 4 are shown as identical machines. When the design is made for a given heat source, heat sink, and cooling load temperature, the optimized heat engine might look quite different from the optimized heat pump.

The machine shown in Fig. 4 employs two floating pistons and a fluid flywheel inertial member to make possible compression before heating. The two venturi tubes perform the function of a speed increaser and a high speed rotary flywheel in a crank-operated Stirling engine. The fluid flywheel stores energy during the cycle so that higher compression ratios are possible. These higher compression ratios produce much more power from the same flow through the regenerator. Power density is increased further by having almost all the gas expand in the hot space of the heat engine and in the cold space of the heat pump. In the VM cycle cooler (ref. 3) and in some earlier machines proposed by the author (ref. 2), gas expansion takes place also in the warm place, since the hot and cold space volumes change very little during expansion. In the bottom part of Fig. 4 the motion of the displacer and piston for both the heat engine and the heat pump are graphed for two engine cycles. Note that for the particular control valve adjustment illustrated, gas is transferred from one end to the other of the engine over 60° phase angle at the ends of a piston stroke. Except for a small dead volume in the regenerator (not shown), all gas is compressed in the warm spaces and is expanded in either the hot or cold spaces. The collapsible packing absorbs the heat of compression and supplies the heat of expansion to give a nearly isothermal process and therefore much more power per cycle. Each packing is totally collapsed one third of the time. During this time the heat of compression is transferred by conduction to either floating piston, and the heat of expansion is transferred by conduction from either the heating loop or the cooling loop.

The fluid flywheel is composed of two venturi tubes and two check valves to attain circulation of all the liquid in the fluid flywheel. Circulation is needed to assure good cooling of the floating pistons and to transfer heat to the warm heat exchanger.

Figure 5 shows an experimental engine which starts itself and is controllable by a valve. Based upon the results of these engine tests, an improved implantable hydraulic power source for powering artificial hearts is being tested (refs. 4, 5). Figure 6 shows a basic schematic of this experimental engine. The heat engine uses a single annulus regenerator which also acts as a gas heater and cooler. The displacer drive piston and the control valve work in hydraulic fluid. The transition from gas to liquid is made at a bellows which sustains no pressure difference. Rubber diaphragms are used at other transition points between gas and liquid. The necked-down portion of the fluid flywheel is the chief mass component because of much higher fluid velocity. The bounce chamber and the engine act as springs. The stepped piston and check valves furnish coulomb damping. Pressure changes in the engine caused by movement of the displacer are the source of the drive energy. When the heat engine is heated and the control valve is closed, no power is produced. As the control valve is opened, action starts smoothly and builds up to the desired power. The fluid flywheel and liquid pump together exhibited a 95% efficiency.

Computed Performance Predictions

Several computer programs have been written to predict the performance of earlier designs of heat operated heat pumps (refs 1, 2). These predictions are generally more applicable to the design presented here than earlier designs since all live spaces are isothermalized instead of just some, as designed previously. Windage losses were estimated. Losses for the inertial component were ignored, but the present gas expansion arrangement results in more power than that assumed previously. Design optimization calculations are planned for the present design. Until they are available the results of previous design optimizations are instructive. Two optimizations were carried through, one for a 500F heat source and one for a 1500F heat source. Both attained 80% of the theoretical efficiency (Eq. 1) for the chosen temperatures.

Key Problems

The key advancement in engine design that has led to the improved performance claimed in this paper is the practical isothermalization of the hot and warm volumes of the heat engine, and the cold and warm volumes of the heat pump. Most key problems relate to actually accomplishing this. If interleaving fins are used there is a problem of fabrication, maintaining proper spacing, and compensating for the lower heat transfer in the nonmeshed portion of the fin. If collapsible packing is used, there is a problem of thermal contact resistance between the foils, having the foils actually make contact when compressed, and fatigue of the foils. Heat transfer from the floating piston to the water may be a problem. In the fluid flywheel, flow friction can be a problem. Vibration caused by acceleration and deceleration of the fluid may be solved by firmly anchoring the machine, freely suspending it, or designing a counterbalanced fluid flywheel. The floating pistons eliminate the friction and expense of seals but they allow the liquid vapors access to the working fluid. If the cold space of the heat pump is allowed to operate below the freezing point, ice will accumulate in the cold space. If needed, diaphragm or wiping seals are possible to largely eliminate this problem.

This machine is not without anticipated problems, but these problems can have a number of satisfactory solutions, so that the chances of attaining a high performance heat engine-heat pump are high.
cial heart shows promise as an efficient, adaptable,

Conclusions

1. A type of free-displacer, free-piston Stirling engine now developed for efficiently operating an artificial heart shows promise as an efficient, adaptable, and inexpensive solar-heated heat pump for buildings.

2. To approach the ultimate performance one should use isothermalized hot, warm, and cold spaces, regenerators made with slots, and expansion entirely from the hot and cold spaces. Also, the overall design must be optimized for efficiency at reasonable power density.

3. This engine can approach within 90% of Carnot efficiency. This heat engine and heat pump can approach within 80% of the limiting Carnot efficiency.

4. The promise of efficiency combined with potential for inexpensive construction and long operating life indicate that this class of machines should be investigated further, with the hope of attaining a breakthrough in our utilization of heat to heat and cool buildings.

REFERENCES

1. Martini, W.R., "The Free-Displacer, Free-Piston Engine—Potential Energy Conserver." IECEC Record 1975 (to be published).
2. Martini, W.R., White, M.A., and DeSteese, J.G., "How Unconventional Stirling Engines Can Help Conserve Energy." IECEC Record 1974, pp. 156-1099.
3. Shelpuk, B., "A Solar Vuilleumier System." Workshop Proceedings, Solar Cooling for Buildings, 6-8 February 1974. NSF-RA-N-74-063, pp. 156-61.
4. Martini, W.R., ét al., "Unconventional Stirling Engines for the Artificial Heart Applications." IECEC Record 1974, pp. 791-98.
5. Johnston, R.P., et al., "A Stirling Engine with Hydraulic Power Output for Powering Artificial Hearts." IECEC Record 1975 (to be published).

DISCUSSION

Q Is your design a kind of Vuilleumier cycle? You use the pressure fluctuation at the low temperature side of the engine to fluctuate the pressure at the high temperature side of the heat pump by means of a connecting fluid. However, it is known in this kind of technology, that you can make a combination of an efficient engine with a bad heat pump or a low efficiency engine with a high efficiency heat pump.

A The type of engine I have designed would be a Vuilleumier cycle cooler if the connecting fluid between the heat pump and heat engine has negligible inertia and the displacer in the engine leads the displacer in the heat pump by 90%. Actually, the connecting fluid in this machine has a large amount of inertia and makes the machine build up to high compression ratios which greatly increase the machine output power without appreciably adding to the heat losses. Also, the phase displacement between the two displacers is not fixed. Because of the higher compression ratio and the more optimum displacer motion, my machine will have a higher efficiency and power output than the Vuilleumier cycle can.

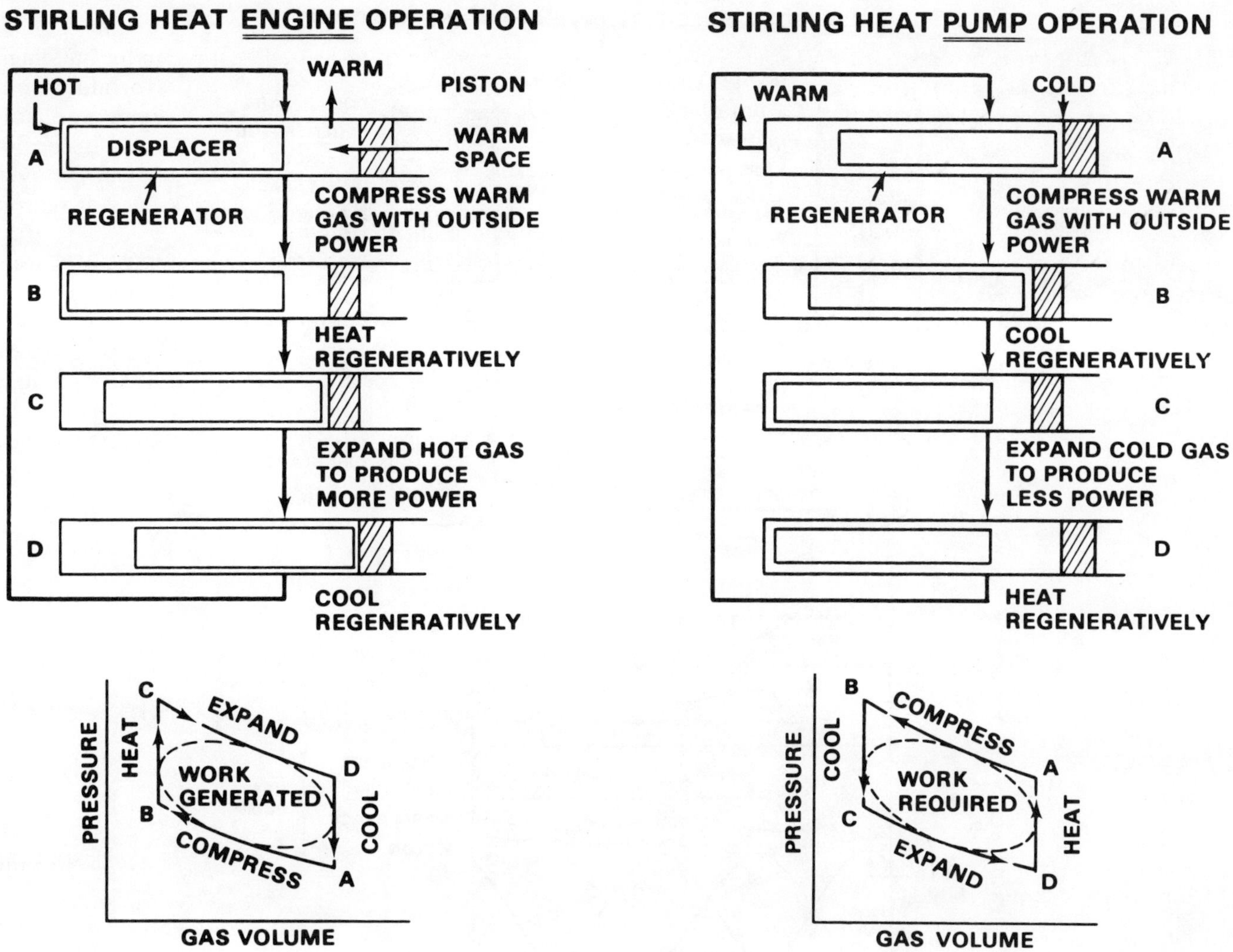

Fig. 1. Stirling engine operation.

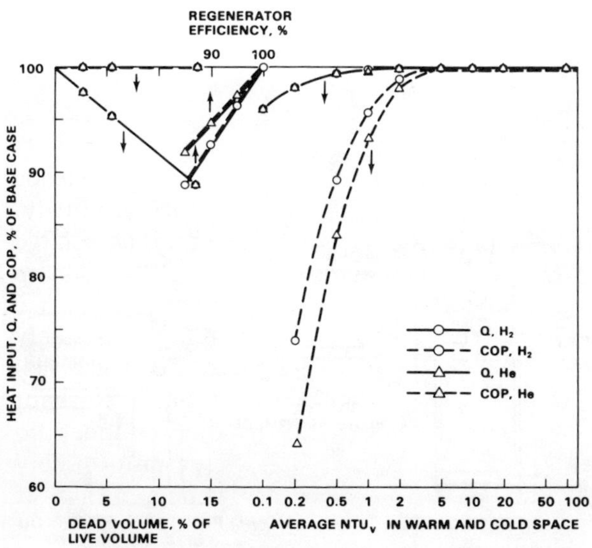

Fig. 2. Effect of imperfections on Stirling heat pump performance.

Fig. 3. Paper model of collapsible packing design.

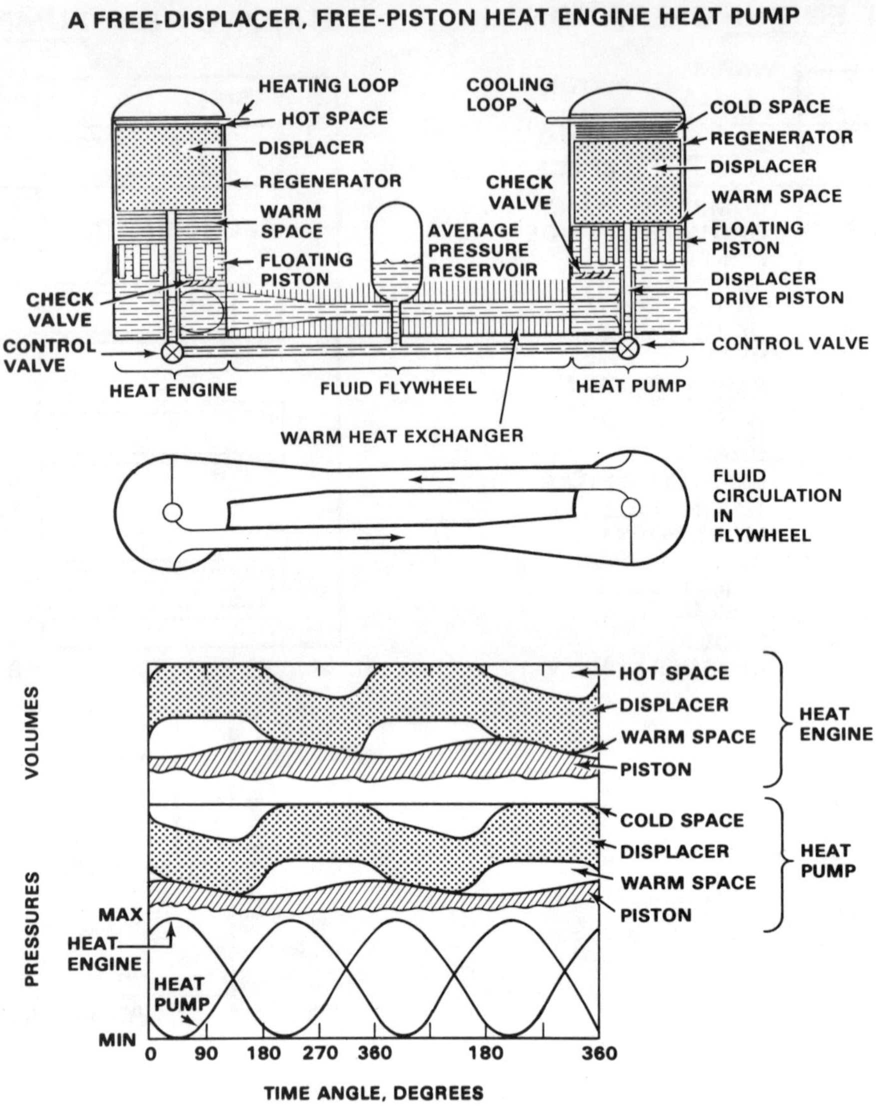

Fig. 4. A free-displacer, free-piston heat engine heat pump.

Fig. 5. Experimental engine with free-displacer drive unit.

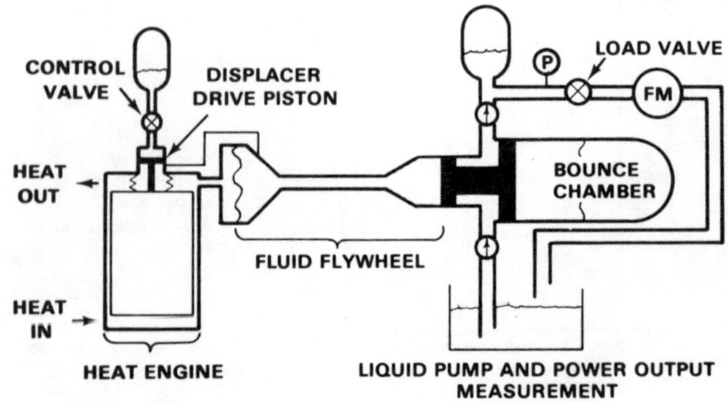

Fig. 6. Simplified schematic of experimental engine.

Part V

Chairman:
Guy H. Cheek

Professional Engineer
Air Conditioning Corporation
Greensboro, N.C.

A Residential Solar Assisted Heat Pump Design

Ved Bansal

Chief Mechanical Engineer
Vosbeck Vosbeck Kendrick Redinger
Alexandria, Va.

Space heating in residential and small commercial buildings accounts for about 12% of this nation's annual energy consumption. This percentage of energy usage is on the rise because all-electric heating is rapidly becoming the favorite of builders and financiers. It is cheaper to install than oil heating and gas heating.

Electric heating is a very inefficient way to heat any building, because two-thirds of oil-, gas-, or coal-generated energy is lost in stacks and in transmission; but since it is cheap to install and easy to maintain, contractors are still installing it in houses across the country.

The only energy that will be available ten years from now at reasonable rates will be electricity. Because gas and oil reserves are on the decline, electricity will be produced by coal, nuclear energy, or hydroelectric power.

Can we achieve both goals of enjoying all electric living and conserving energy? It can be done by using water-to-air heat pumps and solar energy. This combination can put a dent in our country's energy usage.

While we have designed this system with the best information available, we are still in the early stages of technology. More work needs to be done to optimize the use of solar collectors with heat pumps so it can be economically attractive for everyone.

In this context, we are designing a house in Alexandria, Virginia. It will be a three-story house with 1200 sq ft of collector, a floor area of 3955 sq ft, and window area of 680 sq ft. All windows will have insulated type glass. The collector is going to face due south and will be inclined 40° from the horizontal.

Calculated heat loss for the residence is 120,000 Btuh and heat gain 96,000 Btuh. We have selected two heat pumps, each having cooling capacity of 48,000 Btuh and heating capacity of 60,000 Btuh.

As indicated in Fig. 1, during the year Pump P-1 starts to pump water to collector only when the temperature in the collector is 10F greater than the temperature in the storage tank, so that the temperature of water in the tank can go as high as 200F. Domestic water runs through the heat transfer coil in the storage tank to pick up whatever heat is possible, depending on temperature of water in the storage tank. We expect on the average the temperature to rise by 40–50F in winter and up to 80F in summer. This water is further heated by the electric water heater.

Pump P-2 pumps water between 60 and 90F on demand from a room thermostat year-round through the heat pumps. This discharge thermostat is reset by an outdoor sensor. If the outdoor temperature is 0F, water leaving the three-way valve is 60F; if the temperature outdoors is 95F, water leaving the three-way valve is 92F.

When the thermostat calls for heating, the blower F starts and the reversing valve V is energized, as shown in Fig. 2, selecting the path of refrigerant flow. The compressor P starts and pumps hot compressed refrigerant gas to the air side coil A, where the gas is condensed to a liquid by giving up heat to the air. This air returning from the room is warmed and delivered to the room. The liquid refrigerant then passes through a capillary restrictor tube C, causing a pressure drop, and expands (boils) into a gas within the coaxial refrigerant-to-water exchanger E. Heat is absorbed from the circulating water loop which is warmed by the solar collector. The refrigerant gas now passes through the second path of the valve to the suction of the compressor, where the cycle starts over. With this method we can have water stored at 190F and let it go down to 60F in case there is no sun for a week.

Because the heat pump allows the storage water to run at a lower temperature, solar collection efficiencies will increase. The reason, as explained earlier, is that heat pumps can run on water at 60F, whereas for direct heating, the water temperature has to be at a minimum of around 100F.

Because the temperature of water required for heat pumps is low, a solar collector with single glazing can be used.

We are designing this system with an electric heating coil in the heat pump equivalent to heat loss. If for some reason water temperature drops below 60F, the compressor cuts out and electric resistance is energized. In the Washington, D.C., area with a 3000-gal storage tank and a 1200 sq ft panel in this house, this electric heating coil should not come on except possibly two or three times a year.

In summer, as reset by the reset schedule, water temperature leaving the three-way valve maintains a maximum 92F discharge temperature. As described earlier, Pump P-2 runs on demand from thermostat; once the flow is established as shown in the flow diagram, the blower F starts and the reversing valve is deenergized. As shown in Fig. 3, compressor P starts and pumps hot compressed refrigerant gas to the coaxial refrigerant-to-water exchanger, where the gas is condensed to a liquid by giving up heat to the circulating water loop. The liquid refrigerant passes through a capillary restrictor tube causing a pressure drop, and expands into a gas within the air side coil by picking up heat from the warm room return air. The air is cooled and dehumidified and delivered to the room. The refrigerant gas now passes through the second path of valve to the suction of the compressor where the cycle starts over.

During the cooling cycle, water leaving the heat pump may be as high as 103F if the heat pump is designed for 92F water inlet temperature. To reject the heat outdoors, heat exchanger, Pump P-3 and cooling tower will be used as shown in Fig. 1.

As soon as the temperature of water leaving the heat pump reaches 92F, Pump P-3 is energized and in turn the cooling tower is activated. The cooling tower is designed to cool water from 95 to 85F. The 85F water entering the heat exchanger removes heat from 92–103F water leaving the heat pump and cools down to a maximum temperature of 92F. This 92F water is circulated through the three-way valve and through the heat pumps without going through the storage tank.

In the past, air-to-air heat pumps for residential usage have been applied. There has always been a problem of outdoor evaporator coils frosting up. In order to avoid this, heat pumps always have to go in the cooling mode to remove frost from the coils; in the meantime, resistance heating coils have to be energized to heat the house.

We feel in the future heat pumps will be applied to homes with a collector made of corrugated aluminum sheets and a single pane of glass over the aluminum sheet. The advantages of this system are:

1. Economy. This type of collector can be built at about $3.00/sq ft compared to other collectors that are at present selling at the rate of $8.00–10.00/sq ft.
2. Lower water temperatures. As described earlier, this system does not require water above 60F. We know this type of collector can provide water around 120F.
3. This collector can be used in the summertime to reject the heat absorbed during the day to the night time atmosphere.

The last two points need more research so that more data can be made available to designers of these systems.

In future years, heat pumps with hot water heating coils should be made available. The heat pump can act as an air-handling unit as long as the temperature of water in the storage tank maintains the space temperature. If water in the storage tank does not maintain the space temperature, then the compressor will be energized for this purpose.

Regarding the payback period for the solar collector, our calculations indicate that if this house is to be heated with oil or gas, a solar collector is not economically feasible. With electricity rates around 4¢/kwh, the payback period is around 20 years. We believe this period will be reduced as electricity rates keep going up.

DISCUSSION

Q Will you please give data on source and area of collectors, size of tank, fluid used, flow rate, and size of heat pumps?

A Revere. 1200 sq ft. 3000 gal. There will be a separate heat exchanger, and 20% ethylene glycol mixed with water will be pumped through the collector. Flow rate through the heat pumps will be 24 gpm. Heat pumps range from 2 to 4 tons.

Q What is the Btuh you obtained with the collector?

A 300,000 Btuh.

Q Would it not be better to blend domestic hot water from main storage through the auxiliary tank on the output end of the collector circuit, before going to a high grade auxiliary domestic heater?

A I don't see how this will help, since the temperature in storage tank can go as high as 200F.

Q What minimum tank temperature can be obtained with a dry cooling system?

A During summer, tank will be bypassed and in lieu of dry cooler, we have decided to use a cooling tower and a heat exchanger.

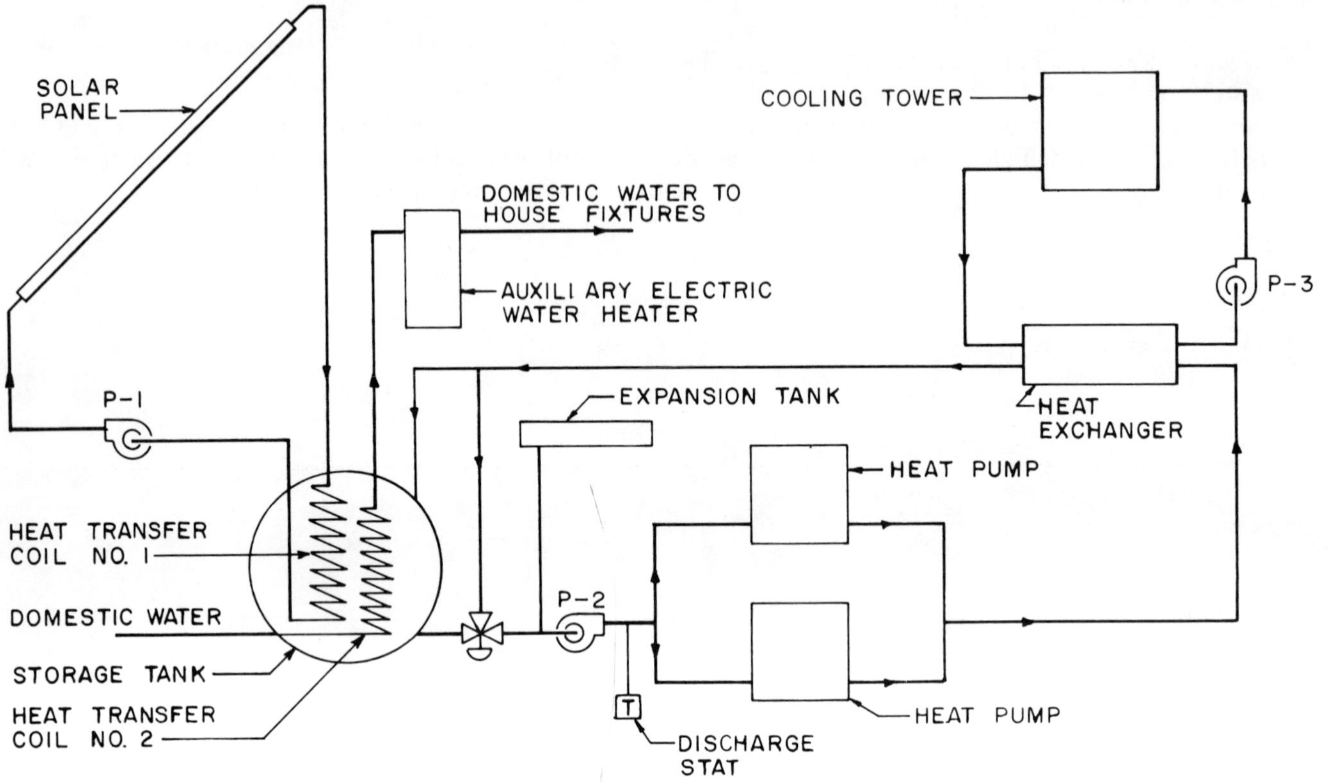

Fig. 1. Schematic flow diagram of house in Alexandria, Virginia.

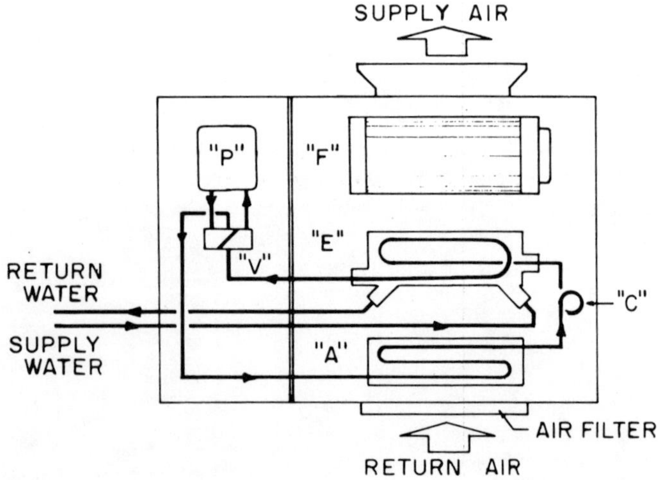

Fig. 2. Flow diagram.

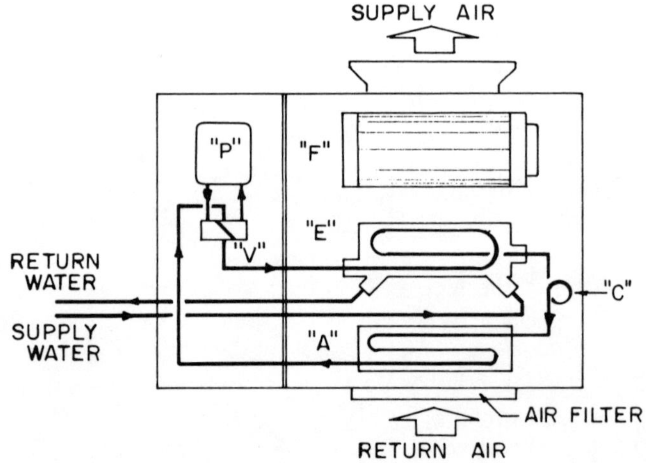

Fig. 3. Flow diagram.

Solar Rankine Cycle (Steam) Drive and Assistance to Heat Pump Systems

Dr. Noam Lior

Assistant Professor of Mechanical Engineering
University of Pennsylvania
Philadelphia, Pa.

This study analyzes various methods for reducing the purchased resource energy consumption of vapor compression heat pumps in both heating and cooling modes, in comparison with conventional systems. One of the methods studied is the application of a novel solar-powered, fuel-superheated Rankine cycle incorporating a steam turbine or rotary expander. System performance is augmented further in the heating mode by utilizing the waste heat from the cycle, as well as by solar energy.

Application of solar energy to augment the performance of heat pumps has been under consideration for a long time (ref. 1). Solar energy collected at sufficiently high temperatures, during periods of adequate insolation and sufficiently high ambient temperature, can be used to heat the building directly (refs. 2, 3). When the ambient conditions cause the solar collector fluid to drop below about 35C, the heat collected can be used as a heat source for the evaporator of a heat pump. Solar heat collection at lower collector temperatures is more efficient, or, alternatively, the collection of the same amount of heat would require a collector which is simpler and less expensive.

An additional purchased-energy saving in the operation of the heat pump can be obtained by driving its compressor by a thermal engine, where the cycle fluid is heated by solar energy. While high collection temperatures result in better cycle efficiency, they require expensive solar concentrators which are also very sensitive to the intensity and direction of the solar radiation. Flat-plate solar collectors provide lower collection temperatures, but are less costly and make use of both the diffuse and direct components of radiation. Resulting from extensive performance studies of flat-plate solar collectors (ref. 4), two of the most viable designs are considered here: a double-glazed collector with a flat-black absorber and a single-glazed collector with a selective-black absorber.

Some of the most promising power generation schemes at present utilize Rankine cycles, employing reciprocating engines (ref. 5), or turbines (refs. 6–8). Turbines, however, have several advantages over reciprocating engines, such as simpler construction, smaller friction losses, no lubricant required in cycle fluid, low maintenance and high reliability, lower noise and vibrations, easy regulation, and smaller size with large expansion to condenser pressure and temperature. Their major disadvantages are that presently available small turbines have low efficiency (below 50%), and their cost. Initial specific studies on small turbines driven by low temperature solar energy have emphasized low turbine cost, without too much concern for efficiency (refs. 9, 10). More recently, a high efficiency (77% isentropic) unit using organic fluids has been built and operated (ref. 7), and a steam turbine of similar efficiency has been designed (ref. 8).

While organic fluids in Rankine cycles have some advantages over water, such as higher molecular weight and possibly positive slopes of temperature versus entropy at the vapor mixture saturation line, they also may have serious disadvantages, such as instability, toxicity, flammability, low thermal conductivity, corrosiveness, and higher cost. They also have low specific energy and usually cannot be superheated to increase the potential enthalpy drop between prespecified pressure limits. Superheating the relatively low temperature steam emerging from the solar collector by another source of energy, such as a fuel-fired superheater, would increase the cycle efficiency, as proposed by Tabor (ref. 11), Heywood (ref. 12), and in the ETI (ref. 8) concept, with relatively small consumption of fuel. Also, the development of small steam turbines can draw upon the larger amount of experience available with steam turbines, as compared to organic fluid Rankine cycles, which are to some extent still in the exploratory stage. For units below about 15 hp, where steam turbines may become impractical, a rotary expander can be used. It has the advantage of being able to operate at lower speeds than the turbine; thus the reducing gear is eliminated. The same design of the expander can be adapted to the heat pump compressor, operating on direct drive. A high efficiency steam expander of this type has been built and operated successfully in fractional hp size.

During the heating season, the application of a thermal engine to drive the heat pump has the added advantage of being able to utilize its waste heat (jacket cooling and exhaust) to augment the output of the heat pump, either by applying the energy directly to the building or as a heat source to the heat pump evaporator.

The following study concerns itself primarily with energy savings in several systems. Further details appear in ref. 13.

Solar Steam Turbine

The combined heat pump-Rankine cycle system is shown in Fig. 1. The Rankine cycle employs superheat by external fuel, a regenerator, and an economizer. Figure 2 is the enthalpy-entropy Mollier Diagram of the cycle.

The major conclusion of this analysis (Fig. 3) is the fact that the proposed solar Rankine steam cycle with fuel-fired superheat has about twice the efficiency when compared to cycles existing at present and utilizing organic fluids with no superheat. The organic fluid systems used for comparison utilize a reciprocating engine (ref. 5) or a turbine (refs. 6, 7).

Cooling

Solar energy utilization for summer cooling is particularly attractive due to the coincidence of high insolation and ambient temperature with high cooling load. Heat flow in the solar Rankine cycle-heat pump system during the cooling period is shown in Fig. 4.

A noteworthy fact is that the effect of reducing the condensing temperature T_{pc} on the system coefficient of performance for cooling, $(COP)_{sc}$, is much larger than on the η_{pc}. This results from the influence of T_{pc} on both power cycle and heat pump performance. For example, decreasing T_{pc} from 37C to 27C for T = 98C and T_{sup} = 550C increases $(COP)_{sc}$ by 74%. Figure 5 also shows η_{el}, the percentile resource energy saving, as compared to a conventional electric drive,

$$\eta_{el} = \frac{[1/0.3(COP)_c] - [E_f/(E_f + E_s)\eta_{pc}(COP)_c]}{1/0.3(COP)_c}$$

$$= \left[1 - \frac{0.3}{(1 + E_s/E_f)\eta_{pc}}\right] \cdot 100 \qquad (1)$$

where the resource energy efficiency of electrical power at the point of consumption is taken to be 30%. The obtained savings are close to 60%.

To be able to determine the payback period relative to added system cost, a criterion, η_a, which expresses the resource energy savings per square meter of solar collector area, was derived. Collector area has been chosen as the normalizing parameter, due to the fact that it constitutes the major part of the cost of the system

$$\eta_a = \eta_{sc} I \left[\frac{\eta_{pc}}{0.3} - \frac{E_f}{E_s}\left(1 - \frac{\eta_{pc}}{0.3}\right)\right] \qquad (2)$$

Finally, to be able to estimate the collector area needed for a specific application, this area per ton of cooling, A, is computed:

$$A = \frac{3516.8}{(E_f/E_s + 1)\,(COP)_{sc}\eta_{sc} I} \qquad (3)$$

The major results for a double-glazed, flat-black collector and a single-glazed, selective-black collector are presented in Fig. 5. The single-glazed selective-black collector is shown to perform better than the double-glazed flat-black one.

Increasing the temperature of the collector improves the efficiency of the power cycle, but it decreases the efficiency of the solar collector. This behavior is expressed in the maxima of the performance curves shown in Fig. 5. The maxima for the double-glazed collector are for about $85 \leq T_{sc} \leq 100$C and for the single-glazed somewhat higher, $95 \leq T_{sc} \leq 115$C.

Heating

Some of the major schemes for heating, outlined in Fig. 6, are described and compared. For the comparison of resource energy consumption by the various schemes, the heat load was calculated using the degree-day method. The hourly heating load E_1 is

$$E_1 = \frac{E_d(65 - T_a)}{T_{bh} - T_{ad}} \quad \text{(all temperatures in °F)} \qquad (4)$$

where E_d is the hourly design load, T_a the average

ambient temperature during that hour, T_{bh} the temperature maintained in the building by the heating system (21C), and T_{ad} the ambient temperature at the design point. Since the example in the next section considers a building in New York City, $T_{ad} = -10C$.

To perform a preliminary, building-independent analysis, Eq. 4 can be rewritten as

$$E_1 = C_b(65 - T_a) = C_b \, \Delta T \qquad (5)$$

where

$$C_b = \frac{E_d}{(T_{bh} - T_{ad})} \quad \text{and} \quad \Delta T = 65 - T_a$$

(temperature in °F).

1. ELECTRIC RESISTANCE HEATING—A (the capital letter refers to the description in Figs. 6 and 7). The resource energy required, E, is simply

$$E = \frac{C_b \, \Delta T}{0.3} \quad \text{(temperature in °C)} \qquad (6)$$

and the specific, building-independent, resource energy E_H is

$$E_H = \frac{E}{C_b} = \frac{\Delta T}{0.3} \quad \text{(temperature in °C)} \qquad (7)$$

shown as line A in Fig. 7.

2. FUEL-FIRED HEATING—B. The efficiency of the furnace is taken to be 60%.

3. ELECTRIC-DRIVEN HEAT PUMP, WITH AUXILIARY ELECTRIC RESISTANCE HEATERS—C. Up to the design point, the resistance heaters are now in operation; thus

$$E_H = \frac{\Delta T}{0.3(COP)_H} \quad \text{(temperature in °C)} \qquad (8)$$

where $(COP)_H$ is the coefficient of performance of the heat pump, for heating.

For temperatures below the design point, electrical resistance heaters augment the heating. The total E_H curve is shown in line C in Fig. 7. The dashed line C_1 represents the heat pump performance without defrosting or electric resistance heaters.

4. ENTIRELY SOLAR HEATING. The specific solar collector area A_H needed to satisfy the entire heat load is shown in Fig. 7. The collector temperature is 37C, possibly the lowest practicably acceptable.

$$A_H = \frac{A}{C_b} = \frac{\Delta T}{\eta_{sc} I} \quad \text{(temperature in °C)} \qquad (9)$$

where A is the total required collector area.

Heating entirely by solar means usually requires prohibitively large collector areas. The areas calculated here serve, however, as a good basis for comparison and as an upper limit which is of interest in the design of such systems.

5. FUEL-ASSISTED SOLAR HEATING. This is the most popular scheme contemplated at present (refs. 2, 3). For I = 500 W/sq m and $T_{sc} = 37C$, it has been chosen to investigate E_H for collector areas of 5 and 25% from that required at the design temperature of −10C. From Fig. 7:

$$0.05 \, A_{Hd} = 0.00602 \, \frac{m^2}{W/°C}$$

and

$$0.25 \, A_{Hd} = 0.031 \, \frac{m^2}{W/°C}$$

E_H for the 5 and 25% collector areas are shown as lines K and L, respectively, in Fig. 7.

6. ELECTRIC-DRIVEN HEAT PUMP WITH SOLAR HEATING. Here, the building is heated partially by the heat pump and partially by solar collectors. For temperatures below the design point, electrical resistance heaters augment the performance of the heat pump. Curve D shows E_H for 0.05 A_{Hd} and curve H shows E_H for 0.25 A_{Hd}.

7. ELECTRIC-DRIVEN HEAT PUMP WITH SOLAR ASSISTANCE TO EVAPORATOR—F. Applying solar heat to the heat pump evaporator has two principal advantages: it makes lower temperature solar collector operation possible, thus improving its efficiency, and it improves performance of the heat pump by increasing its $(COP)_H$ at low ambient temperatures, by eliminating frosting of the evaporator (when the temperature is maintained above freezing) and by stabilizing its operation through the maintenance of a constant evaporator temperature. While the analysis shows that in general the application of solar heat directly to the building is more efficient, it is possible that the solar heated fluid be switched to the heat pump evaporator when ambient conditions do not provide for efficient direct solar heating.

E_H is calculated with $(COP)_H = (COP)_{Hd} = 3.209$ for $T_a = 12C$. It is shown as curve F in Fig. 7.

The specific collector area required in this case can be easily seen to be

$$A_H = \frac{\Delta T[1 - 1/(COP)_H]}{I \eta_{sc}} \quad \text{(temperature in °C)} \qquad (10)$$

approximately half of that required for direct solar heating under above conditions.

8. FUEL-FIRED RANKINE-CYCLE DRIVEN HEAT PUMP—E. In this scheme, the Rankine cycle condenser discharges heat to the building, thus adding to the energy delivered by the heat pump. The exhaust gases are used to defrost the heat pump evaporator. The Rankine cycle used is the one described earlier in this paper, operating at T_3 = 98C and T_{sup} = 550C, with η_T = 75%, and resulting η_{pc} = 18.5%. The combustion efficiency of the Rankine cycle boiler is taken as η_c = 70%.

Below the design temperature, the system is augmented by a fuel-fired furnace. The calculated performance is displayed as curve E in Fig. 7.

9. FUEL-FIRED RANKINE-CYCLE DRIVEN HEAT PUMP WITH SOLAR HEATING. The same Rankine cycle proposed earlier and analyzed for solar cooling can be used for heating. In most geographic locations it would, however, be more effective to use the existing solar collectors to heat the building directly, assisting their performance with the fuel-fired Rankine-cycle driven heat pump. Below the design point, the system is augmented by a fuel-fired furnace. Curve G shows E_H for 0.05 A_{Hd} and curve J shows E_H for 0.25 A_{Hd}.

A Comparative Example of Resource Energy Consumption

The comparative analysis of resource energy consumption is performed on a high-rise office building in New York City. The average insolation during daytime is taken to be 500 W/sq m. Since hour-by-hour calculations are not performed, the effect of energy storage is not considered.

The office building has 30 stories, floor area/story (= roof area) of 1860 sq m, height/story of 3.7 m, wall area of 20,100 sq m and 30% window area. C_b for the occupied period, C_{bo}, was calculated to be 94.2 kw/°C and for the period when it is vacant, C_{bv} = 70.3 kw/°C.

The monthly and annual results for the heating period are summarized in Table 1. The annual results are also displayed in Fig. 8. It can be seen that significant savings can be obtained by the application of solar energy and by the utilization of heat pumps and Rankine cycle drives.

A simple approximation can be made of the compatibility of solar collector area requirements for cooling and heating. The design heating load of this building is 3 MW. Following Eq. 1, the cooling loading for the same building in the New York area would be about ⅔ of the heating load, i.e., 2 MW. The collector area required for cooling is calculated to be about 5600 sq m, slightly larger than the area of one side of the building. Together with the available roof area, the building itself would be sufficient for mounting the collectors. Compared to the area needed for heating, up to about 0.5 A_{Hd} would thus be available for solar energy utilization.

Conclusions

A Rankine cycle utilizing solar energy as the primary source and superheated by an auxiliary fuel source has efficiencies which are about double that of comparable Rankine cycles without auxiliary superheat. The cycle is well suited for driving vapor compression heat pumps for heating and cooling.

TABLE 1. MONTHLY RESOURCE ENERGY CONSUMPTION FOR HEATING, IN 10^2 MW-HRS 30 STORY OFFICE BUILDING IN NEW YORK

Heating System	Gas	Gas + 5% Solar	Heat Pump	Rankine Cycle Heat Pump	Heat Pump + 5% Solar	Solar Assisted Heat Pump	Rankine Cycle Heat Pump + 5% Solar	Gas + 25% Solar	Heat Pump + 25% Solar	Rankine Cycle Heat Pump + 25% Solar
Fig. 7 Curve:	B	K	C	E	D	F	G	L	H	K
Sep.	0.52	0.33	0.30	0.30	0.20	0.30	0.20	0.02	0.01	0.01
Oct.	3.02	2.29	1.87	1.81	1.45	1.81	1.38	0.41	0.32	0.25
Nov.	7.56	6.47	5.09	4.68	4.37	4.67	3.93	2.42	1.91	1.46
Dec.	13.29	12.13	9.54	8.36	8.67	8.29	7.44	6.83	5.36	4.17
Jan.	15.67	14.52	11.44	9.88	10.55	9.77	8.94	9.00	7.06	5.53
Feb.	12.52	11.46	9.03	7.89	8.22	7.83	7.04	6.52	5.14	3.99
Mar.	11.62	10.44	8.20	7.29	7.32	7.26	6.38	5.18	4.10	3.14
Apr.	5.55	4.59	3.59	3.41	2.99	3.41	2.77	1.18	0.92	0.70
May	1.90	1.35	1.13	1.11	0.82	1.11	8.07	0.12	0.09	0.07
Annual	71.66	63.57	50.19	44.73	44.59	44.44	38.88	31.68	24.90	19.33

In cooling it offers a 50–60% resource energy saving as compared to a conventional electric-driven system.

As compared with conventional furnace heating, total resource energy consumption can be reduced threefold to fourfold by employing partial solar heating assisted by a furnace, by a heat pump, or, best of all, by the Rankine-cycle driven heat pump.

ACKNOWLEDGMENT

This study was partially supported by a grant from NSF-RANN to the National Center for Energy Management and Power, University of Pennsylvania.

NOMENCLATURE

(Including values of some of the parameters used in this study)

A = solar collector area, in sq m, needed to provide one ton of cooling
A_H = specific solar collector area for heating (Eq. 9)
$(COP)_c$ = heat pump coefficient of performance for cooling
$(COP)_H$ = heat pump coefficient of performance for heating
$(COP)_{sc}$ = system coefficient of performance for cooling
$(COP)_{sc,tot}$ = overall system coefficient of performance for cooling, including solar collector efficiency
E = resource energy, evaluated before conversion
E_{el} = electrical energy
E_f = fuel (combustion) energy (= $h_5 - h_4$ for superheater)
E_H = E/C_b
E_s = solar energy supplied to Rankine cycle (= $h_3 - h_2$)
E_w = shaft work delivered by turbine
h = enthalpy
I = insolation, W/sq m
T_a = ambient temperature
T_{pc} = temperature of fluid in power-cycle condenser
T_{sc} = temperature of fluid in solar collector
T_{sup} = degrees of superheat
α = absorptivity of solar collector absorber (= 0.95)
ϵ = emissivity of solar collector absorber (= 0.95 for flat-black, = 0.1 for selective-black)
η_e = fuel ratio $= \dfrac{E_f}{E_f + E_s}$
η_e^* = practical fuel ratio $= \dfrac{E_f/\eta_{sup}}{E_f/\eta_{sup} + E_s}$
η_{ec} = efficiency of economizer (= 0.85)
η_{el} = Eq. 1
η_m = mechanical efficiency of turbine (= 0.98)
η_p = efficiency of pump in Rankine cycle (= 0.8)
η_{pc} = Rankine cycle efficiency

$$= \frac{(h_5 - h_6)\eta_m}{h_5 - h_4 + (h_1 - h_{10}) + (h_3 - h_2)}$$

η_R = efficiency of regenerator (= 0.85)
η_{sc} = solar collector efficiency
η_{sup} = efficiency of superheater (= 0.85)
η_T = isentropic efficiency of turbine

Subscript $_d$ refers to load design conditions

REFERENCES

1. Jordan, R.C., and Threlkeld, J.L., *Heating, Piping and Air Conditioning, 26*, 2, 122-30 (February 1954).
2. Tybout, R.A., and Löf, G.O.G., *Natural Re-*

sources Journal, 10, 268-326 (1970).
3. Lorsch, H.G., *Energy Conversion, 13*, 1-5 (1973).
4. Lior, N., and Saunders, A.P., Report No. NSF/RANN/SE/GI27976/TR731/1, University of Pennsylvania, Philadelphia (August 1974).
5. Sargent, S.L., and Teagan, W.P., Paper No. 73-WA/Sol-8, ASME Winter Annual Meeting, Detroit, 11-15 November 1973.
6. Tabor, H., and Bronicki, L.Y., Paper No. 35/5/S4, U.N. Conf. on New Sources of Energy, Rome, 1961.
7. Bronicki, L.Y., Paper No. 729057, 7th IECEC, San Diego, Calif., September 1972, pp. 327-34.
8. ETI (Energy Technology, Inc.), Cleveland, O., private communication, 1973.
9. Leo, B.S., and Hsu, S.T., *Solar Energy, 4*, 2, 16-20 (1960).
10. d'Amelio, C., *Solar Energy, 7*, 3, 138-43 (1963).
11. Tabor, H., *Bull. Res. Council of Israel, 5C*, 5-27 (1955).
12. Heywood, Harold, in *Solar Energy Research*, Daniels, F., and Duffie, J.A., eds., University of Wisconsin Press, Madison, 1955, pp. 227-35.
13. Lior, N., 9th IECEC, San Francisco (also University of Pennsylvania, Department of Mechanical Engineering and Applied Mechanics Report No. MEAM 74-4), 1974.

DISCUSSION

Q Your concept includes several key elements which may be difficult or impossible to achieve. These are as follows:

1. To achieve high COP, you are suggesting using maximum system steam temperatures approaching or exceeding those used in today's large electric power generation systems of 100–1000 MW electric output.
2. You require substantial heat exchange, in regenerations, economizers, and generally at very low pressures (60–100 psia). Regeneration, you suggest, is much greater than used in large steam stations.
3. In small sizes (3–1000 hp) it will be extremely difficult to achieve high performance in a single-stage steam turbine.
4. Pressure at turbine exhaust and condenser will be very low—requiring hard vacuum.
5. System costs will be high due to above complexities.
6. Organic fluids give you high performance at temperatures about 300–350F without any of above complexities.

A 1. We do not foresee any special difficulties due to superheated steam temperature. The proposed operation is at slightly subatmospheric pressure, and the temperatures are comparable or less than these attained in residential combustion heating and cooking appliances utilized in a routine manner.

3. While high-efficiency small-size steam turbines are not common, our initial design indicates that a 15–20 hp turbine with an isentropic efficiency of 75% is feasible. Below 15 hp we are proposing a rotary steam expander of novel design and similar efficiency, which has been designed, built, and successively tested in a small size.

4. 1 in. Hg abs. is not *hard* vacuum and we do not foresee any difficulties in its maintenance.

6. Steam has some distinct advantages over organic fluids, as elaborated in the paper.

Comments 2 and 5 are best answered by the recent Hittman Associates report (HIT-600, NSF-RA-N-75-012, by H.M. Curran et al., April 1975) which analyzes the economics of various solar cooling concepts and finds this system clearly superior to the others, a conclusion which is in agreement with our preliminary design and with ETI's design. The fact that the efficiency of the cycle is about double that of present organic fluid cycles reduces the area of needed solar collectors, and thus the cost.

Q What are the details of the cold storage room you suggest? Does it employ an icebank?

A The cold storage utilized in our off-peak air conditioning experiments consisted at first of sodium sulfate decahydrate phase change salt contained in elongated plastic tubes. Due to the instability of the salt, storage was changed to utilize water in a tank, working between 35 and 53F.

Q Did you try concentrated solar heat for high pressure steam?

A No. We felt this would not be economical for heat pumps utilized in buildings.

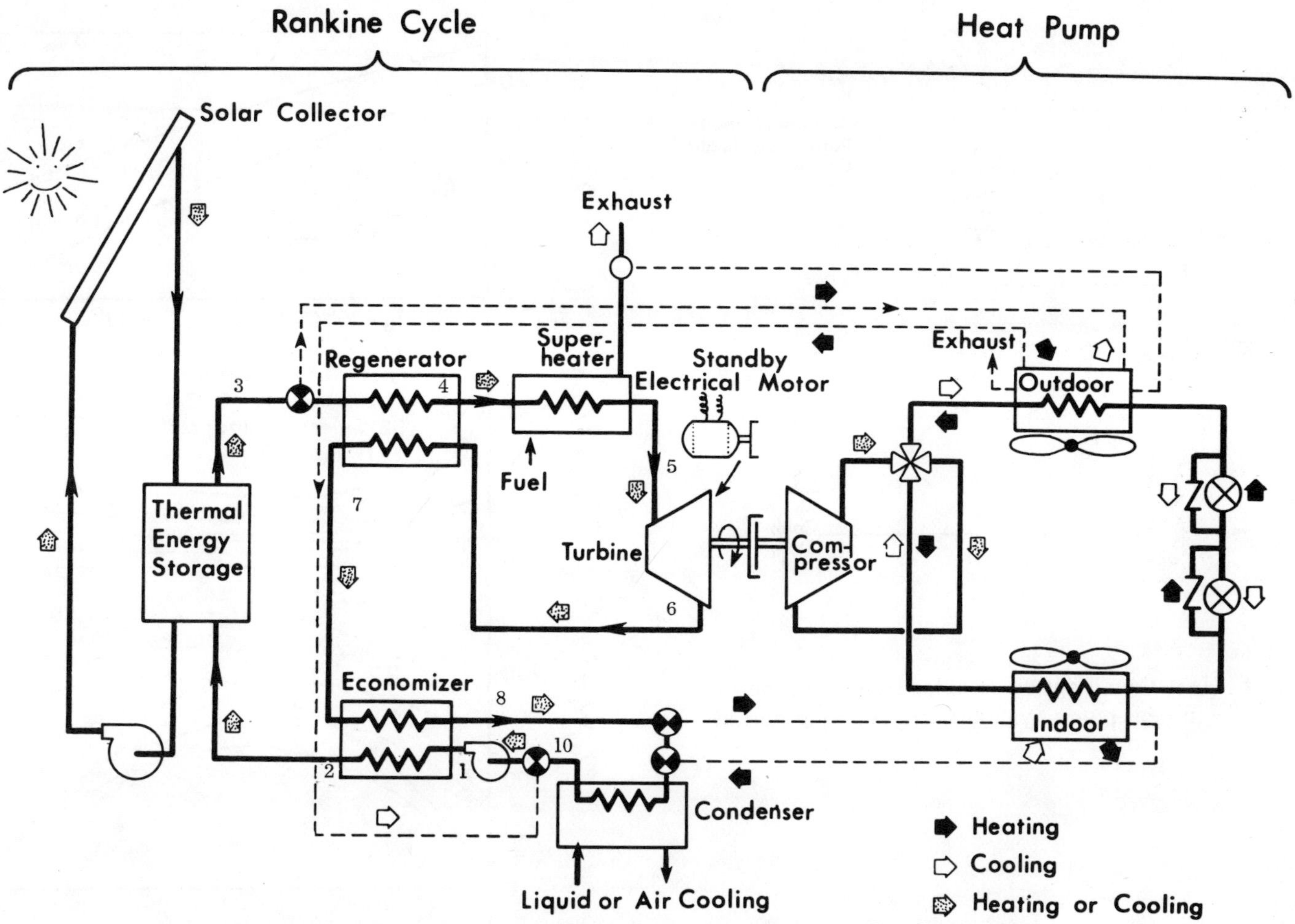

Fig. 1. Solar steam turbine driven heat pump: system diagram.

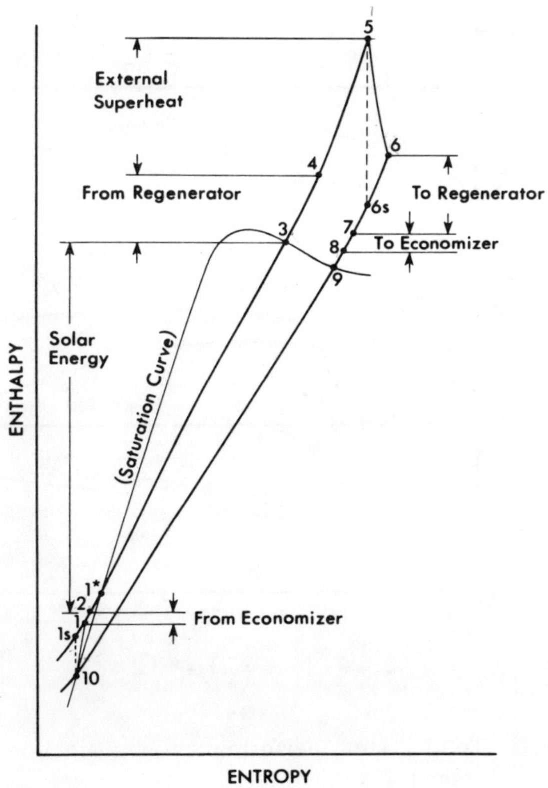

Fig. 2. Solar Rankine cycle Mollier diagram.

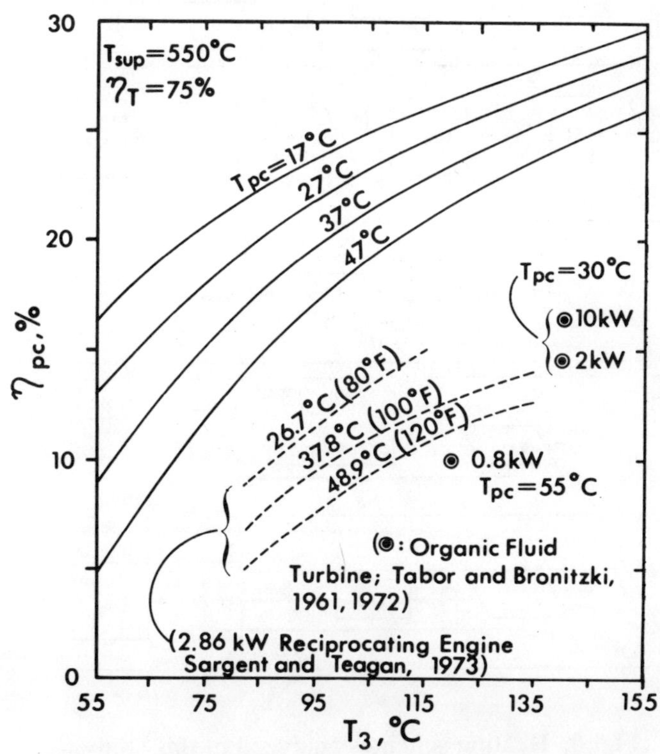

Fig. 3. Rankine cycle efficiency η_{pc} as a function of solar collector temperature, compared to other existing Rankine cycle systems.

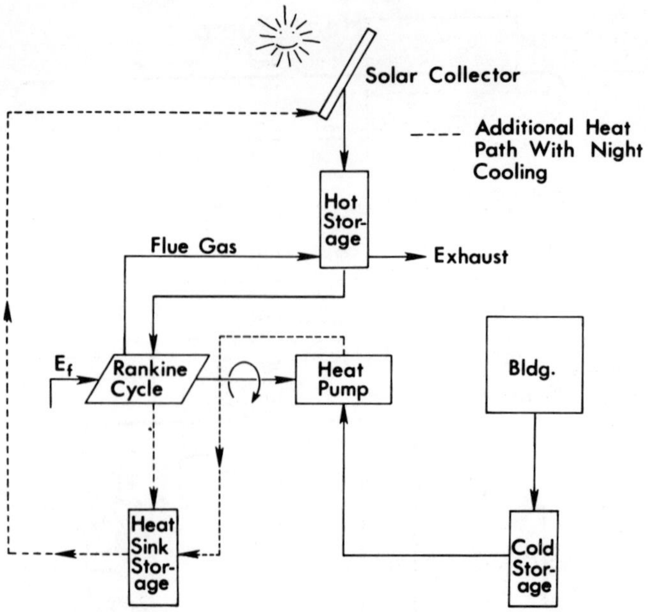

Fig. 4. Heat flow diagram of solar steam turbine driven heat pump system during cooling period.

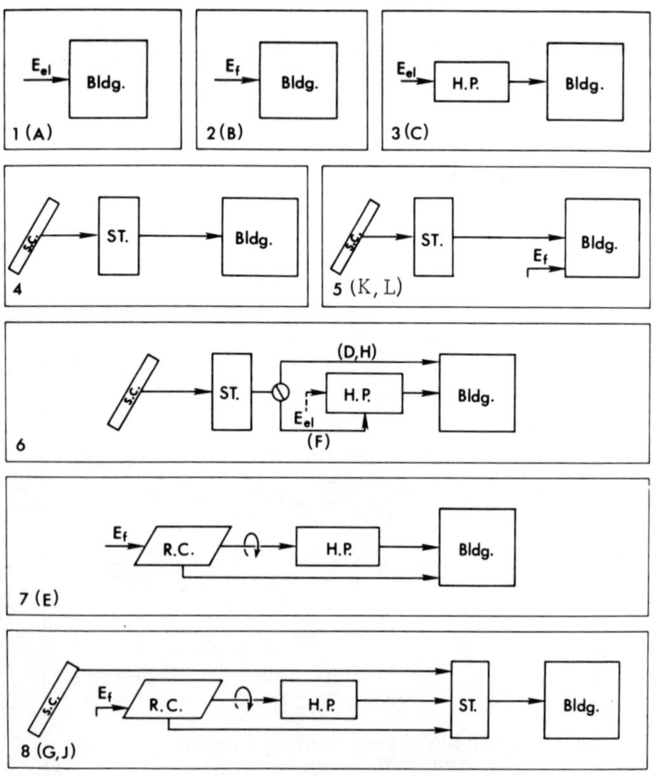

Fig. 6. Heating schemes analyzed in this study.

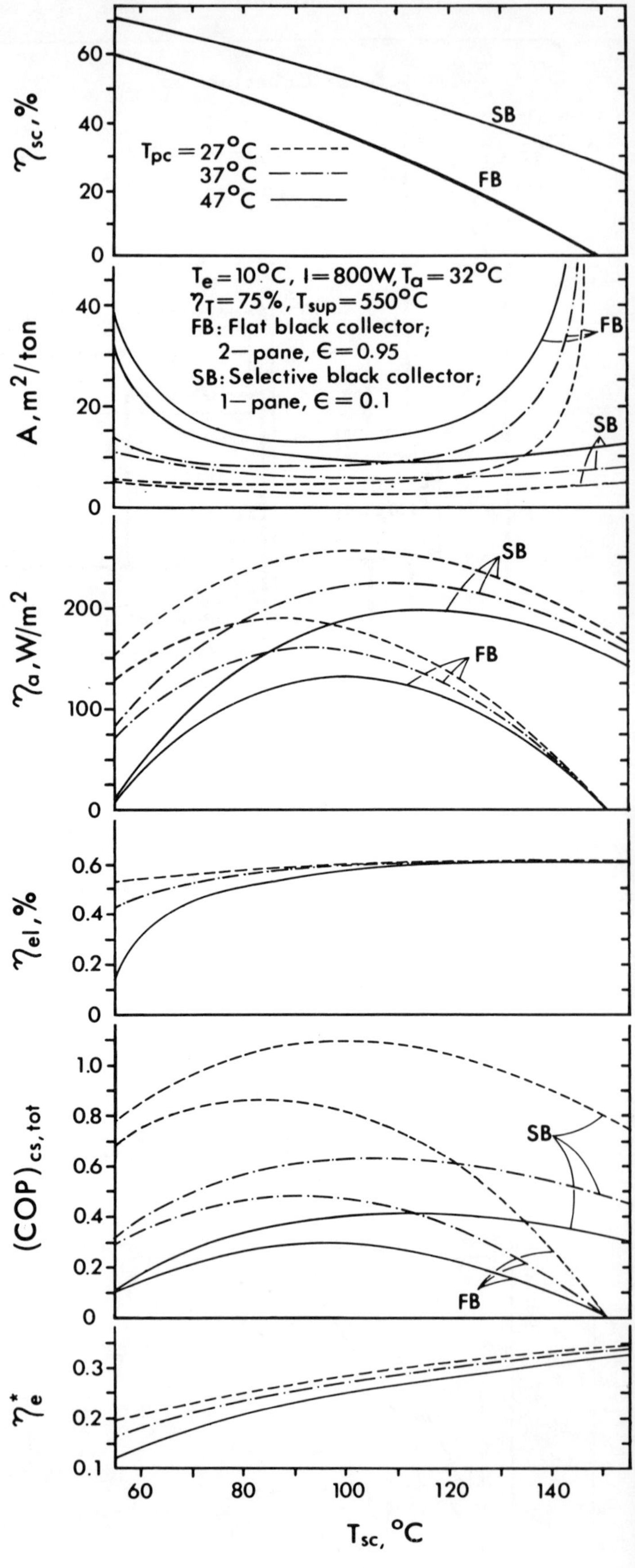

Fig. 5. Total system performance as a function of solar collector temperature.

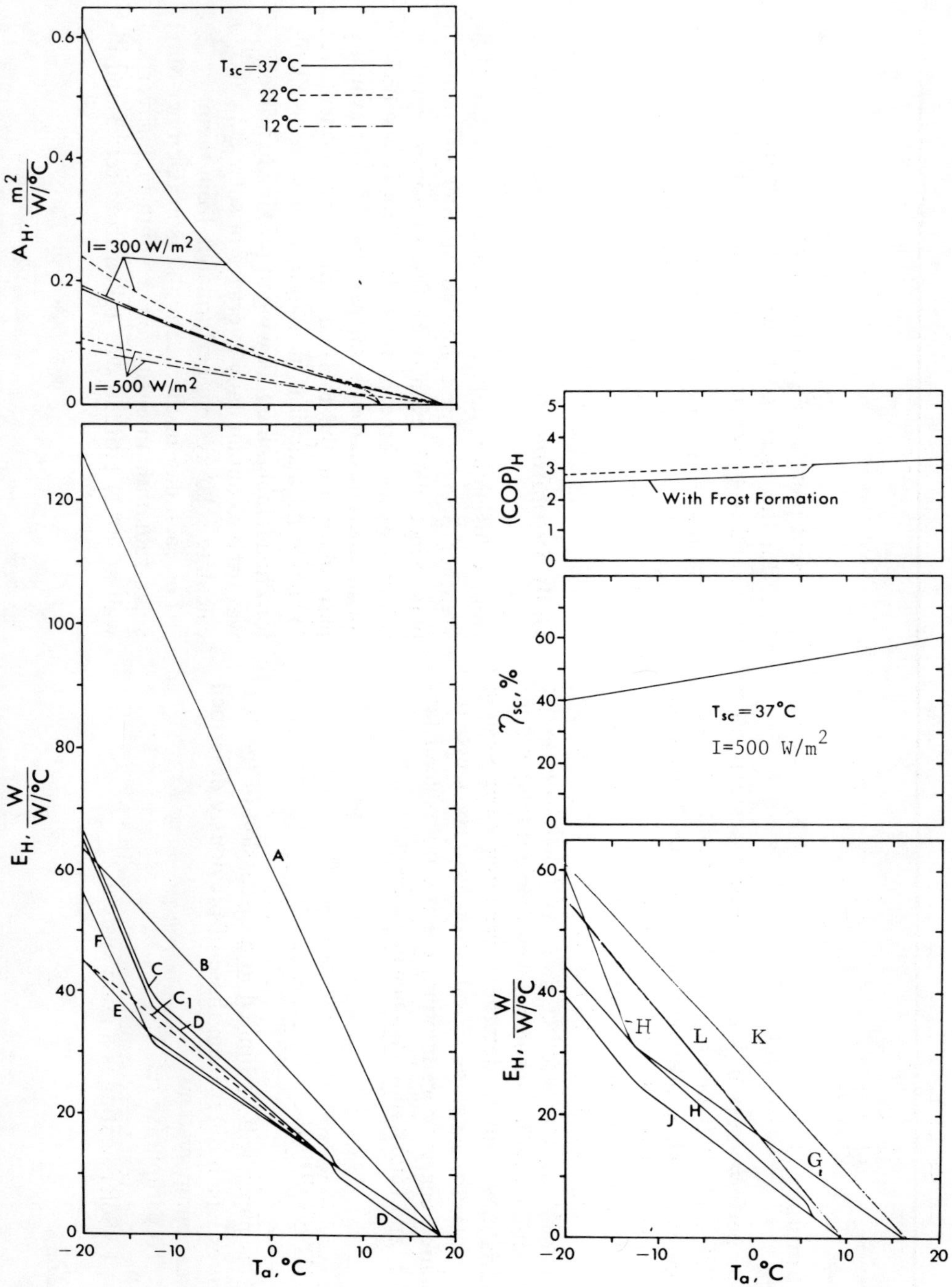

Fig. 7. Total system performance for heating. The solar collector is two-pane, flat-black.

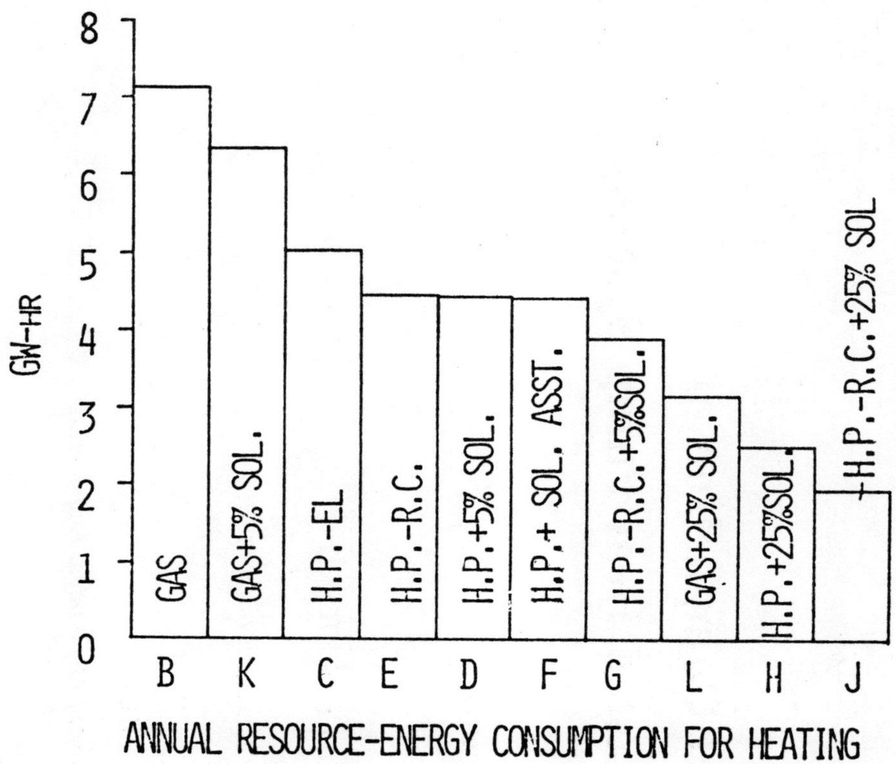

Fig. 8. Annual resource energy consumption for heating a 30-story office building, New York.

Analytical Comparison of Solar Assisted Heat Pump Systems

T.A.V. Cassel

Scientific Development
Bechtel Corporation
San Francisco, Calif.

Dr. Harold G. Lorsch

Principal Engineer
Franklin Institute Research Laboratories
Philadelphia, Pa.

Dr. Noam Lior

Assistant Professor of Mechanical Engineering
University of Pennsylvania
Philadelphia, Pa.

Summary

The performances of six space heating systems for a 1500 sq ft Philadelphia one-family residence were compared. They are: electric resistance heat, combustion furnace, direct solar heating with auxiliary combustion furnace, conventional air-to-air heat pump, and two different solar assisted heat pump systems. Measured weather data were used to reconstruct day-by-day insolation and ambient temperature variations, and to determine resulting solar system performance values. Evaluation criteria were resource energy consumption, electric utility load factor, and return on consumer investment.

Introduction

This paper will be limited to a discussion of the features which differ from those of previously described heat pump systems.

To paraphrase ref. 1:

1. Solar heating is a good system.
2. Heat pumps are good systems.
3. But the marriage between them must be carefully handled to prevent a catastrophic divorce.

This is borne out by the results of a comparative analysis of six space heating systems for a single-family residence in Philadelphia.

Daily Insolation

Hourly insolation data are available from the U.S. Weather Service for a limited number of locations only. Since 1965, even those few locations report insolation every three hours only. This is insufficient for detailed design of solar assisted heating systems, particularly for the thermal storage component (ref. 2). Since no historical insolation data were available for Philadelphia, the following approximate procedure was used to obtain daily averages of solar data and to calculate solar system performance from them.

For each day in the heating season (October-May) solar radiation intensity on a south-facing vertical wall at 40° N latitude was taken from ref. 3 and divided by the possible hours of sunshine for that day during the 1972–73 heating season, reported by NOAA (ref. 4), to obtain the mean hourly insolation. Solar collector efficiency was then computed for each day of the heating season as a function of mean daily values for insolation, ambient temperature (also obtained from ref. 4), angle of incidence of the sun from the collector normal (taken as the altitude angle at solar noon by interpolation from ref. 3), and collector absorber temperature: 75F for heat pump operation and 130F for direct solar heating. A solar collector performance program by Ostad-Hosseini (ref. 5) revised and rewritten for APL (an interactive computer language) was used to compute mean daily collector efficiencies. The daily amount of solar energy available per square foot of collector is then obtained from the relation

$$Q_c = I_0 \times \eta_c \times t_s$$

where I_0 = mean hourly insolation on a vertical, south-facing surface

η_c = collector efficiency

t_s = actual hours of sunshine per day (ref. 4)

Residence and Heating Systems

A 1500 sq ft single-family residence located in Philadelphia at 40° north latitude was chosen as being representative of current construction in the northeastern United States. Its design heat loss and heat gain were determined by the ASHRAE Calculated Heat Loss Method (ref. 6), resulting in a design heating load of 53,000 Btuh and a design cooling load of 32,000 Btuh. In accordance with the usual design practice in that area, a 3-ton heat pump was therefore chosen.

The performances of six different heating systems for this residence were calculated each day and compared for an eight-month heating season from October 1, 1972, through May 31, 1973. These systems were defined as follows:

1. Electric resistance heating, 16 kw capacity.
2. Combustion furnace, 100,000 Btuh input, 60% seasonal efficiency assumed constant throughout.
3. Direct solar heating: vertical, south-facing solar collectors, two glass covers, flat-black absorber ($\alpha = \epsilon = 0.95$), collector areas of 375, 500, and 750 sq ft. Thermal energy storage capacities of 0, 100, 250, and 500 thousand Btu. Auxiliary heat: combustion furnace, 60% efficiency.
4. Conventional air-to-air heat pump, 3-ton. Performance data from ref. 7.
5. Solar assisted heat pump, System A (see Fig. 1).
6. Solar assisted heat pump, System B (see Fig. 2).

Operation of Systems

The conventional systems (1, 2, and 4 above) were operated in their conventional modes based on instantaneous demands. The direct solar heating system (No. 3) was operated, respectively analyzed, as follows:

The daily heat load (ref. 6) is multiplied by the fraction of the day during which the sun shines (ref. 4) to obtain the average heat load which can be supplied directly from the solar collector to the residence. Solar heat collected in excess of this amount is transferred from the collector to thermal energy storage until the storage device capacity is reached. Any solar heat exceeding storage capacity is dumped. Next, the remaining heat load of the building is supplied from the thermal energy storage device until it is depleted or until the heat load is satisfied. If the device is depleted prior to satisfying the heat load, supplementary heat is provided by a 60% efficient auxiliary combustion furnace.

This is an approximate procedure resulting in somewhat conservative results. It was compared with a more accurate, hour-by-hour analysis (ref. 2) for a Washington, D.C., townhouse (hourly insolation data are available for that location), and a difference of 5±% in annual performance was obtained which appears satisfactory.

System No. 5 (Solar Assisted Heat Pump A, Fig. 1) operates as a conventional air-to-air heat pump when the ambient temperature is above 45F. For lower ambient temperatures at which the heat pump COP is severely decreased, the refrigerant circuit receives energy from an evaporator coil inside the thermal storage unit. The temperature of that unit is maintained above 60F, either by solar energy from the 375 sq ft vertical, south-facing solar collector previously described or by 60% efficient auxiliary combustion heat. Supplementary electric resistance heating in the supply air duct is also provided to satisfy conditions of high heating load when the heat pump capacity is exceeded. Thermal storage of 200,000 Btu is effected through the latent heat of fusion of C_{16} paraffin wax which has the following physical properties (ref. 8):

Melting point	58–65F
Density	48 lb/cu ft
Latent heat	86 Btu/lb = 4190 Btu/cu ft

System No. 6 (Solar Assisted Heat Pump B, Fig. 2) permits direct solar heating from the 600 sq ft collector or from the 10,000 lb water/ethylene glycol thermal storage unit provided the storage temperature is above 105F. Below that storage temperature, the heat pump is used to "boost" solar heat from storage to a temperature level sufficient to heat the living space. If the storage temperature drops below the ambient temperature, refrigerant evaporation is switched to a conventional ambient air evaporator coil. Supplementary heating is provided by electric resistance heat.

For all solar systems, the change in collection efficiency with insolation, ambient temperature, and collector temperature was calculated for every day of the heating season. For System 6, the effect of a change in storage temperature was similarly taken into account. It should be pointed out, however, that all calculations were performed on daily average values because hourly insolation data are not available for Philadelphia. In this sense, the calculations are only approximate.

Results

An annual performance comparison of Systems 2 through 5 is given in Fig. 3. The performance is evaluated in terms of resource energy consumption, either at the residence or at the power plant. A conversion/transmission efficiency of 31% was assumed for electrical energy from fuel to the circuit breaker at the residence. Electric resistance heating was taken to be 100% efficient, and combustion furnace operation was assumed to be 60% efficient. The energy consumption of internal fans and pumps was not included.

The authors firmly believe that resource energy is the proper basis of comparison for fundamentally different heating systems; no other single measure appears to be adequate. Coefficient of performance (COP) is suitable for comparing air-source or water-source heat pumps to each other, percentage of heating energy provided by solar collection is suitable for comparing direct heating systems, but neither is satisfactory for use with solar assisted heat pumps. A high-COP solar heat pump requires a larger solar collector or more resistance heating than a low-COP unit, and it may actually consume more purchased energy than a low-COP unit. This is due to the fact that COP is the ratio of heat delivered to space divided by electrical energy input, no account being taken of the air-source or water-source energy because it is supposedly "free." When it is supplied by a solar collector, however, it is apparent that this energy is not free but must be paid for through high initial investment. It is therefore recommended (in accordance with the current draft version of the proposed ASHRAE Standard 90, Energy Conservation in New Building Design) that resource energy be used as a measure of performance for different heating and cooling systems.

The ordinate of Fig. 3 represents total resource energy consumption, the abscissa represents thermal storage, and solar collector size is included as a parameter. The figure shows that solar assistance reduces the resource energy consumption of a conventional heat pump, but the solar assisted heat pump is less energy-conserving than a direct solar heating system except for extremely large storage sizes. It was this somewhat disappointing behavior of heat pump A in System 5 which led to the design of heat pump B in System 6. The annual electricity and oil or gas energy consumption rates for all systems are shown in Table 1. Based on an economic analysis of the data from Fig. 3, the optimum collector and storage sizes shown in the table were used. A graphical comparison of annual resource energy consumption of all six systems investigated is presented in Fig. 4.

Discussion of Results

System No. 5 (Solar Assisted Heat Pump A) consumes 64% less resource energy than electric resistance heating, 30% less than the combustion furnace, and 18% less than the conventional air-to-air heat pump. However, it consumes 15% more than the direct solar heating system No. 3. This behavior is believed to be caused by (*a*) the depletion of thermal storage from heat pump operation, discussed below, and (*b*) the inability of this system to utilize the direct solar heating mode on warm, sunny spring and fall days. Even on those days the heat pump must be operated in System No. 5.

In System No. 6 (Solar Assisted Heat Pump B) the last-mentioned drawback is eliminated, and a direct solar heating mode is possible. As a result the system consumes 71% less resource energy than electric resistance heating, 44% less than the combustion furnace, 34% less than the conventional heat pump, and 15% less than direct solar heating. This relatively small improvement over the direct heating mode can be explained by considering the behavior of the thermal storage unit. If heat pump operation draws the storage temperature down during the night it requires a considerable amount of solar energy to raise that temperature above 105F during the following day before direct heating (the most economical mode) can again occur. In the direct solar heating system, on the other hand, this energy can be used to heat the building during the following day.

In order to examine this effect in more detail, the month of January 1973 was re-run with a 20,000 lb water storage, double that of System No. 6. The result was a 1% drop in direct solar heat supplied and an increase in resource energy consumed. As a consequence it is concluded that solar assisted heat pumps should have two thermal storage units: one for direct heating which is never depleted below 105F, the other for heat pump operation which is allowed to vary over a larger temperature range.

		1	2	3	4	5	6
	Heating System	Electric Resistance	Combustion Furnace	Direct Solar Heating (see note below)	Conventional Heat Pump	Solar Heat Pump A (Fig. 1)	Solar Heat Pump B (Fig.2)
ANNUAL ENERGY CONSUMED	Electrical Energy (kWh)	29,600	-	-	13,000	9,080	8,430
	Fossil Fuel Energy (10^6 Btu)	-	168	110	-	30	- -
	Total Resource Energy (10^6 Btu)	326	168	110	143	130	94
	Electrical Load Factor (Monthly Basis)	0.62	-	-	0.51	0.64	0.57

Note: Values given are for a direct solar heating system with 600 sq ft of collector area and 200,000 Btu of thermal storage. This was considered to be the most cost-effective system for Philadelphia conditions.

Consumer Point of View

With the increasing shortage of gas and oil, combustion furnace heating becomes less attractive and electrical heating modes become more prevalent. Electric resistance heating requires the least capital investment, but has the highest operating cost and resource energy consumption. The additional investment required to go from electric resistance heating to a conventional heat pump can be amortized in approximately five years through savings in electrical heating energy alone; if the cost of an air conditioner is added to System No. 1 in order to compare two systems, both of which have heating and cooling capacities, the break-even point is achieved even sooner.

The additional investment in a solar augmentation system for a conventional heat pump, however, produces an annual saving (return on investment) of only 1.5% for an electricity price of 4¢/kwh. This by itself is not attractive to a homeowner. Either better solar heat pump systems must be designed, or the government must provide financial incentives to reward the homeowner for the reduction in resource energy consumed by solar assistance, or both.

Electric Utility Point of View

The decreased heating capacity of a conventional heat pump during periods of low ambient temperature, when the heating load is large, causes greater proportions of the load to be met by supplemental electric resistance heaters than in the early and late months of the heating season. Figure 5 illustrates this sharply peaking electricity demand during January and February for the conventional heat pump during the 1972–73 heating season. The solar assisted heat pump A (System No. 5) demonstrates a smoother electricity demand curve. Its monthly load factor is 0.64 versus 0.51 for the conventional heat pump and 0.62 for electric resistance heat. The electric load factor for heat pump B (System No. 6) is 0.57, which is 10% better than that of the conventional heat pump (No. 4).

It should be noted that these values are monthly averages which are not directly comparable to the instantaneous (15-min or 30-min) load peak used by electric utilities to calculate load factors and demand charges. But the numbers indicate that the solar assisted heat pump does not automatically have a lower load factor than electric resistance heating—a fear frequently expressed during impact assessment of solar heating on utilities.

Conclusions

1. The addition of solar augmentation to conventional air-to-air heat pumps does not always improve

overall system performance.

2. Care should be taken in matching the two subsystems (heat pump and solar system) to each other and to the heating demands of the building.

3. Resource energy consumption is the proper way of comparing the performance of solar assisted heat pumps to each other and to other heating systems; neither heat pump COP nor percentage of heating supplied by solar is adequate.

4. No solar augmented heat pump system has been designed yet which takes full advantage of the best features of both subsystems.

5. It appears that an optimum marriage between the two subsystems would have to include the following features: a direct heating mode; separate thermal storage systems for direct heating and for heat pump heat source; and a method of using thermal storage to take advantage of off-peak electric power.

6. With current electricity prices and under the present state of development of solar assisted heat pumps, the return on the homeowner's marginal investment of solar agumentation added to a conventional heat pump is approximately 1.5%. Either better designs or government incentives or both are required for the large-scale introduction of solar assisted heat pumps.

ACKNOWLEDGMENT

This paper is primarily based on work performed at the National Center for Energy Management and Power of the University of Pennsylvania under NSF-RANN Grant GI-29729. Additional aspects of this work are scheduled for presentation at the 10th IECEC in August 1975 (ref. 9).

REFERENCES

1. Weinstein, A., "Heat Pumps are Good; Solar-Assisted Heat Pumps are Better." Presented at Solar Heat Pump Workshop, Pennsylvania State University, June 1975.
2. Lorsch, H.G., and Spielvogel, L.G., "Performance Analysis of Solar Augmented Space Heating Systems," *Trans. ASHRAE*, *81*, part 1.
3. Lorsch, H.G., and Niyogi, B., "Influence of Azimuthal Orientation on Collectible Energy in Vertical Solar Collector Building Walls," Report No. NSF/RANN/SE/GI27976/TR72/18, University of Pennsylvania, August 1971.
4. U.S. Dept. of Commerce, *Local Climatological Data, October 1972-May 1973, Philadelphia*, National Oceanic and Atmospheric Administration, Environmental Data Service.
5. Altman, M., Yeh, H., and Lorsch, H.G., "Conservation & Better Utilization of Electric Power by Means of Thermal Energy Storage and Solar Heating," Final Summary Report, No. NSF/RANN/SE/GI27976/PR73/5, University of Pennsylvania, July 1973, sect. 4.4.
6. *ASHRAE Handbook and Product Directory—1973 Systems*, American Society of Heating, Refrigeration and Air Conditioning Engineers, New York, 1973, ch. 11.
7. Carrier Corporation, *Single-Package Heat Pumps-50DQ Series*, Form 50DQ-4P, 1971.
8. Kauffman, K., and Pan, Y., "Congruently Melting Materials for Thermal Energy Storage in Air Conditioning," Report No. NSF/RANN/SE/GI27976/TR73/5, University of Pennsylvania, June 1973.
9. Cassel, T.A.V., Lorsch, H.G., and Lior, N., "Solar Heat Pump Comfort Heating Systems," 10th IECEC, University of Delaware, Newark, August 1975.

DISCUSSION

Q Why apply both the hours of sunshine and the percentage available times insolation? Either should suffice.

A The ratio of actual hours of sunshine to total possible hours of sunshine is not necessarily the same as the ratio of actual to available insolation. For a given number of sunshine hours around noon the actual insolation is higher than for the same number of sunshine hours in the early morning.

Q Resource energy requirement is not as good a common denominator as you implied. A gas or oil furnace needs very high quality and scarce fuel. Power plants don't.

A That is correct.

Q A well-insulated house has a design heat loss of 25,000 Btuh, only half of your assumption. Please comment.

A The calculations were performed for a house which is typical of current construction, not for a specially designed and constructed experimental energy conservation house.

Q A lot of useful data are available on water-to-air heat pumps from about eight manufacturers. Why do you have to fudge?

A To simplify the calculations, which were performed on a computer with a severely limited memory.

Q Did your study take into consideration the increased solar collection efficiency for heat pump due to the lower temperature fluid in the collector, compared to the higher temperature for solar heating with furnace supplementary heat? Also the ability of a low-temperature collector to collect diffuse radiation during cloudy periods?

A Definitely yes.

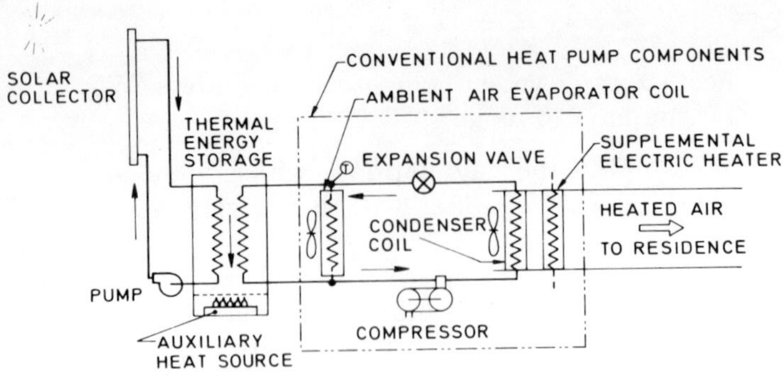

Fig. 1. System No. 5, Solar Assisted Heat Pump A.

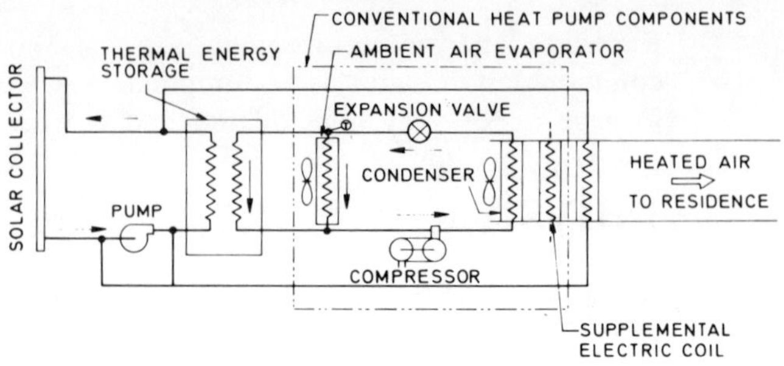

Fig. 2. System No. 6, Solar Assisted Heat Pump B.

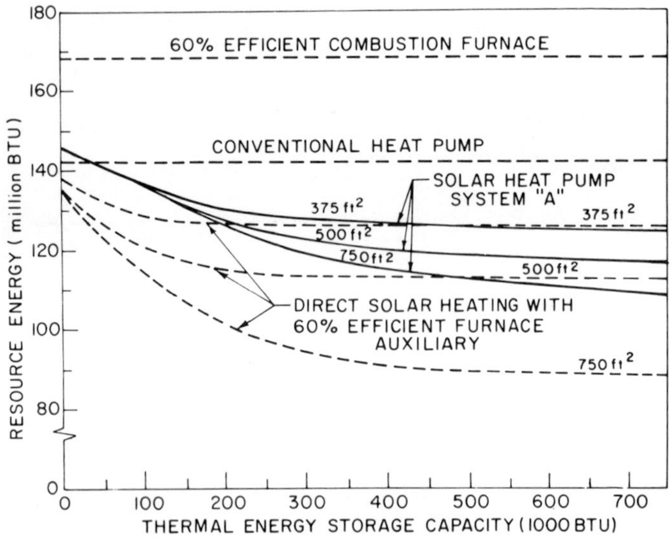

Fig. 3. Variation of resource energy consumption with collector area and thermal storage capacity.

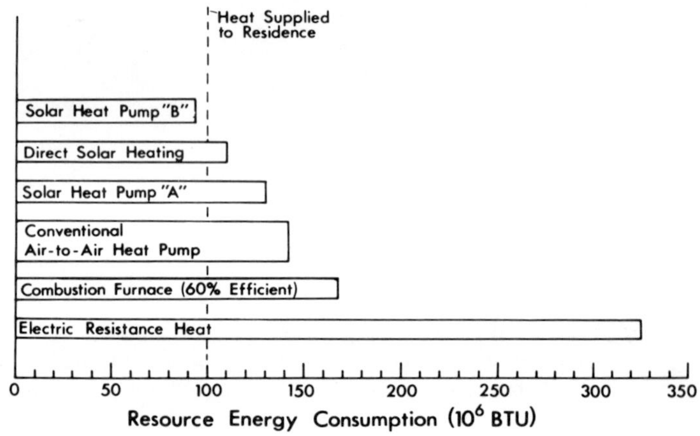

Fig. 4. Annual resource energy consumption for different space heating systems in 1500 sq ft Philadelphia residence.

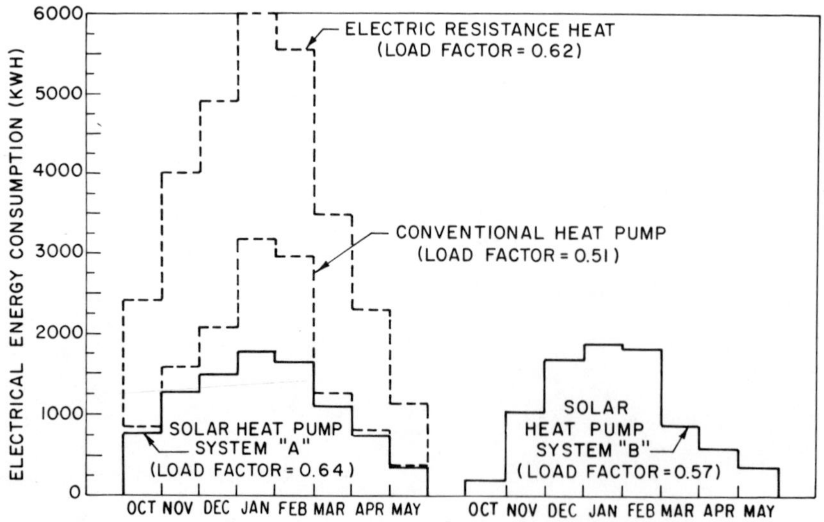

Fig. 5. Monthly electrical energy consumption for different space heating systems in 1500 sq ft Philadelphia residence.

Analysis Approach to Evaluate the Impact of Electric Heating Loads on Utility Operations

James L. Oplinger

Systems Planning
Electric Utility Systems Engineering Department
General Electric Company
Schenectady, N.Y.

It is in the customer's and the utility's best interests to increase off-peak energy uses which do not require significant additional capital expenditures for generation and transmission. These off-peak energy uses increase load factor and maximize revenue/investment ratios, resulting in lowest possible rates for electric service.

Electric heating is an excellent example of off-peak energy use since more than 80% of U.S. utilities are summer peaking. In the future, energy storage could make electric heating an even more desirable off-peak load, even to winter peaking utilities. Electric heating, especially in the form of the heat pump, is becoming increasingly attractive to utilities, consumers, regulatory agencies, and federal energy officials. One of the reasons for this growing interest is the energy-conserving nature of the heat pump coupled with electric heating's potential to conserve scarce fuels (natural gas and oil) through substitution of our more abundant coal and nuclear resources.

The objective of this paper is to present a general analysis approach which can be utilized to evaluate the impact of incremental electric heating loads on utility operations. Although the approach was developed to perform utility-specific economic evaluations and quantify utility impacts, it can also be used to assess consumer and societal impacts. Specifically, the analysis approach can be utilized to quantify the following impacts:

1. Utility impacts: load factor; capital investments; operating costs; annual revenue requirements; and minimum acceptable rates.
2. Customer impacts: annual energy costs.
3. Societal impacts: scarce fuel consumption; and primary energy consumption.

Analysis Approach

Impacts of incremental electric heating loads are evaluated using an analysis approach consisting of the following principal steps:

1. Select a utility to study.
2. Select a time interval to study and identify the number of new electric heated homes (heat pump, electric baseboard, electric furnace) for each year of the study period.
3. Based on connected heating load, load diversity information, weather data, and heating system performance characteristics, calculate hourly incremental loads.
4. Simulate utility operations to determine operating costs and investment costs to serve the additional load.
5. Determine annual revenue requirements and the minimum acceptable rate for the incremental electric heating load.

Incremental Heating Loads

A critical step in this analysis approach is the determination of hour-by-hour incremental heating loads which are determined by a number of factors, including design heat loss, weather data, number of homes, load diversity, heating system performance, and internal heat gains.

Although several approximate techniques based on statistical weather data and the degree-day formula have been developed to approximate hourly incremental loads, the preferred approach is computer simulation using hour-by-hour weather data. Although most energy system simulation programs available were designed for analysis of commercial buildings, many are also applicable for residential analyses. Several simulation programs have specifically been designed for thermal load analysis of residential buildings (refs. 1–3).

Simulation of Utility Operations

The determination of increased operating costs and investment costs required to serve incremental electric heating loads is a rather complex task, because it must take into account generating unit performance characteristics, equipment availability, utility system operating policies, etc. This complex task is accomplished in a rather straightforward manner via computer simulation using the Optimized Generation Planning (OGP) Program (refs. 4, 5). The OGP Program is a sophisticated utility planning tool which incorporates the three major elements of generation planning: reliability evaluation, production costing, and investment costing (refs. 6, 7). The reliability evaluation is performed to establish generation capacity additions required to maintain a prescribed reliability design criterion. Production costing is actually an hour-by-hour simulation of utility operations to determine fuel costs and operating and maintenance costs. The simulation of utility operations consists of maintenance scheduling, unit starting or commitment based on spinning reserve rules, and allocation of hourly demand among individual generating units to minimize fuel costs (economic load dispatch).

Utility Revenue Requirements

By making two computer runs with the OGP Program, a base case run without electric heating and a change case with electric heating, it is possible to establish new generation plant requirements and additional operating costs. After new transmission and distribution plant requirements have been established, it is possible to determine the total cost of electric power for space heating. Total costs include operating costs as well as fixed charges associated with cost of capital equipment. Fixed charges include depreciation, income tax, property tax, and return on investment. Since an appropriate rate of return on capital investment is a part of the fixed charges, total power costs are also referred to as annual revenue requirements (ref. 8). In utility economic analyses, minimum power generation costs and minimum revenue requirements are equivalent and are used interchangeably. The development of per unit annual revenue requirements, based on a levelized fixed charge rate, is illustrated in Fig. 1.

The development of total annual revenue requirements for a typical single family home in the Detroit area is illustrated in Fig. 2. Annual revenue requirements for generation, transmission, and distribution are calculated based on per unit annual revenue requirements ($/kw) and new plant requirements determined by the OGP Program. It is interesting to note that the new generation requirement shown in Fig. 2 is 1.5 kw/home, which is only 10% of the 15 kw/home connected heating load. Off-peak electric heating loads require very little in the way of additional capital expenditures for generation equipment. The reason for the 1.5 kw/home is the fact that additional electric heating does contribute to daily peaks during the winter; as a result, small amounts of generation are required to maintain generation reliability. Total annual revenue requirements are developed by adding fuel costs to new plant revenue requirements. Comparison of annual revenue requirements for incremental electric heating loads with alternative heating costs (natural gas, oil) indicates electric heating's marketability to utility executives.

The minimum acceptable rate for electric space heating is calculated simply by dividing annual revenue requirements by annual kwh sales per home. The minimum acceptable rate is that rate which must be charged to recover investment and fuel costs as well as produce a satisfactory level of return. The difference between this minimum acceptable rate and the present rate schedule is a measure of profitability and indicates to utility executives the desirability of the incremental heating load.

Electric Heating Scarce Fuel Conservation Study

The previous discussion focused on the assessment of utility-specific economic impacts using an incremental analysis approach. This incremental analysis approach is also useful in assessing broader policy type questions.

This approach has been used to perform a study to evaluate residential electric space heating's potential for conserving scarce fuels (natural gas and oil) during the 1975–90 time period (ref. 9). By virtue of utility generation mix and fuel diversity, electric heating offers significant potential to reduce scarce fuel consumption by substituting our more abundant coal and nuclear resources.

Study Scope

To perform the scarce fuels analysis it was necessary to compare electric space heating systems to oil and gas heating systems. In all, five space heating systems were considered: electric resistance (baseboard or ceiling cable); electric furnace (forced warm air); electric heat pump; gas (warm air); and oil (warm air).

A regionalized analysis approach was considered necessary to adequately model climatic variations and regional differences in installed generation capacity, construction starts, and heating equipment mix. The nine Bureau of Census geographic regions were used to conduct the analysis. The analysis approach had three aspects: energy consumption at the home; primary energy requirements at the source; and scarce fuel consumption based on generation mix.

Primary energy requirements at the source were defined as the sum of energy use at the home plus energy requirements associated with the production and delivery of energy for end use. To determine primary energy requirements at the source, it was necessary to conduct fuel cycle analyses for the natural gas, oil, coal, and nuclear fuel cycles. An estimate of overall system efficiency was calculated for each heating system based on energy requirements and losses throughout each fuel cycle including final utilization. Seasonal performance factors for the alternative heating systems were based on information gathered from a number of government, industry, and university sources. In determining energy consumption at the home, space heating requirements were calculated for characteristic single-family and multifamily dwelling units in nine study cities, one from each geographic region. The characteristic single-family home was a two-story colonial of 1700 sq ft while the multifamily home was a garden apartment or townhouse of 1140 sq ft.

The previously discussed incremental analysis approach using the OGP Program was utilized to determine scarce fuel requirements at the power plant for alternative electric heating systems. Study results based on this incremental analysis approach are discussed below.

Study Results

Primary energy consumption at the source, for space heating, is a function of energy consumption at the home, fuel cycle energy requirements and losses, and power generation equipment used to satisfy electric heating loads. A comparison was made of primary energy requirements based on construction forecasts for 1975 and assumed 100% market penetration (new construction) levels for each of the five heating systems. The determination of primary energy requirements was based on the assumption of equal insulation levels in walls and ceilings; however, the analysis did account for differences in ventilation due to combustion air requirements associated with gas/oil heating systems. Based on the previous assumptions, primary energy requirements for heat pumps were approximately 5% less than gas furnaces and 15% less than oil furnaces.

Scarce fuel consumption required for new residential construction during the 1975–90 time period is shown in Fig. 3. Two scenarios are indicated: *(a)* 100% new construction market saturation by natural gas, forced air furnaces, and *(b)* a forecast mix of fossil and electric space heating systems. Annual energy consumption levels for these two scenarios are indicated by the two solid lines in Fig. 3. The difference between these two consumption levels is the annual scarce fuel saving associated with the forecast mix of heating systems. The results indicated by the solid lines are based on the assumption that all new residential construction, electric-heated and gas-heated, will be insulated to the electric heating standard. It should be noted that most homes presently heated with oil or gas have insulation levels significantly less than the electric heating standard. If current practice continues, scarce fuel requirements for gas heating will be substantially greater than the level indicated by the solid line in Fig. 3. The dashed line in Fig. 3 indicates scarce fuel requirements for the first scenario, should current insulation practice for gas-heated homes continue.

Additional insight concerning electric heating scarce fuel requirements can be gained by reviewing the actual regional results used to develop Fig. 3. Regionalized scarce fuel requirements for the previously identified market penetration scenarios are presented in Fig. 4. As in Fig. 3, dashed lines are used in Fig. 4 to indicate scarce fuel requirements for gas heating should current insulation practice continue.

The critical determinant of scarce fuel requirements for electric heating is the mix of installed generating equipment. In this study, forecasts of installed generating capacity for 1975–90 were made for each of the nine Bureau of Census regions. Regionalized forecasts of generation mix, categorized as scarce fueled and nonscarce fueled, are summarized in Fig. 5. As the figure indicates, the amount of generating capacity using scarce fuels varies significantly regionally, but the long term trend is toward significantly less gas and oil fired power generation equipment. Based on FPC statistics for 1974, electric utility gas and oil consumption amounted to less than 35% of their total fuel consumption.

The annual scarce fuel requirements presented in Figs. 3 and 4 are not cumulative totals; only new construction was considered each year. By considering both new construction and all construction since 1975, it is possible to determine cumulative annual scarce fuel requirements. Cumulative results for the two alternative scenarios identified earlier are shown in Fig. 6. Although new construction is only a small fraction of total housing, cumulative scarce fuel savings at the end of the 15-year time period are substantial. Cumulative scarce fuel savings for 1990 fall somewhere between 2 and 4 quadrillion Btu/yr, depending on insulation levels in homes with gas heating. This amounts to an equivalent 1 to 2 million barrels of oil per day, about 16–32% of current daily imports.

Another factor to be considered, although not reflected in the study (ref. 9), is the quality of oil consumed at the power plant. Although electric utilities burn all grades of oil, residual consumption is normally more than 90% of the total. Typical home heating oil, No. 2 distillate, is of much higher quality than the residual burned by utilities. Therefore, electric heating is accomplishing a significant fuel substitution even when utility boilers are oil-fired—low quality residual is being substituted for high quality distillate or natural gas.

Summary

In summary, a generalized analysis approach to evaluate impacts of electric heating loads on utility operations has been presented. The approach utilizes computer simulation of actual utility operations to determine increased capital and operating costs associated with incremental electric heating loads. In addition to assessment of utility-specific economic impacts, the approach has been used to evaluate electric space heating's potential for conserving scarce fuels (natural gas and oil) on a national scale. The results of the evaluation, based on a projected mix of generating equipment and home heating systems, indicate that scarce fuel savings in 15 years could be an equivalent 1 to 2 million barrels of oil per day. This amounts to approximately 16–32% of current daily petroleum imports. Even with today's generation mix, the use of electric heating offers the potential of reducing oil and gas consumption through substitution of our more abundant coal and nuclear resources.

REFERENCES

1. "Solar Heating and Cooling of Buildings" (SHACOB), NSF Contract C-855, Feasibility and Planning Study, Report No. NSF-RANN-74-027B, May 1974.
2. "Residential Energy Consumption Single Family Housing," Final Report HUD Contract No. H-1654, Report No. HUD-HAI-2, Hittman Associates, Inc., Columbia, Md., March 1973.
3. Meriwether, R.F., "Computer Programs for Economic Analysis," Ross F. Meriwether and Associates, Inc., San Antonio, Tex.
4. Marsh, W.D., Moisan, R.W., and Murrell, T.C., "Perspective on the Design and Application of Generation Planning Programs," Nuclear Utility Planning Methods Symposium, Chattanooga, Tenn., February 1974.
5. Adamson, A.M., Kenney, J.F., and Moisan, R.W., "Parametric Sensitivity Method for Establishing Optimum Long Range Generation Mix," *Proc. Am. Power Conf., 35* (1973).
6. Garver, L.L., "Electric Utility Planning Models," Paper presented at the Joint National ORSA/TIMS meeting, Session WP3: Energy Systems Planning Models, 30 April 1975.
7. Kirchmayer, L.K., *Economic Operation of Power Systems*, John Wiley and Sons, New York, 1958.
8. Jeynes, P.H., and VanNimwegen, L., "The Criterion of Economic Choice," *AEEE Trans., Power Apparatus and Systems*, August 1958, p. 606.
9. Oplinger, J.L., et al., "Electric Space Heating: Its Potential for Conserving Scarce Fuels," General Electric Company, Electric Utility Systems Engineering Department, Schenectady, N.Y., July 1974.

DISCUSSION

Q Why don't you consider nuclear fuel as scarce in light of delays, reprocessing limitations, and consistently lower than achieved projections for nuclear plants? Why not consider synthetic fuels from coal by 1985–90?

A The primary objective of the paper was to present a general analysis approach. The scarce fuels study was presented as an example of the many policy-type questions which might be investigated using the approach.

Q Isn't summer versus winter peaking a function of climate and other relatively constant factors? How would you increase use of resistance heating to fill a winter valley in a place where there is little heat demand?

A Many electric utilities with winter valleys are located in areas with substantial heating loads. Future heating loads are a function of both climate and housing starts. As shown in Fig. 4, projected space heating demands in the southern regions are not insignificant.

Q How do you account for an increment of 1.5 kw only for an electrically heated residence?

A Off-peak loads. The new plant requirements in Fig. 2 are based on incremental electric heating loads for a summer peaking utility. The reason for the 1.5 kw/home generation requirement is discussed in the paper.

	TOTAL INSTALLED COST $/KW	FIXED CHARGE RATE*	ANNUAL REVENUE REQUIREMENTS $/KW
BASE LOAD GENERATION	500	.1578	78.91
PEAKING GENERATION	150	.1578	23.67
TRANSMISSION	125	.1592	19.90
DISTRIBUTION	120	.1644	19.73

*Includes 11% Return on Investment

Fig. 1. Per unit annual revenue requirements.

	ANNUAL COST ($/KW)	NEW PLANT REQUIREMENTS (KW/HOME)	ADDED ANNUAL REVENUE REQUIREMENTS $
GENERATION	78.91	1.5	118.36
TRANSMISSION	19.90	1.5	29.85
DISTRIBUTION	19.73	10.0	197.31
CAPITAL INVESTMENT SUBTOTAL			$345.52
FUEL			235.00
TOTAL			$580.52

Fig. 2. Annual revenue requirements for a typical home (1500 sq ft).

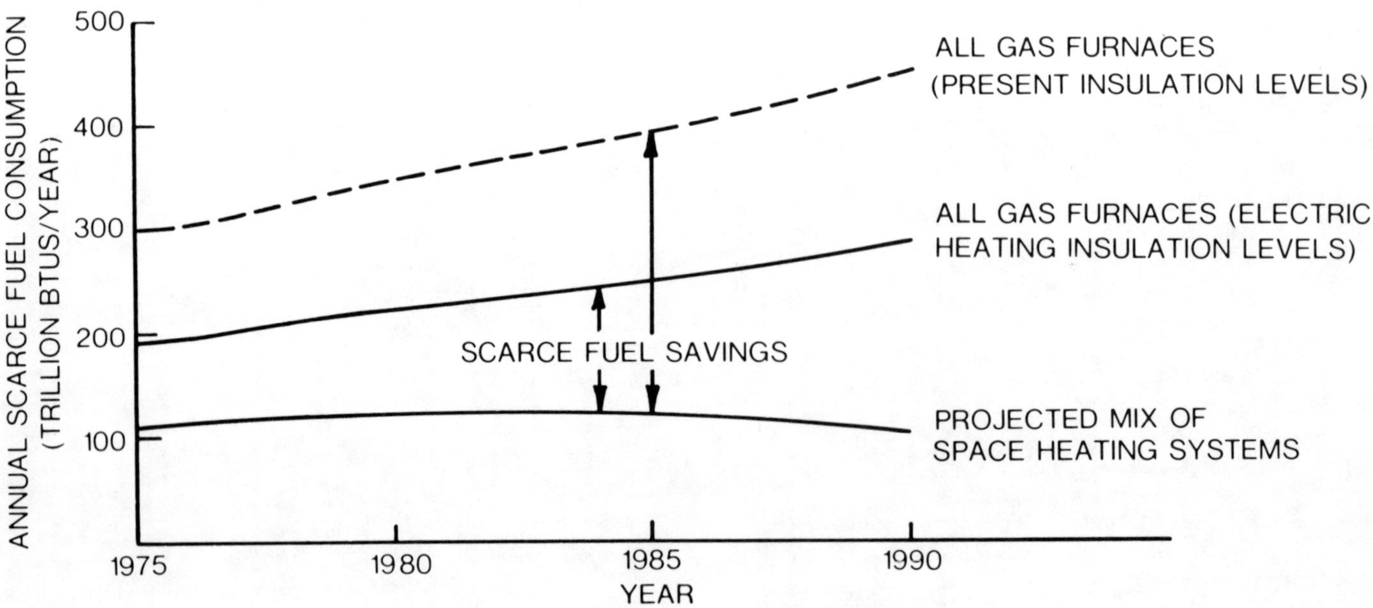

Fig. 3. Annual scarce fuel consumption for new residential construction 1975–90.

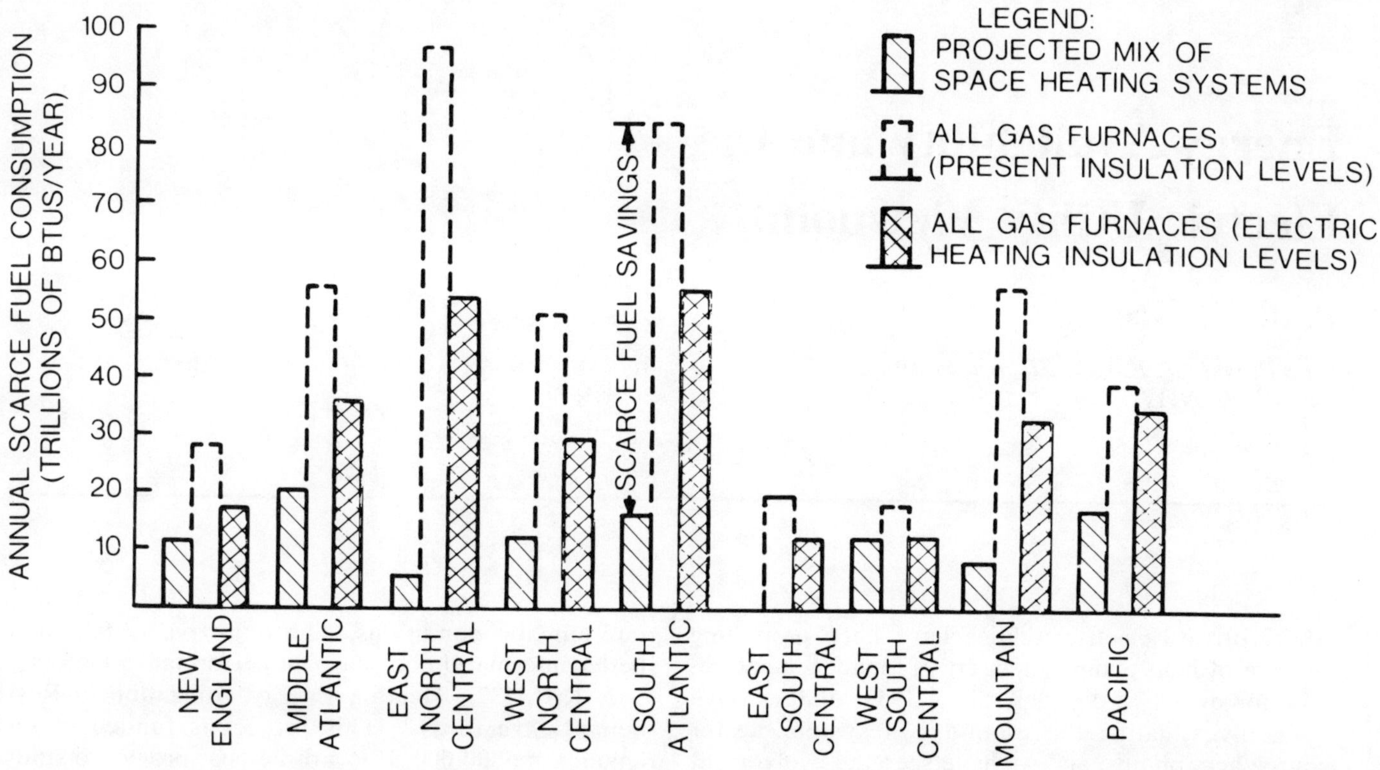

Fig. 4. Regionalized scarce fuel consumption for projected residential construction during 1990.

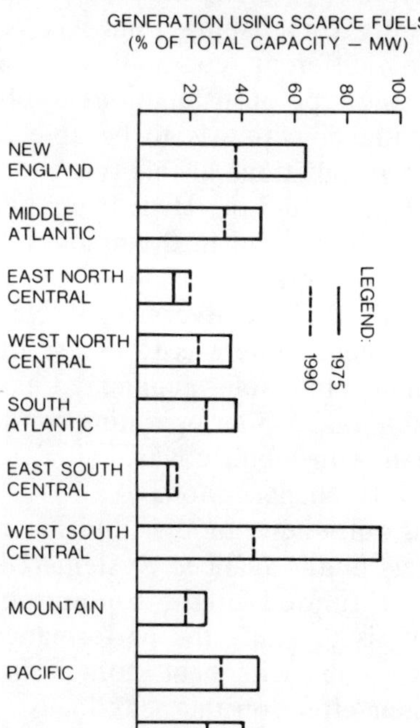

Fig. 5. Installed generating capacity using scarce fuel.

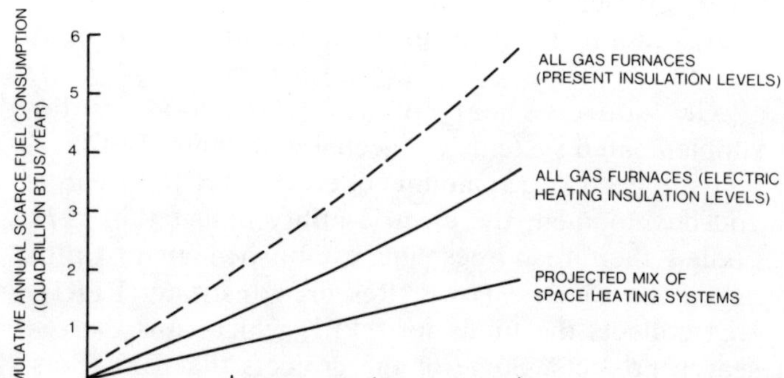

Fig. 6. Cumulative annual scarce fuel consumption for all residential construction since 1975.

Energy Availability and the Electric Utility Viewpoint

Paul C. Greiner

Vice President, Edison Electric Institute
New York, N.Y.

Many private electric utilities have been promoting the use of heat pumps as an energy management device for over two decades. Since 1953, utilities have been investigating the ground coil system, water source heat pumps, and we have seen the evolvement of the air source heat pump, including the open compressor, the sealed cycle, and the various defrost methods.

The electric utility industry has for years talked of the merits of heat recovery systems using heat pumps, double bundle condensers, and energy storage. From data that we have collected, there are at the present time over a million heat pump installations in the United States. This includes residential and commercial types. In research and development the private utilities have worked on their own and with manufacturers to help develop better and more reliable heat pumps. Now, the air source heat pump is a very efficient solar collector. Where does the source energy come from as used in an air source heat pump? It's solar.

Air source heat pumps can have solar supplemented systems connected with them. But they will still need large amounts of electricity. In research and development the electric utility industry in 1973 pooled their resources and established an institute called the Electric Power Research Institute (EPRI). EEI collects the funds for EPRI, which runs the research projects. Some of the projects that EPRI has at present in heat pumps and solar energy are:

1. R.P. 432 "Load and Use Characteristics of Heat Pumps in Single Family Dwellings." This is a three-year project and is funded to the amount of $624,000. It is a joint project with AEIC (Association of Edison Illuminating Companies). In this project 20 utilities will install and monitor some 400 heat pumps in various climatic conditions. The objective is to obtain better information on seasonal performance factors.
2. R.P. 137 "Heating Energy Utilization in Residential Structures." This project is funded in the amount of $800,000. It is a three-year project to study the energy effectiveness of various heating systems. This is an ongoing project at OSU (Ohio State University). At present, we have six houses and three apartments with various systems including heat pumps. We are presently establishing some six remote locations, each in a different geographic area, in order to monitor the same type of installations to obtain better field data; the objective is to be able to predict energy usage in installations of this type.
3. R.P. 385 "Solar Energy and the Heat Pump in a Northern Climate." This is funded in the amount of $140,000. It is a two-year project. It is to be run in connection with SUNY (State University of New York) at Albany. The objective here is to assess the efficiency and performance of a solar augmented heat pump and storage system under poor operating conditions. By this we mean with cloud cover, rain, etc. This is a joint project with Niagara Mohawk.
4. R.P. 1213 is being run under another division of EPRI—The Fossil Fuels and Advanced Systems Division. This project is funded in the amount of $500,000. The objective is to study the performance and reliability of heat pumps with heat storage and solar assistance and their effect on the peak loads of utilities.

These are examples of R & D under way at EPRI. In addition, many individual utilities are conducting research, one such utility project being conducted by Pennsylvania Power & Light Co., included in these proceedings. Other utilities are conducting individual research, such as American Electric Power, Commonwealth Edison, New York State Electric and Gas, Alabama Power, Union Electric, and Mississippi Power.

In order to further understand the electric utility viewpoint on solar supplemented heat pumps, one first needs to understand something about that industry.

First, the electric utility industry historically has been growing (in kwh sales) at the rate of approximately 7%/yr. But in 1974 it had zero growth (primarily due to conservation efforts and a milder than usual year). In 1975 it expects to continue a growth rate of approximately 5.3–5.8%, and this should continue in the future. In 1974 the electric industry consumed 26.9% of the total energy used in the United States (Fig. 1).

The industry is not tied to any one source of energy. The mix of energy consumed in electric generation at the end of 1974 is shown in Fig. 2. Almost two-thirds of this is from noncritical energy sources. For the rest, 17% is produced from natural gas and 17.6% from petroleum. But the petroleum we used in 1974 was almost 88% residual oil (Fig. 3). Residual oil is, as one might expect, the bottom of the barrel. It's what is left over after the cracking process. The mix of source energy varies by region and by utility company. Commonwealth Edison, for instance, in Chicago uses some 97% coal and nuclear. Boston Edison in the northeast is over 30% nuclear. Northeast Utilities is over 40% nuclear.

We are shifting rapidly from the use of critical fuels to what we call noncritical fuels. In Project Independence, a report from the Federal Government, the estimate of known reserves is as follows:

Coal	2000-year supply
Oil	30–40 years (approx.)
Natural gas	50 years (approx.)

We are running short of the critical fuels, oil and natural gas, and the electric utility industry is shifting more to noncritical fuels, coal and nuclear. From a nuclear point of view, at present, the industry is using fission and we have some 50 or more years of known supply of fissionable material. The industry is working on the breeder reactor—the first is being constructed and should go into operation in the 1980s.

Following this, there will be fusion, using seawater (the deuterium in seawater), and it is estimated that 1 cu km of seawater contains the energy equivalent to 1500 billion barrels of oil. That is enough to supply the energy requirements of the United States for 100 years.

We are striving to achieve the goal of some 200 nuclear power plants by 1985. (A 1000-MW nuclear plant saves the equivalent of 10.4 million barrels of oil per year.)

One of the major challenges for the electric utility industry is in load factor and the differential between summer and winter peaks. Over 80% of utilities in the United States have a summer peak. All the electric power pools in the United States have a summer peak, but few utilities have a winter peak; these are the utilities that will be most interested in ways to trim it. Heat storage, supplemented with solar, will help trim these peaks. But with those utilities facing a summer peak, present technology does not help reduce their peak.

At a recent conference (ref. 1), two representatives of Arthur D. Little (Messrs. Gleason and Platt) said they had just completed a project of a hypothetical model prediction for the electric utility industry. This is a project funded by the National Science Foundation. Their findings were these: solar climate control will help level winter peak loads, reduce peak demand, improve load management capability, and have better system security. All of this is due to storage and control, and only on the heating cycle. They reported that as far as the heat actuated air conditioner (present technology) is concerned, from an economic standpoint it is impractical. There are other systems under investigation by various groups, such as NSF and Westinghouse. One system requires a high initial investment and utilizes the freezing of a large tank of water in the winter, using a heat pump; this ice bin is then used for summer cooling. A test installation of this system is being conducted at the University of Tennessee on a small apartment complex.

Summary

From an electric utility standpoint, most utilities have promoted the heat pump as an energy management device for more than 20 years. They have pushed research and development, either on its own or in cooperation with manufacturers. Second, solar supplemented heat pumps really only help winter-peaking companies and these are in the minority. With most utilities and power pools faced with a summer peak, the solar supplemented heat pump really does not help solve this problem. Storage really provides the aid for this kind of a utility or system pool.

It is true that with the Rankine cycle, the absorption and heat actuated air conditioner would help provide relief for summer peaking. But the Arthur D. Little study shows that the heat actuated air conditioner is not economically feasible at present.

The electric utility industry can help solve the energy problem. From a source energy standpoint we see a shift to coal in short term and nuclear in the longer term. The electric heat pump certainly fits into this future.

REFERENCE

1. Load Management Conference, sponsored by Federal Energy Administration, Washington, D.C., June 1975.

DISCUSSION

Q Does EPRI represent only private power companies?

A EPRI represents both private and public utilities.

Q You state that storage will help lower peak demand for winter peaks. Will not *cold storage* assist in lowering the daily peaks during summertime operation? Comment on the value of cold storage reservoirs to the electric utility industry.

A Cold storage would help reduce daily peaks, but solar supplementation does not help storage. To utilize solar supplement you would need a dual system and also a heat exchanger.

Q Does nuclear power from seawater pollute?

A Nuclear generation does not pollute seawater. What I referred to was in the process of fusion of deuterium from seawater used as the source. The cooling water from a nuclear reactor does not create radiation problems.

Q You mentioned the tremendous quantity of energy in 1 cu km of seawater if it could be converted by fusion. When do you think there will be a demonstration fusion reactor? When will fusion power stations exist?

A There are three stages of nuclear generation. The first stage is fission, which is being utilized today. The second stage is the breeder reactor, which, in this country, will be in the demonstration stage in the 1980s (France and the Soviet Union have demonstration projects in operation today). Stage 3 is fusion and that is predicted, depending upon the R & D effort put into it, for demonstration projects approximately in the year 2000.

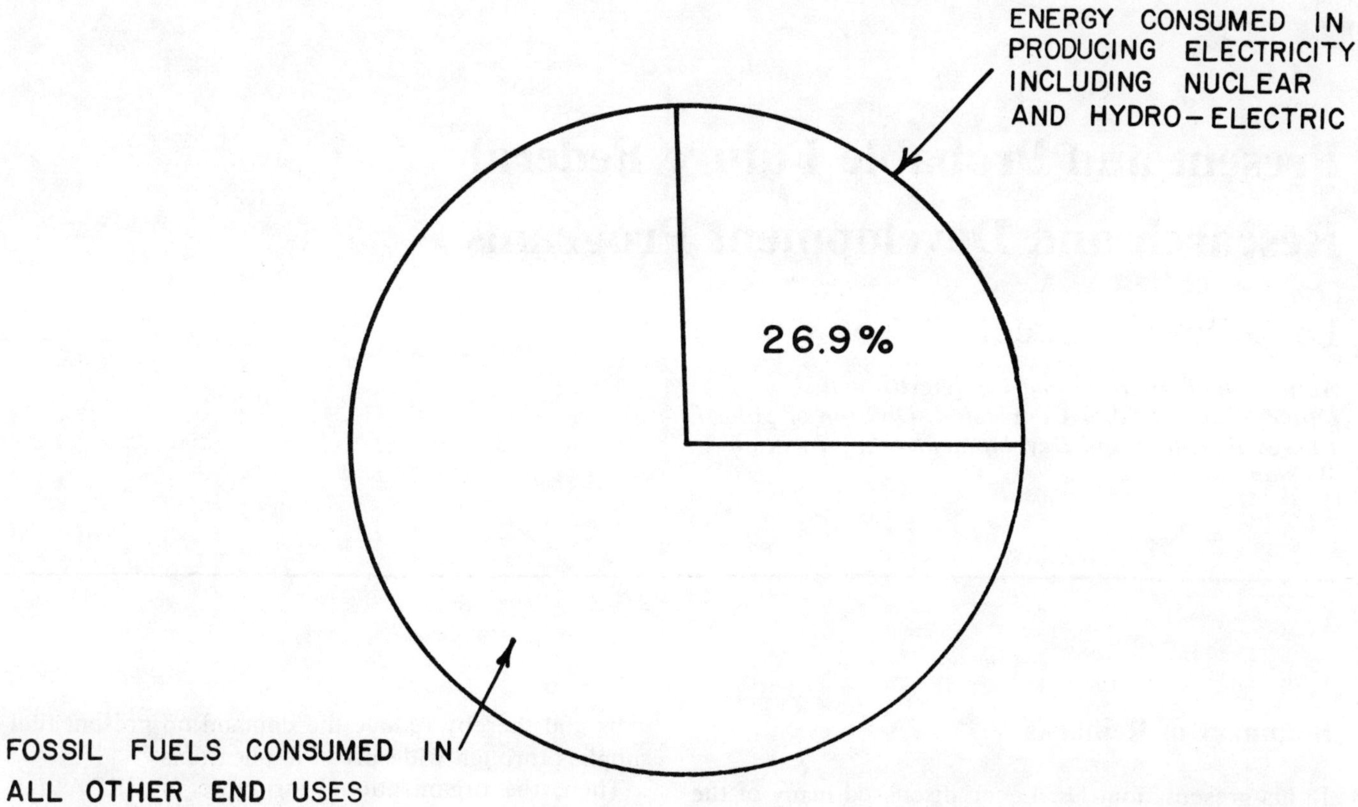

Fig. 1. Energy consumed in generation of electricity in 1974.

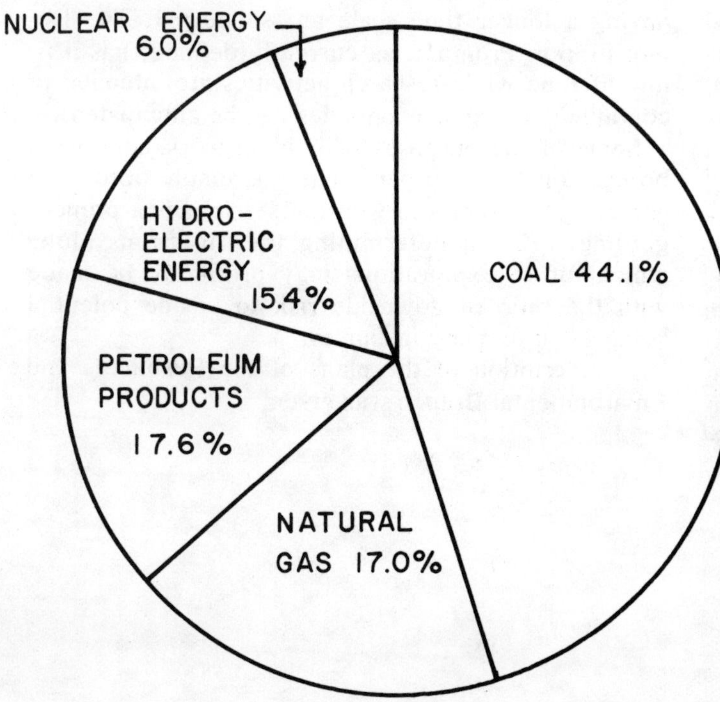

Fig. 2. Sources of energy for electric generation in 1974.

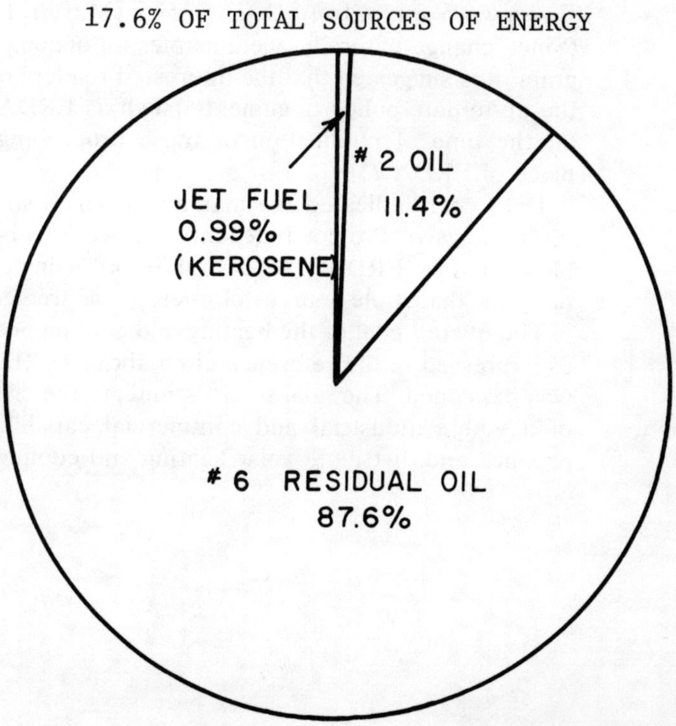

Fig. 3. Petroleum products used in electric generation.

Present and Probable Future Federal Research and Development Programs

Dr. John W. Leech

Acting Chief, Heating and Cooling Branch
Office of Direct Solar Conversion, Division of Solar Energy
Energy Research and Development Administration
Washington, D.C.

Summary of Remarks

In his presentation, Dr. Leech discussed many of the facets of the ERDA program which had not been released to the public until shortly before the Workshop in the form of the document: "Interim Report, National Plan for Solar Heating and Cooling (Residential and Commercial Applications)," Energy Research and Development Administration, Division of Solar Energy, Washington, D.C. 20545 (March 1975). (Since change naturally accompanies an ongoing program, it is suggested that the interested reader consult the up-to-date policy documents [such as ERDA 23A (at the time of publication of these proceedings) in place of ERDA 23].)

Dr. Leech indicated the manner in which some of the findings of Project Independence were to be implemented by ERDA. The desirability of using essentially inexhaustible sources of energy was treated.

The overall goal of the heating and cooling program as expressed in the reference cited above (ERDA 23) was explained. The goal is to "stimulate the creation of a viable industrial and commercial capability to produce and distribute solar heating and cooling systems and thereby reduce the demand on present fuel supplies through widespread applications."

Then, the organizational structure of ERDA and the Division of Solar Energy was explained in order to assist the conference attendees in their future working relations with ERDA personnel. The dual philosophy of the R & D Branch (Heating and Cooling of Buildings Branch) was explained. On the one hand, R & D in direct support of the Demonstration Program is funded, and on the other, explorations having a longer time scale are supported. The Demonstration Program, as currently defined, has a finite lifetime while research activities are intended to continue (as long as economies can be anticipated).

Some of the purposes of holding workshops were brought out. One in particular was emphasized. The gathering of experts in working sessions is a primary guiding force in determining the directions along which future explorations may profitably be made with the ratio of downside risk to upside potential being small in many instances.

A description of the plans of the Resources and Environmental Branch was given.

Part VI: Working Group Reports

Working Group A:
Heat Pump Fluids and Machines Interaction

Moderator:
Frank Biancardi

Chief, Advanced Energy Conversion Systems
United Technologies Research Center
East Hartford, Conn.

A small but spirited group of solar application engineers and researchers met to discuss the work group topic. The discussions concerned a range of topics, which are summarized below along with suggested alternative approaches.

1. The need for a single working fluid capable of providing efficient operation over a wide temperature range was expressed. Although a single working fluid capable of meeting this requirement is not likely to become available due to both thermodynamic and physical limitations, a number of machine approaches were identified which could provide for efficient operation. These approaches include the use of cascaded systems, variable speed options, and/or multiple machine design approaches.

2. The elimination of oil additives and/or development of self-lubricating fluids was expressed. The requirement for oil additives is generally related to the types of machines (compressors) used in heat pump systems. With conventional reciprocating or rotary-vane compressors lubrication will be required. However, in specifically designed turbomachinery it is possible to utilize the working fluid for lubrication and thus eliminate the use of any oil additives.

3. A desire for fluids with improved performance characteristics was expressed. A number of new fluids have been developed by the fluid suppliers which provide small but significant improvements in efficiency, capacity, and/or stability. These fluids are available at reasonable cost today.

4. Replacements for working fluids that may be objectionable for environmental reasons may be required. Feasible, moderate cost replacements for the conventional fluids are available, according to suppliers representatives.

5. The need for new, safe, nontoxic fluids capable of operating at temperatures of 450F and above with good performance characteristics was expressed. Such fluids are being actively studied and may be available at low cost with further development. Finally, it was recommended that an increased dialog be encouraged among fluid suppliers, machine designers, and system installers and specifiers to implement the newest developments in fluids and overcome specific fluid problems.

Working Group B: Heap Pump and Collector Interaction

Moderator:
Alwin B. Newton

Consultant
York, Pa.

Attendance varied, with a maximum of 92 during the 105 minute session. The group had some difficulty in considering only the two elements—the heat pump and collector—since both are so intimately associated with storage. The items on which there was agreement are:

1. Use of the heat pump extends the range of insolation over which heat can be collected from a given collector for building heating purposes.
2. Use of the heat pump increases the total amount of heat collectable and the instantaneous amount per square foot, by making use of lower collector temperature and thus of higher collector efficiency.
3. Discussion of the number of cover glasses as influenced by the use of the heat pump led to no conclusion. 0, 1, or 2 cover glasses have been suggested in the program sessions. It was suggested by the Workshop that ERDA might sponsor analytical work to determine incremental costs per unit of performance, considering effects on square feet of collector.
4. Control technology appeared to the group to be lagging, at least in establishment of simple methods of forcing components to work together economically.
5. Low temperature collection levels which may be made possible via use of heat pumps have lower plumbing costs due to lower insulation needs, or its elimination. It was noted that on some jobs the pipe insulation contract equalled the collector cost.
6. In contrast to some of the above, it was noted that higher temperature collectors may reduce the heat pump hours of operation and thereby extend its operating life, particularly on small heat pumps with reciprocating compressors.
7. The heat pump, collector, and storage can be so operated as to eliminate operation during utility peak periods. It may also be used to fill the utility valleys.
8. It was unanimously agreed among consultants present that the minimum practical hot water temperature for heating at design conditions should be 110F (44C). Otherwise fan power and heat exchanger size become excessive.

The foregoing items need good analysis for maximum energy saving and in order to relate the result to cost effectiveness. ERDA could well sponsor such an analysis.

Working Group C: Heat and Storage System Interaction

Moderator:
Dr. James E. Hill

Mechanical Engineer
National Bureau of Standards
Washington, D. C.

The group met for approximately one and one-half hours and the attendance varied from about 50 to 80 during this period. Three broad subjects were discussed that directly relate to the assigned task. Several miscellaneous topics worth noting were also covered and will be briefly mentioned at the end of this report. The three pertinent topics were:

1. Advantages of different storage media in a solar heat pump system.
2. Discussion of optimum storage capacity in a given solar heat pump system.
3. Discussion of optimum modes of operation or methods of utilizing thermal storage in a solar heat pump system.

1. DIFFERENT STORAGE MEDIA FOR USE IN A SOLAR HEAT PUMP SYSTEM. With respect to the three most often considered storage media, the following points were made by the participants:

1. Water—It is probably the only practical medium for use in large heat pump systems (installed in large commercial buildings) where commercially available heat exchangers and other components are used (ref. 1).
2. Rocks—This medium has possibly the most application with residential air-to-air heat pumps (ref. 2).
3. Phase change materials—The capability of large energy storage per unit volume was noted along with the fact that heat pump performance is improved if it is operated with a constant source temperature which could be obtained by using these materials. Even though these materials have not had wide use in solar heat pump systems to date, there were two applications mentioned. A salt hydrate phase change material has been used successfully for three years with air-to-air heat pumps in "Solar One" at the University of Delaware (ref. 3). In addition, the Oak Ridge National Laboratory is now studying, in conjunction with the University of Tennessee, heat pump use in apartment complexes where the thermal storage medium is ice (ref. 4). This latter application appears feasible in areas where the ambient temperature drops to 10C or lower for a significant number of hours during the heating season.

Conclusion: Each type of storage medium has advantages for specific applications.

Recommendations: A study should be undertaken to determine and document the advantages and disadvantages of available storage media for use with the various types of solar heat pump systems. The study should also include the determination of the best phase change material to be used with each of the various systems.

2. OPTIMUM STORAGE CAPACITY FOR A GIVEN SOLAR HEAT PUMP SYSTEM. There was a divergence of opinion concerning the answer to this question. At least two participants stated that based on computer studies of particular systems, storage capacities large enough to carry a heating system through more than one or two days of sunless weather could not be economically justified. Less vocal opinions were expressed that larger capacity storage systems could be justified under certain conditions.

Conclusion: There is no definitive answer to this question that is universally applicable.

Recommendation: Optimization studies should be undertaken to determine the correct relationship between storage size and heat pump for various types of solar heat pump systems. This recommendation was qualified by several members of the working group who stressed that suboptimization is dangerous. In the optimization study, a range of factors should be included such as what possible action might be taken by utility companies if systems were designed without regard to the impact on the companies' load factor (refs. 5, 6).

3. OPTIMUM MODES OF OPERATION OR METHODS OF UTILIZING THERMAL STORAGE IN A SOLAR HEAT PUMP SYSTEM. Concern was expressed that in designing a solar heat pump system, there is a large number of ways in which components can be combined and controlled, and the performance of the overall system will vary accordingly. Computer study results were cited to indicate vastly different energy consumption for a given collector, storage system, and heat pump depending on how they were interconnected and operated. Bob Tamblyn of Engineering Interface, Ltd., of Toronto has recently discussed the advantages of storing chilled water in large heat pump systems. He cited several specific designs where there has been a significant savings in energy through the use of this technique (ref. 7). Fred Dubin cited similar results at this meeting (ref. 8).

Conclusions: Optimum methods of utilizing thermal storage in solar heat pump systems are not yet completely defined.

Recommendation: Thorough analysis should be made of all types of potential solar heat pump applications, and *system strategies* be determined and documented for the various applications.

4. MISCELLANEOUS TOPICS OF DISCUSSION. First, it was noted that the relevant performance factor for a given solar heat pump system is energy used by the system, not necessarily the heat pump COP. This results from the fact that heat pump systems, and not just heat pumps, are being examined, and the storage-collector-heat pump interaction affects and even governs the system performance.

In addition, it was noted that at this stage of solar heat pump utilization, the percentage of all heat pumps produced that are used in solar systems is extremely small. Therefore, solar heat pump systems must be designed to use commercially available heat pumps, rather than expect any new machines on the market in the forseeable future designed especially for these applications.

Finally, a conventional air-to-air electrically driven heat pump is in reality a solar augmented heat pump since the level of outdoor air temperature indirectly results from the solar energy being absorbed at the surface of the earth. It is recommended that this be acknowledged in some manner by the workshop coordinator in the publication of the proceedings from this meeting. This is consistent with the generally accepted classification of wind and ocean thermal as being solar power generation processes.

REFERENCES

1. Bridgers, F.H., "Comparison of Heat Pump Systems for a Library at Idaho State University," Paper presented in Session III.
2. Barkmann, H.G., "Solar Augmented Air-to-Air Heat Pump Installation," Paper presented in Session III.
3. Boer, K., "A Combined Solar Thermal Electrical House System," Paper No. EH108, presented at the International Congress—The Sun in the Service of Mankind, UNESCO House, Paris, 2–6 July 1973.
4. Hise, E., "Annual Cycle Energy Storage System for Buildings," *Proceedings of the NSF/ERDA Workshop on Solar Energy Storage Subsystems for the Heating and Cooling of Buildings*, held in Charlottesville, Va., 16–18 April 1975, published by the University of Virginia, 1975.
5. Oplinger, J., "Electric Space Heating—Impact on Utility Loads," Paper presented in Session V.
6. Greiner, P.C., "Energy Availability and Electric Utility Viewpoints," Paper presented in Session V.
7. Tamblyn, R.T., "The ABC's of Thermal Storage," *Proceedings of the NSF/ERDA Workshop on Solar Energy Storage Subsystems for the Heating and Cooling of Buildings*, held in Charlottesville, Va., 16–18 April 1975, published by the University of Virginia, 1975.
8. Dubin, F.S., "Solar Assisted Heat Pump Installation in Cold Climates," Paper presented in Session III.

Working Group D:
Solar Energy Heat Pump System Modeling

Moderator:
Dr. John Mitchell

Professor of Mechanical Engineering
University of Wisconsin
Madison, Wisc.

An assessment of (*a*) the current state of the art and (*b*) the areas in which further modeling research is needed were discussed in this workshop. The group consisted of about 50 people and appeared to represent a broad spectrum of interests, ranging from equipment manufacturers to system modelers.

The main modeling building blocks for these systems are:

1. Load models—desired heating and cooling flows.
2. Weather models—representations of realistic weather patterns in desired locations.
3. System models—representations of present and proposed systems.
4. Component models—representations of present and proposed equipment.

The first major goal of system modeling was felt to be the attaining of insight into system operation. This allows an assessment of the performance of a given system and aids further research into the development of better systems. It facilitates the identification of critical parameters and provides a direction to experimental research.

The second major goal was the development of simple calculation procedures for determining system performance and component sizes.

At present, modeling is a research tool used to study the behavior of the system to change in parameters and driving forces. With improved knowledge of system behavior, simplified design procedures will result.

Approximately eight large computer programs for use in system modeling were identified and discussed. These are of varying degrees of sophistication, complexity, and flexibility. In general, these programs allow accurate calculation of energy demands. Various models for the components of a solar heat pump system are available.

The main area of modeling in which further research is needed were felt to be:

1. *Weather models*—Accurate, adequately described solar data are needed on which to base weather models. Modeling efforts into a meaningful statistical analysis of data are needed. These efforts should develop models that allow assessment of the influence of weather without the need to simulate every day. Efforts need to be directed into the number and type of days needed in a simulation in order to size equipment. Methods for determining and evaluating the importance of particular microclimates, including the use of synthesized data, need development.

2. *Transient performance*—Models of the transient behavior of both loads and components are needed to allow better evaluation of system performance. Many components and systems operate under on-off control, and the integrated performance may differ significantly from steady state performance.

3. *Heat pump models*—Data and empirical representations currently exist over limited operating ranges for heat pumps. These are sufficiently accurate for simulations within the test ranges and possibly accurate for performance evaluation over a broad range. Data are needed on reliability, time response, and performance over broader ranges for modeling economic and thermal performance. Fundamental limits on heat pump thermodynamic, thermal, and mechanical performance need to be determined.

4. *Calculation and design procedures*—Continued

efforts are needed to develop calculation procedures for performance that require reduced computer time and, ultimately, are suitable for use with a calculator. Design procedures for equipment selections for a solar heat pump system do not exist. Such procedures are essential in order to transfer these basic system research efforts to practice.

5. *Economic models*—Realistic economic models are needed to accurately assess the costs and benefits of systems. Consistency is currently lacking.